SODIUM-CALCIUM EXCHANGE

PROCEEDINGS OF THE SECOND INTERNATIONAL CONFERENCE

ANNALS OF THE NEW YORK ACADEMY OF SCIENCES
Volume 639

SODIUM-CALCIUM EXCHANGE

PROCEEDINGS OF THE SECOND INTERNATIONAL CONFERENCE

Edited by Mordecai P. Blaustein, Reinaldo DiPolo, and John P. Reeves

The New York Academy of Sciences
New York, New York
1991

Cover art: The illustration on the soft cover edition of this volume shows the effects of sequential exposure to ouabain and low-Na^+ media on intracellular [Ca^{2+}] in a cultured astrocyte. Resting cell is in the upper left. Ouabain induced a small increase in [Ca^{2+}] (upper right). Subsequent exposure to low-Na^+ solution caused Ca^+ to dramatically increase (lower left). The cover photograph was provided by William F. Goldman.

Library of Congress Cataloging-in-Publication Data

Sodium-calcium exchange : proceedings of the second international conference / edited by Mordecal P. Blaustein, Reinaldo DiPolo, and John P. Reeves.

p. cm. — (Annals of the New York Academy of Sciences, ISSN 0077-8923 ; v. 639)

"Result of a conference entitled: Second International Conference on Sodium-Calcium Exchange, which was held by the New York Academy of Sciences on April 7-11, 1991 in Baltimore, Maryland"—Contents p.

Includes bibliographical references and index.

ISBN 0-89766-693-3 (cloth : alk. paper). — ISBN 0-89766-694-1 (paper : alk. paper)

1. Calcium channels—Congresses. 2. Sodium channels—Congresses. I. Blaustein, Mordecai P. II. DiPolo, Reinaldo. III. Reeves, John P., 1942- . IV. New York Academy of Sciences. V. International Conference on Sodium-Calcium Exchange (2nd : 1991 : Baltimore, Md.) VI. Series.

[DNLM: 1. Calcium—metabolism—congresses. 2. Ion Exchange—congresses. 3. Sodium—metabolism—congresses. WI AN626YL v. 639 / / QH 604.5 S6795 1991]

Q11.N5 vol. 639
500 s—dc20
[591.19'214]
DNLM/DLC
for Library of Congress

91-44531
CIP

PCP

Printed in the United States of America

ISBN 0-89766-693-3 (cloth)
ISBN 0-89766-694-1 (paper)
ISSN 0077-8923

ANNALS OF THE NEW YORK ACADEMY OF SCIENCES

Volume 639
December 18, 1991

SODIUM-CALCIUM EXCHANGE: PROCEEDINGS OF THE SECOND INTERNATIONAL CONFERENCE[a]

Editors and Conference Organizers
MORDECAI P. BLAUSTEIN, REINALDO DIPOLO, and JOHN P. REEVES
Advisory Board
T. JEFF A. ALLEN, W. JONATHAN LEDERER, PETER A. MCNAUGHTON, HANNAH RAHAMIMOFF, and HARALD REUTER

CONTENTS

[a] This volume is the result of a conference entitled Second International Conference on Sodium-Calcium Exchange, which was held by the New York Academy of Sciences on April 7-11, 1991 in Baltimore, Maryland.

Financial assistance was received from:

Supporters
- ICI Pharmaceuticals Group
- National Science Foundation

Contributors
- American Cyanamide/Medical Research Division
- CIBA-GEIGY Corporation
- Hoechst-Pharmaceuticals
- Hoffmann-LaRoche, Inc.
- Marrion Merrell Dow Inc.
- Merck Sharp & Dohme Research Laboratories
- Miles Inc./Bayer AG
- Monsanto Company
- Sandoz Research
- The University of Maryland School of Medicine
- The Upjohn Company
- Warner-Lambert Company

Preface

Intracellular Ca^{2+} ions play key second messenger roles in numerous physiological processes in virtually all types of cells. Consequently, considerable interest has been devoted to the elucidation of the mechanisms that help to regulate intracellular free Ca^{2+}, $[Ca^{2+}]_i$. One transport system that plays a critical role in the control of cell Ca^{2+} in many cells is the Na-Ca exchanger that was first identified in the mid-1960s in squid axons, and in mammalian cardiac muscle and intestinal epithelia. The plasma membrane Na-Ca exchanger helps to control cell Ca^{2+} by acting in concert with the plasma membrane ATP-driven Ca^{2+} pump and Ca^{2+} channels, and with intracellular Ca^{2+} sequestration systems. The exchanger appears to be particularly important in cells in which it is required to (i) extrude a large amount of Ca^{2+} rapidly, (ii) regulate the intracellular stores of Ca^{2+}, and/or (iii) modulate $[Ca^{2+}]_i$ when this must remain at relatively elevated levels for long periods of time.

In 1987, the late Dr. Peter F. Baker organized the First International Conference on Sodium-Calcium Exchange to bring investigators from around the world together to discuss this critical transport mechanism.[a] The conference, which was held shortly after Dr. Baker's death, generated renewed enthusiasm for research on this topic. Consequently, during the past few years, novel electrophysiological, fluorescent imaging, and molecular biological methods, as well as many older, established techniques have been used to generate a wealth of new information about this transport system. New controversies have also arisen, for example, with regard to (i) exchanger stoichiometry, potassium dependence, and voltage dependence; (ii) the exchanger's kinetic mechanism (is it consecutive or simultaneous?); (iii) purification and identification of the exchanger protein(s); (iv) the physiological role of Na-Ca exchange (despite new data, there is still controversy about the importance of the exchanger in platelets, in vascular smooth muscle, in skeletal muscle, and in Na^+-transporting epithelia).

Under the auspices of the New York Academy of Sciences, a Second International Conference on Sodium-Calcium Exchange was convened in April of 1991 to discuss these issues and try to resolve some of these controversies and to set an agenda for research in this field for the next several years. The year 1991 is especially appropriate because it marks the silver anniversary of the discovery: Na-Ca exchange was first identified in mammalian cardiac muscle from tracer flux experiments carried out by Reuter and Seitz in Mainz, Germany, in the fall of 1966 and, at the very same time and totally independently, in squid giant axons from tracer flux experiments performed by Baker, Blaustein, Hodgkin, and Steinhardt in Plymouth, England.

The staff of the New York Academy of Sciences provided invaluable assistance with all the administrative aspects of the conference. We are most grateful to Renée Wilkerson and Geri Busacco for their skill and efficiency in navigating us through the rocky shoals of conference organization.

Our Advisory Committee (Drs. Jeff Allen, Jon Lederer, Peter McNaughton, Hannah Rahamimoff, and Harald Reuter) made many very worthwhile suggestions about speakers and session format. The input of the New York Academy of Sciences Conference Committee, under the chairmanship of Dr. Charles Nicholson, was greatly appreciated, as was the very valuable guidance of our Conference Subcommittee (Drs. David Gadsby, Walter Scott, and Eric Simon).

The contributions of the speakers, poster presenters, and all other participants were outstanding. This, combined with the incisive leadership of the session and round-table

[a] ALLEN, T. J. A., D. NOBLE & H. REUTER, 1989. Sodium-Calcium Exchange. Oxford University Press. Oxford, U.K.

chairs (Drs. Ernesto Carafoli, Fredric Fay, Peter McNaughton, Denis Noble, and Harald Reuter) made for a very successful conference that was both lively and memorable.

Excellent secretarial assistance was provided by Arlene L. Reninger and Katherine E. Taylor.

Finally, we thank Linda Mehta for her excellent editorial efforts in helping us to prepare this written record of the conference.

DEDICATION

This volume is dedicated to some of the key figures in the Na-Ca exchange story: Professor Alan L. Hodgkin, who was unable to attend the conference, and to the memory of Peter F. Baker, Brian J. Nunn, and Peter Läuger.

MORDECAI P. BLAUSTEIN
REINALDO DIPOLO
JOHN P. REEVES

Kinetics, Stoichiometry, and Mechanism of Sodium-Calcium Exchange

Introduction to Part I

One of the major findings presented here is the difference between the properties of the Na-Ca exchanger in retinal rods and in other tissues. The exchanger in retinal rods has an absolute requirement for K^+ ions that are transported along with Ca^{2+}: The stoichiometry for this exchanger is 4 Na^+ : (1 Ca^{2+} + 1 K^+). This contrasts with the exchanger in mammalian cardiac muscle and in invertebrate "skeletal" muscle and neurons, in which the stoichiometry is 3 Na^+ : 1 Ca^{2+}. Although K^+ ions activate the exchanger in these other preparations, the K^+ ions are not transported by the exchanger. These observations are consistent with the evidence, presented elsewhere in this volume, that the retinal rod exchanger is structurally different from the cardiac muscle exchanger.

The countertransport mechanism (consecutive exchange of Na^+ and Ca^{2+} versus simultaneous exchange of the two countertransported species) remains controversial in spite of intense efforts to resolve this problem. Very careful studies from several different laboratories have yielded inconclusive results because some observations are most consistent with a consecutive mechanism, while other data appear to fit best with a simultaneous mechanism.

The photolysis of "caged" Ca^{2+}, leading to a very rapid rise in cytosolic Ca^{2+}, is reported to demonstrate intramembrane charge movement associated with the Ca^{2+} translocation step of the exchange cycle. This novel approach is also providing information about the rapid kinetics of the exchanger.

Fundamental Properties of the Na-Ca Exchange

An Overview[a]

P. A. McNAUGHTON

Physiology Group
Biomedical Sciences
King's College London
Strand, London WC2R 2LS, United Kingdom

The three main aspects of the Na-Ca exchange that have attracted the most attention in recent years have been (i) the stoichiometry of the exchange, and the related question of how much charge is transferred per exchange cycle; (ii) the mechanism of ion translocation, and in particular the location of ion-binding sites and the question of whether binding and transport of ions occurs simultaneously or sequentially; and (iii) the molecular structure of the exchange molecule itself. I shall review progress in our understanding of the first two aspects of the exchange; the molecular structure, which is reviewed in more detail elsewhere in this volume, is mentioned only in passing.

STOICHIOMETRY OF THE EXCHANGE

The stoichiometry of the exchange has for some time been widely assumed to be $3Na^+ : 1Ca^{2+}$, mainly on the basis that an exchange of this stoichiometry would be sufficiently powerful to attain the levels of approximately 10^{-8}M free $[Ca^{2+}]$ found in many cells, given the known gradients of $[Na^+]$ and typical values of membrane potential.[1,2] The first clear evidence for a $3Na^+ : 1Ca^{2+}$ stoichiometry was obtained in vesicles isolated from the sarcolemma of cardiac muscle by Reeves and Hale.[3] These authors varied the transmembrane $[Na^+]$ gradient until the net calcium flux across the membrane was zero, and obtained a cubic relationship between $[Ca^{2+}]_o$ and $[Na^+]_o$, as expected for a $3Na^+ : 1Ca^{2+}$ exchange.

Extensive studies on the exchange stoichiometry have been carried out using the outer segments of vertebrate rods. The outer segments of the rods from vertebrate retina may seem an unusual preparation in which to study Na : Ca exchange, but they do in fact offer a number of advantages.[4,5] The pathways for ion flow across the membrane are uniquely simple among higher cells, because instead of the usual zoo of ionic channels, the outer segment contains just one of any significance: the light-sensitive channel itself. The only other current-carrying mechanism is the Na : Ca exchange (or, as we now know, the Na : Ca,K exchange), which extrudes the calcium entering through the light-sensitive channel and which therefore plays an important role in light adaptation. A known amount of calcium can be loaded into the outer segment by simply recording the current that flows through the light-sensitive channel

[a] The parts of this work carried out by the author and colleagues were supported by the Medical Research Council (U.K.) and by NATO.

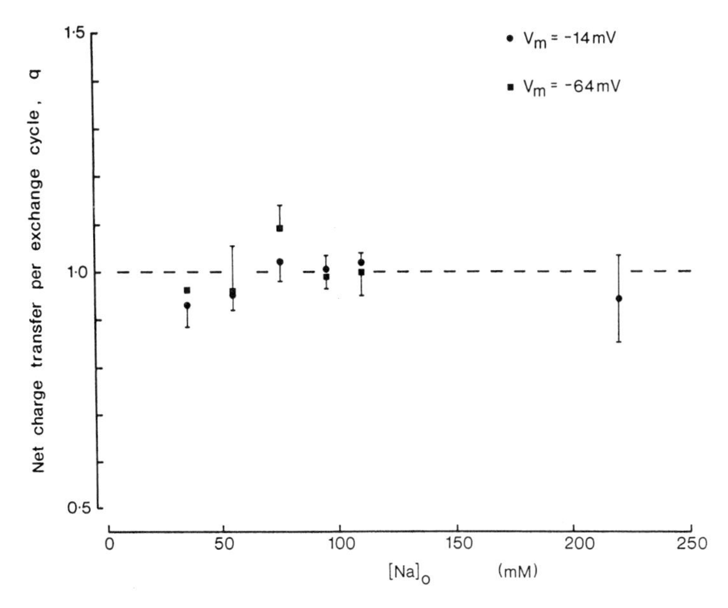

FIGURE 1. The net charge movement through the Na-Ca,K exchange under a variety of conditions. Each point is the value of q, the charge flowing into the cell per Ca^{2+} extruded, at the external $[Na^+]$ indicated, either at −14 mV (●) or −64 mV (■). All the results shown were obtained in solutions containing $0Ca^{2+}$, but the results in the presence and absence of external K^+ (2.5 mM) and Mg^{2+} (1.6 mM) are collected together, as these ions did not appear to affect the value of q. Bars show ± SEM. Over the range of conditions tested, the value of q did not differ significantly from a value of unity. (Taken from Lagnado & McNaughton[5]; used with permission.)

when the cell is exposed to a solution containing only calcium. The current carried by the exchange can then be recorded in isolation after the light-sensitive channels have been closed by a bright flash of light.

Experiments using this basic protocol showed that approximately one charge flows into the cell for every calcium ion that is extruded.[6,7] Recent measurements on voltage-clamped outer segments[8] have shown that the stoichiometry is 1.005 ± 0.01 charges exchanging for every Ca^{2+} over a wide range of ionic conditions and membrane potentials.[8] The result is shown in FIGURE 1.

These results are obviously consistent with a $3Na^+ : 1Ca^{2+}$ exchange, so we were surprised to discover that the exchange obstinately refused to work in the absence of K^+ ions on the same side as Ca^{2+}. From a series of equilibrium measurements similar in principle to those carried out by Reeves and Hale,[3] we were able to show[9] that the exchange in photoreceptors has a stoichiometry of $4\ Na^+ : 1Ca^{2+},1K^+$. The results of these experiments are shown in FIGURE 2.

In an independent series of experiments, Schnetkamp and collaborators showed by a different method that K^+ is cotransported with Ca^{2+} in the photoreceptor exchange.[10–12] They observed the ratio between charge moved in exchange for a Ca^{2+} ion transported by allowing the exchange to operate in isolated rod outer segments in which a current return path had been provided by applying the protonophore FCCP. The number of charges flowing into the outer segment is then equal to the number of

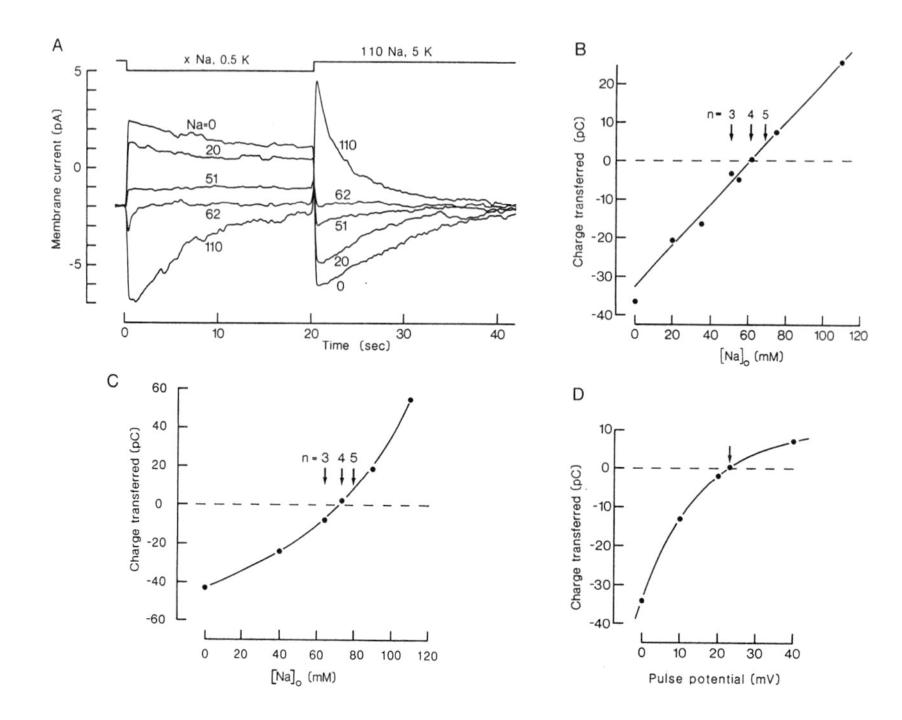

FIGURE 2. Determination of the stoichiometry of the Na-Ca,K exchange in rod outer segments.

(A) Currents observed during a 20-sec exposure to a solution containing 0.5 mM K, 1 mM Ca, and the stated concentration of Na; rod otherwise maintained in 100 mM Na, 5 mM K, and 1 mM Ca.

(B) Relation between $[Na]_o$ during the exposures shown in A and charge flow on return to 110 mM $[Na]_o$. Straight line, drawn by eye, crosses the axis at $[Na]_o = 62$ mM. Arrows show predicted crossing points for exchange stoichiometries of $nNa^+ : 1\ K^+$, where $n = 3$, 4, and 5. Similar results obtained in four experiments, with a mean $[Na^+]_o$ for zero charge transfer of 62.7 ± 0.8 mM (mean ± SEM).

(C) Similar experiment in which $[Ca^{2+}]_o$ and $[Na^+]_o$ were changed simultaneously. Rod was maintained in 110 mM Na, 5 mM Ca, 5 mM K; $[Ca^{2+}]_o$ was reduced to 1 mM, and $[n]$ reduced to 1 mM and $[Na^+]_o$ reduced to values shown on abscissa for 100 sec. Ordinate shows charge transferred on return to control solution. Continuous curve, drawn by eye, crosses the axis at $[Na^+]_o = 72$ mM. Arrows show expected null points for $nNa^+ : 1\ Ca^{2+}$, where $n = 3$, 4, and 5. Similar results obtained in three experiments, with mean $[Na^+]_o$ for zero charge transfer of 73.0 ± 0.6 mM.

(D) Effect of simultaneous change in membrane potential and $[K^+]_o$ on charge transfer. Rod maintained in 110 mM Na, 1 mM Ca, 5 mM K, and $[K^+]_o$ reduced to 2 mM for 40 sec. Depolarizing pulse coincided with the solution change. Ordinate shows the charge transferred on return to control conditions, and abscissa shows the pulse amplitude (from a holding potential of −14 mV). Arrow shows the expected change in V_m to null a 2.5-fold change in $[K^+]_o$ if one charge is countertransported for one K^+. (Taken from Cervetto *et al.*[9]; used with permission.).

H^+ ions flowing out. By repeating the experiment in a solution containing EGTA, which releases $2H^+$ per Ca^{2+} bound, they showed that one charge flows in for every Ca^{2+} extruded, as was found in the electrical measurements. The exchange was also found to depend absolutely on the presence of K^+ on the same side as Ca^{2+}. There is only one major disagreement between these results and those from the electrical measurements outlined above: When a return path is not provided by the addition of FCCP, the exchange still seems able to operate, though at a reduced rate, and Schnetkamp *et al.* propose that the exchange can slip into an electroneutral $3Na^+ : 1Ca^{2+}, 1K^+$ mode when the outer segment is strongly depolarized. Another possible explanation, though, is that endogenous pathways for the return of current exist (leakage channels in the outer segment membrane) and that the exchange continues to operate in an electrogenic mode even when no return path for current is provided by adding FCCP. It would be helpful to measure the membrane potential of these outer segments, because for membrane potentials as depolarized as +40 mV, we know that the exchange continues to operate in electrogenic mode with one charge transferred per Ca^{2+} extruded[8] (see FIG. 1).

The cotransport of Ca^{2+} and K^+ confers a powerful additional boost on the photoreceptor exchange.[9] An important gain (about a factor of 10) comes from the addition of an extra Na^+, but the main gain (about a factor of 50) comes from recruiting the K^+ gradient. A $4Na^+ : 1Ca^{2+}, 1K^+$ stoichiometry is therefore about 500 times more powerful than a $3Na^+ : 1Ca^{2+}$ exchange (the exact figure depends, of course, on the steepness of the transmembrane Na^+ and K^+ gradients).

Why does the photoreceptor exchange cotransport Ca^{2+} and K^+ in exchange for Na^+, while in the cardiac muscle membrane a simple exchange of Na^+ and Ca^{2+} is adequate? The answer probably lies in the uniquely unfavorable conditions in the photoreceptor outer segment. Not only is the resting membrane potential rather depolarized (about −35 mV in dark) but the intracellular $[Na^+]$ is substantially elevated by the large influx of Na^+ through the light-sensitive channels and by the diffusional barrier presented to Na^+ movement to the inner segment (where the $Na^+ : K^+$ pumps are located) by the stacked baffles of the discs.[13] The solution adopted by red cells in the blood—to extrude Ca^{2+} by an ATP-driven Ca^{2+} pump—may be unacceptable because there is no biochemical machinery for producing ATP within the outer segment; the mitochondria, together with most of the other functions of a normal neuron, are located in the inner segment.

The discovery that K^+ is cotransported with Ca^{2+} in photoreceptors has stimulated a number of groups to reexamine the exchange in other tissues (see Perry & McNaughton[14] for a review). In cardiac muscle Noma and coworkers[15] had found a reversal potential consistent with a $3Na^+ : 1Ca^{2+}$ exchange; further experiments by Yasui and Kimura, specifically designed to examine the possibility of a K^+ dependence, found no evidence for such an effect.[16] Crespo, Grantham, and Cannell[17] examined the reversal potential of the exchange in cardiac muscle by measuring intracellular $[Ca^{2+}]$. Once again, they found the reversal potential to be well predicted by a $3Na^+ : 1Ca^{2+}$ exchange, and in an explicit test for K^+ dependence they found no effect. Finally, in squid giant axon Condrescu *et al.*[18] found that when the reversed exchange was stimulated by the application of external Rb^+ and Ca^{2+} influx rose but that the Rb^+ influx did not. The well-known stimulatory effect[1] of monovalent cations applied on the same side as Ca^{2+} does not, therefore, seem to involve transport of the stimulating ion, a situation that is quite different from that in photoreceptors.

One study has, however, uncovered evidence for an involvement of K^+ transport in the Na-Ca exchange from brain synaptomsomes. Rahamimoff and collaborators[19] found that part of the Ca^{2+} influx from synaptomsomes, under conditions when the exchange was operating in the reversed mode, depended on Rb^+ (they used Rb^+ as a

K^+ surrogate), and that Rb^+ was cotransported with Ca^{2+}. The experiments showed that one Rb^+ was transported per five to eight Ca^{2+}. The most likely explanation of this result is that in the majority of cells in the brain the Na-Ca exchange molecules operate independently of the K^+ gradient with a $3Na^+ : 1Ca^{2+}$ stoichiometry, while in some cells, perhaps those operating with especially unfavorable Na^+ gradients or with a requirement for a particularly powerful Ca^{2+} extrusion mechanism, the mode of $4Na^+ : 1Ca^{2+},1K^+$ is used. It will be most interesting to discover which particular cells require the 4Na : 1Ca,K mode; this should be easy to determine once antibodies to both exchange molecules have been produced.

The conclusion arrived at in these studies is that there are two quite distinct Na-Ca exchange mechanisms: One, found in heart, squid giant axon, and throughout much of the brain, has a stoichiometry of $3Na^+ : 1Ca^{2+}$; another, properly called the Na-Ca,K exchange,[9] is found in photoreceptor outer segments and in other (as yet unidentified) cells of the central nervous system, and has a stoichiometry of $4Na^+ : 1Ca^{2+},1K^+$. The idea that these are two quite distinct molecules has recently been confirmed by determination of their primary sequence: The sequence of the photoreceptor Na-Ca,K exchange (reported elsewhere in this volume in work from the laboratory of N. J. Cook) is quite different from the published sequence of the cardiac Na-Ca exchange.[20] Although there is little or no sequence homology between these two molecules, a considerable degree of structural homology exists, and it is possible that the two exchange molecules diverged from a common progenitor many millions of years ago. Of one thing there can now be no doubt, though: There are at least two quite distinct Na^+-linked Ca^{2+} transport mechanisms, a conclusion that was not at all apparent two years ago.

LOCATION OF ION-BINDING SITES

In the most intuitively simple models of the exchange, the ion-binding sites are located at the membrane surface, and after binding has taken place the ion is translocated across the full width of the membrane and of its transmembrane potential difference. The ion is often imagined as moving attached to its binding site, but a physical movement across the whole membrane of the binding site—which is after all part of a large protein—seems unlikely, and more plausible models would involve diffusion of the ion from one binding site to another through a channel within the exchange protein.

There is, however, no reason why the binding site should not be located partway across the membrane, so that the ion has access to it through a short channel.[21] Some evidence supports this idea in the case of the Na-binding sites of the photoreceptor Na-Ca,K exchange[22] (see FIG. 3A). In this experiment the voltage dependence of the exchange was measured at various external sodium concentrations, and it was found that the exchange rate became independent of membrane potential when the external Na-binding site was saturated. One possible explanation for these results is shown in FIGURE 3B: The voltage sensitivity of the exchange may be conferred by a voltage dependence of Na binding, which could arise if the binding site was located in a channel and therefore sensed a fraction of the transmembrane voltage. The fractional occupancy of the binding site would therefore be increased by hyperpolarization at low $[Na]_o$, while at high $[Na]_o$ the binding site would be saturated irrespective of membrane potential, and the voltage sensitivity would therefore disappear.

The evidence for a channel associated with the Na-binding site prompted us to examine other binding sites to see if a similar voltage dependence could be observed.[23]

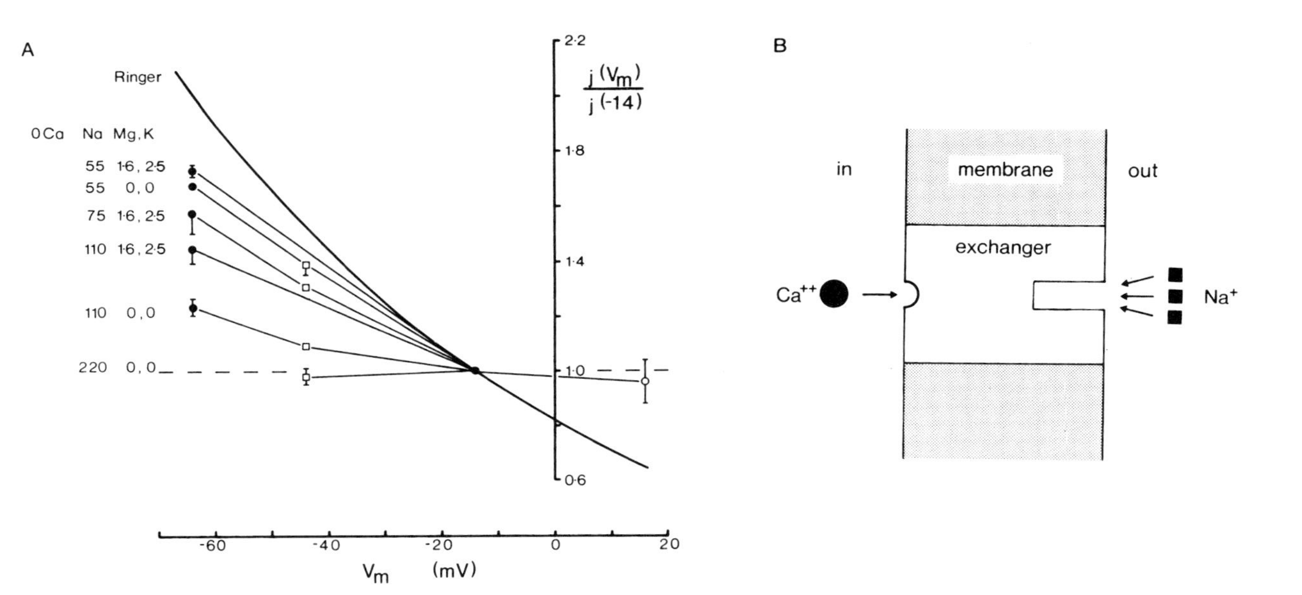

FIGURE 3. Voltage dependence of Na-Ca exchange current under various external ionic conditions. **(A)** Exchange current at a potential V_m is plotted relative to its magnitude at −14 mV. The curve is that fitted to the data in Ringer's solution obtained in separate experiments. External ionic concentrations (mM) are shown beside each set of points, which have been connected by straight lines. Bars show ±SEM. **(B)** Schematic diagram of the Na-Ca exchange molecule showing the proposed locations of the Na^+- and Ca^{2+}-binding sites of the unloaded carrier. Some or all of the Na-binding sites are located within the membrane electric field. The internal Ca^{2+}-binding site is outside the membrane electric field. This diagram does not, for simplicity, show the binding of the K^+ ion at the internal membrane surface. (Modified from Lagnado *et al.*[22])

One particularly interesting possibility is that there might be only a single binding site for each ion, located a fixed distance across the membrane. If this was true the voltage sensitivity for ion binding at one side should be predictable from that observed for binding at the other. For instance, in the case of the binding of calcium, the affinity at the internal membrane surface is known to be independent of membrane potential,[22] which would predict a strong dependence of the external binding if external calcium must diffuse down a long channel in order to access its binding site near the internal membrane surface. In the event, no voltage dependence of binding of external calcium was observed,[23] and a similar result was obtained for external K. We conclude that the simple model of a single, fixed binding site fails, at least for the binding of Ca and K, and there must therefore be two separate binding sites for each of these ions, one at the external surface and one at the internal surface. Whether these two sites are connected by a channel or whether a single site moves physically between two locations has, of course, yet to be determined.

SIMULTANEOUS OR SEQUENTIAL?

Two main classes of models for the exchange have been considered: ones in which ions bind independently to both sides of the membrane before translocation occurs (simultaneous models) and ones in which binding occurs first on one side and then the other (sequential models).[7,24,25] There are, of course, a number of other permutations and combinations, and these two models simply represent the extremes.[26,27] One prediction of a simple simultaneous model is that the affinity of an ion-binding site should be unaffected by ion concentrations at the opposite membrane surface. This seems to be true for the internal Ca-binding site: The affinity is not greatly affected by changes in $[Na^+]_o$ in squid axons,[24,25] while in photoreceptors the affinity is independent of external $[Na^+]$, $[K^+]$, $[Mg^{2+}]$, $[Ca^{2+}]$, and of changes in membrane potential.[22,28]

In the cardiac muscle exchange, though, Li and Kimura[29] have recently shown that the external Ca^{2+} affinity of the exchange operating in the reversed direction is strongly affected by changes in $[Na^+]_i$, with elevations in $[Na^+]_i$ reducing the affinity for external Ca^{2+}. This is exactly what one would expect for a sequential mechanism: Raising $[Na^+]_i$ tends to pull the exchange binding sites across to the internal membrane surface, thereby reducing the availability of Ca-binding sites at the external membrane surface and consequently reducing the apparent affinity for external Ca. This result seems clear enough, but how it can be reconciled with the earlier evidence suggestive of a simultaneous mechanism has yet to be resolved.

REFERENCES

1. Baker, P. F., M. P. Blaustein, A. L. Hodgkin & R. A. Steinhardt. 1969. The influence of calcium on sodium efflux in squid axons. J. Physiol. **200:** 431-458.
2. Blaustein, M. P. & A. L. Hodgkin. The effect of cyanide on the efflux of calcium from squid axons. J. Physiol. **200:** 497-527.
3. Reeves, J. P. & C. C. Hale. 1984. The stoichiometry of the cardiac sodium-calcium exchange system. J. Biol. Chem. **259:** 7733-7739.
4. Lagnado, L. & P. A. McNaughton. 1989. The sodium : calcium exchange in photoreceptors. *In* The Sodium-Calcium Exchange. T. J. A. Allen, D. Noble & H. Reuter, Eds. Oxford University Press. Oxford, England.
5. Lagnado, L. & P. A. McNaughton. 1990. The electrogenic properties of the Na : Ca exchange. J. Membr. Biol. **113:** 177-191.

6. YAU, K.-W. & K. NAKATANI. 1984. Electrogenic Na-Ca exchange in retinal rod outer segment. Nature **311:** 661-663.
7. HODGKIN, A. L., P. A. MCNAUGHTON & B. J. NUNN. 1987. Measurement of sodium-calcium exchange in salamander rods. J. Physiol. **391:** 347-370.
8. LAGNADO, L. & P. A. MCNAUGHTON. 1991. Net charge transport during sodium-dependent calcium extrusion in isolated salamander rod outer segments. J. Gen. Physiol. In press.
9. CERVETTO, L., L. LAGNADO, R. J. PERRY, D. W. ROBINSON & P. A. MCNAUGHTON. 1989. Extrusion of calcium from rod outer segments is driven by both sodium and potassium gradients. Nature **337:** 740-743.
10. SCHNETKAMP, P. P. M., D. K. BASU & R. T. SZERENCSEI. 1989. Na^+-Ca^{2+} exchange in bovine rod outer segments requires and transports K^+. Am. J. Physiol. **257:** C153-C157.
11. SCHNETKAMP, P. P. M. & R. T. SZERENCSEI. 1991. Effect of potassium ions and membrane potential on the Na-Ca-K exchanger in isolated intact bovine outer segments. J. Biol. Chem. **266:** 189-197.
12. SCHNETKAMP, P. P. M., R. T. SZEERENCSEI & D. K. BASI. 1991. Unidirectional Na^+, Ca^{2+} and K^+ fluxes through the bovine rod outer segment Na-Ca-K exchanger. J. Biol. Chem. **266:** 198-206.
13. LAMB, T. D., P. A. MCNAUGHTON & K.-W. YAU. 1981. Longitudinal spread of activation and background desensitization in toad rod outer segments. J. Physiol. **319:** 463-496.
14. PERRY, R. J. & P. A. MCNAUGHTON. 1991. Calcium regulation in neurones: Transport processes. Curr. Opinion Neurobiol. **1:** 98-104.
15. EHARA, T., S. MATSUOKA & A. NOMA. 1989. Measurement of reversal potential of Na^+-Ca^{2+} exchange current in single guinea-pig ventricular cells. J. Physiol. **410:** 227-249.
16. YASUI, K. & J. KIMURA. 1990. Is potassium co-transported by the cardiac Na-Ca exchanger? Pfluegers Arch. **415:** 513-515.
17. CRESPO, L. M., C. J. & M. B. CANNELL. 1990. Kinetics, stoichiometry and role of the Na-Ca exchange mechanism in isolated cardiac myocytes. Nature **345:** 618-621.
18. CONDRESCU, M., H. ROJAS, A. GERARDI, R. DIPOLO & L. BEAUGÉ. 1990. In squid nerve fibres, monovalent activating cations are not co-transported during Na^+/Ca^{2+} exchange. Biochim. Biophys. Acta **1024:** 198-202.
19. DAHAN, D., R. SPANIER & H. RAHAMIMOFF. 1991. The modulation of rat brain Na^+-Ca^{2+} exchange by K^+. J. Biol. Chem. **266:** 2067-2075.
20. NICOLL, D. A., S. LONGONI & K. D. PHILIPSON. 1990. Molecular cloning and functional expression of the cardiac sarcolemmal Na^+-Ca^{2+} exchanger. Science **250:** 562-656.
21. LAUGER, P. 1987. Voltage dependence of the sodium : calcium exchange: Predictions from kinetic models. J. Membr. Biol. **99:** 1-11.
22. LAGNADO, L., L. CERVETTO & P. A. MCNAUGHTON. 1988. Ion transport by the Na : Ca exchange in isolated rod outer segments. Proc. Natl. Acad. Sci. USA **85:** 4548-4552.
23. PERRY, R. J. & P. A. MCNAUGHTON. 1991. Characteristics of the Ca_o and K_o binding sites of the Na-Ca,K exchange in isolated salamander rod outer segments. J. Physiol. **434:** 70P.
24. BAKER, P. F. & P. A. MCNAUGHTON. 1976. Kinetics and energetics of calcium efflux from intact squid giant axons. J. Physiol. **259:** 103-144.
25. BLAUSTEIN, M. P. 1977. Effects of internal and external cations and of ATP on sodium-calcium and calcium-calcium exchange in squid axons. Biophys. J. **20:** 79-110.
26. HILGEMANN, D. W. 1989. Numerical probes of Na : Ca exchange. *In* The Sodium-Calcium Exchange. T. J. A. Allen, D. Noble & H. Reuter, Eds. Oxford University Press. Oxford, England.
27. HILGEMANN, D. W. 1988. Numerical probes of sodium-calcium exchange. Progr. Biophys. **51:** 1-45.
28. HODGKIN, A. L. & B. J. NUNN. 1987. The effects of ions on sodium-calcium exchange in salamander rods. J. Physiol. **391:** 371-398.
29. LI, J. & J. KIMURA. 1990. Translocation mechanism of Na : Ca exchange in single cardiac cells of guinea pig. J. Gen. Physiol. **96:** 777-788.

The Stoichiometry of Na-Ca+K Exchange in Rod Outer Segments Isolated from Bovine Retinas[a]

PAUL P. M. SCHNETKAMP,[b] DEBESH K. BASU, AND ROBERT T. SZERENCSEI

Department of Medical Biochemistry
University of Calgary
Calgary, Alberta T2N 4N1, Canada

INTRODUCTION

The outer segments of retinal rods (ROS) exhibit dynamic Ca^{2+} fluxes; in the dark, Ca^{2+} enters ROS via the light-sensitive and cGMP-gated channels and Ca^{2+} is extruded via the Na-Ca exchanger.[1] The presence of Na-Ca exchange in retinal rods was first described by measurements of Na^+- and Ca^{2+}-stimulated ^{45}Ca fluxes in bovine ROS.[2-4] Subsequently, the suction electrode technique was used to measure Na-Ca exchange currents in amphibian ROS and to determine the electrogenicity of Na-Ca exchange.[5] ROS proved to be a useful preparation for the study of Na-Ca exchange, because the density of the Na-Ca exchanger protein in the ROS plasma membrane is high, as judged from Na-Ca exchange fluxes[6,7] and from purification of the exchanger protein.[8,9] The Na-Ca exchanger predominantly resides in the ROS plasma membrane and not in the internal disk membranes, since disruption of intact ROS greatly reduces Na-Ca exchange fluxes[3,10] and since antibodies against the Na-Ca exchange protein specifically label the ROS plasma membrane.[11] The latter study also demonstrated that the Na-Ca exchanger is localized to the outer segment and is not found in other parts of the rod cell. A curious observation is that large Na-Ca exchange fluxes can be measured in ROS with a plasma membrane that appears to be leaky to small solutes such as ATP[2] or didansylcysteine.[10]

On a functional level, the ROS Na-Ca exchanger initially appeared to be quite similar to those observed in other tissues except for the effects of other alkali cations, most noticeably K^+, on Na-Ca and Ca-Ca exchange fluxes (reviewed in Schnetkamp[7]). A more profound difference was revealed recently when it was reported that the ROS Na-Ca exchanger requires and transports K^+ in the Na-Ca exchange mode.[12-14] This conclusion was based on observations that:

1. Outward Na-Ca exchange current[12] as well as Ca^{2+} influx,[14] both indicating reverse Na-Ca exchange, were only observed when external K^+ was present.

[a]This research was supported through a grant from the Medical Research Council of Canada. PPMS is a Medical Scholar of the Alberta Heritage Foundation for Medical Research. DKB is a recipient of a studentship from the Alberta Heritage Foundation for Medical Research.

[b]Address for correspondence: Department of Medical Biochemistry, University of Calgary, 3330 Hospital Dr. N.W., Calgary, Alberta T2N 4N1, Canada.

2. External K^+ concentration influenced the direction of Na-Ca exchange current.[12]
3. Na^+-stimulated Ca^{2+} release was accompanied by a stoichiometric K^+ release.[14]

Both studies[12,14] concluded that Na-Ca exchange in ROS most likely operates at a 4Na : (1Ca + 1K) stoichiometry. Subsequent, more detailed, studies in bovine ROS described the effects of K^+ ions on Na-Ca exchange and on K^+ transport via Na-Ca and Ca-Ca exchange: Considerable flexibility in the Na : Ca and Ca : K transport ratios was observed and K^+-independent Na-Ca and Ca-Ca exchange fluxes were observed as well.[15-17] In this paper we will discuss our experiments on the stoichiometry and electrogenicity of Na-Ca exchange in bovine ROS.

METHODOLOGICAL CONSIDERATIONS

Preparations

Stoichiometry determinations require measurements of large Na-Ca exchange fluxes, unlike the small net fluxes observed under physiological conditions when the exchanger operates close to equilibrium and is exposed to the low and subsaturating free cytosolic Ca^{2+} concentration. In our studies we have used two different preparations of intact ROS isolated from bovine retinas[7,15,18] : Ca^{2+}-rich ROS are greatly enriched in Ca^{2+} and contain about 24 mM total Ca^{2+} at a free -Ca^{2+} concentration of about 20-30 μM, whereas Ca^{2+}-depleted ROS contain 40-50 mM total Na^+ but no detectable Ca^{2+}. Ca^{2+}-rich ROS were used to study Na^+-stimulated Ca^{2+} release or inward Na-Ca exchange current as well as Ca-Ca and K-K exchange, whereas Ca^{2+}-depleted ROS were used to study Ca^{2+} influx or outward Na-Ca exchange current as well as Na-Na exchange.

Flux Measurements

A number of different analytical techniques were used to measure Na^+, K^+, and Ca^{2+} fluxes via Na-Ca+K exchange in bovine ROS; detailed descriptions can be found in the indicated references:

1. Atomic absorption spectroscopy is the most direct analytical technique to measure changes in Na^+, Ca^{2+}, and K^+ concentration with as major drawbacks the large amount of ROS required and limitations in kinetic resolution and sensitivity.[3,14,17]
2. ^{45}Ca, ^{86}Rb, and ^{22}Na fluxes were measured with a rapid filtration technique over borosilicate glass fiber filters.[3,4,17] Radioisotope fluxes offer adequate kinetic resolution, excellent sensitivity, and are the only tool to measure self-exchange fluxes; quantitative measurements of ion fluxes, however, require precise knowledge of the "cold" concentrations of the cations involved as well as their compartmentalization within ROS.
3. Ca^{2+} influx in and efflux from ROS was measured with arsenazo III as Ca^{2+}-indicating dye in the external medium.[7,16] Arsenazo III and dual-wavelength spectrophotometry offer excellent kinetic resolution and very good sensitivity; changes in total Ca^{2+} concentration in the cuvette by as little as 0.01 μM can be detected.

4. The equivalent of Na-Ca exchange currents were measured in suspensions of bovine ROS as Na^+- or Ca^{2+}-induced proton fluxes carried by the electrogenic protonophore FCCP (carbonyl cyanide *p*-trifluoromethoxyphenylhydrazone). Proton fluxes via FCCP formed a current loop with the Na-Ca exchange current (see below) and were measured with the pH-indicating dye phenol red and dual-wavelength spectrophotometry.[14,15]
5. Another optical method to measure the equivalent of ionic currents in an ensemble of cells or vesicles in suspension utilizes the dye neutral red and dual-wavelength spectrophotometry.[18-20] Applied to intact bovine ROS, electrogenic Na-Ca+K exchange fluxes were measured with a resolution equivalent to a current of 0.01 pA.[21]
6. Free intracellular Ca^{2+} concentration in bovine ROS and its control via Na-Ca+K exchange can be measured with the fluorescent Ca^{2+}-indicating dye fluo-3 (manuscripts submitted).

RESULTS

In order to address the stoichiometry of the Na-Ca+K exchanger in bovine ROS, we will discuss three different experimental aspects of this problem separately. First, we will discuss the K^+ requirement of both Ca^{2+}- and Na^+-stimulated Ca^{2+} fluxes, then we will discuss Na^+- and Ca^{2+}-dependent K^+ transport, and, finally, we will discuss our measurements of the electrogenicity of Na-Ca+K exchange and present our conclusion on the overall stoichiometry of Na-Ca+K exchange.

Internal K^+ Requirement of Na^+-Stimulated Ca^{2+} Release

If Na-Ca+K exchange in bovine ROS can only operate at a fixed stoichiometry and with the simultaneous release of both Ca^{2+} and K^+, Na^+-stimulated Ca^{2+} release should be abolished when internal K^+ is completely removed. Internal K^+ can be removed from ROS by treatment with the ionophore gramicidin followed by washing. When K^+-free ROS were assayed for Na-Ca exchange, Na^+-stimulated Ca^{2+} release was still observed, albeit at a much reduced rate; Na^+-stimulated Ca^{2+} release could be reactivated by addition of K^+, although the degree of activation by K^+ decreased as the Na^+ concentration decreased.[16] Na^+-stimulated Ca^{2+} release in the absence of internal K^+ implies exchange stoichiometries without participation of K^+.

The use of the nonselective ionophore gramicidin to remove intracellular K^+ compromises the ability of the ROS plasma membrane to maintain alkali cation gradients. Here, we illustrate the internal K^+ requirement of Na^+-stimulated Ca^{2+} release using a less invasive way to remove internal K^+. The ion selectivity of the K^+ transport site of the exchanger enables us to substitute intracellular K^+ for Cs^+. Cs^+ is a very poor substitute for K^+ in activating Na^+-stimulated Ca^{2+} release,[16] but it can cause the slow release of internal K^+ (half-time of several minutes) via Cs-K exchange that accompanies Ca-Ca self-exchange.[17] In the experiment illustrated in FIGURE 1, we exposed Ca^{2+}-rich ROS to 10 mM CsCl for different time intervals. Subsequently, 50 mM NaCl was added to assay for rapid and slow components of Ca^{2+} release. In the absence of CsCl, the typical kinetics of Na^+-stimulated Ca^{2+} release in Ca^{2+}-rich

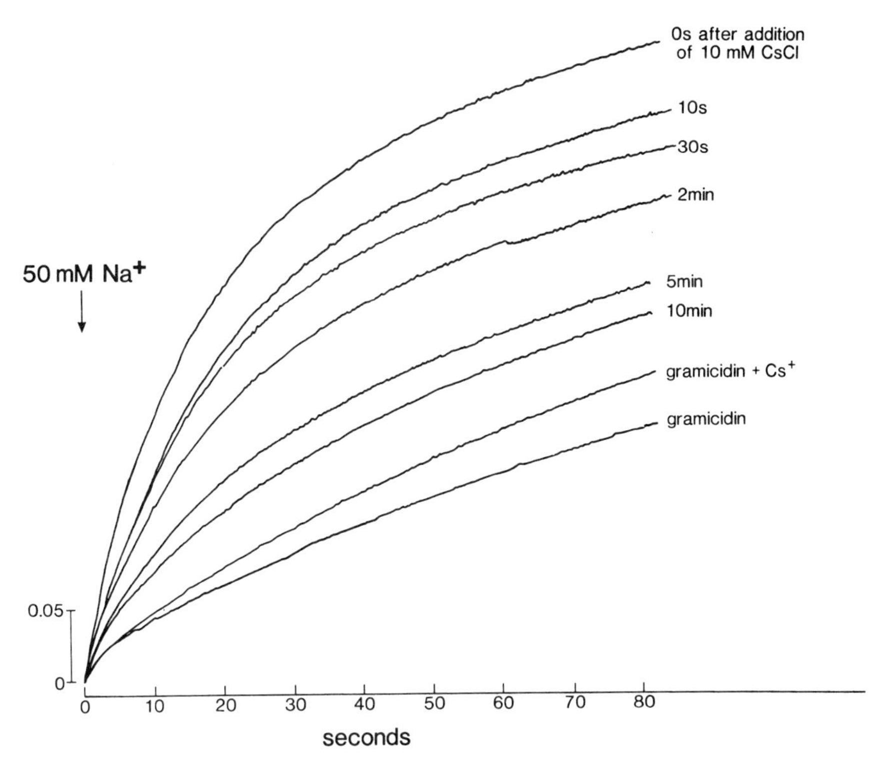

FIGURE 1. Internal K^+ requirement of Na^+-stimulated Ca^{2+} release. Ca^{2+}-rich ROS were incubated in 600 mM sucrose, 20 mM Hepes [4-(2-hydroxyethyl)-1-piperazineethanesulfonic acid] adjusted to pH 7.4 with arginine, 100 μM arsenazo III, 1 μM gramicidin (when indicated), and 10 mM CsCl (as indicated). The cuvette contained ROS to a final rhodopsin concentration of 4 μM. Na^+-stimulated Ca^{2+} release was initiated at time zero by addition of 50 mM NaCl and was indicated by the increase in light absorption (A_{650}-A_{750}) as described in detail elsewhere.[7] The calibration bar of 0.05 absorbance units represents a Ca^{2+} release of 0.54 mol Ca^{2+}/mol rhodopsin or a change in total internal Ca^{2+} concentration in ROS by 1.6 mM. Temperature: 25°C.

bovine ROS were observed: A rapid component lasted 20-30 sec and was followed by a persistent slow component. Exposure of ROS to 10 mM CsCl for time periods of increasing duration caused a progressive reduction in the amount of Ca^{2+} released by the rapid component of Na-Ca+K exchange, but had little effect on the persistent slow component of Na^+-stimulated Ca^{2+} release. After a ten-minute exposure to CsCl, Na^+-stimulated Ca^{2+} release was similar to that observed with gramicidin and CsCl present, indicating that at that point most internal K^+ had been replaced by Cs^+. Na^+-stimulated Ca^{2+} release in the presence of gramicidin alone represented most likely K^+-independent Na-Ca exchange. From the Cs^+-induced and time-dependent reduction in the amplitude of Na^+-stimulated Ca^{2+} release, we conclude that the rapid component of Na^+-stimulated Ca^{2+} release in bovine ROS requires intracellular K^+.

K^+ Transport via Na-Ca+K Exchange

We have used three different methods to demonstrate K^+ efflux associated with Na^+-stimulated Ca^{2+} release. First, atomic absorption measurements showed that addition of Na^+ to Ca^{2+}-rich ROS caused the stoichiometric release of both Ca^{2+} and K^+ with very similar kinetics and very similar Na^+ dependence.[14,17] A second method used ROS preloaded with ^{45}Ca and ^{86}Rb (the effects of K^+ and Rb^+ on Na-Ca exchange in ROS are very similar): Na^+ stimulated the release of both radioisotopes with very similar kinetics and a very similar sensitivity to inhibitors of Na-Ca exchange.[15] Finally, we measured the equivalent of inward Na-Ca exchange current and correlated the amount of charge transported with the amount of K^+ lost (see later FIG. 3).

A different approach to demonstrate K^+ transport via the Na-Ca+K exchange protein takes advantage of the observation that the ROS Na-Ca+K exchanger (like Na-Ca exchangers in other tissues) also performs Ca-Ca self-exchange.[3,4] Simultaneous release of Ca^{2+} and K^+ during the Na-Ca+K exchange cycle suggests that Ca-Ca self-exchange may actually be a combined Ca-Ca and K-K exchange. We measured influx of ^{45}Ca and ^{86}Rb as a function of external free Ca^{2+} concentration[17] and as a function of the presence of different inhibitors of Na-Ca exchange[15]; in both sets of experiments, ^{86}Rb influx closely followed the pattern of ^{45}Ca influx. Here, we present the kinetic analysis of the Ca^{2+}-dependence of ^{86}Rb and ^{45}Ca influx (FIG. 2). Equilibration of externally added "hot" radioisotope with the "cold" internal isotope of the same cation can be described by two opposing first-order reactions if no net cation flux occurs.[3] In the case of a single homogeneous pool of internal cations, the graphic presentation shown in FIGURE 2 is expected to yield straight lines with slopes that are proportional to the rate constant. When ^{86}Rb and ^{45}Ca influx into Ca^{2+}-rich ROS were analyzed this way, straight lines were observed with slopes that increased with increasing free external Ca^{2+} concentration. The above kinetic analysis yields several useful conclusions.

1. The rate constants for ^{86}Rb and ^{45}Ca influx, respectively, depended in a very similar manner on the external free-Ca^{2+} concentration, suggesting that influx for both isotopes is controlled by the same Ca^{2+}-dependent process, most likely binding of Ca^{2+} to the Ca^{2+} transport site of the Na-Ca+K exchange protein.
2. The straight lines indicate that a single rate constant (i.e., most likely transport via a single protein species) controls rapid entry of both ^{45}Ca and ^{86}Rb into a single and homogeneous internal pool containing both Ca^{2+} and K^+. A much slower entry into a second pool (most likely the intradiskal space) can be observed on a time scale of one hour (not illustrated).
3. The absence of ^{86}Rb influx in the presence of EDTA (ethylenediaminetetraacetic acid) suggests that the ROS plasma membrane does not contain functional K^+ channels or carriers other than Na-Ca+K exchange.

Electrogenicity of Na-Ca+K Exchange in Bovine ROS

Small cell particles such as mammalian ROS do not offer sufficient collecting area for electrophysiological measurements of the exchange current. Despite the fact that the exchanger can change total intracellular Ca^{2+} by as much as 0.5 mM/sec, maximal exchange currents of only 1-2 pA can be anticipated. Essential to our procedure for obtaining the stoichiometry of the Na-Ca+K exchange process is a protocol to measure the equivalent of Na-Ca+K exchange current in intact bovine ROS in suspension. Na^+-stimulated Ca^{2+} release of the magnitude observed in bovine ROS would carry a 1-2-pA inward exchange current, which, if not compensated, would depolarize the plasma membrane of bovine ROS by 2-3 V/sec. We added to the ROS suspension the electrogenic protonophore FCCP with the objective of forming a current loop between inward Na-Ca+K exchange current and an outward current of protons, carried by

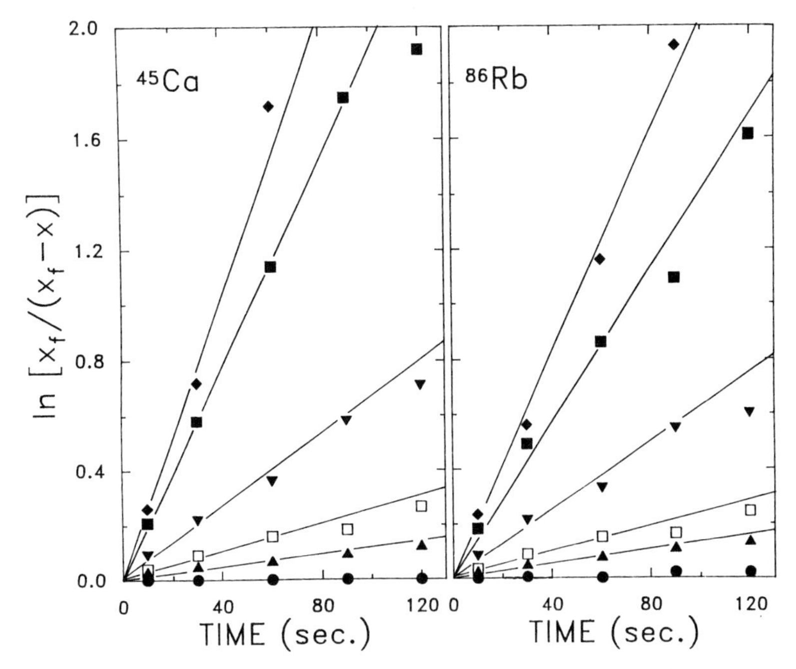

FIGURE 2. Kinetic analysis of Ca^{2+}-stimulated ^{45}Ca and ^{86}Rb influx. Ca^{2+}-rich ROS were incubated in 600 mM sucrose, 20 mM Hepes (pH 7.4), and 200 μM $CaCl_2$. Rhodopsin concentration was 25 μM. Radioisotope uptake was initiated by addition of ^{45}Ca and ^{86}Rb and measured by filtration through borosilicate glass fiber filters as described in detail in Schnetkamp *et al.* [17] Free-Ca^{2+} concentration was adjusted through application of the Ca^{2+} chelators EDTA, HEDTA (*N*-hydroxyethylethylenediaminetetraacetic acid), and NTA (nitrilotriacetic acid) to: <5 nM (filled circles); 0.17 μM (triangles); 0.38 μM (open squares); 1.0 μM (inverted triangles); 6.1 μM (filled squares); 200 μM (diamonds). X_f represents equilibrium radioisotope uptake after 5 minutes at 200 μM Ca^{2+}; X represents radioisotope uptake for each given time point and each given free-Ca^{2+} concentration. Values for X_f were 30% (^{45}Ca) and 24% (^{86}Rb) of the total amount of radioisotope added. Temperature: 25°C.

FCCP and measured with the pH-indicating dye phenol red and dual-wavelength spectrophotometry.[14,15] The central assumption in this protocol is that charge neutrality (at least to a very good approximation) is maintained for overall ion transport across the ROS plasma membrane. The capacitor equation dictates that for a cylindrical structure of 1 × 20 μm, a net inward current of 1 pA maintained for 1 sec would cause a depolarization of the ROS plasma membrane by 1.6 V. Charge neutrality of overall cation transport was tested when we compared gramicidin-induced Na^+ and K^+ release from ROS with gramicidin-induced proton influx measured in different aliquots of the same ROS suspension: We observed that the coupling ratio between the combined efflux of Na^+ and K^+ and influx of protons was 1.0, validating the principle of charge neutrality.[21] Na^+-stimulated efflux of either Ca^{2+}, K^+, or protons was not dependent on the anion used (chloride versus sulphate) in accordance with our earlier conclusion that the ROS plasma membrane is impermeable to these anions.[4,19]

The above-described procedure for measuring Na-Ca+K exchange current via a counter current of protons carried by FCCP is conceptually very similar to the procedure adopted to measure the light-sensitive current[22] or the Na-Ca exchange current[5,23] in ROS with the suction electrode technique. The suction electrode is positioned over the rod inner segment leaving the rod outer segment exposed to the experimental solution. The electrode measures K^+ currents across the rod inner segment plasma membrane that electrically compensate the currents of interest across the ROS plasma membrane.

Na^+-induced proton efflux carried by FCCP and measured with phenol red displayed very similar kinetics and a very similar sigmoidal dependence on the external Na^+ concentration when compared with Na^+-stimulated Ca^{2+} efflux measured with arsenazo III.[15] This result indicates that an outward proton current carried by FCCP electrically compensated for inward Na-Ca+K exchange current and that the Na-Ca+K exchanger was the exclusive carrier of inward current in bovine ROS under conditions that the cGMP-gated channel was closed. The above proton fluxes were measured in bovine ROS with a resolution equivalent to a current of 0.01 pA. To determine the electrogenicity of Na-Ca+K exchange, it is highly desirable to measure Na^+-induced proton release and Na^+-induced Ca^{2+} release under identical experimental conditions. At physiological pH, the Ca^{2+} chelator EGTA [ethylene glycol-bis (β-aminoethyl ether)-*N,N,N',N'*-tetraacetic acid] releases two protons for each Ca^{2+} it binds, whereas the Ca^{2+} chelator BAPTA [1,2-bis(2-aminophenoxy)ethane-*N,N,N',N'*-tetraacetic acid] does not. This circumstance can be used to measure Na^+-stimulated Ca^{2+} release with a pH-indicating dye when Ca^{2+} released to the external medium binds to EGTA and releases two protons. From a comparison between Na^+-induced acidification of the external medium with EGTA present and the Na^+-induced acidification observed with BAPTA present, we determined that the electrogenicity of Na-Ca+K exchange in bovine ROS was one positive charge carried by Na^+ for each Ca^{2+} released.[14,15]

Stoichiometry of Na-Ca+K Exchange in Bovine ROS

The principle of electroneutrality of overall ion transport in isolated bovine ROS not only suggested a protocol for measuring the electrogenicity of Na-Ca+K exchange in ROS, it also provided the cornerstone for our assessment of the overall transport stoichiometry. The kinetic and quantitative correlation between Na^+-induced proton fluxes (carried via FCCP) and Na^+-induced Ca^{2+} release indicates that Na-Ca+K exchange is the only electrogenic Na^+ transporter in the ROS plasma membrane.[14] Electroneutrality of overall ion transport implies that the stoichiometry of Na-Ca+K exchange can be obtained by a simple summation of the positive charges (Ca^{2+}, K^+, and protons) released during Na^+-stimulated Ca^{2+} release.[14,15] First, we discuss more direct evidence that the Na-Ca+K exchanger is the only functional ion transporter present in the plasma membrane of our preparation, isolated bovine ROS. We have used the radioisotopes ^{86}Rb, ^{22}Na, and ^{45}Ca to examine cation fluxes in bovine ROS. Both ^{86}Rb and ^{45}Ca fluxes were absolutely dependent on the presence of either Na^+ or Ca^{2+} in a manner suggesting that fluxes of these radioisotopes mediated by Na-Ca+K exchange exceeded those carried by other putative transporters by at least 50-fold.[17] Likewise, ^{22}Na fluxes were strongly stimulated by addition of "cold" Na^+, indicating that Na^+ transport in isolated bovine ROS is exclusively mediated by a cooperative mechanism such as Na-Ca exchange,[15] while complete inhibition of ^{22}Na influx by external Ca^{2+} is consistent with exclusive Na^+ transport via Na-Ca exchange as well.[17]

Na^+-stimulated Ca^{2+} efflux in the absence of ionophores caused the release of one Ca^{2+} ion plus one K^+ ion with no release of protons or Mg^{2+}, leading to a stoichiometry of 3Na : (1Ca + 1K). In the presence of FCCP, Na^+-stimulated Ca^{2+} efflux caused

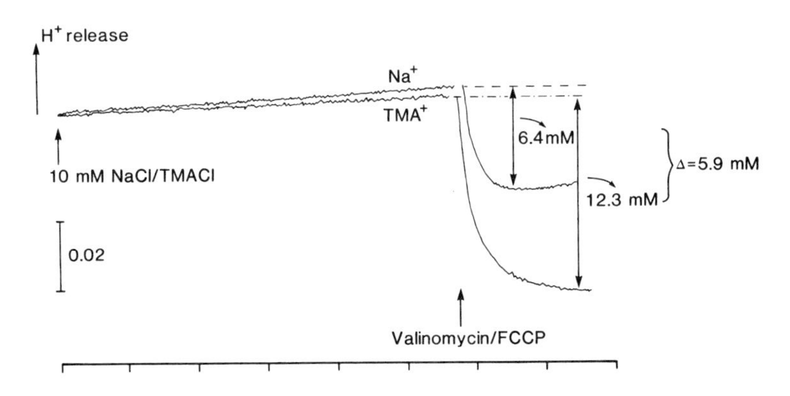

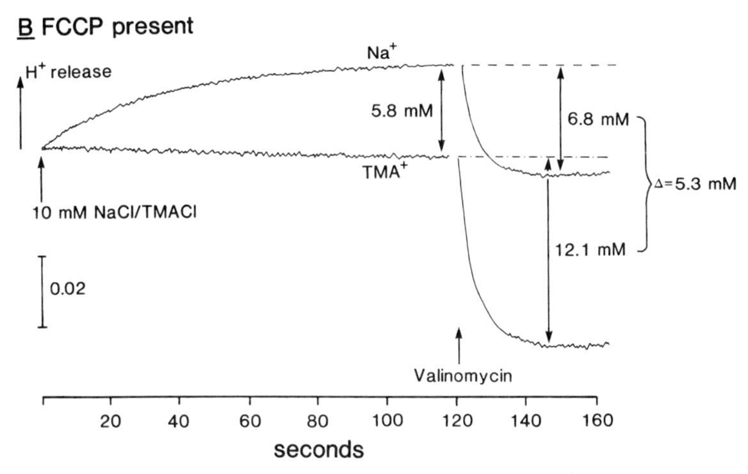

FIGURE 3. The effect of FCCP on Na^+-induced proton and Na^+-induced K^+ release. Ca^{2+}-rich ROS were incubated in 600 mM sucrose, 0.5 mM Hepes (pH 7.4), 20 mM tetramethylammonium chloride (TMACl), 200 μM BAPTA, 1 μM FCCP (panel B only) and 40 μM phenol red. The rhodopsin concentration in the cuvette amounted to 6.7 μM. Proton release was initiated at time zero by addition of 10 mM NaCl or 10 mM TMACl. Proton release was measured as A_{650}-A_{570} as described in detail in Schnetkamp *et al.*[14] and Schnetkamp.[15] Absorption changes were calibrated by addition of HCl. At the second arrow at about 2 minutes, 1 μM valinomycin was added (panel B) or 1 μM FCCP plus 1 μM valinomycin (panel A). Temperature: 25°C.

the release of one Ca^{2+} ion, one K^+ ion, and one proton (the latter via FCCP), resulting in a stoichiometry of 4Na : (1Ca + 1K). A key observation is that addition of FCCP does not uncouple K^+ from Ca^{2+} release, but adds the release of one proton and raises the number of positive charges released for each Ca^{2+} released from three to four.[14,15] The experiment shown in FIGURE 3 illustrates Na^+-induced proton release and compares it with the Na^+-induced reduction in total internal K^+. The internal K^+ content was measured by the alkalinization of the external medium upon addition of the K^+-selective ionophore valinomycin (FCCP present). Valinomycin-induced release of internal K^+ is electrically coupled to an equally large and oppositely directed proton influx via FCCP.[15] Ca^{2+}-rich ROS were incubated in a slightly buffered sucrose solution containing the dye phenol red to monitor proton fluxes. Na^+-stimulated Ca^{2+} release was initiated at time zero by addition of 10 mM NaCl, which caused a proton efflux only when FCCP was present (in control traces 10 mM tetramethylammonium chloride was added to indicate nonspecific drift in pH; TMACl did not cause the release of either Ca^{2+}, K^+, or protons). Proton efflux after two minutes of exposure to 10 mM NaCl in the presence of FCCP indicated a decrease in total internal proton concentration by 5.8 mM, accompanying a decrease in total internal Ca^{2+} by the same amount (not illustrated). At the end of the two-minute exposure to either NaCl or TMACl, addition of the K^+-selective ionophore valinomycin (FCCP present) caused an alkalinization, indicating the influx of protons that accompanied the release of internal K^+. A two-minute exposure of ROS to 10 mM NaCl caused a smaller alkalinization of the external medium when compared with ROS exposed to TMACl, indicating that less K^+ was released. Most likely, Na^+-stimulated Ca^{2+} efflux was accompanied by an equimolar release of K^+, leading to a significant reduction in total internal K^+ to be released by the subsequent addition of valinomycin. Two important observations can be made:

1. Na^+-induced K^+ release in the presence of FCCP (5.3 mM) was comparable to that observed in the absence of FCCP (5.9 mM).
2. Na^+-induced proton release (5.8 mM) and Na^+-induced K^+ release (5.3 mM) were equally large.

The observations discussed above suggest that the Na : Ca coupling ratio of Na-Ca+K exchange in bovine ROS can be either 3 or 4. Corroborative evidence for this conclusion came from the observation that FCCP increased the ratio of ^{22}Na uptake/^{45}Ca release by 31%, consistent with a switch in Na : Ca coupling ratio from 3 to 4.[17]

The Ca : K Coupling Ratio of Na-Ca+K Exchange

Na^+-stimulated Ca^{2+} release was accompanied by an equally large K^+ release when Na^+ was the only external cation present.[14,17] When Na^+-stimulated Ca^{2+} release was measured with both Na^+ and K^+ present in the external medium, the ratio of $^{86}Rb/^{45}Ca$ release was not constant and, thus, the Ca : K coupling ratio of Na-Ca+K exchange transport was not constant.[17] Similarly, the coupling ratio of Ca^{2+} and K^+ fluxes via the Ca-Ca self-exchange mode was not constant: Complete depletion of ROS from internal K^+ had little effect on the rate of Ca-Ca exchange fluxes.[15] On a mechanistic level, the above observations can be accounted for by a model in which the Ca^{2+} and K^+ site of the exchanger can be translocated independently of each other.[17] Fixed stoichiometric coupling ratios of Na-Ca+K exchange transport observed when Na^+ is the only cation present in the external medium are due to the fact that the exchanger protein transports Na^+ only in a cooperative fashion, requiring that all Na^+ transport sites are occupied, leading to simultaneous transport of all Na^+ ions bound.

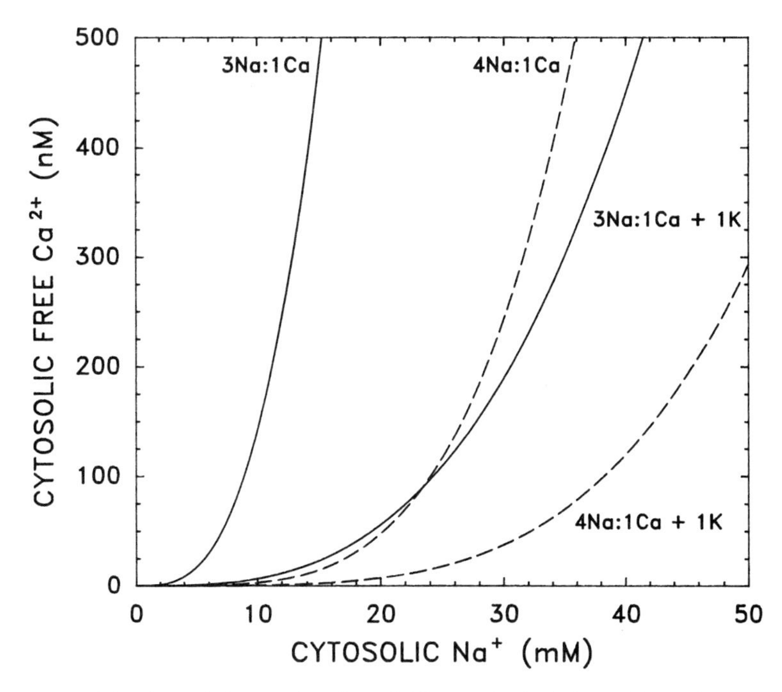

FIGURE 4. Model calculations of the effect of different Na-Ca-(K) stoichiometries on equilibrium free cytosolic Ca^{2+} as function of internal Na^+ concentration. Calculations were done as described in the text. External Na^+ concentration was taken as 150 mM, external Ca^{2+} as 1.5 mM, K_o/K_i as 1/20 and membrane potential was assumed to be −30 mV. Solid lines indicate stoichiometries of 3Na : 1Ca and 4Na : 1Ca as indicated; dashed lines indicate stoichiometries of 3Na : (1Ca + 1K) and 4Na : (1Ca + 1K) as indicated.

Physiological Relevance of Na-Ca+K Exchange and Its Flexible Coupling Ratios

Na-Ca exchange has been found in several tissues. To date, it is clear that the cardiac Na-Ca exchanger neither requires nor translocates K^+ and operates at a 3Na : 1Ca stoichiometry.[24–27] A recent study suggests that a subpopulation of the Na-Ca exchange activity in brain synaptosomes requires and transports K^+.[28] The physiology of retinal rods provides perhaps some clues for the necessity of K^+ transport via Na-Ca exchange. In darkness, a constant influx of Na^+ across the rod plasma membrane results in a depolarized membrane potential and quite possibly an elevated internal Na^+ concentration; both depolarized membrane potential and reduced transmembrane Na^+ gradient compromise the thermodynamic competence of the exchanger to maintain cytosolic Ca^{2+} concentration below 200 nM. FIGURE 4 illustrates a few calculations on the equilibrium free-Ca^{2+} concentration as a function of the cytosolic Na^+ concentration for four hypothetical Na-Ca-(K) exchangers with different transport stoichiometries. The equilibrium free-Ca^{2+} concentration is determined by the transport stoichiometry according to the equation:

$$Ca_i/Ca_o = (Na_i/Na_o)^n \cdot (K_o/K_i) \cdot \exp(V_m zF/RT)$$

where the subscripts "i" and "o" refer to intracellular and extracellular concentrations, respectively, V_m is the membrane potential, z is net positive charge transported for each transport cycle, F is Faraday's constant, R is the gas constant, and T the temperature. The Na-Ca+K exchanger is capable of maintaining a dark free cytosolic Ca^{2+} concentration in ROS of about 200 nM.[29] The 3Na : 1Ca exchanger would be adequate for this purpose if cytosolic Na^+ never exceeds 10 mM, the 4Na : 1Ca and 3Na : (1Ca + 1K) exchangers would still be adequate if cytosolic Na^+ gets as high as 28 mM, whereas the 4Na : (1Ca + 1K) exchanger would allow cytosolic Na^+ concentrations of up to 44 mM. The above calculations suggest that cells with sustained depolarized membrane potentials and elevated internal Na^+ concentrations require a Na-Ca+K exchanger for proper maintenance of cytosolic free Ca^{2+}. This description certainly fits retinal rod cells in the dark. However, rod cells are exposed for prolonged periods of time to sufficient light to saturate the cell. Under these conditions, a more negative membrane potential is maintained, and, more importantly, the transmembrane Na^+ gradient is expected to be maintained at normal levels. Thus, under constant illumination the 4Na : (1Ca + 1K) exchanger is expected to lower free cytosolic Ca^{2+} to undesirably low levels of < 1 nM. We believe that the flexible Ca : K coupling ratios observed in our experiments under nonphysiological conditions are likely to be used for adapting the Na-Ca+K exchanger to operate in retinal rods in the light.

SUMMARY

Ca^{2+} extrusion in the outer segments of retinal rods (ROS) is mediated by a protein that couples both the inward Na^+ gradient and the outward K^+ gradient to Ca^{2+} extrusion. Na^+-stimulated Ca^{2+} release from ROS requires internal K^+ and is accompanied by release of internal K^+, whereas a slow component of Na^+-stimulated Ca^{2+} release does not require K^+. In this paper we discuss our observations on the K^+ transport via Na-Ca+K exchange in bovine ROS, on the electrogenicity and stoichiometry of the ROS Na-Ca+K exchanger, and on the mechanism on coupling Ca^{2+} to K^+ via this protein. Finally, we discuss briefly the physiological implications of Na-Ca+K exchange.

REFERENCES

1. YAU, K.-W. & K. NAKATANI. 1985. Nature **313:** 579-582.
2. SCHNETKAMP, P. P. M., F. J. M. DAEMEN & S. L. BONTING. 1977. Biochim. Biophys. Acta **468:** 259-270.
3. SCHNETKAMP, P. P. M. 1979. Biochim. Biophys. Acta **554:** 441-459.
4. SCHNETKAMP, P. P. M. 1980. Biochim. Biophys. Acta **598:** 66-90.
5. YAU, K.-W. & K. NAKATANI. 1984. Nature **311:** 661-663.
6. SCHNETKAMP, P. P. M., A. A. KLOMPMAKERS & F. J. M. DAEMEN. 1979. Biochim. Biophys. Acta **552:** 379-389.
7. SCHNETKAMP, P. P. M. 1986. J. Physiol. **373:** 25-45.
8. COOK, N. J. & U. B. KAUPP. 1988. J. Biol. Chem. **263:** 11382-11388.
9. NICOLL, D. A. & M. L. APPLEBURY. 1989. J. Biol. Chem. **264:** 16207-16213.
10. SCHNETKAMP, P. P. M. & M. D. BOWNDS. 1987. J. Gen. Physiol. **89:** 481-500.
11. HAASE, W., W. FRIESE, R. D. GORDON, H. MULLER & N. J. COOK. 1990. J. Neurosci. **10:** 1486-1494.
12. CERVETTO, L., L. LAGNADO, R. J. PERRY, D. W. ROBINSON & P. A. MCNAUGHTON. 1989. Nature **337:** 740-743.
13. SCHNETKAMP, P. P. M., R. T. SZERENCSEI & D. K. BASU. 1988. Biophys. J. **53:** 389a.
14. SCHNETKAMP, P. P. M., D. K. BASU & R. T. SZERENCSEI. 1989. Am. J. Physiol. **257:** C153-C157.

15. SCHNETKAMP, P. P. M. 1989. Prog. Biophys. Mol. Biol. **54:** 1-29.
16. SCHNETKAMP, P. P. M. & R. T. SZERENCSEI. 1991. J. Biol. Chem. **266:** 189-197.
17. SCHNETKAMP, P. P. M., R. T. SZERENCSEI & D. K. BASU. 1991. J. Biol. Chem. **266:** 198-206.
18. SCHNETKAMP, P. P. M. 1985. J. Membr. Biol. **88:** 249-262.
19. SCHNETKAMP, P. P. M. 1985. J. Membr. Biol. **88:** 263-275.
20. SCHNETKAMP, P. P. M. 1990. J. Gen. Physiol. **96:** 517-534.
21. SCHNETKAMP, P. P. M. 1991. J. Gen. Physiol. In press.
22. BAYLOR, D. A., T. D. LAMB & K.-W. YAU. 1979. J. Physiol. **288:** 589-611.
23. HODGKIN, A. L., P. A. MCNAUGHTON & B. J. NUNN. 1987. J. Physiol. **391:** 347-370.
24. SLAUGHTER, R. S., J. L. SUTKO & J. P. REEVES. 1983. J. Biol. Chem. **258:** 3183-3190.
25. REEVES, J. P. & C. C. HALE. 1984. J. Biol. Chem. **259:** 7733-7739.
26. EHARA, T. , S. MATSUOKA & A. NOMA. 1989. J. Physiol. **410:** 227-249.
27. YASUI, K. & J. KIMURA. 1990. Eur. J. Physiol. **415:** 513-515.
28. DAHAN, D., R. SPANIER & H. RAHAMIMOFF. 1991. J. Biol. Chem. **266:** 2067-2075.
29. RATTO, G. M., R. PAYNE, W. G. OWEN & R. Y. TSIEN. 1988. J. Neurosci. **8:** 3240-3246.

Stoichiometry and Regulation of the Na-Ca Exchanger in Barnacle Muscle Cells[a]

HECTOR RASGADO-FLORES, JAIME DESANTIAGO,
AND RICARDO ESPINOSA-TANGUMA

Department of Physiology and Biophysics
University of Health Sciences/
The Chicago Medical School
North Chicago, Illinois 60064

INTRODUCTION

Intracellular Ca^{2+} plays a critical second messenger role in animal cells. This role is regulated by changes in the intracellular free-Ca^{2+} concentration ($[Ca^{2+}]_i$). Changes in $[Ca^{2+}]_i$ can be achieved either by the release or sequestration of Ca^{2+} by intracellular stores and/or by an increase in Ca^{2+} influx or efflux across the plasmalemma. Ca^{2+} influx can be induced by the activation of voltage-gated or receptor-operated Ca^{2+} channels; Ca^{2+} efflux can be induced by the activation of an ATP-driven Ca^{2+} pump. In addition to these mechanisms, the Na-Ca exchanger can transport Ca^{2+} bidirectionally across the plasmalemma in exchange for Na^+.[1] The coupled Ca^{2+} efflux and Na^+ influx mediated by this exchanger are operationally defined as the "Ca^{2+} efflux" mode of exchange; the coupled Ca^{2+} influx and Na^+ efflux have been termed the "Ca^{2+} influx" mode.

The ability of the exchanger to regulate $[Ca^{2+}]_i$ under physiological conditions is determined by two factors: the *direction* of transport and its *magnitude.* The direction of (net) Ca^{2+} transport is regulated by the chemical gradients of Na^+ and Ca^{2+} and the membrane potential (i.e., thermodynamic factors). The magnitude of the exchange is determined by kinetic factors. The relative importance of each of these factors in determining the exchanger's direction of net Ca^{2+} transport is dependent on the exchanger's stoichiometry. Kinetic factors that determine exchange rate can be divided into primary (i.e., partial saturation of the exchanger's transporting sites) and secondary categories (e.g., modulation by ATP, pH_i, K_i, and Mg_i; and activation by Ca_i). Secondary kinetic factors are critical; they can determine whether or not the exchanger mediates a net Ca^{2+} transport under a given set of conditions. For example, operation of the Na-Ca exchanger in any of its modes (e.g., Na-Ca or Na-Na exchange)[2-4] has an absolute requirement for Ca_i. To understand how the exchanger regulates $[Ca^{2+}]_i$ in parallel with the Ca^{2+}-buffering and other Ca^{2+}-transporting mechanisms, it is essential to know its stoichiometry and to characterize its secondary kinetic factors.

Barnacle muscle cells are a valuable model for the study of the Na-Ca exchanger. The exchanger in these cells has similar properties to its counterpart in other animal cells, including cardiac,[5] skeletal,[6] and smooth muscle[7] cells, as well as squid axons[3]

[a] H.R.-F. is an established investigator of the American Heart Association. This work was supported by the AHA-Metropolitan Chicago Affiliate.

and synaptosomes.[8] Barnacle muscle cells, because of their large size (one cell is about 2 cm in length and 2 mm in diameter) can be internally perfused, and both extracellular and intracellular solution composition can be controlled. The efflux and influx of labeled ions can be measured, and the cells can be voltage-clamped.

Recently, we showed that it is possible to determine the coupling ratio of the exchange directly from measurements of Ca^{2+} influx and Na^{+} efflux.[4,9] These results demonstrate that three Na^{+} are exchanged for one Ca^{2+}. Recently, however, Cerveto *et al.*[10] reported the unexpected observation that in salamander retinal rods, the Na-Ca exchanger cotransports both Ca^{2+} and K^{+} in exchange for Na^{+}. The proposed stoichiometry for this exchange is 4 Na : 1 Ca, 1K. In view of the fact that in squid axons[11] and in barnacle muscle (DeSantiago, Espinosa-Tanguma, Rasgado-Flores, unpublished observations) intracellular K^{+} activates Na_o-dependent Ca^{2+} efflux, the question arises whether K^{+} is transported by these exchangers or whether it simply activates them. One way to address this issue is to determine if the barnacle muscle exchanger can operate in the complete absence of intra- and extracellular K^{+}.

In this study we have also attempted to activate the Na-Ca exchanger without increasing $[Ca^{2+}]_i$. This manipulation would be very beneficial for studying the exchanger's properties under voltage-clamp conditions. In barnacle muscles, a rise in $[Ca^{2+}]_i$, besides activating the exchanger, also opens nonselective cation channels.[12] These channels increase the membrane conductance and cannot be blocked because of the lack of a specific inhibitor. In vesicles[13] and excised patches[14] from cardiac cells, mild treatment with α-chymotrypsin produces an activated exchanger that no longer requires Ca_i to operate. In the present work we examined the ability of a mild treatment with α-chymotrypsin to activate the exchanger and to eliminate the need to raise $[Ca^{2+}]_i$.

The results provide new insight into the stoichiometry of the exchanger in barnacle muscle cells as well as its regulation by secondary kinetic factors.

METHODS

The data in FIGURE 1 and the methods used to obtain these results are published.[9] As described below, these methods were modified for more recent experiments.

Experimental Setup

Internally perfused, single giant muscle cells from the barnacle, *Balanus nubilus*[15] were dissected, cannulated, and mounted in the experimental chamber. A double-barrel capillary tube was inserted axially through the cut basal end of the muscle fiber. The open tip was positioned close to the uncut tendon end of the fiber and was used to perfuse the cell with the desired intracellular solutions. The other barrel was shorter and ended about midway along the length of the muscle fiber. It was filled with 3 M KCl (or 3 M NaCl for experiments performed under K_i-free conditions) and was used to monitor the membrane potential. With this arrangement, the internal perfusion fluid flows out of the end of the longer barrel, into the myoplasmic space near the tendon end of the cell, and then back through the myoplasmic space until it exits through the glass end-cannula. Although the normal contraction threshold of barnacle muscle cells is about 1 μM,[16] internally perfused cells do not contract when $[Ca^{2+}]_i$ is raised above this level.[17]

Measurement of Tracer Efflux

To measure Na^+ or Ca^{2+} efflux, ^{22}Na or ^{45}Ca was added to the internal perfusate (0.6 mCi/mmol Na; 6-60 mCi/mmol Ca) and the respective tracer appearance in the superfusion fluid was monitored with an on-line liquid scintillation counter. The extracellular fluid was also collected and counted (counting error $\leq 3\%$) using standard liquid scintillation techniques. The fibers were perfused at a rate of 1.7 μl/min. Because the volume of these cells was about 50 μl, the intracellular fluid was changed in the fibers approximately once every 30 minutes. The superfusion rate was 4 ml/min. In experiments measuring Ca^{2+} efflux, the external solutions were Ca^{2+}-free to prevent Ca-Ca exchange. When Na^+ efflux was measured, 0.1 mM ouabain was added to the external solution to prevent Na^+ efflux via the Na-K pump; 10 μM bumetanide was also added to prevent any possible Na^+ efflux mediated by Na-K-Cl cotransport.[18] The Ca_o-dependent Na^+ efflux was determined by comparing the ^{22}Na efflux in the presence and absence of Ca_o (replaced by Mg). Na_o-dependent flux was determined by comparing the ^{45}Ca (for Na_o/Ca_i exchange) or ^{22}Na (for Na-Na exchange) efflux in the presence and absence of Na_o (replaced by Tris).

External (Superfusion) Solutions

The Na seawater used for superfusion (artificial sea water, NaSW) contained (in mM): 455 NaCl; 10 KCl; 25 $MgCl_2$; 11 $CaCl_2$; 3 tris(hydroxymethyl) aminomethane (Tris) base (pH 7.8, adjusted at room temperature with Tris-HCl). In most experiments the external solutions contained Tris (TrisSW) as the main extracellular cation. These solutions were prepared by completely replacing the NaCl by an isosmotic Tris-Base/Tris-HCl mixture (pH 7.8). Some solutions were also Ca^+-free; in these instances, the $CaCl_2$ was replaced with $MgCl_2$. Experiments in which Ca^{2+} efflux was measured were performed under K^+-free conditions. The external K^+-free solutions were prepared by replacing KCl by Tris-HCl.

Internal (Perfusion) Solutions

Two internal solutions were used depending on whether Ca^{2+} or Na^+ efflux was measured.

The solution for Ca^{2+}-efflux experiments contained a high free $[Ca^{2+}]$ (20 μM), a K^+ channel blocker (phenylpropyltriethylammonium bromide, PPTEA) and were K^+-free. The composition of this solution was (in mM): Na-Hepes, 6; Tris-HCl, 38; ATP-Mg, 4; PPTEA, 20; HEDTA, 10; $CaCl_2$, 8; caffeine, 3.5; FCCP, 0.025; glycine, 200; Hepes, 60 (pH = 7.3); and an ATP-regenerating system (1.5 mM phosphoenol pyruvate [Na-salt] and 0.08 mg/ml pyruvate kinase). The free-Ca^{2+} concentration was calculated assuming a Ca-HEDTA dissociation constant of 5 μM.[19]

The solution for Na^+-efflux experiments (standard Na^+-efflux perfusion solution) contained a low $[Ca^{2+}]$ (10^{-8}M), high $[K^+]$ (200 mM), and high $[Na^+]$ (46 mM). The composition of this solution was (in mM): Na-Hepes, 46; K-aspartate, 172; KCl, 38; $MgCl_2$, 7; Tris-HCl, 38; ATP-Mg, 4; EGTA, 8; $CaCl_2$, 0.56; caffeine, 3.5; FCCP, 0.025; glycine, 200; Hepes, 60 (pH = 7.3); 1.5 mM phosphoenol pyruvate; and 0.08 mg/ml pyruvate kinase. The free $[Ca^{2+}]$ was calculated assuming a Ca-EGTA stability constant of 7.54×10^6 M^{-1}.[20] Osmolarity was adjusted with sucrose.

The osmolarity of all the internal and external solutions was 1000 ± 10 mOsM. The temperature of all experiments was 16°C.

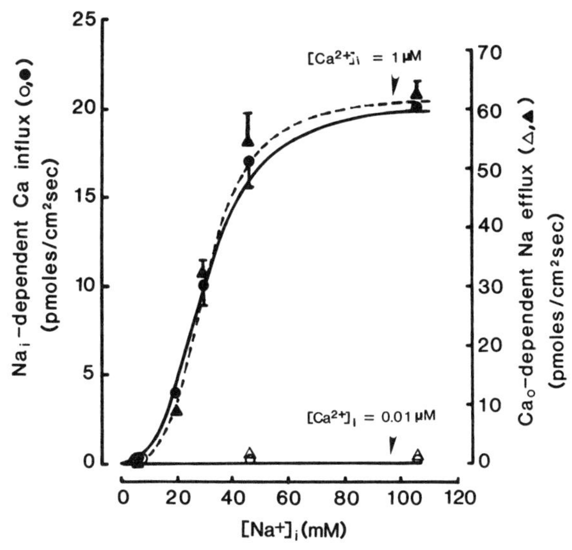

FIGURE 1. Effects of $[Na^+]_i$ and $[Ca^{2+}]_i$ on the Ca_o-dependent Na^+ efflux (△, ▲ ; right-hand ordinate scale) and the Na_i-dependent Ca^{2+} influx (○ , ●; left-hand ordinate scale) in internally perfused barnacle muscle cells. Data from 27 cells are summarized in the figure. Each symbol represents the mean of at least three flux measurements; the bars indicate ± SEM. The external solution in all experiments was Tris-SW containing 0.1 mM ouabain; $[Ca^{2+}]_i$ was either 0.01 mM (open symbols) or 1.0 mM (solid symbols). Note that the ordinate scale for the Ca^{2+} influx is expanded threefold, relative to the scale for the Na^+ efflux. The solid line fits the Hill equation with a Hill coefficient of 3.0, a K_{Nai} of 30 mM, and maximal fluxes of 20.5 pmol/cm²·sec for Na_i-dependent Ca^{2+} influx and 61.5 pmol/cm²·sec for Ca_o-dependent Na^+ efflux. The discontinuous line is the best fit to the data and has a Hill coefficient of 3.7 ± 0.4 and calculated $J_{Na(Max)}$ = 62.4 ± 1.8 pmol/cm² sec and K_{Nai} = 30.0 ± 0.8 mM. All the data in this figure were obtained with V_M between −33 and −43 mV; more than 90% of the data were obtained with V_M = −37 ± 3 mV. (Reprinted from Rasgado-Flores *et al.*[9] with permission from the *Journal of General Physiology.*)

RESULTS AND DISCUSSION

The Coupling Ratio of Na-Ca Exchange in Barnacle Muscle Cells Is 3 Na : 1 Ca

Direct measurement of the stoichiometry of Na-Ca exchange can be made by measuring the counterion dependence of the fluxes mediated by the exchanger. In barnacle muscle cells, however, only the Na_o-dependent ^{45}Ca-efflux component of the Ca^{2+} efflux mode of exchange can be reliably measured: the Ca_i-dependent ^{22}Na-influx component of exchange is accompanied by ^{22}Na influx through a Ca_i-activated nonselective cation channel.[12] Consequently, attempts to measure the stoichiometry of the exchanger in this way give overestimated coupling ratios.[21] We decided,therefore, to measure the stoichiometry of the exchanger by studying the Ca^{2+} influx mode of exchange. The experimental conditions were chosen to eliminate Ca-Ca and Na-Na exchange so that the ratio of the coupled Ca^{2+} influx and Na^+ efflux could be measured directly. The fact that both fluxes were activated by intracellular (nontransported) Ca^{2+} indicated that the fluxes were coupled. These studies provided a kinetic description of the coupled Na^+-efflux/Ca^{2+}-influx exchange as well as the first direct measurements of the exchanger's stoichiometry.[4,9] FIGURE 1 shows the relationship between the

Na_i-dependent Ca^{2+} influx (circles, left-hand ordinate) and Ca_o-dependent Na^+ efflux (triangles, right-hand ordinate) at various $[Na^+]_i$, at $[Ca^{2+}]_i = 10^{-8}$ M (open symbols) and $[Ca^{2+}]_i = 10^{-6}$ M (closed symbols) in an internally perfused barnacle muscle cell. Data from 27 cells are summarized in the figure. To prevent Na-Na and Ca-Ca exchange, Na^+ was replaced in all external solutions with Tris and 0.1 mM ouabain was added. The Ca^{2+} influx and Na^+ efflux curves at $[Ca^{2+}]_i = 10^{-6}$ M are virtually superimposable when the Na^+ efflux scale is made one-third as large as the Ca^{2+} influx scale. The ratio of the Ca_o-dependent Na^+ efflux to the Na_i-dependent Ca^{2+} influx, over this entire range of $[Na^+]_i$, was 2.8-3.2 : 1. The solid line is a least-squares fit to the Hill equation with a Hill coefficient of 3.0. When $[Ca^{2+}]_i$ was 10^{-8} M, there was no measurable Ca_o-dependent Na^+ efflux or Na_i-dependent Ca^{2+} influx at any of the $[Na^+]_i$ tested. These results indicate that the coupling ratio of the Na-Ca exchanger in barnacle muscle cells is 3 Na^+ : 1 Ca^{2+}, and that intracellular Na^+ and Ca^{2+} are both required to activate the Ca^{2+}-influx mode of exchange.

With this coupling ratio, the reversal potential ($E_{Na/Ca}$) of the exchanger is given by:

$$E_{Na/Ca} = 3\ E_{Na} - 2\ E_{Ca} \quad (1)$$

where E_{Na} and E_{Ca} are, respectively, the equilibrium potentials for Na^+ and Ca^{2+}. The reversal potential of the exchanger is the potential at which the transporter cannot mediate net movement of Na^+ or Ca^{2+}. The direction of net Ca^{2+} transport mediated by the Na-Ca exchanger is determined by the net driving force ($V_M - E_{Na/Ca}$). However, the rate of transport depends upon kinetic factors. Under normal resting conditions, V_m may be more negative than $E_{Na/Ca}$ so that the Na-Ca exchanger drives Ca^{2+} out of cells and contributes to keep $[Ca^{2+}]_i$ low. During activation, however, when the membrane is depolarized and $[Ca^{2+}]_i$ and $[Na^+]_i$ rise, the exchanger may contribute to Ca^{2+} entry; upon repolarization, the exchanger should drive Ca^{2+} out of the cell. Indeed, many observations indicate that the Na-Ca exchange is voltage-sensitive and electrogenic. However, an analysis of the effects of membrane potential on the exchanger-mediated Ca^{2+} transport is needed to clarify the physiological role of this transport mechanism.

In Barnacle Muscle Cells, K^+ Is Not Required for Operation of the Ca^{2+}-Efflux Mode of Exchange

In vertebrate retinal cells, operation of this transporter has an absolute requirement for K^+.[10] To determine if the barnacle muscle exchanger absolutely requires K^+ to operate, we measured Na_o-dependent Ca^{2+} efflux in cells dialyzed and superfused with K^+-free solutions. Extracellular Ca^{2+} was also removed to prevent unidirectional Ca^{2+} efflux via Ca-Ca exchange. In barnacle muscle, complete removal of extracellular K^+ (K_o) is made difficult by the fact that leakage of intracellular K^+ may result in K^+ entrapment in the deep, branching invaginations of the cell's surface membrane. To prevent loss of K_i and facilitate its removal from the extracellular space, a K^+-channel blocker was added to the perfusion fluid. PPTEA was chosen because this substance was found to be the most effective tetraethylammonium derivative that blocks K^+ channels in squid giant axons[22] and has also proven to be effective in barnacle muscle (DeSantiago, Espinosa-Tanguma, Rakowski & Rasgado-Flores, unpublished observa-

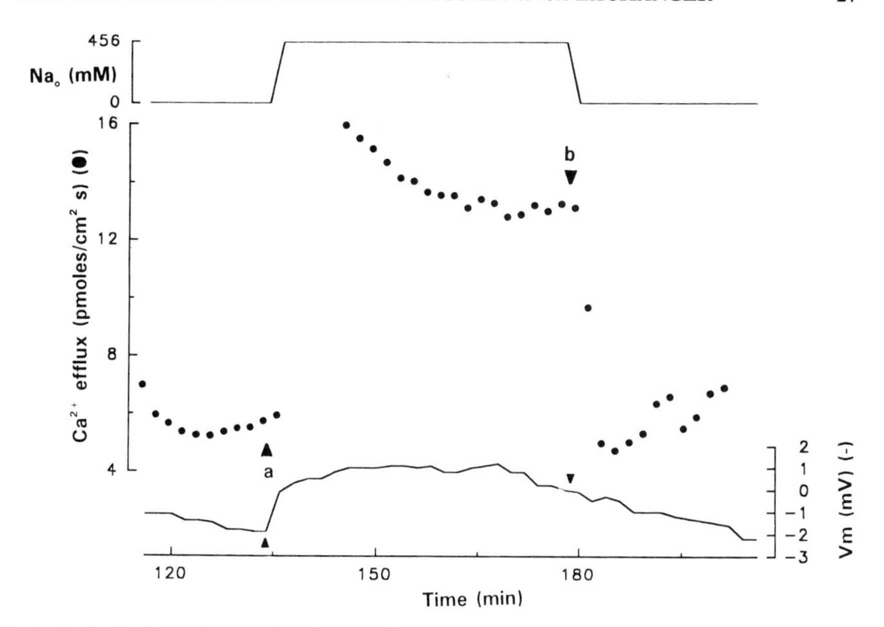

FIGURE 2. Effect of external Na^+ on Ca^{2+} efflux and membrane potential (V_m) from a barnacle muscle cell perfused and superfused with solutions containing 0 K^+. The internal fluid was the Ca^{2+}-efflux perfusion solution containing 0 K^+, 20 mM PPTEA, and 20 μM $[Ca^{2+}]_i$. The external solutions were K^+- and Ca^{2+}-free TrisSW and NaSW. At **a,** Na_o was added (replacing Tris); at **b,** Na_o was replaced by Tris.

tions). FIGURE 2 shows that in Ca_o-, K_o-, and K_i-free conditions, addition of Na^+ to the extracellular solution (at *a*) activated the Ca^{2+} efflux by about 7 $pmol \cdot cm^{-2} \cdot sec^{-1}$. This increase in Ca^{2+} efflux was accompanied by a 2.5-mV depolarization. Likewise, replacement of Na_o by Tris (*b*) produced a reduction of Ca^{2+} efflux of about 7 $pmol \cdot cm^{-2} \cdot sec^{-1}$. This indicates that, in contrast to retinal rods, the barnacle muscle Na-Ca exchanger can operate in the absence of K^+. Although this is reasonable evidence that the exchanger does not transport K^+ if this ion is not available, the possibility that the exchanger can operate in different stoichiometries depending on whether K^+ is available can be raised. This possibility, however, is not consistent with the direct measurement of stoichiometry. A definitive answer to the question of whether K^+ is transported could be provided by assessing the effect of Na_o removal on the simultaneous effluxes of Ca^{2+} and K^+. A lack of effect on K^+ efflux while Ca^{2+} efflux is reduced in response to Na_o removal would demonstrate that the exchanger only transports Ca^{2+} in the presence of K^+. This double-label experiment can be performed by perfusing the cell with a solution containing both ^{45}Ca and ^{42}K since the half-lives of these isotopes are very different. A complication, however, is that in barnacle muscle, the steady-state unidirectional K^+ efflux is much larger (5- to 10-fold) than the Ca^{2+} efflux. Small changes in unidirectional K^+ efflux can only be measured if most of the K^+ flux pathways are blocked (e.g., resting channels, voltage-activated channels, Ca^{2+}-activated K^+ channels). At present we are testing various blockers to reduce the unidirectional steady-state K^+ efflux level.

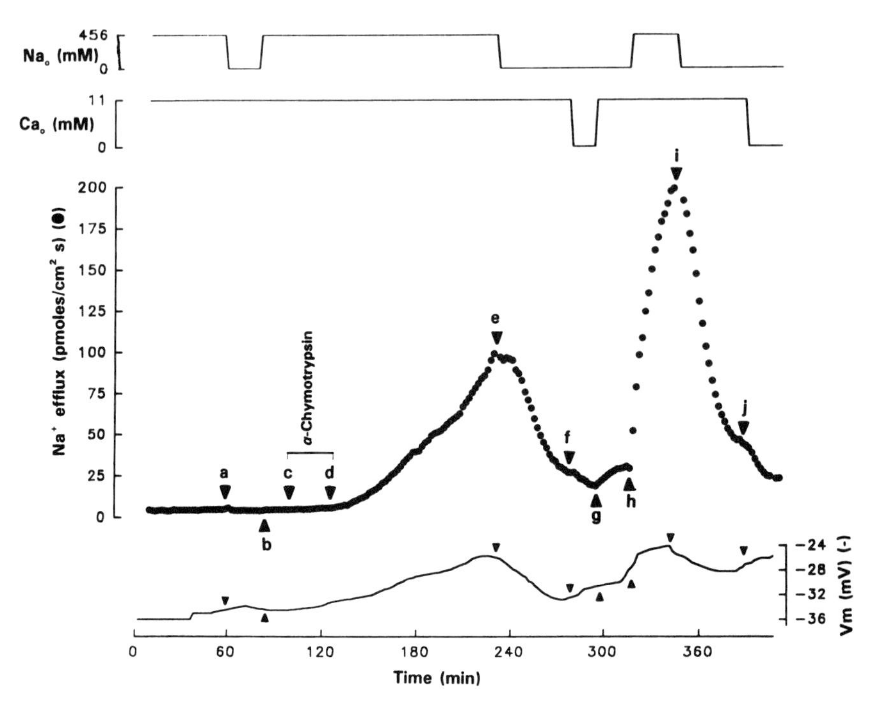

FIGURE 3. Effects of external Na^+, Ca^{2+}, and α-chymotrypsin on Na^+ efflux and membrane potential in a barnacle muscle cell perfused with the standard Na^+-efflux perfusate. The external solutions used were the normal NaSW or TrisSW containing 11 mM or 0 Ca^{2+}; 0.1 mM ouabain and 10 μM bumetanide were added to these solutions. From **a** to **b**, Na_o was replaced by Tris; from **c** to **d**, the cell was perfused with 0.5 mg/ml α-chymotrypsin; at **e**, Na_o was replaced by Tris; from **f** to **g**, Ca_o was replaced with Mg_o; at **h**, normal Na_o was restored; at **i**, Na_o was replaced by Tris; at **j**, Ca_o was replaced with Mg_o.

A Mild Treatment with α-Chymotrypsin Activates the Barnacle Muscle Na-Ca Exchanger

Intracellular Ca^{2+} is a critical secondary kinetic factor whose presence is normally required for operation of the exchanger. FIGURE 1 shows that even a very large driving force favoring the Na^+-efflux/Ca^{2+}-influx mode of exchange is ineffective in promoting exchange in the absence of sufficient Ca_i (0.01 μM). Half-maximal activation for the Ca_o-dependent Na^+ efflux into a Tris-SW solution containing Ca^{2+} is attained with $[Ca^{2+}]_i = 0.7$ μM.[9] There are three likely means (or combinations thereof) by which Ca_i could activate the exchanger: by increasing (i) the exchanger's binding affinities for Ca^{2+} and Na^+, (ii) the exchanger's voltage sensitivity, and (iii) the exchanger's turnover rate. It is not clear which of these mechanisms is responsible for the activating role of Ca_i. The internally perfused, voltage-clamped barnacle muscle cell offers the opportunity to perform the appropriate experiments. Activation of the barnacle muscle's exchanger by raising $[Ca^{2+}]_i$, however, increases the membrane conductance by activation of nonselective cation channels and limits the voltage excursion attainable under voltage-clamp conditions. We therefore attempted to activate the exchanger by mild treatment with α-chymotrypsin[13,14] instead of raising $[Ca^{2+}]_i$. FIGURE 3 shows an experiment in which Na^+ efflux was measured in a barnacle muscle perfused with the standard Na^+-efflux perfusion solution (containing 46 mM Na^+ and low $[Ca^{2+}]_i$, 10^{-8} M). The external solution contained 0.1 mM ouabain and 10 μM bumetanide. Under these low Ca_i conditions, the exchanger should not be activated. Indeed, removal of Na_o (from *a* to *b*) to test for the presence of Na-Na exchange showed no appreciable change in Na^+ efflux. Perfusion for 30 min with a solution containing 0.5 mg/ml of α-chymotrypsin (from *c* to *d*) produced an immediate increase in Na^+ efflux. This increase was reversed (by 74 $pmol \cdot cm^{-2} \cdot sec^{-1}$) when Na_o was replaced by Tris (at *e*). Na^+ efflux was further reduced (by 12 $pmol \cdot cm^{-2} \cdot sec^{-1}$) when Ca_o was removed (at *f*). This indicates that the exchanger was activated and able to engage in Na-Na and Na_i-Ca_o exchange after the α-chymotrypsin treatment. Subsequent restoration of normal Ca_o (at *g*) and Na_o (*h*) produced even larger fluxes that were reversible on removal of these ions (*i, j,* respectively). This shows that the α-chymotrypsin treatment was not complete after the initial 30 min of perfusion.

Activation of the Exchanger by α-Chymotrypsin Is Not Due to an Increase in $[Ca^{2+}]_i$

FIGURE 3 shows that α-chymotrypsin activates the exchanger in the presence of Ca_o. This effect could be due to a direct action of the protease on the exchanger and/or to an indirect effect by increasing the membrane permeability to Ca^{2+} and causing an increase in $[Ca^{2+}]_i$. To rule out this second possibility, we carried out a similar protocol to that shown in FIGURE 3, but in this case the external solution was Ca^{2+} free. Under these conditions Ca_o-dependent Na^+ efflux cannot be measured and the activity of the exchanger had to be limited to testing the presence of Na-Na exchange. FIGURE 4 shows a representative experiment of this kind. As expected, the initial removal of Na_o (from *a* to *b*) produced no effect on Na^+ efflux. The α-chymotrypsin treatment (*c* to *d*) produced an increase in Na^+ efflux. Under these conditions, removal of Na_o (at *e*) produced a reduction of Na^+ efflux. This indicates that the activation of the exchanger by α-chymotrypsin can take place in the absence of extracellular Ca^{2+}.

Na^+ Efflux Activated by α-Chymotrypsin Is Not Mediated by Ca^{2+} Channels

Barnacle muscle cells, like other invertebrate muscles, lack Na^+ channels: Their electrical excitability is mediated instead by Ca^{2+} channels.[23] The fact that K_o depolarization (in the absence of Ca_o) promotes Na^+ efflux in these cells[24] suggests that Na^+ can permeate through Ca^{2+} channels. FIGURES 3 to 6 all show that replacement of Na_o by Tris produces a (4-8-mV) hyperpolarization. This effect may be a consequence of a lower membrane permeability for Tris than for Na^+. If α-chymotrypsin activates Ca^{2+} channels and Na^+ can pass through these channels, the reduction in Na^+ efflux resulting from replacing Na_o by Tris could result from the accompanying membrane hyperpolarization. This possibility can be tested by measuring the effect of Ca^{2+} channel blockers on α-chymotrypsin-activated Na-Na exchange. Because verapamil is a potent Ca^{2+} channel blocker in barnacle muscle,[25] we measured its effect on the Na_o-dependent Na^+ efflux activated by α-chymotrypsin. FIGURE 5 shows an experiment of this type. The internal fluid was the standard Na^+-efflux perfusion solution, the external solution was Ca^{2+} free, and the cell was treated with α-chymotrypsin 96 min before the first data point plotted on the graph. At *a,* replacement of Na_o by Tris induced a large reduction in Na^+ efflux (96 $pmol \cdot cm^{-2} \cdot sec^{-1}$); at *b,* 100 μM verapamil was added and was accompanied by an increase in Na^+ efflux that could be due to a drug-induced

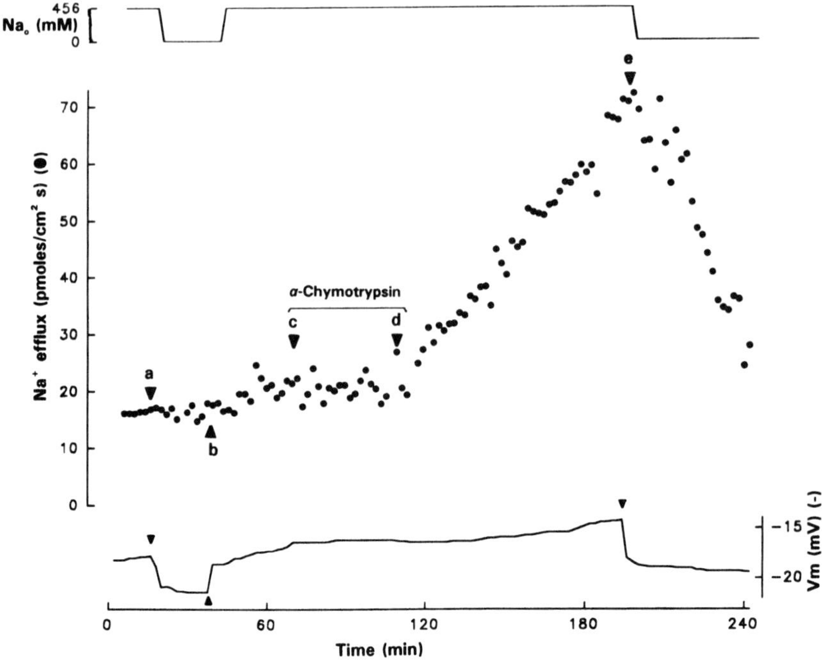

FIGURE 4. Effects of Na$_o$ and α-chymotrypsin on Na$^+$ efflux and membrane potential in a barnacle muscle cell perfused with the standard Na$^+$-efflux perfusate. From **a** to **b,** Na$_o$ was replaced by Tris; from **c** to **d** the cell was perfused with 0.5 mg/ml α-chymotrypsin; at **e,** Na$_o$ was replaced by Tris. The external solutions contained 0.1 mM ouabain and 10 μM bumetanide.

increase in cell leakiness. At *c* normal Na$_o$ was restored, and this induced an increase in Na$^+$ efflux of the same magnitude as when Na$_o$ was removed (95 pmol·cm^{-2}·sec^{-1}). In the presence of verapamil, removal of Na$_o$ (from *d* to *e*) produced a transient decrease of Na$^+$ efflux (80 pmol·cm^{-2}·sec^{-1}) indicating that the α-chymotrypsin-activated Na$^+$ efflux is not mediated by Ca^{2+} channels.

Na-Na Exchange Activated by α-Chymotrypsin Is Mediated by the Na-Ca Exchanger

The evidence presented above demonstrates that α-chymotrypsin activates a Na-Na exchange, presumably mediated by the Na-Ca exchanger. However, the fact that other co- and countertransport mechanisms present in the barnacle muscle can also mediate Na-Na exchange makes it necessary to consider whether any of these other transport mechanisms are activated by α-chymotrypsin. In the presence of ouabain and bumetanide to eliminate Na-Na exchange by the Na-K pump or by Na-K-Cl cotransport, two other possible mechanisms merit consideration: (i) the Na-dependent pH$_i$-regulating system (i.e., Na + HCO$_3$/Cl + H exchange)[26]; and (ii) the Na-Mg exchanger.[27,28] The first possibility can be dismissed because 0.1 mM DIDS (4,4'-diisothiocyanatostilbene-2,2' disulfonic acid), an effective inhibitor of the pH$_i$ regulating system[26] did not affect the Na-Na exchange (data not shown). Possible participation

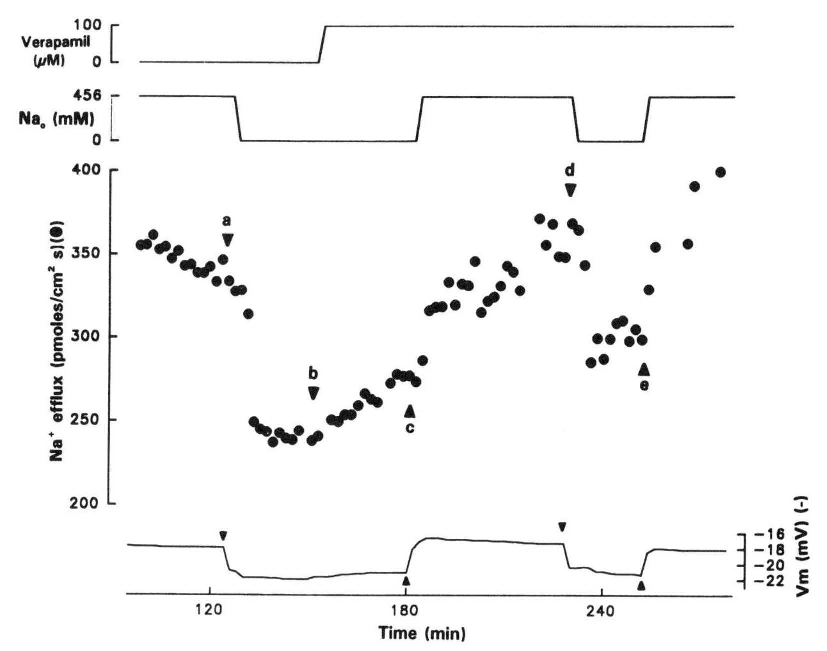

FIGURE 5. Effects of Na_o and verapamil on Na^+ efflux and membrane potential in a barnacle muscle cell treated with α-chymotrypsin. The cell was perfused with α-chymotrypsin (0.5 mg/ml·30 min) 96 min before the first data point plotted in the graph. The external solutions were Ca^{2+}-free. At **a,** Na_o was replaced by Tris; at **b,** 100 μM verapamil was added; at **c,** normal Na_o was restored; from **d** to **e,** Na_o was replaced by Tris.

of the Na-Mg exchanger, however, has not been ruled out. In squid axons[3,28] and barnacle muscle,[4] a Na-Na exchange that does not require Ca_i to be activated can be readily measured. This exchange may be mediated by the Na-Mg exchanger.[28] Separation of the Na-Na exchanges mediated by either the Na-Ca and the Na-Mg exchangers is possible.[3] The exchange mediated by the Na-Ca exchanger requires Ca_i, is activated by alkalinization and ATP_i, and is inhibited by Ca_o. The Ca_o-mediated inhibition results from switching the exchange mode from the (fast) Na-Na exchange (in the absence of Ca_o) to the (slow) Na_i/Ca_o exchange.[3] The exchange mediated by the (presumably) Na-Mg exchanger has an absolute requirement for ATP_i, does not require Ca_i, and is not affected by pH_i or Ca_o.

To test whether the α-chymotrypsin-activated Na-Na exchange is indeed mediated by the Na-Ca exchanger, we tested the effect of Ca_o on this exchange. FIGURE 6 shows such an experiment. The cell was perfused with the standard Na^+-efflux perfusion fluid and was treated with α-chymotrypsin two hours before time "zero" in the graph. The external solutions contained 0.1 mM ouabain, 10 μM bumetanide, and 100 μM verapamil. At *a,* Na_o was replaced by Tris, producing a reduction in Na^+ efflux of 49 $pmol \cdot cm^{-2} \cdot sec^{-1}$. Removal of Ca_o (*b*) induced a further reduction in the flux (29 $pmol \cdot cm^{-2} \cdot sec^{-1}$), indicating the presence of Ca_o-Na_i exchange. Readmission of normal Na_o when Ca_o was absent (at *c*) induced a large increase in Na^+ efflux (234 $pmol \cdot cm^{-2} \cdot sec^{-1}$. The greater magnitude of this flux as compared to that when Na_o was removed (at *a*) may result from the time-dependent exchanger activation by α-chymotrypsin and/or to a faster Na-Na exchange manifested in the absence of Ca_o.

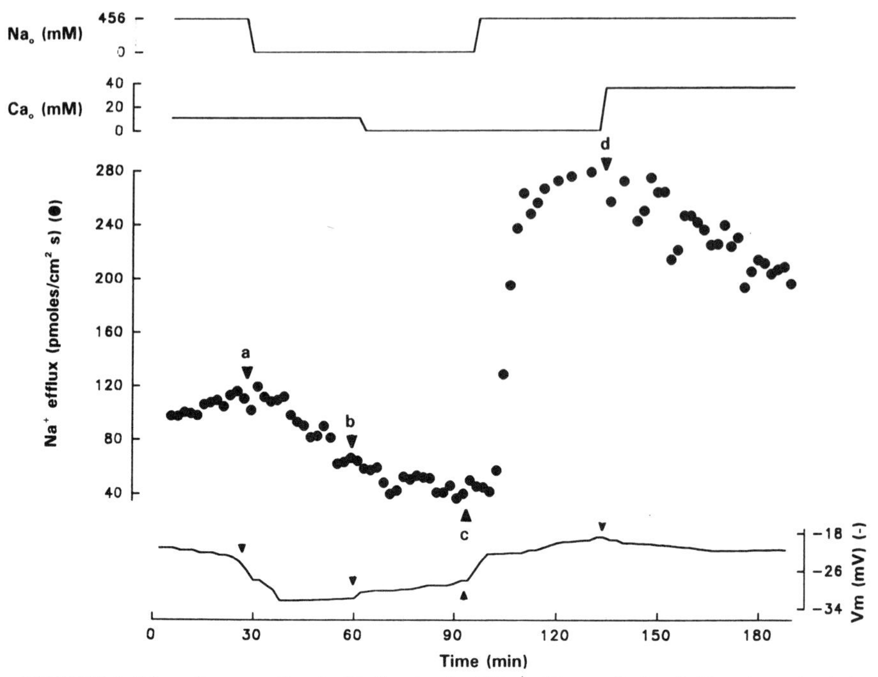

FIGURE 6. The cell was perfused with the standard Na^+ efflux perfusion fluid and was treated with α-chymotrypsin two hours before the time zero in the graph. The external solutions contained 0.1 mM ouabain, 10μM bumetanide, and 100 μM verapamil. At **a,** Na_o was replaced by Tris; at **b** Ca_o was a replaced by Mg_o; at **c,** normal Na_o was restored; and at **d** Ca_o was increased to 40 mM.

In any event, addition of 40 mM Ca_o (*d*) produced a pronounced inhibition of the rate of Na^+ efflux (by 89 pmol$\cdot$cm^{-2} $\cdot$ sec^{-1} c^{-1} in 1 hour) towards the value obtained in the absence of Na_o. This must result from inhibition of the (faster) Na-Na exchanger as the (slower) Na_i-Ca_o exchange is activated. That the fluxes activated by α-chymotrypsin are mediated by the Na-Ca exchanger is also supported by the observation that this treatment simultaneously activates a fast Na-Na exchange and a slow Na_i-Ca_o exchange (FIGS. 3 and 6).

CONCLUSIONS

The stoichiometry of the Na-Ca exchanger in barnacle muscle cells is 3 Na : 1 Ca. Consistent with this result we find that K^+ is not required for operation of the exchanger. In addition, mild treatment with α-chymotrypsin modifies the Na-Ca exchanger so that it no longer requires Ca_i to operate. This treatment will be very helpful for studying the exchanger's properties under voltage-clamp conditions. Some of the interesting questions that need to be addressed are: (i) What is the voltage dependence of the various modes of exchange mediated by the Na-Ca exchanger?; (ii) How does Ca_i activate the exchanger?; (iii) Why is it that Na-Na (and Ca-Ca)[29] modes of exchange are faster than Na-Ca exchange?. Answers to these questions may be obtained using internally perfused, voltage-clamped barnacle muscle cells.

ACKNOWLEDGMENTS

We wish to thank Drs. M. P. Blaustein and R. F. Rakowski for helpful comments on the manuscript.

REFERENCES

1. BLAUSTEIN, M. P. 1974. Rev. Physiol. Biochem. Pharmacol. **70:** 32-82.
2. DIPOLO, R. & L. BEAUGÉ. 1986. Biochim. Biophys. Acta **894:** 298-306.
3. DIPOLO, R. & L. BEAUGÉ. 1987. J. Gen. Physiol. **90:** 505-525.
4. RASGADO-FLORES, H. & M. P. BLAUSTEIN. 1987. Am. J Physiol. (Cell Physiol.) **252:** C499-C504.
5. KIMURA, J., A. NOMA & H. IRISAWA. 1986. Nature **319:** 596-597.
6. CASTILLO, E., H. GONZALEZ-SERRATOS, H. RASGADO-FLORES & M. ROZYCKA. 1991. Ann. N.Y. Acad. Sci. This volume.
7. ASHIDA, T. & M. P. BLAUSTEIN. 1987. J. Physiol. (London) **392:** 617-635.
8. NACHSHEN, D. A., S. SANCHEZ-ARMASS & A. M. WEINSTEIN. 1986. J. Physiol. (London) **381:** 17-28.
9. RASGADO-FLORES, H., E. M. SANTIAGO & M. P. BLAUSTEIN. 1988. J. Gen. Phsyiol. **93:** 1219-1241.
10. CERVETO, L., L. LAGNADO, R. J. PERRY, D. W. ROBINSON & P. A. MCNAUGHTON. 1989. Nature **337:** 740-743.
11. DIPOLO, R. & E. ROJAS. 1984. Biochim. Biophys. Acta **776:** 313-316.10.
12. SHEU, S.-S. & M. P. BLAUSTEIN. 1983. Am. J. Physiol. **244**(Cell Physiol. 13): C297-C302.
13. PHILIPSON, K. & A. Y. NISHIMOTO. 1982. Am. J. Physiol. **243**(Cell Physiol. 12): C191-C195.
14. HILGEMANN, D. W. 1990. Nature **344:** 242-245.
15. NELSON, M. T. & M. P. BLAUSTEIN. 1980. J. Gen. Physiol. **75:** 183-206.
16. HAGIWARA, S. & S. NAKAJIMA. 1966. J. Gen. Physiol. **49:** 807-818.
17. NELSON, M. T. & M. P. BLAUSTEIN. 1981. Nature **289:** 314-316.
18. ALTAMIRANO, A. & J. RUSSELL. 1978. J. Gen. Physiol. **89:** 669-686.
19. DIPOLO, R., J. REQUENA, F. J. BRINLEY, L. J. MULLINS, A. SCARPA & T. TIFFERT. 1976. J. Gen. Physiol. **67:** 433-467.
20. BLINKS, J. R., W. G. WIER, P. HESS & F. G. PRENDERGAST. 1982. Prog. Biophys. Molec. Biophys. **40:** 1-114.
21. LEDERER, W. J. & M. T. NELSON. 1983. J. Physiol. (London) **341:** 325-339.
22. ARMSTRONG, C. M. 1971. J. Gen. Physiol. **58:** 413-437.
23. HAGIWARA, S. & K. NAKA. 1964. J. Gen. Physiol. **48:** 141-162.
24. BITTAR, E. E., S. CHEN, B. G. DANIELDSON, H. A. HARTMANN & E. Y. TONG. 1972. J. Physiol. (London) **221:** 389-414.
25. XIE, H. & E. BITTAR. 1989. Biochim. Biophys. Acta **1014:** 207-209.
26. RUSSELL, J. M., W. F. BORON & M. S. BRODWICK. 1983. J. Gen. Physiol. **82:** 47-78.
27. DIPOLO, R. & L. BEAUGÉ. 1988. Biochim. Biophys. Acta **946:** 424-428.
28. GONZALEZ-SERRATOS, H. & H. RASGADO-FLORES. 1990. Am. J. Physiol. **259**(Cell Physiol. 28): C541-C548.
29. BLAUSTEIN, M. P. 1977. Biophys. J. **20:** 79-111.

Voltage Dependence of Sodium-Calcium Exchange and the Control of Calcium Extrusion in the Heart[a]

JOHN H. B. BRIDGE,[b] JOHN SMOLLEY,
KENNETH W. SPITZER, AND THOMAS K. CHIN

Nora Eccles Harrison Cardiovascular Research and Training Institute
and
Division of Cardiology
University of Utah
Salt Lake City, Utah 84112

INTRODUCTION

It has recently become possible to measure inward currents ($I_{Na\text{-}Ca}$) associated with calcium (Ca) extrusion that are attributable to sodium-calcium (Na-Ca) exchange.[1–6] However, there is little information on how Ca extrusion is controlled during the course of a cardiac action potential. When sarcoplasmic reticular (SR) Ca is released and intracellular Ca begins to rise, both the SR Ca pump and the Na-Ca exchange will compete to remove this Ca from the cytosol. It seems reasonably certain that two factors will profoundly influence the time course of the Na-Ca exchange. These are first the concentration of intracellular free Ca at various times during the action potential and second (since the Na-Ca exchange is voltage sensitive)[7] the trajectory of membrane potential. If, for instance, the Ca transient occurs early during the plateau of the action potential when membrane potential is changing only very slowly, one would expect the time course of Na-Ca exchange to be largely determined by the time course of the Ca transient. That is, as Ca rises the exchange will activate and as it falls the exchange will inactivate. Because $I_{Na\text{-}Ca}$ is believed to be linearly related to intracellular Ca concentration,[8] the exchange should reflect the time course of the Ca transient.[9] If on the other hand the action potential completely repolarizes before the Ca transient is over, the time course will be more difficult to predict. It is reasonable to assume that at a particular concentration of Ca exchange current will increase during membrane repolarization.[7] If at all Ca concentrations the exchange current-voltage relationship is steep, one would expect the inactivating effect of declining Ca to be offset by the activating effect of repolarization.[10] Under these circumstances the trajectory of membrane potential will significantly affect the time course of exchange. Thus, exchange could be sustained during the decline of the Ca transient.

[a] This work was supported by Grants HL42357 and HL42873 from the National Institutes of Health (Heart, Lung, and Blood Institute) and by awards from the Nora Eccles Treadwell Foundation and the Richard A. and Nora Eccles Harrison Fund for Cardiovascular Research.

[b] Address for correspondence: John H. B. Bridge, Nora Eccles Harrison CVRTI, University of Utah, Building 500, Salt Lake City, UT 84112.

In this report we first provide evidence in heart cells that the Na-Ca exchange can extrude sufficient Ca to produce mechanical relaxation. We also describe methods to isolate inward Na-Ca current on an interval that is similar to contraction. We measured a similar current in SF21 cells infected with Baculo virus containing DNA coding for the canine Na-Ca exchanger. This confirms the reliability of our method for measuring exchange current. We further investigate the Na-Ca exchange current-voltage relationship under circumstances where intracellular Ca is elevated with a view toward understanding the extent to which $I_{Na\text{-}Ca}$ is activated over the range of potentials exhibited by the cardiac action potential. Finally we investigate the time course of $I_{Na\text{-}Ca}$ during a cardiac action potential. Our preliminary results suggest that the Ca transient (inferred from the time course of $I_{Na\text{-}Ca}$) decays significantly before the action potential has repolarized. In our studies on ventricular cells $I_{Na\text{-}Ca}$ seems to be transiently activated during the early part of the action potential plateau, although this could be quite variable.

METHODS

The methods used in this study (including rapid cooling of rabbit papillary muscles) have mainly been described elsewhere.[5,11,12] Briefly guinea pig ventricular myocytes were obtained by enzymatic dispersion. These cells were voltage clamped[13] at 30°C using either a discontinuous voltage-clamp circuit (Axoclamp 2A, Axon Instruments, Burlingame, California) or, in later experiments, a current-to-voltage converter (Axopatch 1C, Axon Instruments, Burlingame, California). Cell shortening was measured using a video-based motion detector.[14] In action-potential clamp experiments, an action potential was recorded using the Axopatch 1C in current-clamp mode. This action potential was digitized and then used to command an action-potential clamp. In general $I_{Na\text{-}Ca}$ was activated by stimulating a ventricular cell to contract in the absence of extracellular Na (in the presence of caffeine or ryanodine this raises intracellular Ca) and then rapidly applying extracellular Na. Rapid application of Na was accomplished by means of a rapid switching device that has already been described.[15] Exchange currents were also recorded in SF21 cells. These were infected with recombinant Baculo virus containing DNA coding for the canine Na-Ca exchanger. Recombinant virus was provided by Dr. Ken Philipson. Reverse Na-Ca exchange was activated by rapidly reducing extracellular Na to zero. This raises intracellular Ca. Forward exchange was activated by reapplying extracellular Na. The pipette solutions used in ventricular cell experiments with ryanodine were similar to those described for the measurement of $I_{Na\text{-}Ca}$ in caffeine-treated cells.[5]

Pipette solutions for the experiments on SF21 cells contained 120 mM Cs (CsCl and CsOH), 20 mM NaCl, 0.2 mM $MgCl_2$, 10 μM EGTA, 3 mM MgATP, and 10 mM Hepes adjusted to pH 7.1 with CsOH. For experiments using action-potential clamps, the pipette solutions contained 140 mM K (KCl and KOH), 1.3 mM MgATP, and 10 mM Hepes adjusted to pH 7.1 with KOH. Control external solutions for ventricular and SF21 cell experiments contained 144.4 mM Na (NaCl and NaOH), 1.0 mM $MgCl_2$, 2.7 mM $CaCl_2$, 11.0 mM dextrose, and 10 mM Hepes adjusted to pH 7.4 with NaOH. In the zero-Na solution Li salts were used to replace Na. The external solutions for the action-potential clamp experiments (both zero and normal Na) were slightly modified and contained 4.4 mM KCl and either 140 mM Na or Li. Ryanodine was used at a concentration of 1 μM and in later experiments at a concentration of 10 μM.

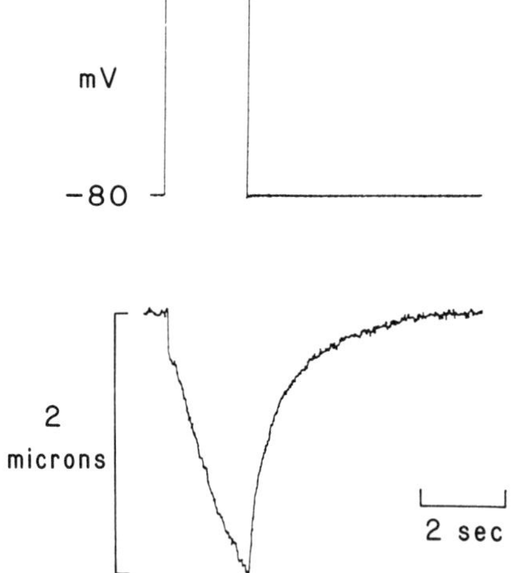

FIGURE 1. Cell contraction (measured as shortening in the presence of 10 mM caffeine. The cell was voltage clamped from −80 to 0 mV for 2 seconds (From Bridge *et al.*[11] *Science,* **241:** 823-825, 1988. Reproduced with permission from *Science* and the authors. Copyright 1988 by AAAS.)

RESULTS

Influence of Na-Ca Exchange on Mechanical Relaxation

We have conducted several different types of experiments on mammalian heart muscle all of which indicate that under appropriate conditions the Na-Ca exchange provides a powerful voltage-sensitive process that can extrude sufficient Ca to bring about mechanical relaxation. Guinea pig ventricular myocytes were treated with 10 mM caffeine and then subjected to a 2-sec voltage-clamp pulse from a holding of −80 mV to 0 mV. This produced a contracture that relaxed upon electrical repolarization[11] (FIG. 1). Since the SR cannot sequester Ca in the presence of caffeine, it seems likely that some process other than SR Ca sequestration was responsible for mechanical relaxation. This process shows a striking dependence on voltage. Thus if relaxation takes place at more depolarized potentials, it is slowed (FIG. 2). In addition mechanical relaxation in the presence of caffeine shows a significant dependence on extracellular Na. Contractures were first obtained with 2-sec voltage-clamp pulses in the absence of Na. These contractures did not relax rapidly upon electrical repolarization. However, application of 144.4 mM Na produced a prompt relaxation (FIG. 3). This voltage- and Na-dependent relaxation is likely to be produced by Na-Ca exchange. Under these circumstances the Na-Ca exchange can remove sufficient Ca (approximately 60 μmoles/liter cell water within 1 sec)[5] from the cytosol to produce mechanical relaxation. Another example of the relaxing effect of Na-Ca exchange was obtained by an examination of cooling contractures. When caffeine-treated rabbit papillary muscles relax from cooling contractures, the relaxation rate is also dependent on extracellular Na and membrane potential.[16] Additional analysis of relaxation from cooling contractures suggests that in the presence of a functioning SR Ca pump the Na-Ca

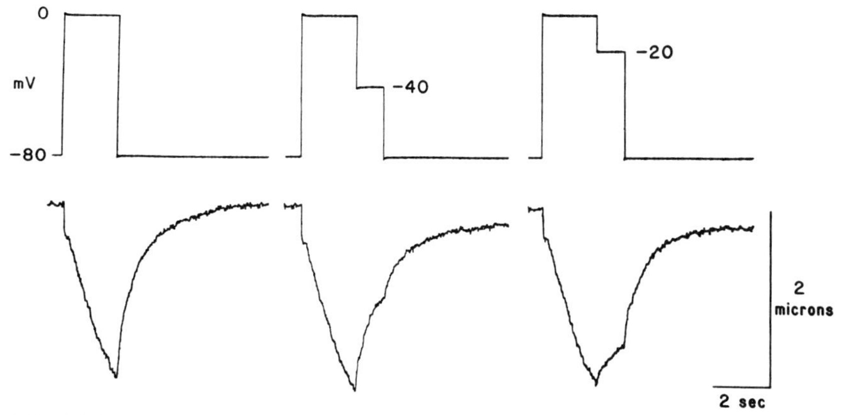

FIGURE 2. Tonic contractions elicited in the presence of 10 mM caffeine by voltage-clamp pulses from −80 mV to 0 mV. The dependence of relaxation on voltage was measured by repolarizing to different potentials (in this example −80, −40, and −20 mV). (From Bridge *et al.*,[11] *Science,* **241:** 823-825, 1988. Reproduced with permission from *Science* and the authors. Copyright 1988 by AAAS.)

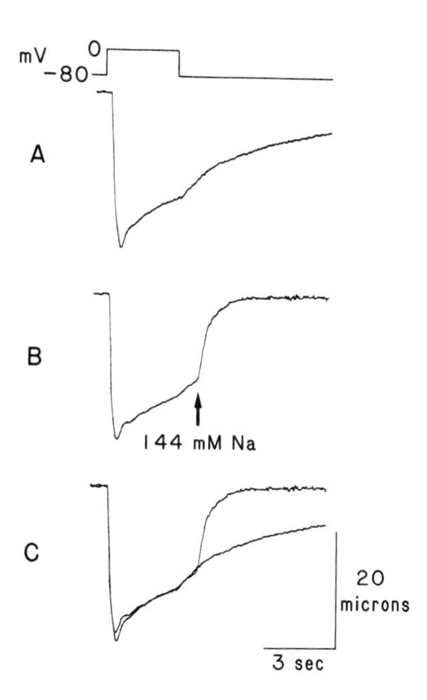

FIGURE 3. Tonic contractions elicited in the absence of a Na^+ gradient with voltage-clamp pulses from −80 mV to 0 mV. The contractions were either maintained or (in this example) relaxed slowly during the clamp pulse. **(A)** This slow relaxation accelerated on repolarization. **(B)** Sudden application of 144.4 mM external Na^+ after the return to −80 mV caused an eightfold increase in relaxation rate. **(C)** Contractions in (A) and (B) are superimposed to emphasize the effect of external Na^+ on relaxation rate. All solutions contained 10.0 mM caffeine. (From Bridge *et al.*,[11] *Science,* **241:** 823-825, 1988. Reproduced with permission from *Science* and the authors. Copyright 1988 by AAAS.)

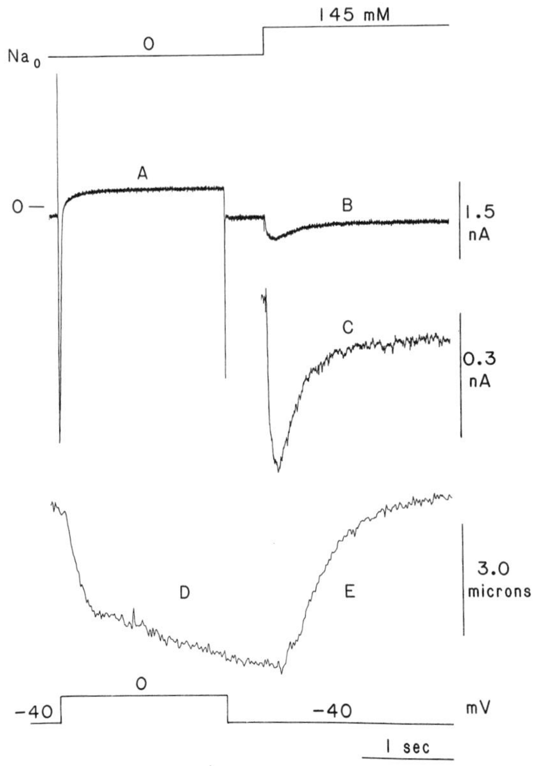

FIGURE 4. Two-second voltage-clamp pulses in the absence of Na^+ and presence of 10.0 mM caffeine cause contraction. Rapid application of Na 500 msec after repolarization causes relaxation. **(A)** I_{Ca} elicited by membrane depolarization. **(B)** The application of Na produces putative transient inward $I_{Na\text{-}Ca}$. This current is displayed on an expanded scale in **(C)**. **(D)** Contraction (cell shortening) activated by I_{Ca} recorded in (A). **(E)** After repolarization to −40 mV, relaxation does not occur until 500 msec after the clamp pulse when Na is suddenly applied. (From Bridge *et al.*,[5] *Science,* **248:** 376-378, 1990. Reproduced with permission from *Science* and the authors. Copyright 1990 by AAAS.)

exchange is capable of removing up to 30% of the cytosolic Ca (see D. Bers, this volume). This indicates that the exchange is capable of strenuous competition with the SR Ca pump for cytosolic Ca.

Membrane Currents Underlying Na-Dependent Relaxation in Heart

If Na-Ca exchange is responsible for Na- and voltage-dependent relaxation, then because the Na-Ca exchange is electrogenic, it should be possible to record a transient inward Na-dependent current that underlies Ca extrusion and accompanies relaxation. In studies that have already been reported,[5] guinea pig ventricular myocytes were treated with 10 mM caffeine and then voltage clamped and held at −40 mV in 0 mM Na replaced with 145 mM lithium. A two-second clamp step to 0 mV produced an inward Ca current I_{Ca} and a contracture (FIG. 4). Upon repolarization the cell did not

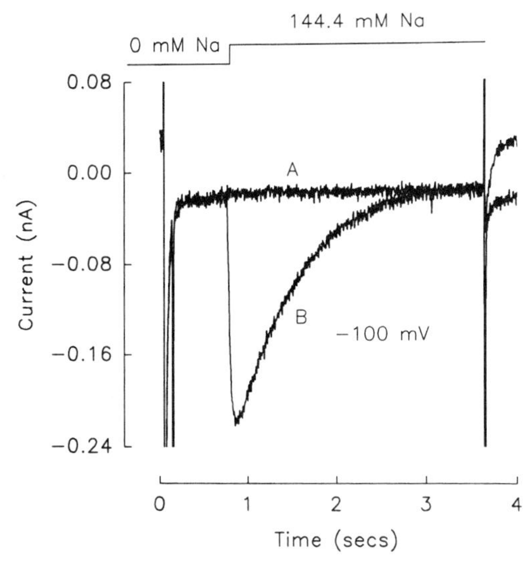

FIGURE 5. $I_{Na\text{-}Ca}$ activated in the presence of ryanodine. The cell was tetanized in the absence of Na with 500-msec voltage-clamp pulses from −40 to +10 mV. After the last pulse the cell was clamped to −100 mV and Na was rapidly applied producing transient inward exchange current (**B**). The current measured in the absence of Na is also displayed (**A**).

relax, presumably because in the presence of caffeine and in the absence of Na no processes exist that can rapidly reduce cytosolic free Ca (see D. Bers, this volume). Rapid application of Na 500 msec after the onset of repolarization produced a prompt mechanical relaxation and a transient inward current.[15] This transient current could not be elicited if the cell was clamped using pipettes containing 14 mM EGTA. This indicates that the current is activated by intracellular Ca. Thus we measured a Na-dependent transient inward current that requires elevated intracellular Ca for its activation and that accompanies mechanical relaxation. It seems therefore likely that this current is $I_{Na\text{-}Ca}$. This conclusion is strengthened by the observation that the total charge movement associated with I_{Ca} was twice that associated with the transient inward putative $I_{Na\text{-}Ca}$. This relationship is expected if (as is now widely believed[17–19]) three Na^+ exchange with a single Ca^{2+} and if all the Ca that entered the cell during the initial depolarization was removed by the Na-Ca exchange.

Na-Ca Exchange in Heart and SF21 Cells

Because it is difficult to maintain voltage clamps in the presence of caffeine, we used ryanodine to tetanize heart cells. Ten 100-msec clamp pulses from −40 to +10 mV produced a tetanic contracture. After the last pulse we repolarized the cell to −100 mV and rapidly applied Na. This activated a transient inward current and prompt relaxation of the cell. Currents recorded in the presence and absence of Na are displayed in FIGURE 5. We also measured $I_{Na\text{-}Ca}$ in SF21 cells infected with recombinant Baculo virus containing DNA coding for the canine Na-Ca exchanger. These cells were voltage clamped and held at 0 mV for several seconds to load them with Ca via reverse exchange. The holding potential was then set to −100 mV and Na was rapidly applied.

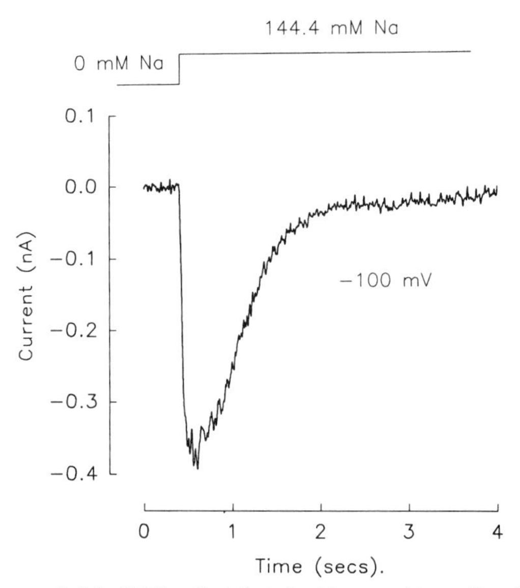

FIGURE 6. $I_{Na\text{-}Ca}$ recorded in SF21 cells infected with recombinant Baculo virus containing DNA coding the canine Na-Ca exchanger. The cell was voltage clamped at −40 mV and superfused with Na-deficient Tyrode solution. This caused Ca to enter the cell by reverse exchange. The membrane was then clamped to −100 mV and Na was rapidly applied. This activated transient exchange current. We substracted stable resting current obtained in the absence of Na.

This resulted in a large transient inward current that exhibited a striking resemblance to the exchange current in the ventricular cells. This current is displayed in FIGURE 6 after subtraction of a stable background current that is present in the absence of Na. This current could not be detected in uninfected cells. It is, however, noteworthy that the current measured in the SF21 cell was approximately twice those typically measured in ryanodine-treated heart cells. We do not know the extent to which Ca is elevated in SF21 cells under these circumstances. However, if the level of free Ca is similar in these cells and ventricular cells and since the SF21 cells are smaller than ventricular cells, this finding suggests that the density of exchange sites is at least twice that in ventricular cells. This is in accord with findings by Dr. Ken Philipson (personal communication). The rate of activation and inactivation of $I_{Na\text{-}Ca}$ in both cell types is similar. These results indicate that qualitatively similar currents can be obtained from ventricular cells and SF21 cells known to express the mammalian cardiac Na-Ca exchanger. This strengthens the view that the methods we employed to measure $I_{Na\text{-}Ca}$ are reliable. Moreover unless the nonspecific current of mammalian ventricular cells closely resembled $I_{Na\text{-}Ca}$ recorded in insect cells, it seems unlikely that putative $I_{Na\text{-}Ca}$ recorded in guinea pig ventricular cells is a nonspecific current of the type reported by Ono *et al.*[20]

Voltage Dependence of Na-Ca Exchange Current

We have investigated the voltage dependence of $I_{Na\text{-}Ca}$ under circumstances where intracellular Ca is assumed to be elevated. Guinea pig ventricular myocytes were

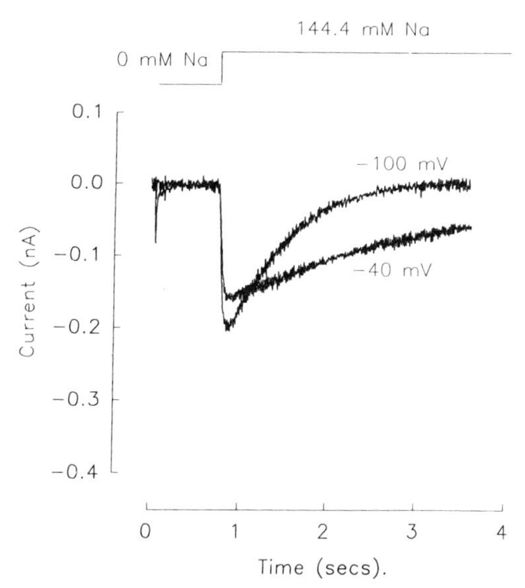

FIGURE 7. By subtracting Na-dependent current from Na-independent current, we obtained difference currents at −40 mV and −100 mV.

tetanized with ten brief voltage-clamp pulses from −40 to +10 mV in a solution containing no added Na and 1-10 μM ryanodine. After the last clamp pulse the cell was held at potentials that ranged between −100 mV to +60 mV. Five-hundred milliseconds after the onset of this holding potential, 144.4 mM Na was rapidly applied to the cell. This resulted in the activation of $I_{Na\text{-}Ca}$. Current was also measured at the same potential but without applying Na. Examples of difference currents measured at −100 mV and at −40 mV are displayed in FIGURE 7. These are mainly attributable to Na-Ca exchange. It is noteworthy that depolarization not only suppresses peak current but also slows the decay of current. Presumably at more positive potentials intracellular Ca cannot be extruded so rapidly so that it falls more slowly and slows the inactivation of exchange current. From current records like these measured over a whole range of potentials, we have been able to construct current-voltage relationships for peak $I_{Na\text{-}Ca}$ (FIG. 8). It should be emphasized that only the transient current is considered to be associated with Ca extrusion and that when this current settles we assume that Ca extrusion is negligible. Often, a stable current remains after settling. This disappears in the absence of Na. The possible origin of this current has been discussed.[5] It is apparent from this current-voltage relationship that despite the elevated intracellular Ca, which we assume would favor Ca extrusion over the whole range of potentials, most of the current is activated between +20 and −80 mV. From this we might infer that most of the current is activated during the steeply repolarizing phase of the cardiac action potential.

When Is Na-Ca Exchange Current Activated During the Ventricular Action Potential?

$I_{Na\text{-}Ca}$ will only be activated during the steeply repolarizing phase of the action potential if intracellular Ca is elevated to some extent; that is, if intracellular Ca has

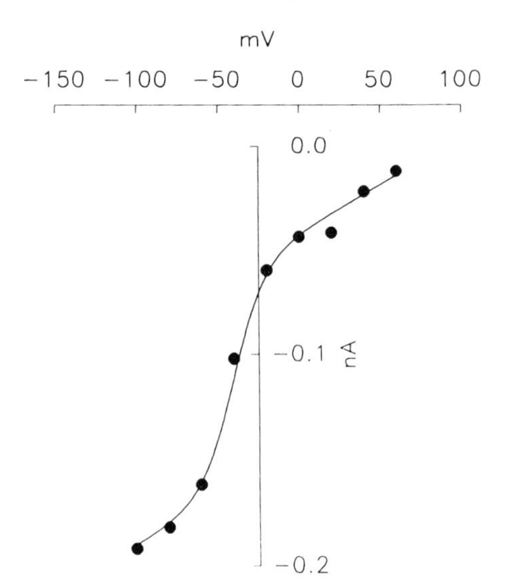

FIGURE 8. Plot peak transient difference current ($I_{Na\text{-}Ca}$) against membrane potential.

not been reduced by the sarcoplasmic reticulum before complete repolarization has taken place. To obtain some idea of the time course of $I_{Na\text{-}Ca}$ during the cardiac action potential, we subjected a cell to voltage-clamp pulses that were of the configuration of a cardiac action potential. This procedure has already been used to detect $I_{Na\text{-}Ca}$ in heart cells.[21] To accomplish this, we used a digitized action potential as the command pulse. The external solution was a modified standard Tyrode solution and the pipette solution contained 140 mM K, 3 mM MgATP, and no Na. Thus, in these cells the SR was functioning. We first applied an action potential shaped clamp pulse in the absence of Na. After rapidly switching to a Na-containing solution, we applied a second clamp. The difference currents revealed a significant transient current that peaked within 12 msec and activated early during the plateau of the action potential (FIG. 9). When this experiment was repeated with pipettes containing 14 mM EGTA, this transient current was abolished as expected of $I_{Na\text{-}Ca}$. This result suggests that at least in this cell the Ca transient was over well before the steeply repolarizing phase of the action potential. Thus in this case $I_{Na\text{-}Ca}$ did not occur during the late phase of repolarization as we initially supposed, and $I_{Na\text{-}Ca}$ was largely controlled by intracellular Ca and not voltage.

DISCUSSION

These results indicate that under circumstances where the SR cannot sequester Ca (e.g., after caffeine treatment) a voltage- and Na-dependent process can extrude sufficient Ca to produce mechanical relaxation. By using a rapid perfusion technique, it is possible to isolate a transient inward exchange current that underlies sodium- and voltage-dependent relaxation. At least in caffeine-treated cells, this current can quantitatively account for mechanical relaxation. It is also possible to tetanize ryanodine-treated cells in the absence of extracellular Na and by rapidly applying Na to activate $I_{Na\text{-}Ca}$ transiently. Peak values of $I_{Na\text{-}Ca}$ activated in this way revealed a sigmoid current-voltage relationship. Finally we used action-potential clamps and measured mem-

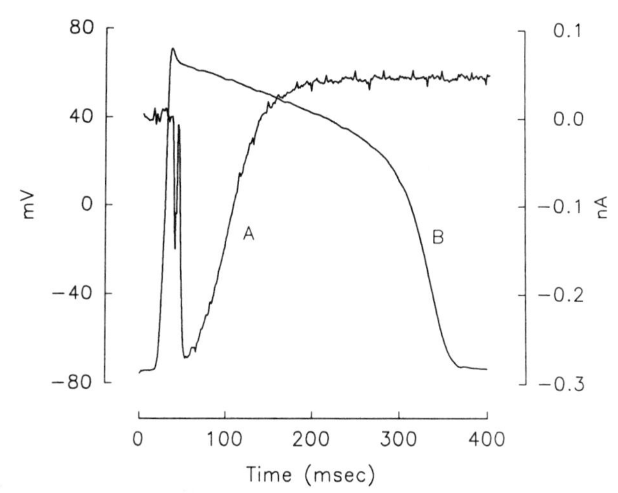

FIGURE 9. A voltage-clamp pulse in the configuration of an action potential is displayed in (**B**). By subjecting the cell to action-potential clamps in the presence and absence of 140 mM Na, we obtained the difference current displayed in (**A**). This is putative $I_{Na\text{-}Ca}$.

brane currents in the presence and absence of Na. The difference current obtained by subtracting Na-dependent from Na-independent current revealed transient behavior. This current could not be obtained if 14 mM EGTA was included in the pipette. From this we infer that it was $I_{Na\text{-}Ca}$ and that it activated early during the action-potential plateau and was over before repolarization was complete.

In this study $I_{Na\text{-}Ca}$ was activated in two different ways. In the first it was activated by applying Na after first elevating intracellular Ca by tetanizing the cell. The transient nature of this current can be explained by supposing that extracellular Na reaches a peak value before intracellular Ca has significantly declined. At this point the current is maximal. As intracellular Ca declines the current inactivates. The second method involved stimulating the cell to contract by subjecting it to a voltage-clamp pulse that is the same configuration as an action potential. In the first case the clamp pulse is applied in the absence of extracellular Na, in which case there is no $I_{Na\text{-}Ca}$. Then the clamp pulse is applied in the presence of 144.4 mM Na, in which case $I_{Na\text{-}Ca}$ is present. The difference in membrane current in the presence and absence of Na reveals the time course of $I_{Na\text{-}Ca}$ during this simulated action potential. Under these circumstances $I_{Na\text{-}Ca}$ is activated and then inactivated by the rise and fall of intracellular Ca as it is released and resequestered by the SR (FIG. 9). This probably explains the fact that $I_{Na\text{-}Ca}$ is more rapidly activated by this method than by external application of Na, which involves diffusion delays as Na penetrates the T-tubules.

When $I_{Na\text{-}Ca}$ is activated by rapid application of Na, we have assumed that the transient current settles when intracellular Ca is pumped down to sufficiently low levels that inward $I_{Na\text{-}Ca}$ is negligible.[22] At more negative potentials where Ca is probably reduced to about 100 nM or so, this assumption is probably valid. Thus the difference between peak and settled current is a measure of the magnitude of $I_{Na\text{-}Ca}$. These currents were measured in the absence of external potassium and with Cs replacing potassium in the pipette. This was done to eliminate as much contaminating potassium current as possible.[23] These experiments were also conducted on a 10-sec cycle length in which the cell was tetanized in the absence of Na and then Na was applied for the remaining

80% of the cycle. Since potassium was absent and therefore the Na pump was presumably blocked, one would expect intracellular Na accumulation. Upon removing extracellular Na to tetanize the cell, outward exchange current is stimulated. When Na is reapplied this outward current would be suppressed, giving the appearance of an activated inward current. There is probably also a background Na current which is inward upon Na application and appears even in the presence of large quantities of intracellular EGTA when inward $I_{Na\text{-}Ca}$ cannot occur to any significant extent. The existence of these currents can explain why $I_{Na\text{-}Ca}$ does not settle to values that existed before the application of Na. We also observed that during a long experiment when the cell spent a lot of time in the absence of extracellular potassium these background currents tended to enlarge. This is consistent with continued intracellular Na accumulation leading to ever-increasing reverse exchange.

Our view that the transient inward current activated by Na application is $I_{Na\text{-}Ca}$ is strengthened by recent preliminary results obtained with SF21 cells. The recently cloned cardiac Na-Ca exchanger[24] has been expressed using the Baculo virus/SF21 expression system. Inward $I_{Na\text{-}Ca}$ could be activated by rapid Na application to a cell that had been loaded with Ca by reverse exchange. This current qualitatively resembled the exchange current measured in ventricular cells. In the example shown, $I_{Na\text{-}Ca}$ in the SF21 was twice that typically measured in ventricular cells but qualitatively resembled the ventricular exchange current. These results strengthen our conclusion that we have in fact isolated $I_{Na\text{-}Ca}$ in ventricular cells. Moreover $I_{Na\text{-}Ca}$ in heart cells is unlikely to be explained by the existence of a Ca-activated nonspecific current because no such current appears to exists in the SF21 cells. If one concludes that putative $I_{Na\text{-}Ca}$ measured in ventricular cells is in fact Ca-activated, nonspecific current, then one must also conclude that the similarity between the ventricular cell current and $I_{Na\text{-}Ca}$ measured in SF21 cells is fortuitous. This seems unlikely.

The form of the Na-Ca exchange current-voltage relationship measured at what is assumed to be elevated intracellular Ca is sigmoid. About 75% of the exchange current is activated between 20 and −80 mV. There appear to be signs of saturation of the current at both positive and negative potentials, and Hilgemann has offered an explanation for this.[25] In the vertebrate rod outer segment, the form of the exchange current-voltage relationship is independent of intracellular Ca.[26] There is also some evidence that this is the case in guinea pig ventricular cells.[7] Moreover, we might expect the current-voltage relationship to shift to the left as intracellular Ca is reduced and to the right as it increases.[7] Finally it is reasonable to assume that the current-voltage relationship will move to the right as intracellular Na is reduced and to the left as it is increased. Thus, as intracellular Na increases at fixed intracellular Ca and membrane potential, less inward current can be activated and less Ca extrusion will occur. These facts will certainly have a considerable bearing on the time course of Ca extrusion during a cardiac action potential.

With an intact SR the rise and fall of intracellular Ca during a twitch will be largely controlled by the release and sequestration of intracellular Ca by the SR. Thus the activation of $I_{Na\text{-}Ca}$ and the interval over which it is activated will be largely determined by this process. However, predicting the time course is more complicated. For example, the peak of $I_{Na\text{-}Ca}$ will not necessarily coincide with the peak of the Ca transient.[6] The results with the action-potential clamp (FIG. 9) provide perhaps the simplest case to consider. These results were obtained with pipette solutions containing no Na and in the presence of extracellular potassium. Thus it is likely that intracellular Na was very low. The current-voltage relationship for $I_{Na\text{-}Ca}$ is likely to be shifted to the right so that at positive potentials larger currents than those measured at higher intracellular Na might be expected. It is clear from FIGURE 9 that $I_{Na\text{-}Ca}$ activates within 12 msec and is completely over within 200 msec during the early plateau of the action potential when voltage is not changing significantly. Under these circumstances the time course of the exchange current will largely be determined by and will also reflect the time course of the Ca transient. In particular the magnitude of $I_{Na\text{-}Ca}$ will be approximately proportional to the concentration of intracellular Ca.[8] These results are in close agreement with those of Terrar and White,[9] who demonstrated an envelope of tail currents of similar time course in ventricular cells.

What sort of behavior might we expect if repolarization is more rapid in relationship to the Ca transient? Although Ca is falling during the plateau of the action potential, provided repolarization is rapid, the inactivating effect of declining Ca may be more than offset by the activating effect of repolarization. Under these circumstances $I_{Na\text{-}Ca}$ will not be proportional to Ca concentration, and the peak value of $I_{Na\text{-}Ca}$ may occur after the peak of the Ca transient. An example of this sort of behavior is clearly described by Egan *et al.*[6] where peak $I_{Na\text{-}Ca}$ occurs after the peak of the Ca transient. In this case (their FIG. 11) the action potential is shorter and repolarizes much more rapidly during the Ca transient than occurred in our study depicted in FIGURE 9. Others have reported that Na-Ca exchange is activated during late repolarization of the action potential.[21,27] There are also reports of Ca transients that considerably outlast the action potential.[28,29] Under these circumstances we might expect exchange to be sustained or even reactivated to some extent during the period of rapid repolarization.[30] With suitable assumptions as to the extent of Ca-dependent activation, Noble (see this volume) has proposed secondary activation of the exchange, which appears as a second peak in the exchange current. In relatively short action potentials produced by rat ventricular cells at room temperature, Schouten and ter Keurs[31] have inferred that Na-Ca exchange contributes to the slow phase of repolarization.[31] In atrial cells Earm *et al.*[32] demonstrate that $I_{Na\text{-}Ca}$ contributes to the repolarization of atrial cell action potentials. Presumably in these cases repolarization is occurring during the Ca transient. In general the precise trajectory of $I_{Na\text{-}Ca}$ will depend on the temporal relationship between the time course of the Ca transient and the time course of membrane repolarization, and this might be quite variable.

As soon as Ca rises in the cytosol SR, Ca sequestration will begin. At the same time the Na-Ca exchange will extrude Ca. Thus two processes will compete for cytosolic Ca. During the steady state the relative quantities of Ca sequestered and extruded will remain unchanged. If, however, the action potential suddenly shortens in relationship to the Ca transient, then relatively more Ca will be extruded with the result that less Ca will enter the SR. The next contraction should be appropriately diminished. Recent results from cells with intact SR support this idea.[33] Conversely if the action potential is suddenly prolonged, the SR Ca pool should enlarge with attendant positive inotropy. This suggests that the Na-Ca exchange might be capable of regulating SR Ca content in ways that are extremely dependent on intracellular Ca, intracellular Na, and the trajectory of membrane potential.

In summary a number of studies indicate that in a variety of heart cell types $I_{Na\text{-}Ca}$ occurs during systole rather than after it. The precise timing of exchange and, among other things, the regulation of SR Ca content will vary between cell types and within a given cell depending on the condition of that cell.

ACKNOWLEDGMENTS

We thank Dr. K. Philipson and Dr. Xhaoping Li for providing recombinant Baculo virus. We also thank Ms. Virginia Hill and Prof. Donald Summers for their help and advice with the SF21 expression system. Dr. Loess Miller kindly provided the SF21 cell line. Mr. Dale Anderson provided indispensable technical assistance.

REFERENCES

1. KIMURA, J., A. NOMA & H. IRISAWA. 1986. Na-Ca exchange current in mammalian heart cells. Nature **319:** 596-599.
2. HUME, J. R. & A. UEHARA. 1986. Properties of "creep currents" in single frog atrial cells. J. Gen. Physiol. **87:** 833-855.
3. HUME, J. R. & A. UEHARA. 1986. "Creep currents" in single frog atrial cells may be generated by electrogenic Na-Ca exchange. J. Gen. Physiol. **87:** 857-884.
4. LIPP, P. & L. POTT. 1988. Transient inward current in guinea-pig atrial myocytes reflects a change of sodium-calcium exchange current. J. Physiol. **397:** 601-630.
5. BRIDGE, J. H. B., J. R. SMOLLEY & K. W. SPITZER. 1990. The relationship between charge movements associated with I_{Ca} and $I_{Na\text{-}Ca}$ in cardiac myocytes. Science **248:** 376-378.
6. EGAN, T. M., D. NOBLE, T. POWELL, A. J. SPINDLER & V. W. TWIST. 1989. Sodium-calcium exchange during the action potential in guinea-pig ventricular cells. J. Physiol. **411:** 630-661.
7. MIURA, Y. & J. KIMURA. 1989. Sodium-calcium exchange current. J. Gen. Physiol. **93:** 1129-1145.
8. BARCENAS-RUIZ, L., D. J. BEUCKELMANN & W. G. WIER. 1987. Sodium-calcium exchange in heart: Membrane currents and changes in $[Ca^{2+}]_i$. Science **238:** 1720-1722.
9. TERRAR, D. A. & E. WHITE. 1989. Changes in cytosolic calcium monitored by inward currents during action potentials in guinea-pig ventricular cells. Proc. R. Soc. London [Biol.] **238:** 171-188.
10. POWELL, T. & D. NOBLE. 1989. Calcium movements during each heart beat. Mol. Cell. Biochem. **89:** 103-108.
11. BRIDGE, J. H. B., K. W. SPITZER & P. R. ERSHLER. 1988. Relaxation of isolated ventricular cardiomyocytes by a voltage-dependent process. Science **241:** 823-825.
12. BRIDGE, J. H. B. 1986. Relationships between the sarcoplasmic reticulum and calcium transport revealed by rapidly cooling rabbit ventricular muscle. J. Gen. Physiol. **88:** 437-473.
13. HAMILL, O., E. MARTY, E. NEHER, B. SAKMANN & F. SIGWORTH. 1981. Improved patch-clamp techniques for high-resolution current recording from cells and cell-free membrane patches. Pfluegers Arch. **391:** 85-100.
14. STEADMAN, B. W., K. B. MOORE, K. W. SPITZER & J. H. B. BRIDGE. 1988. A video system for measuring motion in contracting heart cells. IEE Trans. Biomed. Eng. **35:** 264-272.
15. SPITZER, K. W. & J. H. B. BRIDGE. 1989. A simple device for rapidly exchanging solution surrounding a single cardiac cell. Am. J. Physiol. Cell Physiol. **256:** C441-C447.
16. BERS, D. M. & J. H. B. BRIDGE. 1989. Relaxation of rabbit ventricular muscle by Na-Ca exchange and sarcoplasmic reticulum. Circ. Res. **65:** 334-342.
17. REEVES, J. P. & C. C. HALE. 1984. The stoichiometry of the cardiac sodium-calcium exchange system. J. Biol. Chem. **259:** 7733-7739.
18. BRIDGE, J. H. B.& J. B. BASSINGTHWAIGHTE. 1983. Uphill sodium transport driven by an inward calcium gradient in heart muscle. Science **219:** 178-180.
19. RASGADO-FLORES, H., E. M. SANTIAGO & M. P. BLAUSTEIN. 1989. Kinetics and stoichiometry of coupled Na efflux and Ca influx (Na-Ca exchange) in barnacle muscle. J. Gen. Physiol. **93:** 1219-1241.
20. EHARA, T., A. NOMA & K. ONO. 1988. Calcium-activated non-selective cation channel in ventricular cells isolated from adult guinea-pig hearts. J. Physiol. **403:** 117-133.
21. DOERR, T., R. DENGER, A. DOERR & W. TRAUTWEIN. 1990. Ionic currents contributing to the action potential in single ventricular myocytes of the guinea pig studied with action potential clamp. Pfluegers Arch. **416:** 230-237.
22. EHARA, T., S. MATSUOKA & A. NOMA. 1989. Measurement of reversal potential of Na^{+}-Ca^{2+} exchange current in single guinea-pig ventricular cells. J. Physiol. **410:** 227-249.

23. MATSUDA, H. & A. NOMA. 1984. Isolation of calcium current and its sensitivity to monovalent cations in dialysed ventricular cells of guinea-pig. J. Physiol. **357:** 553-573.
24. NICOLL, D. A., S. LONGONI & K. D. PHILIPSON. 1990. Molecular cloning and functional expression of the cardiac sarcolemmal Na^{+}-Ca^{2+} exchanger. Science **250:** 562-565.
25. HILGEMANN, D. W. 1988. Numerical approximations of sodium-calcium exchange. Prog. Biophys. Molec. Biol. **51:** 1-45.
26. LAGNADO, L., L. CERVETTO & P. A. MCNAUGHTON. 1988. Ion transport by the Na-Ca exchange in isolated rod outer segments. Proc. Natl. Acad. Sci. USA **85:** 4548-4552.
27. SCHOUTEN, V. J. A., H. E. D. J. TER KEURS & J. M. QUAEGEBEUR. 1990. Influence of electrogenic Na-Ca exchange on the action potential in human heart muscle. Cardiov. Res. **24:** 758-767.
28. BARRY, W. H. & J. H. B. BRIDGE. 1988. The relative importance of calcium influx and efflux via Na-Ca exchange in cultured myocardial cells. *In* Cellular Calcium Regulation. D. R. Pfeiffer, Ed. Plenum. New York.
29. BEUCKELMANN, D. J. & W. G. WIER. 1988. Mechanism of release of calcium from sarcoplasmic reticulum of guinea-pig cardiac cells. J. Physiol. **405:** 233-255.
30. RYDER, K. O., S. M. BRYANT & G. HART. 1990. Calcium and calcium-activated currents in hypertrophied left ventricular myocytes isolated from guinea-pig. J. Physiol. **430:** 71P.
31. SCHOUTEN, V. J. A. & H. E. D. J. TER KEURS. 1985. The slow repolarization phase of the action potential in rat heart. J. Physiol. **360:** 13-25.
32. EARM, Y. E., W. K. HO & I. S. SO. 1990. Inward current generated by Na-Ca exchange during the action potential in single atrial cells of the rabbit. Proc. R. Soc. London [Biol.] **240:** 61-81.
33. CHIN, T. K., K. W. SPITZER & J. H. B. BRIDGE. 1990. Activation of electrogenic Na-Ca exchange during twitches in isolated guinea pig ventricular cell. J. Gen. Physiol. **96:** 1a-101a.

Translocation Mechanism of Cardiac Na-Ca Exchange[a]

JINMING LI

Department of Pharmacology
China Medical University
5-3 Nanjing Street Heping District
Shenyang, People's Republic of China

JUNKO KIMURA[b]

Department of Pharmacology
Yamagata University School of Medicine
2-2-2 Iida-nishi
Yamagata 990-23, Japan

INTRODUCTION

In the cardiac myocyte, the electrogenic Na-Ca exchange is a major sarcolemmal calcium efflux system.[1-3] The mechanism of the exchange, whether it is simultaneous or consecutive, has not been resolved.[4,5] In the simultaneous exchange scheme, both Na and Ca bind to the exchanger and then the translocation takes place, while in the consecutive scheme Na (for example) binds to the exchanger and is translocated first and then Ca binds and is translocated, so that Na and Ca cross the membrane in two steps with "ping-pong" kinetics. The method of examining these different modes of reactions has been well established in enzyme kinetics[6] and has been applied to the membrane transport mechanism.[4,5] Recently the membrane current generated by the Na-Ca exchange has been isolated by loading Na and Ca on the opposite side of the membrane while blocking almost all other membrane currents in single cardiac myocytes using the patch-clamp and intracellular dialysis techniques.[7,8] This method allows detailed kinetic studies on the Na-Ca exchange.[9,10] In the present study, we examined the transport mechanism of the Na-Ca exchange by measuring the outward exchange current under the condition that inward exchange current could not flow.

METHODS

The methods of cell preparation, whole-cell voltage clamp and intracellular dialysis, were described in detail previously.[9] In brief, single ventricular cells are isolated from

[a]This work was supported by a grant from the Ministry of Education, Science and Culture and by the Japan Heart Foundation Research Grant for 1989.

[b]Address for correspondence: Junko Kimura; Department of Pharmacology, Yamagata University School of Medicine, 2-2-2 Iida-nishi, Yamagata 990-23, Japan.

TABLE 1. Composition of the External Solution

	Concentration (mM)
Li	150
Mg	0.5
Cl	150
EGTA	5, 1, or 0
Free Ca	0.1
	0.25
	0.5
	1
	2
	3
HEPES	10
Ouabain	0.02
Verapamil	0.002
pH	7.4

NOTE: Reproduced from Li & Kimura[22] with permission from the *Journal of General Physiology* (Rockefeller University Press).

guinea pig heart by perfusing collagenase (Yakult, Japan; 8 mg/50 ml) in Ca-free Tyrode solution on the Langendorff apparatus. The patch pipette for the whole-cell voltage clamp had a tip diameter of 4-5 μm (resistance of 1-2 MΩ).

When a Giga-ohm seal was established with normal Tyrode solution in a glass pipette, the pipette solution was changed to the test internal solution, which contained a high concentration of BAPTA (Ca-chelating agent) and various concentrations of Na using a thin polyethylene tubing in the pipette. The patch membrane was then ruptured to perform the whole-cell clamp technique. At least 5 min was required for the equilibration of the solution in the cell interior before giving test clamp pulses. The current-voltage relation was obtained by ramp pulses from the holding potential of -10 mV to $\pm$ 100 mV at the rate of 0.4 V/sec, which was recorded on line by a personal computer (NEC PC9801-VM). The descending portion of the ramp was used for display.[9]

Solutions

The compositions of external and internal solutions are described in TABLES 1 and 2, respectively. The concentration of free Ca was calculated by the equations of Fabiato and Fabiato[11] with the correction of Tsien and Rink[12] using the binding constants of BAPTA.[13] All the experiments were performed at about 36°C.

Data

The numerical data are shown as the mean $\pm$ standard error unless otherwise stated.

TABLE 2. Composition of the Internal Solutions[a]

	150 mM Na	30 mM Na	20 mM Na
	(mM)	(mM)	(mM)
Na	150	30	20
Cs	0	70	120
Tris	0	10	0
Mg	5	5	5
Aspartate	45	60	45
Cl	30	30	30
ATP	10	10	10
CrP	5	5	5
BAPTA	30	30	30
Ca	14.5	14.5	14.5
HEPES	10	10	10

NOTE: Reproduced from Li & Kimura[22] with permission from the *Journal of General Physiology* (Rockefeller University Press).
[a]pCa 7, pH 7.4.

Theory

To distinguish the two types of transport kinetics, consecutive (ping pong) or simultaneous (sequential), it is necessary to obtain the values of apparent K_m (Michaelis constant) of one of the transported ions and the ratio of the apparent K_m/apparent I_{max} (maximum current density) according to the following theory.

1. The consecutive model is called the "ping-pong, bi-bi system" in Segel's enzyme kinetics (Segal,[6] p. 606), and the scheme of a one-way reaction; that is, Na_i-Ca_o exchange in the presence of Na only inside and Ca outside, is as follows:

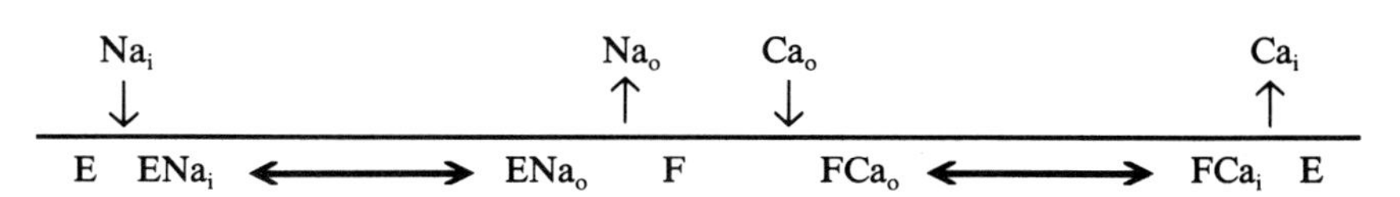

where E is the form of the exchanger that has a higher affinity to Na and F has a higher affinity to Ca. ENa is the exchanger-Na complex and ECa is the exchanger-Ca complex (subscript "o" or "i" indicates external or internal, respectively). The velocity equation for Hanes-Woolf linear plot is as follows (arranged from Segel,[6] Chapter IX, p. 142):

$$\frac{[Ca]_o}{i} = \frac{K_m Ca_o}{I_{max}} + \frac{\frac{(K_m Na_i)^n}{([Na]_i)^n} + 1}{I_{max}} [Ca]_o \tag{1}$$

where $K_m Ca_o$ is the Michaelis constant or the concentration of Ca_o that yields half-maximal velocity (or current, i) at saturating $[Na]_i$, $K_m Na_i$ is that for $[Na]_i$ at saturating $[Ca]_o$, I_{max} is the maximum current magnitude, and n is the Hill coefficient for $[Na]_i$.[6] The equation is the form of y = a + bx where y = $[Ca]_o$/i and x = $[Ca]_o$. The Y-intercept is constant regardless of $[Na]_i$, while the slope decreases as $[Na]_i$ increases,

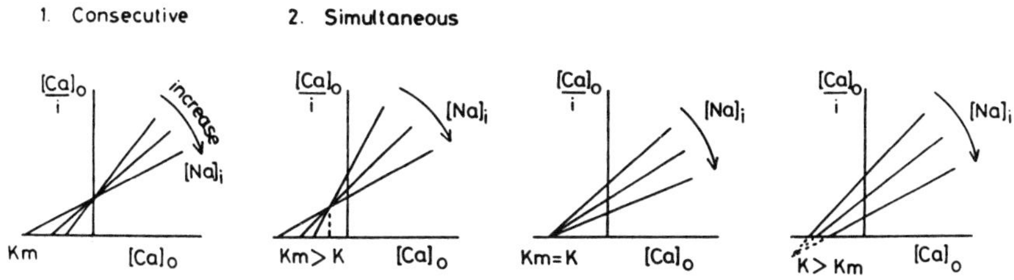

FIGURE 1. Expected patterns of a Hanes-Woolf plot of the consecutive model and three distinct simultaneous models. K_m represents the Michaelis constant and K represents the dissociation constant. See text for explanations. (Reproduced from Li & Kimura[22] with permission from the *Journal of General Physiology* [Rockefeller University Press].)

and the absolute value of X-intercept (apparent $K_m Ca_o$) increases as $[Na]_i$ increases. The scheme of the graph is shown in FIGURE 1.

2. The simultaneous (also called sequential) exchange model corresponds to "random bi-bi" or "ordered bi-bi system" of Segel[6] (see pp. 274 and 560), where both Na and Ca bind the exchange molecule and are translocated at the same time. Although various types of "bi-bi" reactions exist, the simplest models are considered in the present study. In the "random" bi-bi system, Na and Ca bind and dissociate at random, while in the "ordered" bi-bi system, Na and Ca bind and dissociate in a fixed order. It has been known that Na and Ca compete with each other in the external binding site.[9,14] This evidence limits the order of binding, if it is an "ordered bi-bi," to a reaction illustrated by the following scheme.

$$\begin{array}{ccccccc} Na_o & Ca_i & & & Na_i & Ca_o & \\ \downarrow & \downarrow & & & \uparrow & \uparrow & \\ \hline E \quad ENa_o & ENa_oCa_i & \longleftrightarrow & ENa_iCa_o & & ECa_i & E \end{array}$$

ENaCa is the exchanger-Na-Ca complex. The consequent velocity equations of the two types of simultaneous reactions are identical. Thus for the $[Na]_i$-$[Ca]_o$ exchange mode, the velocity equation for the Hanes-Woolf linear plot is as follows (derived from Segel,[6] Chapter IX, p. 89):

$$\frac{[Ca]_o}{i} = \frac{\dfrac{(K_m Na_i)^n \, KCa_o}{([Na]_i)^n} + K_m Ca_o}{I_{max}} + \frac{\dfrac{(K_m Na_i)^n}{([Na]_i)^n} + 1}{I_{max}} [Ca]_o \qquad (2)$$

KCa_o is the dissociation constant of $[Ca]_o$ and the other symbols are the same as in Equation (1). This equation is also the form of y = a + bx where y = $[Ca]_o/i$ and x = $[Ca]_o$. The difference between Equations 1 and 2 are the value of "a" or Y-intercept, which is constant in Eq. (1) while it is the function of $[Na]_i$ in Eq. (2).

The plot of Equation (2) varies further according to the ratio of $K_m Ca_o$ and KCa_o (see FIG. 1): (i) If $K_m Ca_o > KCa_o$, the lines obtained at different $[Na]_i$ will cross on the left of Y-axis and the absolute value of x-intercept (apparent $K_m Ca_o$) increases as $[Na]_i$ increases. (ii) If $K_m Ca_o = KCa_o$, the apparent $K_m Ca_o$ is constant regardless of $[Na]_i$. (iii) If $K_m Ca_o < KCa_o$, the apparent $K_m Ca_o$ decreases as $[Na]_i$ increases.

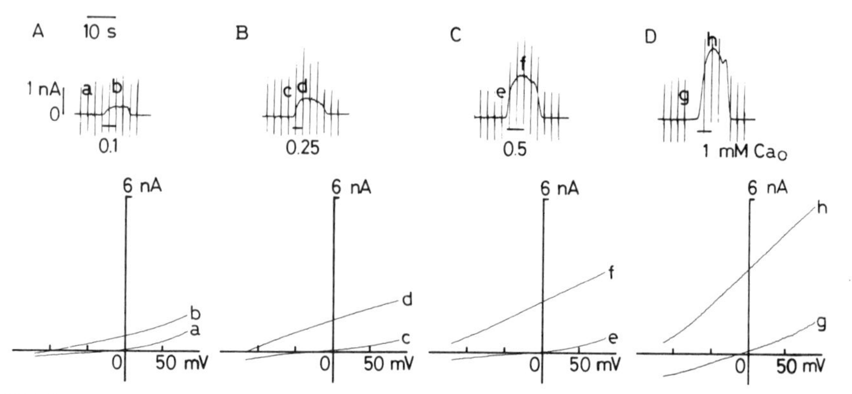

FIGURE 2. *Top:* Chart of the outward shift of the holding current at −10 mM induced by a brief application of $[Ca]_o$ indicated by the bar below each trace. $[Na]_i$ is 150 mM. The vertical lines indicate the current in response to ramp pulses of ± 100 mV. *Bottom:* I-V relations taken at the corresponding labels in the inset. The I-V curves of a, c, e, and g are the control, and b, d, f, and h are during $[Ca]_o$ superfusion. The data of A-D are taken from different cells. The capacitance of the cells was 181 pF, 121 pF, 172 pF, and 184 pF in A-D, respectively. (Reproduced from Li & Kimura[22] with permission from the *Journal of General Physiology* [Rockefeller University Press].)

Thus in order to distinguish the Na-Ca exchange reaction from among the above four models, the value of the Y-intercept and the direction of the shift of apparent K_m values of $[Ca]_o$ were examined at various $[Na]_i$ in the following experiment.

RESULTS

It has been suggested that it is advantageous to study the translocation mechanism under "zero trans" conditions.[4] Therefore we set the ionic conditions to develop only an outward exchange current by loading Na only inside and Ca outside. Ca_i cannot be eliminated because it is necessary for the exchange to operate,[7,15-17] so that a minimum amount of 0.1 μM free Ca_i was present in all the internal solutions. Although Na-Ca exchange is known to operate in the Ca-Ca exchange mode, Ca-Ca exchange is inhibited in the presence of a high concentration of Na.[18,19] Therefore we assumed that the fraction of Ca-Ca exchange is insignificant even in the presence of a 100 nM $[Ca]_i$ and that it is possible to measure the translocation kinetics.

As shown at the top of FIGURE 2, a brief superfusion of various $[Ca]_o$ at the holding potential of −10 mV induced an outward exchange current in the presence of 150 mM $[Na]_i$. The control external solution is Ca-free (1 or 5 mM EGTA and no added Ca) and the test solution contains either 0.1, 0.25, 0.5, 1, 2, or 3 mM free Ca. The current-voltage (I-V) relations obtained by the ramp pulses of ± 100 mV from the holding potential before and during the $[Ca]_o$ superfusion are shown below in FIGURE 2A to D. The net exchange current was obtained as the difference between the two

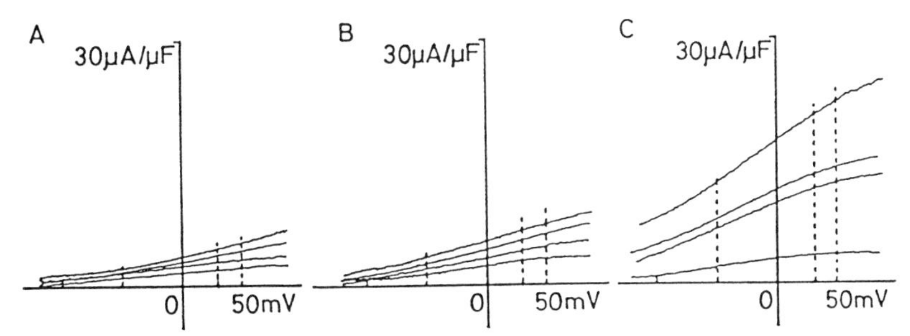

FIGURE 3. The superimposed difference currents between the control and at the peak response during $[Ca]_o$ superfusion. **A,** 20 mM $[Na]_i$; **B,** 30 mM $[Na]_i$; **C,** 150 mM $[Na]_i$ (obtained from FIG. 2). The current magnitudes are expressed as the current density. The average capacitance of the cells is 156 ± 39 pF (n = 33) at 20 mM $[Na]_i$, 153 ± 49 pF (n = 20) at 30 mM $[Na]_i$, and 129 ± 37 pF (n = 28) at 150 mM. The current increases as $[Na]_i$ and $[Ca]_o$ get higher. The vertical dotted lines indicate the potentials where the current magnitudes are measured for the dose-response relation. (Reproduced from Li & Kimura[22] with permission from the *Journal of General Physiology* [Rockefeller University Press].)

traces, and the representative I-V curves are superimposed in FIGURE 3C. Because one concentration of Ca was tested in each cell in most of the experiments, the current density was obtained by dividing the current magnitude with the capacitance of the cell to compare in different cells.

The I-V curves of the exchange current were obtained not only at 150 mM $[Na]_i$ but also at 20 and 30 mM $[Na]_i$, and the representative traces of each set of difference currents are superimposed in FIGURE 3. As $[Na]_i$ increased, the magnitude of the current density increased. The current density was measured at four different potentials, that is, +50, +30, 0, and −50 mV, and was plotted against $[Ca]_o$ to obtain the dose-response relations at three different $[Na]_i$.

TABLE 3 summarizes the numerical values of the current density obtained from the above experiments. FIGURE 4 shows the four sets of dose-response relations at different potentials plotted from TABLE 3. The curves are drawn by nonlinear regression analysis using the SAS computer program (Marquardt Methods, SAS Inc. Ltd., Cary, NC). All the curves are fitted well by hyperbolae, indicating that the reaction of $[Ca]_o$ binding to the exchanger is of the Michaelis-Menten type. We estimated the K_m and I_{max} values according to Sakoda and Hiromi[20] who demonstrated that the direct nonlinear fitting method is better than any linear plot analysis to estimate K_m and I_{max} values of the Michaelis-Menten type of reaction. This method, however, does not give values of the standard error of K_m/I_{max} straightforwardly. Therefore we employed Hanes-Woolf linear plot to estimate the K_m/I_{max} values with their standard errors directly (FIG. 5). This plot is known to be the most reliable among three different linear plots of Michaelis-Menten type kinetics.[21]

FIGURE 5 shows the Hanes-Woolf plot of the corresponding data shown in FIGURE 4. The lines were drawn by the least-squares method. TABLE 4 summarizes the numerical results obtained by the Hanes-Woolf plot (K'_m/I'_{max}) and the nonlinear regression curves (K'_m and I'_{max}). The apparent K_m values of $[Ca]_o$ obtained by the Hanes-Woolf plot are 0.51-0.62 mM at 150 mM $[Na]_i$, 0.16-0.21 mM at 30 mM, and 0.12-0.20 mM at 20 mM $[Na]_i$, which are similar to those obtained by the nonlinear regression shown in TABLE 4. The apparent K_mCa_o increases as $[Na]_i$ increases at all the potentials measured. This result is consistent with either the consecutive model or the simultaneous one of the $K_m > K_d$ type and excludes two other types of simultaneous models.

TABLE 3. Value of the Current Density Obtained by the Experiment Representatively Shown by FIGURES 2 and 3

$[Na]_i$ (mM)	$[Ca]_o$ (mM)	I_{NaCa} (μA/μF)				n
		−50 mV	0 mV	30 mV	50 mV	
20	0.1	0.84 ± 0.27	1.53 ± 0.19	1.93 ± 0.30	2.09 ± 0.35	5
	0.25	2.01 ± 0.45	2.92 ± 0.46	3.63 ± 0.52	4.08 ± 0.60	7
	0.5	2.27 ± 0.75	3.33 ± 1.16	4.38 ± 1.55	5.07 ± 1.75	6
	1	2.79 ± 0.83	4.41 ± 1.21	5.51 ± 1.46	5.58 ± 2.60	7
	2	2.80 ± 0.66	4.33 ± 1.00	5.15 ± 1.35	5.79 ± 1.59	8
30	0.1	1.59 ± 0.19	2.61 ± 0.34	3.17 ± 0.49	3.53 ± 0.53	5
	0.25	1.83 ± 0.12	3.15 ± 0.31	3.96 ± 0.34	4.42 ± 0.48	5
	0.5	3.67 ± 0.60	5.59 ± 0.76	6.99 ± 1.06	7.86 ± 1.21	5
	1	3.47 ± 0.47	5.37 ± 0.62	6.72 ± 0.89	7.25 ± 0.79	5
150	0.1	2.60 ± 1.01	3.57 ± 1.13	4.13 ± 1.47	4.48 ± 1.95	5
	0.25	6.75 ± 0.49	8.99 ± 1.41	10.01 ± 2.10	10.57 ± 2.66	5
	0.5	9.24 ± 2.12	11.99 ± 1.77	13.18 ± 1.71	13.71 ± 1.64	4
	1	11.90 ± 1.68	16.83 ± 1.37	19.70 ± 1.41	22.41 ± 2.43	5
	2	11.93 ± 1.27	18.58 ± 2.32	22.55 ± 3.39	24.86 ± 3.80	4
	3	15.19 ± 2.20	21.05 ± 1.82	24.59 ± 1.93	26.75 ± 2.25	5

NOTE: I_{NaCa} indicates the Na-Ca exchange current; n is the number of cells.

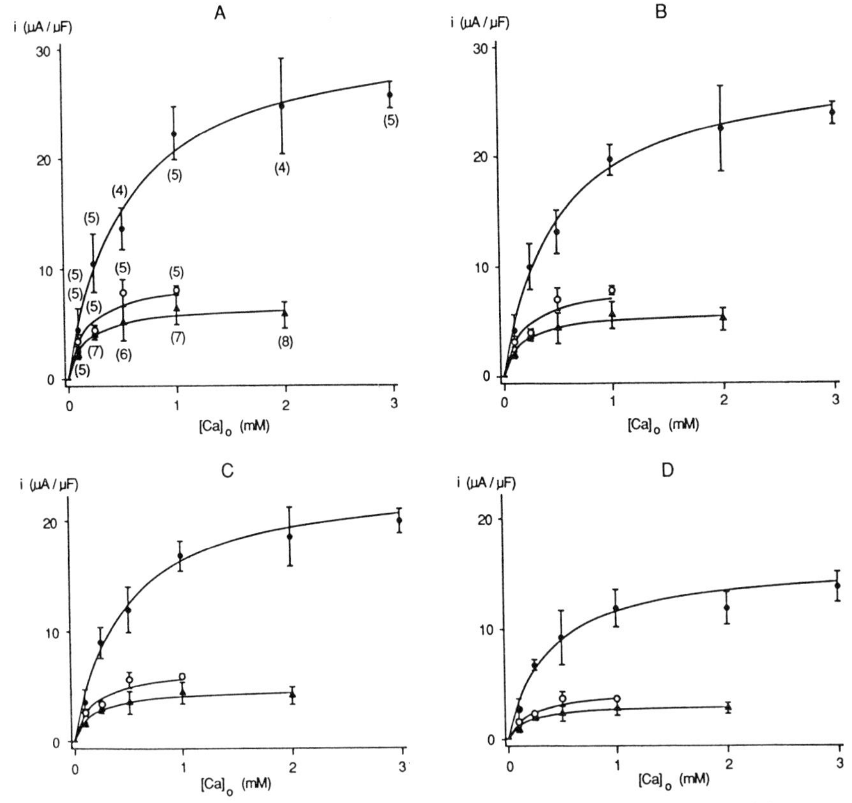

FIGURE 4. The dose-response curves between $[Ca]_o$ (abscissa) and the magnitude of the current density (ordinate) measured at four different voltages as indicated in FIGURE 3. **A,** +50 mV; **B,** +30 mV; **C,** 0 mV; **D,** −50 mV. Solid circles indicate the currents at 150 mM $[Na]_i$; open circles, 30 mM $[Na]_i$; and solid triangles, 20 mM $[Na]_i$. The number of cells at each point in B-D are the same as in A. The curves were drawn by the nonlinear regression (NLIN) procedure using a SAS computer program (Marquardt method). (Reproduced from Li & Kimura[22] with permission from the *Journal of General Physiology* [Rockefeller University Press].)

The ratio between K'_mCa_o and the apparent maximum current density or the value of Y-intercept is similar at three different $[Na]_i$ at +50 mV but tends to increase as $[Na]_o$ decreases at all the potentials. FIGURE 6 shows 95% confident intervals of K'_m/I'_{max} drawn by the computer program SAS. It can be seen that the overlapping area of the 95% confidence intervals includes the Y axis at all the potentials measured.

DISCUSSION

We have previously reported in the *Journal of General Physiology*[22] that our results are consistent with the simultaneous model, since the K'_m/I'_{max} values appeared to increase as $[Na]_i$ decreased, although we could not completely exclude the possibility of the consecutive mechanism because the values of K_m/I_{max} were so small and so close

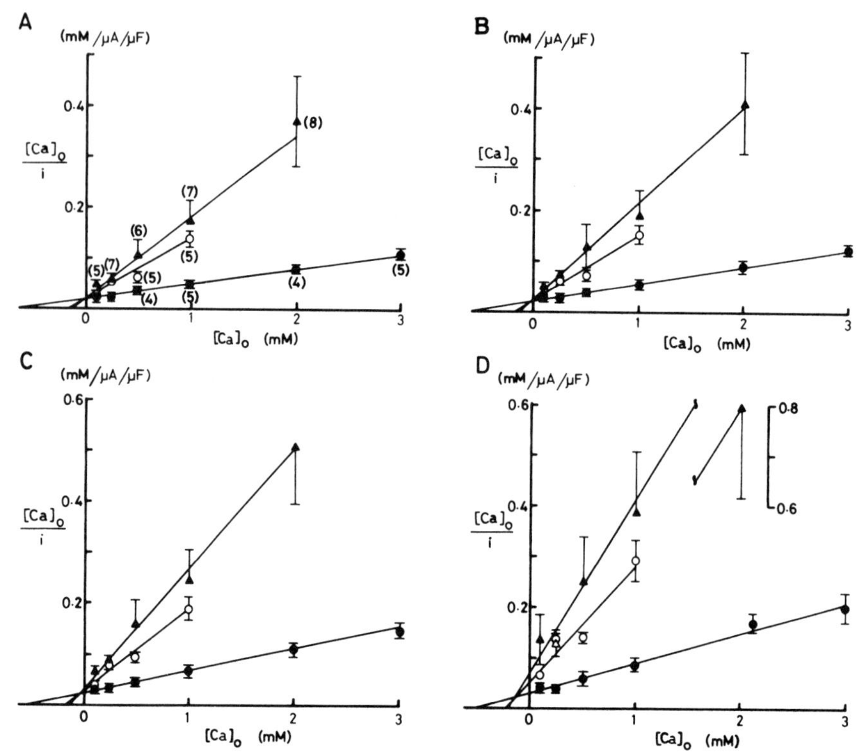

FIGURE 5. Hanes-Woolf plot of the data shown in FIGURE 3 and TABLE 3. **A,** +50 mV; **B,** +30 mV; **C,** 0 mV; and **D,** −50 mV. The lines were drawn by the least squares fit method. All three lines cross at the same point close to Y-axis but deviate to the left as the potential shifts in the negative direction.

to each other. In that article, howver, the analysis was satistically insufficient because the K'_m/I'_{max} values were compared without testing the range of experimental error. In the present analysis, we directly obtained the K'_m/I'_{max} values wih the standard error from the Hanes-Woolf plot. Student's t test was then performed on the difference between each possible pair of the K'_m/I'_{max} values in each potential group. It indicated that the difference was insignificant at +50, +30, and partly at 0 mV ($p < 0.05$). Furthermore, the 95% confidence intervals of the Hanes-Woolf linear regression revealed that the overlapping area at the three $[Na]_i$ included Y axis at all the four potentials (FIG. 6). These results cannot reject a null hypothesis that the K'_m/I'_{max} values are constant. This analysis therefore indicates that there is a possibility that the Na-Ca exchange is a consecutive mechanism.

The overlapping area of 95% confidence intervals shifted toward the left at more negative potentials. The Y-axis lies at the center of the overlapping area at 50 mV, but at −50 mV the Y axis lies on the right edge. The reason for this may be that the experimental error becomes relatively and progressively larger at more negative potentials where the current density is smaller, while the K_m values hardly change at the different potentials. Therefore our data must be more reliable at more positive potentials. Outward exchange current could have accumulated Ca and depeleted Na under

TABLE 4. Data Obtained by Nonlinear Regression (K'_m and I'_{max}) and by Hanes-Woolf Plot (K'_m/I'_{max})

E_m (mV)	$[Na]_i$ (mM)	$K'_m Ca_o \pm$ SE (mM)	$I'_{max} \pm$ SE (μA/μF)	$K'_m Ca_o/I'_{max} \pm$ SE (mM/μA/μF)	n
50	20	0.17 ± 0.062	6.74 ± 0.58	0.020 ± 0.015	33
	30	0.18 ± 0.055	9.11 ± 0.87	0.018 ± 0.005	20
	150	0.54 ± 0.095	32.05 ± 1.82	0.019 ± 0.002	28
30	20	0.17 ± 0.061	5.94 ± 0.50	0.024 ± 0.016	33
	30	0.19 ± 0.058	8.43 ± 0.81	0.022 ± 0.006	20
	150	0.52 ± 0.079	28.88 ± 1.41	0.020 ± 0.002	28
0	20	0.17 ± 0.059	4.77 ± 0.40	0.031 ± 0.020	33
	30	0.19 ± 0.053	6.68 ± 0.60	0.027 ± 0.007	20
	150	0.46 ± 0.067	23.92 ± 1.07	0.022 ± 0.003	28
−50	20	0.19 ± 0.067	3.17 ± 0.28	0.069 ± 0.033	33
	30	0.23 ± 0.076	4.51 ± 0.52	0.049 ± 0.013	20
	150	0.39 ± 0.082	16.21 ± 0.98	0.030 ± 0.006	28

NOTE: E_m indicates the membrane potential; $K_m Ca_o$, the apparent Michaelis constant for $[Ca]_o$; I'_{max}, the apparent maximum current density; and $K'_m Ca_o/I'_{max}$ indicates the ratio between the two values.

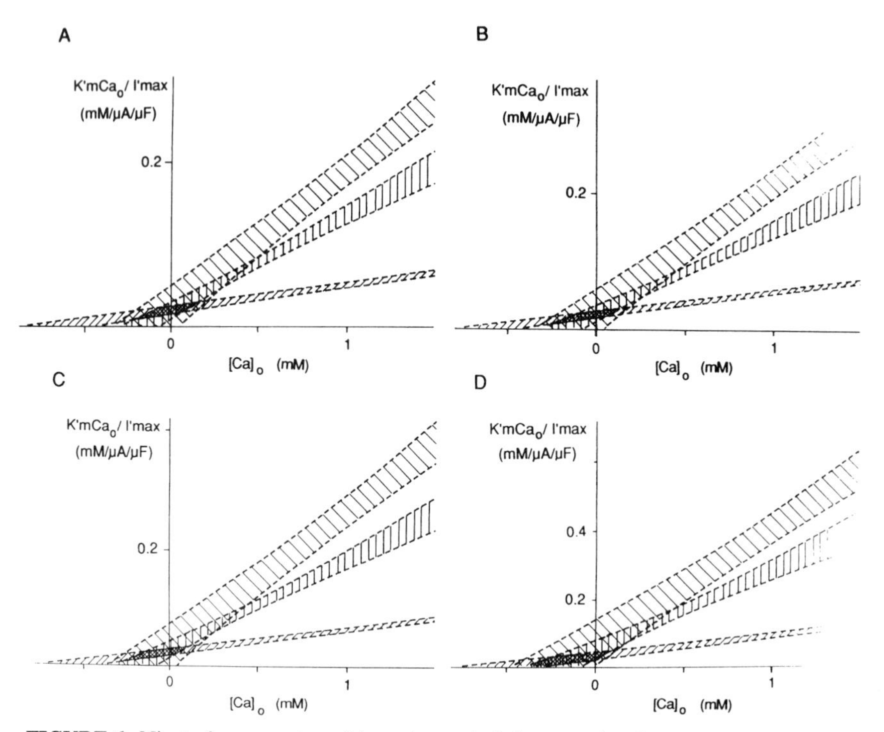

FIGURE 6. Ninety-five percent confidence interval of the regression lines shown in FIGURE 5. **A,** +50 mV; **B,** +30 mV; **C,** 0 mV; **D,** −50 mV. Note that the scales of the ordinate are all different. The darkest area is the overlapping area which includes the Y axis at all the potentials.

the membrane, and as a result it might have distorted $[Ca]_i$ and $[Na]_i$, although the external solution was changed as quickly as possible, and the current was measured at the earliest possible phase of the peak to avoid such artifacts. This effect may have additionally contributed to the apparent shift of the overlapping range.

So far various workers have suggested the simultaneous mechanism for the Na-Ca exchange because they did not see the shift of K_m values of the counterion when changing the ionic concentration on one side. Blaustein[23] measured $[Na]_o$-dependent ^{45}Ca efflux by changing $[Ca]_i$ to three different concentrations, and it did not change the K_mNa_o in perfused squid axon. Philipson and Nishimoto[24] examined $[Na]_i$-dependent ^{45}Ca uptake in dog ventricular vesicles, and at $[Na]_i$ of 140, 56, and 28 mM the same K_mCa_o of 28 μM was found. Hodgkin and Nunn[25] observed in the salamander rod that the fractional saturation by internal Ca did not affect the external Na-activation kinetics. Ledvora and Hegyvary[19] reported that the K_m for $[Ca]_o$ did not depend on $[Na]_i$ in cardiac vesicles. Their observation was carried out in the presence of ionic flux in both directions. All other workers' experiments were done under nominally "zero-*trans*" conditions. The reason why these workers did not see the shift of K_m values in either direction might be that the range of investigation was too small and/or that the range of shift of the K_m value was too small to detect. Also if the experiment was performed without buffering Ca, micromolar amounts of Ca contamination might have distorted the nominal "zero-*trans*" condition. In our result, the shift of K_mCa_o was only from about 0.17 to 0.5 mM for a 7.5-fold change of $[Na]_i$.

A piece of evidence supporting this argument was recently obtained by Khananshvilli.[26] He performed an experiment similar to ours with flux measurement in reconstituted proteoliposomes under zero-*trans* conditions using a Ca chelator. His result obtained clearly constant K'_m/V_{max} values with an approximately 3.5-fold increase in K_mCa_o with an eightfold increase in $[Na]_i$, and thus he concluded that the Na-Ca exchange is consecutive. The range of K_m shift he obtained is similar to our data.

More experimental data have been obtained to support a consecutive mechanism for the Na-Ca exchanger. Hilgemann observed in his giant patch method[27] that as counterion (*trans*) concentration is reduced toward zero, the apparent K_d of the transported ion (*cis*) decreases toward zero (personal) communication). Niggli and Lederer[28] recorded a transient current component preceding a Na-Ca exchange current upon photorelease of Ca, which they attributed to a conformation change of the exchanger and claimed that it suggests a consecutive mechanism. In addition, Phillipson and his colleagues[29] succeeded in cloning the amino acid sequence of the dog heart Na-Ca exchanger, which revealed a similar primary structure to the Na-K pump, although the amino acid sequence itself was largely different. This result also supports the consecutive mechanism, since the Na-K pump is known to be consecutive.[30]

The consecutive mechanism is also known in other systems, such as the anion exchanger[31] and Na-Li exchange of red blood cells.[32] For the Na-H exchanger, the possibility of a consecutive mechanism still remains, although a limited experimental result was consistent with a simultaneous mechanism.[33]

The maximum density of Na-Ca exchange current was surprisingly large, that is, 32 $\mu A/\mu F$ at 50 mV at 150 mM $[Na]_i$, in our experiment. Using this value, we calculated the density of the carrier based on the estimate of the turnover number for the exchanger of 1000/sec.[34] The number became $20 \times 10^{10}/\mu F$ or 40×10^6 per cell assuming the membrane capacity of a ventricular cell at 200 pF. This value is similar to the estimated density of the Na-K pump molecule (26×10^6 per cell).[35] Our value is about 10 times larger than the lower limit estimate of $250/\mu m^2$ by Niggli and Lederer,[28] who calculated the turnover of the exchanger to be 2500/sec.

The Na-Ca exchange current is voltage dependent and so either the Ca or the Na translocation step must be voltage dependent. In the present study, apparent K_mCa_o values were not significantly different among the four different potentials at any $[Na]_i$ ($p < 0.05$). This evidence may indicate that voltage dependence is not at the Ca-translocation step and so it may be at the Na-translocation step.

SUMMARY

In single cardiac ventricular cells of guinea pig, we have studied the ionic translocation mechanism of the electrogenic Na-Ca exchange, that is, whether Na and Ca ions countercross the membrane simultaneously or consecutively. The dose-response relations between the external Ca ($[Ca]_o$) and the outward Na-Ca exchange current were measured at three different internal Na concentrations ($[Na]_i$) in the absence of external Na. Hyperbolic regression curves and Hanes-Woolf linear plots of the dose-response relation revealed that apparent K_m values for external Ca (K'_mCa_o) decrease progressively as $[Na]_i$ decreases. The ratio of K'_mCa_o to apparent I_{max} value (I'_{max}) showed a slight increasing tendency as $[Na]_i$ decreased. We previously interpreted the data as consistent with the simultaneous mechanism[22] but without statistical analysis. Here we performed careful statistical analysis, which indicated that the K'_{max}/I'_{max} values were not significantly different among the different $[Na]_i$ at most of the potentials. This result suggests that Na-Ca exchange is likely to be a consecutive mechanism.

ACKNOWLEDGMENT

We are grateful to Dr. Donald Hilgemann for comments on the manuscript.

REFERENCES

1. Mullins, L. J. 1981. Ion Transport in Heart.: 20-43. Raven Press. New York.
2. Langer, P. 1982. Annu. Rev. Physiol. **44:** 435-439.
3. Carafoli, E. 1987. Annu. Rev. Biochem. **56:** 395-433.
4. Läuger, P. 1987. J. Membr. Biol. **99:** 1-11.
5. Hilgemann, D. W. 1988. Prog. Biophys. Molec. Biol. **51:** 1-45.
6. Segel, I. H. 1975. Enzyme kinetics. Wiley. New York.
7. Kimura, J., A. Noma & H. Irisawa. 1986. Nature **319:** 596-597.
8. Kimura, J., S. Miyamae & A. Noma. 1987. J. Physiol. **384:** 199-222.
9. Miura, Y. & J. Kimura. 1989. J. Gen. Physiol. **93:** 1129-1145.
10. Gadsby, D. C., M. Nakao, M. Noda & R. N. Shepherd. 1988. J. Physiol. **407:** 135P.
11. Fabiato, A. & F. Fabiato. 1979. J. Physiol. **75:** 463-505.
12. Tsien, R. Y. & T. J. Rink. 1980. Biochim. Biophys. Acta **599:** 623-638.
13. Tsien, R. Y. 1980. Biochemistry **19:** 2396-2404.
14. Reeves, J. P. & J. L. Stukuo. 1983. J. Biol. Chem. **258:** 3178-3182.
15. Baker, P. F. 1972. Prog. Biophys. Molec. Biol. **24:** 177-223.
16. Baker, P. F. & P. A. McNaughton. 1976. J. Physiol. **259:** 103-144.
17. Allen, T. J. A. & P. F. Baker. 1985. Nature **315:** 755-756.
18. Ledvora, R. & C. Hegyvary. 1983. Biochim. Biophys. Acta **729:** 123-136.
19. Slaughter, R. S. & J. L. Sutko. 1983. J. Biol. Chem. **258:** 3178-3182.
20. Sakoda, M. & K. Hiromi. 1976. J. Biochem. **80:** 547-555.
21. Atkins, G. L. & I. A. Nimmo. 1975. Biochem. J. **149:** 775-777.
22. Li, J. & J. Kimura. 1990. J. Gen. Physiol. **96:** 777-788.
23. Blaustein, M. P. 1977. Biophys. J. **20:** 79-111.
24. Philipson, K. D. & A. Y. Nishimoto. 1982. J. Biol. Chem. **257:** 5111-5117.
25. Hodgkin, A. L. & B. J. Nunn. 1987. J. Physiol. **391:** 371-398.
26. Khananshvili, D. 1990. Biochemistry. **29:** 2437-2442.
27. Hilgemann, D. W. 1989. Pflugers Arch. **415:** 247-249.
28. Niggli, E. & W. J. Lederer. 1991. Nature **349:** 621-624.
29. Nicoll, D. A., S. Longoni & K. D. Philipson. 1991. Science **250:** 562-565.
30. De Weer, P., D. C., Gadsby & R. F. Rakowski. 1988. Annu. Rev. Physiol. **50:** 225-241.
31. Gunn, R. & O. Fröhlich. 1979. J. Gen. Physiol. **74:** 351-374.
32. Ryu, K. H., N. C. Adragna & P. K. Lauf. 1989. Am. J. Physiol. **257:** C58-64.
33. Green, J., D. T. Yamaguchi, C. R. Kleeman & S. Muallem. 1988. J. Gen. Physiol. **92:** 239-261.
34. Cheon, J. & J. P. Reeves. 1988. J. Biol. Chem. **263:** 2309-2315.
35. Bahinski, A., M. Nakao & D. C. Gadsby. 1988. Proc. Natl. Acad. Sci. USA **85:** 3412-3416.

Photorelease of Ca^{2+} Produces Na-Ca Exchange Currents and Na-Ca Exchange "Gating" Currents[a]

ERNST NIGGLI

Department of Physiology
University of Bern
3012 Bern, Switzerland

W. JONATHAN LEDERER

Department of Physiology
University of Maryland, School of Medicine
Baltimore, Maryland 21201

INTRODUCTION

A large amount of information about the Na-Ca exchanger has been acquired during the last 24 years after its identification as a countertransport system. Transport properties of the exchanger have been studied in a variety of tissues including heart, smooth and skeletal muscle, nerve, and rod outer segments.[1] Experimentally, the activity of the Na-Ca exchanger has usually been assessed by measuring isotopic fluxes, changes of ionic concentrations or membrane current while changes of intra- and extracellular ionic concentrations have been imposed to activate or inhibit the Na-Ca exchanger. However, knowledge about the molecular operations of the Na-Ca exchanger has lagged far behind the accumulated information about the cellular functions of this transporter.

The recent development of photosensitive, biologically inactive precursor molecules ("caged compounds") allows rapid concentration jumps of the biologically active substance by photolysis with a flash of ultraviolet light. This technique has opened the door for kinetic studies of biochemical and physiological processes on a microsecond to millisecond time-scale (for reviews see Kaplan[2] and Gurney & Lester[3]). Here we present results that have been obtained in isolated guinea-pig ventricular myocytes by rapidly (within < 1 msec) increasing the intracellular Ca^{2+} concentration $[Ca^{2+}]_i$ using flash-photolysis of a light-sensitive Ca^{2+} buffer ("caged" calcium, DM-nitrophen[4]) while simultaneously measuring membrane current with the whole-cell patch-clamp technique. Concentration jumps of $[Ca^{2+}]_i$ were produced in a variety of experimental settings. Under experimental conditions designed to slow down or inhibit the exchanger, we were able to resolve the Na-Ca exchange current (I_{NaCa}) and a transient current

[a] This work was supported by grants from the National Institutes of Health to W.J.L. (HL25675 and HL36974) and grants from the Swiss National Science Foundation to E. N. (31-28545.90 and 87-30-BE).

component possibly reflecting a conformational rearrangement of the Na-Ca exchanger molecules after Ca^{2+} binding (I_{conf}).[5]

METHODS

All experiments were performed with enzymatically isolated cardiac myocytes in a chamber mounted on the stage of an inverted microscope. Whole-cell membrane currents were recorded with a patch-clamp amplifier and digitized at appropriate sampling rates. Cells were held at 0 mV to elevate resting $[Ca^{2+}]_i$ and to saturate DM-nitrophen with Ca^{2+}. This resulted in a stepwise increase of $[Ca^{2+}]_i$ upon photolysis of DM-nitrophen.[2] A flash unit that discharges electrical energy up to 230 Wsec through a short-arc Xenon-flashlamp was used to produce a pulse of light ~400-μsec duration. The light was collected with an ellipsoidal mirror and a segment of the spectrum (330 nm to 390 nm) was directed onto the preparation using an epi-illumination arrangement. Special care was taken to eliminate the electromagnetic interference (EMI) artifact. A combination of fiber-optic and shielding techniques was used to reduce coupling between the flash unit and the voltage-clamp amplifier. Cell length was measured with optical methods. The electrode filling solution had the following composition (in mM); Cs-gluconate, 120; TEA-Cl, 20; HEPES, 20; Na_4-DM-nitrophen, 2; reduced glutathione (GSH), 2; $CaCl_2$, 0.5; K-ATP, 5; pH 7.2. Superfusing solution: NaCl, 145; KCl, 4; CsCl, 1; $BaCl_2$, 0.5; $CaCl_2$, 1; HEPES, 10; glucose, 10; verapamil, 0.01; ryanodine, 0.01; pH 7.4. To reduce the mechanical activity associated with the large Ca^{2+} transient, 20 mM, 2,3-butanedione-monoxime (BDM), which had no significant effect on the current components, was added. Contractions indicative of successful Ca^{2+} release could still be observed.

RESULTS AND DISCUSSION

Rapid concentration jumps of $[Ca^{2+}]_i$ were produced to investigate properties of the Na-Ca exchanger in single cardiac myocytes. Activity of the electrogenic Na-Ca exchanger was measured as whole-cell membrane currents recorded under experimental conditions designed to minimize currents through ionic channels of the cell membrane. Interference from Ca^{2+} release and uptake by the sarcoplasmic reticulum (SR) was pharmacologically eliminated with ryanodine.[6] Before photolytic flashes were elicited, the cells were preloaded with DM-nitrophen and Ca^{2+} by dialysis through the patch-clamp electrode for at least 10 minutes. DM-nitrophen is a light-sensitive Ca^{2+}-buffer which increases the apparent K_d for Ca^{2+} from 5 nM to 3 mM after photolysis (at pH 7 and zero Mg^{2+}) and rapidly releases Ca^{2+} with a $t_{1/2}$ of less than 200 μsec.[7] FIGURE 1A illustrates the change of whole-cell current and cell-length produced by photolysis of DM-nitrophen at a fixed membrane potential of -40 mV. Immediately after the flash (arrow) an inward current was fully activated and the cell started to shorten. The inward current subsequently decayed with a $\tau \approx 550$ msec while $[Ca^{2+}]_i$ was declining as a result of the cycling Na-Ca exchanger and as indicated by relengthening of the cell. Inward current activated by photorelease of Ca^{2+} is consistent with electrogenic Na-Ca exchange current (I_{NaCa}), since the Na-Ca exchanger extrudes one Ca^{2+} in exchange for 3 Na^+. However, other Ca^{2+}-activated changes of the membrane conductance could contribute to the observed inward current as well (e.g., the Ca^{2+}-activated nonspecific cation channels responsible for I_{NS}).[8-10]

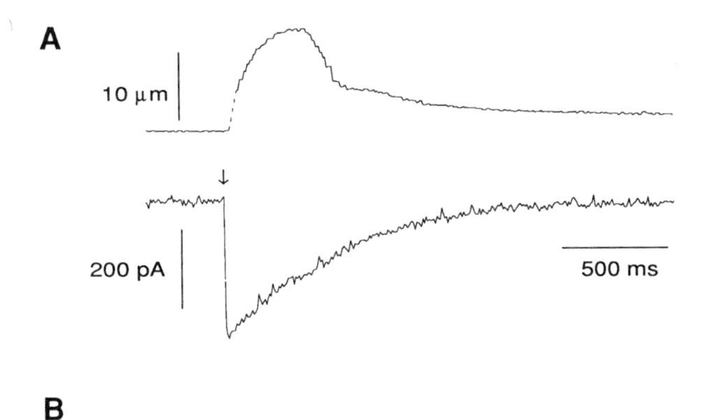

FIGURE 1. Time-course and voltage dependence of Na-Ca exchange current (I_{NaCa}). Changes of cell-length and whole-cell membrane current were produced by flash photolysis of caged Ca^{2+} at a fixed membrane potential of -40 mV (**A**). After the flash (arrow) an inward current was generated, and cell shortening was initiated. A monoexponential decay of the inward current ($\tau \approx 550$ msec) was observed while $[Ca^{2+}]_i$ declined as indicated by relengthening of the cell. The voltage-dependence of peak I_{NaCa} is plotted in (**B**). Currents are normalized for photolytic consumption of DM-nitrophen. The solid line represent the best fit of an Eyring rate theory model to the data (used with permission from *Nature*[5]).

To test for this possibility, we examined the voltage dependence of the Ca^{2+}-activated currents and the sensitivity of these currents towards inhibitors of the Na-Ca exchanger (e.g, Ni^{2+}, La^{3+}, zero $[Na^+]_o$). In the tested range of membrane potentials (between -100 mV and $+100$ mV) the current was always inwards suggesting that it represents I_{NaCa} (FIG. 1B and other experiments). This finding still does not rigorously exclude a small contribution from Ca^{2+}-dependent channels. However, in the absence of extracellular Na^+ (replaced by Li^+) and in the presence of 4 mM Ni^{2+} the current activated by flash-photolysis of caged Ca^{2+} was completely abolished. Because Li^+ permeates I_{NS} channels as Na^+ does[10] and because Ni^{2+} does not block I_{NS},[8] these maneuvers would have revealed any contaminating I_{NS} activated by photorelease of Ca^{2+}. Taken together, these findings strongly suggest that the currents activated by jumps of $[Ca^{2+}]_i$ arose from I_{NaCa} alone and that no other membrane currents were activated (or inhibited) by the increase of $[Ca^{2+}]_i$. We do not know, at present, why the Ca^{2+}-activated nonspecific cation current (I_{NS}) is not manifest under our experimental conditions.

Although many cellular properties of the Na-Ca exchanger are well characterized, very little is known about its function as a molecule. Nevertheless, several models for a transport mechanism have been proposed and discussed.[1,11-15] FIGURE 2A shows three possible simple models for a Na-Ca exchange mechanism. In principle, the exchanger may either transport the Ca^{2+} and Na^+ ions in two consecutive membrane-crossing molecular transitions or may move the ions simultaneously during one transition. If a simultaneous exchanger has to rearrange itself before it can undergo the next transport cycle (and thus also undergoes two consecutive membrane-crossing transitions only one of which moves ions), it is called a two-step simultaneous transporter, whereas a one-step simultaneous carrier has identical inside and outside configurations and does not require such an additional step before it can enter the next transport cycle.

A current-transient during rapid activation of the Na-Ca exchangers is predicted for several models of the transporter if the equilibrium distribution of carrier states can be disturbed more rapidly than the transporters are able to cycle. In our experiments, we rapidly ($t_{1/2}$ < 200 μsec) increased $[Ca^{2+}]_i$ from an estimated concentration of around 100 nM to concentrations in the micromolar range. Assuming diffusion-limited Ca^{2+} association to the binding site on the exchanger, the rate of Ca^{2+} binding would increase in parallel with $[Ca^{2+}]_i$. Depending on the rate of other partial reactions of the transport cycle, Ca^{2+} binding may thus be rate-limiting for cycling in the Na_o-Ca_i exchange mode (Ca^{2+} removal or "forward mode") before the flash, but another partial reaction may become rate-limiting at the elevated $[Ca^{2+}]_i$ prevailing after photolysis of DM-nitrophen. To simulate and qualitatively predict the effects of Ca^{2+} concentration jumps, we implemented a simple mathematical model of a consecutive Na-Ca exchange mechanism using the simulation software Facsimile.[16] All ion-binding steps were modeled as diffusion-limited, second-order association reactions, dissociation rates were derived using K_d values set equal to published apparent K_M values.[17,18]

FIGURE 2B shows membrane currents predicted by our own computer model during the early activation phase with a temporal resolution 100 times higher than in FIGURE 1A. An instantaneous increase of $[Ca^{2+}]_i$ from 100 nM to 10 μM is simulated for two different cases. In the first case, the rate-limiting step for cycling (associated with Na^+ translocation in our consecutive model) is faster than the rate at which Ca^{2+}-bound exchanger states can be formed. This results in a monotonically increasing inward current, as expected for a process limited by the diffusion of Ca^{2+} ions to the binding site on the exchanger (dashed line). In the second simulation, Ca^{2+}-bound exchanger states are formed more rapidly than the exchanger is able to cycle (due to a slow and rate-limiting Na^+ translocating step), resulting in a current transient before a new steady-state current is reached. In the consecutive model this transient current is the consequence of a temporary accumulation of E*-Ca_i states (see FIG. 2A). This would in turn lead to a current transient if charge is moved during the Ca^{2+}-translocating step. At present the data appear to favor this type of model because the current transient is not abolished by zero $[Na^+]_o$ (see below) and because the consecutive model is consistent with Na-Na and Ca-Ca exchange without requiring additional complexity. In a two-step simultaneous model, a current transient is also expected, resulting from the rapid decrease of E-states (as Ca^{2+} binds to form E-Ca_i states, see FIG. 2A). The rapid coversion of the ion-free E-states to the E-Ca-states is paralleled by a net conver-

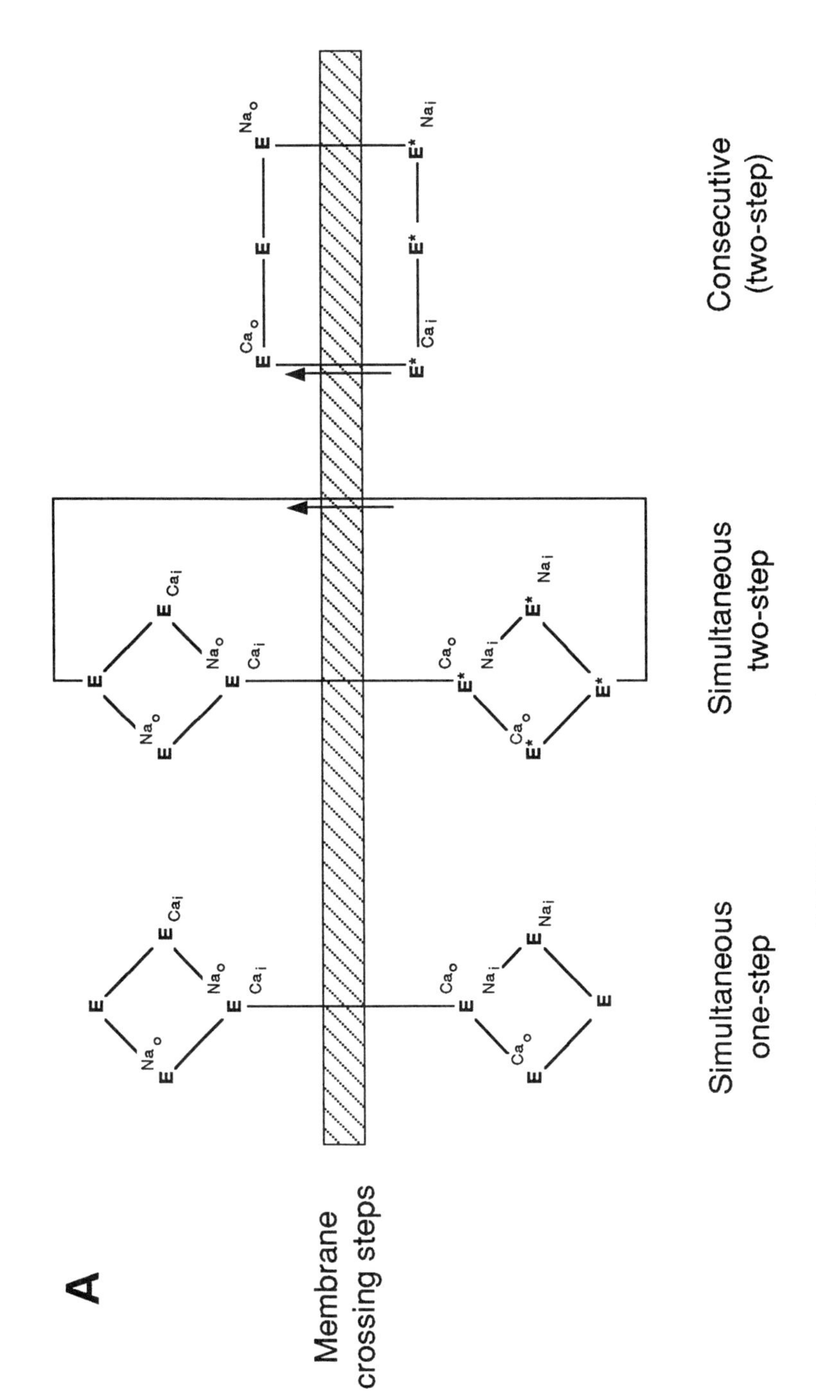

FIGURE 2. (Legend on following page.)

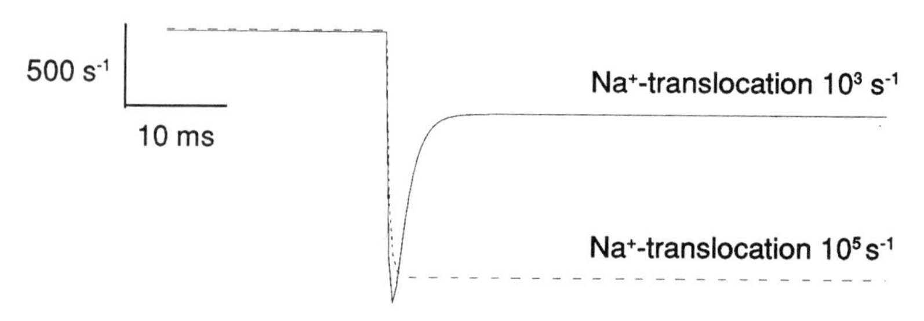

FIGURE 2. Models of the Na-Ca exchange transport mechanism predict I_{NaCa} and I_{conf}. Three possible simple diagrams for the Na-Ca exchanger are shown in **A** (facing page) including a one-step simultaneous, a two-step simultaneous, and a consecutive transport mechanism. For details refer to the text. An instantaneous increase in $[Ca^{2+}]_i$ may produce a current transient, if Ca^{2+}-bound exchanger states can be produced faster than the exchanger is able to cycle, as shown in the results of a computer simulation (**B**). (Panel A used with permission from *Nature.*[5])

sion of ion-free E*-states into ion-free E-states, a molecular transition generating current if charge is moved. Such current transients, once resolvable experimentally, would be a source of information about the molecular operations of the Na-Ca exchanger. Furthermore, the charge moved during the transients would allow estimates of site densities and turnover rates of the Na-Ca exchanger molecules.

As suggested by the computer simulations, the activation of the Na-Ca exchanger was examined on a rapid time scale. The activation of I_{NaCa} at high temporal resolution is shown in FIGURE 3A. After the flash we observed a monotonically increasing inward current ($\tau \approx 400$ μsec) without any appreciable current transient. The decay of inward current seen in FIGURE 1A is too slow to be noticeable at this temporal resolution. From this record we learned that under the given experimental conditions the exchanger molecules were apparently turning over more rapidly than we were able to produce Ca^{2+}-bound carrier states. However, in independent experiments we have observed a large effect of temperature on the Na-Ca exchanger with a 3.5-fold reduction of peak I_{NaCa} after lowering the temperature from 35°C to 25°C. Therefore, cooling appeared to be a possible approach to slow down the turnover of the exchanger in order to reveal current transients. As we had anticipated, the current recorded at 12°C (FIG. 3B) showed a small initial transient that could clearly be distinguished from the reduced (after cooling), more maintained current. We have called this transient current I_{conf} because it arises from a conformational rearrangement of the exchanger molecules due to Ca^{2+} binding. The maintained current component was again identified as the Na-Ca exchange current I_{NaCa} based on its voltage dependence (FIG. 1B) and its sensitivity to extracellular inhibitors of the Na-Ca exchanger (e.g., DCB (FIG. 4A), Ni^{2+}, La^{3+}, or zero $[Na^+]_o$). The current transient (I_{conf}) appears not to be greatly affected by voltage and the application of these inhibitors (see, for example, FIG. 4A). This is consistent with I_{conf} arising from a conformational change of the Na-Ca exchanger, because it represents the charge moved during a partial reaction of transport and does not require complete cycling of the exchangers. Minor effects of membrane voltage or inhibitors on the charge moved cannot be excluded because the two currents

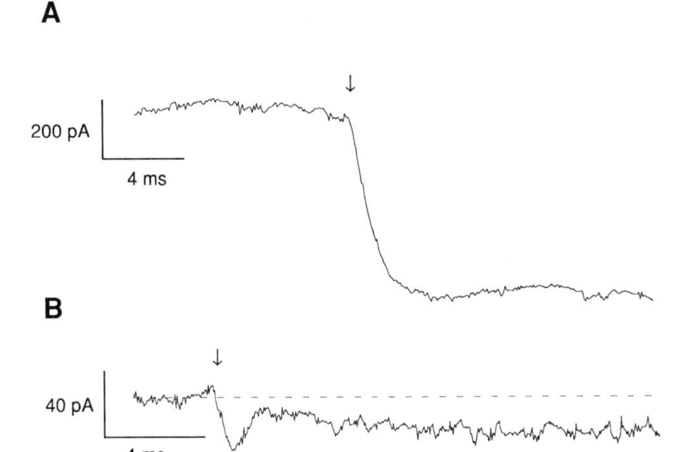

FIGURE 3. Activation of I_{NaCa} at high temporal resolution. (**A**) shows the activation phase of the exchange current at 35°C. The current rises in a monotonic fashion, as predicted for a transporter cycling more rapidly than Ca^{2+}-bound exchanger states can be generated experimentally. Note that on this time scale the slow decay of I_{NaCa} is not noticeable. At 12°C (**B**) the exchange current is drastically reduced, indicating a slower turnover rate. A small initial current transient can now clearly be distinguished from I_{NaCa} (used with permission from *Nature*[5]).

overlap temporally and I_{conf} cannot be accurately resolved before inhibitors are applied. Several properties of I_{conf} nevertheless indicate that it may also be produced by the Na-Ca exchangers. First, a transient current like I_{conf} is expected if Ca^{2+}-bound states can be formed faster than the exchanger is able to cycle and if charge is moved during a molecular transition following Ca^{2+} binding (see above). Second, it is possible to exclude all other known Ca^{2+}-dependent current sources like the Ca^{2+}-activated nonspecific cation channels and the Ca^{2+}-activated K^+ channels. Both types of channels would not produce a transient current, because they continue to have high open probabilities as long as $[Ca^{2+}]_i$ remains high.[10] Furthermore, both channel types would exhibit a reversal potential thus generating outward current at positive potentials. It is still conceivable, that Ca^{2+}-dependent gating of these channels generates a current that is voltage-independent over some range of membrane potentials. But the charge movement that we observe after photorelease of Ca^{2+} is quite large (up to 2 nC μF^{-1}) and corresponds to $\sim$125 charges per μm^{-2}, inconsistent with the observed low density of these channels (below 1 per μm^2).[10,19]

Ni^{2+} or La^{3+} ions cannot be used to examine the effects of inhibition of the carriers from the cytosolic side of the membrane. The affinity of EDTA-based chelators like DM-nitrophen for Ni^{2+} is orders of magnitude larger than for Ca^{2+} and these ions would essentially displace Ca^{2+} from DM-nitrophen, rendering any photorelease of Ca^{2+} impossible. We have thus carried out experiments with an organic Na-Ca exchange inhibitor, the amiloride derivative dichlorobenzamil (DCB). In studies with cardiac sarcolemmal vesicles, DCB had properties of a competitive inhibitor for Na^+ binding, thus effectively blocking the transport cycle, but it only weakly affected Ca-Ca exchange in the presence of K^+.[20] In our experiments I_{NaCa} induced by photorelease of Ca^{2+} was reduced markedly by extracellular application of 100 μM DCB ($\sim$90% inhibition, see FIG. 4A). This finding is similar to the effect of DCB observed on currents induced by spontaneous and caffeine-induced Ca^{2+}-release from the SR.[21,22]

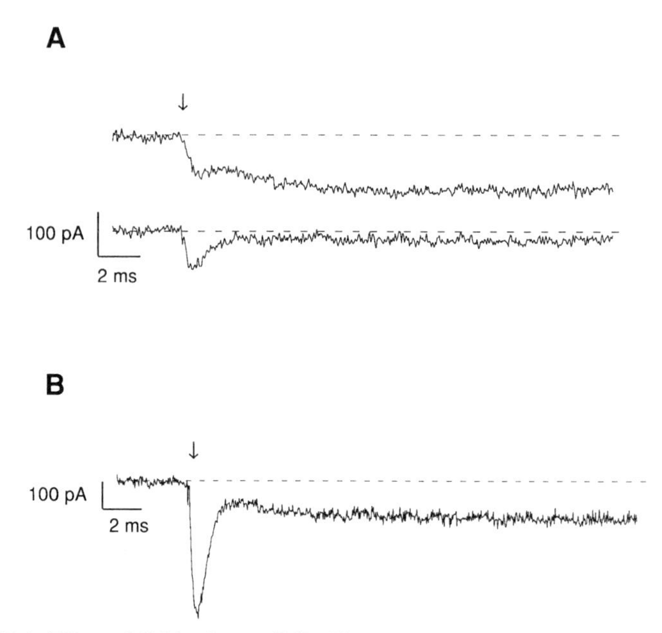

FIGURE 4. Effects of dichlorobenzamil (DCB) on I_{NaCa} and I_{conf}. The control experiment (**A**, upper trace) shows both I_{NaCa} and I_{conf} activated by photorelease of Ca^{2+}. The organic Na-Ca exchange inhibitor DCB, added to the superfusion solution at a concentration of 100 μM, inhibited the more maintained current by ~90% (**A**, lower trace). Effects on I_{conf} cannot be excluded, but are likely to be small. A representative current recorded after intracellular application of DCB is shown in (**B**). I_{NaCa} was inhibited slightly, but the charge moved during I_{conf} was dramatically increased (~6-fold) (used with permission from *Nature*[5]).

This inhibition is expected from DCB interfering with extracellular Na^+ binding to the transporter while the exchanger is extruding intracellular Ca^{2+}, thus operating in the "forward-mode" (see FIG. 2A, consecutive transporter). Similar to the effect of ionic inhibitors, I_{conf} seems little affected by extracellular DCB since it is generated by partial reactions of the exchanger cycle that do not involve Na^+ binding. In contrast, when DCB was dialyzed into the cytosol through the patch-clamp pipette, I_{NaCa} was not significantly inhibited. However, the charge moved during I_{conf} was dramatically increased by intracellular DCB (~6-fold, see FIG. 4B). For the interpretation of this result we need to consider the following experimental details. During our experiments we are measuring a difference-current activated by jumps of $[Ca^{2+}]_i$. This difference current (I_{NaCa}) reflects a change of the exchanger turnover rate in the "forward mode." At the same time, the "backward" reactions are also continuously present but are not affected by photorelease of $[Ca^{2+}]_i$ since $[Na^+]_i$ and $[Ca^{2+}]_o$ remain largely unchanged (but see Lederer *et al.*[23]). Because the "foward mode" we stimulate does not involve Na^+ binding on the intracellular side, the turover rate and thus I_{NaCa} is not greatly affected by intracellular DCB. However, intracellular DCB presumably interferes with a partial reaction of the "reverse mode," that is, intracellular Na^+ binding and subsequent translocation of Na^+-loaded exchangers to the outside. Thereby DCB is able to change the distribution of inside and outside states and increases the population of intracellular exchanger states ready to bind Ca^{2+} and to produce I_{conf} upon a flash. In summary, the effects of DCB are in agreement with models of the Na-Ca exchanger but are clearly inconsistent with the known nonspecific blocking effects of DCB on other transporters and ionic currents,[24] again suggesting that I_{conf} is generated by a conformational rearrangement of Na-Ca exchanger molecules after Ca^{2+} binding.

To deduce the amount of charge moved during the two current-producing transitions, an Eyring rate theory model was fitted to the voltage dependence of I_{NaCa}. The voltage dependence of I_{NaCa} represents the rate-limiting step of cycling and is presumably associated with Na^+ translocation in a consecutive transporter. This voltage dependence can be modeled by a charge moving over a symmetric energy barrier. The observed shallow current/voltage relationship (*e*-fold change for 114 mV) indicates that less than one charge is moved during this transition. Therefore, we have introduced the concept of apparent fractional charges into our model. The data can now be fitted by +0.44 elementary charges moving inwards with the Na^+-translocating step. Apparent fractional charges are representations of real elementary charges that only move through a fraction of the membrane potential.[25] There is no real justification to assume *a priori* that the apparent moved charge is an integral multiple of the elementary charge.

Once the charge on the Na^+-bound state of the exchanger was determined, the charge on the unloaded (−2.56) and on the Ca^{2+}-loaded forms of the exchanger (−0.56) could be calculated and used to estimate a lower limit for the site density of carriers in the cell membrane. Given the charge moved in the presence of intracellular DCB when most (but probably not all) transporters were in the "inside" states before the flash, we estimated a lower limit of 250 exchangers per μm^2. Using the established 3 : 1 stoichiometry for full cycling and the observed peak I_{NaCa}, the upper limit for the turnover rate under these conditions would be 2500 sec^{-1} (at −80 mV and 35°C, with $[Ca^{2+}]_i$ below saturation of the Na-Ca exchangers).

We anticipate that rapid concentration jumps of ions under different conditions may provide additional details about the molecular operations of the Na-Ca exchanger. Furthermore, the combined application of this biophysical technique with molecular biology methods[26] may provide insight into the structure-function relationship of this important transport protein.

REFERENCES

1. ALLEN, T. J. A., D. NOBLE & H. REUTER. 1989. Sodium-Calcium Exchange. Oxford University Press. Oxford, U.K.
2. KAPLAN, J. H. 1990. Photochemical manipulation of divalent cation levels. Annu. Rev. Physiol. **52:** 897-914.
3. GURNEY, A. M. & H. A. LESTER. 1987. Light-flash physiology with synthetic photosensitive compounds. Physiol. Rev. **67:** 583-617.
4. KAPLAN, J. H. & G. C. R. ELLIS-DAVIES. 1988. Photolabile chelators for the rapid photorelease of divalent cations. Proc. Natl. Acad. Sci. USA **85:** 6571-6575.
5. NIGGLI, E. & W. J. LEDERER. 1991. Molecular operations of the sodium-calcium exchanger revealed by conformation currents. Nature **349:** 621-624.
6. NIGGLI, E. & W. J. LEDERER. 1990. Voltage-independent calcium release in heart muscle. Science **250:** 565-568.
7. FIDLER, N., G. ELLIS-DAVIES, J. H. KAPLAN & J. A. MCCRAY. 1988. Rate of Ca^{2+} release following laser photolysis of a new caged Ca^{2+}. Biophys. J.**53:** 599a.
8. NIGGLI, E. 1989. Strontium-induced creep currents associated with tonic contractions in cardiac myocytes isolated from guinea-pigs. J. Physiol. **414:** 549-568.
9. COLQUHOUN, D., E. NEHER, H. REUTER & C. F. STEVENS. 1981. Inward current channel activated by intracellular Ca in cultured cardiac cells. Nature **294:** 752-754.

10. EHARA, T., A. NOMA & K. ONO. 1988. Calcium-activated non-selective cation channel in ventricular cells isolated from adult guinea-pig hearts. J. Physiol. **403:** 117-133.
11. MULLINS, L. J. 1977. A mechanism for Na/Ca transport. J. Gen. Physiol. **70:** 681-695.
12. DIFRANCESCO, D. & D. NOBLE. 1985. A model of cardiac electrical activity incorporating ionic pumps and concentration changes. Phil. Trans. R. Soc. (London) **307:** 353-398.
13. LÄUGER, P. 1987. Voltage dependence of sodium-calcium exchange: Predictions from kinetic models. J. Membr. Biol. **99:** 1-11.
14. EISNER, D. A. & W. J. LEDERER. 1985. Na-Ca exchange: Stoichiometry and electrogenicity. Am. J. Physiol. **248:** C189-202.
15. HILGEMANN, D. W. 1988. Numerical approximations of sodium-calcium exchange. Prog. Biophys. Molec. Biol. **51:** 1-45.
16. CHANCE, E. M., A. R. CURTIS, I. P. JONES & C. R. KIRBY. 1977. Facsimile: A computer program for flow and chemistry simulation, and general initial value problems. UK Atomic Energy Authority, Harwell, UK.
17. KIMURA, J., S. MIYAMAE & A. NOMA. 1987. Identificatin of sodium-calcium exchange current in single ventricular cells of guinea-pig. J. Physiol. **384:** 199-222.
18. MIURA, Y. & J. KIMURA. 1989. Sodium calcium exchange current: Dependence on internal Ca and Na and competitive binding of external Na and Ca. J. Gen. Physiol. **93:** 1129-1241.
19. CALLEWAERT, G., J. VEREECKE & E. CARMELIET. 1986. Existence of a calcium-dependent potassium channel in the membrane of cow cardiac Purkinje cells. Pflügers Arch. **406:** 424-426.
20. SLAUGHTER, R. S., M. L. GARCIA, E. J. CRAGOE, J. P. REEVES & G. J. KACZOROWSKI. 1988. Inhibition of sodium-calcium exchange in cardiac sarcolemmal membrane vesicles. 1. Mechanism of inhibition by amilorids analogues. Biochemistry **27:** 2403-2409.
21. LIPP, P. & L. POTT. 1988. Voltage dependence of sodium-calcium exchange currents in guinea-pig atrial myocytes determined by means of an inhibitor. J. Physiol. **403:** 355n366.
22. CALLEWAERT, G., L. CLEEMANN & M. MORAD. Caffeine-induced Ca^{2+} release activates Ca^{2+} extrusion via Na^{+}-Ca^{2+} exchanger in cardiac myocytes. Am. J. Physiol. **257:** C147-152.
23. LEDERER, W. J., E. NIGGLI & R. W. HADLEY. Sodium-calcium exchange in excitable cells: Fuzzy space. Science **248:** 283.
24. BIELEFELD, D. R., R. W. HADLEY, P. M. VASSILEV & J. R. HUME. 1986. Membrane electrical properties of vesicular Na-Ca exchange inhibitors in single atrial myocytes. Circ. Res. **59:** 381-389.
25. LÄUGER, P. 1984. Thermodynamic and kinetic properties of electrogenic ion pumps. Biochim. Biophys. Acta **779:** 307-341.
26. NICOLL, D. A., S. LONGONI & K. D. PHILIPSON. 1990. Molecular cloning and functional expression of the cardiac sarcolemmal Na^{+}-Ca^{2+} exchanger. Science **250:** 562-565.

The Exchange in Intact Squid Axons[a]

T. JEFF A. ALLEN

Department of Physiology
University of Bristol
BS1 5LS, Bristol, United Kingdom

INTRODUCTION

The introduction of the squid giant axon to studies on Ca homeostasis has proved very useful in laying the foundations in the study of Ca regulation in other less tractable systems. The Na-Ca exchange mechanism in the intact squid axon can be easily monitored by a number of techniques; these allow the study of the various transporting exchange modes across a plasma membrane little affected by experimental preparation. The main aim of this article is to present features of the exchange fluxes measured after microinjection of $^{22}NaCl$ or $^{45}CaCl_2$ into the axon interior. These observations study the exchange at a basic level since control of the internal environment is limited, but results can be readily compared to studies using other, more complex techniques.

The methods used in these studies have been described in detail by Baker *et al.*[1] and Allen and Baker[2]; the squid used, *Loligo forbesi*, were supplied by the Marine Biological Association, Plymouth, between the months of September and January and had diameters ranging between 500 and 1100 μM. After removal of adhering connective tissue and small nerve fibers, axons were usually placed in a horizontal chamber that permitted axial microinjection of substances. Additionally, injectors could be reloaded and repositioned inside the axon to allow injection of substances during the course of the experiment. Temperature was held close to 18 °C. The chamber also allowed electrical control of membrane potential, V_m, by voltage clamping and could be covered by a light tight box to allow aequorin glow to be monitored. Artificial sea waters of differing compositions superfused the axon at rates between 1-2 ml/min.

ACTIVATION BY MONOVALENT CATIONS

Axons axially injected with ^{22}Na and exposed to artificial seawaters containing various amounts of calcium show a clear Ca_o-dependent ^{22}Na efflux (Ca_o-Na_i exchange); however, the nature and magnitude of this flux is dependent on the ionic composition

[a]This work was supported by the Medical Research Council.

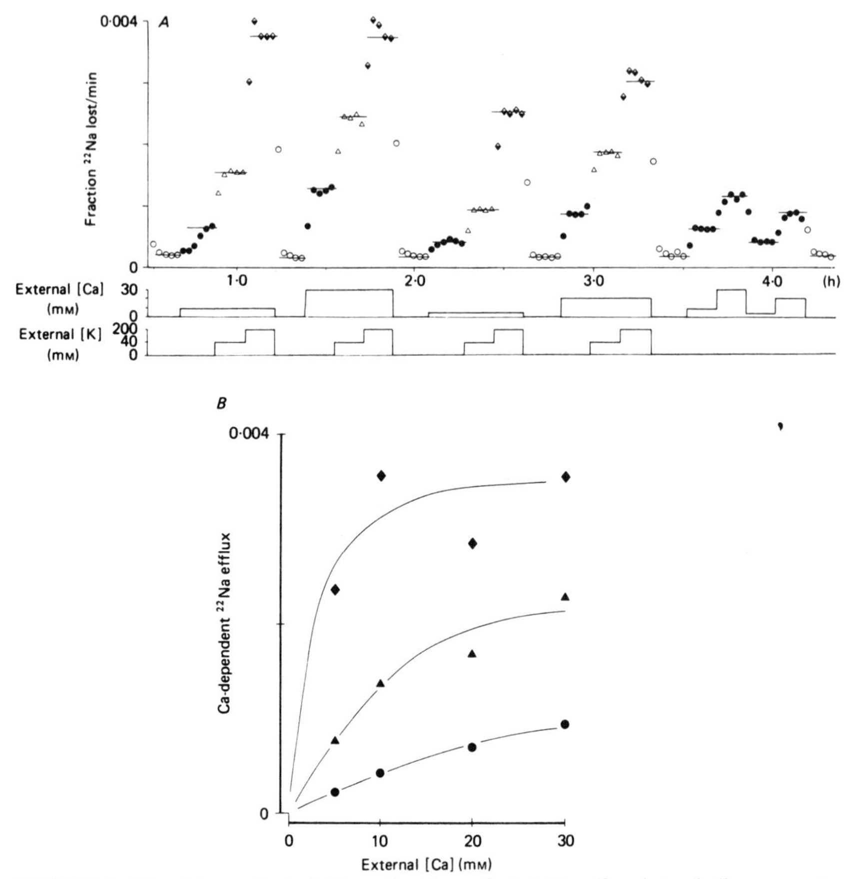

FIGURE 1. Stimulatory effect of K on the Ca_o-activated Na efflux into choline sea water. Isosmotic replacement of choline by K. **Part a.** Raw data. **Part b.** Ca_o-dependent Na efflux at 0 mM K_o (circles), 40 mM K_o (triangles) and 200 mM K_o (diamonds) as a function of Ca_o. The curves have been drawn with K_m and V_{max}, respectively; 53.2 mM and 0.0025 in 0 mM K; 17.0 mM and 0.0034 in 40 mM K; 2.1 mM and 0.0037 in 200 mM K. (From Allen & Baker[2]; used with permission.)

of the medium.[1,2] The experiment shown in FIGURE 1 demonstrates that the apparent K_m for Ca_o is increased by raising the K^+ concentration of the sea water. The raw data of Part *A* show that the efflux of ^{22}Na, after inhibition of the sodium pump by 10^{-5} M ouabain, can be increased by addition of calcium and reversibly activated to differing levels by K^+ by substitution for choline; in control experiments addition of potassium in the absence of calcium fails to activate ^{22}Na efflux, demonstrating that the Na pump is fully inhibited. The curves drawn in Part *B* show that the Ca concentration required for half-maximal activation of the Ca_o-Na_i exchange flux decreases from 53 mM in nominally K-free choline sea water to 2 mM in a mixture of 200 mM K and 200 mM choline sea water. This strong activating effect of K could be readily considered to reflect the electrogenic nature of the exchange mechanism, in which 3 Na^+ are transported for 1 Ca^{2+}; however, a similar activating effect on the Ca_o-dependent ^{22}Na efflux is seen when Li^+ is used as the substituting cation, and also with Na^+ at low concentrations: These cations have little effect on V_m. If, however, K^+ depolarization is performed in the presence of lithium in place of choline or Tris, different results are obtained. An increase in K^+ from 10 mM to 100 mM has little effect on the K_m for Ca_o (~2 mM), but increases V_{max} by some 30%.

The clear suggestion from these and similar experiments is that the activating effects of K^+ on the Ca_o-Na_i exchange may be a mixture of both chemical and electrical actions. When the forward (Na_o-Ca_i) mode of Na-Ca exchange is examined, the activating effect of Na_o on ^{45}Ca efflux is strongly sigmoidal in the presence of ATP but becomes more like a rectangular hyperbola in its absence.[3-5] In axons that have been preinjected with ^{45}Ca, there is little difference between the Na_o-dependent flux in choline, Tris, or Li^+ mixtures, suggesting that there is no additional activating cation effect at the Na-binding site during Na_o-Ca_i exchange. The action of K^+ depolarization during this forward exchange mode is inhibition of the flux, as expected for an electrogenic process.

EXPERIMENTS COMBINING FLUX STUDIES WITH VOLTAGE CLAMP

Data presented in the previous section showed that experiments that use K^+ to examine the effects of V_m on Na-Ca exchange fluxes may include a chemical activation component, the result of which may overestimate the voltage sensitivity of the exchange process. To examine this possibility further, axons were voltage clamped to effect electrical changes in V_m while measuring unidirectional exchange fluxes.

FIGURE 2 is one such experiment in which the Ca_o-dependent ^{22}Na fluxes can be monitored during changes in V_m effected by K^+ or electrical means. In the first half of the experiment K^+ brings about a clear large activation of the ^{22}Na efflux into choline sea water containing 10 mM Ca; in the second half, electrical depolarization by voltage clamp fails to activate the flux to a similar extent. In other experiments, hyperpolarization during a K^+ challenge only removes part of the activated flux into choline sea water. Similar experiments performed in Li^+ sea water, in which the Ca_o-Na_i exchange has high affinity for Ca_o, show that the smaller activation effected by K^+ can be mimicked by electrical depolarization.

During forward exchange, the inhibition of the Na_o-dependent ^{45}Ca efflux by K^+ depolarization in intact axons can be removed and mimicked by voltage clamp.[5] This effect is best studied during cyanide poisoning, which raises Ca_i and attenuates Ca pump activity due to the low levels of ATP. At alkaline pH (pH 9) it is possible to isolate Na_o-Ca_i exchange fluxes from Ca pump activity and examine the flux in the presence of ATP.[6,5] The effects of K^+ depolarization on Na_o-Ca_i exchange are generally similar at pH 9 to those in poisoned axons.

A problem found in combining isotope flux studies with voltage clamp was possible intracellular acidification caused during steady depolarizations; indeed recent studies[7] have shown that these small pH_i changes can influence measurements of Ca_i by aequorin. In all experiments combining isotope fluxes with voltage clamp,[2,5] axons were routinely preinjected with a mixture of 3-[-*N*-morpholino]-propane-sulphonic acid (MOPS) buffer and tetraethylammonium chloride (TEA) to aid the pH-buffering capacity of axoplasm and reduce the currents necessary to hold potential constant. The possibility raised runs as follows: Since K^+ depolarization activates the Ca_o-Na_i exchange flux to a greater extent when the affinity to Ca_o is low (i.e., in choline sea water), the lack of the voltage clamp to mimic K^+ activation could be due to intracellular

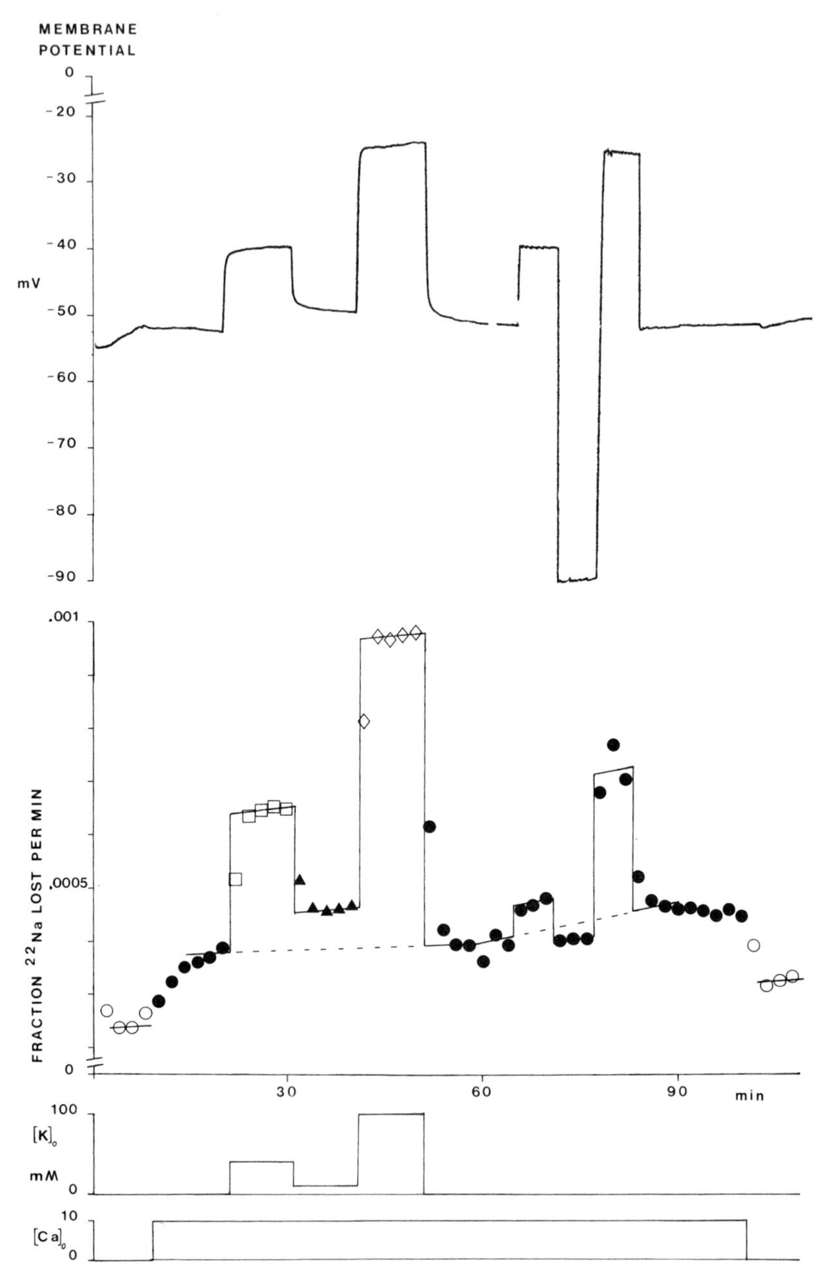

FIGURE 2. Equivalent electrical depolarizations fail to mimic the effects of high K on the Ca_o-dependent Na efflux into choline sea water. 0 Ca (open circles), 10 mM Ca (filled circles), 10 mM Ca-10 mM K (filled triangles), 10 mM Ca-40 mM K (open squares), 10 mM Ca-100 mM K (open diamonds). Axon prepared by injection of (final concentrations in mM); MOPS, 14.7; HEPES, 14.7; TEA, 29; pH 7.3. (From Allen & Baker[2]; used with permission.)

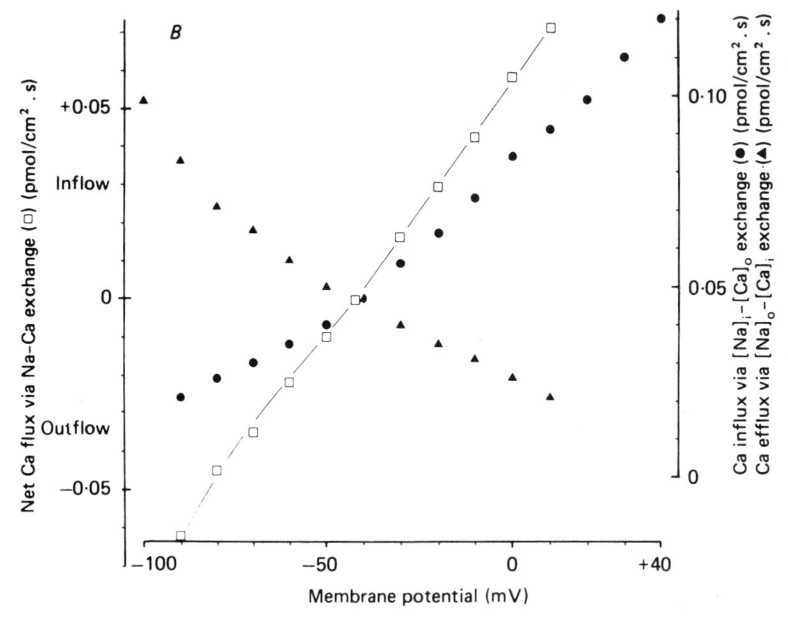

FIGURE 3. Comparison of Ca fluxes via both Na_o-Ca_i and Ca_o-Na_i components of Na-Ca exchange and the voltage dependence of the net Ca flux through the exchanger. Na_o-dependent Ca efflux (triangles), Ca_o-dependent Na efflux (circles). Net Ca flux (squares) is calculated as the difference between the Na_o-Ca_i and Ca_o-Na_i exchange fluxes, assuming a Ca_o-dependent Na flux of 0.04 pmol·cm^{-2}·sec^{-1} and Na_o-dependent Ca flux of 0.05 pmol·cm^{-2}·sec^{-1} at −50 mV. (From Allen & Baker[5]; used with permission.)

acidification taking place. If this were so, Ca_o-Na_i exchange displays a much greater sensitivity to potential under these conditions, depolarization not only increasing V_{max}, but also increasing the K_m for Ca_o. This is a less likely explanation for the strong K^+ activation of reverse exchange that occurs into choline seawater. The data on Ca_o-Na_i exchange in lithium seawater[2] shows that depolarization does not strongly affect the affinity for Ca_o, and this is supported by the recent findings of DiPolo and Beaugé[8] using dialyzed axons; there was no change in K_m for Ca_o during Ca_o-dependent Ca efflux (a mode of Na-Ca exchange) in dialyzed axons measured at −50 and 0 mV.

The activation of the reverse exchange by monovalent cations has received much interest because it has been unclear whether the activating cation is also transported. This possibility will be addressed in later sections.

The contribution of Na-Ca exchange fluxes to voltage-sensitive Ca movements remain a central question in a number of studies. In the intact axon the voltage-sensitivity of Ca transport through the exchange can be simply estimated by combining the unidirectional flux data obtained from intact axons during excursions in membrane potential effected by K^+ and voltage clamp. This is summarized in FIGURE 3. The data are obtained in choline seawater containing 100-200 mM Na—conditions under which both forward and reverse exchange modes operate. The open squares in the figure represent net Ca flux via the exchange; this is zero near −40 mV and inward at positive potentials, displaying no bell-shaped relationship as seen for Ca channels.

EFFECTS OF INHIBITORS ON THE EXCHANGE

A clear advantage of studying Na-Ca exchange fluxes in the squid axon is the capacity to examine the effects of putative exchange blockers on unidirectional fluxes via its transport modes. By far the best alternative system has been plasma membrane vesicles, and their use in the development of amiloride-based inhibitors has been reviewed.[9] However, studies using vesicles do have limitations since a potent Na-Ca exchange blocker, 3,4-dichlorobenzamil (DCB) which has a K_I of $\sim 20\ \mu M$, proves less selective in intact cells; it blocks Ca channel currents with equal potency. Studies using intact squid axons injected with ^{22}Na or ^{45}Ca show that DCB (50-100 μM) had no effect on Ca_o-Na_i or Na_o-Ca_i exchange fluxes, once activated by low concentrations of sodium due to the proposed competitive nature of DCB inhibition.[10] These concentrations of DCB block calcium-channel currents in barnacle muscle fibers (unpublished observations). Na-Ca exchange in cardiac vesicles can also be effectively inhibited by the alkaline earth metals.[11] A few reports have successfully used these cations to block either currents (e.g., Kimura *et al.*[12]) or changes in a^i_{Na} (e.g., Aikin *et al.*[13]) mediated by the exchange.

The intact squid axon has been used to study the effectiveness of these inhibitors on both forward and reverse modes of the exchange mechanism.[14] FIGURE 4 compares the relative effects of Mn, Co, and Ni on the Ca_o-dependent flux when the Ca-binding site is activated to differing degrees in choline sea water (i.e., in 10 and 40 mM K). The raw data (Part a) shows the large stimulation of ^{22}Na efflux into 10 mM calcium by 40 mM K, and this is almost completely removed by increasing the inhibiting cation from 1 to 10 mM. At 10 mM these cations also clearly inhibit the flux at low K^+ and further reduced ^{22}Na efflux in the nominal absence of Ca. The curves drawn in Part b show that the concentration required for half-inhibition of Ca_o-Na_i exchange lies close to 1-2 mM, with Ni displaying the greater potency. At 10 mM they failed to activate ^{22}Na efflux.

The effects of Mn^{2+} on the Ca_o-Na_i and Na_o-Ca_i exchange modes have been studied in greater detail owing to its more reversible nature. The findings indicate that Mn is a more effective inhibitor of the Ca_o-Na_i exchange mode when the Ca_o affinity is higher and is also less effective during Na_o-Ca_i exchange.[14] Efflux experiments were therefore conducted into choline sea waters containing 100 mM sodium, which activates both fluxes and allows easier comparison. The data from two axons are shown in FIGURE 5. The Mn concentration required to effect half-maximal inhibition of the exchange fluxes was 30 mM (Na_o-Ca_i mode) and 7 mM (Ca_o-Na_i mode), with apparent K_I of 10 mM and 1 mM, respectively. These differences in apparent K_I are dependent on the levels of ATP in the axon; in the presence of ATP, Na_o-Ca_i exchange has greater affinity for Na_o and the apparent affinity for Mn changes to 3-4 mM. These parameters deserve further exploration under more carefully controlled conditions.

EFFECTS OF MANIPULATIONS IN CA_i ON NA-CA EXCHANGE FLUXES

One of the most interesting features of Na-Ca exchange is the observation that Ca_i can further stimulate Ca_o-Na_i exchange. This positive feedback system, in which raising internal calcium promotes further Ca influx on the exchange, has been carefully studied in dialyzed axons,[15-17] nerve membrane vesicles,[18] and has been observed in barnacle muscle cells[19] and when measuring currents carried by the exchange (e.g., Miura & Kimura[20] and Hilgemann[21]). The first indications for this calcium requirement were

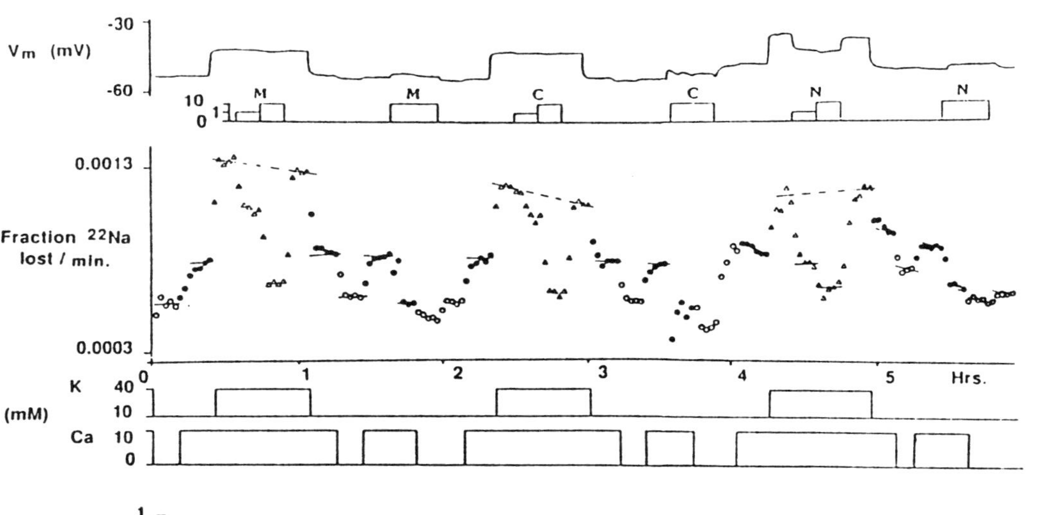

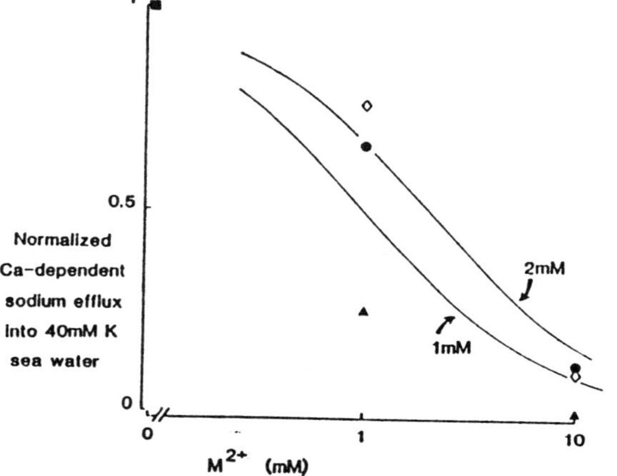

FIGURE 4. Comparison of the effects of Mn, Co, and Ni on the Ca_o-dependent ^{22}Na efflux into choline and choline-potassium sea water mixtures. **Top** shows the raw data of the effects of 1 mM and 10 mM of the inhibitory cation on the components of sodium efflux in the presence and nominal absence of calcium. M = Mn^{2+}, C = Co^{2+}, N = Ni^{2+}, scale is mM. Open circles: Ca-free media, Closed circles: 10 mM Ca, Open triangles: 10 mM Ca + 40 mM K. **Bottom** (left): Concentration-dependent inhibition of the Ca-dependent flux by divalent cations (M^{2+}). Normalized data from Part a. Curves drawn with concentration for 50% inhibition of 1 mM and 2 mM. (From Allen[14]; used with permission.)

after injections of Ca chelators into intact axons[22,23] to lower Ca_i. The nature of Ca_o-Na_i exchange stimulation exerted by Ca_i is dependent on the levels of ATP in dialyzed axons; in the nominal absence of ATP the K_m for internal Ca increases from 2 μM to close to 10 μM.[16]

The stimulatory effects of Ca_i on Ca_o-Na_i exchange, by further increasing Ca influx, should oppose Ca efflux via the Na_o-Ca_i exchange. This interesting interaction could be addressed in measurements of net Ca fluxes in intact axons, if it were possible to selectively isolate the forward and reverse exchange modes. Studies using Ca indicators as a measure of Na-Ca exchange activity (e.g., Mullins *et al.*[24] and Requena *et al.*[25,26]) show that half-activation of Ca influx is 0.67 mM and that this Ca entry is a steep function of Na_i, suggesting a coupling of $4Na^+$ in exchange for $1Ca^{2+}$. Comparable studies using radioisotope fluxes reflect a K_m for Ca_o of ~2 mM to 50 mM (intact axons[2]) and ~5-10 mM (dialyzed axons[16,17]), depending on the extent of activation at the monovalent cation binding site. A possible explanation for the higher affinity of Ca_o-Na_i exchange for Ca_o when measured by Ca indicators is the contribution of other Ca fluxes, such as Ca pump and Ca-Ca exchange, and also intracellular buffering.

Attempts to examine the stimulatory effects of Ca_i on Ca_o-Na_i exchange in intact axons by simply injecting Ca chelators, such as Quin-2, BAPTA, and EGTA, with levels of calcium to control free Ca in the μM range, invariably lead to inhibition of the Ca_o-dependent ^{22}Na flux and corresponding Ca uptakes.[27,14] This failure to observe an increase in Ca_o-Na_i exchange activity raises the possibility that axoplasm could rapidly strip Ca from the chelator, before the exchange could see a prolonged increase in Ca_i. Another means to effect increases in Ca_i in the absence of chelators is by blockage of ruthenium red sensitive mitochondrial Ca uptakes.[28] Ruthenium red alone does not affect the exchange process in dialyzed squid axons.[29]

FIGURE 6 is one such experiment measuring the components of calcium efflux from an intact axon. Axial injection of ruthenium red has little effect on the Na-dependent and uncoupled ^{45}Ca efflux until after imposition of a Ca_i load by removal of Na_o.

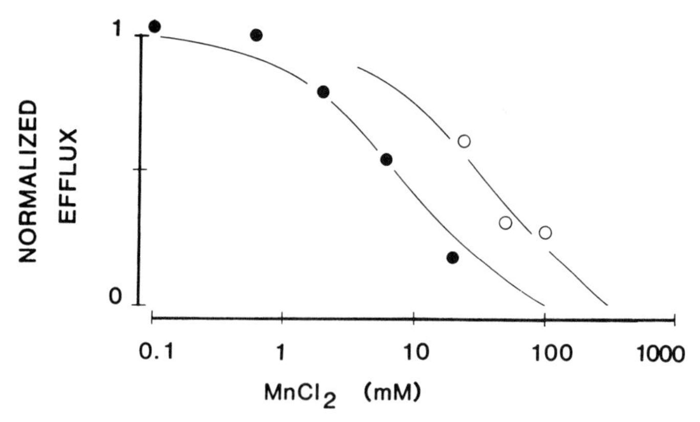

FIGURE 5. Mn inhibition of the Na-dependent Ca efflux (Na_o-Ca_i exchange) and Ca-dependent Na efflux (Ca_o-Na_i exchange) into choline sea waters containing 100 mM sodium. *Closed circles*: normalized Ca_o-dependent ^{22}Na efflux into 10 mM calcium. *Open circles*: normalized Na_o-dependent ^{45}Ca efflux (50 μM Ca_o; CN^- poisoned axon). Curves drawn with concentration for 50% inhibition of 7 mM and 30 mM.

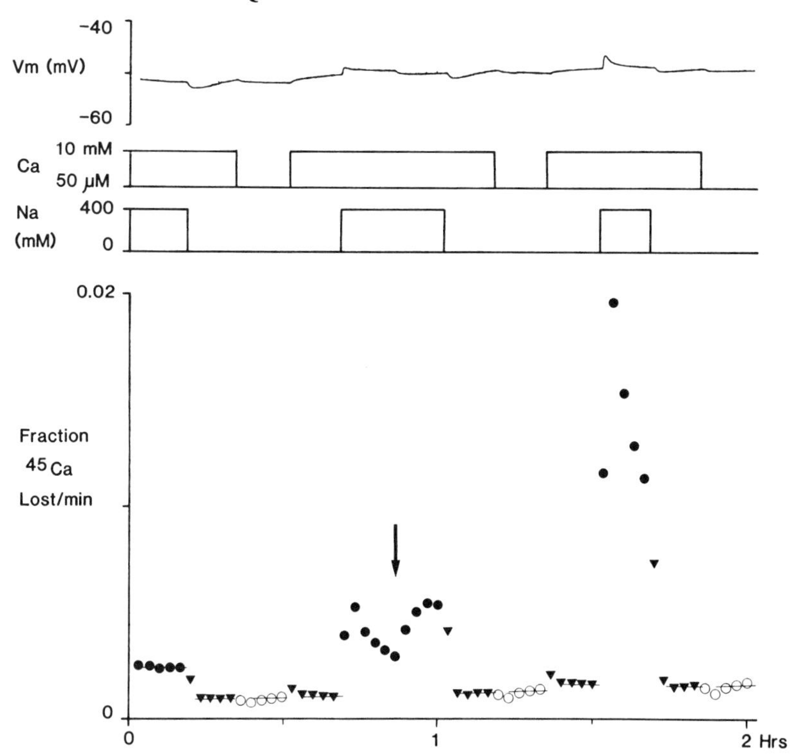

FIGURE 6. The effects of intracellular ruthenium red on the sodium-dependent calcium efflux. Arrow on figure denotes period during which axon was injected with a final concentration of 210 μM ruthenium red and 4.2 mM MOPS (pH 7.3). *Closed circles*: Na-dependent calcium efflux. *Open circles*: uncoupled calcium efflux (Ca pump). *Triangles*: Ca-dependent calcium efflux. Choline replaced sodium. (From Allen[14]; used with permission.)

Readdition of Na_o leads to a clear activation of the Na-dependent efflux but not the uncoupled flux (Ca pump). As Na_o-Ca_i exchange is half-activated by Ca_i of ~2 μM and the Ca pump ~0.2 μM,[30] it appears that by inhibiting mitochondrial buffering, Ca inflow leads to an increase in Ca_i and stimulation of Na_o-Ca_i exchange flux. The transient nature of the Na-activated ^{45}Ca efflux in FIGURE 6 suggests that its high V_{max} allows it to rapidly lower Ca_i levels.

The increase in Ca_i effected by ruthenium red is seen more clearly in axons injected with aequorin. FIGURE 7 is an experiment in which both aequorin glow and ^{22}Na efflux are monitored. Although the experiment monitors the Ca_o-dependent flux into both choline and lithium sea waters, the flux into 10 mM Ca lithium sea water is easier to follow (due to the large differences in K_m for Ca_o). At the first arrow, ruthenium red is axially injected; readdition of calcium leads to a marked increase in Ca_i, but not Ca_o-Na_i exchange flux as measured by the Ca_o-dependent ^{22}Na flux. Although there is no absolute measure of Ca_i in this experiment, it is probable that Ca_i rises from just below 1 μM to perhaps 2-5 μM.

In dialyzed axons, the K_m for Ca_i activation is close to 2 μM during both Na_o-Ca_i[30,31] and Ca_o-Na_i exchange.[17] In intact axons an increase in Ca_i by ruthenium red stimulates Na_o-Ca_i exchange but not Ca_o-Na_i exchange.[14] Because buffering Ca_i above micromolar levels with calcium chelators also invariably inhibits Ca_o-Na_i exchange in intact axons, it has therefore proved very difficult to observe the expected catalytic stimulation of Ca_o-Na_i exchange by increases in Ca_i.

It is important to account for these observations. It is unlikely that Ca_o-Na_i exchange is already maximally activated in studies on intact axons: in dialyzed axons, to see the stimulatory effect of Ca_i during Ca_o-Na_i exchange in the absence of EGTA, 200 μM calcium was used[29] (see also FIG. 11a of DiPolo & Beaugé[17]). Another interesting feature of Ca_o-Na_i exchange in intact axons when compared to dialyzed axons are the changes in K_m for Ca_o. Raising Ca_i has little effect on the Ca_o affinity during Ca_o-Na_i exchange in dialyzed axons, and this K_m of ~3-5 mM is not strongly dependent on the activating cation.[16,17,32] However, in intact axons, the affinity for Ca_o can change markedly from ~50 mM to ~2 mM depending on the extent of Ca_o-Na_i exchange activation at the monovalent cation-binding site.[1,2]

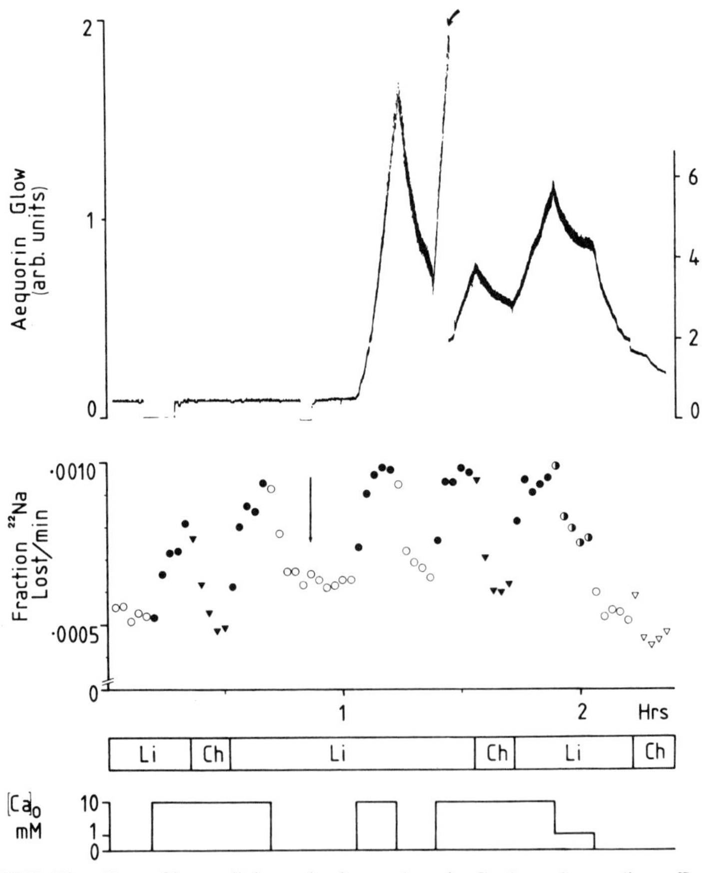

FIGURE 7. The effects of intracellular ruthenium red on the Ca-dependent sodium efflux. Axon was preinjected with aequorin and then 25 min later with ^{22}Na. At first arrow, a final concentration of 170 μM ruthenium red with 3.4 mM MOPS (pH 7.3) was injected. Readdition of calcium activated the Ca-dependent sodium efflux and led to a large increase in Ca_i. Second arrow denotes point at which amplifier gain was reduced. The photomultiplier tube was switched off for two periods during the experiment. (From Allen[14]; used with permission.)

TABLE 1. Do Ca and Li Transport Together during Na_i-Ca_o Exchange in Intact Squid Axons?[a]

		0 Ca	10 mM Ca	Net Lithium Uptake	
Part 1	(a) Li uptake[b]	2.09	3.69	1.60	
	(b) Li uptake	1.80	2.93	1.13	
		10 mM	10 mM Ca + Mn	Mn-Sensitive Uptake	Percent Reduction
Part 2	(c) Ca uptake	4.11,5.53	2.12	2.7	56%
	Li uptake	18.8,18.8	13.3	5.5	29%
	(d) Ca uptake	2.51	0.86	1.6	66%
	Li uptake	10.7	7.2	3.5	33%
	(e) Ca uptake	1.77	1.22	0.6	31%
	Li uptake	7.3	9.3	−2.0	

[a] Paired thermos axons were incubated in 400 mM lithium sea water with TTX (1 μM) and ouabain (10^{-5} M) for 30 min, and washed in 10 mM Ca choline sea water (5 min) at end of this period. To determine lithium uptake, axoplasm was extruded and dissolved in nitric acid for atomic absorption spectrophotometry. In Part 2 axoplasm was first liquified in 1 M $CaCl_2$ and then halved to permit scintilation counting of ^{45}Ca. Axon pair (c) shows two determinations of Ca and Li uptake in 10 mM Ca as one axon was divided into half (unpublished data).

[b] All uptakes are in pmole · cm^{-2} · see^{-1}.

It has been previously suggested that exchange kinetics are subtly altered during the preparation of membrane vesicles[2,8,14]; there is also a possibility that subtle changes also occur in dialyzed axons, and this deserves further investigation. Perhaps introduction of EGTA, which reduces ^{22}Na leak, thus prolonging dialysis experiments, may influence the exchange at the cytoplasmic face during experiments on dialyzed axons. Testing this possibility is very difficult but could be done at constant Ca_i with varying amounts of chelator—but retaining tight control at Ca_i just under the membrane. It would be very interesting if EGTA does cause a shift in the Ca_i or Ca_o affinity during Ca_o-Na_i exchange.

The half-activating effect of Ca_i on Ca_o-Na_i exchange, as performed by EGTA or EDTA in the presence of ATP, ranges from 0.02 μM in cardiac ventricular myocytes,[20] to 0.3 μM or 0.7 μM in barnacle muscle cells,[19] to 2 μM in dialyzed squid axons[17]: It remains uncertain how this reflects on the relative importance of the exchange mechanism in mediating Ca influx in these systems.

DURING REVERSE EXCHANGE, DOES THE ACTIVATING CATION GET TRANSPORTED?

Much interest has been focused on the potassium stimulation of Na-Ca exchange in rod outer segments in which K^+ is transported alongside Ca, with a stoichiometry of $4Na^+ : 1\ Ca^{2+} + 1K^+$ and K_m for $K_o \sim 1$ mM.[33-35] As the reverse exchange in squid axons displays a clear activation by monovalent cations, Condrescu *et al.*[36] have investigated whether the activating cation is also transported in dialyzed axons. Their results, using ^{86}Rb as tracer for K^+, show very clearly that the activating Rb cation is not transported during Ca_o-Na_i exchange. It appears that K^+ is also not transported with currents carried by the exchange in cardiac myocytes[37]; a caveat to this study, however, is that no activating effect of K^+ was observed.

One approach to examine this effect in intact squid axons is to simply monitor lithium uptakes under conditions that activate or inhibit Ca_o-Na_i exchange (TABLE 1). The preliminary data from two paired axons (Part 1) show that lithium uptake is increased by addition of 10 mM calcium to the sea water. In Part 2 both ^{45}Ca uptake and lithium uptakes are measured on the same axons with or without 25 mM Mn to inhibit the exchange.[14] The data from three pairs of axons show that the average inhibition effected by Mn on ^{45}Ca uptakes was 51%, while that for lithium uptakes was 20%. Furthermore, lithium uptakes were highly variable, ranging from 2.93 to 18.8 $pmole \cdot cm^{-2} \cdot sec^{-1}$ (Parts 1 & 2). Measuring lithium uptakes under these conditions does not allow a good estimate of other possible routes of influx, and in Part 2 one of the three pairs of axons actually shows an increase in lithium uptake in the presence of Mn. Lithium uptakes did not match those expected for Ca uptakes via Ca_o-Na_i exchange and these preliminary data tend to support the findings that the monovalent cation is not transported,[36,37] and the exchange in photoreceptors has a different stoichiometry.[33–35] It would be very interesting, however, to examine lithium uptakes into intact axons with high and low Na_i.

SUMMARY

The exchange in intact axons displays a number of features in common with other systems, but a number of interesting points remain to be examined. Both forward (Na_o-Ca_i) and reverse (Ca_o-Na_i) exchange are sensitive to changes in membrane potential, but potassium depolarization can also stimulate Ca_o-Na_i exchange by chemical activation at a monovalent cation-binding site. By monitoring lithium uptakes into intact axons, activating cations do not appear to be transported on the exchange, but this deserves further examination under more stringent conditions. Ca_o-Na_i exchange in intact axons appears activated by monovalent cations to a greater extent compared to dialyzed axons that exhibit little, if any, shift in the K_m for Ca_o. The catalytic effect of Ca_i on Ca_o-Na_i exchange seen in dialyzed axons proves elusive to study in intact axons, with or without introduction of Ca chelators. Experiments using ruthenium red suggest that free calcium can be dissociated from Ca_o-Na_i exchange fluxes; this finding is also important to those studies monitoring exchange activity using Ca indicators. The possibility that Ca chelators may effect changes in the kinetics of Na-Ca exchange is a subject that needs further investigation.

ACKNOWLEDGMENTS

I am deeply indebted to P. F. Baker for his guidance and enthusiasm during the stages of this study.

REFERENCES

1. BAKER, P. F., M. P. BLAUSTEIN, A. L. HODGKIN & R. A. STEINHARDT. 1969. The influence of calcium on sodium efflux in squid axons. J. Physiol. **200:** 431-458.
2. ALLEN, T. J. A. & P. F. BAKER. 1986. Comparison of the effects of potassium and membrane potential on the Ca-dependent sodium efflux in squid axons. J. Physiol. **378:** 53-76.
3. BLAUSTEIN, M. P. & A. L. HODGKIN. 1969. The effect of cyanide on the efflux of calcium from squid axons. J. Physiol. **200:** 497-527.

4. BAKER, P. F. & P. A. MCNAUGHTON. 1976. Kinetics and energetics of calcium efflux from intact squid axons. J. Physiol. **259:** 103-144.
5. ALLEN, T. J. A. & P. F. BAKER. 1986. The influence of membrane potential on calcium efflux from giant axons of *Loligo*. J. Physiol. **378:** 77-96.
6. DIPOLO, R. & L. BEAUGÉ. 1982. The effect of pH on Ca^{2+} extrusion mechanisms in dialysed squid axons. Biochim. Biophys. Acta **688:** 237-245.
7. MULLINS, L. J., J. WHITTEMBURY & J. REQUENA. 1989. Changes in internal ionized Ca and H in voltage clamped squid axons. Cell Calcium **10:** 401-412.
8. DIPOLO, R. & L. BEAUGÉ. 1990. Asymmetrical properties of the Na-Ca exchanger in voltage-clamped, internally dialyzed squid axons under symmetrical ionic conditions. J. Gen. Physiol. **95:** 819-835.
9. KACZOROWSKI, G. J., M. L. GARCIA, F. KING & S. SLAUGHTER. 1989. Development and use of inhibitors to study Na-Ca exchange. *In* Sodium-Calcium Exchange. T. J. A. Allen, D. Noble & H. Reuter, Eds. Oxford Press. London.
10. ALLEN, T. J. A. & P. F. BAKER. 1986. Effects of amiloride and 3′,4′-dichlorobenzamil on Na-Ca exchange in squid axons. J. Physiol. **378:** 107P.
11. TROSPER, T. L. & K. D. PHILIPSON. 1983. Effects of divalent and trivalent cations on sodium-calcium exchange in cardiac sarcolemmal vesicles. Biochim. Biophys. Acta **731:** 63-68.
12. KIMURA, J., S. MIYAMAE & A. NOMA. 1987. Identification of sodium-calcium exchange current in single ventricular cells of guinea pig. J. Physiol. **384:** 199-222.
13. AIKIN, C. C., A. F. BRADING & D. WARMSLEY. 1987. An investigation of sodium-calcium exchange in the smooth muscle of guinea-pig ureter. J. Physiol. **391:** 325-346.
14. ALLEN, T. J. A. 1990. The effects of manganese and changes in internal free calcium on Na-Ca exchange fluxes in the squid axon. Biochem. Biophys. Acta **1030:** 101-110.
15. DIPOLO, R. 1979. Calcium influx in internally dialyzed squid giant axons. J. Gen. Physiol. **73:** 91-113.
16. DIPOLO, R. & L. BEAUGÉ. 1986. Reverse Na-Ca exchange requires internal Ca and/or ATP in squid axons. Biochim. Biophys. Acta **854:** 298-306.
17. DIPOLO, R. & L. BEAUGÉ. 1987. Characterization of the reverse Na/Ca exchange in squid axons and its modulation by Ca_i and ATP. J. Gen. Physiol. **90:** 505-525.
18. CONDRESCU, M., A. GERARDI & R. DIPOLO. 1988. Na^+-Ca^{2+} exchange in squid optic nerve membrane vesicles is activated by internal calcium. Biochim. Biophys. Acta **946:** 289-298.
19. RASGADO-FLORES, H., E. M. SANTIAGO & M. P. BLAUSTEIN. 1989. Kinetics and stoichiometry of coupled Na efflux and Ca influx (Na/Ca exchange) in barnacle muscle cells. J. Gen. Physiol. **93:** 1219-1241.
20. MIURA, Y. & J. KIMURA. 1989. Sodium-calcium exchange current. Dependence on internal Ca and Na and competitive binding of external Na and Ca. J. Gen. Physiol. **93:** 1129-1145.
21. HILGEMAN, D. W. 1989. Giant excised cardiac sarcolemmal membrane patches: Sodium and sodium-calcium exchange currents. Pflügers Arch. **415:** 247-249.
22. BAKER, P. F. 1970. Sodium-calcium exchange across nerve cell membranes. *In* Calcium and Cellular Function. A. W. Cuthbert, Ed.: 96-107. Macmillan. New York.
23. BAKER, P. F. 1972. Transport and metabolism of calcium ions in nerve. Prog. Biophys. Mol. Biol. **24:** 177-223.
24. MULLINS, L. J., T. TIFFERT, G. VASSORT & J. WITTEMBURY. 1983. Effects of internal sodium and hydrogen ions and of external calcium ions and membrane potential on calcium entry in squid axons. J. Physiol. **338:** 295-319.
25. REQUENA, J., L. J. MULLINS, J. WHITTEMBURY & F. J. BRINLEY, JR. 1986. Dependence of ionized and total Ca^{2+} in squid axons on Na^+_o-free or high-K^+_o conditions. J. Gen. Physiol. **87:** 143-423.
26. REQUENA, J., J. WHITTEMBURY & L. J. MULLINS. 1989. Calcium entry in squid axons during voltage clamp pulses. Cell Calcium **10:** 413-423.
27. ALLEN, T. J. A. & P. F. BAKER. 1985. Intracellular Ca indicator Quin-2 inhibits Ca^{2+} inflow via Na_i-Ca_o exchange in squid axon. Nature **316:** 755-756.

28. BAKER, P. F. & J. A. UMBACH. 1987. Calcium buffering in axons and axoplasm of *Loligo*. J. Physiol. **383:** 369-394.
29. DIPOLO, R., L. BEAUGÉ & H. ROJAS. 1989. In dialyzed squid axons Ca_i^{2+} activates Ca_o^{2+}-Na_i^+ and Na_o^+-Na_i^+ exchanges in the absence of Ca chelating agents. Biochim. Biophys. Acta **978:** 328-332.
30. DIPOLO, R., H. ROJAS & L. BEAUGÉ. 1980. Mechanisms of Ca transport in the giant axon of the squid and their physiological role. Cell Calcium **1:** 147-169.
31. BLAUSTEIN, M. P. 1977. Effects of internal and external cations and of ATP on sodium-calcium and calcium-calcium exchange in squid axons. Biophys. J. **20:** 79-110.
32. ALLEN, T. J. A., R. DIPOLO & H. ROJAS. 1988. Effects of external lithium and internal calcium on the Ca_o affinity of reverse sodium-calcium exchange in the dialysed squid axon. J. Physiol. **407:** 136P.
33. CERVETTO, L., L. LAGNADO, R. J. PERRY & P. A. MCNAUGHTON. 1989. Extrusion of calcium from rod outer segments is driven by both sodium and potassium gradients. Nature **337:** 740-743.
34. SCHNETKAMP, P. P. M., D. K. BASU & R. T. SZERENCSEI. 1989. Na^+-Ca^{2+} exchange in bovine rod outer segments requires and transports K^+. Am. J. Physiol. **257:** C153-C157.
35. FRIEDEL, U., G. WOLBRING, P. WOHLFART & N. J. COOK. 1991. The sodium-calcium exchanger of bovine rod photoreceptors: K-dependence of the purified and reconstituted protein. Biochim. Biophys. Acta **1061:** 247-252.
36. CONDRESCU, M., H. ROJAS, A. GERARDI, R. DIPOLO & L. BEAUGÉ. 1990. In squid nerve fibres monovalent activating cations are not cotransported during Na/Ca exchange. Biochim. Biophys. Acta **1024:** 198-202.
37. YASUI, K. & J. KIMURA. 1990. Is potassium co-transported by the cardiac Na-Ca exchange? Pfluegers Arch. Eur. J. Physiol. **415:** 513-515.

Mechanism of Partial Reactions in the Cardiac Na^+-Ca^{2+} Exchange System[a]

DANIEL KHANANSHVILI

Department of Biochemistry
Weizmann Institute of Science
Rehovot, 76100, POB 26, Israel

The Na^+-Ca^{2+} exchange system is a typical representative of a large family of antiporters (otherwise known as "exchange-only" systems).[1,2] In general, antiporters modulate primary ion gradients, created by ion pumps, and can be regulated by numerous regulatory modes, including membrane potential.[1-4] However, an understanding of cellular regulatory mechanisms is seriously hampered because very little is known about the basic mechanism of ion transport. Most antiporters (especially cation antiporters), are low-abundance proteins ($<$ 0.1-0.2% of total membrane protein) exhibiting a high rate of catalytic turnover ($>$ 1000 sec^{-1}).[1-3] This presents serious problems for obtaining reasonable quantities of purified proteins for structural studies and for observing pre-steady-state kinetics on the level of a single catalytic cycle.

The exchange-only systems can expose ion-binding sites to the opposite sides of the membrane at different stages of the catalytic cycle.[1,2,5] The ion pumps can also exhibit an alternative exposure of ion-binding sites, but in order to provide this, they have to undergo additional steps of phosphorylation and dephosphorylation.[1,2] Two different basic mechanisms of ion transport can be considered for the Na^+-Ca^{2+} exchanger; the ping-pong (consecutive) and the sequential (simultaneous) mechanisms.[1,5,6] In the ping-pong mechanism, translocation of Na^+ and Ca^{2+} ions are separate events, while in the sequential mechanism the exchanger binds Na^+ and Ca^{2+}, and translocates both ions in a single step. A distinction between the two basic mechanisms was a difficult issue for many years (for review see Refs. 3-5). By using a semi-rapid mixing device, we have recently measured the V_{max}, K_m, and V_{max}/K_m parameters of Na_i-dependent ^{45}Ca uptake in EGTA-entrapped proteoliposomes (zero-*trans* conditions).[6] On the basis of these measurements, we have concluded that the experimental data are consistent with the ping-pong mechanism. A strong support for the ping-pong mechanism has been very recently obtained in heart cells by using a combination of "caged-Ca" and patch-clamp techniques.[7] Additional evidence for the ping-pong mechanism has come from the measurement of transient charge movements of Na^+ half-cycle via the exchanger (Hilgemann & Philipson, personal communication).

[a] This work was supported by the Miron Bantrel Foundation.

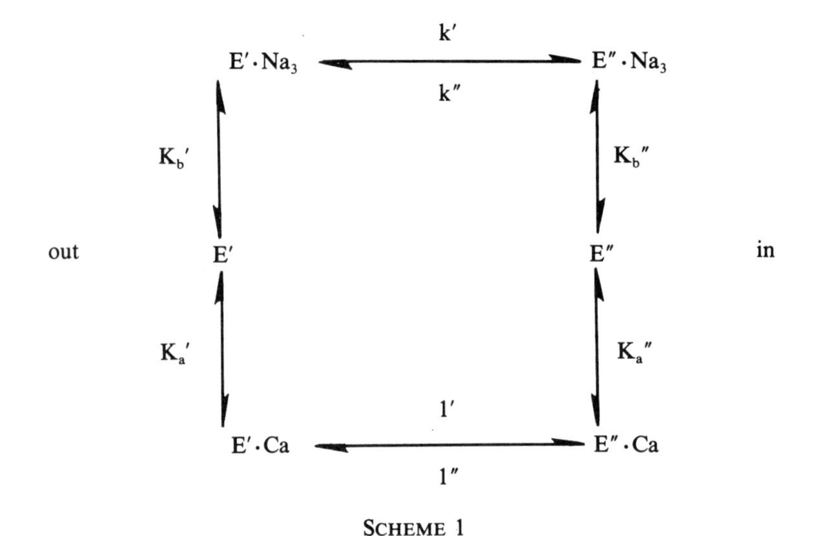

SCHEME 1

However, the real mechanism of ion transport in the Na^+-Ca^{2+} exchanger could be a much more complicated process than a simple ping-pong mechanism, such as the one described in SCHEME 1. For example, the "simple" ping-pong mechanism can be modified by operation of putative "regulatory" ion-binding sites, possible interaction of two (or more) catalytic units, cooperative interaction of catalytic Na^+-ion-binding sites, and so forth.

Reconstitution of functionally active Na^+-Ca^{2+} exchange protein(s) in proteoliposomes[8-10] permits the study of ion fluxes under controlled experimental conditions.[6,11,20] This provides wide opportunities to apply a steady-state kinetic approach in order to analyze ion-transport mechanisms.[1,6] The main goal of this paper is to analyze different modes of the Na^+-Ca^{2+} exchanger (Na^+-Ca^{2+}, Ca^{2+}-Ca^{2+}, and Na^+-Na^+ exchanges) under the zero-*trans,* infinite-*trans,* and equilibrium exchange conditions.

EVIDENCE FOR INVOLVEMENT OF CA-CA EXCHANGE IN A CATALYTIC TURNOVER OF THE NA-CA EXCHANGE CYCLE

According to the ping-pong mechanism, the Ca^{2+}-Ca^{2+} and Na^+-Na^+ exchange reactions are reversible partial reactions of the Na^+-Ca^{2+} exchange cycle, while according to the sequential mechanism the Ca^{2+}-Ca^{2+} and Na^+-Na^+ exchange reactions do not represent elementary steps of the catalytic cycle.[5-7] In order to distinguish 1between these two possibilities, it is useful to consider the kinetic parameters of the Na^+-Ca^{2+} and Ca^{2+}-Ca^{2+} exchange reactions under the infinite-*trans* conditions (TABLE 1). A simple ping-pong mechanism (SCHEME 1) predicts that the ratio R^{it} between the observed K_m values of the Na^+-Ca^{2+} and Ca^{2+}-Ca^{2+} exchange reactions is equal to the ratio of the V_{max} values (Eq. 1):

$$R^{it} = \frac{K_m{}^{it}(Na/Ca)}{K_m{}^{it}(Ca/Ca)} = \frac{V^{it}{}_{max}(Na/Ca)}{V^{it}{}_{max}(Ca/Ca)} = \frac{1 + 1'/1''}{1 + 1'/k''} \qquad (1)$$

In this series of experiments, the same preparations of vesicles (sarcolemma membranes or proteoliposomes) were loaded with a high, fixed concentration of unlabeled Ca^{2+}

TABLE 1. Kinetic Parameters of Partial Reactions under the Infinite-*trans*, Infinite-*cis*, and Equilibrium Exchange Conditions

Reaction	Rate Constant	Equilibrium Constant
Na-Ca exchange (infinite-*trans*)	$K_m^{it}\,(Na/Ca) = K_a' \dfrac{k''}{l' + k''}$	$V^{it}_{max}\,(Na/Ca) = \dfrac{k''l'\,[E]_t}{k'' + l'}$
Ca-Ca exchange (infinite-*trans*)	$K_m^{it}\,(Ca/Ca) = K_a' \dfrac{l''}{l' + l''}$	$V^{it}_{max}\,(Ca/Ca) = \dfrac{l''l'\,[E]_t}{l'' + l'}$
Na-Ca exchange (infinite-*cis*)	$K_m^{ic}\,(Na/Ca) = K_a'' \dfrac{k'}{l'' + k'}$	$V^{ic}_{max}\,(Na/Ca) = \dfrac{k'l''\,[E]_t}{l'' + k'}$
Ca-Ca exchange (infinite-*cis*)	$K_m^{ic}\,(Ca/Ca) = K_a'' \dfrac{l'}{l'' + l'}$	$V^{ic}_{max}\,(Ca/Ca) = \dfrac{l'l''\,[E]_t}{l'' + l'}$
Na-Na exchange (equilibrium)		$V^{ee}_{max}\,(Na/Na) = \dfrac{k''k'\,[E]_t}{k'' + k'}$
Ca-Ca exchange (equilibrium)		$V^{ee}_{max}\,(Ca/Ca) = \dfrac{l''l'\,[E]_t}{l'' + l'}$

NOTE: The rate and equilibrium constants have been written according to the simple ping-pong mechanism described in SCHEME 1.

(for assay of Ca^{2+}-Ca^{2+}) or Na^+ (for assay of Na^+-Ca^{2+} exchange) and diluted (100-200-fold) in assay medium containing varying $[^{45}Ca]_o$. In order to avoid any side effects of monovalent ions, the initial rates (t = 1 sec) of the ^{45}Ca uptake were measured in sucrose medium, in the absence of monovalent ions. As can be seen from TABLE 2, under the infinite-*trans* conditions the observed ratios of the kinetic parameters were $[K_m^{it}(Na/Ca)]/[K_m^{it}(Ca/Ca)] = 2.1 \pm 0.1$ and $[V^{it}_{max}(Na/Ca)]/[V^{it}_{max}(Na/Ca)] = 2.4 \pm 0.1$, suggesting that within the experimental error the R^{it} is a constant value for a given preparation of vesicles. This was reproducible either for the sarcolemmal membrane vesicles or proteoliposomes ($n = 6$). These results provide independent support of the ping-pong mechanism. It is important to note that in different preparations of sarcolemma vesicles the absolute values of R^{it} varied from 2.1 to 5.3. This means that, although in different preparations the absolute values of kinetic parameters can vary, their ratios are compatible with Eq. 1, as predicted by a ping-pong mechanism.

IS ANY PARTICULAR ION-TRANSPORT STEP RATE-LIMITING FOR NA-CA EXCHANGE CYCLE?

A rate-limiting step of the catalytic turnover can be identified by direct measurement of rate constants involved in Na^+ and Ca^{2+} binding and transport. At this moment no experimental setup is available to measure absolute rate constants of elementary reactions on the level of single turnover. On the other hand the equilibrium exchange experiments of homogeneous Na^+-Na^+ and Ca^{2+}-Ca^{2+} exchange reactions may provide useful information about the separate steps of ion-transport mechanisms.[1,5] It is

TABLE 2. The Observed Kinetic Values of the Na^+-Ca^{2+} and Ca^{2+}-Ca^{2+} Exchanges under the Infinite-*trans* Conditions

Reaction	Observed K_m (μM)	Observed V_{max} (nmol Ca · mg^{-1} · sec^{-1})
Na-Ca exchange	K_m^{it} (Na/Ca) = 45 ± 9	V^{it}_{max} (Na/Ca) = 19 ± 2
Ca-Ca exchange	K_m^{it} (Ca/Ca) = 21 ± 4	V^{it}_{max} (Ca/Ca) = 8 ± 1
	$\frac{K_m^{it}\,(Na/Ca)}{K_m^{it}\,(Ca/Ca)} = 2.1 \pm 0.1$	$\frac{V^{it}_{max}\,(Na/Ca)}{V^{it}_{max}\,(Ca/Ca)} = 2.4 \pm 0.1$

NOTE: Proteoliposomes were obtained[6,20] and equilibrated in 20 mM MOPS/Tris, pH 7.4, and 0.25 M sucrose either with 160 mM NaCl or 250 mM $CaCl_2$. Na or Ca-loaded vesicles were diluted 100-fold in MOPS/Tris/sucrose medium containing 2 μM endogenous Ca and 1-120 μM added ^{45}Ca. The initial rates (t = 1 sec) of the Na_i-dependent ^{45}Ca uptake were measured and the specific radioactivity for each $[Ca]_o$ was corrected as described before.[6] The lines were computed according to $v = V_{max}\,[Ca]_o/(K_m + [Ca]_o)$ to give an optimal fit of the K_m and V_{max} values to the experimental points (Enzfitter PC-IBM program).

convenient to do these experiments at saturating concentrations of ions on both sides of the membrane, assuming that under these conditions the ion binding is not rate limiting (TABLE 1). In the same preparations of proteoliposomes (n = 5), the initial rates of the equilibrium Na^+-Na^+ exchange were at least two- to threefold slower than the initial rates of the Ca^{2+}-Ca^{2+} exchange (TABLE 3). If one takes into account a probable stoichiometry of the equilibrium exchange reactions, that is, 1Ca : 1Ca (Ca^{2+}-Ca^{2+} exchange) and $3Na^+$: $3Na^+$ (Na^+-Na^+ exchange) the rate constant of the limiting step of Na^+-Na^+ exchange is six- to ninefold slower than that of Ca^{2+}-Ca^{2+} exchange. If we propose that inward-outward rate constants are symmetric for Na^+ ($k' \sim k''$) and Ca^{2+} ($l' \sim l''$) transport, the rate constants involved in Na^+ transport could be rate limiting. However, this conclusion has certain limitations, because of possible asymmetry of the rate constants involved in each ion-transport step. For example, if $k' \sim k''$ and $l' >> l''$, a slowest rate constant of the catalytic cycle could be an outward transport of Ca^{2+} (l''). This means that in the case of unidirectional Ca^{2+} influx [$V^{12}_{max} = l'\,k''/(l'' + k'')$], the outward Na^+-transport step (k'') might be rate limiting. On the other hand in the case of unidirectional Ca^{2+}-efflux [$V^{21}_{max} = l''\,k'/(l'' + k'')$], the outward Ca^{2+} transport (l'') may control the overall turnover of the catalytic cycle.

RELATIONSHIP BETWEEN THE HOMOGENEOUS AND HETEROGENEOUS EXCHANGE REACTIONS

Although the Na^+-Na^+ and Ca^{2+}-Ca^{2+}exchange reactions reflect partial steps of the Na^+-Ca^{2+} exchanges cycle, the heterogeneous exchange is not a simple sum of homogeneous ion-exchange reactions. This can be seen by putting rate expressions for V_{max} values (see TABLE 1): $V^{it}_{max}(Na/Ca) + V^{ic}_{max}(Na/Ca) \neq V^{ee}_{max}(Na/Na) + V^{ee}_{max}(Ca/Ca)$. This means that even in the absence of any regulatory modes the heterogeneous exchange reactions cannot be just a "simple sum" of homogeneous partial reactions. Likewise, this situation may become even more complicated, if we take into account a possible regulation of elementary reactions by different modes. For example, the rate-limiting step during the Na^+-Ca^{2+} exchange may be controlled by conditions, which do not operate during the homogeneous Na^+-Na^+ or Ca^{2+}-Ca^{2+} exchanges. Conceivably, during Na^+-Ca^{2+} exchange, there are additional interactions of Na^+ and Ca^{2+} with the exchanger, which do not exist in Na^+-Na^+ and Ca^{2+}-Ca^{2+} exchange conditions. It is possible, for example, that during the Na^+-Ca^{2+} exchange a Na^+-binding to the "regulatory" site may accelerate a Ca^{2+} transport step and/or the Ca^{2+} binding may control a Na^+-transport step. This kind of regulatory mechanism seems to be in good agreement with the experimental fact that monovalent ions, including Na^+, accelerate the Ca^{2+}-Ca^{2+} exchange.[3,4]

POSSIBLE ASYMMETRY OF Ca^{2+} MOVEMENTS

It has already been established that in whole cells the extracellular and intracellular K_m values are very asymmetric for Ca^{2+}.[3,4,12] The reasons for this asymmetry are not known. Likewise, this kind of asymmetry has not been demonstrated in plasma membrane vesicles or reconstituted proteoliposomes.[3,4] The apparent binding affinity of an ion (K_m) is a function of the intrinsic binding constant and rate constants and depends on the chosen experimental setup (TABLE 1). An advantage of Equation 1 is that it does not contain any equilibrium binding constants, while the R^{it} values reflect only

TABLE 3. Observed V_{max} Values of the Ca^{2+}-Ca^{2+}, Na^+-Na^+, and Na^+-Ca^{2+} Exchanges under the Equilibrium Exchange, Infinite-*trans* and Infinite-*cis* Conditions

Exchange	
Na-Na (equilibrium)	V^{ee}_{max} (Na/Na) = 3.0 ± 0.2 nmol Na · mg^{-1} · sec^{-1}
Ca-Ca (equilibrium)	V^{ee}_{max} (Ca/Ca) = 6.7 ± 0.3 nmol Ca · mg^{-1} · sec^{-1}
Na-Ca (infinite-*trans*)	V^{it}_{max} (Na/Ca) = 19 ± 2 nmol Ca · mg^{-1} · sec^{-1}
Na-Ca (infinite-*cis*)	V^{ic}_{max} (Na/Ca) = 12 ± 3 nmol Ca · mg^{-1} · sec^{-1}
	$R > 3$ $R^{it} > 2$ $R^{ee} < 0.5$ $\frac{R^{it}}{R^{ee}} > 4$

NOTE: Proteoliposomes (see legend in TABLE 1) were equilibrated with 160 mM NaCl or 200 μM $CaCl_2$. The initial rates of the equilibrium Na^+-Na^+ ($[^{22}Na]_o$ = $[Na]_i$ = 160 mM) and Ca^{2+}-Ca^{2+} ($[^{45}Ca]_o$ = $[Ca]_i$ = 200 μM); exchanges were measured under the conditions outlined before.[20] The assay of infinite-*trans* Na^+-Ca^{2+} exchange was carried out as described in TABLE 2. The infinite-*cis* Na^+-Ca^{2+} exchange was measured with 160 μM $^{45}Ca_i$ in assay medium containing 20 mM MOPS/Tris, pH 7.4, and 160 mM NaCl.

the relationship between the rate constants (l′/l″ and l″/k″) involved in ion transport. As can be seen from TABLE 2, the observed value of $R^{it} > 2$, indicating that $l' > l''$ and $k'' > l''$ (see Eq. 1). This may reflect a situation in which the Ca^{2+} movement in opposite directions of the membrane could be an asymmetric process. This means that the $K_{eq} = l'/l''$ for Ca^{2+}-bound species could be different from unity. Variability of the absolute R^{it} values (2.2-5.3) in different preparations of sarcolemmal vesicles may suggest the presence of some "regulatory" modes controlling an asymmetry of Ca^{2+} movements in the Na^+-Ca^{2+} exchange protein. It is possible that in intact cells the absolute R^{it} values are even larger than the absolute values of R^{it} observed here in sarcolemmal membranes or reconstituted proteoliposomes. This may contribute to a 200-500-fold asymmetry of intracellular versus extracellular K_m values for Ca^{2+} observed in intact cells.[3,4,12]

RELATIVE RATE AND EQUILIBRIUM CONSTANTS

The V_{max} values of the Na^+-Na^+ and Ca^{2+}-Ca^{2+} exchange rates measured under the equilibrium, infinite-*trans*, and infinite-*cis* conditions may provide additional information about the relationship of rate constants. TABLE 3 shows that the sum of V_{max} values of unidirectional heterogeneous reactions is at least threefold larger than the sum of V_{max} values of homogeneous reactions (equilibrium exchange conditions).

$$R = \frac{V^{it}_{max}(Na/Ca) + V^{ic}_{max}(Na/Ca)}{V^{ee}_{max}(Na/Na) + V^{ee}_{max}(Ca/Ca)} = \frac{l'k' + l'k'' + l''k' + l''k''}{l'l'' + l'k' + l''k'' + k''k''} \quad (2)$$

Equation 2 shows that if $R > 3$, then $l' > k'$ and $k'' > l''$. As can be seen from TABLE 3, at saturating concentrations of ions on both sides of the membrane, the ratio between the V_{max} values of the Na^+-Na^+ and Ca^{2+}-Ca^{2+} exchanges (Eq. 3) is less than 0.5.

$$R^{ee} = \frac{V^{ee}_{max}(Na/Na)}{V^{ee}_{max}(Ca/Ca)} = \frac{k'(1 + l'/l'')}{l'(1 + k'/k'')} \quad (3)$$

If one takes into account a stoichiometry of 3Na : 1Ca for the cardiac Na^+-Ca^{2+} exchange,[13,14] a value of $R^{ee} < 0.5$ can be considered as an upper limit (R^{ee} can be even less than 0.15, because of the possible stoichiometry of 3Na : 3Na for the Na^+-Na^+ exchange). Assuming that $k'' > l''$, $l' > k'$, $l' > l''$ (see above), we may come to the conclusion that the $R^{ee} < 0.5$ suggests that $(l'/l'') > (k'/k'')$ (see Eq. 2). This means that $K_{eq} = l'/l''$ involving Ca-bound species might be more asymmetric than the $K_{eq} = k'/k''$ for Na-bound species. These estimated relationships between the rate constants are compatible with the observed value of R^{it}/R^{ee} ratio (Eq. 4) being larger than four (TABLE 3):

$$\frac{R^{it}}{R^{ee}} = \frac{l'(1 + k'/k'')}{k'(1 + l'/k'')} \quad (4)$$

Putting together all available information, we propose that in the absence of monovalent ions a sequence of relative rate constants can be described as $l' > k' \sim k'' > l''$. Conceivably, this "intrinsic" relationship between the rate constants can be altered or regulated by different modes.

TABLE 4. Effect of Voltage on the Kinetic Parameters of the Unidirectional Na_i-dependent ^{45}Ca Uptake under the Zero-*trans* and Infinite-*trans* Conditions

$\Delta\Psi$ (mV)	K_m^{12} (μM)	V^{12}_{max} (nmol Ca · mg^{-1} · sec^{-1})	V^{12}_{max}/K_m^{12}
+90	40 ± 5	27 ± 2	0.7 ± 0.1
0	26 ± 6	19 ± 2	0.7 ± 0.2
−95	18 ± 3	10 ± 1	0.5 ± 0.1

NOTE: EGTA-entrapped proteoliposomes were prepared in 20 mM MOPS/Tris, pH 7.4, 160 mM NaCl with different concentrations of potassium (KCl + choline-Cl = 140 mM) and treated with 1 μM valinomycin before the experiment.[6,20] Proteoliposomes were diluted 100-fold in assay medium 20 mM MOPS/Tris, pH 7.4, 300 mM choline-Cl, 2-120 μM of added $^{45}CaCl_2$ and 2μM of endogenous Ca^{2+}. The kinetic parameters were measured as described in TABLE 1.

EFFECT OF MEMBRANE POTENTIAL ON THE INTRINSIC AFFINITY OF ION BINDING

The voltage dependence of the Na^+-Ca^{2+} exchange has been extensively studied[15-18]; however, the mechanism of voltage-dependent modulation of partial reactions is not understood yet. The electrogenic nature of the cardiac Na^+-Ca^{2+} exchange system is a reflection of equilibrium properties of the system. The kinetic counterpart can be reflected in voltage sensitivity of the rates of partial reactions involving charge movements in the electric field.[1,5,19] The voltage dependence of the rate constants may reflect charge-carrying, conformational, and dielectric properties of ligand-protein interactions.[1,5] Charge translocation can result, for example, from conformational transitions associated with movements of polar amino acid residues or from migration of ions in an "ion-well" connecting a binding site with the aqueous medium.[1,5,19]

The effect of voltage on the partial reactions may provide useful information about the mechanisms and regulation of Ca^{2+} and Na^+-transport steps. For example, the rates of nonelectrogenic modes (1 : 1 exchange), such as Na^+-Na^+ and Ca^{2+}-Ca^{2+} exchanges, can be voltage-sensitive.[1,5] Recently we have tested an effect of diffusion potentials (K^+-valinomycin) in reconstituted proteoliposomes by measuring voltage-sensitive ion fluxes of the Na^+-Ca^{2+}, Ca^{2+}-Ca^{2+}, and Na^+-Na^+ exchange reactions under the zero-*trans*, infinite-*trans*, and equilibrium-exchange conditions.[20]

In order to test a possible effect of voltage on K_a' and l′, the effect of varying voltage on the K_m, V_{max}, and V_{max}/K_m has been tested in EGTA-entrapped proteoliposomes (zero-*trans* conditions) and fixed saturating $[Na]_i$ (infinite-*trans* conditions). Under these conditions, when the ion fluxes are unidirectional and Na^+-binding sites are saturated by Na^+, voltage changes from −95 mV to +90 mV increases both the V_{max} (from 10 ± 1 to 27 ± 2 nmol·mg^{-1}·sec^{-1}) and the K_m (from 18 ± 3 to 40 ± 5 μM) values (TABLE 4). Within the same range of voltage, the V_{max}/K_m values range between 0.5 and 0.7 nmol·mg^{-1}·sec^{-1}·μM^{-1}. A voltage-sensitive V_{max} of Na^+-Ca^{2+} exchange suggests that at least one of the rate constants (k″ or l′) is voltage sensitive $\{V^{12}_{max} = l'k''[E]_T/(l' + k'')\}$. A voltage-sensitive increase of K_m for Ca^{2+} $[K_m^{12} = K_a' k''/(l' + k'')]$ by imposition of inside-positive charge may reflect a voltage dependence of K_a' (dissociation constant) and the rate constants k″ and/or l′. The fact that the V_{max}/K_m values are voltage insensitive within experimental error is compatible with the ping-pong mechanism. This also suggests that the K_a' and l′ terms are voltage insensitive or

TABLE 5. Effect of Voltage on the Initial Rates of the Equilibrium Na^+-Na^+ and Ca^{2+}-Ca^{2+} Exchanges at the Saturating and Nonsaturating Ionic Conditions

Experiment	Saturating Ionic Conditions	
	$[Na]_o = [Na]_i$ = 160 mM (nmol Na · mg^{-1} · sec^{-1})	$[Ca]_o = [Ca]_i$ = 200 μM (nmol Ca · mg^{-1} · sec^{-1})
− Val. + KCl	2.8 ± 0.2	7.7 ± 0.5
+ Val. + KCl	2.9 ± 0.3	7.3 ± 0.3
	Nonsaturating Ionic Conditions	
	$[Na]_o = [Na]_i$ = 5 mM (nmol Na · mg^{-1} · sec^{-1})	$[Ca]_o = [Ca]_i$ = 2 μM (nmol Ca · mg^{-1} · sec^{-1})
− Val. + KCl	0.5 ± 0.1	0.04 ± 0.01
+ Val. + KCl	0.3 ± 0.1	0.02 ± 0.01

NOTE: The Na-loaded proteoliposomes were equilibrated with 160 mM or 5 mM NaCl ([Na] + [Ch-Cl] = 300 mM). The Ca-loaded proteoliposomes were obtained with 200 μM or 2 μM $CaCl_2$ (in 300 mM choline-Cl). All preparations were buffered with 20 mM MOPS-Tris, pH 7.4. The different preparations of proteoliposomes were treated with or without valinomycin (see legend in TABLE 4) and diluted in the same buffer plus 20 mM KCl containing ^{22}Na ($[^{22}Na]_o = [Na]_i$) or ^{45}Ca ($[^{45}Ca]_o = [Ca]_i$). The initial rates of the equilibrium Na^+-Na^+ and Ca^{2+}-Ca^{2+} exchanges were measured as described in TABLE 3.

the membrane potential affects both the K_a' and l' proportionally $\{V^{12}_{max}/K_m^{12} = l'[E]_T/K_a'\}$.

The rate of the equilibrium Ca^{2+}-Ca^{2+} exchange can be described as

$$v = \frac{l'l''\,[E]_T}{l''(1 + K_a'/[Ca]_o) + l'(1 + K_a''/[Ca]_i)} \quad (5)$$

At nonsaturating $[Ca]_o = [Ca]_i$ the Ca^{2+}-Ca^{2+} exchange is voltage sensitive (TABLE 5), suggesting a possible effect of voltage on the intrinsic binding affinity (K_a' and/or K_a'') for Ca^{2+} (i.e., if Ca^{2+} enters a narrow vestibule). A voltage-dependent effect was also observed on the equilibrium Na^+-Na^+ exchange at nonsaturating $[Na]_o = [Na]_i$ = 5 mM (TABLE 5), suggesting that Na^+ binding to the protein (K_b' and/or K_b'' are voltage sensitive) is also a voltage-dependent process.

HOW A MEMBRANE POTENTIAL CONTROLS THE Na^+-Ca^{2+} EXCHANGE

As can be seen from TABLE 5 the initial rate of the equilibrium Ca^{2+}-Ca^{2+} exchange is voltage-insensitive at saturating $[Ca]_o = [Ca]_i$ = 200 μM $\{V_{max} = l'l''[E]_T/(l' + l'')\}$. The initial rate of the Na^+-Na^+ exchange is also not affected by membrane potential with saturating $[Na]_o = [Na]_i$ = 160 mM $\{V_{max} = k'k''[E]_T/(k' + k'')\}$ (TABLE 5). Under saturating ionic conditions it is expected that the exchange rates will decline in a bell-shaped manner upon varying a voltage if the involved rate constants are voltage-dependent and rate-limiting. As can be seen from TABLE 5, this is not the case. Under saturating ionic concentrations, both the Ca^{2+}-Ca^{2+} and Na^+-Na^+

exchanges appear to be voltage insensitive. There are two possible explanations for this effect: (a) the Ca^{2+}-Ca^{2+} and Na^+-Na^+ exchanges are rate limited by a voltage-independent step; (b) the exchange rates may become voltage independent at any value of z (charge on the ion-binding site), if the dielectric distances over which the charge moves vanish.[5,19]

An apparent paradox is that both partial reactions are "voltage-insensitive" at saturating ionic conditions, while the Na^+-Ca^{2+} exchange reaction exhibits a voltage-sensitive V_{max} (FIGS. 2 and 3). Even if we assume that the k″ is voltage-dependent and the Na^+ transport is involved in a rate-limiting step of Na^+-Ca^{2+} exchange ($k'' < l'$), this may or may not account for a voltage-dependent modulation of V_{max} during the Na^+-Ca^{2+} exchange. In order to be affected by voltage, either a voltage-independent step (presumably a Ca^{2+} transport) must be accelerated (l′ is accelerated) or the voltage-dependent step (presumably a Na^+ transport) must operate much more slowly (k″ is decelerated) as compared to the homogeneous-exchange conditions. As suggested above, during Na^+-Ca^{2+} exchange some additional interactions of Na^+ and Ca^{2+} with the exchanger may exist (presumably these interactions do not operate under the Na^+-Na^+ and Ca^{2+}-Ca^{2+} exchange conditions) and may affect the voltage sensitivity of V_{max}.

CHARGE-CARRYING PROPERTIES OF ION-PROTEIN INTERMEDIATES

Voltage-dependent effects on the Na^+-Ca^{2+} exchange were tested in the whole cardiac cell[12,17] and sarcolemmal vesicles[21,22]; however, the experimental conditions were not "zero-*trans*." By using EGTA-entrapped proteoliposomes, we have found recently that under the zero-*trans* conditions and saturating $[Ca]_o$ and $[Na]_i$ a linear and neither exponential nor bell-shaped flux-voltage curve is observed over a wide range of voltage (from −104 mV to +135).[20] This makes unlikely the charge (z) values of 0, −1, or −4 for unliganded protein, since in this case both Na^+- and Ca^{2+}-transport steps would carry either a positive charge (for $z = 0$ or -1) or negative charge (for $z = -4$) leading to a bell-shaped voltage curve.[5,19] This is because the extreme values of voltage (negative or positive) will inhibit a rate-limiting partial reaction (either Na^+ or Ca^{2+} transport) causing a bell-shaped voltage dependence of Na^+-Ca^{2+} exchange. A model with a single charge-carrying step should reproduce a monotonic flux-voltage relation that leads to a plateau value at extremes of voltage. The observed linear voltage curve is most simply explained as being a part of such a curve in the range of tested voltage.[1,5,19] This would indicate a charge value of unliganded protein as $z = -2$ ($E \cdot Na_3$ species are positively charged and $E \cdot Ca$ are electroneutral) or of $z = -3$ ($E \cdot Ca$ species carry a negative charge and $E \cdot Na_3$ are electroneutral). We prefer the possibility that the cation-binding domain contains $z = -2$ rather than, say, $z = -3$, because independent evidence described above supports the possibility that in the unidirectional Ca-influx mode of Na^+-Ca^{2+} exchange the Na^+-transport step might be rate-limiting ($l' > k''$).

The binding of three Na^+ ions to the cation-binding domain carrying −2 charges has been proposed before for Na^+, K^+-ATPase.[23] This kind of model has been supported for the Na^+, K^+-ATPase by using electrophysiological[25] and biophysical[26] approaches. Similarity between the Na^+, K^+-ATPase and the Na^+-Ca^{2+} exchangers seems likely, because both proteins bind three Na^+. Likewise, it has been shown recently that the 23-amino-acid sequence of a putative transmembrane cation-binding domain contains two carboxyl residues in both proteins.[24]

ION WELLS AND ION-BINDING DOMAIN

Ion migration through a putative "ion well" might be a complicated process involving partial dehydration of ions, specific conformational changes of protein, modification of dielectric properties of the environment, and so forth.[1,5,19] These collected effects could be reflected in a modification of intrinsic binding affinity of ion-protein interaction induced by membrane potential[5,19]:

$$K_a' = K_a \exp\left(\frac{n\Delta\Psi\alpha'}{RT/F}\right) \quad \textbf{(6)}$$

in which n is the number of elementary charges, K_a is a dissociation constant at zero voltage and α' is the fraction of the field traversed (dimensionless dielectric distance) by the ion in reaching its site. Our observations do not support the possibility that there are large changes in V_{max}/K_m (see above), suggesting that K_a' is not extensively modified by a membrane potential (Eq. 2). If we assume that K_a' is voltage insensitive ($K_a'/K_a = 1$), $\alpha' = 0$ (Eq. 6), suggesting that the Ca^{2+}-binding site is outside of the electric field. One can put an upper limit for the voltage-dependent modulation of the K_a' (membrane potential is varied over 180 mV) as $K_a'/K_a = 2$. In this case $\alpha' = 0.05$, suggesting that even under these limiting conditions the present experimental data are consistent with very small values of dielectric distance α' for the Ca^{2+}-binding site. For a membrane thickness of 40-50 Å, this corresponds to a well of about 2-3 Å in depth. If this kind of well for Ca^{2+} exists at all, it can be on the order of the diameter of an unhydrated Ca^{2+} ion (x-radius is 0.95 Å). It is possible that the α' is smaller than the α'' (the dielectric distance for Ca^{2+}-binding on the other side of the membrane). Assuming that K_a'' is affected by membrane potential more extensively than K_a' would reflect a voltage-sensitive Ca^{2+}-Ca^{2+} exchange at very low $[Ca]_o = [Ca]_i$ (TABLE 5).

REFERENCES

1. STEIN, W. D. 1986. *In* Transport and Diffusion across Cell Membranes. Academic Press. New York.
2. KHANANSHVILI, D. 1990. Curr. Opin. Cell Biol. **2:** 731-734.
3. PHILIPSON, K. D. 1990. *In* Calcium and the Heart. G. A. Langer, Ed.: 85-108. Raven Press. New York.
4. REEVES, J. P. 1985. Curr. Top. Membr. Trans. **25:** 77-127.
5. LAUGER, P. 1984. Biochem. Biophys. Acta **779:** 307-341.
6. KHANANSHVILI, D. 1990. Biochemistry **29:** 2437-2442.
7. NIGGLI, E. & W. J. LEDERER. 1991. Nature **349:** 621-624.
8. MIAYMOTO, H. & E. RACKER. 1980. J. Biol. Chem. **255:** 2656-2658.
9. SOLDATI, L., S. LONGONI & E. CARAFOLI. 1985. J. Biol. Chem. **260:** 13321-1327.
10. VEMURI, R. & K. D. PHILIPSON. 1987. Biochem. Biophys. Acta **937:** 258-268.
11. CHEON, J. & J. P. REEVES. 1988. J. Biol. Chem. **263:** 2309-2315.
12. LI, J. & J. KIMURA. 1990. **29:** 777-788.
13. PITTS, B. J. R. 1979. J. Biol. Chem. **254:** 6232-6235.
14. REEVES, J. P. & C. C. HALE. 1984. J. Biol. Chem. **259:** 7733-7739.
15. LEGNADO, L. & P. A. MCNAUGHTON. 1990. J. Membr. Biol. **117:** 177-191.
16. LEGNADO, L., L. CERVETTO & P. A. MCNAUGHTON. 1988. Proc. Natl. Acad. Sci. USA, **85:** 4548-4552.
17. KIMURA, I., S. MIYAMAE & A. NOMA. 1987. J. Physiol. (London) **384:** 199-222.
18. EGAN, T. M., D. NOBEL, S. J. NOBEL, T. POWELL & A. J. SPINDLER. 1989. J. Physiol. London **411:** 639-662.
19. LAUGER, P. 1984. Biochem. Biophys. Acta **779:** 307-341.

20. KHANANSHVILI, D. 1991. J. Biol. Chem. Submitted.
21. PHILIPSON, K. D. & A. Y. NISHIMOTO. 1980. J. Biol. Chem. **255:** 6880-6882.
22. LEDVORA, R. F. & C. HEGYVARY. 1983. Biochem. Biophys. Acta **729:** 123-136.
23. GOLDSHLEGGER, R., S. J. D. KARLISH, A. REPHAELI & W. D. STEIN. 1987. J. Physiol. (London) **387:** 331-355.
24. NICOLL, D. A., S. LONGONI & K. D. PHILIPSON. 1990. Science **250:** 562-564.
25. RAKOWSKI, R. F., D. C. GADSBY & P. DE WEER. 1989. J. Gen. Physiol. **93:** 903-941.
26. APELL, H.-Y. 1989. J. Membr. Biol. **110:** 103-114.

Is Stoichiometry Constant in Na-Ca Exchange?

L. J. MULLINS

Department of Biophysics
University of Maryland School of Medicine
Baltimore, Maryland 21201

The operation of Na-Ca exchange has turned out to be a very complex process. My purpose here is not to write a review or scientific paper but rather to outline some of the problems as I see them and to propose that one variable that has not been seriously considered, namely, a variable coupling ratio for Na-Ca, might help in understanding some of the problems.

Variable coupling has not been seriously considered because the Na-K pump, which has been studied in detail, does not show such an effect, at least not when it is operated in a Na-K transport mode. However, I find it unlikely that Na-K and Na-Ca exchange operate in anything like similar modes. For Na-K a fixed amount of energy is transferred to the protein; this energy is represented by the use of gamma-P to phosphorylate the pump protein. By contrast, the Na-Ca mechanism is energized by three or four Na ions binding to the protein, and, under certain conditions, one could imagine that two Na ions would suffice to energize transport—under others, four Na ions might be required.

MODELS

Everyone seems to have a model, and my complaint about them is that they usually have two or three variables instead of ten or twelve, which would be necessary if the model were actually to describe the behavior of the exchange.

The last model paper I published was in 1977. At that time it seemed that a quite simple exchange system would serve to account for the known phenomena. Since that time, however, observations have led to an appreciation of the complexity of the system and have made model construction very difficult. The following attempts to categorize some of the difficulties. My advice, as mentioned above, is to forget about the Na pump as a model of Na-Ca exchange; I am certain that Na-Ca exchange doesn't work that way.

THE EVOLUTION OF IDEAS ABOUT NA-CA EXCHANGE

a. It was a 2-Na or 1-Ca exchanger that could move Ca in one direction using the energy of the Na gradient.

b. The difficulty was that Ca in the cell would be far too high compared with measurements that had been made.

c. An early variant of this scheme was a proposal by Baker, Blaustein, and Hodgkin that perhaps the carrier carried 3 Na and 1 Ca + 1 K. This idea preserved an electroneutral exchange but was not pursued with any experimental studies, and it still gave equilibrium values for internal Ca that were far too high.

d. Membrane vesicle studies and physiological measurements on nerve found a sensitivity of the exchange fluxes to membrane potential. The simplest explanation of such findings were that three or more Na were transported per Ca so that an electric current would be generated by the operation of Na-Ca exchange. My original suggestion was that four Na were transported per Ca, and I still think that under some conditions this may be true. If so, why does everyone find three Na per Ca? Experiments are often done at relatively high internal Ca (because of the difficulty of measurement in the range of tens of nM); at high Ca there is no need to couple more than three Na per Ca.

e. It has proved possible to measure such a current in cardiac cell patches and in intact cells.

f. In some cells the loading of the Na-Ca carrier is four Na per Ca + K. This gives a far lower limiting Ca in the cell but still produces an electric current.

g. Proteases applied to the inside of the membrane cause the Na-Ca exchanger to behave much like simple models predict—there is no requirement of internal Ca for the reverse operations of the exchanger, nor is there a requirement for internal sensitivity for Na to affect voltage sensitivity of the exchange.

PRESENT STATE OF EXPERIMENTAL FINDINGS ABOUT THE CARRIER

a. In addition to the conventional expectations that Na-Ca requires Na outside to produce Ca efflux, and the reverse, Na inside to activate Ca entry, a number of findings make it apparent that the carrier is a very complex entity.

b. The voltage sensitivity of Ca efflux in squid axons disappears if internal Na is zero. A simple explanation of this finding might be that the Na gradient is now sufficiently large to extrude Ca in an electroneutral (2 Na/Ca) mode and, hence, that coupling can be variable.

c. The inverse ion movement, Ca influx in exchange for internal Na, requires physiological levels of internal Ca. At first I didn't believe that this was more than an artifact of isotopic measurement, but the evidence is convincing that exchange currents in cardiac cells disappear if internal Ca is made very low. The most obvious explanation of this effect is that internal Ca is an activator of the exchange system in addition to being transported by it, but clearly, more work needs to be done to understand this.

d. Activation of Na-Ca exchange by monovalent cations also needs study because in addition to Na, there is an activation of the exchange by K (in addition to any role it may play in affecting membrane potential), but the effect of K is complicated by its possible involvement in being moved by the Na-Ca exchange itself. Lithium is capable of producing exchange fluxes of Ca many-fold larger than the maximum Na-Ca flux, but its role in energizing Li-Ca exchange is small. Choline and other organic cations change the affinity of the carrier for Ca at the outside of the membrane, often in quite dramatic ways, and thus lead to confusing ideas about the Ca affinity of the carrier.

e. Na-Ca exchange is very sensitive to pH inside with H ions inhibiting the exchange.

f. ATP enhances both influx and efflux of Ca via the exchanger—there is no evidence that it produces a net flux.

g. Net flux measured as Na-Ca current has transient and steady-state values—that is, carrier current inactivates. This suggests that part of the exchange that initially gave current has reverted to an electroneutral mode.

WHAT NEEDS TO BE DONE

a. How many different kinds of Na-Ca exchange are there? Are there mixtures of these in cells?

b. Look for the classic two Na per Ca, does it exist?

c. Look carefully at coupling and at voltage sensitivity of fluxes—compare with net fluxes.

Regulation of Sodium-Calcium Exchange

Introduction to Part II

Much new information about the regulation of the Na-Ca exchanger is presented in this section. The main regulatory ligands appear to be intracellular ATP and Ca^{2+} and extracellular alkali metal cations. In addition, the lipid composition of the plasmalemma appears to modulate exchanger activity. An important methodological advance described here is the use of "giant" excised membrane patches; this technique appears to be especially useful for exploring the regulation of exchanger-mediated currents by intracellular Ca^{2+} and ATP.

The mechanism by which ATP stimulates exchange activity is still unknown. The evidence favoring a phosphorylation process is discussed, but other data from cardiac myocyte patches suggest that a redistribution of charged membrane phospholipids might be involved instead.

In invertebrate nerve and muscle cells, activation of the exchanger by intracellular Ca^{2+} occurs at a concentration that is significantly higher than the resting Ca^{2+} level, but within the dynamic range of cytosolic Ca^{2+} concentrations during cell activity. In cardiac muscle, however, the Ca^{2+} regulatory site may have a higher affinity for Ca^{2+}, so that this site might be close to saturation even during diastole.

An external monovalent cation-activating site has been well documented in cardiac muscle and in neurons. The alkali metal ions, K^+, Rb^+, Li^+, and Na^+, all appear to be effective activating ions. In cardiac myocytes, but not in squid axons, the activating external monoavlent ions affect the voltage dependence of the exchanger.

Regulation of Na-Ca Exchange

An Overview[a]

REINALDO DIPOLO AND LUIS BEAUGÉ

Centro de Biofiísica y Bioquímica
IVIC, Apartado 21827
Caracas, 1020A, Venezuela

Instituto M. y M. Ferreyra
Casilla de Correo 389
5000 Córdoba, Argentina

INTRODUCTION

The regulation of the Na-Ca exchanger is a subject that has received much attention in the past decade. In general, the importance of regulatory processes is that they are able to modify the activity of carriers, pumps, and channels, without necessarily changing the driving force of the transported ions.

Several physiological ligands and experimental procedures are known to influence the exchange activity. They include nucleotides, intracellular Ca^{2+} and Mg^{2+}, hydrogen ions, monovalent cations, orthophosphate, redox potential, and modifications of the lipid and protein composition of the plasma membrane (see Reeves[1] and DiPolo & Beaugé[2] for references). At present we do not have a clear picture of how the Na-Ca exchange is actually regulated, nor the physiological role, if any, of these modifiers. An additional complexity is given by the fact that the response to several of the aforementioned ligands varies with the preparation used. Particularly illustrative are the different effects found in intact cells or intact plasma membrane (injected, perfused, and dialyzed cells and giant excised patches) compared with isolated membrane vesicles.[1,3,4] Because many of the contributors to this publication will deal with several regulatory features of the Na-Ca exchanger, we will restrict ourselves to the two major *in vivo* modulators: ATP and intracellular ionized Ca.

THE EFFECTS OF ATP AND OTHER NUCLEOTIDES

In 1973, Baker and Glitsch,[5] working on injected squid axons, found that if the intracellular ionized Ca was kept constant with EGTA, cyanide poisoning and apyrase injection caused an inhibition of the Na-Ca exchange. Poisoning was also accompanied by changes in the kinetic parameters of the antiporter. These experiments suggested that ATP affects the Na-Ca exchanger. That ATP was indeed the responsible for this effect was later shown in dialyzed nerve fibers.[6,7] A typical ATP effect in a squid axon

[a] This work was supported by grants from the National Institutes of Health (R01 HL-39243-03); the National Science Foundation (BNS-8817299); CONICIT, Venezuela (S1-2231); and CONICET, Argentina (30000800/88).

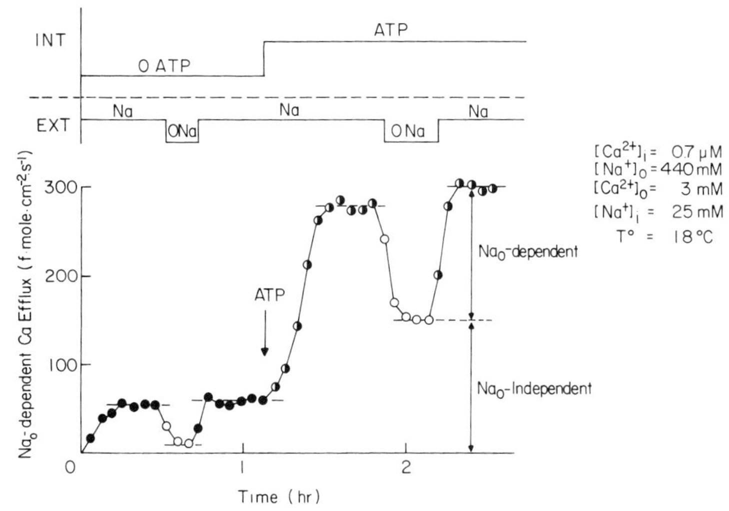

FIGURE 1. The effect of ATP on the Na_o-dependent Ca efflux in an axon dialyzed with submicromolar Ca_i concentration (0.7 μM). Note the existence of a sizable Ca efflux in the complete absence of the nucleotide. At this particular $[Ca^{2+}]_i$, ATP induces a similar activation of both the Na-Ca exchange and the Ca pump.

dialyzed with a 0.7 μM Ca_i concentration is illustrated in FIGURE 1. The experiment shows that even in the complete absence of intracellular ATP, a sizable Na_o-dependent Ca efflux exists, indicating that the requirement for ATP is not absolute. On the other hand, ATP causes an increase in the Na_o-dependent Ca efflux as well as in the Na_o-independent component now known to correspond to the Ca pump.[8]

The kinetics of the ATP effect show that the nucleotide stimulates the Na-Ca exchange, acting with low apparent affinity ($K_{1/2}$ = 250 μM); this contrasts with a much higher affinity of the Ca pump (see FIG. 2). Modulation of the exchanger by ATP can manifest itself in several ways. Among the most striking effects are: (i) an increase in the apparent affinity of the transporting sites for external Na and internal Ca without changes in the maximal rate of transport (see FIG. 3); (ii) an increase in the affinity of the Ca_i regulatory site (see FIG. 4), and (iii) even in the complete absence of Ca_i^{2+} ATP can induce a distinct although small reverse Na-Ca exchange (see FIG. 2 from DiPolo and Beaugé[9]). These kinetic consequences are seen in all transport modes of the carrier (forward, reverse, Ca-Ca, and Na-Na exchanges).

When considering the mechanism by which ATP stimulates the Na-Ca exchange, the fact that V_{max} remains unchanged indicates that a direct coupling of ATP hydrolysis with transport is unlikely. On the other hand, there is much indirect evidence suggesting that phosphorylation of either the carrier or other structures is required. These are the following: (i) *Need for Mg*: As is the case for most phosphorylating reactions, Mg ions are also essential for ATP stimulation of the Na-Ca exchange. This is illustrated in FIGURE 5 which shows the Na_i-dependent Ca influx measured in a dialyzed axon under zero *trans* conditions (Na_o-free, Ca_o-containing medium). Without ATP the influx of Ca is low and remains unaffected when the nucleotide is added in the absence of Mg_i; the inclusion of $MgCl_2$ together with ATP results in a fivefold increase in the flux levels. (ii) *Selectivity*: Stimulation by ATP is highly selective because other phosphate compounds (GTP, GDP, cGTP, UTP, ITP, ADP, AMP, cAMP, acetyl phosphate and phosphoarginine) are ineffective.[10,11] An additional important feature in favor of the phosphorylation hypothesis (see TABLE 1 and FIG. 6) is that in squid axons two phosphorylating ATP analogues can also mimic the ATP effect: these are ATPγS and AMP-CPP. ATPγS, an analogue that is known to act as a substrate for kinases but not for ATPases, is a much better substrate than ATP in stimulating the Na-Ca exchange. In the experiment shown in FIGURE 6, it can be seen that the activation by ATPγS doubles that obtained with ATP (see FIG. 1) without affecting the Ca pump component. All the above considerations agree with the suggestion by Caroni and Carafoli[12] that a balance between phosphorylation and dephosphorylation reactions controls the activity of the carrier. (iii) *Vanadate and phosphate stimulation*: These compounds stimulate Na-Ca exchange *only* in the presence of ATP. FIGURE 7 shows the consequences of adding 100 μM internal vanadate with or without 1 mM ATP. In the axon dialyzed without ATP (FIG. 7A) vanadate has no effect. In the presence of MgATP (FIG. 7B) addition of vanadate in the absence of external Na completely blocks the Ca pump component. Readmission of external Na results in an Ca efflux that is much higher than the Na_o-dependent component observed with ATP and no vanadate. ATPases and phosphatases seem to be more sensitive to vanadate than kinases (Beaugé, unpublished). Therefore, a plausible explanation for these results is that this compound shifts the kinase to phosphatase activity ratio in favor of kinases; in turn this would lead to an increase in the level of phosphorylated structure. In support of this idea is the fact that inorganic phosphate also stimulates the exchanger (in the presence of ATP) at a concentration that coincides with that at which it acts as product inhibitor of phosphatase activity (Beaugé[13]).

A problem with accepting the idea that a protein kinase-phosphatase system is involved in the process of phosphorylation is that in cardiac myocytes and squid axons

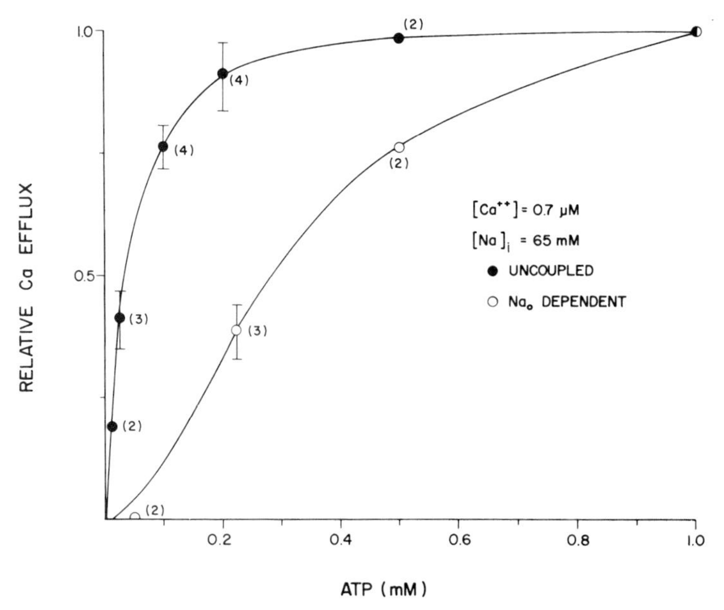

FIGURE 2. Effect of internal ATP on the Ca pump and the Na-Ca exchange components of Ca efflux in dialyzed squid giant axons. The values have been normalized to the maximal flux attained in each condition. Note the low apparent affinity of the exchanger ($K_{1/2}$ = 250 μM) compared to the Ca pump ($K_{1/2}$ = 20 μM). (Redrawn from DiPolo and Beaugé.[8])

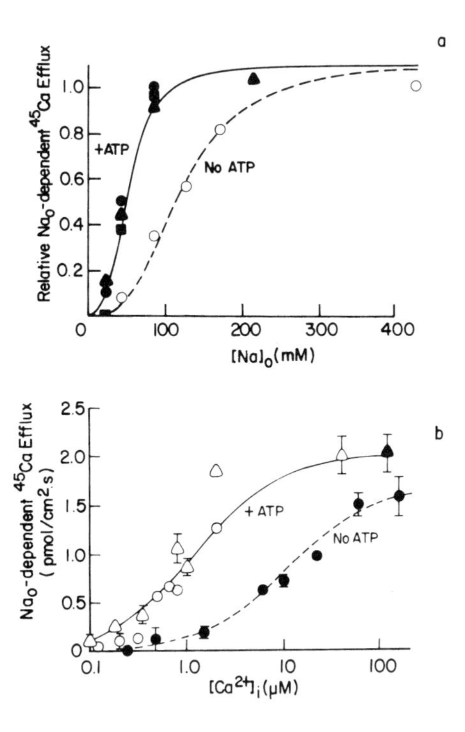

FIGURE 3. The effect of ATP on the kinetic of activation of the Na-Ca exchange by Na_o ions and Ca_i ions in dialyzed squid axons. Observe the increase in the apparent affinity for both cations when ATP is added to the dialysis solution. (Taken from Blaustein[21] used with permission.)

the ATP effect is insensitive to a variety of calmodulin antagonists and phorbol esters[1] (also, DiPolo and Beaugé, unpublished). So far, the only positive report is one by Vigne *et al.*[14] in smooth muscle, which suggests that a protein kinase C is involved. An interesting alternative to account for the ATP effect has been recently put forward by Collins and Hilgemann (this volume). These authors propose that a modification of the lipid environment of the exchanger (perhaps through a aminophospholipid translocase) might be implicated in this effect.

REGULATION BY INTRACELLULAR CA

The other regulatory process we would like to discuss is the activation of the Na-Ca exchange by calcium acting on a cytoplasmic *regulatory* site. In 1970, Peter Baker[15] found that injection of EGTA caused a marked inhibition of the reverse Na-Ca exchange. Later work in dialyzed axons demonstrated that this inhibition was secondary to a decrease in the intracellular ionized Ca concentration and not a direct effect of the exogenous Ca chelator.[9,16] The existence of the internal regulatory Ca site can be unambiguously shown in those modes of operation of the exchanger that intrinsically do not require Ca_i; that is, reverse exchange and Na-Na exchange. FIGURE 8 shows

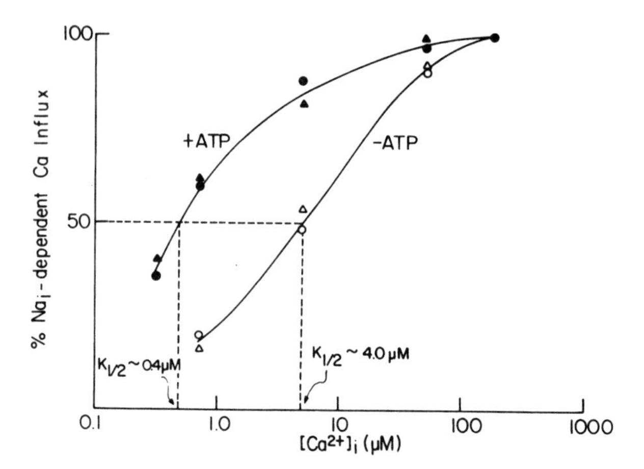

FIGURE 4. This figure summarizes the results of several axons in which the Na_i-dependent Ca influx was measured as a function of the $[Ca^{2+}]_i$ in the presence and absence of MgATP. Note that the $K_{1/2}$ for Ca_i^{2+} ions increases 10-fold when MgATP is added.

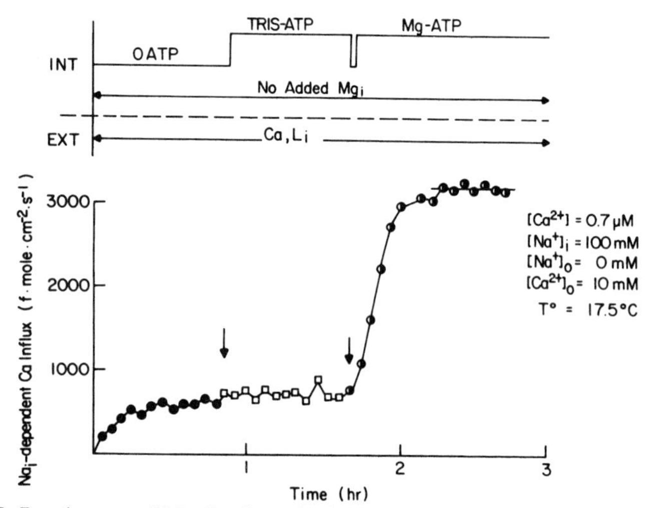

FIGURE 5. Requirement of Mg_i for the activation of Na-Ca exchange by ATP in a dialyzed squid axon. Note the absence of an ATP effect when added as a TRIS salt in comparison with a large increase in the presence of Mg.

the need for cytoplasmic Ca ions to promote Na_i-Ca_o reverse exchange in an axon dialyzed without ATP. In the absence Ca_i the efflux of Na into 10 mM external Ca sodium-free seawater is near the leak values. Upon addition of Ca_i, there is a sharp increase in Na efflux, which is brought back to leak values when the external Ca is removed. FIGURE 9 illustrates the Ca_i activation curves for both the reverse exchange and Na-Na exchange fluxes through the Na-Ca antiporter. This cytoplasmic Ca regula-

TABLE 1. The Effect of Nucleotides and High-Energy Phosphate Compounds on the Na-Ca Exchanger in Dialyzed Squid Axons

Compound	Effect (%)
ATP	100
ATPγS	200
AMP-CPP, AMP-NPP	50
AMP-PCP, AMP-PNP	0
ADP, AMP, c-AMP	0
GTP, UTP, c-GTP	0
UTP, UDP	0
Acetyl phosphate	0

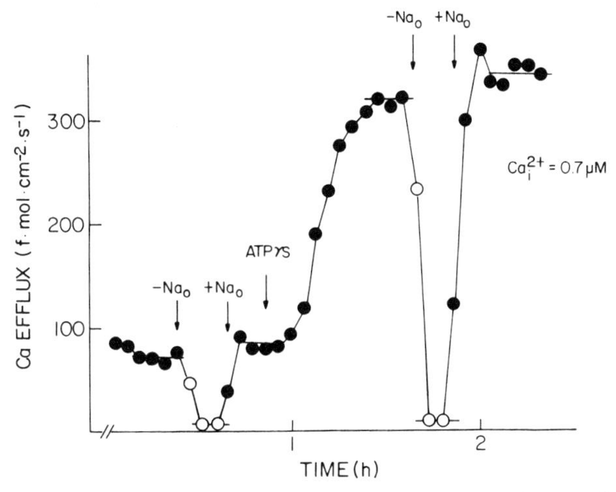

FIGURE 6. Effect of ATPγS on the Na_o-dependent Ca efflux in a squid axon dialyzed with a submicromolar ionized Ca_i concentration (0.7 μM). Note that the activation of Ca efflux by the ATP analogue is completely dependent on Na_o since no activation of the Ca pump component was observed. (Redrawn from DiPolo and Beaugé[23]; used with permission.)

tory site is not a unique property of the squid axon, for it has been found in several other preparations. A summary of the apparent affinities of the Ca_i regulatory site for Ca_i is presented in TABLE 2.

Intracellular Na is known to displace Ca from its internal transporting site. This in principle could give a tool to separate Ca transport from Ca regulatory events, provided Na ions do not displace Ca from the regulatory site as well. In order to explore this possibility, as a first step we have determined the Na_i inhibition of the forward and Ca-Ca exchange at saturating $[Ca_i]$. Internal Na inhibits both forward and Ca-Ca components with the same apparent affinity; at about 100 mM Na_i both components are very low in magnitude (FIG. 10; see also Beaugé & DiPolo, this volume). Accordingly, any activation of the reverse exchange by Ca_i at 100 mM or higher Na_i, should indicate a direct effect on the regulatory Ca_i site. Furthermore, even at the highest levels of Na_i-dependent Ca influx (360 mM Na_i), there was no detectable

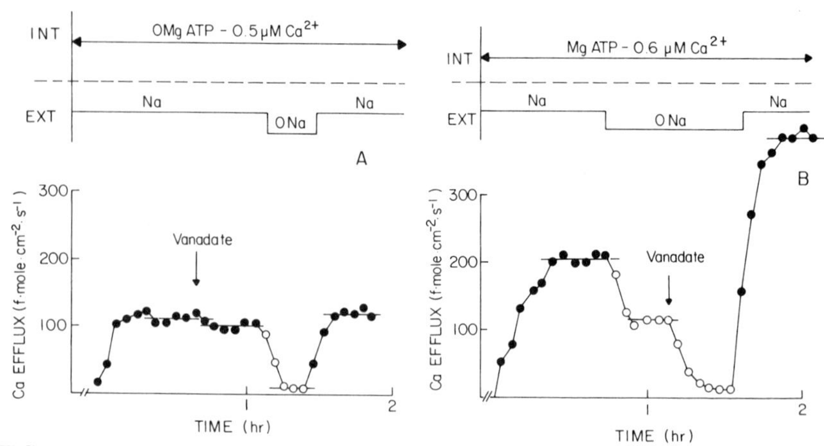

FIGURE 7. Effect of vanadate on the Na_o-dependent Ca efflux in squid axons dialyzed without and with ATP. (**A**) in the absence of ATP, there is no effect of vanadate on Ca efflux. (**B**) In the presence of ATP, vanadate inhibits the Ca pump component and markedly increases the Ca efflux.

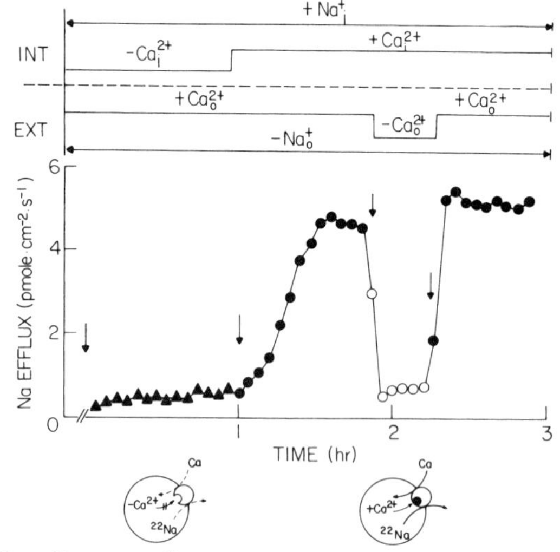

FIGURE 8. The effect of internal Ca^{2+} on the Ca_o-dependent Na efflux (reverse Na-Ca exchange) in an axon dialyzed without ATP. Notice that in the absence of ATP, reverse exchange can only be detected in the presence of internal ionized Ca.

efflux of Ca above the leak levels. This shows that the Ca_i that binds to the regulatory site is not transported. The Ca_i activation of the Na_i-dependent Ca influx under zero *trans* conditions in the presence of 100 mM Na is shown in FIGURE 11. The relationship between the different steady-state Ca influx and the $[Ca^{2+}]_i$ follows a hyperbolic curve with a $K_{1/2}$ for Ca_i^{2+} of about 5 μM in the absence of ATP (note that the plot is on semilogarithmic scale). As it was pointed out before (see FIG. 4), the presence of MgATP increases by 10 fold the Ca_i affinity.

The results just described indicate that, at 100 mM concentration, Na_i displace Ca_i from the transporting but not from the regulatory site. To further investigate any possible Na_i-Ca_i antagonism at the regulatory locus, we performed experiments similar to those of FIGURE 10 in which the $[Na]_i$ was increased to levels similar to those found in seawater. A monotonic stimulation of Na_i up to 440 mM was observed. These results are very important since they demonstrate for the first time that the Ca_i transport and Ca_i regulatory sites behave differently with respect to at least one ligand, Na ions. This strongly indicates that both sites correspond to different entities.

Using the same approach we have tested whether membrane potential changes have any effect on the apparent affinity of the Ca_i regulatory site. To that aim, we measured Ca_i activation curves of the reverse Na-Ca exchange in axons whose membrane potentials were clamped at zero and −50 mV. As shown in FIGURE 12 no differences in the apparent Ca_i affinity were detected at these two membrane potentials.

A key point to be addressed concerns the physiological relevance of the positive Ca_i feedback described in nerve cells as a general regulatory mechanism for the Na-Ca exchanger. In order that Ca_i can be an effective regulator, the K_m of the site for Ca_i must be close to the resting ionized Ca_i level. In the case of cardiac muscle cells with a K_m significantly lower than the resting ionized Ca_i, its role as modulator of the

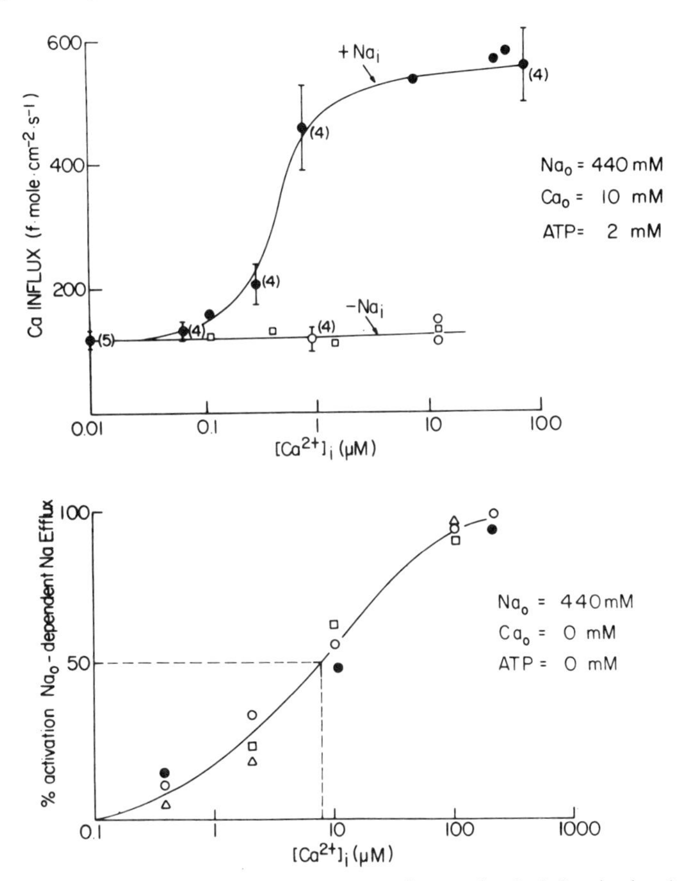

FIGURE 9. Top: the effect of ionized Ca_i concentration on the Ca influx in the absence and presence of internal Na in squid axons dialyzed with 2 mM MgATP and bathed in Na seawater. **Bottom**: the effect of ionized Ca_i concentration on Na-Na exchange in the absence of ATP. (Taken from DiPolo and Beaugé[2]; used with permission.)

TABLE 2. Apparent Affinity for Ca_i ($K_{0.5}$) of the Ca_i Regulatory Site in Different Preparations

System	$K_{0.5}$	Method	References
Barnacle	0.4-0.6 μM (+ATP)	Ca_o-dep Na efflux	Rasgado-Flores *et al.*[8] (1989)
Squid	0.4-0.8 μM (+ATP) 4-8 μM (−ATP)	Na_i-dep Ca influx Zero *trans*	DiPolo & Beaugé[2] (1990)
Myocyte	22-50 nM	Reverse $I_{Na/Ca}$	Noda *et al.*[22] (1988) Miura & Kimura[26] (1989)
Sarcolemma vesicles	0.1-0.5 mM	Na_i-Ca uptake	Reeves & Poronnik[24] (1987)
Squid optic vesicles	12 μM	Na_i-Ca uptake EGTA-loaded vesicles	Condrescu *et al.*[25] (1988)

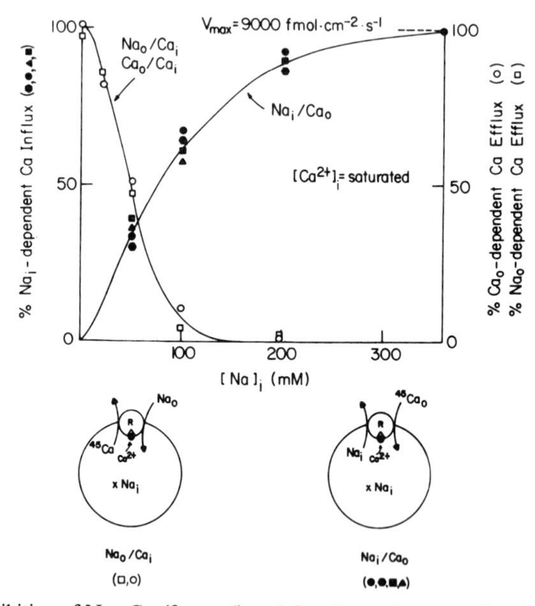

FIGURE 10. Inhibition of Na_o-Ca_i (forward) and Ca_o-Ca_i exchanges and activation of Ca_o-Na_i (reverse) exchange by Na_i at saturating $[Ca^{2+}]_i$. Observe that at 100 mM or higher, forward and Ca-Ca exchange are very small in contrast with a still-increasing reverse exchange with Na_i.

exchanger becomes unclear (however, see Hilgemann,[17]). A different situation appears to exist in nerve cells and barnacle muscle where the K_m for Ca_i is much higher than the resting Ca_i concentration (DiPolo and Beaugé,[3,4] Rasgado-Flores *et al.*[18] 1989). In these two preparations, as well as in any other with similar kinetic characteristics, $[Ca^{2+}]_i$ increase due to Ca^{2+} entry through voltage-sensitive Ca channels during membrane depolarization will enhance Ca influx via the Na-Ca exchanger.[9,19] It is likely

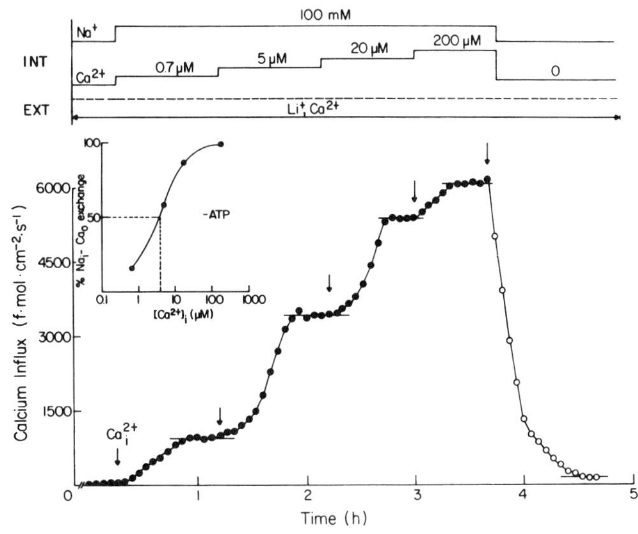

FIGURE 11. Apparent affinity of the regulatory Ca_i site from steady-state measurements of Na_i-dependent Ca influx in the presence of 100 mM Na_i.

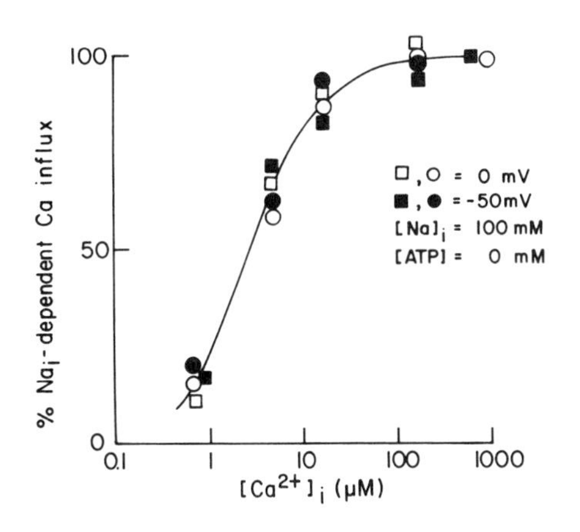

FIGURE 12. Lack of effect of membrane potential on the apparent affinity of the intracellular regulatory site for Ca_i.

that this extra Ca entry via the Na-Ca exchanger could be used to refill the endoplasmic reticulum[20] as well as to activated different Ca_i-dependent processes.

ACKNOWLEDGMENTS

We wish to thank Dhuwya Otero for the drawings and Lillian Palacios for the secretarial help.

REFERENCES

1. REEVES, J. P. 1990. Sodium-calcium exchange. *In* Intracellular Calcium Regulation. F. Bronner, Ed.: 305-347. Liss. New York.
2. DIPOLO, R. & L. BEAUGÉ. 1990. Calcium transport in excitable cells. *In* Intracellular Calcium Regulation. F. Bronner, Ed.: 381-413. Liss. New York.
3. DIPOLO, R. & L. BEAUGÉ. 1990. Asymmetrical properties of the Na-Ca exchanger in voltage-clamped, internally dialyzed squid axons under symmetrical ionic conditions. J. Gen. Physiol. **95:** 819-835.
4. BAKER, P. F. & T. J. A. ALLEN. 1986. Sodium-calcium (Na-Ca) Exchange. *In* Intracellular Calcium Regulation. H. Bader, K. Gietzen, J. Rosenthal, R. Rudel & N. U. Wolf, Eds.: 35-47. Manchester Univ. Press. Manchester, England.
5. BAKER, P. F. & H. G. GLITSCH. 1973. Does metabolic energy participate directly in the Na-dependent extrusion of Ca from squid axons? J. Physiol. **233:** 44-46.
6. DIPOLO, R. 1973. Calcium efflux from internally dialyzed squid axons. J. Gen. Physiol. **62:** 575-589.
7. DIPOLO, R. 1974. The effect of ATP on Ca efflux in dialyzed squid giant axons. J. Gen. Physiol. **64:** 503-517.
8. DIPOLO, R. & L. BEAUGÉ. 1979. Physiological role of ATP-driven calcium pump in squid axons. Nature **278:** 271-273.
9. DIPOLO, R. & L. BEAUGÉ. 1986. In squid axons reverse Na/Ca exchange requires internal Ca and/or ATP. Biochim. Biophys. Acta **854:** 298-306.
10. DIPOLO, R. 1976. The influence of nucleotides upon Ca fluxes. Fed. Proc. **35:** 2579-2582.
11. DIPOLO, R. 1977. Characterization of the ATP-dependent Ca efflux in dialyzed squid axons. J. Gen. Physiol. **69:** 795-814.
12. CARONI, P. & E. CARAFOLI. 1983. The regulation of the Na^+-Ca^{2+} exchanger of heart sarcolemma. Eur. J. Biochem. **132:** 451-460.
13. BEAUGÉ, L. 1979. Vanadate-potassium interactions in the inhibition of Na,K-ATPase, *In* Na,K-ATPase. Structure and Kinetics. J. Norby & J. Skou, Eds.: 373-388. Academic Press. New York.
14. VIGNE, P., J. P. BREITTMAYER, D. DUVAL, C. FRELIN & M. LASDUNZKI. 1988. The Na^+-Ca^{2+} antiporter in aortic smooth muscle cells. Characterization and demonstration of an activation by phorbol ester. J. Biol. Chem. **263:** 8078-8083.
15. BAKER, P. F. 1970. Sodium-calcium exchange across the nerve cell membrane. *In* Calcium and Cellular Function. A. W. Cuthbert, Ed.: 96-107. MacMillian. New York.
16. DIPOLO, R. 1979. Calcium influx in internally dialyzed squid giant axons. J. Gen. Physiol. **73:** 91-113.
17. HILGEMANN, D. W. 1988. Numerical approximations of Na/Ca exchange. Prog. Biophys. Mol. Biol. **51:** 1-45.
18. RASGADO-FLORES, H., E. M. SANTIAGO & M. P. BLAUSTEIN. 1989. Kinetic and stoichiometry of coupled Na efflux and Ca influx by Na-Ca exchange in barnacle muscle cells. J. Gen. Physiol. **93:** 1219-1241.
19. DIPOLO, R., H. ROJAS & L. BEAUGÉ. 1982. Ca entry at rest and during prolonged depolarization in dialyzed squid axons. Cell Calcium **3:** 19-41.
20. LEDERER, J. W., E. NIGGLI & R. W. HADLEY. 1990. Sodium-calcium exchange in excitable cells. Fuzzy space. Nature **342:** 283.
21. BLAUSTEIN, M. P. 1977. Effects of internal and external cations and of ATP on sodium-calcium and calcium-calcium exchange in squid axons. Biophys. J. **20:** 79-110.
22. NODA, M., R. N. SHEPHERD & D. C. GASBY. 1988. Activation by $(Ca^{2+})_i$ and block by 3′,4′-dichlorobenzamil of outward Na-Ca exchange current in guinea-pig ventricular myocytes. Biophys. J. **53:** 342a.
23. DIPOLO, R. & L. BEAUGÉ. 1987. In squid axons ATP modulates Na/Ca exchange by a Ca-dependent phosphorylation. Biochim. Biophys. Acta. **897:** 347-353.

24. REEVES, J. P. & P. PORONNIK. 1987. Modulation of Na^+-Ca^{2+} exchange in sarcolemmal vesicles by intracellular Ca^{2+}. Am. J. Physiol. **252:** C17-C23.
25. CONDRESCU, M., A. GERARDI & R. DIPOLO. 1988. Na^+-Ca^{2+} exchange in squid optic nerve membrane vesicles is activated by internal calcium. Biochim. Biophys. Acta **946:** 289-298.
26. MIURA, Y. & J. KIMURA. 1989. Sodium-calcium exchange currents; dependence of internal Ca and Na and competition binding of external Na and Ca. J. Gen. Physiol. **93:** 1129-1145.

Intracellular Ionized Calcium Changes in Squid Giant Axons Monitored by Fura-2 and Aequorin[a]

J. REQUENA,[b] J. WHITTEMBURY,[c] A. SCARPA,[c]
J. F. BRINLEY, JR.,[d] AND L. J. MULLINS[d]

[b]*Centro de Biociencias*
Instituto Internacional de Estudios Avanzados (IDEA)
Apartado 17606-Parque Central
Caracas 1015A, Venezuela

[c]*Department of Physiology and Biophysics*
Case Western Reserve University School of Medicine
Cleveland, Ohio

[d]*Department of Biophysics*
University of Maryland School of Medicine
Baltimore, Maryland

INTRODUCTION

Understanding Na-Ca exchange requires a knowledge of the intracellular concentration of free Ca ($[Ca^{2+}]_i$) inside the cells. Since measurements of $[Ca^{2+}]_i$ tend to depend on the method used for its determination, it seemed critical to compare two commonly used techniques, namely aequorin and the dye Fura-2. This paper gives the details of such a comparison.

Over the years there has been discussion on the value of the $[Ca^{2+}]_i$ in squid giant axons. Apart from the physiological importance of this parameter, which establishes the relevance of the various Ca extrusion mechanisms, the measurement in itself has proved to be remarkably difficult.

The first measurement was that of Baker *et al.*[1] where an upper limit of 300 nM was set up for $[Ca^{2+}]_i$ with the aid of aequorin injected into axons and an *in vitro* calibration of the photoprotein. Later, DiPolo *et al.*[2] measured some 30 nM of $[Ca^{2+}]_i$ using aequorin confined to a dialysis capillary placed in the axon center. In the same study, arsenazo III injected into the axon was used as a Ca probe, and since the dye absorbance measures the free Ca averaged over the axon cross section, a higher value of 60 nM was reported. To complement this series of studies, Requena *et al.*[3] injected axons with aequorin and phenol red. The luminescence measurement yielded a value for $[Ca^{2+}]_i$ of 75 nM in the axon periphery, because phenol red acted as a screen to absorb aequorin light from the axon core.[4] These measurements relied on a null point *in vivo* calibration method based upon Ca EGTA/EGTA buffer mixtures.

Besides these optical-based measurements, Ca-sensitive microelectrodes have also been used in axons to determine intracellular Ca activity and values of 79 nM and 106

[a]This work was partially supported by Grant No. S1-1147 and S1-1197 from CONICIT (Caracas, Venezuela) and RO1 NS 17718 from the National Institutes of Health, Bethesda, MD.

nM have been reported.[5,6] When an optical indicator (aequorin or arsenazo III) and a Ca-sensing electrode were used simultaneously and their results compared, however, Requena *et al.*[7] concluded that small changes in $[Ca^{2+}]_i$, which are easily detectable by an optical Ca probe, are not registered by the electrode. It would appear, then, that the microelectrodes are not very sensitive at the resting level of free Ca encountered within the axoplasm.

The methods so far outlined are difficult in two areas: first, the extremely low levels of $[Ca^{2+}]_i$ to be detected, coupled to the presence of other ionic species at much higher concentrations that act as potential interfering cations; second, the need to have an *in vivo* calibration using a buffer system, powerful enough to overcome endogenous buffers, with dissociation constant not only adequate for the cytoplasmic Ca concentration range but with a value that is accurately known for the ionic environment present inside the cell.

An important advance in Ca detection was achieved by Tsien and collaborators,[8,9] who designed and synthesized a series of Ca chelators and fluorescence indicators with a simple 1 to 1 stoichiometry, high selectivity for Ca, and low affinity for Mg^{2+} and H^+. This new generation of Ca fluorescence probes offers a number of advantages, including sensitivity, reduced interference from light scattering, and straightforward calibration which may not require a specific buffer system. Besides, indicators such as Quin-2 and Fura-2 have shown that their properties are much the same in calibrating solutions as in cytoplasm with very little evidence of biological toxicity. Notwithstanding this last point, Allen and Baker[10] reported that an axon injected with the fluorescence indicator Quin-2 showed no Ca influx depending upon internal Na (Na_i), a manifestation of Na-Ca exchange.

In the present paper the measurement of free Ca level inside squid giant axon was studied using both aequorin and Fura-2 in an attempt to calibrate the luminescent photoprotein and to verify the adequacy of a widely used fluorescence indicator as a probe suitable to analyze Na-Ca exchange. A preliminary report of this work has appeared elsewhere.[11]

METHODS

Axons

The experiments were performed at the Marine Biological Laboratory, Woods Hole, MA. The hindmost giant axon from the stellate ganglion was dissected from live specimens of *Loligo pealei*. After careful cleaning for connective tissue, the axon, while being mounted in a chamber, was positioned on pedestals at either end. Through slits opened in the axon, electrodes and the microinjector were inserted as described elsewhere.[12,13]

Microinjection and Experimental Arrangement

A microinjector was used to introduce the aequorin and Fura-2. It also served, simultaneously, as the current and voltage electrode for voltage clamping.[14] For injection purposes a tiny bubble of air was loaded at the tip of the capillary and then about 0.15 μl of aequorin (plus TEA and/or Fura-2 solution was sucked into the capillary. Once the capillary was properly placed into position, this solution was injected inside the axon (final concentrations of Fura-2 and of aequorin were about 10 μM). Extracellular solution changes in the experimental chamber were completed in about 10 sec and axons were always immersed in seawater, which was flowing at a rate of 1 ml/min. Experiments were done at 15-16 °C. The voltage-clamp experimental setup has been described elsewhere.[15]

Fura-2 and Aequorin Measurement

For Fura-2 measurement a Time Sharing Multiple Wavelength Spectrophotometer[16] (Johnson Foundation, University of Pennsylvania, Philadelphia, PA) was adapted for fluorescence measurements. A Xe (75 W) arc lamp with a voltage-stabilized source was used to excite the Fura-2 via a custom-made quartz optical guide (Dolan Jenner, MA) fitted with an ultraviolet narrow band filtered at 340 nm, 360 nm, and 380 nm. Ninety-degree fluorescence from the axon filtered at 510 nm was collected over a selected phototube with a glass optical guide. Aequorin luminescence was measured through this optical guide with photon-counting equipment described elsewhere.[17]

Calibration

Fura-2 fluorescence (F) and $[Ca^{2+}]_i$ were related by means of the calibration equation for a fluorescent species tested at a single wavelength:

$$[Ca^{2+}]_i = K_d \cdot (F - F_{min})/(F_{max} - F) \quad (1)$$

where K_d is the dissociation constant of the complex Ca: Fura-2 at the prevailing ionic conditions of the reaction medium and F_{min} and F_{max} represent the extreme limiting values for the fluorescence signal, taken at a given excitation wavelength, of free Fura-2 and dye saturated with Ca^{2+}, respectively. Another method to calculate $[Ca^{2+}]_i$ is based upon the electronically computed ratio (R) of fluorescence signals elicited at two wavelengths (F_{340} over F_{380}) according to the following equation[9]:

$$[Ca^{2+}]_i = K_d \cdot \beta \cdot (R - R_{min})/(R_{max} - R) \quad (2)$$

where β is a constant that depends on the geometry and nature of the experimental preparation and is related to the ratio of fluorescence of Fura-2 to Ca : Fura-2 at a given wavelength as shown elsewhere.[9] R_{min} and R_{max} represent the extreme limiting values for the ratio of fluorescence signals taken at excitation wavelength 340 nm over 380 nm of free Fura-2 and dye saturated with Ca^{2+}, respectively. For each experiment, axon autofluorescence was electronically subtracted before calculating the ratio.

Solutions

Regular artificial seawater used in these experiments had the following composition (mM): 440 NaCl, 10 KCl, 10 $CaCl_2$, 50 $MgCl_2$, 10 TRIS-TES (Tris[hydroxymethy] aminomethane hydroxide *N*-Tris[hydroxymethyl]methyl-2-amino-ethane sulfonic acid) pH 7.8 buffer and 0.1 EDTA (ethylene-dyamine-tetraacetic acid) and 10 mM $NaHCO_3$. Bicarbonate was added to all seawaters in order to ensure proton pumping and steady pH_i levels.[18,19] The voltage clamp experiments were done in depolarizing artificial seawater containing 150 mM of the chloride salt of each of Na, K, and of an inert cation, either Li, *N*-methylglucamine (NMG), or Tris. When needed Na was replaced by either Li, Tris, or NMG as mentioned in the text. To prepare low-Ca seawater, Ca_o was replaced by Mg_o while for high-Ca seawater (112 mM), Na_o and Mg_o were replaced by Ca_o. Fura-2 and aequorin microinjection solutions were made by dissolving 1 mg of salt-free aequorin (kindly supplied by Dr. John Blinks, Mayo Clinic, Rochester, MN) or K-Fura (Molecular Probes, Eugene, OR), into 80 μl of 5 μM K_2EGTA (ethyleneglycol-bis [β-aminoethyl ether]-*N,N,N',N'*-tetraacetic di-K salt) plus 3 μM K_2EDTA solution at neutral pH (made from glass distilled water). TEA-MOPS solution (2.4 M) was prepared by neutralizing in cold a 40% solution of tetraetylammonia hydroxide (TEA-OH, Alfa Products, Danvers MA) with enough MOPS (4-morpholine-propanesulfonic acid, Sigma Chemical, St. Louis, MO) to yield a pH 8.0 ± 0.1 solution. Fura-2, aequorin, and the MOPS solutions were made in 0.6 M KCl. Usually equal parts of the aequorin and TEA-MOPS solutions were injected; however, when aequorin alone was injected, K currents were suppressed by the addition of 3,4-diaminopyridine (1 to 5 mM) to the seawater. All seawaters were freshly made and adjusted to 1010 mOsm. Unless otherwise specified, all experiments were done in the presence of 100 nM tetrodotoxin (TTX; Sigma Chem., St. Louis, MO). KCl solution for the reference electrode was either made from recrystallized pure salt (in 500 μM EGTA, 300 μM EDTA solution) or made from a suprapure sample of KCl (E. G. Merck, Darmstadt, Germany) to avoid contamination of the aequorin with Ca carried by the salt.

RESULTS

In Vitro Calibration of Fura-2

FIGURE 1 shows a plot of Fura-2 fluorescence as function of Ca concentration in a high-ionic-strength medium containing the chloride salt of 350 mM K, 25 mM Na, 5 mM Mg, and 10 mM MOPS buffer at pH 7.30. The intercept of the linear regression of the reciprocal of the fluorescence on the reciprocal of the Ca concentration yields an apparent dissociation constant for the dye:Ca complex of 870 nM. This high K_d value, for a medium closely resembling the ionic environment of axoplasm, reflects the fact that the Fura-2 binding constant strongly depends on ionic strength as shown in Grynkiewicz *et al.*,[9] and Poenie *et al.*,[20] where it was found that in 0.25 M ionic strength medium, the K_d of Fura-2 is 760 nM while at 0.125M its value is 224 nM.

In Vivo Calibration of Fura-2

A typical protocol for calibration of Fura-2 fluorescence in a squid giant axon consists of allowing the axon to gain or to lose $[Ca]_i$ by using various experimental procedures until the two extreme limiting levels of dye fluorescence were detected. In the case of FIGURE 2, Ca_i rose following exposure of the axon to a mixture of high-Ca (60 mM) seawater and depolarization induced by high-K artificial seawater (450 mM K) dosed with CN (2 mM to block mitochondrial respiration). Under this condition of massive Ca load into the axon, the fluorescence from a 380-nm excitation wavelength reaches a minimum value while fluorescence from 340 nm (and the ratio of fluorescences at 340 nm/380 nm) reaches a maximum. Toward the end of the experiment, the axon was injected with EGTA (final concentration about 20 mM heavily buffered with MOPS at pH 7.3) to sequester all of the intracellular Ca. This procedure is thought to guarantee a fluorescence minimum for the 340-nm excitation wavelength (and the ratio of fluorescences at 340 nm/380 nm) and a maximum for the fluorescence at 380 nm.

Through experimental protocol, the fluorescence parameters needed for an *in vivo* calibration of Fura-2 in squid giant axons were obtained and their values are summarized in TABLE 1. For all the experiments listed, the 380-nm elicited fluorescence and the ratio of fluorescences show good reproducibility for the different experimental conditions tested. From the limiting maximum and minimum fluorescence observed at the 380-nm excitation wavelength, the value of factor β of Eq. 2 could be calculated as follows: First the K_d value was obtained for the ionic media resembling axoplasm, then the resting or initial $[Ca^{2+}]_i$ was determined from Eq. 1 for several experiments, which was then substituted into Eq. 2 to calculate the value β. A value of 3.55 ± 0.49 ($n = 15$) was, thus, obtained for β.

The average resting $[Ca^{2+}]_i$ determined from an initial ratio of fluorescences of 0.884 ± 0.029 corresponds to 184 ± 18 nM ($n = 27$) for fresh axons exposed to 3 mM Ca seawater. Finally, for axons injected with equimolar quantities of CaEGTA and EGTA (pH 7.3 and final concentration 5 mM) a ratio of fluorescences of 0.818 was measured which corresponded to a $[Ca^{2+}]_i$ of 140 nM.

Ca Entry during Stimulation

In the next series of experiments, the effect of electrical stimulation on Ca levels measured from Fura-2 and from aequorin were recorded alternately on the same axon.

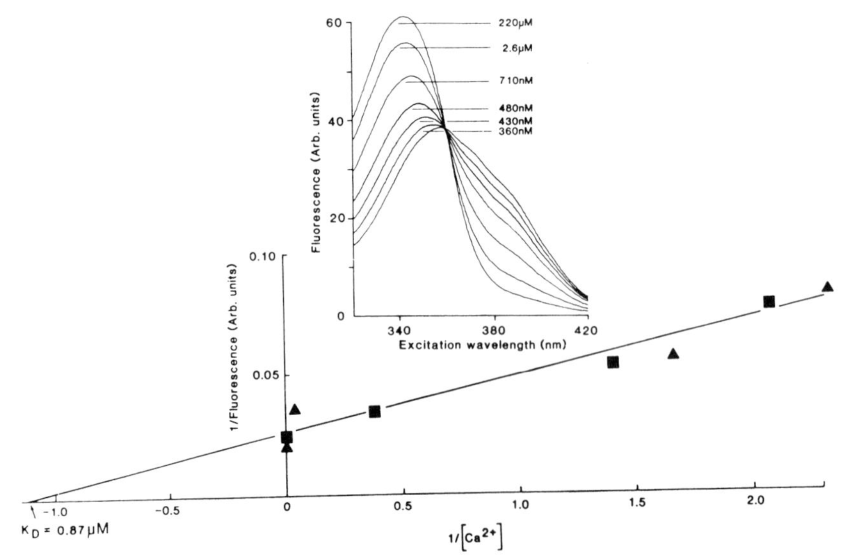

FIGURE 1. *In vitro* Lineweaver-Burk linear transformation plot of the reciprocal Fura-2 fluorescence measured at 510 nm (arbitrary units, excitation wavelength 340 nm) on the reciprocal of the EGTA-buffered Ca concentration (µM) in a high ionic strength medium. The concentration of Ca in the cuvette was adjusted with a EGTA/CaEGTA buffer mixture (pH 7.3 and EGTA total concentration 5 mM) and read simultaneously with a Ca electrode. The electrode was calibrated by means of a set of high ionic strength solutions without EGTA and with Ca concentration standardized above 10 µM Ca. The electrode response was linear down to 10^{-7} M Ca with a slope of 27-28 mV. Temperature was 17 °C. Insert shows fluorescence as a function of excitation wavelength (nm) for various Ca concentrations (µM) as measured in an Aminco Spectrofluorometer.

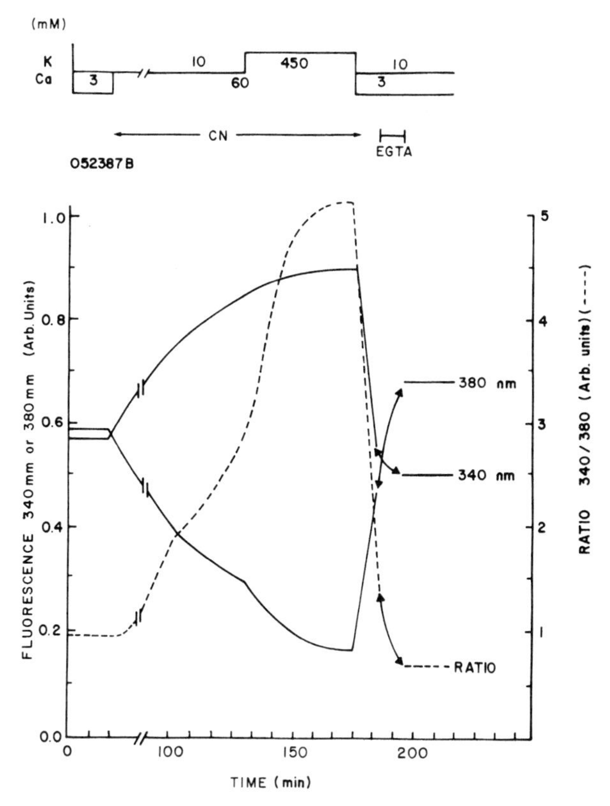

FIGURE 2. Time course (in minutes) of Fura-2 fluorescence (left ordinate corresponds to excitation wavelengths 340 nm and 380 nm, in arbitrary units, solid lines) (right ordinate corresponds to fluorescence ratio 340 nm/380nm, in arbitrary units, dashed line) in an axon that was allowed to gain intracellular Ca until a saturation level of fluorescence was detected. After this, the axon was injected with EGTA buffered with MOPS at pH 7.3 to sequester all intracellular Ca and yield the free fluorescence value (final EGTA concentration in axoplasm 5 mM).

FIGURE 3 is one example of this series and shows the time course of luminescence or fluorescence in an unpoisoned axon exposed to high-Ca (112 mM) artificial seawater and stimulated at 90 impulses/sec. The left ordinate of the figure corresponds to the $[Ca^{2+}]_i$ as determined from the measurement of the Fura-2 ratio of fluorescences at 340 nm/ 380 nm, but it only applies to the aequorin luminescence trace. From the record it can be calculated that there would be a net change in intracellular Ca^{2+} concentration on the order of 0.7 pM Ca/impulse if the axon had been exposed to regular seawater (Ca_o = 10 mM) during the stimulation episode. This value ought to be compared with the reported value of 2.4 nM Ca/impulse measured as the total Ca net gain.[21,22] These numbers mean that only one in some 3400 Ca^{2+} ions that cross the axolemma is left available by the powerful axoplasm-buffering systems detectable by the ion probe.

Extracellular Cations

In FIGURE 4, aequorin luminescence and Fura-2 fluorescence measurements were carried out in a squid giant axon exposed repetitively to high-K seawater (450 mM)

TABLE 1. Initial, Minimum, Maximum, and CaEGTA/EGTA-Buffered[a] Fluorescence (F_{380}), the 340/380 Ratio (R) and the Calculated Correction Factor (β) in Squid Giant Axon Injected with Fura-2

Axon No.	F_o (initial)	F_{min}	F_{max}	F_{EGTA}	R_o (initial)	R_{min}	R_{max}	R_{EGTA}	β
050787A	0.400	0.980	—	—	0.775	0.680	—	—	—
050987A	0.480	0.520	0.128	—	0.760	0.561	3.35	—	2.64
050987B	0.420	0.470	0.034	—	0.862	0.690	5.63	—	3.49
051087B	0.335	—	0.100	—	0.520	—	5.88	—	—
051187A	0.340	0.445	0.065	—	1.090	0.600	6.00	—	3.75
051187B	0.555	—	0.070	0.650	0.950	—	5.45	0.920	—
051287A	0.525	0.600	0.100	—	1.095	0.635	7.04	—	2.13
051287B	0.605	0.650	—	—	0.890	0.470	—	—	—
051287D	0.735	0.785	0.045	—	0.890	0.540	4.93	—	1.59
051387A	0.310	—	0.035	—	0.790	—	3.80	—	—
051387B	0.345	—	0.010	—	0.710	—	6.40	—	—
051587A	0.265	0.385	0.055	—	0.940	0.720	3.92	—	7.74
051587B	0.130	0.260	—	—	0.950	0.560	—	—	—
051687A	0.540	—	—	0.590	1.045	—	—	0.805	—
051687B	0.600	—	—	0.610	1.070	—	—	0.865	—
051687C	0.585	—	—	0.655	0.950	—	—	0.775	—
051687D	0.565	—	—	0.605	0.785	—	—	0.725	—
051887A	0.685	0.630	0.025	—	1.226	0.819	9.00	—	1.30
051887B	0.510	0.690	0.180	—	0.706	0.566	3.05	—	3.54
051987A	0.410	0.655	0.030	—	0.781	0.545	7.05	—	—
051987B	0.510	0.530	0.055	—	0.773	0.562	8.00	—	1.12
052087A	0.470	0.540	0.060	—	0.829	0.578	7.20	—	4.41
052187A	0.310	0.375	0.055	—	0.843	0.552	4.40	—	3.11
052187B	0.550	0.680	0.085	—	0.807	0.583	6.20	—	6.99
052287A	0.210	0.275	0.040	—	0.865	0.545	4.70	—	4.79
052387A	0.470	0.515	0.045	—	1.008	0.590	5.40	—	2.53
052387B	0.495	0.605	0.115	—	0.960	0.640	5.15	—	4.12
MEAN					0.884	0.601	5.62	0.818	3.55
(SEM)					0.029	0.020	0.35	0.034	0.49

[a] Total EGTA = 5 mM equimolar mixture.

and high-Li or regular artificial seawater (450 mM Li or Na, respectively) containing, throughout, 3 mM Ca_o. In this example, depolarization by high-K seawater induces a rise in ionized Ca from 300 nM to 550 nM in about 100 sec. This change in Ca_i level through the reversal operation of Na-Ca exchange[23,24] corresponds to a net rate of Ca inflow on the order of 25 $fmol \cdot cm^{-2} \cdot sec^{-1}$. Similarly, near the end of the record, when the main extracellular cation was changed from Li to Na, ionized Ca dropped quickly from 375 nM to 325 nM, reflecting the activation of the Na-Ca exchange mechanism in its forward mode of operation. This change in Ca levels corresponds to a net rate of extrusion on the order of 10 $fmol \cdot cm^{-1} \cdot sec^{-1}$.

In Vivo *Calibration of Aequorin*

The injection into one axon of the two Ca probes, aequorin and Fura-2 and the power to switch very quickly from one optical recording system to the other enabled

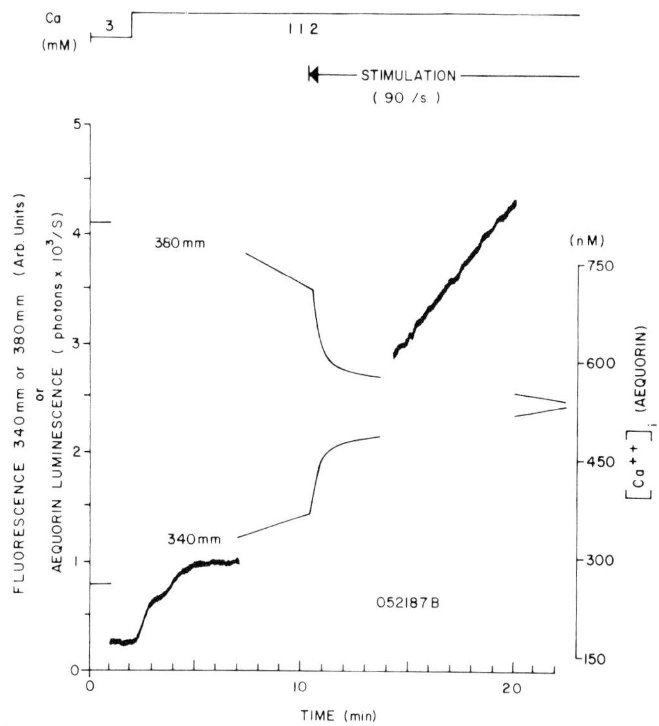

FIGURE 3. Time course (in minutes) of aequorin luminescence (rough trace; photons/sec $\times 10^3$) or Fura-2 fluorescence (line trace; arbitrary units) in an axon. The left ordinate shows the intracellular Ca concentration (in nM) corresponding to the aequorin trace. The axon was exposed two minutes after the onset of the experiment to 112 mM Ca artificial seawater and 8 minutes later it was stimulated at 90 impules/sec.

us to relate the luminescence from the photoprotein to a stable Fura-2 fluorescence reading and, through it, to a Ca concentration level. By carrying out these dual measurements over a very large range of Ca concentrations, it was possible to calibrate the *in vivo* aequorin light response. FIGURE 5 summarizes data available from seven axons. Aequorin luminescence is plotted as a function of the $[Ca^{2+}]_i$ derived from the measurement of the ratio of fluorescences recorded at 340 nm/380 nm of Fura-2. It is observed that at the low range of Ca concentrations, 75 to 750 nM, the experimental points are well fitted by a straight line with a slope of one while, in the high range of Ca concentration, 750 nM to 10,000 nM, they can be fitted to a line with a slope equal to 2.

Voltage-Clamp Experiments

Another series of experiments were designed to address the question of the quantitative determination of intracellular Ca level changes following pulses of membrane potential under voltage-clamp conditions. While the use of this technique is possible in squid axons injected with aequorin,[15,25] it does not seem feasible to carry it out with

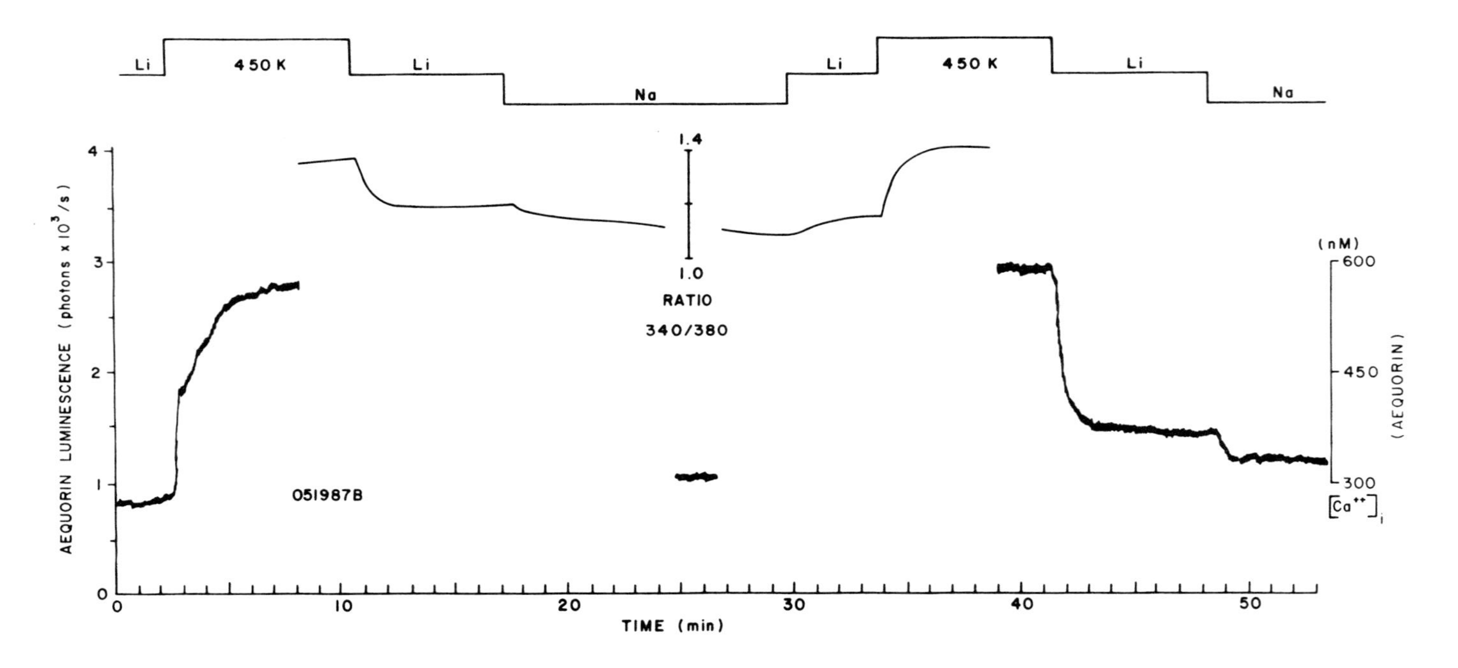

FIGURE 4. Time course (in min) of aequorin luminescence (rough trace; photons/sec $\times$ 10^3), or in the inset the Fura-2 fluorescence (line trace; ratio 340 nm/380 nm in arbitrary units) and the corresponding intracellular Ca concentration (in nM) in an axon initially in Li artificial seawater, which was exposed three minutes after the onset of the experiment to 450 mM K and 3 mM Ca artificial seawater. Seven minutes later, the extracellular K was changed for Li and then to Na artificial seawater. Further ionic changes of the artificial seawater are shown.

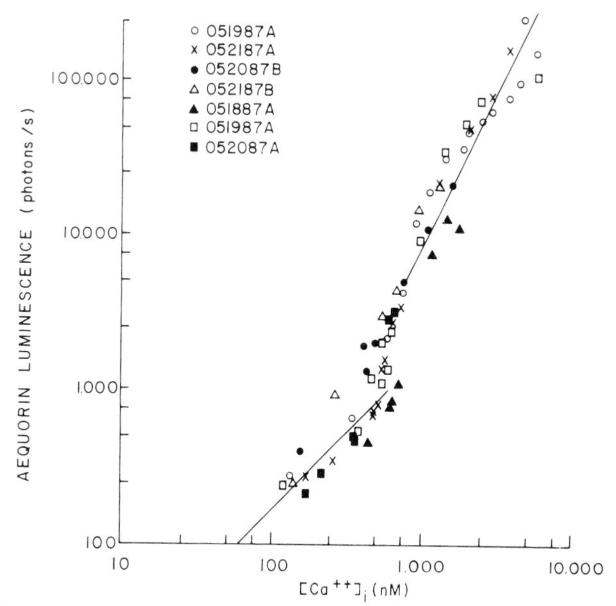

FIGURE 5. *In vivo* calibration of aequorin luminescence (log scale in photons/sec) as function of intracellular Ca concentration (in nM) as computed from Fura-2 fluorescence ratio (340 nm/380 nm) also in a log scale. Data from seven different axons.

Fura-2 treated axons, at least using the conventional Pt/Pt black internal electrode arrangement. Indeed, as shown in FIGURE 6, while there is an apparent increase in the ratio of fluorescences with depolarization, a closer examination of the emission corresponding to the individual excitation wavelength used, namely 340 nm, 360 nm, and 380 nm, reveals a loss of signal as a consequence of the train of depolarizing voltage-clamp pulses. The plain fact that concomitant with the onset of the train of voltage-clamp pulses the fluorescence elicited from the isosbestic wavelength (360 nm) shows a decay constitutes a clear indication that the apparent increase in $[Ca^{2+}]_i$ reflected from the ratio measurement is an artifact caused by the decomposition of Fura-2 molecules, most probably through an electrochemical reaction catalyzed over the Pt surface of the current-passing electrode.

DISCUSSION

The experiments reported in this study show that in squid axons injected with Fura-2 it is possible to study intracellular Ca homeostasis using either the individual excitation wavelengths (340 nm or 380 nm) or, better still, the ratio of the fluorescence signals at excitation wavelengths 340 nm over 380 nm. The preferential use of the electronically computed ratio of fluorescences as a measuring device has been discussed elsewhere.[9] It is clear that the choice of this parameter compensates for variations in dye content inside the cell, for shifts in instrumental sensitivity, and for a lack of calibration in any particular experiment.

The values of the dissociation constant for Fura-2 in high ionic strength media and for the limiting ratio of fluorescences levels for free and Ca-saturated Fura-2 obtained

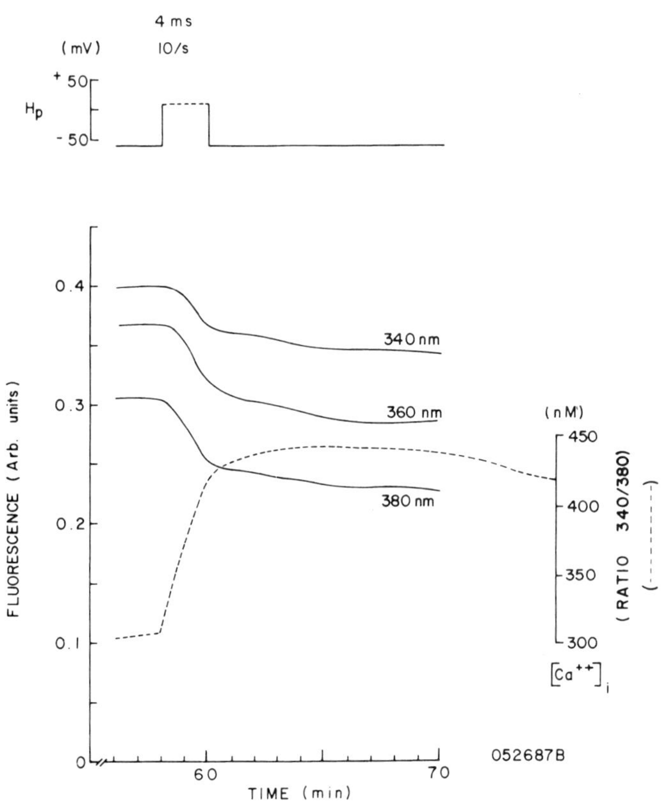

FIGURE 6. Time course (in minutes) of the effect of a train of depolarizing voltage-clamp pulses on Fura-2 fluorescence (left ordinate, arbitrary units) for the three wavelengths employed to excite the dye (340 nm, 360 nm, and 380 nm) and the Ca concentration (in nM) as calculated for the fluorescence ratio (340 nm/380 nm, left ordinate, dotted line). A train of depolarizing voltage-clamp pulses (70-mV amplitude, 4-msec duration at 10/sec for 2 min) was delivered 3 minutes after the start of the record.

here allowed us to obtain an expression of fluorescence as a function of $[Ca^{2+}]_i$ in squid giant axon. An average resting free Ca level of 184 nM in fresh axons exposed to 3 mM Ca_o was obtained. Although this value is higher than that reported previously from other optical indicators in squid axon, it is not an unreasonable figure. In fact, most marine animals cells analyzed with Fura-2 have consistently shown values for $[Ca^{2+}]_i$ on the order of 100 nM to 200 nM.[20] That this value is correct is further supported by the fact that in axons injected with equimolar amounts of CaEGTA/EGTA buffer mixture, the free Ca level measured with Fura-2 was 140 nM. This apparent dissociation constant for the CaEGTA complex agrees with the 150 nM value listed in the literature.[2]

As discussed elsewhere,[27] methods to determine $[Ca^{2+}]$ based on optical indicators and used to study Ca homeostasis in cells suffer from the inability of these methods to deal with calcium buffering, which represents a sizable compartment where Ca ions can exchange while invisible to the indicator. Moreover, such a compartment could be, at least partially, under metabolic control.[28] Thus, the findings reported herein must be analyzed with these limitations in mind. In fact, during Ca entry with electrical stimulation, only one ion in 3,400 that cross the axolemma shows up as ionized Ca in the axoplasm, while at rest, only one Ca out of 400 is present as ionized calcium. This last number arises from knowledge of the total content for Ca in axoplasm, which is in the order of 70 μmol/kg,[21] and the average value for resting $[Ca^{2+}]_i$ in axoplasm determined herein. This difference between the buffering activity in axons at rest and in activity, tends to confirm the notion[29] that Ca buffering is a dynamic variable in Ca homeostasis such that during a Ca load, buffers of different capacity and affinity are sequentially activated to deal with more of the incoming ions.

The nature of the response of aequorin to Ca inside a cell has been the subject of controversy.[26] Therefore, a most interesting finding of this study is that the fluorescent Ca indicator Fura-2 allowed the calibration of the photoluminescent protein aequorin in squid giant axons in a manner relatively independent of the presence of exogenous Ca-buffer systems such as EGTA. Our results should be analyzed keeping in mind the data of Allen, Blinks, and Prendergast,[30] who showed that light emission from aequorin for change in $[Ca^{2+}]$ was zero at pCa-9, reached slightly more than 2 at pCa-6, and declined once again to zero at pCa-3. This finding implies that there exists a region of the curve where the slope of light emission versus Ca concentration is 1, and we find that this occurs in the range pCa-7 to pCa-8. Similarly there exists a much larger region of Ca concentration (pCa-4.5 to pCa-6.5) in which the slope is 2 or more. This observation is reflected in FIGURE 5, which shows that, in the range of intracellular Ca concentration 100 nM to 1000 nM, aequorin responds, for all practical purposes, linearly to Ca concentration. Above 1000 nM, and up to 10 μM, the luminescent response changes to a square law relationship with $[Ca^{2+}]_i$.

Finally, since the two optical Ca indicators used, namely Fura-2 and aequorin, showed responses very similar in time course to physiological experiments such as stimulation, depolarization by high-K_o seawater, or exposure to CN, it would appear that both indicators are equally good in determining intracellular Ca changes in squid axon. However, the finding reported herein that Fura-2 is not a suitable probe for conventional voltage-clamp experiments in squid giant axon reduces its usefulness as a tool to study Ca homeostasis in this classical model of excitable cells.

SUMMARY

Squid giant axons were injected simultaneously with Ca indicators Fura-2 and aequorin. Fura-2 was calibrated *in situ* by measuring fluorescence at 510 nm upon UV excitation at 340 nm, 360 nm, and 380 nm with a time-sharing multiple wavelength spectrofluorimeter. Limiting values for dye fluorescence were obtained by allowing a massive load of Ca to enter the axon with the aid of procedures such as prolonged depolarization in the presence of CN (for saturation) and by sequestration of all Ca present in the axoplasm accomplished with injection of EGTA into the axon (for a zero-Ca signal). The average intracellular Ca concentration obtained with Fura-2 was 184 nM. The sensitivity of Fura-2 to intracellular Ca is at least as great as that of aequorin, thus permitting its use in the characterization of Ca homeostasis mechanisms such as Na-Ca exchange. It was found, however, that for voltage-clamp experiments requiring an internal current electrode, Fura-2 is not a convenient Ca probe because electrode reactions in the axoplasm denature the dye, thereby restricting its use in characterization of Ca movements associated with electrically induced changes in membrane potential. A comparison of aequorin luminescence with Fura-2 fluorescence demonstrated that light output by aequorin is linear with intracellular Ca concentrations up to values of 750 nM, changing to a square law relationship from 750 nM up to 10 μM Ca.

ACKNOWLEDGMENTS

The authors wish to thank Dr. Guillermo Whittembury for critical reading of the manuscript and to the director and staff of the Marine Biological Laboratory, Woods Hole, Massachusetts for facilities placed at their disposal.

REFERENCES

1. BAKER, P. F., A. L. HODGKIN & E. G. RIDGWAY. 1971. Depolarization and Ca entry in Squid Axons. J. Physiol. (London) **218:** 709-755.
2. DIPOLO, R., J. REQUENA, J. F. BRINLEY, JR., L. J. MULLINS, A. SCARPA & T. TIFFERT. 1976. Ionized Ca concentration in squid Axons. J. Gen. Physiol. **67:** 433-467.
3. REQUENA, J., L. J. MULLINS, J. WHITTEMBURY & F. J. BRINLEY, JR. 1986. Dependence of ionized and total Ca in squid axons on Na_o-free or high K_o conditions. J. Gen. Physiol. **87:** 143-159.
4. MULLINS, L. J. & J. REQUENA. 1979. Ca measurement in the periphery of an axon. J. Gen. Physiol. **74:** 393-413.
5. BAKER, P. F. & G. E. UMBACH. 1985. Calcium buffering in axons and axoplasm of *Loligo.* J. Physiol. **383:** 369-394.
6. DIPOLO, R., H. ROJAS, J. VERGARA, R. LÓPEZ & C. CAPUTO. 1983. Biochim. Biophys. Acta **738:** 311-318.
7. REQUENA, J., J. WHITTEMBURY, T. TIFFERT, D. A. EISNER & L. J. MULLINS. 1984. A comparison of measurements of intracellular Ca by Ca electrode and optical indicators. Biochim. Biophys. Acta **805:** 393-404.
8. TSIEN, R. Y. 1983. Intracellular measurements of ion activities. Annu. Rev. Biophys. Bioeng. **12:** 91-116.
9. GRYNKIEWICZ, G., M. POENIE & R. TSIEN. 1985. A new generation of Ca indicators with greatly improved fluorescence properties. J. Biol. Chem. **260:** 3340-3350.
10. ALLEN, T. J. A. & P. F. BAKER. 1985. Intracellular Ca indicator Quin-2 inhibits Ca inflow via Na_i/Ca_o exchange in squid axon. Nature **316:** 755-756.
11. REQUENA, J., J. WHITTEMBURY, A. SCARPA & L. J. MULLINS. 1988. Fura 2 and aequorin as indicators of intracellular Ca in squid giant axons. FASEB J. **2(4):** A322.
12. REQUENA, J., R. DIPOLO, F. J. BRINLEY, JR. & L. J. MULLINS. 1977. The control of ionized calcium in squid axons. J. Gen. Physiol. **70:** 329-353.
13. DIPOLO, R., F. BEZANILLA, C. CAPUTO & H. ROJAS. 1985. Voltage dependence of the Na/Ca exchange in voltage-clamp, dialyzed squid axons. Na-dependent Ca efflux. J. Gen. Physiol. **86:** 457-478.
14. VANWAGONER, D. & J. WHITTEMBURY. 1985. An improved current electrode/injection capillary for large cells voltage clamp. J. Physiol. **365:** 8p.
15. MULLINS, L. J., J. REQUENA & J. WHITTEMBURY. 1985. Ca entry in squid axons during voltage clamp pulses in mainly Na/Ca exchange. Proc. Natl. Acad. Sci. USA **82:** 1847-1851.
16. SCARPA, A. 1972. Spectrophotometric measurement of Ca by murexide. Methods Enzymol. **24:** 343-351.
17. MULLINS, L. J. & J. REQUENA. 1981. The "late" Ca channel in squid axons. J. Gen. Physiol. **78:** 683-700.
18. BORON, W. F. & J. M. RUSSELL. 1983. Stoichiometry and ion dependencies of the intracellular-pH-regulating mechanism in squid giant axons. J. Gen. Physiol. **81:** 373-399.
19. BORON, W. F. 1985. Intracellular pH regulation mechanism of the squid axon. Relation between external Na^+ and HCO^{3-} dependences. J. Gen. Physiol. **85:** 325-346.

20. POENIE, M., J. ALDERTON, R. Y. TSIEN & R. STEINHARDT. 1985. Changes in free Ca levels with stages of the cell division cycle. Nature **315:** 147-149.
21. REQUENA, J., L. J. MULLINS & F. J. BRINLEY, JR.. 1979. Calcium content and net fluxes in squid giant axons. J. Gen. Physiol. **73:** 327-342.
22. REQUENA, J. 1983. Ca transport and movement in nerve fibers. Ann. Rev. Biophys. Bioeng. **12:** 237-257.
23. BAKER, P. F., A. L. HODGKIN & E. G. RIDGWAY. 1971. Depolarization and Ca entry in squid axons. J. Physiol. (London) **218:** 709-755.
24. REQUENA, J., L. J. MULLINS, J. WHITTEMBURY & F. J. BRINLEY, JR. 1986. Dependence of ionized and total Ca in squid axons on Na_o-free or high-K_o conditions. J. Gen. Physiol. **87:** 143-159.
25. REQUENA, J., J. WHITTEMBURY & L. J. MULLINS. 1989. Ca entry in squid axons during voltage clamp pulses. Cell Calcium **10:** 413-423.
26. REQUENA, J. & L. J. MULLINS. 1979. Ca movement in nerve fibers. Q. Rev. Biophys. **12:** 371-460.
27. MULLINS, L. J. & J. REQUENA. 1989. Comparing sodium-calcium exchange as studied with isotopes and measurements of $[Ca]_i$. *In* Na/Ca Exchange. J. Allen, D. Noble & H. Reuter, Eds. Oxford University Press. Oxford, England.
28. BRINLEY, JR., F. J., T. TIFFERT, A. SCARPA & L. J. MULLINS. 1977. Intracellular calcium buffering capacity in isolated squid axons. J. Gen. Physiol. **70:** 355-384.
29. BRINLEY, JR., F. J. 1978. Calcium buffering in squid axons. Annu. Rev. Biophys. Bioeng. **7:** 363-92.
30. ALLEN, D. G., J. R. BLINKS & F. G. PRENDERGAST. 1979. Aequorin luminescence: Relation of light emissions to Ca concentration—A Ca independent component. Science **196:** 996-998.

Cardiac Na^{+}-Ca^{2+} Exchange System in Giant Membrane Patches

DONALD W. HILGEMANN, ANTHONY COLLINS,
AND DAVID P. CASH

Department of Physiology
Southwestern Medical Center at Dallas
Dallas, Texas 75235

GEORG A. NAGEL

Cardiovascular Laboratories
Rockefeller University
New York, New York 10021

INTRODUCTION

This article presents an overview of recent work on cardiac Na-Ca exchange in giant, excised membrane patches.[1] The need to develop this experimental model became pressing in the period of 1985-87. On the one hand, many properties of Na-Ca exchange currents in cardiac myocytes[2,3] were consistent with previous data from cardiac vesicles (for review, see Reeves & Philipson[4]) and converged neatly with data from intact tissue.[5] Current densities of the inward exchange current, for example, fit remarkably well with rates of sodium-dependent calcium efflux measured during the late action potential.[6] On the other hand, many questions about the catalytic cycle of the exchanger remained unanswered and new enigmas emerged (for overview, see Hilgemann[7]). The exchanger in intact cells appeared to be regulated secondarily by cytoplasmic calcium and ATP,[8] similar to the squid axon system. Ion dependencies of the exchanger in intact cells appeared to be asymmetrical (Kimura *et al.*[3]; more recently, Li & Kimura[9]), similar to the axon exchanger and in contrast to conclusions from cardiac vesicle studies. Although internal perfusion of cardiac myocytes[10] represented a great improvement over previous electrophysiological myocyte techniques, control of the cytoplasmic environment remained inadequate for many purposes.

METHODS

Methods to obtain giant, inside-out excised patches have been described previously.[1,11] Unless indicated otherwise, results described are from guinea-pig myocytes. Pipette perfusion was carried out with the method of Soejima and Noma.[10] *Xenopus* oocytes expressing the cardiac exchanger were provided by Drs. K. D. Philipson and D. A. Nicoll (see Nicoll *et al.*[12]). Before seal formation, oocytes were shrunk in 3× hypertonic solution and the vitellin layer was removed mechanically. Giant patches (20-35 μm diameter) were formed and excised using bullet-shaped pipettes.

TABLE 1. Composition of Standard Solutions for Inward and Outward Current Measurements[a]

	Standard Superfusion Solution	Pipette Solution A	Pipette Solution B
EGTA	9 mM	20 μM	2 mM
$CaCl_2$	0-10 mM	4-8 mM	—
(Na + Cs)-MES	100 mM	—	—
(NMG or Li)-MES	—	100 mM	—
(Na + Li)-MES	—	—	140 mM
(Li + K)-Cl	—	20 mM	—
$MgCl_2$	1 mM	2 mM	2-6 mM
TEA-MES	18 mM	20 mM	20 mM
Cs-MES	18 mM	20 mM	20 mM
HEPES	25 mM	10 mM	10 mM
Ouabain	—	200 μM	200 μM
Verapamil	—	5 μM	5 μM
$BaCl_2$	—	0.5 mM	—
4-AP	—	1 mM	1 mM
pH	7.0	7.0-7.2	7.0-7.2

[a] Chemical abbreviations employed: ethylene glycol-bis (β-aminoethyl ether) *N,N,N',N'*-tetraacetic acid, *EGTA; 2*-[*N*-morpholino] ethanesulfonic acid, MES; tetraethylammonium, TEA; *N*- methylglucamine, NMG; *N*-[2-hydroxyethyl] piperazine-*N'*-[2-ethanesulfonic acid], HEPES; 4-aminopyridine, 4AP.

Compositions of the standard solutions employed for inward and outward current measurements were as presented in TABLE 1. Pipette (extracellular) Solution A was used for outward exchange current. Pipette Solution B was used for inward exchange current.

THE CATALYTIC MECHANISM

Limited proteolysis of the cytoplasmic surface by chymotrypsin (1 mg/ml for ~30 sec) functionally removes secondary modulations of the exchanger by cytoplasmic calcium and sodium and leaves the exchanger highly stimulated.[11] Because the exchanger then appears to function in a simple fashion and the "deregulated" exchange currents are remarkably stable, studies of the catalytic mechanism have been performed on patches pretreated with chymotrypsin.

Ion dependencies of exchange current were studied in more than 80 stable, 15-25-μm diameter cardiac patches with 1 to 8 gigaohm seals. In the great majority of patches, calcium-activated nonspecific conductance was evidently absent or was negligible. The dependencies of inward current on cytoplasmic calcium concentration and of outward current on cytoplasmic sodium concentration were determined at different extracellular (pipette) concentrations of the counterion, at different membrane potentials, and with different concentrations of the counterion on the cytoplasmic side. At 0 mV, accurate concentration-current relations could be obtained when experimental conditions had reduced the "maximum" exchange current by more than 95% from the fully activated current. Current-voltage (I-V) relations were determined under selected conditions, usually by activating the exchange current with ion jumps at each potential of interest.

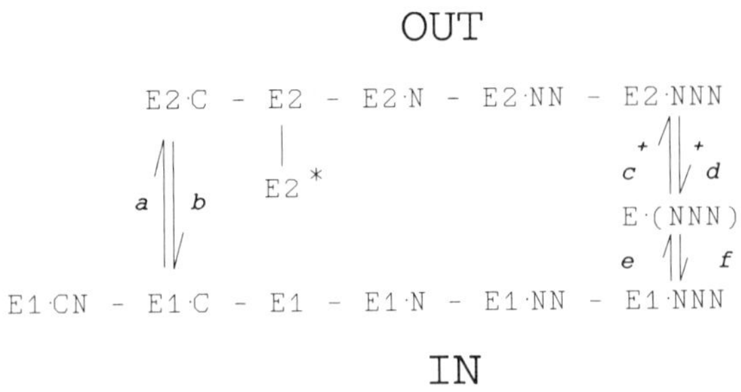

OUT

E2·C - E2 - E2·N - E2·NN - E2·NNN

E2*

a b c d e f

E·(NNN)

E1·CN - E1·C - E1 - E1·N - E1·NN - E1·NNN

IN

FIGURE 1. Simple reaction scheme of the catalytic mechanism of cardiac sodium-calcium exchange, which was fitted to all data on "deregulated" exchange current. Arrows indicate ion translocation reactions simulated kinetically. Bars indicate ion-binding reactions, which were usually assumed to be instantaneous. "E1" is the empty carrier with binding sites facing the cytoplasmic side and "E2" with binding sites facing the extracellular side. Other states correspond to states with ions bound, whereby "C" indicates a bound calcium ion, and "N" indicates a bound sodium ion. On the cytoplasmic side one sodium ion can bind to the calcium-loaded carrier, giving the E1-CN state. The extracellular E2* state adequately accounts for lower calcium and sodium affinities on the extracellular side than on the cytoplasmic side. The "c" and "d" reactions correspond to deocclusion and occlusion of sodium ions from the extracellular side. Effects of voltage on cytoplasmic ion dependencies of the exchanger are very well fitted when this reaction is assumed to move one charge through the membrane field. Reactions "e" and "f" represent occlusion and deocclusion of sodium from the cytoplasmic side. Reactions "a" and "b" represent the entire calcium translocation pathway.

The entire data base from these experiments was fitted simultaneously to different models of the catalytic mechanism using a least-squares method. For some purposes, results of others on the extracellular ion and voltage dependencies of exchange currents were added to the data base. Data fitted included interactions of sodium and calcium on opposite membrane sides and on the same membrane side. (FIG. 2 shows examples.)

FIGURE 1 shows schematically the simplest exchange model that accounted well for the data base from "deregulated" patches. This model is presented with the explicit expectation that individual assumptions will be changed, as more detailed information becomes available on specific aspects of the cycle. The model is of the consecutive type, moving 3 Na and 1 Ca in separate steps, rather than one of the different possible simultaneous models (e.g. Hilgemann[7]). Binding of sodium ions is sequential. On the cytoplasmic side it must be assumed that one sodium ion can bind when the carrier is calcium loaded, forming the E1-CN state. Most of the voltage dependence is assumed to reside in the sodium translocation pathway. Similar to the sodium pump, the major charge movement in the cycle takes place at the extracellular end of the translocation pathway.

Reasons for the choice of a consecutive model are based on the ion concentration dependencies of exchange current as counter ion concentration is decreased toward zero. The important pattern has already been described for the outward exchange current in cardiac myocytes[9] and for calcium uptake into proteoliposomes.[13] Our work extends findings to the inward exchange current and through a greater concentration range. First, the apparent K_d for cytoplasmic calcium of inward current appears to decrease toward zero as the concentration of extracellular sodium is decreased toward zero (namely, from about 4 μM at 150 mM extracellular sodium to less than 1 μM with 20 mM extracellular sodium). FIGURE 2A shows results of a relevant experiment with pipette perfusion of three different sodium concentrations. Second, the apparent K_d for cytoplasmic sodium of outward current appears to decrease toward zero as the concentration of extracellular calcium is decreased toward zero (namely, from about 18 mM with 5 mM extracellular calcium to about 5 mM with 10 μM extracellular calcium). Results of a relevant experiment with pipette perfusion are given in FIGURE 2C. The ratio of the apparent K_d to the maximum current becomes nearly constant only with less than 100 μM extracellular calcium.

On the cytoplasmic side, sodium and calcium were found to interact in surprisingly different fashions with respect to the inward and outward exchange currents (FIGS. 2B and D). As shown in 2D, cytoplasmic calcium inhibits the outward current in an apparently competitive fashion with sodium. As shown in 2B, cytoplasmic sodium

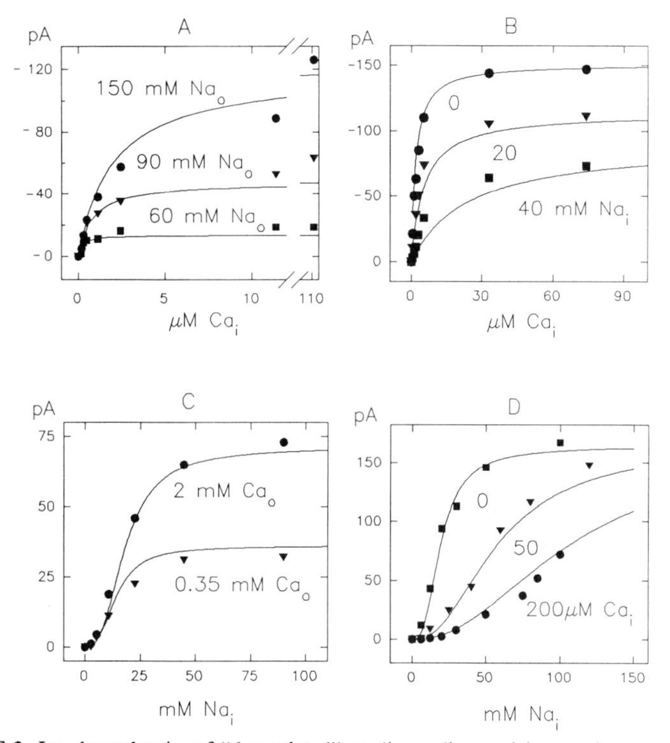

FIGURE 2. Ion dependencies of "deregulated" cardiac sodium-calcium exchange current. All results at 0 mV, 34 °C. These results and results from > 20 other experiments are fitted simultaneously to the model of FIGURE 1. The model requires only 6 parameters to be fitted, apart from a scaler for data from each experiment. **A.** Cytoplasmic free-calcium dependence of inward exchange current at 150, 80, and 40 mM extracellular sodium but simulated as 150, 90, and 60 mM as indicated. Pipette perfusion experiment. Note shifts of calcium dependency to lower free calcium as sodium is reduced. **B.** Cytoplasmic calcium dependence of inward exchange current at 0, 20, and 40 mM cytoplasmic free sodium; 150 mM extracellular sodium. Note the drop of maximum current with increasing sodium concentration. **C.** Cytoplasmic sodium dependence of outward exchange current at 2 and 0.35 mM extracellular calcium. Pipette perfusion experiment. **D.** Cytoplasmic sodium dependence of outward exchange current at 0, 50, and 200 μM cytoplasmic free calcium. See text for further explanations.

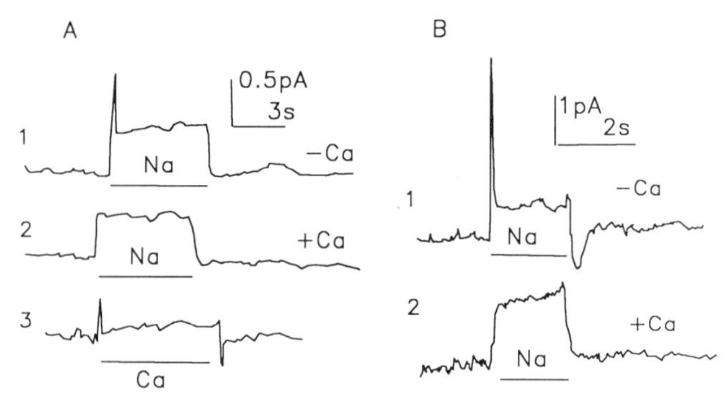

FIGURE 3. Probable "half-cycle" charge movement of sodium-calcium exchange. **A.** Cardiac membrane patch, pretreated with chymotrypsin; 22-μm diameter; 10 μM extracellular free calcium and no sodium. Cytoplasmic concentration jumps are made to 80% completion in 150 msec. Application of 40 mM cytoplasmic sodium in the absence of cytoplasmic calcium induces a current transient followed by steady-state current (1, "−Ca"). In the presence of 0.2 μM free calcium (2, "+Ca") the transient is abolished. **B.** Patch from oocyte expressing the cardiac exchanger; 32 μm diameter; Pretreated with chymotrypsin; 5 mM extracellular sodium and no extracellular calcium. Application of 40 mM cytoplasmic sodium in the absence of cytoplasmic calcium induces a current transient followed by steady-state current ("−Ca"). In the presence of 0.2 μM free calcium ("+Ca") the transient is abolished. In the oocyte, the steady-state current is probably a genuine background sodium current. See text for further explanations.

inhibits the inward current in a mixed fashion with respect to cytoplasmic calcium. The maximum inward current is strongly reduced by cytoplasmic sodium, which is similar to results published for squid axon.[14] These behaviors are well accounted for by the binding scheme of FIGURE 1 in which one sodium ion can bind after calcium has bound on the cytoplasmic side.

A further important reason for assuming a consecutive model is that it appears possible to isolate an electrogenic ion translocation step of the cycle via cytoplasmic ion jumps. To attempt to isolate the individual ion translocation reactions of the exchanger, either a small concentration of calcium (3-20 μM; see FIG. 3) or sodium (2-20 mM) is included in the pipette, and solutions on the cytoplasmic side are initially sodium- and calcium-free. In a consecutive exchange model, this insures that the ion-translocation reactions move binding sites to face the cytoplasmic side. From that configuration, a rapid step of the cytoplasmic sodium or calcium concentration should move binding sites back to the extracellular (pipette) side and generate a current transient if the corresponding translocation pathway is electrogenic. When low calcium is included in the pipette, as in FIGURE 3A, a small amount of steady-state exchange current is activated by application of 40 mM sodium. A current transient precedes the steady-state current. The time course of the transient current is quite certainly determined by the speed of solution switches, not the kinetics of exchange reactions. The total amount of charge moved would quantitate the site density of exchanger in the patch, if all charge movement were related to sodium translocation. With this assumption, exchanger site density is about 200 to 400 per μm^2, which indicates maximum turnover rates of about 5000 sec^{-1} to account for maximum current densities.

The sodium-induced current transients are identified with the Na-Ca exchanger by multiple criteria: (1) They are blocked by preapplication of low cytoplasmic calcium concentrations, which should shift the binding sites back to the extracellular side. (2) They are blocked by nickel, which blocks steady-state exchange current. (3) They are blocked by dichlorobenzamil, an organic inhibitor of the exchanger. (4) They are blocked by exchanger inhibitory peptide.[15] They are shifted in their ion dependencies by changes of extracellular ion concentrations, as predicted by the model of FIGURE 1. (6) They are present in the membranes of *Xenopus* oocytes expressing the cardiac exchanger (see FIG. 3B). By all these same criteria, the possibility of negative charge movement in the translocation of calcium has not been supported by the equivalent experiments with cytoplasmic calcium pulses. In contrast, under some conditions current transients consistent with positive charge movement in calciuni translocation are found (see third record in 3A), both in cardiac patches and oocyte patches expressing the cardiac exchanger. These results would be consistent with a possibility that calcium moves through a small fraction of membrane field on approach to its binding sites on one membrane side or the other (or both).

In summary, data obtained on the chymotrypsin-deregulated catalytic mechanism is consistent with a consecutive mechanism with positive charge movement through the membrane field during sodium translocation. Strong caution remains indicated until further data are available on voltage sensitivity of the concentration dependencies from the extracellular side and until charge movements are resolved kinetically by voltage-pulse experiments.

SECONDARY MODULATIONS BY CYTOPLASMIC Na^+ AND Ca^{2+}

As already mentioned, secondary processes modulate the sodium-calcium exchange function when cytoplasmic calcium and sodium concentrations are changed. As a starting point to understand the origins of these mechanisms, time-dependent changes of both inward and outward exchange currents were analyzed with step changes of cytoplasmic sodium and calcium. For reasons of brevity, details cannot be presented in this article, except to demonstrate the minimum complexities involved in measurements of outward current.

FIGURE 4A demonstrates the existence of at least three inactive exchanger states connected with the modulatory processes. Part A of the figure shows the usual results of five possible cytoplasmic concentration jumps of sodium (to activate the current directly) and/or calcium (to activate the current indirectly). As shown in the first result of FIGURE 4A, when sodium is increased from 0 to 100 mM in the presence of 2 μM free calcium, outward current activates and then inactivates with a time constant of a few seconds to an intermediate steady-state level. From the steady-state level in the presence of 100 mM cytoplasmic sodium (second result), outward current turns off and on upon removal and readdition of calcium with a similar time course of a few seconds. For the third result, the 2 μM calcium was applied before sodium (as in the first result), and cytoplasmic solution was then switched to one with sodium, but without calcium. The outward current activates nearly to the level of the first result, and the current turns off with the slow time course of the previous results. As shown in the fourth result, no current can be activated if sodium is applied in the complete absence of calcium. However, if calcium is applied together with sodium after a calcium-free/sodium-free period, as in the fifth record, most of the current is available within about 0.4 sec. The current then inactivates slowly as in the first result. Since no current can be activated without addition of calcium, a fast calcium-dependent activation process must be operative during the "upstroke" of current in this sequence.

FIGURE 4B shows a simulation of the results assuming two parallel inactivation processes, described in part C of FIGURE 4. The simulation assumes for simplicity that the exchanger exists in only one active state (A), functioning via the catalytic cycle

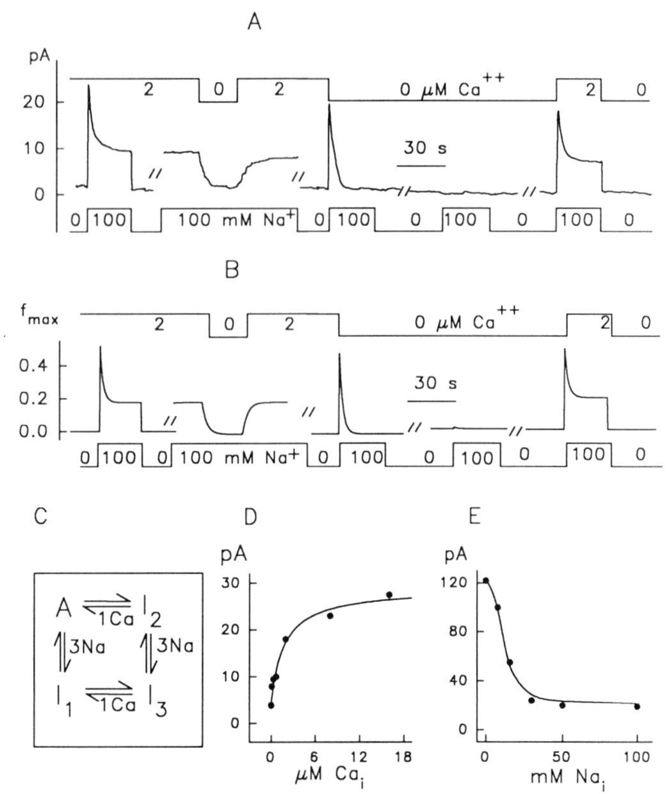

FIGURE 4. Secondary modulation of outward exchange current: sodium-dependent inactivation and calcium-dependent activation. **A.** Typical outward exchange current transients in cardiac membrane patch. Concentration jumps are indicated above and below the records. Note that no current can be activated by sodium in the absence of calcium. Calcium acts over the time course of many seconds when applied after equilibration in high cytoplasmic sodium (second record), but acts within the solution switch time (<0.4 sec) when applied together with sodium after a sodium-free period (fifth record). **B.** Simulation of the responses in A. **C.** Scheme for simulation of results. "A" indicates active state of the exchanger. "I" states are inactive states. Transitions to I_1 and I_3 states occur only when catalytic sites are fully loaded with sodium. In the absence of cytoplasmic sodium and calcium, the exchanger accumulates in the I_2 state. The transition from I_2 to A is very fast in the presence of calcium and accounts for the rapid effect of calcium in the fifth current records. **D.** Cytoplasmic calcium dependence of steady-state outward exchange current. The relationship is well fit with a Hill coefficient of 1. **E.** Cytoplasmic sodium-dependence of sodium-dependent inactivation. Data points give the peak (total) exchange current activated by 100 mM cytoplasmic sodium after equilibration with the given sodium concentrations. The relationship is fit with a Hill coefficient of 2.8.

described in FIGURE 1. The exchanger is assumed to undergo two parallel inactivation reactions, resulting in three possible inactive states ($I_{1,2,\text{and }3}$). Calcium-dependent activation (transitions from I_2 to A and from I_3 to I_1) is assumed for the time being to reflect a slow binding of a single calcium to a regulatory site. The reverse transitions are assumed to be slow dissociation reactions. That binding of only one calcium is needed is suggested by the fact that steady-state activation of the outward exchange current by cytoplasmic calcium is not sigmoid (FIG. 4D). In ATP-depleted, excised patches the half-maximum calcium concentration is about 2 μM, a log unit higher than described in intact cells.[3]

For sodium-dependent inactivation (transitions from A to I_1 and from I_2 to I_3), three sodium ions must be bound to catalytic sites. Evidence for this assumption came from analysis of steady-state, sodium-dependent inactivation, exactly equivalent to steady-state inactivation of channels with voltage. Fractional inactivation of the exchanger at different steady-state sodium concentrations was determined from the total current that could be activated upon stepping to a saturating sodium concentration. As shown in FIGURE 4E, the steady-state inactivation curve has a Hill coefficient very close to 3 and was always as large as the steady-state outward exchange current determined in the same experiments ($n = 5$). Recovery from inactivation is left in the model as a simple time-dependent transition.

Much further experimentation along the lines of FIGURE 4 is needed to develop an adequate analytical understanding of the modulatory processes. In particular, the necessary conditions for entrance into the inactive states with calcium-dependent recovery must be defined (I_2 and I_3). The assumption of a single active catalytic state of the exchanger is a likely heel of Achilles for the simple scheme.

The presently available experimental tools, in particular molecular biological manipulations of the cloned cardiac exchanger, should allow mechanistic questions about the modulatory processes to be answered hand-in-hand with the analytical questions. The following results appear to answer a central question as to whether the secondary modulatory processes are inherent to the exchanger protein itself, or whether additional proteins such as calmodulin and/or cytoskeletal proteins might be involved.[a]

In giant patches from *Xenopus* oocytes optimally expressing cardiac exchange activity (4 to 6 days after RNA injection), outward exchange current was found to behave indistinguishably from the chymotrypsin-deregulated cardiac system. There was no secondary activation of outward current by cytoplasmic calcium, no sodium-dependent inactivation, and the cytoplasmic sodium dependencies of outward current over a range of 120 mV superimposed after scaling to the maximum currents in each type of patch. This remarkable functional similarity suggested that endogenous proteases in the oocyte might "deregulate" the expressed exchanger, as is known to happen when cardiac myocytes are "aged" at room temperature for long periods in the presence of calcium (unpublished results). For this reason, care was taken to study oocytes as soon as expression could be detected after RNA injection. Experimental results were then obtained with remarkable similarity to results obtained in patches from freshly prepared cardiac myocytes (see Hilgemann[1]). Outward exchange current turned off and on slowly upon removal and reapplication of cytoplasmic calcium. The outward current transient on application of cytoplasmic sodium was either small or absent. Secondary calcium dependence could be effectively removed by chymotrypsin.

Finally, expression results with a drastically mutated exchanger protein deserve mention: From the fact that chymotrypsin treatment removes secondary modulations of the exchanger, a critical involvement of peripheral portions of the exchanger protein on the cytoplasmic side was expected. A direct test of this assumption became possible because nearly all of the putative cytoplasmic domain of the exchanger, including about 400 amino acids, could be deleted without loss of exchange activity. An experimental

[a] This work was carried out in collaboration with Drs. K. D. Philipson and D. A. Nicoll (UCLA School of Medicine, Los Angeles, CA).

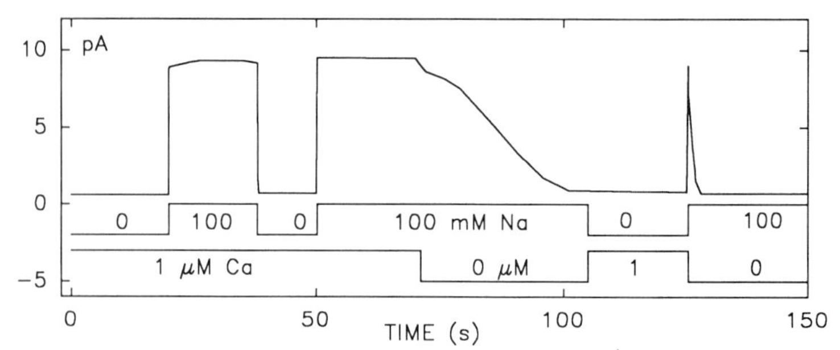

FIGURE 5. Excised membrane patch from an oocyte expressing a deletion mutant cardiac exchanger protein, which lacks nearly all of the putative cytoplasmic domain of the exchanger. Na (100 mM) activates outward exchanger current in the usual manner, and the current turns off slowly when cytoplasmic calcium is reduced from 1 to 0 μM. Reactivation (not shown) is also slow. Solution switches from Ca-containing Cs solution to Na-containing/Ca-free solution are accompanied by complete activation of the current and very rapid decay.

record of outward current is shown in FIGURE 5. Remarkably, outward exchange current generated by the mutated exchanger showed secondary activation by cytoplasmic calcium with qualitatively normal kinetic properties and dependence on calcium. The secondary calcium activation could *not* be removed by chymotrypsin. These results strongly suggest that the secondary modulation by cytoplasmic calcium may arise from a membrane-spanning portion(s) of the exchanger. Evidently, chymotrypsin does not remove a regulatory domain, but rather disrupts "communication" between regulatory and catalytic domains.

STIMULATION BY MG-ATP

Outward Na-Ca exchange current in giant, excised cardiac sarcolemmal patches is stimulated by Mg-ATP, unless the cytoplasmic surface of the patch has been previously treated with chymotrypsin.[11] Upon removal of Mg-ATP the current typically remains elevated or declines slowly (t_{50} = 10 to 25 minutes), although particularly with rabbit and rat sarcolemma a much faster reversal has been observed (t_{50} = 10 to 40 seconds). After the outward current is allowed to "run down" in Mg-ATP-free solution, a second Mg-ATP effect can be evoked, precluding the possibility that Mg-ATP acts by causing the dissociation of an inhibitory factor.

Stimulation of Na-Ca exchange in excised patches by Mg-ATP does not require internal Ca. Application of Mg-ATP to the cytoplasmic surface of a patch under Ca-free conditions in the presence of 9 mM EGTA had a stimulatory effect on the outward exchange current. This effect was not apparent until Ca was replaced after Mg-ATP removal, since the current is dependent upon internal Ca[11] independent from Mg-ATP.

We have used several different approaches to address the possibility that the stimulation by Mg-ATP of Na-Ca exchange current in excised patches may be mediated by a protein kinase, and we have been unable to obtain any evidence consistent with such a mechanism. Specifically, the effect of Mg-ATP was not mimicked by Mg-ATP-γ-S; was not reversed by any of three different phosphatases; was not blocked by the protein kinase inhibitor H7 (200 μM); and was not blocked by peptide inhibitors of protein kinase C, cAMP-dependent protein kinase, and Ca-calmodulin-dependent protein kinase II. In addition, a protein kinase-mediated effect would be expected to have micromolar dependence on Mg-ATP (also if tyrosine kinases were involved; see for example, Morris & Kahn,[17] p. 265), whereas the effect observed in excised patches requires millimolar concentrations of Mg-ATP (half-maximum initial rates, > 2 mM Mg-ATP). These results led us to investigate alternative mechanisms, although the possibility remains that Na-Ca exchange may be stimulated by more than one ATP-dependent mechanism, one of which might be a protein kinase that is absent in excised sarcolemmal patches.

In accord with the results of Philipson and associates,[18-20] the Na-Ca exchange current in excised patches is stimulated by phospholipase D and by the negatively charged detergent SDS and is inhibited by positively charged detergents. Thus, possible candidates for the mechanism of the Mg-ATP effect would include systems that increase or unmask negative charge on phospholipids in the cytoplasmic leaflet of the sarcolemma in a Mg-ATP-dependent manner. Exchanger activity is known to be strongly dependent on phosphatidylserine (PS; see Reeves & Philipson[4] for review), giving rise to speculation that phospholipid asymmetry could be a determinant of exchange function. An aminophospholipid translocase ("flippase"), as described in erythrocytes,[21] then becomes a candidate for the ATP mechanism (together with other processes that might increase negative membrane surface charge).

The following results are consistent with phospholipid flippase involvement:

1. The concentration dependence of initial rates of exchange current stimulation by Mg-ATP is very similar to that of PS translocase activity in resealed red cell ghosts (K_a > 1 mM; Seigneuret & Devaux[21]; A. J. Schroit, personal communication of unpublished data).

2. Pharmacological agents known to block PS translocase also block the stimulation of exchange current by Mg-ATP. Our best example is the sulfhydryl reagent pyridyldithioethylamine (PDA), the most potent published inhibitor of flippase activity.[22] When 0.2 mM PDA is included in the patch electrode, the Mg-ATP effect is abolished, although exchange current and sodium pump currents are not significantly reduced.

3. Exchange stimulation by ATP is insensitive to vanadate. Although vanadate was originally reported to inhibit flippase activity in intact red cells via the spin-label technique, recent detailed studies in resealed ghosts using fluorescent PS analogues demonstrate that the system is indeed insensitive to vanadate (A. J. Schroit, personal communication).

5. Pentalysine in micromolar concentrations binds (cross-links) reversibly to PS head groups,[23] and, as shown in FIGURE 6, pentalysine potently inhibits outward exchange current after stimulation by Mg-ATP. Pentalysine is without effect on the "deregulated" outward exchange current.

6. The cross-linking oxidizer, diamide, induces randomization of phospholipids in ATP-depleted red cells.[24] In giant patches, diamide can greatly accelerate reversal of the ATP effect in the absence of ATP. Like pentalysine, diamide was without effect after chymotrypsin treatment.

7. Channel-forming antibiotics, whose conductivity is sensitive to membrane surface charge, have been used to test whether negative charge on the cytoplasmic membrane surface is actually increased during application of Mg-ATP. As shown in FIGURE 7, potassium current carried by nonactin is roughly doubled when ATP is applied under conditions where external negative surface charge is eliminated by high Mg and Ca on the pipette side. This effect would indicate an increase of internal negative surface potential of about −25 mV. The effect is completely masked by 20 μM pentalysine.

8. The ATP effect is closely mimicked by addition of PS to the electrode hydrocarbon coat used to promote seal formation.

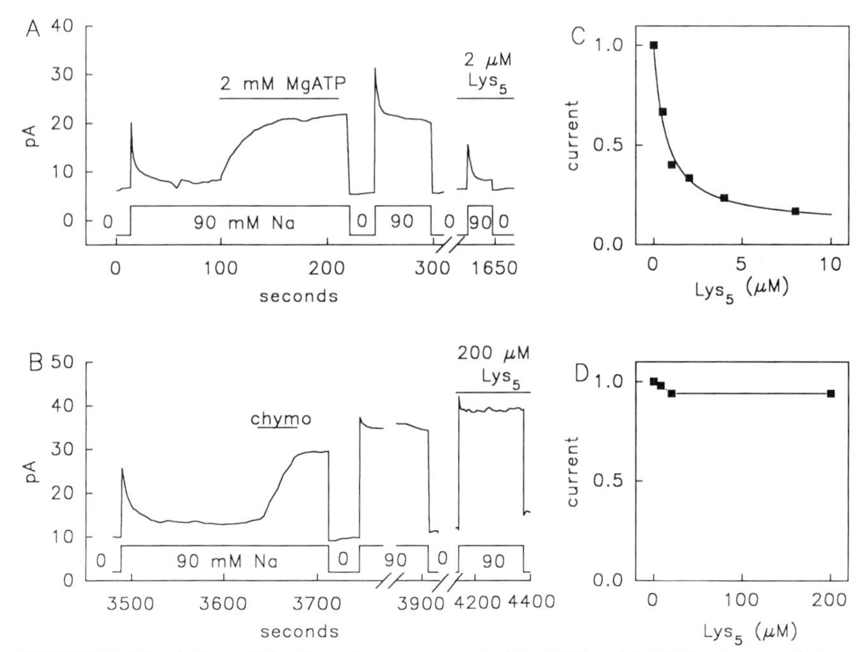

FIGURE 6. Stimulation of Na-Ca exchange current by Mg-ATP and inhibition by pentalysine. **A.** Switching from zero Na to 90 mM Na (at 20 sec) on the cytoplasmic surface of the patch produced an outward current transient. Addition of 2 mM Mg-ATP to the superfusion solution (at 100 sec; indicated by the bar above the current record) increased the steady-state current. Switching to zero Na (at 220 sec) and then back to 90 mM Na (at 240 sec) revealed that there was no change in the leak current and that the instantaneous peak current was also increased. Between 300 seconds and 1600 seconds the current ran down slowly and was twice restimulated with Mg-ATP. Just before pentalysine treatment the instantaneous peak current was 26 pA and the steady-state current was 12.5 pA (without leak subtraction). Addition of 2 μM pentalysine (Lys_5; indicated by the bar above the current record) to the superfusion solution inhibited the peak current by 50% and the steady-state current by 60%. **B.** From the same patch as in (A). Treatment of the cytoplasmic surface of the patch (at 3640 sec) with 3 mg/ml chymotrypsin (chymo) increased the steady-state current without affecting the leak, as shown by switching to zero Na (at 3710 sec). Switching back to 90 mM Na (at 3740 sec) revealed that the transient was almost eliminated. The current continued to increase slowly between 3900 seconds and 4400 seconds in the presence of 8, 20, and then 200 μM pentalysine. This increase was not due to pentalysine, since the current continued to increase slowly after pentalysine removal. **C.** Concentration dependence of inhibition of steady-state outward Na-Ca exchange current by pentalysine. Current is expressed as the fraction of the steady-state current in the absence of pentalysine. Half-maximal inhibition is at 0.68 μM pentalysine. **D.** Effect of pentalysine on outward Na-Ca exchange current after treatment of the patch with chymotrypsin.

DISCUSSION

The results described have been discussed with their presentation so that only a few summarizing comments are appropriate. The overall impression from data on the catalytic mechanism in giant patches is that the exchanger functions with some analogy to the Na-K pump.[25] The cycle appears to be consecutive, and positive charge appears to move through membrane field during sodium translocation. Many reasons for cau-

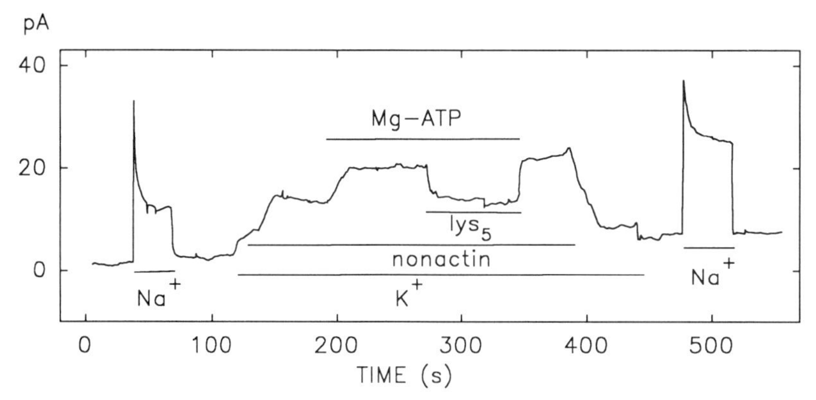

FIGURE 7. ATP-dependent stimulation of nonactin-carrying K current in giant, excised patch from rabbit myocyte. Pipette solution is Pipette Solution A with 20 mM additional TEA-MES instead of 20 mM Cs-MES. "Na+" marks activation of outward exchange with usual substitution of 100 mM Cs-MES for Na-MES. During the period marked "K+", 50 mM K-MES was superfused, and during the period marked, "nonactin," 5 μM of the K-specific ionophore was superfused. At approximately the 200-sec time, 2 mM Mg-ATP was superfused, and the nonactin-induced K current is nearly doubled. As indicated by "lys_5", pentalysine masked completely and reversibly the effect of ATP. Note that after removal of pentalysine and ATP the nonactin current remains stimulated, similar to the effect of ATP on exchange current. After return to the usual cytoplasmic solutions, reactivation of exchange current indicates that it was stimulated in the usual manner by the Mg-ATP exposure.

tion remain until these conclusions are supported by other approaches (e.g., ion self-exchange studies) and until the nature of the secondary modulatory processes is better understood. It is attractive to think that the secondary modulatory processes could be physiological regulators of exchange activity, but details remain a matter of speculation.

It is presently unclear whether Mg-ATP acts by the same or different mechanisms to stimulate Na-Ca exchange in giant patches and in squid giant axon. We do not obtain stimulatory effects with the ATP analogue, ATP-γ-S, which directly conflicts with reports from the axon.[26] The hypothesis that negatively charged lipids mediate the Mg-ATP mechanism fits well with previous work (and speculations) on the interactions of lipids with the exchanger (for review, see Reeves & Philipson[4]). On the one hand, it is known that phosphatidic acid, negatively charged amphiphiles, and Mg-ATP stimulate exchange activity in a qualitatively similar way, whose common denominator is logically an increase of negative surface charge. On the other hand, it is known from experiments in giant patches (unpublished observations) and in cardiac vesicles[27] that screening of negative surface charge with high concentrations of spermine, spermidine, and dimethonium has almost no effect on exchange activity. Thus, three different types of negative charges, when anchored appropriately at the membrane surface, appear to interact directly with an inhibitory domain of the exchanger. The positively charged cytoplasmic region corresponding to exchanger inhibitory peptide, XIP,[15] is an attractive candidate for the inhibitory domain.

REFERENCES

1. HILGEMANN, D. W. 1989. Giant excised cardiac sarcolemmal membrane patches: Sodium and sodium-calcium exchange currents. Pfluegers Arch. **415:** 247-249.

2. Kimura, J., A. Noma & H. Irisawa. 1986. Sodium-calcium exchange current in mammalian heart cells. Nature **319:** 596-597.
3. Kimura, J., S. Miyamae & A. Noma. 1987. Identification of sodium-calcium exchange current in single ventricular cells of guinea-pig. J. Physiol. **384:** 199-222.
4. Reeves, J. P. & K. D. Philipson. 1989. Sodium-calcium exchange activity in plasma membrane vesicles. *In* Sodium-Calcium Exchange. T. J. A. Allen, D. Noble & H. Reuter, Eds.: 27-53. Oxford University Press. Oxford, England.
5. Hilgemann, D. W. 1986. Extracellular calcium transients and action potential configuration changes related to post-stimulatory potentiation in rabbit atrium. J. Gen. Physiol. **87:** 675-706.
6. Hilgemann, D. W. & D. Noble. 1987. Excitation-contraction coupling and extracellular calcium transients in rabbit atrium. Proc. R. Soc. London B **230:** 163-205.
7. Hilgemann, D. W. 1988. Numerical approximation of sodium-calcium exchange. Prog. Biophys. Molec. Biol. **51:** 1-45.
8. Haworth, R. A. *et al.* 1987. Inhibition of calcium influx in isolated adult rat heart cells by ATP depletion. Circ. Res. **60:** 586-594.
9. Li, J. & J. Kimura. 1990. Translocation mechanism of Na-Ca exchange in single cardiac cells of guinea pig. J. Gen. Physiol. **96:** 777-788.
10. Soejima, M. & A. Noma. 1984. Mode of regulation of the ACh-sensitive K-channel by the muscarinic receptor in rabbit atrial cells. Pfluegers Arch. **400:** 424-431.
11. Hilgemann, D. W. 1990. Regulation and deregulation of cardiac Na-Ca exchange in giant excised sarcolemmal membrane patches. Nature **344:** 242-245.
12. Nicoll, D. A., S. Longoni & K. D. Philipson. 1990. Molecular cloning and functional expression of the cardiac sarcolemmal Na-Ca exchanger. Science **250:** 562-565.
13. Khananshvili, D. 1990. Distinction between the two basic mechanisms of cation transport in the cardiac Na-Ca exchange system. Biochemistry **29:** 2437-2442.
14. Blaustein, M. P. & J. M. Russell. 1975. Sodium-calcium exchange and calcium-calcium exchange in internally dialyzed squid axons. Biophys. J. **20:** 79-111.
15. Li, Z. *et al.* 1991. Identification of a peptide inhibitor of the cardiac sarcolemmal Na-Ca exchanger. J. Biol. Chem. **266:** 1014-1020.
16. Niggli, E. & W. J. Lederer. 1991. Molecular operations of the sodium-calcium exchanger revealed by conformation currents. Nature **349:** 621-624.
17. Morris, F. W. & C. R. Kahn. 1986. The insulin receptor and tyrosine phosphorylation. *In* The Enzymes. Vol. **27:** 247-305. P. D. Boyer & E. G. Krebs, Eds.
18. Philipson, K. D. 1984. Interaction of charged amphiphiles with Na-Ca exchange in cardiac sarcolemmal vesicles. J. Biol. Chem. **259:** 13999-14002.
19. Philipson, K. D. & A. Y. Nishimoto. 1984. Stimulation of Na-Ca exchange in cardiac sarcolemmal vesicles by phospholipase D. J. Biol. Chem. **259:** 16-19.
20. Vemuri, R. & K. D. Philipson. 1988. Phospholipid composition modulates the Na-Ca exchange activity of cardiac sarcolemma in reconstituted vesicles. Biochim. Biophys. Acta **939:** 503-508.
21. Seigneuret, M. & P. F. Devaux. 1984. ATP-dependent asymmetric distribution of spin-labeled phospholipids in the erythrocyte membrane: Relation to shape changes. Proc. Natl. Acad. Sci. USA **81:** 3751-3755.
22. Connor, J. & A. J. Schroit. 1989. Transbilayer movement of phosphatidylserine in nonhuman erythrocytes: Evidence that the aminophospholipid transporter is a ubiquitous membrane protein. Biochemistry **28:** 9680-9685.
23. Kim, J., M. Mosior, L. A. Chung, H. Wu & S. McLaughlin. 1991. Binding of peptides with basic residues to membranes containing acidic phospholipids. Biophys. J. **80:** 135–148.
24. Middlekoop, E. *et al.* 1989. Involvement of ATP-dependent aminophospholipid translocation in maintaining phospholipid asymmetry in diamide-treated human erythrocytes. Biochim. Biophys. Acta **981:** 151-160.

25. GADSBY, D. C., A. BAHINSHI & M. NAKAO. 1989. Voltage dependence of Na/K pump current. Curr. Topics Membr. Trans. **34:** 269-288.
26. DIPOLO, R. & L. BEAUGE. 1987. In squid axons, ATP modulates Na-Ca exchange by a Ca_i-dependent phosphorylation. Biochim. Biophys. Acta **897:** 347-354.
27. BERS, D. M. 1985. Calcium at the surface of cardiac plasma membrane vesicles: Cation binding, surface charge screening, and Na-Ca exchange. J. Membr. Biol. **85:** 251-61.

Influence of External Monovalent Cations on Na-Ca Exchange Current-Voltage Relationships in Cardiac Myocytes[a]

DAVID C. GADSBY,[b] MAMI NODA,[c]
R. NEAL SHEPHERD,[d] AND MASAKAZU NAKAO[e]

Laboratory of Cardiac Physiology
Rockefeller University
New York, New York 10021

INTRODUCTION

The well-known effect of a reduction in extracellular Na (Na_o) concentration to increase cardiac contractility, and the offsetting effect of a simultaneous reduction in extracellular Ca (Ca_o) concentration, were first given a modern theoretical basis by Lüttgau and Niedergerke[1] who proposed a specific competition between these two ions for an extracellular binding site that ultimately regulated contractility. They also suggested that the binding site might be a carrier that facilitated the transport of Na and/or Ca ions and that, if the ion-binding site complex were charged, the transport rate would be voltage dependent. Subsequent flux studies did indeed support an interaction between extracellular Na and Ca ions.[2] Later, more direct evidence consistent with our present concept of Na-Ca countertransport came, almost simultaneously, from two sources. Reuter and Seitz[3] demonstrated that Ca efflux from guinea pig atria was markedly slowed by removing extracellular Na, regardless of the Na substitute used (Li, K, or sucrose), and Baker *et al.*[4] demonstrated a similar countertransport of intracellular Na (Na_i) for Ca_o in the squid giant axon, and suggested[5] that three or more Na ions might be exchanged for each Ca ion and, thus, net charge transferred.

The charge-transporting, or electrogenic, nature of the exchange was subsequently supported by the observation, in both squid axon (e.g., Blaustein *et al.*,[6] Mullins & Brinley,[7] Baker & McNaughton[8]) and cardiac sarcolemmal vesicles (e.g., Bers *et al.*,[9] Caroni *et al.*,[10] Reeves & Sutko[11]), that the Na-Ca exchange rate was voltage sensitive. This dependence on voltage was exploited by Reeves and Hale[12] who used a thermodynamic equilibrium approach to show that the transport stoichiometry for Na-Ca

[a] This work was supported by National Institutes of Health Grant HL-14899 and the Irma T. Hirschl Trust.

[b] Address for correspondence: Dr. David C. Gadsby, Laboratory of Cardiac Physiology, The Rockefeller University, 1230 York Avenue, New York, NY 10021.

[c] Present address: Department of Biophysics, Neuroinformation Research Institute, Kanazawa University School of Medicine, Kanazawa 920, Japan.

[d] Present address: VA Medical Center, Durham, NC.

[e] Present Address: Department of Anesthesiology, Hiroshima University School of Medicine, Hiroshima 734, Japan.

exchange in sarcolemmal vesicles was in fact 3 Na : 1 Ca, the now accepted stoichiometry for the cardiac exchanger. Likewise, the reversal potential of the membrane current associated with Na-Ca exchange in isolated, internally dialyzed myocytes[13] has been shown to be consistent with an exchange ratio of 3 Na : 1 Ca.[14]

Apart from the transport ratio, however, little is known about the reaction mechanism of Na-Ca exchange, and some of what is known remains controversial. One notable feature of the exchange, observed early on in the work with squid axons,[4] was the apparent requirement of Ca_o-dependent Na efflux for the presence extracellularly of small monovalent cations such as Li, K, or Na. Subsequent work on squid showed that Ca_o-dependent Ca efflux, but not Na_o-dependent Ca efflux, was also sensitive to the presence of monovalent cations (e.g. Blaustein[15]), and a similar distinction between their influence on Ca-Ca exchange and on Na-Ca exchange was noted in cardiac sarcolemmal vesicles.[16]

It was of interest to see whether the activating effect of small monovalent cations on Na_i-Ca_o exchange could be observed in cardiac myocytes. Somewhat surprisingly, Miura and Kimura[17] had found in internally dialyzed myocytes that, as $[Na]_o$ was lowered below 50 mM, outward Na-Ca exchange current seemed inhibited to the same degree whether Li ions or the much larger *N*-methylglucamine (NMG) ions were used to replace Na. We have extended the conditions to the complete replacement of external Na by Li or NMG, taking care to keep the intracellular Na sites saturated. We find a striking difference between the outward Na-Ca exchange current-voltage relationship obtained in Li-containing, and that in NMG-containing, Na-free solution. The difference suggests that both the voltage sensitivity and the magnitude of Na_i-Ca_o exchange depend critically on the presence in the extracellular solution of certain small monovalent cations.

METHODS

In principle, the methods used to examine Na-Ca exchange currents are similar to those developed for studying voltage- and ion-dependent characteristics of currents generated by the electrogenic Na-K exchange pump in cardiac myocytes (e.g., Gadsby *et al.*,[18] Gadsby & Nakao[19]), except for the additional precautions needed to buffer Na-Ca exchange-mediated changes in intracellular concentrations of Ca and H ions.[13,14,20] Myocytes were isolated from adult guinea pig ventricles following treatment with collagenase and were stored in a high [K], Ca-free solution.[21,22]

Wide-tipped pipettes (resistance ~ 1 MΩ) fitted with a pipette perfusion device[23,24] were used to voltage-clamp and internally dialyze the myocytes. Whole-cell currents were elicited by 100-msec voltage pulses from the holding potential, -40 mV, to potentials between -140 and $+60$ mV. The holding potential and compositions of internal and external solutions were selected to minimize ion-channel and Na-K pump currents. For the experiments described here, the internal solution included 10 mM MgATP, 40 mM HEPES/CsOH (pH 7.4), 50 mM EGTA/20.6 mM Ca (pCa 7.3), 20 mM TEACl, and 100 mM NaOH. Nominally Ca-free external solutions included a total of 145 mM NaCl, LiCl, and/or NMGCl, plus 10 μM verapamil, 50 μM ouabain, 4 mM $MgCl_2$, and usually 2 mM $BaCl_2$; solutions with up to 4 mM [Ca] were made by replacing $MgCl_2$ with $CaCl_2$. External solutions were preheated to 35-36° C before they flowed into the chamber via a two-position valve.

Whole-cell currents[25] and voltage signals were filtered at 2 kHz (6-pole Bessel), digitized on-line at 6 kHz (12-bit resolution), and stored in a microcomputer for later analysis. The voltage-clamp amplifier was connected to the solution in the pipette and

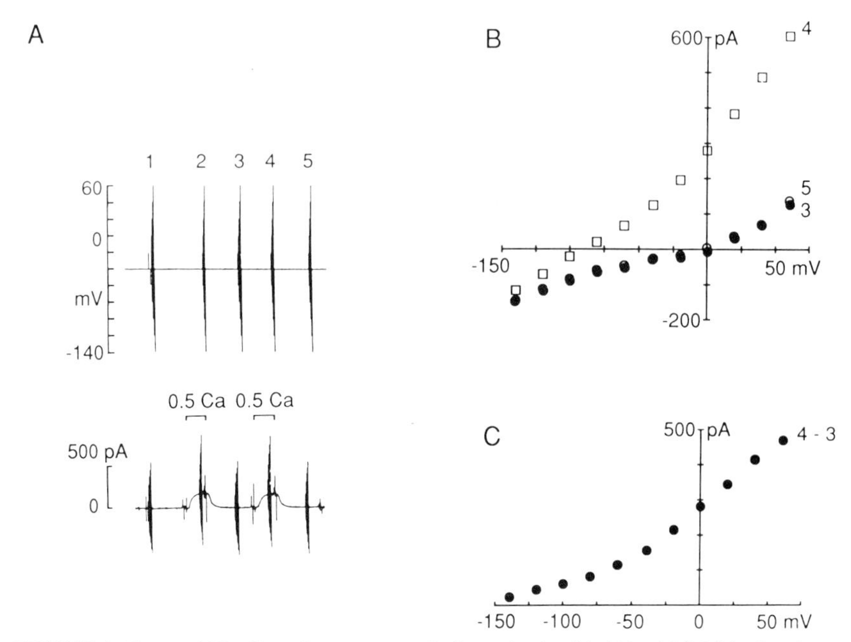

FIGURE 1. Outward Na-Ca exchange current is determined, with 100 mM [Na] in the pipette, as current activated upon sudden exposure to extracellular Ca. **A.** Chart record of the membrane potential (upper trace) and whole-cell current (lower trace); holding potential = −40 mV. The bars above the current record mark 60-sec periods of exposure to 0.5 mM extracellular Ca. Numbered groups of vertical lines on voltage and current traces indicate collection of current-voltage (I-V) data using 100-msec voltage-clamp pulses to membrane potentials between +60 mV and −140 mV before, during, and after the exposures to calcium. Free $[Ca]_{pip}$ = 50 nM; $[Na]_o$ = 145 mM; cell capacitance = 164 pF. **B.** Whole-cell I-V relationships before (3), during (4), and after (5) exposure to 0.5 mM Ca. Steady-state current levels were determined at each voltage by averaging digitized data points over a 30-msec period near the end of each 100-msec clamp pulse. **C.** Ca_o-activated current (4-3), obtained from the data in (B) by subtracting steady current levels before the second exposure to Ca_o (3) from those determined during that exposure to 0.5 mM $[Ca]_o$ (4).

to the experimental chamber by 3 M KCl-filled half cells to minimize voltage errors from changes in liquid junction potentials.

RESULTS AND DISCUSSION

As illustrated in FIGURE 1A, outward Na-Ca exchange current was elicited by briefly raising $[Ca]_o$ from nominally zero to millimolar levels (Ca replacing Mg). The size and voltage dependence of the Ca_o-activated outward current (FIG. 1C) were determined by subtracting steady-state, whole-cell current levels attained during voltage pulses in the nominal absence of external Ca from those recorded in its presence (FIG. 1B). In 145 mM [Na] external solution, Ca_o-activated outward current increased in a roughly exponential manner with voltage from an extremely small value at large negative potentials. Noma, Kimura, and colleagues have presented evidence[13,20] that, under these conditions, the Ca_o-activated outward current reflects Na-Ca exchanger-mediated extrusion of 3 Na ions per Ca ion imported into the cell,[14] the same transport stoichiometry as previously established from thermodynamic equilibrium experiments on cardiac sarcolemmal vesicles.[12]

The Ca_o-activated outward current was enhanced by moderate experimental reduction of the external [Na]. At 50% $[Na]_o$, for example, we found the Ca_o-activated outward current to be increased ~2-fold (at 0 mV), and the augmented Ca_o-activated outward current-voltage relationship was the same whether Li ions or NMG ions were substituted for Na ions. As shown by Miura and Kimura[17] for replacement of Na by Li, this increase in outward Na-Ca exchange current can be attributed to an increase in the apparent affinity of the Na-Ca exchanger for external Ca ions. Those authors also reported that, although the Ca_o-activated outward current was progressively increased as $[Na]_o$ was lowered to 50 mM, further replacement of Na by Li was associated with a decline of outward Na-Ca exchange current.[17] In contrast to those results, however, as shown in FIGURE 2, our experiments demonstrated that full replacement of external Na by Li resulted in a several-fold increase in Ca_o-activated outward current at 0 mV, whereas complete replacement of external Na by NMG caused a reduction of the current at 0 mV to about three-quarters of its level at 145 mM $[Na]_o$.

FIGURE 2 also illustrates a second striking difference between the effects of these two Na ion substitutes. The enlarged Ca_o-activated outward current into 145 mM Li solution showed strong voltage dependence with a steep positive slope, while the much smaller Ca_o-activated outward current into 145 mM NMG was only weakly, if at all, voltage dependent and, indeed, often showed a variable negative slope at the largest positive voltages tested. Both the voltage dependence and the magnitude of outward Na-Ca exchange current appear to depend critically on the extracellular presence of certain small monovalent cations. Judging from the shapes of the outward current-voltage relationships (FIG. 2), both Na and Li, but not NMG, ions seem effective in this role. Furthermore, since replacement of 50% of external Na yields the same current-voltage relationship irrespective of whether Li or NMG ions are used as Na substitute, the monovalent cation site would seem to be saturated by 72.5 mM Na and, hence, must have a relatively high affinity for Na ions.

Presumably, the monovalent cation site remains saturated throughout experiments in which Na is progressively replaced with Li, and the apparent competition between external Na (but not Li) and Ca ions reflects interaction of Na ions with a relatively low affinity site. Miura and Kimura,[17] for example, estimated that the inhibitory constant for external Na was ~44 mM.

Recent detailed kinetic analysis of radiotracer Ca uptake by reconstituted cardiac Na-Ca exchanger into lipid vesicles has indicated that transport occurs via a consecutive, ping-pong cycle. (Kananshvili,[26] but *cf.* Li & Kimura[27]). Simplified kinetic schemes of either the consecutive or simultaneous type (e.g., Hansen *et al.*,[28] Johnson & Kootsey,[29] Läuger,[30] Hilgemann[31]) can easily account for outward Na-Ca exchange current-voltage relationships in high external [Na] solutions, like those illustrated in FIGURES 1 and 2. If external Na, but not Li, ions are assumed to compete with Ca for transport sites on the exchanger, such schemes can also account for the observed increase in exchange current on substituting Li for Na (FIG. 2B), no matter which step in the cycle is assumed to be voltage dependent. However, unloading of the monovalent cation site, as occurs in NMG solution (FIG. 2A), results in an attenuated and relatively voltage-independent outward Na-Ca exchange current, which suggests that a slow voltage-independent step then rate-limits the exchange cycle.

The activation of Na_i-Ca_o exchange by external movovalent cations in our preparation recalls that previously observed in squid[4,32] and *Myxicola*[33] giant axons, and in synaptic membranes,[34] but voltage dependence of that activation was not analyzed in

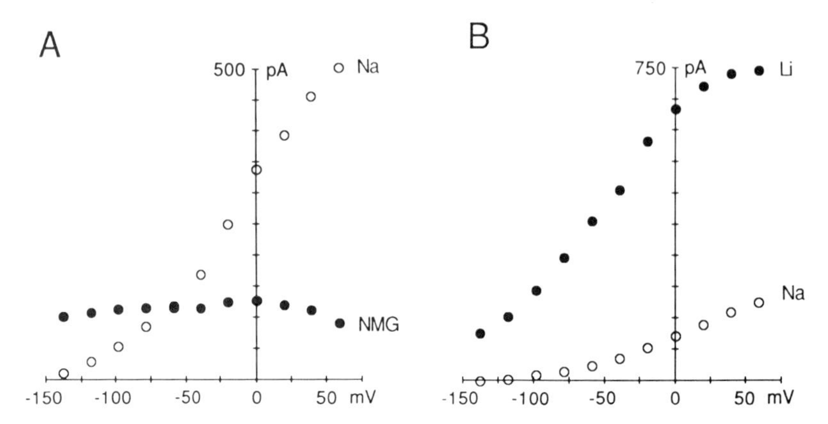

FIGURE 2. Relative size and voltage dependence of outward Na-Ca exchange current in the absence of extracellular Na depends importantly on the Na substitute. **A.** 0.5 mM Ca_o-activated current in the presence of 145 mM $[Na]_o$ (open circles), and after replacing all of the Na with 145 mM *N*-methylglucamine (NMG, filled circles). Free $[Ca]_{pip}$ = 50 nM; $[Na]_{pip}$ = 100 mM; cell capacitance = 189 pF. **B.** 0.25 mM Ca_o-activated current in the presence of 145 mM $[Na]_o$ (open circles), and after replacing all of the Na with 145 mM Li (filled circles). Free $[Ca]_{pip}$ = 50 nM; $[Na]_{pip}$ = 100 mM; cell capacitance = 102 pF.

those studies. Although the relationship between the cardiac and squid axon Na-Ca exchangers is not yet clear, the finding of Allen and Baker[35] that Ca-Ca exchange seems voltage insensitive in the absence of monovalent cations but acquires voltage sensitivity in their presence echoes our results on Na-Ca exchange. On the other hand, DiPolo and Beaugé,[36] also studying squid axons, have reported that Ca-Ca exchange shows the same voltage sensitivity in the presence or absence of the monovalent cation, Li. In the case of cardiac myocytes, the striking difference between the outward Na-Ca exchange current-voltage relationships in Li and in NMG solution suggests that close examination of the mechanism of the activating effect of monovalent cations should provide further insight into the mechanism of the overall exchange process.

ACKNOWLEDGMENTS

We thank Dr. John H. B. Bridge for help with preliminary experiments, Dr. Paul F. Cranefield for constant support and encouragement, and Peter Hoff for technical assistance.

REFERENCES

1. LÜTTGAU, H. C. & R. NIEDERGERKE. 1958. The antagonism between Ca and Na ions on the frog heart. J. Physiol. **143:** 486-505.
2. NIEDERGERKE, R. 1963. Movements of Ca in frog heart ventricles at rest and during contractures. J. Physiol. **167:** 155-550.
3. REUTER, H. & N. SEITZ. 1968. The dependence of calcium efflux from cardiac muscle on temperature and external ion composition. J. Physiol. **195:** 451-470.

4. BAKER, P. F., M. P. BLAUSTEIN, A. L. HODGKIN & R. A. STEINHARDT. 1969. The influence of calcium on sodium efflux in squid axons. J. Physiol. **200:** 431-458.
5. BLAUSTEIN, M. P. & A. L. HODGKIN. 1969. The effect of cyanide on the efflux of calcium from squid axons. J. Physiol. **200:** 497-527.
6. BLAUSTEIN, M. P., J. M. RUSSELL & P. DE WEER. 1974. Calcium efflux from internally dialyzed squid axons: The influence of external and internal cations. J. Supramol. Struct. **2:** 558-581.
7. MULLINS, L. J. & F. J. BRINLEY, JR. 1975. Sensitivity of calcium efflux from squid axons to changes in membrane potential. J. Gen. Physiol. **65:** 135-152.
8. BAKER, P. F. & P. A. MCNAUGHTON. 1976. Kinetics and energetics of calcium efflux from intact squid giant axons. J. Physiol. **259:** 103-144.
9. BERS, D. M., K. D. PHILIPSON & A. Y. NISHIMOTO. 1980. Sodium-calcium exchange and sidedness of isolated cardiac sarcolemmal vesicles. Biochim. Biophys. Acta **601:** 358-371.
10. CARONI, P., L. REINLIB & E. CARAFOLI. 1980. Charge movements during the Na-Ca exchange in heart vesicles. Proc. Natl. Acad. Sci. USA **77:** 6354-6358.
11. REEVES, J. P. & J. L. SUTKO. 1980. Sodium-calcium exchange activity generates a current in cardiac membrane vesicles. Science **208:** 1461-1464.
12. REEVES, J. P. & C. C. HALE. 1984. The stoichiometry of the cardiac sodium-calcium exchange system. J. Biol. Chem. **259:** 7733-7739.
13. KIMURA, J., S. MIYAMAE & A. NOMA 1987. Identification of sodium-calcium exchange current in single ventricular cells of guinea-pig. J. Physiol. **384:** 199-222.
14. EHARA, T., S. MATSUOKA & A. NOMA. 1989. Measurement of reversal potential of Na^+-Ca^+ exchange current in single guinea-pig ventricular cells J. Physiol. **410:** 227-249.
15. BLAUSTEIN, M. P. 1977. Effects of internal and external cations and of ATP on sodium-calcium and calcium-calcium exchange in squid axons. Biophys. J. **20:** 79-111.
16. SLAUGHTER, R. S., J. L. SUTKO & J. P. REEVES. 1983. Equilibrium calcium-calcium exchange in cardiac sarcolemmal vesicles. J. Biol. Chem. **208:** 3183-3190.
17. MIURA, Y. & J. KIMURA. 1989. Dependence on Internal Ca and Na and competitive binding of external Na and Ca. J. Gen. Physiol. **93:** 1129-1145.
18. GADSBY, D. C., J. KIMURA & A. NOMA. 1985. Voltage dependence of Na/K pump current in isolated heart cells. Nature **315:** 63-65.
19. GADSBY, D. & M. NAKAO. 1989. Steady-state current-voltage relationship of the Na/K pump in guinea-pig ventricular myocytes. J. Gen. Physiol. **94:** 511-537.
20. KIMURA, I., A. NOMA & H. IRISAWA. 1986. Na-Ca exchange current in mammalian hearts cells. Nature **319:** 596-597.
21. ISENBERG, G. & U. KLOCKNER. 1982. Calcium tolerant ventricular myocytes prepared by preincubation in a "KB Medium." Pfluegers Arch. **395:** 6-18.
22. MATSUDA, H., A. NOMA, Y. KURACHI & H. IRISAWA. 1982. Transient depolarization and spontaneous voltage fluctuations in isolated single cells from guinea pig ventricles. Circ. Res. **51:** 142-151.
23. SOEJIMA, M. & A. NOMA. 1984. Mode of regulation of the ACH-sensitive K-channel by the muscarinic receptor in rabbit atrial cells. Pflügers Arch. Physiol. **400:** 424-31.
24. SATO, R., A. NOMA, Y. KURACHI & H. IRISAWA. 1985. Effects of intracellular acidification on membrane currents in ventricular cells of the guinea pig. Circ. Res. **57:** 553-561.
25. HAMILL, O. P., A. MARTY, E. NETHER, B. SAKMANN & F. J. SIGWORTH. 1981. Improved patch-clamp techniques for high resolution current recordings from cells and cell-free membrane patches. Pfluegers Arch. **391:** 85-100.
26. KANANSHVILI, D. 1990. Distinction between the two basic mechanisms of cation transport in the cardiac Na^+-Ca^{2+} exchange system. Biochemistry **29:** 2437-2442.
27. LI, J. & J. KIMURA. 1990. Translocation mechanism of Na-Ca exchange in single cardiac cells of guinea pig. J. Physiol. **384:** 199-222.
28. HANSEN, U.-P., D. GRADMANN, D. SANDERS & C. L. SLAYMAN. 1981. Interpretation of current-voltage relationships for "active" ion transport systems: I. Steady-state reaction kinetic analysis of Class-I mechanisms. J. Membr. Biol. **63:** 165-190.
29. JOHNSON, E. A. & J. M. KOOTSEY. 1985. A minimum mechansim for Na^+-Ca^{++} exchange: Net and unidirectional Ca^{++} fluxes as functions of ion composition and membrane potential. J. Membr. Biol. **86:** 167-187.

30. LÄUGER, P. 1987. Voltage dependence of sodium-calcium exchange: Predictions from kinetic models. J. Membr. Biol. **99:** 1-11.
31. HILGEMANN, D. 1988. Numerical approximations of sodium-calcium exchange. Prog. Biophys. Molec. Biol. **51:** 1-45.
32. DIPOLO, R. & H. ROJAS. 1984. Effect of internal and external K^+ on Na^+-Ca^{2+} exchange in dialyzed squid axons under voltage clamp conditions. Biochem. Biophys. Acta **776:** 313-316.
33. SJODIN, R. A. & R. F. ABERCROMBIE. 1978. The influence of external cations and membrane potential on Ca-activated Na efflux in *Myxicola* giant axons. J. Gen. Physiol. **71:** 453-466.
34. COUTINHO, O. P., A. P. CARVALHO & C. A. M. CARVALHO. 1983. Effect of monovalent cations on Na^+/Ca^{2+} exchange and ATP-dependent Ca^{2+} transport in synaptic plasma membranes. J. Neurochem. **41:** 670-676.
35. ALLEN, T. J. A. & P. F. BAKER. 1986. Influence of membrane potential on calcium eflux from giant axons of *Loligo.* J. Physiol. **378:** 77-96.
36. DIPOLO, R. & L. BEAUGÉ. 1990. Asymmetrical properties of the Na-Ca exchanger in voltage-clamped, internally dialyzed squid axons under symmetrical ionic conditions. J. Gen. Physiol. **95:** 819-835.

Effects of Monovalent Cations on Na-Ca Exchange in Nerve Cells[a]

LUIS BEAUGÉ AND REINALDO DIPOLO

Instituto M. y M. Ferreyra
Casilla de Correo 389
5000 Córdoba, Argentina
Centro de Biofísica y Bioquímica
IVIC, Apartado 21827
Caracas 1020A, Venezuela

INTRODUCTION

A facilitated diffusion system in which two substrates compete for the same site in the carrier can give rise to a countertransport mechanism. This was the idea first put forward to account for the Na-Ca exchange across plasma membrane.[1-4] Under this conception the roles of the driven and the driving species are interchangeable, the role being a function of their concentrations and affinities for the carrier. The information obtained over the last decade about the properties of the Na-Ca exchange has led to the conclusion that the system is much more complicated than originally thought: In addition to the basic properties of an exchanger (Na-Ca competition), the function of the carrier is regulated by several ligands such as Mg, monovalent cations, ATP, protons, and inorganic phosphate.[5-7] Furthermore, there are conspicuous asymmetries in the interactions of the carrier with transported and regulatory ligands.[8,9] In this contribution we will deal with the interactions of monovalent cations with the Na-Ca exchanger.

EXTERNAL MONOVALENT CATIONS

In nerve cells, an increase in the external K concentration inhibits the Na_o-dependent Ca efflux. However, this is not due to a chemical effect of K_o for that inhibition can be fully explained by the simultaneous membrane depolarization.[10] Moreover, extracellular Li ions, which do not alter the electrical membrane potential, show no effect on the Na_o-dependent efflux of Ca.[11,12] On the other hand, in those partial reactions in which external Na can be removed (reversal and Ca-Ca exchange) there is a marked stimulation by external monovalent cations. This has been observed in intact, injected, and dialyzed axons. The order of effectiveness follows the sequence Li $>$ K $=$ Rb.[7,12-14] This K_o activation of the Na_i-Ca_o exchange must be chemical because it cannot be abolished when the membrane depolarization is prevented by electrical means.[10,15] Since the forward exchange is not activated by any external monovalent cations other than Na, it is possible that the site(s) we are considering is (are) generated

[a]This work was supported by grants from the National Institutes of Health (ROI HL-39243-02); the National Science Foundation (BNS-8817299); CONICIT, Venezuela (SI-1924); and CONICET, Argentina (30000800/88).

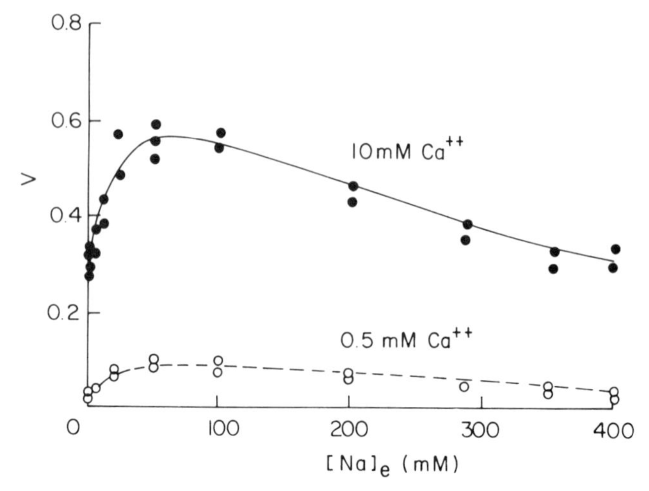

FIGURE 1. Steady-state levels of calcium influx at 0.5 mM and 10 mM external Ca as a function of the external Na concentration in squid axons dialyzed with ATP-free solutions. The experimental points correspond to data obtained from several axons expressed as a fraction of the maximal influx. The lines through the points are the simultaneous best fit to the equations resulting from the scheme in FIGURE 2. Note: (i) the intracellular ionized Ca concentration was 200 μM in all cases; (ii) external NaCl was replaced with equiosmolar amounts of Tris-HCl.

when the external side of the carrier binds Ca (see also Reeves[16]). On the other hand, Na exclusion from these sites may not be absolute because, although high external Na inhibits Ca-Ca exchange, low Na_o promotes Ca entry. These data could be accounted for by considering two actions of Na_o: At low concentrations Na_o binds to the activator monovalent site; at high concentrations it displaces Ca from the external transporting site.[1,17] In other words, if Na ions also have an activating role, any stimulation that could be brought about by other external monovalent cations would be masked in Na_o-containing solutions.

We have performed experiments to analyze the Na and Ca interactions on the external side of the carrier. In doing so, we measured the influx of Ca at 0.5 mM and 10 mM Ca_o as a function of the external Na concentration in axons dialyzed without ATP. As shown in FIGURE 1 the curves are biphasic with a peak stimulation at about 50-70 mM Na followed by a decline in the influx of Ca at higher Na concentrations. The data points of several experiments were fitted to a model shown in FIGURE 2. The

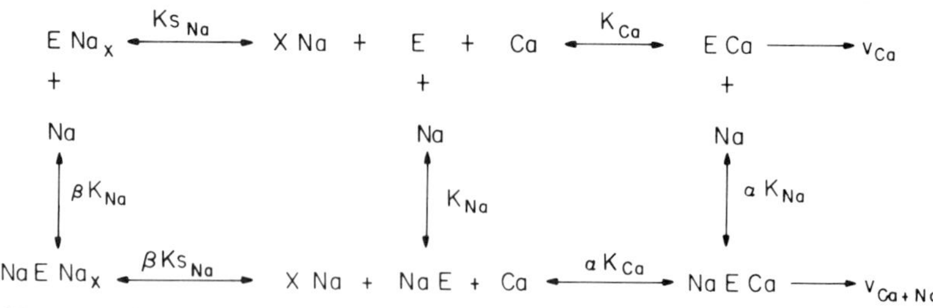

FIGURE 2. Kinetic scheme for the Na and Ca interactions with the external transport and monovalent cation stimulatory site. E represents the Na-Ca exchange carrier. All other expressions are defined in the text.

basic features of the model are: (i) Na and Ca ions compete with each other for the transporting site; (ii) there is an additional site for Na (activating monovalent cation site) where this cation exerts an activating role; (iii) the total influx of Ca is the sum of two components, one with the carrier binding just Ca_o and the other with the carrier binding Ca_o plus Na_o at the activating site; (iv) the equation was obtained assuming rapid random equilibrium binding; (v) the possibility of cross reactivity between transport and stimulating sites was allowed. The resulting equations are given in FIGURE 3.

$$V_{(Ca)} = \frac{1}{1 + \frac{K_{Ca}}{[Ca]} + \frac{K_{Ca}[Na]}{Ks_{Na}[Ca]} + \frac{K_{Ca}[Na]^{x}}{Ks_{Na}\, K_{Na}[Ca]} + \frac{[Na]}{\alpha\, Ks_{Na}} + \frac{K_{Ca}[Na]^{x+1}}{\beta\, Ks_{Na}\, K_{Na}[Ca]}}$$

$$V_{(Ca+Na)} = \frac{[Na]}{[Na] + \frac{\alpha K_{Ca}[Na]}{[Ca]} + \alpha\, Ks_{Na} + \frac{\alpha K_{Ca} Ks_{Na}}{[Ca]} + \frac{\alpha K_{Ca} Ks_{Na}[Na]^{x}}{K_{Na}[Ca]} + \frac{\alpha K_{Ca}[Na]^{x+1}}{\beta K_{Na}[Ca]}}$$

$$V = V_{(Ca)} + V_{(Ca+Na)}$$

FIGURE 3. Rapid random equilibrium solution of the kinetic scheme given in FIGURE 2. V_{Ca} = Ca influx via the carrier with the activating Na site empty, taken as a fraction of the maximal influx. V_{Ca+Na} = Ca influx via the carrier with Na bound to the activating monovalent cation site, taken as a fraction of the maximal influx. K_{Ca} = affinity constant of the external transport site for Ca ions. K_{Na} = coefficient related to the affinity of the external transport site(s) for Na ions. Ks_{Na} = affinity constant of the external monovalent cation activating site for Na ions. x = number of Na ions bound to the external transport site(s). α = cross-reactivity coefficient for the interaction of Na ions with the activating sites and Ca ions with the transport sites. β = cross-reactivity coefficient for the interaction of Na ions with the activating and transport sites. [Na] = extracellular Na ion concentration. [Ca] = extracellular Ca ion concentration.

When this scheme was left free to choose the best fitting parameters, it produced the following results:

(a) The number of Na-transporting sites was close to two.

(b) Only one Na was bound to the stimulatory site with a Ks_{Na} of about 45 mM.

(c) The binding of Ca to its transporting site increased the affinity of the regulatory site for Na about fivefold. Likewise, the binding of Na to the monovalent activating site increased the affinity of the transport site for Ca by the same magnitude; in addition it produced a twofold increase in the maximal rate of translocation (V_{max}).

(d) The binding of Na to the external transporting site increased the affinity of the activating site for this cation by twofold.

In summary, in squid axons it is possible to account for all external Na and Ca interactions on the Na-Ca exchanger on the basis of Na-Ca competition at the transporting site together with a positive cross reactivity between the monovalent activating and the transporting site. In addition, the occupancy of the monovalent activating locus by Na increases the maximal rate of exchange.

Other monovalent cations (K, Li, Rb, NH_4) do not displace Ca from the external transport site. Since Li has the additional advantage of not affecting the membrane potential, we have chosen this cation to study other characteristics of the external monovalent cation site. FIGURE 4 shows that Li activates Ca-Ca exchange following a monotonic function with $K_{1/2}$ of about 150 mM. Although ATP activates the maximal rate of Ca-Ca exchange,[7,8] it is clear from FIGURE 4 that the apparent affinity for Li remains unchanged. This contrasts with the ATP-induced increase in the apparent affinity of the external transport site for Na.[14,18,19]

Na-Ca exchange is electrogenic. In squid axons, forward, reverse and Ca-Ca exchange are voltage sensitive. The presence of an additional external monovalent cation-binding site could be involved in the voltage dependence of the exchange reactions (see Allen & Baker[10]). Therefore, we have explored whether or not that voltage sensitivity was dependent on the presence of an external monovalent cation. FIGURE 5 shows that this is not the case, since similar sensitivity to membrane potential was observed in K-free Tris and K-free lithium seawaters.

INTERNAL MONOVALENT CATIONS

In squid nerve bathed in normal artificial seawater (440 mM Na, 10 mM K) intracellular potassium, at constant membrane potential, is a powerful activator of the forward Na-Ca exchange.[14,20] The $K_{1/2}$ for potassium is about 90 mM and is independent of Na_i, ruling out any K_i-Na_i competition. If an internal monovalent activating site similar to that found on the external membrane side is responsible for this potassium effect, we should be able to reproduce at least some of the results described above for the external site. The experimental protocol was designed in order to have the monovalent and the Ca transport loci on the same side. FIGURE 6 shows that 200 mM external Li is a powerful stimulator of Ca-Ca exchange even in the absence of any intracellular monovalent cations. However, the same experiment demonstrates that internal Li fails to induce any stimulation of Ca-Ca exchange unless there is a simultaneous presence

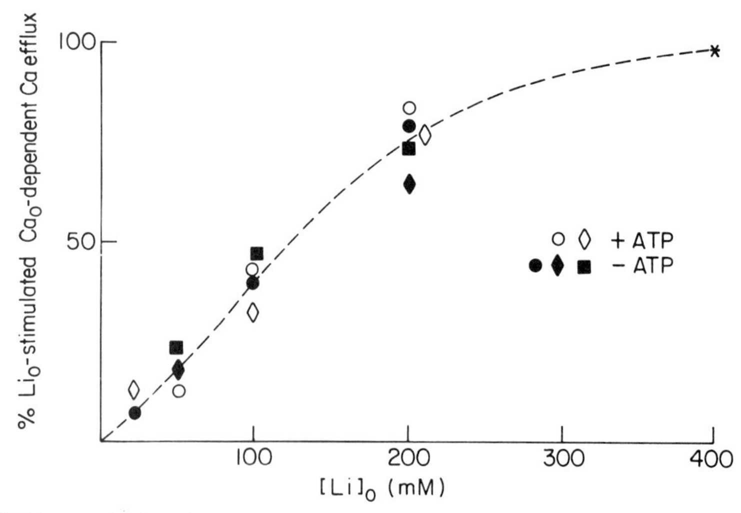

FIGURE 4. Activation of Ca_o-dependent Ca efflux by external Li ions in squid axons dialyzed with and without ATP. Different symbols represent different axons. Notice that ATP does not change the apparent affinity for the monovalent cation-activating site. (Reproduced from DiPolo & Beaugé[8] with permission from the Rockefeller University Press.)

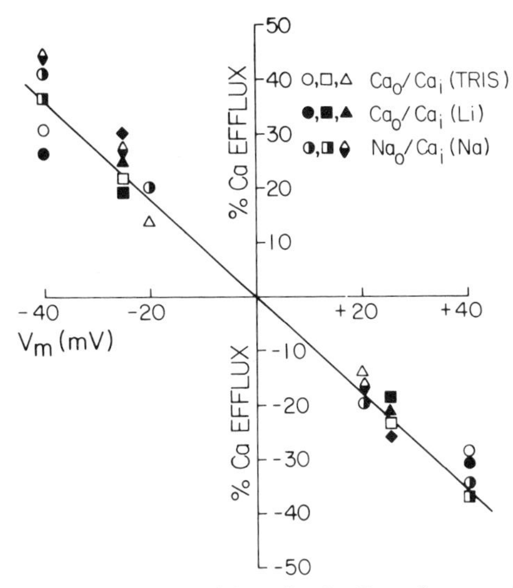

FIGURE 5. Influence of membrane potential on the Ca-Ca exchange and forward Na-Ca exchange modes of the Na-Ca exchange system in dialyzed squid axons. Note: (i) Ca-Ca exchange was explored under complete symmetric ionic conditions; (ii) the sensitivity to membrane potential of the Ca-Ca exchange does not depend on the presence of an external monovalent cation (compare Tris with Li); (iii) similar voltage sensitivity was found for the forward and Ca-Ca exchange modes. (Reproduced from DiPolo & Beaugé[8] with permission from the Rockefeller University Press.)

of Li in the external medium. This may reflect a real differential response to internal K and Li ions; alternatively, the K_i stimulation reported before may have occurred because the external monovalent cation site was occupied by the K or Na present in the seawater. The latter explanation implies that, for internal K or Li to stimulate the exchanger, the external monovalent cation site must be loaded. Regardless of the mechanism involved, this observation constitutes one of the most striking asymmetries of this system.

If there is an intracellular cation monovalent site that accepts Na in a fashion similar to the external one, a biphasic response to Na_i should be expected. To test for this possibility, we elected conditions that would have the monovalent external site largely occupied; that is, forward exchange into Na seawater and Ca-Ca exchange into Li seawater. FIGURE 7 shows that intracellular Na monotonically inhibits both modes at 0.7 and 200 μM Ca_i. This constitutes a drastic difference from the results observed with external Na. Possible explanations are: (i) There is no intracellular monovalent cation site; (ii) the site exists but it does not accept Na ions or it binds Na with very low affinity.

The interaction of Na and Ca with the intracellular transport site can be adequately explained by the simple competitive kinetic model shown in FIGURE 8. The fitting of the data point to that equation was carried out given a K_{Ca} of 25 μM.[6] The results of several experiments indicates that 3 Na ions bind for each Ca with an apparent affinity of 23 mM.

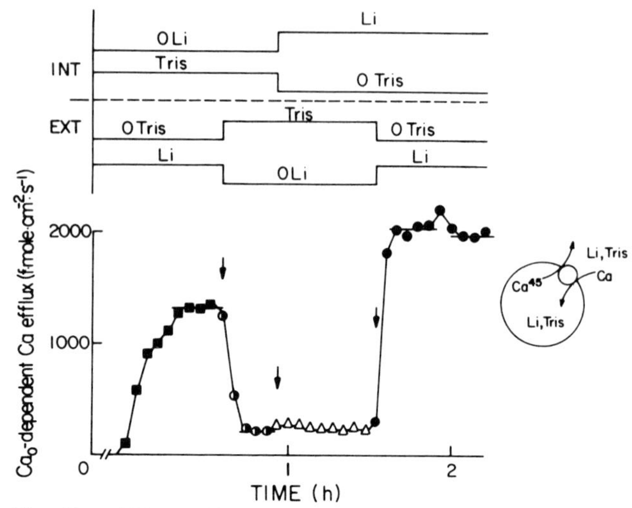

FIGURE 6. The effect of Li_o and Li_i on the Ca_o-dependent Ca efflux in dialyzed squid axons. The correspondence between symbols and ionic composition is the following: filled squares, $Tris_i$-Li_o; half-filled circles, $Tris_i$-$Tris_o$; empty triangles, Li_i-$Tris_o$; filled circles, Li_i-Li_o in the presence of 20 mM Ca_o; empty circles, Li_i-Li_o in the absence of extracellular Ca ions. Note: (i) All concentrations are expressed in mM; (ii) the arrows indicate changes in the internal or external medium. (Partial reproduction from DiPolo & Beaugé[8] with permission from the Rockefeller University Press.)

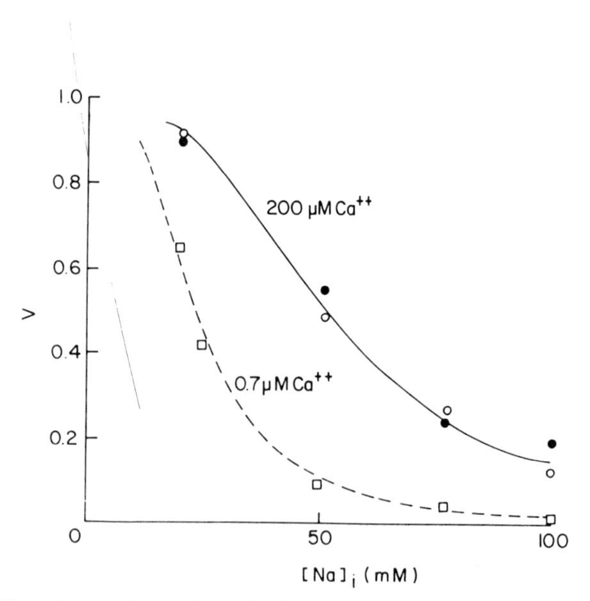

FIGURE 7. Effect of intracellular Na on the forward Na-Ca (filled circles) and Ca-Ca (empty circles) exchanges in the presence of 200 μM ionized Ca_i and Ca-Ca exchange (solid squares) in the presence of 0.7 μM ionized Ca_i. The data points correspond to several axons dialyzed with ATP-free solutions and were simultaneously fitted to the equation resulting from the kinetic scheme of FIGURE 8.

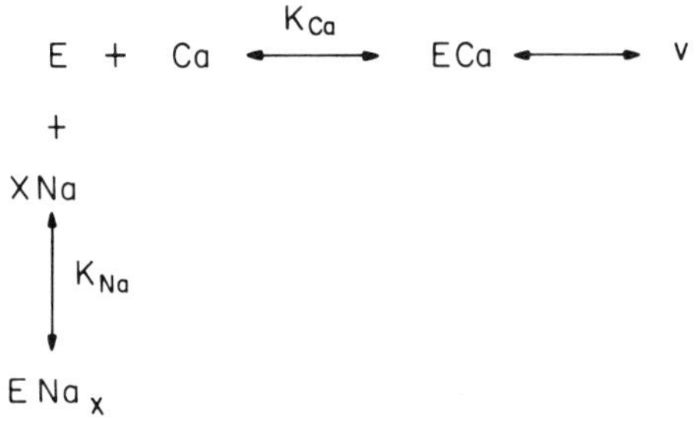

$$E + Ca \overset{K_{Ca}}{\longleftrightarrow} ECa \longleftrightarrow v$$

$$+$$

$$XNa$$

$$\updownarrow K_{Na}$$

$$ENa_x$$

$$v = \frac{v_i}{v_o} = \frac{1 + \frac{K_{Ca}}{[Ca]}}{1 + \frac{K_{Ca}}{[Ca]} \left(1 + \frac{[Na]^x}{K_{Na}}\right)}$$

FIGURE 8. Kinetic scheme and random rapid equilibrium solution for competitive Na and Ca interactions with the intracellular transport site of the Na-Ca exchange carrier. v_i = rate of transport in the presence of Na_i. v_o = rate of transport in the absence of Na_i. K_{Ca} = affinity constant of the internal transporting site for Ca ions. K_{Na} = coefficient related to the affinity of the internal transporting site(s) for Na. x = Hill coefficient for Na ions. [Na] = intracellular Na concentration. [Ca] = intracellular Ca concentration.

IS THE ACTIVATING MONOVALENT CATION COTRANSPORTED?

Recent evidence gathered from rod outer segments indicates that K ions are indeed cotransported with a stoichiometry ratio of 4 Na^+ : 1 Ca^{2+} + 1 K^+.[21,22] Therefore, a decisive point to settle is whether this is also the case in nerve cells. In order to explore this possibility, we measured in dialyzed squid axons the influx of [^{86}Rb]K during full activation of the reverse Na-Ca exchange by internal Ca in the presence of 50 mM K_o. An introductory experiment (FIG. 9A) shows that activation of the reverse Na-Ca by 20 mM external K amounts to about 4 pmole · cm^{-2} · sec^{-1}. FIGURE 9B illustrates another experiment under similar conditions in which [^{86}Rb]K influx at 50 mM K_o was measured in the absence and presence of a saturating Ca_i concentration. This figure demonstrates that addition to 100 μM Ca_i has no effect on the steady-state level of [^{86}Rb]K influx obtained in the complete absence of internal Ca. If this reversed exchange were linked to a cotransport of potassium with a 1 K^+ : 1 Ca^{2+} ratio, an increase in 4 pmole·cm^{-2}·sec^{-1} would be expected following addition of Ca_i. Within the resolution of the method, an increase of that magnitude should have been detected. Other experiments[23] in which we measured the efflux of [^{86}Rb]K in dialyzed axons or the [^{86}Rb]K uptake in isolated squid optic nerve vesicles also failed in showing any increase in K transport coexisting with the maximal activation of the Na-Ca exchange. Consequently we must conclude that in squid nerve there is no evidence that the activating monovalent cation is cotransported with Ca.

ACKNOWLEDGMENTS

We wish to thank Dhuwya Otero for the drawings and Lilian Palacios for secretarial help.

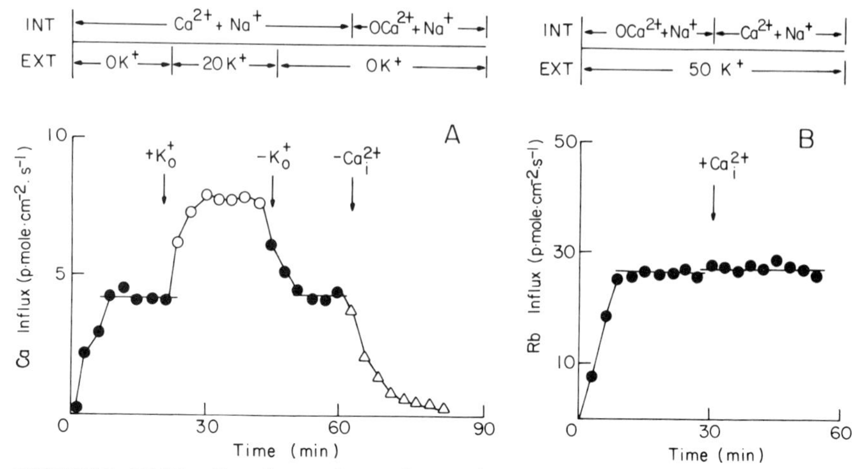

FIGURE 9. (**A**) The effect of external potassium on the Na_i-dependent Ca influx (reverse Na-Ca exchange) at constant zero millivolts membrane potential. (**B**) The effect of internal Ca^{2+} on the [^{86}Rb]K influx in the presence of 50 mM external K. (Taken from Condrescu *et al.*[23] with permission from Elsevier Press.)

REFERENCES

1. BAKER, P. F., M. P. BLAUSTEIN, A. L. HODGKIN & R. A. STEINHARDT. 1969. The influence of calcium on sodium efflux in squid axons. J. Physiol. **200:** 431-458.
2. MULLINS, L. J. 1977. A mechanism of Na/Ca transport. J. Gen. Physiol. **70:** 681-696.
3. WONG, A. Y. K. & J. B. BASSINTHWAIGHTE. 1981. The kinetics of Na-Ca exchange in excitable tissue. Math. Biosci. **53:** 275-310.
4. REEVES, J. P. & C. C. HALES. 1984. The stoichiometry of the cardiac sodium-calcium exchange system. J. Biol. Chem. **259:** 7733-7739.
5. DIPOLO, R. & L. BEAUGÉ. 1984. Interaction of physiological ligands with the Ca pump and the Na/Ca exchange in squid axons. J. Gen. Physiol. **84:** 895-914.
6. DIPOLO, R. & L. BEAUGÉ. 1986. In squid axons reverse Na/Ca exchange requires internal Ca and/or ATP. Biochim. Biophys. Acta **854:** 298-306.
7. BAKER, P. F. & T. J. A. ALLEN. 1986. Sodium-Calcium (Na-Ca) Exchange in Intracellular Calcium Regulation. H. Bader, K. Gietzen, J. Rosenthal, R. Rudel & N. U. Wolf, Eds.: 35-47. Manchester Univ. Press.
8. DIPOLO, R. & L. BEAUGÉ. 1990. Asymmetrical properties of the Na-Ca exchanger in voltage-clamped, internally dialyzed squid axons under symmetrical ionic conditions. J. Gen. Physiol. **95:** 819-885.
9. DIPOLO, R. & L. BEAUGÉ. 1990. Calcium transport in excitable cells. *In* Intracellular Calcium Regulation. F. Bronner, Ed.: 381-413. Alan R. Liss. New York.
10. ALLEN, T. J. A. & P. F. BAKER. 1986. Comparison on the effects of potassium and membrane potential on the calcium-dependent sodium efflux in squid axons. J. Physiol. **378:** 53-76.
11. BAKER, P. F. 1978. The regulation of intracellular calcium in squid giant axons of *Loligo* and *Myxicola.* Ann. N.Y. Acad. Sci. **307:** 250-268.
12. BAKER, P. F. & P. A. MCNAUGHTON. 1976. Kinetics and energetics of calcium efflux from intact squid axons. J. Physiol. **259:** 103-144.
13. ALLEN, T. J. A. & P. F. BAKER. 1986. Influence of membrane potential on calcium efflux from giant axons of *Loligo.* J. Physiol. **378:** 77-96.
14. BLAUSTEIN, M. P. 1977. Effects of internal and external cations and of ATP on sodium-calcium and calcium-calcium exchange in squid axons. Biophys. J. **20:** 79-110.

15. BAKER, P. F. & R. DIPOLO. 1984. Axonal calcium and magnesium homeostasis. Curr. Topics Membr. Transp. **22:** 195-242.
16. REEVES, J. P. 1985. The sarcolemma sodium-calcium exchange system. Curr. Topics Membr. Transport **25:** 77-119.
17. BAKER, P. F. 1970. Sodium-calcium exchange across the nerve cell membrane. *In* Calcium and Cellular Function. A. W. Cuthbert, Ed.: 96-107. MacMillan. New York.
18. DIPOLO, R. 1974. The effect of ATP on Ca efflux in dialyzed squid giant axons. J. Gen. Physiol. **64:** 503-517.
19. BAKER, P. F. & H. G. GLITSCH. 1973. Does metabolic energy participate directly in the Na-dependent extrusion of Ca from squid axons? J. Physiol. **233:** 44-46P.
20. DIPOLO, R. & H. ROJAS. 1984. Effect of internal and external K^+ on Na^+-Ca^{2+} exchange in dialyzed squid axons under voltage-clamp conditions. Biochim. Biophys. Acta **776:** 313-316.
21. CERVETTO, L., L. LAGNADO, R. J. PERRY & P. A. MCNAUGHTON. 1989. Extrusion of calcium from rod outer segments is driven by both sodium and potassium gradients. Nature **337:** 740-743.
22. SCHNETKAMP, P. P. M., D. K. BASU & R. T. SZERENCSEI. 1989. Na^+-Ca^{2+} exchange in bovine rod outer segments requires and transports K^+. Am. J. Physiol. **257:** C153-C157.
23. CONDRESCU, M., H. ROJAS, A. GERARDI, R. DIPOLO & L. BEAUGE. 1990. In squid nerve fibers monovalent activating cations are not cotransported during Na-Ca exchange. Biochim. Biophys. Acta **1024:** 198-202.

Modulation of Sodium-Calcium Exchange by Lipids[a]

SISTO LUCIANI,[b] SERGIO BOVA,
GABRIELLA CARGNELLI, FEDERICO CUSINATO,
AND PATRIZIA DEBETTO

Department of Pharmacology
University of Padova School of Medicine
Padova, Italy

INTRODUCTION

Sodium-calcium exchange catalyzes electrogenic counter transport of Na^+ for Ca^{2+} across the plasma membrane of many excitable cells and some epithelia. It can operate in both directions depending upon Na^+ and Ca^{2+} gradients and membrane potential. Na^+-Ca^{2+} exchange plays a critical role in regulating intracellular Ca^{2+} concentration and therefore contractility in heart cells. Its activity varies during the course of the cardiac cycle: It supports Ca^{2+} uptake during the early part of the cardiac cycle and contributes, together with SR, to heart relaxation by extruding Ca^{2+} from the cell.[1] It has been suggested that Na^+-Ca^{2+} exchange is involved in many aspects of cardiac function[2] including beat-to-beat regulation of cardiac contraction, force-frequency relationship, β-adrenergic relaxation, the calcium paradox and reperfusion injury, the positive inotropic effect of cardiac glycosides,[2] and the positive inotropic and antihypertensive effects of amiloride.[3–6]

In view of its physiological relevance, the regulation of Na^+-Ca^{2+} exchange activity has been the subject of a considerable number of studies. Among the factors affecting Na^+-Ca^{2+} exchange, one of the most thoroughly investigated has been the membrane lipid environment. The effects of the lipid environment have been studied both in native sarcolemmal vesicles and in proteoliposomes reconstituted in a defined lipid composition. Stimulation of Na^+-Ca^{2+} exchange by anionic amphiphiles has been observed by using phospholipases.[7] Phospholipase treatment induces a relative enrichment of negatively charged lipids that stimulate the Na^+-Ca^{2+} exchanger,[7] presumably by increasing its affinity for Ca^{2+}. Somewhat more direct evidence of lipid modulation of the Na^+-Ca^{2+} exchanger has been obtained from experiments with proteoliposomes reconstituted in lipid mixtures of defined composition.[8–11] Reconstitution of the Na^+-Ca^{2+} exchanger, first achieved with beef heart sarcolemma,[12] has been obtained from various sources using different experimental protocols, as recently reviewed.[13] The effects of phosphatidylserine (PS), cholesterol, and polyphosphoinositides will be reported below together with a note on a possible role of Ca-phospholipid binding proteins on the Na^+-Ca^{2+} exchanger.

[a] Work in the authors' laboratory was supported by the Ministry of University and Research, (60% and 40%) and CNR Italia-USA grants.

[b] Address for correspondence: Department of Pharmacology, University of Padova, Largo E. Meneghetti, 2, 35131, Padova, Italy.

TABLE 1. Na^{+}-Ca^{2+} Exchange Activity of Beef Heart Sarcolemmal Vesicles and Reconstituted Proteoliposomes

	Na^{+}-Ca^{2+} Exchange (nmoles Ca^{2+} uptake/mg protein)	
	$+Na^{+}$ Gradient	$-Na^{+}$ Gradient
Sarcolemmal vesicles	30	2
Proteoliposomes[a]	158	12

[a] Proteoliposomes were reconstituted with the following lipid mixture: PC : PS (90 : 10).

PHOSPHATIDYLSERINE

The effect of phospholipids on Na^{+}-Ca^{2+} exchange activity was first studied in proteoliposomes reconstituted from dog heart sarcolemma by a cholate-dyalisis procedure. It was shown[8] that reconstituted proteoliposomes consisting of phosphatidylcholine (PC) as the sole lipid had no detectable Na^{+}-Ca^{2+} exchange activity, whereas proteoliposomes containing a mixture of PC and PS (90 : 10, weight : weight) elicited significant exchange activity. The same result was obtained in cholate-extracted beef heart sarcolemmal vesicles reconstituted in the same phospholipid mixture (TABLE 1). The stimulating effect of acidic phospholipids on the Na^{+}-Ca^{2+} exchanger was also observed in proteoliposomes reconstituted in the presence of mixtures of PC and PS or PC and phosphatidylinositol[9] (PI) and in proteoliposomes containing PS or phosphatidic acid (PA) or cardiolipin but not PI or phosphatidylglycerol.[10] These findings are in good agreement with the results of Barzilai *et al.*,[14] which shows that PC was unable to reconstitute Na^{+}-Ca^{2+} exchange activity in synaptic plasma membranes, whereas brain phospholipids, which have a high PS content, are optimal for reconstitution of the synaptic membrane's Na^{+}-Ca^{2+} exchanger. Stimulation of Na^{+}-Ca^{2+} exchange activity by negatively charged fatty acids has also been shown.[15]

These results indicate that the lipid composition of the mixture is important to achieve a proper reconstitution; the presence of negatively charged phospholipids in the membrane environment of the Na^{+}-Ca^{2+} exchanger may induce an expanded bilayer more suitable to accommodating protein conformational changes and thus Na^{+}-Ca^{2+} exchange activity. The importance of negatively charged phospholipids for the reconstitution of membrane-bound enzymes has been observed in several cases.[16]

CHOLESTEROL

Cholesterol is one of the main lipid components in most eukaryotic plasma membranes. In cardiac sarcolemma cholesterol represents 35% of the bilayer lipid. Cholesterol is known to modify the physical properties of the bilayer by inducing a reorganization of phospholipid packing. It can affect the catalytic properties of several membrane transport systems by modification of membrane fluidity and thickness, but it may also directly interact with membrane-bound proteins.[17] It has been shown that in reconstituted proteoliposomes, in addition to anionic phospholipids, cholesterol is required for high rates of Na^{+}-Ca^{2+} exchange.[10] An increase of Na^{+}-Ca^{2+} exchange activity has also been observed in cardiac sarcolemmal vesicles enriched in cholesterol.[18] It was found that the sterol requirement was specific for cholesterol: Cholesterol

TABLE 2. Time-Course of Na^+-Ca^{2+} Exchange Activity of Cardiac Sarcolemmal Vesicles and Reconstituted Proteoliposomes

		Na^+-Ca^{2+} Exchange (nmole Ca^{2+} uptake/mg protein) Proteoliposomes[a]		
Time (sec)	Vesicles	PC : PS (52 : 48)	PC : PS : Chol[b] (25 : 39 : 36)	PC : PS : DSPC (31 : 48 : 21)
2	2.9 ± 0.2	10.8 ± 1.1	16.2 ± 3.4	3.1 ± 0.2
5	7.3 ± 0.2	27.9 ± 2.0	41.7 ± 3.2	7.9 ± 0.8
10	14.3 ± 1.2	56.5 ± 1.4	82.8 ± 5.7	15.8 ± 0.6
15	21.4 ± 1.3	82.6 ± 4.0	122.8 ± 8.2	23.4 ± 1.0
30	35.8 ± 2.5	92.2 ± 6.4	193.5 ± 9.4	36.7 ± 2.0

NOTE: Reproduced from Debetto *et al.*[11] with permission from the *Archives of Biochemistry and Biophysics.*

[a]Reconstituted from cardiac sarcolemmal vesicles with indicated lipid composition expressed as molar ratio. Data are means ± SE of at least four different experiments.

[b]Chol., cholesterol.

analogues in the medium for reconstitution of the Na^+-Ca^{2+} exchanger were unable to sustain an elevated Na^+-Ca^{2+} exchange activity.[19] Likewise, the oxidation of membrane cholesterol *in situ* completely eliminated Na^+-Ca^{2+} exchange activity in cardiac sarcolemmal vesicles.[18] In either case the modification of Na^+-Ca^{2+} exchange activity was not due to changes in Ca^{2+} permeability. However, it remained to be ascertained whether the effect of cholesterol could be ascribed to a physical modification of the lipid environment surrounding the exchanger or to direct interaction with the exchanger protein.

In order to clarify this point, the influence of temperature on the interaction of the Na^+-Ca^{2+} exchanger with the membrane lipid environment was studied in proteoliposomes reconstituted in different lipid mixtures.[11] The time course of Na^+-Ca^{2+} exchange activity reconstituted in three different lipid mixtures containing PC : PS (52 : 48, mol : mol), PC : PS : cholesterol (25 : 39 : 36, mol : mol), and PC : PS : distearoylphosphatidylcholine (DSPC) (31 : 48 : 21, mol : mol) is shown in TABLE 2. Proteoliposomes reconstituted with acidic phospholipids such as PS exhibited substantially higher exchange activity than the native vesicles, in agreement with previous observations.[8–10] Na^+-Ca^{2+} exchange activity was further stimulated when cholesterol was included in the lipid mixture.[11] However, when DSPC was used in place of cholesterol, the reconstituted proteoliposomes showed a very low exchange activity. The Arrhenius plots in FIGURE 1 illustrate the temperature dependence of Na^+-Ca^{2+} exchange in proteoliposomes reconstituted into the three different lipid mixtures. A discontinuity in the Arrhenius plots occurred at 23°C (PC : PS, FIG. 1A), 33°C (PC : PS : cholesterol, FIG. 1B), and 23°C (PC : PS : DSPC, FIG. 1C). The slopes of the curves became steeper for temperatures below the break points, indicating a substantial increase in the apparent energies of activation at low temperature. Both the increase in the thermotropic transition temperature in PC : PS : cholesterol proteoliposomes and the stimulation of the exchanger activity by cholesterol may result from the known condensing effect of this compound. However, the cholesterol stimulation could also be ascribed to a specific direct interaction of this compound with the exchanger protein. These results support the latter hypothesis since DSPC in PC : PS : DSPC proteoliposomes, although conferring a higher rigidity to the membrane than PC in PC : PS

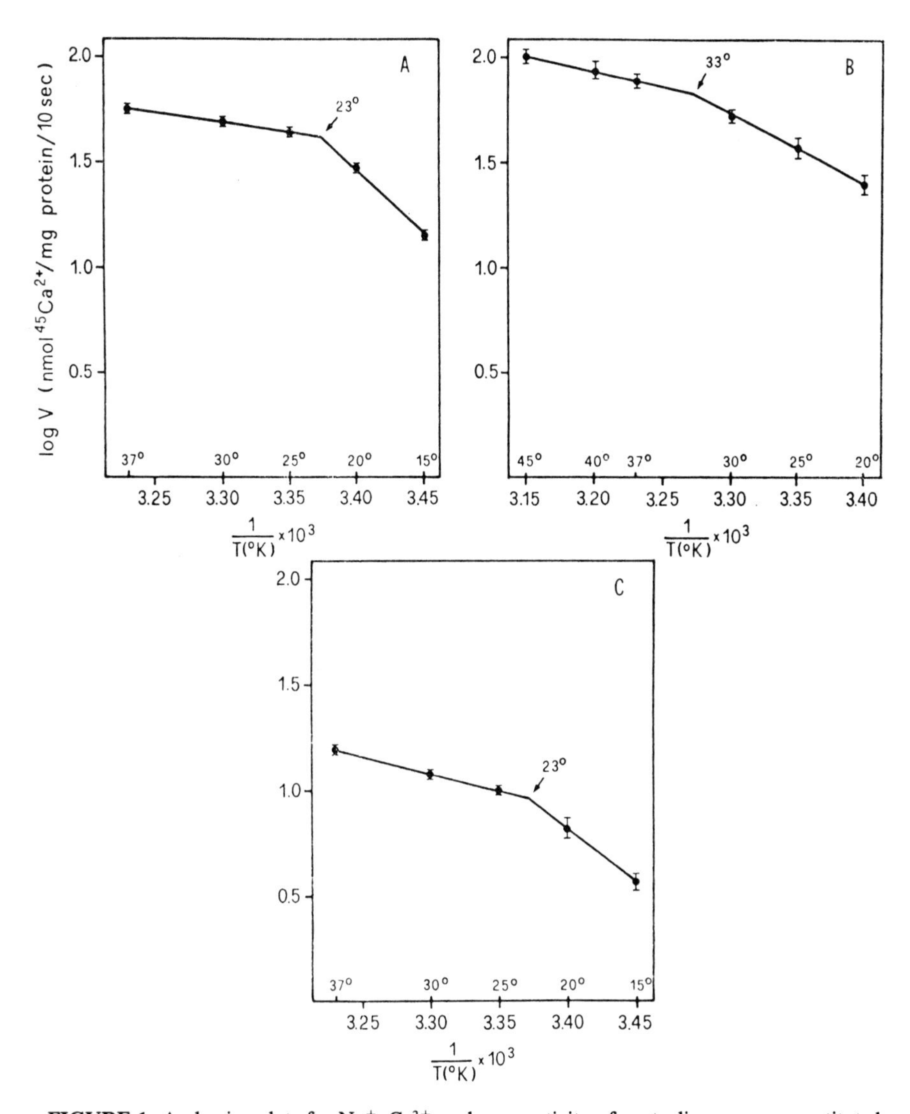

FIGURE 1. Arrhenius plots for Na^+-Ca^{2+} exchange activity of proteoliposomes reconstituted into different lipid mixtures. Proteoliposomes were reconstituted from cardiac sarcolemma with the following lipid mixtures: (**A**) PC : PS (52 : 48, mol : mol); (**B**) PC : PS : cholesterol (25 : 39 : 36, mol : mol); (**C**) PC : PS : DSPC (31 : 48 : 21, mol : mol). Data are means ± SE of four different experiments. The straight lines were fitted by the least-squares method. The arrows indicate the break-point temperatures in ° C. (Reproduced from Debetto *et al.*[11] with permission from the *Archives of Biochemistry and Biophysics.*)

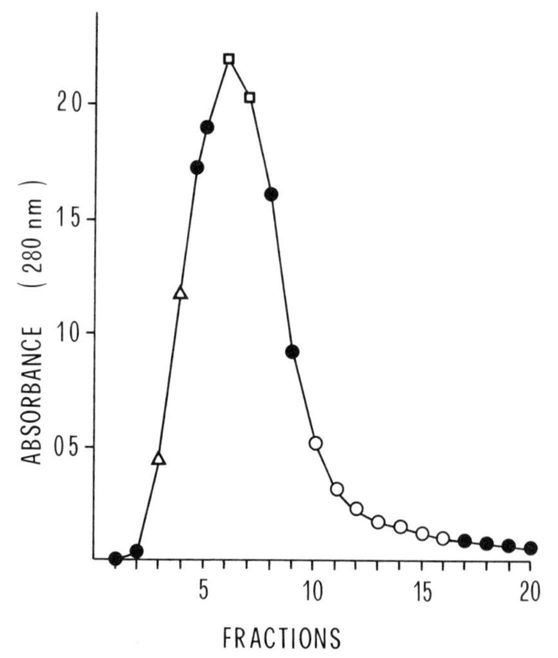

FIGURE 2. Gel filtration (Sepharose-4B) of proteoliposomes reconstituted with the following lipid mixture: PC : PS : cholesterol (25 : 39 : 36, mol : mol). Front (△), peak (□) and tail (○) fractions.

proteoliposomes, gives rise to a lower reconstituted exchange activity. In order to further investigate whether the thermotropic transition was due to some inherent property of the Na^+-Ca^{2+} exchanger protein or to an alteration in the ordering of lipids, a study[20] to assess the lipid fluidity by steady-state fluorescence polarization of reconstituted PC : PS : cholesterol proteoliposomes was undertaken by using diphenylhexatriene (DPH) as fluorescent probe.[21] The temperature dependence of anisotropy of PC : PS : cholesterol proteoliposomes showed a considerable variability in different preparations. To clarify this observation, a fractionation of proteoliposomes was obtained by gel filtration on a Sepharose-4B column. It is shown in FIGURE 2 that three main populations of proteoliposomes can be identified by absorbance at 280 nm: a front, a peak, and a tail. A detailed analysis of the three populations showed differences in fluorescence anisotropy as well as in Na^+-Ca^{2+} exchange activity. Na^+-Ca^{2+} exchange activity decreased, moving from the initial fractions toward the end, and this effect was parallel to an increase of the lipid-protein ratio. Furthermore, it was observed that the lipid composition of the peak fraction was significantly different from the original mixture used for reconstitution: The percentage of cholesterol was 18% by molar ratio (36% in the original mixture), whereas the ratio between PC and PS remained constant. Therefore, proteoliposomes with the same composition as the peak population were prepared and after fractionation used for anisotropy measurements as a function of the temperature. FIGURE 3 shows the temperature dependence of fluorescence anisotropy in PC : PS : cholesterol (35 : 47 : 18, mol : mol) liposomes and PC : PS : cholesterol (35 : 47 : 18, mol : mol) proteoliposomes. A linear decrease in anisotropy with an increase in the temperature but without phase transition was observed for both liposomes and proteoliposomes. No significant difference was observed between liposomes and proteoliposomes either in the absolute values of anisotropy or in the slope of the curves. These results indicate that the break in the Arrhenius plots of the Na^{+}-Ca^{2+} exchange activity in PC : PS : cholesterol proteoliposomes cannot be ascribed to a phase transition of this lipid mixture. A coincidence between the break in the Arrhenius plot and phase transition has been observed only in a limited number of membrane-bound enzymes.[22]

It can be therefore concluded that the fluidity of the membrane bilayer is not playing a major role in the Na^{+}-Ca^{2+} exchange activity in this temperature range and with this lipid mixture. The discontinuity in the Arrhenius plots can be ascribed to a different physicochemical modification of the bilayer induced by cholesterol or to a direct effect of this compound on the Na^{+}-Ca^{2+} exchange protein.

POLYPHOSPHOINOSITIDES

Polyphosphoinositides represent a minor fraction (< 1%) of total cellular lipids. However, their distribution among cellular membranes, their main localization at the

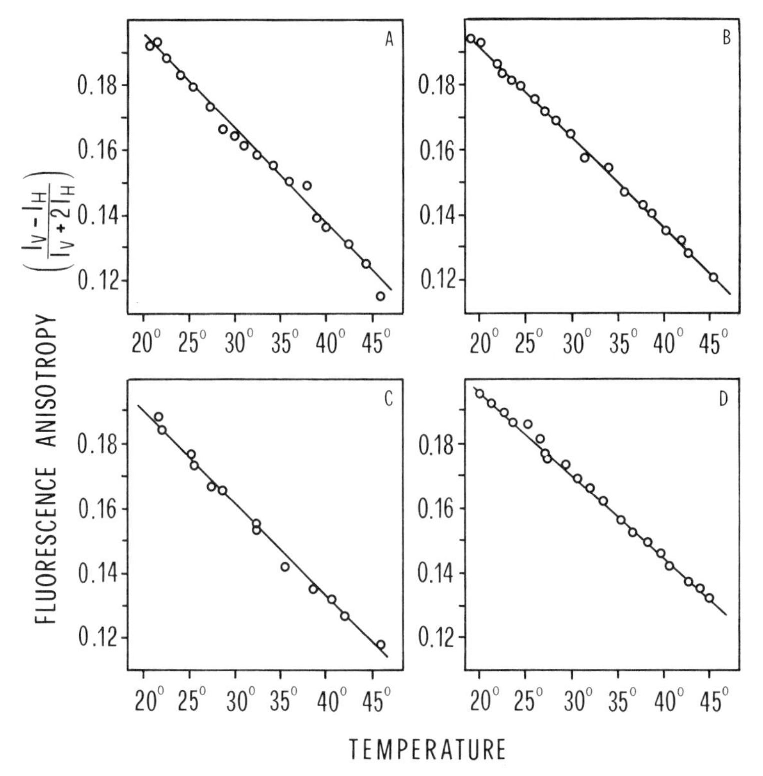

FIGURE 3. Steady-state fluorescence anisotropy in PC : PS : cholesterol (35 : 47 : 18, mol : mol) liposomes: 3A, bulk preparation; 3B, peak fraction and in PC : PS : cholesterol (35 : 47 : 18, mol : mol) proteoliposomes: 3C, bulk preparation; 3D, peak fraction. Fractionation as in FIGURE 2. Steady-state fluorescence anisotropy (A_s) is expressed as $A_s = (I_v - I_h)/(I_v + 2I_h)$ where I_v and I_h are the intensities of the emitted light passing polarizers oriented vertically (I_v) or horizontally (I_h) with respect to the vertical polarization of the exciting light.

TABLE 3. Effect of Phospholipid Composition on Reconstitution of Na^+-Ca^{2+} Exchange Activity[a]

Phospholipids	Na^+-Ca^{2+} Exchange (nmole Ca^{2+} uptake · mg protein · 90 sec)
PC : PI (90 : 10)	2.1 ± 0.5
PC : PI-4-P (90 : 10)	12.7 ± 1.5

[a] Data are means ± SE of 6-10 different experiments.

TABLE 4. Na^+-Ca^{2+} Exchange Activity of Reconstituted Proteoliposomes as a Function of PI-4-P Content

Phospholipids	Na^+-Ca^{2+} Exchange (nmole Ca^{2+} uptake · mg protein · 90 sec)
PC : PI (90 : 10)	1.7
PC : PI : PI-4-P (90 : 7.5 : 2.5)	6.8
PC : PI : PI-4-P (90 : 5 : 5)	11.2
PC : PI-4-P (90 : 10)	27.0

cytoplasmic side of the plasma membrane, the complexity of their metabolism, and their role in intracellular signaling has catalyzed a great number of studies in recent years.[23] Recently, the discovery of phosphatidylinositol 3-kinase and its products, phosphoinositides phosphorylated at the D-3 position of the inositol ring, has revealed a new group of potential signals.[24] The existence of multiple cellular localizations of phosphatidylinositol 4-phosphate (PI-4-P) and phosphatidylinositol 4,5-bisphosphate (PI-4,5-P_2) indicate that these lipids have other functions in addition to PI turnover. Several cytoskeletal actin-binding proteins have been found to bind PI-4-P and PI-4,5-P_2.[25] Furthermore, it has been shown[26] that low concentrations of PI-4-P or PI-4,5-P_2 stimulated Ca^{2+}-ATPase purified from rat brain synaptosomes. Phosphatidylinositol is a negatively charged phospholipid but, as discussed before, is unable to elicit significant Na^+-Ca^{2+} exchange activity when present in the lipid mixture used for reconstitution in proteoliposomes.[10] Furthermore, it has been shown recently that treatment of cardiac sarcolemmal membrane with phospholipase C results in a specific hydrolysis of phosphatidylinositol and significant stimulation of Na^+-Ca^{2+} exchange activity.[27] In this context it was considered worth investigating the role of polyphosphoinositides in the modulation of Na^+-Ca^{2+} exchange activity in cardiac sarcolemma. The Na^+-Ca^{2+} exchanger was reconstituted by the cholate-dyalysis procedure in a lipid mixture containing PC : PI-4-P or PC : PI-4,5-P_2 with a lipid/protein dilution of 1 to 20. As shown in TABLE 3, the presence of PI-4-P gives rise to a significant Na^+-Ca^{2+} exchange activity (about sixfold higher than that in the presence of PI). In TABLE 4 it is shown that increasing concentrations of PI-4-P in a mixture containing PC : PI : PI-4-P induce a consistent increase of Na^+-Ca^{2+} exchange activity. A higher activation of Na^+-Ca^{2+} exchange activity was obtained when PI-4-P was substituted with PI-4,5-P_2. Preliminary results[28] also showed that dephosphorylation of PI-4-P incorporated into proteoliposomes by potato acid phosphatase reduce the exchange activity. Although the concentration of PI-4-P used in these experiments is relatively high with respect to the

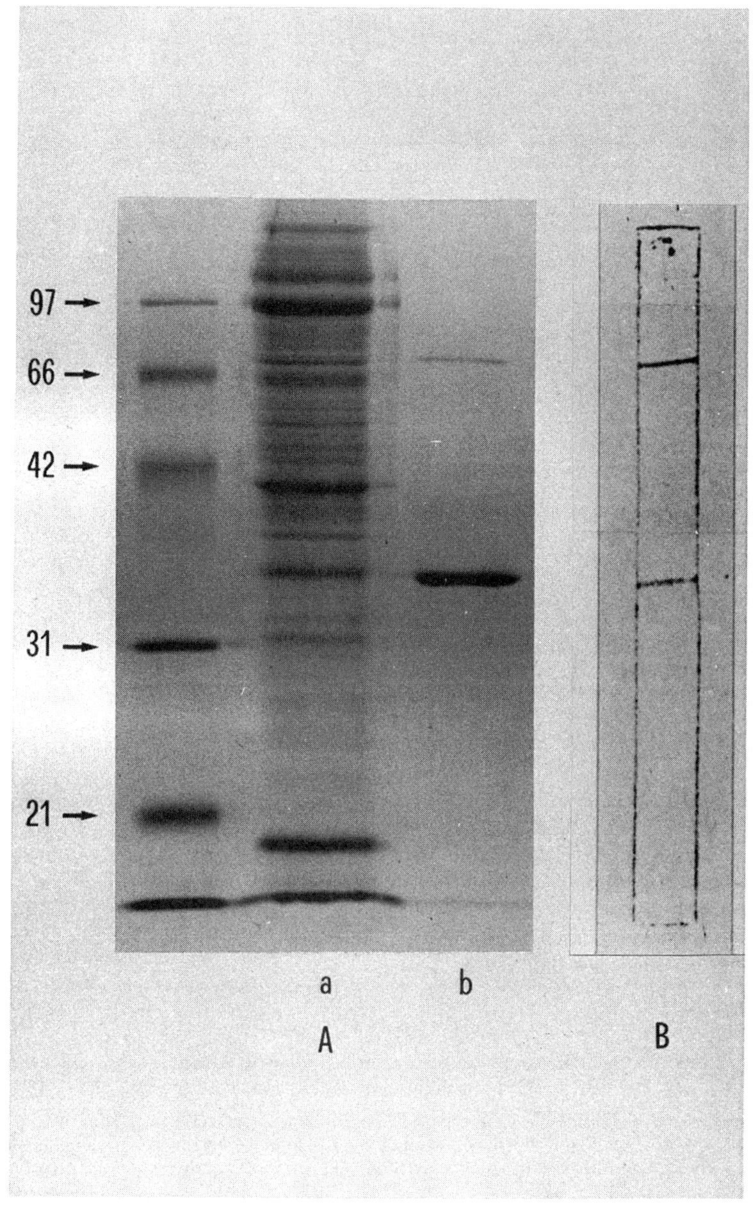

FIGURE 4. (**A**) Sodium-dodecylsulfate polyacrylamide gel electrophoresis (10%) of cardiac sarcolemmal vesicles (*lane a*) and of calpactins purified from cardiac muscle (*lane b*). Molecular mass markers, expressed in kDa are shown on the left. (**B**) Immunoblot of cardiac sarcolemmal vesicles stained with p36-kDa antibody and visualized with peroxidase-protein-A-diaminobenzidine. (Reproduced from Sobota *et al.*[30] with permission from *Biochemical and Biophysical Research Research Communications.*)

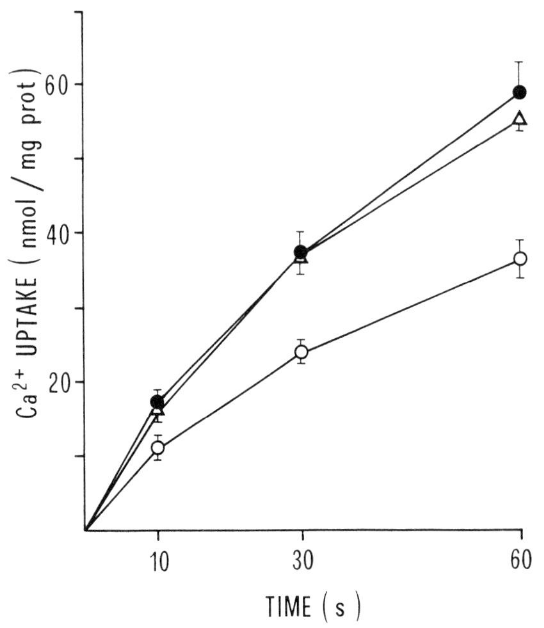

FIGURE 5. Na^+-Ca^{2+} exchange activity in vesicles treated with 2 mM EGTA (○), with 2 mM EGTA and, subsequently, with 1.5 μg calpactins plus 2 μM $CaCl_2$ (●) and in control vesicles (△). (Reproduced from Sobota *et al.*[30] with permission from *Biochemical and Biophysical Research Communications.*)

physiological concentration of this lipid in the native membrane, it should be pointed out that these lipids are concentrated in the inner half of the membrane bilayer, and it is probable that only the plasma membrane contains PI-4,5-P_2. Therefore these results may provide useful information about the physiological regulation of the Na^+-Ca^{2+} exchanger by lipids of the PI cycle.

CA-PHOSPHOLIPID BINDING PROTEINS

Annexins (also called calpactins) are a family of membrane-associated Ca-binding proteins widely diffused in all mammalian tissues that share the ability to bind phospholipids in a calcium-dependent manner.[29] The binding to phospholipids greatly increases their affinity for calcium. Recently in our laboratory a calpactin (calpactin 1) has been isolated and immunologically identified (FIG. 4) from cardiac tissue.[30] The effect of calpactin 1 has been studied on Na^+-Ca^{2+} exchange activity in cardiac sarcolemmal vesicles. Treatment of cardiac sarcolemmal vesicles with ethylene glycol-bis[2-aminoethylether] *N,N*-tetraacetic acid (EGTA) significantly reduces (30-40%) Na^+-Ca^{2+} exchange activity. Even though the exact nature of this effect is not known, it can be assumed that in these conditions calpactins are released from the membrane. Addition of calpactin 1 (1.5 μg) in the presence of 2 μM $CaCl_2$ to EGTA-treated vesicles completely restores the Na^+-Ca^{2+} exchange activity (FIG. 5). Neither Ca^{2+} nor calpactins alone are able to restore Na^+-Ca^{2+} exchange activity. Addition of calpactins to untreated vesicles was without any effect on Na^+-Ca^{2+} exchange. The effect of calpactins on Na^+-Ca^{2+} exchange activity needs to be clarified, but at least two hypotheses can be put forward: (a) Calpactins may induce an increase of Ca^{2+} concentration in the microenvironment around the exchanger; and (b) physicochemical modification of membrane properties induced by the binding of calpactins to phospholipids may cause a variation of lipid composition in the vicinity of the exchanger.

ACKNOWLEDGMENTS

We are indebted to Lewis Cantley (Tufts University, Department of Physiology) for his kind hospitality to one of us (S.L.) on sabbatical leave from the University of Padova and for helpful discussions.

REFERENCES

1. PHILIPSON, K. D. 1985. Annu. Rev. Physiol. **47:** 561-571.
2. BLAUSTEIN, M. P. 1988. J. Cardiovasc. Pharmacol. **12**(Suppl.5): 56-68.
3. FLOREANI, M. & S. LUCIANI. 1984. Eur. J. Pharmacol. **105:** 317-322.
4. LUCIANI, S. & M. FLOREANI. 1985. Trends Pharmacol. Sci. **6:** 316.
5. BOVA, S., G. CARGNELLI & S. LUCIANI. 1988. Br. J. Pharmacol. **93:** 601-608.
6. CARGNELLI, G., S. BOVA & S. LUCIANI. 1989. Br. J. Pharmacol. **97:** 533-541.
7. PHILIPSON, K. D., J. S. FRANK & A. Y. NISHIMOTO. 1983. J. Biol. Chem. **258:** 5905-5910.
8. LUCIANI, S. 1984. Biochim. Biophys. Acta **772:** 127-134.
9. SOLDATI, L., S. LONGONI & E. CARAFOLI. 1985. J. Biol. Chem. **260:** 13321-13327.
10. VEMURI, R. & K. D. PHILIPSON. 1987. Biochim. Biophys. Acta **937:** 258-268.
11. DEBETTO, P., F. CUSINATO & S. LUCIANI. 1990. Arch. Biochem. Biophys. **278:** 205-210.
12. MIYAMOTO, H. & E. RACKER. 1980. J. Biol. Chem. **255:** 2656-2658.
13. RAHAMIMOFF, H. 1989. *In* Sodium-Calcium Exchange. T. J. A. Allen, D. Noble & H. Reuter, Eds.: 153-177. Oxford University Press. Oxford, England.
14. BARZILAI, A., R. SPANIER & H. RAHAMIMOFF. 1984. Proc. Natl. Acad. Sci. USA **81:** 6521-6525.
15. PHILIPSON, K. D. & R. WARD. 1985. J. Biol. Chem. **260:** 9666-9671.
16. SANDERMANN, H. 1978. Biochim. Biophys. Acta **515:** 209-237.
17. YEAGLE, P. L. 1985. Biochim. Biophys. Acta **822:** 267-287.
18. KUTRYK, M. J. B. & G. N. PIERCE. 1988. J. Biol. Chem. **263:** 13167-13172.
19. VEMURI, R. & K. D. PHILIPSON. 1989. J. Biol. Chem. **264:** 8680-8685.
20. CUSINATO, F. 1991. Ph.D. Thesis, University of Padova, Padova, Italy.
21. SHIMITZKY, M. & Y. BARENHOLZ. 1978. Biochim. Biophys. Acta **515:** 367-394.
22. CARRUTHERS, A. & D. L. MELCHIOR. 1988. *In* Methods for Studying Membrane Fluidity. R. C. Aloia, C. C. Curtain & L. M. Gordon, Eds.: 201-225. Alan R. Liss. New York.
23. CARPENTER, C. L. & L. C. CANTLEY. 1990. Biochemistry **29:** 11147-11156.
24. WHITMAN, M., C. P. DOWNES, M. KEELER, T. KELLER & L. CANTLEY. 1988. Nature **332:** 644-646.
25. JANMEY, P. A. & T. P. STOSSEL. 1987. Nature **325:** 362-364.
26. PENNISTON, J. R. 1982. Ann. N.Y. Acad. Sci. **402:** 296-303.
27. PIERCE, G. M. & V. PANAGIA. 1989. J. Biol. Chem. **264:** 15344-15350.
28. LUCIANI, S. & L. CANTLEY. Unpublished results.
29. KLEE, C. B. 1988. Biochemistry **27:** 6645-6653.
30. SOBOTA, A., F. CUSINATO & S. LUCIANI. 1990. Biochem. Biophys. Res. Commun. **172:** 1067-1072.

Adenosine Receptors Modulate the Na^+-Ca^{2+} Exchanger in Cerebral Nerve Endings[a]

M. TAGLIALATELA,[b] L. M. T. CANZONIERO,[b]
A. M. ROSSI,[b] G. MITA,[c] G. F. DIRENZO,[b] AND
L. ANNUNZIATO[b,d]

[b]*Institute of Pharmacology*
2nd School of Medicine
University of Naples

[c]*I.I.G.B., C.N.R.*
80131 Naples, Italy

It is widely accepted that adenosine acts as a neuromodulator in the central nervous system.[1] In fact, adenosine prevents depolarization-induced release of various neurotransmitters from cerebral nerve endings.[2] This action has been attributed to its ability to interfere with transmembrane Ca^{2+} movements.

Because different pathways for Ca^{2+} entry into presynaptic nerve terminals have been described,[3] we studied the effect of the addition of the adenosine-catabolizing enzyme, adenosine deaminase (ADA) and the receptor antagonist theophylline on $^{45}Ca^{2+}$ entrance through voltage-operated calcium channels (VOCCs) and the Na^+-Ca^{2+} antiporter.

RESULTS AND DISCUSSION

ADA caused a dose-dependent enhancement of $^{45}Ca^{2+}$ uptake elicited by extracellular Na^+ removal (145 mM choline), whereas it failed to modify depolarization (55 mM K^+) induced $^{45}Ca^{2+}$ uptake (FIG. 1).

The A_1 selective agonist, phenylisopropyladenosine (L-PIA, 100 μM), but not the rather selective A_2 agonist, *N*-ethylcarboxamido adenosine (NECA,[4] 100 μM), was able to counteract the stimulatory effect exerted by ADA (10 IU/ml) on Na^+-dependent $^{45}Ca^{2+}$ uptake.

Theophylline (0.3-10 mM) dose-dependently and selectively reinforced $^{45}Ca^{2+}$ uptake through the Na^+-Ca^{2+} antiporter, whereas $^{45}Ca^{2+}$ uptake through VOCCs was unaffected (FIG. 2). Furthermore, if both theophylline and ADA were added at the same time, their stimulatory effect on Na^+-dependent $^{45}Ca^{2+}$ uptake displayed no additive effect.

The results of the present study suggest that adenosine, once released from presynaptic nerve terminals, can exert an inhibitory effect on the Na^+-Ca^{2+} antiporter, due

[a]This work was supported by C.N.R. Grants 88.3416 and 89.04676.CT04.

[d]Address for correspondence: Prof. Lucio Annunziato, M.D., Unit of Pharmacology, 2nd School of Medicine, University of Naples, Via Sergio Pansini 5, 80131 Naples, Italy.

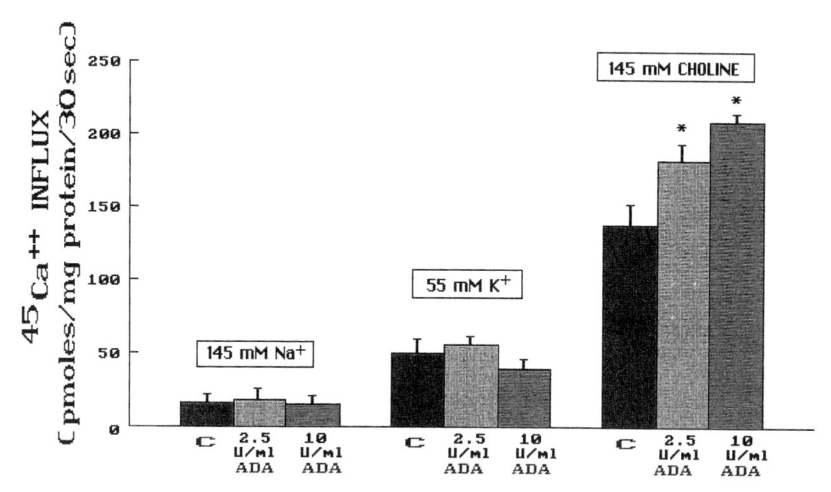

FIGURE 1. Effect of ADA on $^{45}Ca^{2+}$ uptake elicited by depolarization (55 mM K^+) or by extracellular Na^+ removal (145 mM choline). Synaptosomes were preincubated for 30 min in a 145 mM Na^+-containing medium. After this period, ADA (2.5, 10 IU) was added to the synaptosomes for 30 min. After the preincubation phase, synaptosomes were exposed to $^{45}Ca^{2+}$ in an extracellular medium containing either 145 mM NaCl, 55 mM K^+, or 145 mM choline for 30 sec. Data are the mean ± SE of at least three separate determinations.

*Values statistically different ($p < 0.01$) versus respective control value.

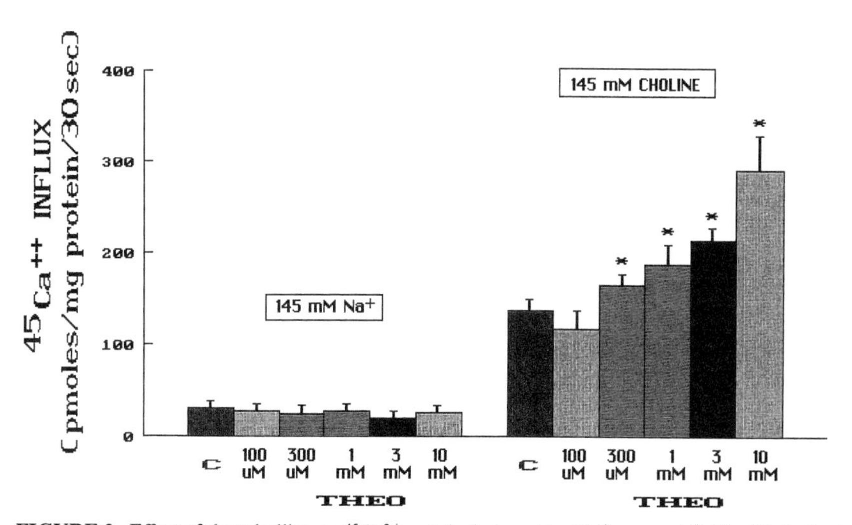

FIGURE 2. Effect of theophylline on $^{45}Ca^{2+}$ uptake induced by Na^+ removal (145 mM choline). Whole brain synaptosomes were preincubated for 60 min in a 145 mM NaCl-containing medium. Theophylline (0.1-10 mM) was added to the synaptosomes during the last 30 min of the preincubation period. After the preincubation phase synaptosomes of each group were exposed to $^{45}Ca^{2+}$ in an extracellular medium containing either 145 mM Na^+ or 145 mM choline. After 30 sec, $^{45}Ca^{2+}$ trapped into synaptosomes was determined by liquid scintillation counting. Data are the mean ± SE of at least three separate determinations.

*Values statistically different ($p < 0.01$) versus respective control value.

to an interaction of the nucleoside with A_1 receptor subtype, since L-PIA counteracted the stimulatory effect exerted by ADA on the synaptosomal Na^+-Ca^{2+} antiporter activity.

The existence of this purinergic inhibitory tone on the activity of the synaptosomal Na^+-Ca^{2+} exchanger is further supported by the specific enhancement of Ca^{2+} entrance through the exchanger induced by theophylline, which antagonizes the effects exerted by endogenous adenosine on adenosine receptors. Furthermore, the fact that the stimulatory action of ADA and theophylline on Na^+-dependent $^{45}Ca^{2+}$ uptake was not additive suggests that both act through a common mechanism.

REFERENCES

1. SNYDER, S. 1985. Adenosine as a neuromodulator. Ann. Rev. Neurosci. **8:** 104-123.
2. HARMS, H., G. WARDEH & A. H. MULDER. 1979. Effects of adenosine on depolarization-induced release of various radiolabeled neurotransmitters from slices of rat corpus striatum. Neuropharmacology **18:** 577-580.
3. TAGLIALATELA, M., G. F. DIRENZO & L. ANNUNZIATO. 1990. Na^+-Ca^{++} exchange activity in central nerve endings. I. Ionic conditions that discriminate $^{45}Ca^{++}$ uptake through the exchanger from that occurring through voltage-operated Ca^{++} channels. Mol. Pharmacol. **38:** 385-392.
4. WILLIAMS M. 1987. Purine receptors in mammalian tissues: Pharmacology and functional significance. Ann. Rev. Pharmacol. Toxicol. **27:** 315-345.

Activation of Ca^{2+}-Na^+ Exchange by Platelet-Derived Growth Factor in Vascular Smooth Muscle Cells

MASSIMO CIRILLO AND MITZY L. CANESSA

Endocrine-Hypertension Division
Brigham & Women's Hospital
Harvard Medical School
Boston, Massachusetts 02115

Regulation of ion transport by growth factors may play an important role in the enhanced vascular smooth muscle cell (VSMC) proliferation observed in genetically hypertensive rats.[1] Platelet-derived growth factor (PDGF) initiates a proliferative response through a complex biochemical cascade involving the receptor tyrosine kinase activity, activation of phospholipase C, generation of diacylglycerol (DAG) and inositol 1,4,5-trisphosphate (IP_3), and subsequent Ca^{2+} mobilization from intracellular stores.[2] The present study was designed to investigate Ca^{2+} transport in quiescent VSMC and at the initiation of the cell cycle induced by PDGF BB, which is the most effective isoform to stimulate growth and contraction.[3]

Enzymatically dissociated VSMC from thoracic aorta of normotensive adult rats were cultured (passages 5-10)[1] and made quiescent (48 hours, 0.3% serum) to study Ca^{2+} efflux; VSMC were loaded to equilibrium with ^{45}Ca and washed out with ^{45}Ca-free balanced salt solution every 0.5 min. Cellular Ca^{2+} per milligram of protein was determined from the cell-associated radioactivity and the specific activity of the loading solution. Ca^{2+} efflux kinetics were best fitted by a two-compartment exponential model (FIG. 1). The effect of low temperature (FIG. 1) and Ca^{2+} (data not shown) indicated that the rate constant (k_1) of the first compartment reflected Ca^{2+} release from extracellular sites while the second compartment rate constant (k_2) reflected active Ca^{2+} efflux from an intracellular pool (FIG. 1). Under basal conditions, k_2 was not affected by external Na^+ (Na^+_o) (TABLE 1), indicating that Ca^{2+}-Na^+ exchange is silent in quiescent VSMC.

PDGF (50 ng/ml) transiently increased cytosolic Ca^{2+} (Ca_i), which peaked between 2-3 min and returned to baseline by 6-8 min. The early (0 to 3 min) response of Ca^{2+} efflux to PDGF (50 ng/ml), associated with the peak elevation of Ca_i, showed a fourfold increase of k_2 (TABLE 1), mediated by a Na_o-independent pathway. The late response of Ca^{2+} efflux to PDGF (10 min) showed a 2.3-fold increase in k_2 (TABLE 1) despite Ca_i levels having already returned to baseline. The stimulation of Ca^{2+} efflux by PDGF occurred through Na_o-independent (1.7-fold) and Na_o-dependent pathways (TABLE 1).

We hypothesize that activation of Ca^{2+}-Na^+ exchange and the Ca^{2+} pump by PDGF might be due to receptor-induced activation of protein kinase C[2,3] with phosphorylation of both transporters leading to an increased turnover rate and/or Ca_i affinity. Our results suggest that PDGF modulation of Ca^{2+}-Na^+ exchange may play an important role in regulation of Ca_i in proliferating hypertensive VSMC.

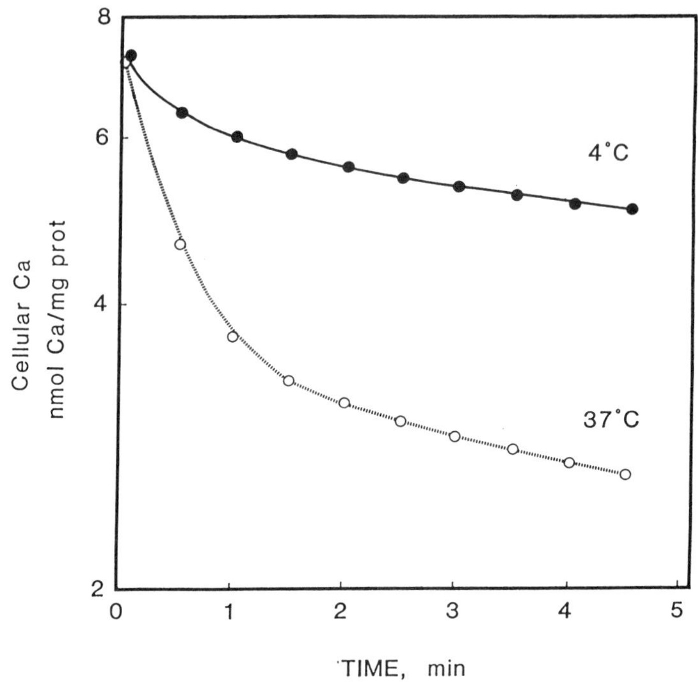

FIGURE 1. Time course of cell-associated Ca^{2+} in quiescent VSMC incubated at 37 and 4°C. VSMC were loaded to equilibrium with ^{45}Ca and washed out with ^{45}Ca-free balanced salt solution (mM: 140 NaCl, 4 KCl, 1.2 $CaCl_2$, 1.0 $MgCl_2$, 1.0 phosphate, 10 glucose, and 20 MOPS-Tris, pH 7.4). Cellular Ca^{2+} per milligram of protein was determined from the cell-associated radioactivity and the specific activity of the loading solution. Ca^{2+} efflux kinetics were best fitted by a two-compartment exponential model. Symbols represent experimental data and lines are the theoretical curves obtained by nonlinear fitting of the data to a two-compartment model. The kinetic parameters of this model were the following at 37°C: A_1 (size of the first compartment): 3.9 nmoles/mg protein; k_1 (rate constant of the first compartment): 2.1 min^{-1}; A_2 (size of the second compartment): 3.4 nmoles/mg of protein; k_2 (rate constant of the second compartment): 6.3×10^{-2} min^{-1}. At 4°C, the parameters were A_1: 1.2 nmoles/mg protein; k_1: 1.95 min^{-1}, A_2: 6.1 nmoles/mg protein; k_2: 4.2×10^{-2} min^{-1}.

TABLE 1. Early and Late Effects of PDGF BB on Calcium Efflux from Rat Cultured Vascular Smooth Muscle Cells

	Na_o-Independent[a]	Na_o-Dependent[a]
Basal	6.2 ± 0.7	none
Early PDGF effect	25.3 ± 0.3	none
Δ PDGF (n = 6)	19.1 ± 0.3	none
Basal	5.9 ± 0.6	none
Late PDGF effect	10.1 ± 0.6	4.0 ± 0.7
Δ PDGF (n = 12)	4.2 ± 0.8	4.0 ± 0.7

NOTE: Recombinant PDGF BB 50 ng/ml. Early effect 0-3 min and late effect 10-15 min.
[a] K_2 rate constant, $10^{-2} \times min^{-1}$.

REFERENCES

1. Socorro, L., G. Vallega, A. Nunn, T. J. Moore & M. Canessa. 1990. Vascular smooth muscle cells from the Milan hypertensive rat exhibit decreased functional angiotensin II receptors. Hypertension **15:** 591-599.
2. Williams, L. T. 1989. Signal transduction by the platelet-derived growth factor receptor. Science **243:** 1564-1570.
3. Block, L. H., L. R. Emmons, E. Vogt, A. Sachinidis, W. Vetter & J. Hoppe. 1989. Ca^{2+} channel blockers inhibit the action of recombinant platelet-derived growth factor in vascular smooth muscle cells. Proc. Natl. Acad. Sci. USA **86:** 2388-2392.
4. Vigne, P., J. P. Breittmayer, D. Duval, C. Frelin & M. Lazdunski. 1988. The Na^{+}/Ca^{2+} antiporter in aortic smooth muscle cells. J. Biol. Chem. **263:** 8078-8083.

Voltage-Dependent Block of the Na-Ca Exchanger in Heart Muscle Examined Using Giant Excised Patches from Guinea Pig Cardiac Myocytes

A. E. DOERING AND W. J. LEDERER

Department of Physiology
University of Maryland School of Medicine
Baltimore, Maryland 21201

The sodium-calcium exchanger is abundant in the sarcolemmal membrane of mammalian heart muscle. Approximately 250 exchangers are found per square micron of sarcolemma, and they turn over at about 2500 sec^{-1} under physiological conditions.[1] Recent experiments have shown that this sodium-calcium exchanger is responsible for extruding most of the calcium that enters the intracellular compartment during the cardiac cycle.[2-4] Additional investigations have revealed the amino acid sequence of the cardiac sodium-calcium exchanger protein[5] and have reported on many of the transport and regulatory properties of the sodium-calcium exchanger.[6-9] The clear physiologic importance of sodium-calcium exchange coupled with the development of new methods[1,10] has led to our investigation of characteristics of the exchanger. In this brief report we use Hilgemann's giant patch-clamp technique to provide preliminary evidence that the blockade of Na-Ca exchange function by intracellular lanthanum may be voltage dependent.

We measured outward sodium-calcium exchange current in giant membrane patches[9,11,12] and used a modified oil-gate bath[13,14] with five chambers to change solutions. Acutely dissociated adult guinea pig ventricular cells were incubated overnight in a depolarizing medium containing 135 mM KCl, 0 Ca at 4°C. These cells spontaneously form blebs of sarcolemmal membrane that are separated from the contractile proteins. It is possible to seal a 10-μm tip diameter patch pipette to a membrane bleb, and to produce an inside-out giant excised patch on the tip of the pipette by rapidly moving the pipette away from the cell. Once the giant patch is formed, solutions at the cytoplasmic face of the patch can be changed by moving the pipette tip from one bath chamber to another through an oil gate. The pipette solution (i.e. extracellular) contained (in mM) 160 *N*-methyl glucamine (NMG), 10 TEACl, 20 PIPES, 0.020 EGTA, 7.5 $CaCl_2$, 0.025 ouabain, 0.0025 D-600, and 1 4-aminopyridine. The bath solution (i.e. cytoplasmic) contained 140 NMG, 10 TEACl, 5 ATP, 1 $MgCl_2$, 20 PIPES, 20 PAPTA, and 0.3 μM free Ca^+. When 140 mM sodium was added to a solution, it replaced an equimolar amount of NMG. All blocking agents were added to the bath solution. EGTA and $CaCl_2$ were not included in the bath solutions with metal blockers since they are chelated by EGTA. All experiments were done at 35°C. "Sodium-activated" current was measured after subtracting "leak" current, which was characterized in each experiment as the current activated by 140 mM potassium. Interestingly, in some experiments the leak current was slightly reduced by the blockers.

TABLE 1. Sodium-Calcium Exchange Blockers

	Percent Blockade					
	This Study		Other Studies			
Blocker	%	mM	%	mM	Method	References
La^{3+}	90	5.0	100	0.1	Whole cell[a]	Kimura *et al.*[7]
Ni^{2+}	73	5.0	80	1.0	Whole cell[a]	Ehara *et al.*[17] (Fig. 6)
			100	1.0	Whole cell[a]	Niggli & Lederer[1]
Co^{2+}	100	5.0	100	1.0	Excised patch	Hilgemann[11] (Fig. 2)
Amiloride	100	5.0	25	3.0	Whole cell[a]	Kimura *et al.*[18]
Lidocaine	60	5.0	—	—	—	—
H^+	60	pH 6.4	100	pH 6.0	Perfused axon	Baker & McNaughton[19]
			50	pH 6.7	Vesicle	Philipson *et al.*[20]

[a] Extracellular application of blocker.

Virtually all studies of sodium-calcium exchange utilize blockers of the exchanger, the blockers usually being applied in the extracellular solution. In this study, we examine cytoplasmic blockade of sodium-calcium exchanger function by several widely used antagonists of the sodium-calcium exchanger. Because it has been shown that the calcium affinity of the exchanger is higher on the intracellular side of the membrane than on the extracellular side,[15] we anticipated that blockers may also have different affinities. We recorded blockade of the sodium-activated current by the cytoplasmic application of lanthanum, nickel, cobalt, lidocaine, amiloride, and protons. TABLE 1 shows the efficacy of blockade of the sodium-activated outward current measured as described above and illustrated in panel A of FIGURE 1. The values in the table are percent reduction in steady-state current measured with the membrane potential held at 0 mV. It appears that lanthanum is a less effective blocker from the cytoplasmic side of the membrane than from the extracellular side. It is interesting to note that lidocaine, the sodium-channel blocker, also acts as a poor blocker of the sodium-activated current. Although we presume that its actions are nonspecific because this concentration also reduces the calcium current in heart,[16] we do not have direct evidence of its mode of action. All of the antagonists studied are only effective at relatively high concentrations, except for protons, for which a 0.3 μM increase in concentration blocks the current 60%.

The sodium-calcium exchanger is known to be voltage-dependent, and evidence has been presented that suggests the voltage dependence is associated with charge translocation in the Na-transport step[1] (see also Niggli & Lederer, this volume). There remains some uncertainty about the best kinetic model to describe sodium-calcium exchange transport, although there is increasing support for a consecutive transport model[1,21] (see Niggli & Lederer, also Hilgemann & Nagel, this volume). We hoped that some insight would develop from an examination of the voltage dependence of sodium-calcium exchanger block by various agents. Consequently, we investigated the sensitivity of the sodium-activated outward current to membrane potential. The current-voltage relation is shown in panel A of FIGURE 1. It has a shallow slope (39 mV/50% change or 131 mV/e-fold change). We also measured the effect of blockers on the sodium-activated current at different membrane potentials. In data not shown,

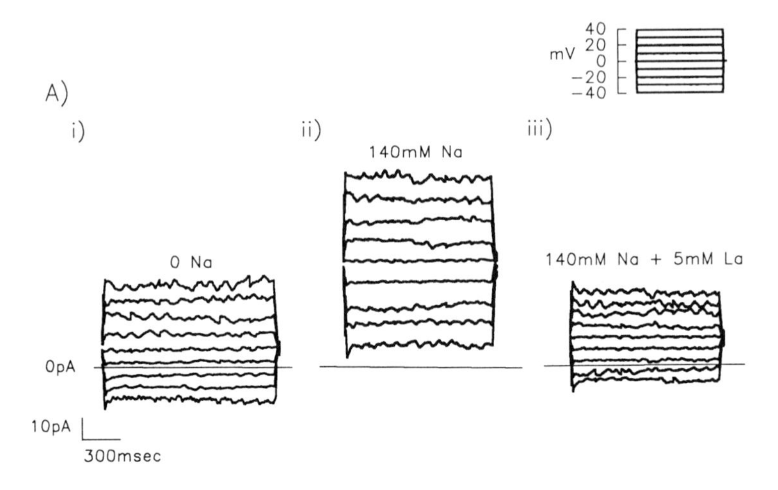

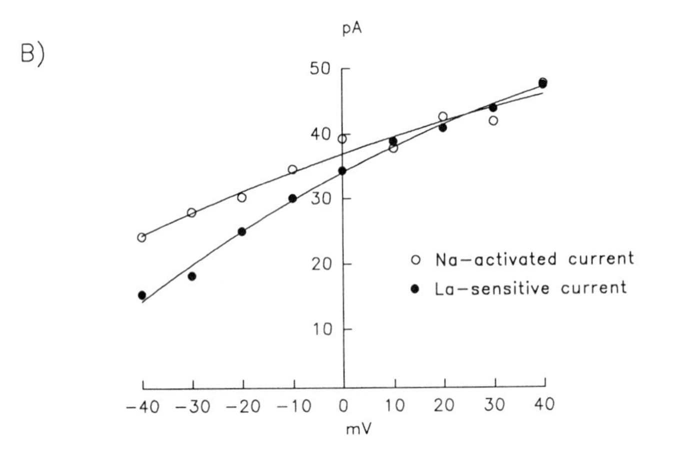

FIGURE 1. Voltage-dependent block of sodium-calcium exchange current by lanthanum. *Panel A:* Membrane potential is stepped to voltages ranging from −40mV to +40mV (inset), and the resulting current is recorded. The current record **(i)** shows the background voltage-dependent current in the absence of cytoplasmic sodium. When sodium replaces NMG in the cytoplasmic solution, an outward current is activated, as shown in **(ii)**. With sodium still present, 5 mM lanthanum is added to the cytoplasmic solution, and the outward current is blocked as shown in record **(iii)**. *Panel B:* Difference currents are plotted versus voltage. The points are fit with first-order polynomials. (○) Sodium-activated currents (140 mM Na), after background currents are subtracted. (●) Sodium-activated currents (140 mM Na) minus sodium-activated current in the presence of lanthanum. Before subtraction, the background cuurents were removed from each record.

the exchanger was partly blocked by 5 mM $NiCl_2$ in a voltage-independent manner, consistent with reports by Ehara *et al.*[20] Most other blockers also acted in a voltage-independent manner. However, both cobalt and lanthanum blocked more effectively at depolarized potentials.

FIGURE 1 shows that the current blocked by lanthanum has a steeper voltage dependence than the sodium-activated current. The apparent voltage-dependent blockade of the sodium-activated current could be due to several different factors.

The sodium-activated current could contain a component that was not generated by sodium-calcium exchanger activity. If this were true, it would be possible to produce voltage-dependent block several ways. The component that is not sodium-calcium exchange could be voltage dependent and/or it could itself be altered by lanthanum in a voltage-dependent manner.

Alternatively, the sodium-activated current may be purely due to activity of the sodium-calcium exchanger. In this case, the action of lanthanum could alter a voltage-sensitive step in the transport mechanism of the sodium-calcium exchanger. However, more information is needed to identify which step of the sodium-calcium exchange reaction scheme is the one that is affected by lanthanum. It is also possible that the affinity of the exchanger for lanthanum is voltage dependent or that lanthanum interacts with the exchanger within the membrane, that is, within the electrical field. Membrane hyperpolarization would be expected to electrostatically draw the trivalent cation lanthanum to the cytoplasmic side of the membrane, reducing the likelihood that it would reach an interaction site in the membrane voltage field.

Recently, Li *et al.* have reported that a 20-amino-acid peptide, XIP, acts as a potent and specific blocker of cardiac sodium-calcium exchange current.[22] It would be interesting to compare the voltage dependence of the XIP-sensitive current with the voltage dependence of the lanthanum-sensitive current. This new blocker may prove a useful tool to identify the origin of the apparent voltage dependence of the lanthanum effect on sodium-calcium exchange.

REFERENCES

1. NIGGLI, E. & W. J. LEDERER. 1991. Molecular operations of the sodium-calcium exchanger revealed by conformation currents. Nature **349:** 621-624.
2. CRESPO, L. M., C. J. GRANTHAM & M. B. CANNELL. 1990. Kinetics, stoichiometry, and role of the Na-Ca exchange mechanism in isolated cardiac myocytes. Nature **345:** 618-621.
3. BERS, D. M. & J. H. B. BRIDGE. 1989. Relaxation of rabbit ventricular muscle by Na-Ca exchange and sarcoplasmic reticulum calcium pump: Ryanodine and voltage sensitivity. Circ. Res. **65:** 334-342.
4. BERS, D. M., W. J. LEDERER & J. R. BERLIN. 1990. Intracellular Ca transients in rat cardiac myocytes: Role of Na-Ca exchange in excitation-contraction coupling. Am. J. Physiol. **258:** C944-C954.
5. NICOLL, D. A., S. LONGONI & K. D. PHILIPSON. 1990. Molecular cloning and functional expression of the cardiac Na^+-Ca^{2+} exchanger. Science **250:** 562-565.
6. REEVES, J. P. & C. C. HALE. 1984. The stoichiometry of the cardiac sodium-calcium exchange system. J. Biol. Chem. **259:** 7733-7739.
7. KIMURA, J., A. NOMA & H. IRISAWA. 1986. Na-Ca exchange current in mammalian heart cells. Nature **319:** 596-597.
8. PHILIPSON, K. D., M. M. BERSOHN & A. Y. NISHIMOTO. 1982. Effects of pH on Na^+-Ca^{2+} exchange in canine cardiac sarcolemmal vesicles. Circ. Res. **50:** 287-293.
9. DOERING, A. E. & W. J. LEDERER. 1991. Cytoplasmic acidity inhibits sodium-calcium exchange in cardiac cells. Biophys. J. **59:** 544a.
10. HILGEMANN, D. W. 1990. Regulation and deregulation of cardiac Na^+-Ca^{2+} exchange in giant excised sarcolemmal membrane patches. Nature **344:** 242-245.

11. HILGEMANN, D. W. 1989. Giant excised cardiac sarcolemmal membrane patches: Sodium and sodium-calcium exchange currents. Pflugers Arch. **415:** 247-249.
12. DOERING, A. E. & W. J. LEDERER. 1991. Single channels are activated along with sodium-calcium exchange in cardiac giant excised patches. Biophys. J. **59:** 462a.
13. QIN, D. & A. NOMA. 1988. A new oil-gate concentration jump technique applied to inside-out patch clamp recording. Am. J. Physiol. **255:** H980-H984.
14. LEDERER, W. J. & C. G. NICHOLS. 1989. Nucleotide modulation of the activity of rat heart ATP-sensitive K^+ channels in isolated membrane patches. J. Physiol. **419:** 193-211.
15. DIPOLO, R. & L. BEAUGE. 1990. Asymmetrical properties of the Na-Ca exchanger in voltage clamped, internally dialyzed squid axons under symmetrical ionic conditions. J. Gen. Physiol. **95:** 819-835.
16. DOERING, A. E. & W. J. LEDERER. 1990. Lidocaine reduces mammalian cardiac calcium currents. Biophys. J. **57:** 133a.
17. EHARA, T., S. MATSUOKA & A. NOMA. 1989. Measurement of reversal potential of Na^+-Ca^{2+} exchange current in single guinea-pig ventricular cells. J. Physiol. **410:** 227-249.
18. KIMURA, J., S. MIYAMAE & A. NOMA. 1987. Identification of sodium-calcium exchange current in single ventricular cells of guinea-pig. J. Physiol. **384:** 199-222.
19. BAKER, P. F. & P. A. MCNAUGHTON. 1977. Selective inhibition of the Ca-dependent Na efflux from intact squid axons by a fall in intracellular pH. J. Physiol. **269:** 78P.
20. PHILIPSON, K. D., M. M. BERSOHN & A. Y. NISHIMOTO. 1982. Effects of pH on Na^+-Ca^{2+} exchange in canine cardiac sarcolemmal vesicles. Circ. Res. **50(2):** 287-293.
21. KHANANSHVILI, D. 1990. Distinction between the two basic mechanisms of cation transport in the cardiac Na^+-Ca^{2+} exchange system. Biochemistry **29:** 2437-2442.
22. LI, Z., D. A. NICOLL, A. COLLINS, D. W. HILGEMANN, A. G. FILOTEO, J. T. PENNISTON, J. N. WEISS, J. M. TOMICH & K. D. PHILIPSON. 1991. Identification of a peptide inhibitor of the cardiac Na^+-Ca^{2+} exchanger. J. Biol. Chem. **266(2):** 1014-1020.

Sodium-Calcium Exchange in Crude Plasma Membrane Vesicles from Aortic Myocytes

Proteolysis Partially Restores Exchange Activity Lost During Vesicle Preparation

RONG-MING LYU

Department of Pharmacology
University of Alabama at Birmingham
Birmingham, Alabama 35294

The present findings suggest that endogenous proteolysis stimulates Na^+-Ca^{2+} exchange activity in membrane vesicles from aortic myocytes. A crude plasma membrane fraction was prepared from rat aortic myocytes in the presence or absence of protease inhibitors (40 KIU/ml solid aprotinin, 4 μg/ml leupeptin, 4 μg/ml pepstatin A, 0.8 mM phenylmethylsulfonylfluoride, and 1 mM benzamidine) as described in FIGURE 1. Na^+-Ca^{2+} exchange activity was taken as the difference between $^{45}Ca^{2+}$ uptake in assay solutions containing 160 mM KCl or NaCl as previously described.[1] Exchange activity was plentiful in "Na vesicles" from aortic myocytes, but no activity was detected in vesicles from human skin fibroblasts (FIG. 1A and 1B). The following results confirmed that $^{45}Ca^{2+}$ uptake depended on the Na^+ gradient. "K vesicles," which were prepared in the presence of 160 mM KCl instead of NaCl, almost eliminated the difference in $^{45}Ca^{2+}$ between the KCl and NaCl assay solutions (FIG. 1C). Moreover, passive loading of the "K vesicles" with Na^+ largely restored exchange activity (FIG. 1D).

Exchange activity in vesicles from aortic myocytes has the following additional properties: (a) *N*-(2,4-dimethylbenzyl)amiloride (DMB) (6-50 μM) inhibited Na^+ gradient-dependent $^{45}Ca^{2+}$ uptake by the vesicles. DMB had no effect on $^{45}Ca^{2+}$ uptake assayed in the presence of 160 mM external Na^+. The concentration of DMB that inhibited exchange activity by 50% was similar for vesicles and intact cells (8 versus 11 μM, respectively). (b) Diltiazem (100 μM), which blocks mitochondrial Na^+-Ca^{2+} exchange,[2] had no effect on $^{45}Ca^{2+}$ uptake by crude vesicles. (c) The rate of $^{45}Ca^{2+}$ uptake by the vesicles was somewhat less when 160 mM choline or *N*-methyl-D-glucamine chloride was used instead of KCl in the assay solution, as was also found for $^{45}Ca^{2+}$uptake by the intact cells (see Smith *et al.*, this volume). (d) Neither the depletion of cellular ATP before vesicle preparation, nor *in vitro* treatment with MgATP under a variety of conditions, affected Na^+-Ca^{2+} exchange activity in the vesicles. In the intact myocytes, however, ATP depletion with rotenone inhibits exchange activity by ~80%, and ATP repletion by glucose restores exchange activity.[3]

FIGURE 2 shows the time course of $^{45}Ca^{2+}$ uptake by vesicles prepared in the presence or absence of the protease inhibitors. Crude vesicles prepared in the presence of the inhibitors had about half as much exchange activity as those prepared in the absence of the protease inhibitors (FIG. 2). Addition of the protease inhibitors to vesicles that were prepared in their absence had no effect on the time course of $^{45}Ca^{2+}$ uptake

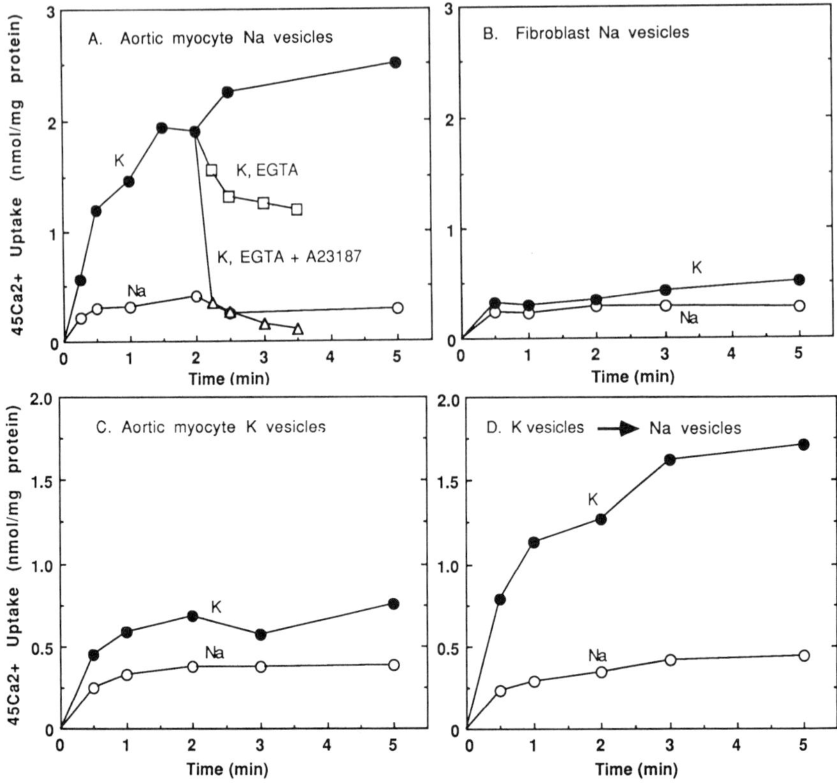

FIGURE 1. An outwardly directed Na^+ gradient drives $^{45}Ca^{2+}$ uptake by membrane vesicles from rat aortic myocytes, but not those from human skin fibroblasts. Cultures of rat aortic myocytes and human skin fibroblasts were rinsed with saline containing 10 mM HEPES/Tris, pH 7.4, scraped from culture dishes and suspended in a solution containing the protease inhibitors, 20 mM MOPS/Tris, pH 7.4, and 160 mM NaCl ("Na vesicles," **A** and **D**) or 160 mM KCl ("K vesicles," **C**), and disrupted with a Dounce homogenizer. After removing unbroken cells, nuclei, and mitochondria by differential centrifugation, crude vesicles were sedimented and suspended in a small volume of homogenization buffer. $^{45}Ca^{2+}$ uptake was measured (in duplicate or triplicate) in solutions containing 1.5 μCi $^{45}Ca^{2+}$, 20 mM MOPS/Tris, pH 7.4, and either 160 mM NaCl ("Na") or KCl ("K"). Panel **A** shows $^{45}Ca^{2+}$ efflux, which was started at 2 min by adding 0.9 ml of a solution containing 160 mM KCl, 0.1 mM K_2H_2EGTA, and 20 mM MOPS/ Tris, pH 7.4, with or without 2 μM A23187 as indicated. Panel **D** shows $^{45}Ca^{2+}$ uptake by "K vesicles" that were diluted with homogenization buffer containing 160 mM NaCl and incubated on ice overnight to load them with Na^+.

in the NaCl or KCl assay solutions, indicating that the inhibitors do not directly affect exchange activity. These results were confirmed by similar experiments on plasma membrane vesicles that were purified by centrifugation in a sucrose gradient.[4] Additionally, removing the protease inhibitors and adding trypsin to the vesicles increased exchange activity approximately twofold. Preparation of the plasma membranes in the presence of the protease inhibitors decreased the V_{max} of exchange by ~70% and had no effect on the K_m for Ca^{2+}. Large losses of exchange activity occur during vesicle preparation. The V_{max} of exchange activity in intact cells is several-fold higher than the

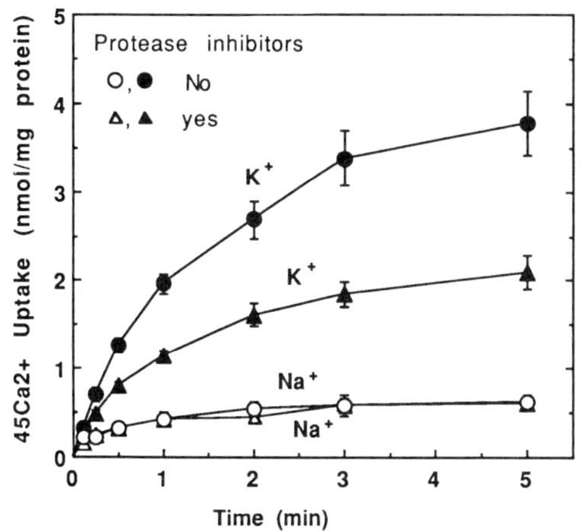

FIGURE 2. Time course of $^{45}Ca^{2+}$ uptake by crude membrane vesicles prepared in the presence (triangles) or absence (circles) of protease inhibitors. $^{45}Ca^{2+}$ uptake was measured with 17 μM $^{45}Ca^{2+}$ in "Na" (unfilled markers) or "K" (filled markers) solutions as indicated in the legend to FIGURE 1. Three days after seeding, half of the cultures were used to prepare vesicles in the absence of protease inhibitors and half to prepare vesicles with the inhibitors as described in the legend to FIGURE 1 for "Na vesicles." Values are mean ± SE for five experiments.

V_{max} of plasma membranes, which were purified several-fold in the presence of the proteases inhibitors. Endogenous proteolysis during membrane preparation apparently compensates for some of the decrease in the V_{max} of vesicles compared to intact cells.

ACKNOWLEDGMENTS

This work was done in Dr. J. B. Smith's laboratory in collaboration with Dr. J. P. Reeves, Roche Institute of Molecular Biology, Nutley, New Jersey.

REFERENCES

1. REEVES, J. P. 1988. Methods Enzymol. **157:** 505-510.
2. MATLIB, M. A. & A. SCHWARTZ. 1983. Life Sci. **32:** 2837-2842.
3. SMITH, J. B. & L. SMITH. 1990. Am. J. Physiol. **259:** C302-C309.
4. LYU, R. M., J. P. REEVES & J. B. SMITH. Biochim. Biophys. Acta **106A:** 97-104.

Molecular Studies

Introduction to Part III

The major findings reported in the MOLECULAR STUDIES section of the conference were the purification and cloning of the cardiac and retinal rod exchangers. The purified proteins have different molecular weights and different functional characteristics: Unlike the cardiac exchanger, which does not require K^+, the retinal rod exchanger depends upon the presence of K^+ for activity. Both exchangers are similar in their membrane topography: In both cases a large cytoplasmic domain is flanked on both sides by multiple transmembrane segments. Nevertheless, these two exchangers appear to have little or no homology in their amino acid sequences. The cardiac and retinal rod exchangers exhibited limited immunological cross-reactivity in some studies, but other antibodies were found to be distinct for one exchanger or the other. Thus, there appear to be at least two types of exchangers that are distinct in both their molecular structure and functional behavior. Studies of the K^+-dependence of exchange activity in synaptosomal membranes raise the possibility that both types of exchanger may be present in brain tissue.

Molecular Studies of the Cardiac Sarcolemmal Sodium-Calcium Exchanger

D. A. NICOLL AND K. D. PHILIPSON

Departments of Medicine and Physiology
and the
Cardiovascular Research Laboratories
University of California, Los Angeles
Center for Health Sciences
Los Angeles, California 90024-1760

INTRODUCTION

The Na^+-Ca^{2+} exchanger of the cardiac sarcolemma is important in returning the cardiac myocyte to its resting state after excitation. During excitation, Ca^{2+} enters the myocyte through voltage-dependent Ca^{2+} channels and induces release of Ca^{2+} from the sarcoplasmic reticulum. The Ca^{2+} interacts with the contractile proteins to cause contraction of the cell. During relaxation, Ca^{2+} is resequestered within the sarcoplasmic reticulum. In addition, that Ca^{2+} which had entered the cell via the Ca^{2+} channels must be removed from the cell. The Na^+-Ca^{2+}exchanger is the primary means for extrusion of the excess Ca^{2+}. This chapter reviews our purification and molecular cloning of the cardiac sarcolemmal Na^+-Ca^{2+} exchanger.

PURIFICATION OF THE CANINE CARDIAC SARCOLEMMAL EXCHANGER

The Na^+-Ca^{2+} exchanger has been purified in a series of steps that include alkaline extraction, solubilization with the detergent decylmaltoside, ion-exchange chromatography on DEAE-Sepharose, and affinity chromatography with wheat-germ agglutinin.[1] Analysis of the purified exchanger by SDS-PAGE under reducing conditions and visualization by silver staining shows three major polypeptides migrating at 160, 120, and 70 kDa. Western blot analysis demonstrates that the 70-kDa band is actually composed of more than one distinct band. Under nonreducing conditions only one prominent polypeptide at 160 kDa is observed.

The relationship between the three polypeptides was further investigated by proteolytic and immunological analysis.[1] Treatment of the purified exchanger with the protease chymotrypsin under mild conditions resulted in a decrease in the amount of the 120-kDa and an increase in the amount of 70-kDa polypeptide as visualized by silver staining or by a reaction with an antibody raised to the purified exchanger. This indicates that the 70-kDa polypeptides probably arise from proteolysis of the 120-kDa polypeptide.

Additionally, antibodies specific for the 160-, 120-, or 70-kDa polypeptides were obtained by affinity purification.[1] Sarcolemmal proteins were separated by SDS-PAGE, transferred to nitrocellulose, and reacted with antibody to the exchanger. Antibodies specific to each of the three polypeptides were prepared by excising each of the three bands and eluting the specific antibody. Using the affinity-purified antibodies on Western blots against sarcolemmal proteins verified that the three polypeptides were immunogically related: Antibody eluted from either the 70-, 120-, or 160-kDa polypeptide reacted with all three of the 160-, 120-, and 70-kDa polypeptides.

These data indicate that the exchanger is a single polypeptide. The nonreduced exchanger appears to migrate as a 160-kDa species. Reduction shifts the protein to a 120-kDa species, and proteolysis reduces the protein to a smaller 70-kDa form.

The identity of the exchanger polypeptides was further confirmed following the successful cloning and sequencing of the exchanger (see below). On the basis of the deduced amino acid sequence of the exchanger, a short peptide corresponding to residues 648-662 was synthesized.[2] Antibodies to the peptide were raised in rabbits and the antibody was then used to detect the exchanger protein on Western blots of sarcolemmal membrane proteins. The antibody to the peptide gave the same reaction on Western blots as an antibody raised against the purified exchanger, thereby verifying the molecular identity of the exchanger.

CLONING OF THE CANINE CARDIAC SARCOLEMMAL EXCHANGER

A partial cDNA clone for the exchanger was obtained by screening a λgt11 expression library with an antibody to the exchanger, and a clone containing the entire coding region was obtained by screening with the partial clone.[2] The exchanger clone was greater than 6.5 kb and hybridized with mRNA of 7 kb on Northern blots. The exchanger clone maps to human chromosome 2 as determined by somatic cell hybridization (Lusis, A. J., D. A. Nicoll & K. D. Philipson, unpublished observation). The coding region of the exchanger clone was sequenced and found to specify a protein of 970 amino acids with a molecular weight of 110 kDa.

The identity of the exchanger clone was confirmed by expression of exchanger activity in *Xenopus* oocytes.[2] RNA was synthesized from the exchanger clone and injected into oocytes. Na^+-gradient-dependent Ca^{2+} uptake into the cells was measured by incubating Na^+-loaded cells in Na^+-free medium containing $^{45}Ca^{2+}$. Cells injected with exchanger cRNA had high levels of exchanger activity as compared to cells injected with water.

Analysis of the exchanger by *in vitro* translation indicates that the exchanger is glycosylated only at the first of its six potential N-linked glycosylation sites and that the exchanger has a cleaved leader peptide (unpublished observations). The existence of a cleaved leader peptide is also supported by the presence of a consensus leader peptide cleavage site[2,3] and by the work of Durkin, *et al.*[4] who have determined the NH_2-terminal amino acid sequence of the exchanger to be ETEMEG. To reflect this information, the amino acid sequence of the mature protein has been renumbered such that the leader sequence is numbered -32 to -1, and the first residues of the mature protein are ETEMEG (FIG. 1). Following cleavage of the leader peptide, the deduced molecular weight of the polypeptide portion of the exchanger is 105 kDa.

MODEL FOR THE EXCHANGER

A model for the exchanger protein is necessary for the design of experiments examining the structure-function relationships of the exchanger. On the basis of the

```
              -32  ML QLRLLPTFSM GCHLLAVVAL LFSHVDLISA
           *                                             1
  1  ETEMEGEGNE TGECTGSYYC KKGVILPIWE PQDPSFGDKI ARATVYFVAM

 51  VYMFLGVSII ADRFMSSIEV ITSQEKEITI KKPNGETTKT TVRIWNETVS
                       2                                  3
101  NLTLMALGSS APEILLSVIE VCGHNFTAGD LGPSTIVGSA AFNMFIIIAL
                                       4
151  CVYVVPDGET RKIKHLRVFF VTAAWSIFAY TWLYIILSVI SPGVVEVWEG
          5
201  LLTFFFFPIC VVFAWVADRR LLFYKYVYKR YRAGKQRGMI IEHEGDRPSS

251  KTEIEMDGKV VNSHVDNFLD GALVLEVDER DQDDEEARRE MARILKELKQ

301  KHPEKEIEQL IELANYQVLS QQQKSRAFYR IQATRLMTGA GNILKRHAAD

351  QARKAVSMHE VNTEVAENDP VSKIFFEQGT YQCLENCGTV ALTIIRRGGD

401  LTNTVFVDFR TEDGTANAGS DYEFTEGTVV FKPGETQKEI RVGIIDDDIF

451  EEDENFLVHL SNVKVSSEAS EDGILEANHV SALACLGSPS TATVTIFDDD

501  HAGIFTFEEP VTHVSESIGI MEVKVLRTSG ARGNVIVPYK TIEGTARGGG

551  EDFEDTCGEL EFQNDEIVKT ISVKVIDDEE YEKNKTFFLE IGEPRLVEMS

601  EKKALLLNEL GGFTITGKYL YGQPVFRKVH AREHPIPSTV ITIAEEYDDK

651  QPLTSKEEEE RRIAEMGRPI LGEHTKLEVI IEESYEFKST VDKLIKKTNL
                                                         6
701  ALVVGTNSWR EQFIEAITVS AGEDDDDDEC GEEKLPSCFD YVMHFLTVFW
                                     7                   8
751  KVLFAFVPPT EYWNGWACFI VSILMIGILT AFIGDLASHF GCTIGLKDSV
                                                         9
801  TAVVFVALGT SVPDTFASKV AATQDQYADA SIGNVTGSNA VNVFLGIGVA
                                          10
851  WSIAAIYHAA NGEQFKVSPG TLAFSVTLFT IFAFINVGVL LYRRRPEIGG
                           11
901  ELGGPRTAKL LTSCLFVLLW LLYIFFSSLE AYCHIKGF
```

FIGURE 1. Deduced amino acid sequence of the sarcolemmal exchanger. The amino acid sequence was deduced from the nucleotide sequence of the exchanger cDNA clone (GenBank accession number M36119). Residues −32 to −1 correspond to the cleaved leader peptide. Note that the numbering has been changed from Nicoll *et al.*[2] Potential membrane-spanning segments are underlined and numbered. The glycosylation site is indicated with an asterisk.

primary structure of the exchanger, we propose the tentative model shown in FIGURE 2. Potential functionally important regions are discussed below.

Hydropathy analysis of the exchanger indicates 12 potential membrane-spanning segments, the first of which corresponds to the cleaved leader peptide. This leaves 11 potential membrane-spanning segments. The first five membrane-spanning segments are in the amino-terminal portion of the protein. Because only the asparagine residue at position 9 is glycosylated, and therefore most likely to be at the extracellular surface, the amino terminus is modeled to be extracellular. This places the region between

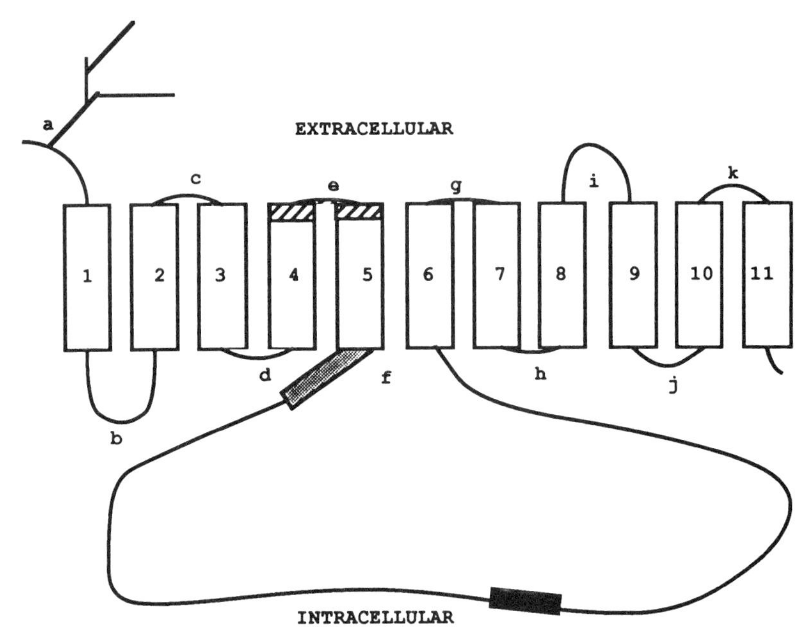

FIGURE 2. Model for the exchanger. The model is based on hydropathy analysis. Note that the model has one less putative transmembrane segment than our previously published model[2] to reflect the removal of a cleaved leader peptide during protein processing. Specific regions of interest are indicated: The single glycosylation site at N9; ▨ The region of similarity to Na^+,K^+-ATPase; ⊛ the exchanger inhibitory peptide region; and ● the potential calmodulin-like Ca^{2+}-binding site.

membrane-spanning segments 4 and 5 at the extracellular surface. This region is 48% similar to an analogous region in the Na^+, K^+ ATPase (FIG. 3), which has also been modeled to an extracellular surface.

Following membrane-spanning segment 5, there is a long hydrophilic, presumably cytoplasmic, segment. The amino-terminal portion (residues 219-238) of this segment contains a number of basic residues interspersed with hydrophobic residues. This motif is common among the calmodulin-binding domains of a number of proteins. For some calmodulin-binding proteins, the calmodulin-binding region has been found to be autoinhibitory. That is, a peptide corresponding to the calmodulin-binding domain can inhibit protein function.[5] With this in mind, a peptide corresponding to residues 219-238 was synthesized and subsequently found to inhibit the exchanger.[6] This peptide was named XIP (eXchanger **I**nhibitory **P**eptide). XIP inhibited the exchanger only at the intracellular surface, thereby supporting the model that the XIP region and the portion of the exchanger with which XIP interacts is on the intracellular surface. The domain with which XIP interacts has not yet been defined. Because XIP has an overall net positive charge, it is anticipated that the interaction site might contain a net negative charge. The exchanger has a number of acidic regions that might serve as an XIP interaction site, most of which are in the long hydrophilic region.

The role of the large hydrophilic domain in exchanger function is not yet clear. A mutant exchanger ("-Bcl") with a deletion encompassing most of the hydrophilic domain (residues 240-679 were deleted, see FIG. 4a) was constructed and examined for exchanger activity. The −Bcl exchanger was capable of expressing Na^+-gradient-dependent Ca^{2+} uptake (FIG. 4b). Therefore, the deleted region is not essential for ion translocation. Alternative roles for the deleted region may include regulatory functions and/or attachment sites for cytoskeletal elements.

The 200 amino acids at the carboxyl terminus of the exchanger contain hydrophobic regions and are modeled to contain an additional six membrane-spanning segments. With a total of 11 membrane-spanning segments, the amino and carboxyl termini of the exchanger would be expected to be on opposite sides of the membrane; Hence, the carboxyl terminus would be located on the intracellular surface. It should be noted, however, that the region between membrane-spanning segments 10 and 11 contains a number of basic residues. Others have noted that membrane-spanning segments generally have positive charges on the cytoplasmic side and few basic residues in the adjoining extracellular segments.[7] Perhaps the number of membrane-spanning segments in the carboxyl end of the exchanger is not six. Either more or fewer membrane-spanning segments in this part of the exchanger is possible. For example, putative membrane-spanning segment 8 is one of the shortest (19 residues) and has the lowest hydropathy value of the six membrane-spanning segments of the carboxyl region. If this segment does not span the membrane, then the basic region between membrane-spanning segments 10 and 11 would be intracellular and the carboxyl terminus would be extracellular. Likewise, the same would be seen if there was an additional membrane-spanning segment in the exchanger that is not detected by hydropathy analysis.

The exchanger's primary structure was examined for regions of possible functional importance. A potential EF-hand type Ca^{2+}-binding site is located at position 498-509 (FIG. 3a). This site could be the high-affinity Ca^{2+} regulatory site that has been observed by others.[8] However, preliminary data obtained from giant membrane patches of

A

	*		*		*		*		*		*	
CaM	D	-	D	G	D	G	-	I	-	-	E	E
NaCaX	D	D	D	H	A	G	I	F	T	F	E	E

B

FIGURE 3. Two regions of the exchanger which may interact with calcium. **A.** Residues 498-509 of the exchanger (NaCaX) are similar to the consensus EF-hand calcium-binding site (CaM). The residues in this site that chelate calcium are indicated with asterisks. Dashed lines indicate nonconserved amino acid residues. **B.** Residues 180-202 of the exchanger are 48% identical to residues 308-330 of the Na^+-K^+ATPase. A similar region in the sarcoplasmic reticular Ca^{2+} pump is also shown.

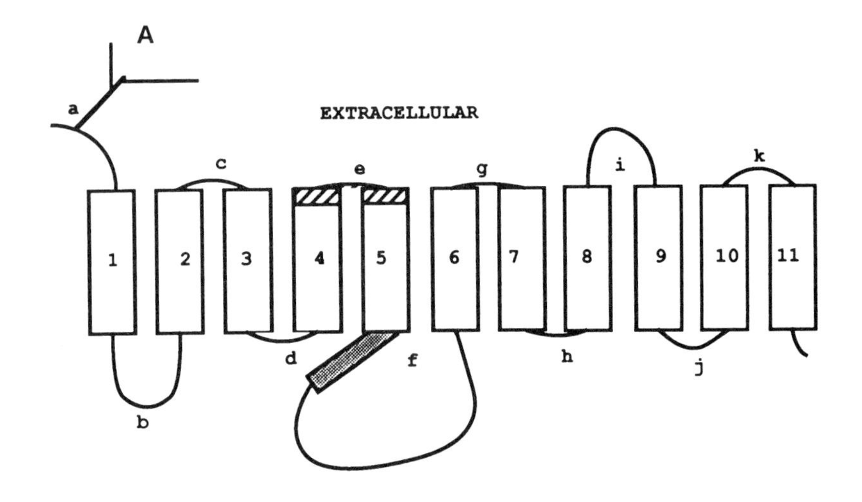

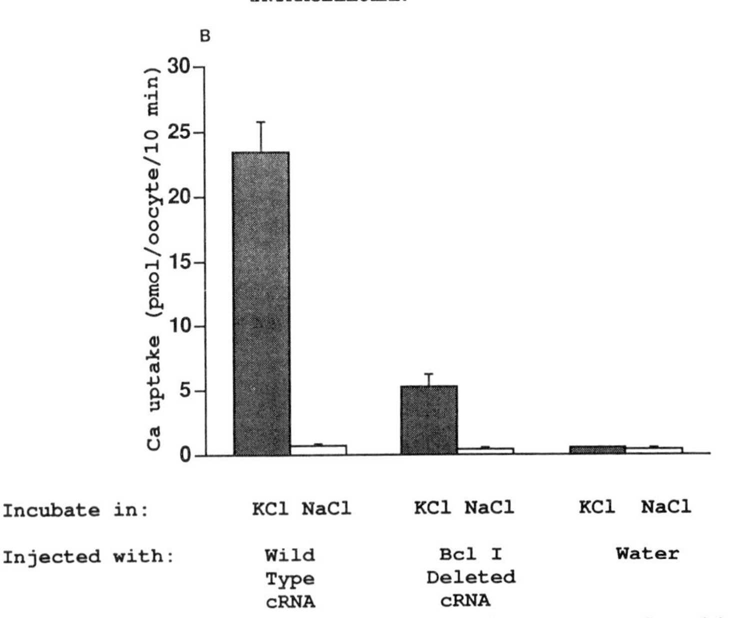

FIGURE 4. The large cytoplasmic region of the exchanger is not necessary for activity. To examine the role of the large cytoplasmic loop in exchanger activity, a mutant with a deletion of residues 240-679 was produced by digesting the wild-type clone with the restriction enzyme Bcl I, gel purifying the resulting fragment and re-ligating to form the deleted clone. **A.** Model for the −Bcl I deleted exchanger. **B.** Water or cRNA that was synthesized from either the wild-type or −Bcl, deleted clone and injected into oocytes. Exchanger activity was measured as described previously.[2]

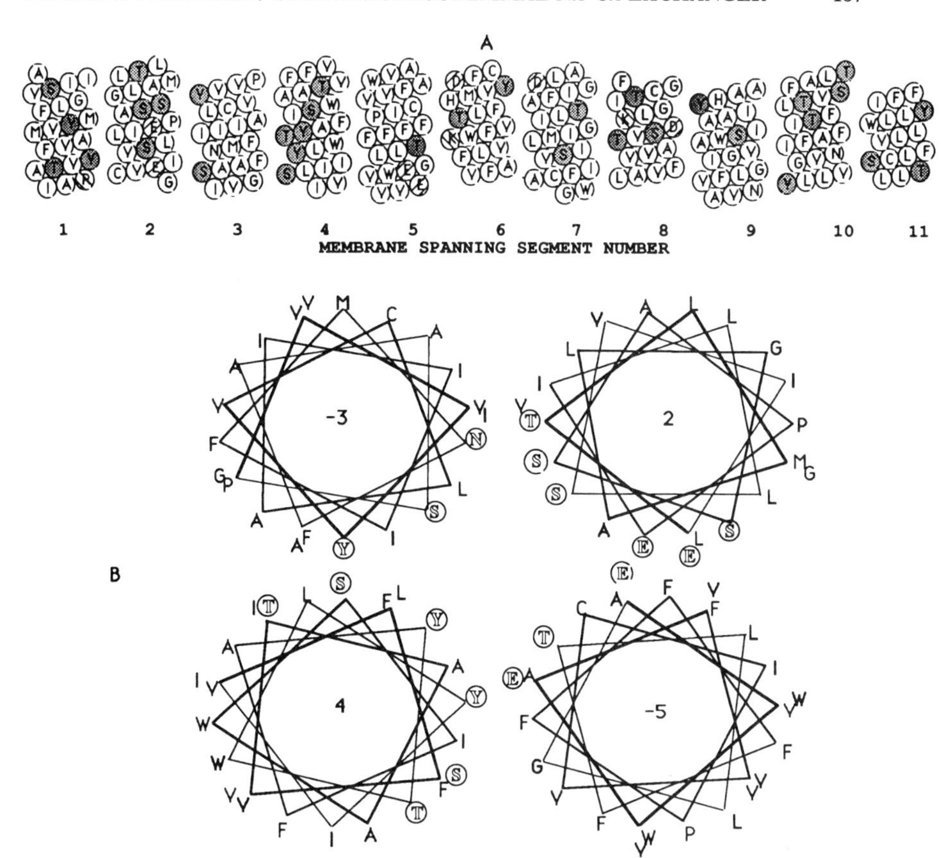

FIGURE 5. Hydrophilic residues in membrane-spanning segments may be involved in ion translocation. **A.** Membrane-spanning segments contain a number of ◌ hydroxyl-containing residues and also some acidic⊘ and basic ⊘ residues. **B.** Helical wheel representations of four of the membrane-spanning segments show that these helices are amphipathic and could form an ion-translocation pathway lined with hydroxyl-containing or charged residues.

oocytes expressing the −Bcl exchanger, which has this site deleted, indicate that the −Bcl exchanger is still regulated by intracellular Ca^{2+} (D. Hilgemann, D. Nicoll, K. Philipson, unpublished observation).

Another potential Ca^{2+}-interaction site is in membrane-spanning segment 5. This is the region that is homologous to the Na^+,K^+-ATPase (FIG. 3b). While the role of this region in the Na^+,K^+-ATPase is not known, a similar region in the sarcoplasmic reticular Ca^{2+} pump appears to be involved in Ca^{2+} binding.[9]

During the Na^+-Ca^{2+} exchange cycle, ions must cross the membrane. The conduction pathway across the membrane is likely to be lined with hydrophilic and/or charged residues to provide a compatible environment for ion translocation. An examination of the putative membrane-spanning segments shows that there are a number of charged residues and many hydroxyl-containing residues in these segments (FIG. 5a). Most of the membrane-spanning segments are amphipathic, with hydrophobic residues on one face and hydrophilic residues on the other. These amphipathic helices may be arranged in the membrane to form an ion-translocation pore (FIG. 5b).

SUMMARY

The molecular nature of the canine cardiac sarcolemmal Na^{+}-Ca^{2+} exchanger has been investigated by purification of the protein and by sequencing and expression of an exchanger cDNA clone. The mature exchanger protein is apparently 120 kDa, with glycosylation at a single asparagine residue near the amino terminus. A proposed model for the exchanger protein includes 11 transmembrane segments, a large cytoplasmic domain that is not involved in ion translocation, an exchanger inhibitory site, two Ca^{2+} interaction sites and an ion-translocation pathway. Experiments are now under way to test the proposed model.

REFERENCES

1. PHILIPSON, K. D., S. L. LONGONI & R. WARD. 1988. Biochim. Biophys. Acta **945:** 298-306.
2. NICOLL, D. N., S. LONGONI & K. D. PHILIPSON. 1990. Science **250:** 562-565.
3. VON HEIJNE, G. 1986. Nucleic Acids Res. **14:** 4683.
4. DURKIN, J. T., D. C. AHRENS, J. D. HULMES, Y.-C. E. PAN & J. P. REEVES. 1991. Biophys. J. **59:** 137a.
5. ENYEDI, A., T. VORHERR, P. JAMES, D. J. MCCORMICK, A. G. FILOTEO, E. CARAFOLI & J. T. PENNISTON. 1989. J. Biol. Chem. **264:** 12313-12321.
6. LI, Z., D. A. NICOLL, A. COLLINS, D. W. HILGEMANN, A. G. FILOTEO, J. T. PENNISTON, J. N. WEISS, J. M. TOMICH & K. D. PHILIPSON. 1991. J. Biol. Chem. **266:** 1014-1020.
7. BOYD, D. & J. BECKWITH. 1990. Cell. **62:** 1031-1033.
8. MIURA, Y. & J. KIMURA. 1989. J. Gen. Physiol. **93:** 1129-1145.
9. CLARKE, D. M., T. W. LOO, G. INESI & D. H. MACLENNAN. 1989. Nature **339:** 476-478.

Molecular and Functional Studies of the Cardiac Sodium-Calcium Exchanger

JOHN T. DURKIN, DIANE C. AHRENS, JOSEPH F. ACETO, MADALINA CONDRESCU, AND JOHN P. REEVES[a]

Roche Institute of Molecular Biology
Roche Research Center
Nutley, New Jersey 07110

In heart cells, the Na^+-Ca^{2+} exchange system is the predominant mechanism for mediating Ca^{2+} efflux.[1,2] Therefore, understanding the Na^+-Ca^{2+} exchange reaction at a molecular level has obvious importance for understanding the regulation of cardiac contractility and perhaps in devising new therapeutic approaches to the treatment of heart disease as well. Recently the cardiac exchanger has been purified, cloned, and sequenced.[3] The cDNA sequence yields a predicted molecular weight of 108 kDa for the exchange protein. This represents a major breakthrough in our understanding of Na^+-Ca^{2+} exchange, and, because of the power and precision of molecular biological techniques, it is certain to stimulate a great deal of new research in this area. In this contribution, we will describe a new method for the purification of the bovine cardiac Na^+-Ca^{2+} exchanger[4] and its use in obtaining sequence information on the mature exchange protein and in examining its functional activity under pre-steady-state conditions.

PURIFICATION OF THE CARDIAC EXCHANGER

The procedure we have developed for purifying the exchange carrier follows the scheme depicted in FIGURE 1. Silver-stained SDS gels of samples taken at various stages of the purification procedure are presented in FIGURE 2. The details of this procedure will be described elsewhere,[4] and only the broad outline and basic rationale of the various steps will be presented here. The starting material is bovine cardiac sarcolemmal vesicles (FIG. 2, lane A), prepared as described by Cheon and Reeves[5] in the presence of a cocktail of protease inhibitors.[4] The vesicles are extracted with an alkaline buffer as originally described by Philipson *et al.*[6] to remove peripheral membrane proteins; the extracted vesicles are then solubilized with 2% Na cholate, 100 mM NaCl (pH 7.4) containing soybean phospholipids (2 mg/ml asolectin). The exogenous lipids are necessary to preserve the activity of the solubilized protein.[7] After removing the insoluble debris by centrifugation, the solubilized material (FIG. 2, lane B) is applied to a column of Affi-Gel 102 for anion-exchange chromatography. Affi-Gel

[a] Address for correspondence: John P. Reeves, Ph.D., Roche Institute of Molecular Biology, Roche Research Center, 340 Kingsland St., Nutley, NJ 07110-1199.

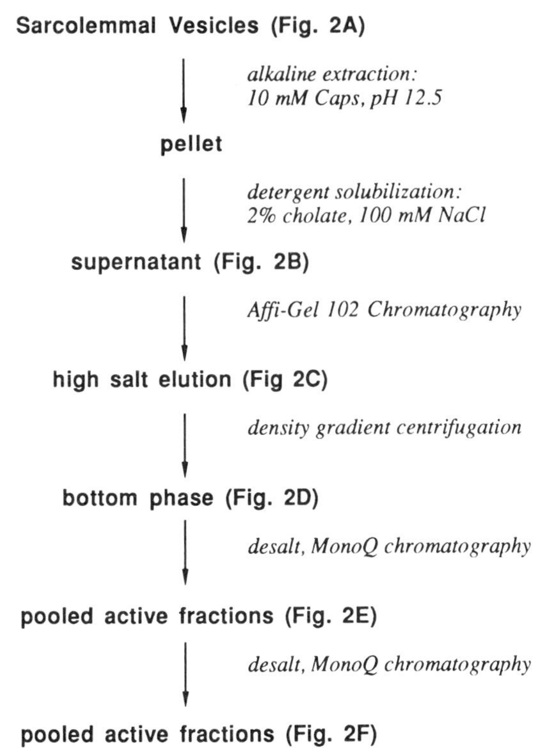

FIGURE 1. Outline of the procedure for purifying the cardiac Na^{+}-Ca^{2+} exchange carrier. The references in parenthesis denote the lanes in the SDS gel in FIGURE 2 for the appropriate stage of purification.

102 (Bio-Rad) contains terminal amino groups coupled to an agarose matrix at a relatively low site density. The low site density of amino groups may be an important factor in the effectiveness of this step because the use of exchange resins with a higher site density (e.g. DEAE-Sepharose) led to substantial losses in activity and increased binding of other proteins to the column. The bound material is eluted with 500 mM NaCl containing 2% cholate (FIG. 2, lane C) and concentrated by density gradient centrifugation.

At this stage of the purification procedure (FIG. 2, lane D), the exchanger is 30-50% pure and the total recovery of activity is 60-80%. The V_{max} for exchange activity in reconstituted preparations[b] of this material is 1-2 μmole/mg protein·sec as compared to 0.02-0.05 μmol/mg protein·sec for reconstituted preparations of detergent-solubilized sarcolemmal vesicles without any purification. The procedure up to this point is remarkably simple and can readily be scaled up to produce large quantities of

[b]Reconstitution involves adding exogenous phospholipids (asolectin) to the detergent-solubilized material and then removing the detergent with adsorbent beads (Bio-Beads SM-2; Bio-Rad). Vesicles, called proteoliposomes, reform spontaneously as the detergent is removed and exchange activity is measured following standard vesicle assay techniques.

highly active exchanger in a concentrated form. This partially purified preparation is very useful for studies of functional activity where a high site density of exchanger is required (cf. below).

Further purification of the exchanger involves two additional rounds of anion exchange chromatography on a Mono-Q column using the FPLC (fast protein liquid chromatography) system developed by Pharmacia (FIG. 2, lanes E & F). Unfortunately, some exchange activity is lost on the Mono-Q column, and, despite the increase in purity, the specific activity of the final product is approximately the same as the partially purified preparations described above.

CHARACTERISTICS OF THE PURIFIED EXCHANGER

Two principal bands, at 120 kDa and 160 kDa, are observed in SDS gels of the purified exchanger (FIG. 2F). Some preparations also show traces of a band at 70 kDa,

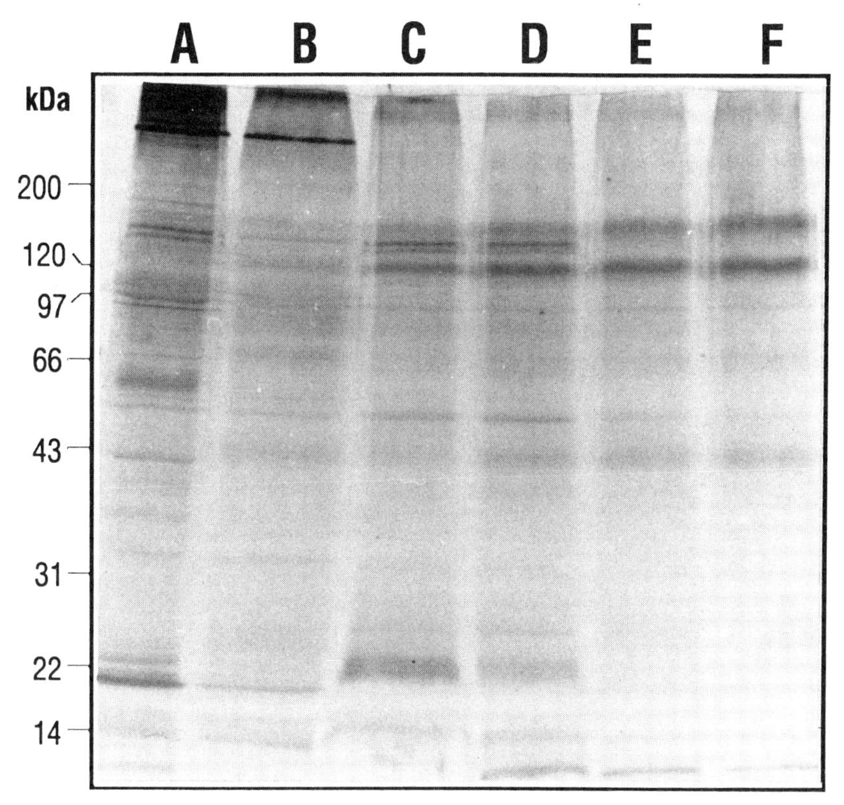

FIGURE 2. SDS polyacrylamide gel of samples taken at various stages of purification (cf. FIG. 1). **A**, native vesicles; **B**, cholate-solubilized extract; **C**, high-salt elution from Affi-Gel 102 column; **D**, Affi-gel elution after density gradient centrifugation; **E**, pooled active fractions after first Mono-Q column; **F**, pooled active fractions after second run on Mono-Q. The protein bands were visualized by silver staining (Bio-Rad). Reprinted from Durkin *et al.*[4] with permission from the *Archives of Biochemistry and Biophysics.*

as originally reported by Philipson *et al.*[8]; this appears to be a proteolytic fragment of the intact exchanger, and the low amounts of this band in these preparations can probably be attributed to the use of protease inhibitors at every stage of the procedure. Additional minor bands are also observed in some preparations, but their intensity shows no correlation during purification with Na^{+}-Ca^{2+} exchange activity (FIG. 2).

The two bands at 160 and 120 kDa do not appear to be different subunits of the exchanger. The NH_2-terminal amino acid sequences of the two bands are identical (cf. below). The 120-kDa band does not appear to be a proteolytic fragment of the 160-kDa band because treatment of the exchanger with trypsin or chymotrypsin does not produce more of the 120-kDa form at the expense of the 160-kDa form. The 120-kDa form can be converted to the 160-kDa form by a variety of procedures, that is, boiling the samples,[4] or preparing the samples for electrophoresis under nonreducing conditions.[8] This suggests that the 160-kDa band is not simply glycosylated more heavily than the 120-kDa band. While the origin of these two bands is rather mysterious, the simplest explanation is that the two bands represent alternate conformational states of the same protein that differ in their ability to bind SDS, and therefore migrate with different mobilities during electrophoresis. The difference in conformation, however, would have to be extremely stable to withstand the denaturing effects of SDS in the electrophoresis sample buffer. Because oxidizing conditions appear to cause all the bands (including the 70-kDa band, when present) to migrate at 160 kDa,[8] it is tempting to suggest that disulfide bond formation is involved. It is unknown whether there is any physiological significance to the presence of these two putative conformational states of the exchanger. It is interesting to note, however, that previous studies with native vesicles had shown that exchange activity could be markedly stimulated by a combination of redox reagents that would be expected to produce sulfhydryl-disulfide interchange.[9]

IMMUNOLOGICAL CROSS-REACTIVITY WITH RETINAL ROD EXCHANGER

The bovine cardiac exchanger cross-reacts with polyclonal antibodies to the canine cardiac exchanger and the bovine retinal rod exchanger (FIG. 3). Bands at 160, 120, and a doublet at approximately 70 kDa show strong reactivity with the canine cardiac antibodies in Western blots of partially purified preparations of the exchanger (FIG. 3, right lane). This pattern is also observed in Western blots of native vesicles (data not shown). The doublet at 70 kDa shows a surprisingly high reactivity with the antibodies given the low staining intensity of the bands in these preparations (FIG. 2F). The most likely explanation for this result is that there is a highly reactive epitope that becomes much more accessible for interaction with the antibodies after proteolytic cleavage of the exchanger.

The immunological cross-reactivity of the canine and bovine cardiac exchangers is not surprising, since there appears to be a high degree of sequence homology between the two proteins (cf. below). The bovine retinal rod exchanger, however, has a higher molecular weight (220 kDa)[10] and shows little sequence homology with the cardiac exchanger (N. Cook, personal communication). These differences may reflect the functional differences between the two exchangers, since the exchange reaction in retinal rods involves cotransport of Ca^{2+} and K^{+},[11,12] whereas the cardiac exchange reaction does not involve K^{+}. Despite these dissimilarities, the cardiac exchanger cross-reacts weakly with antibodies to the retinal rod exchanger (FIG. 3, left lane), suggesting that some degree of homology must exist between the two proteins.

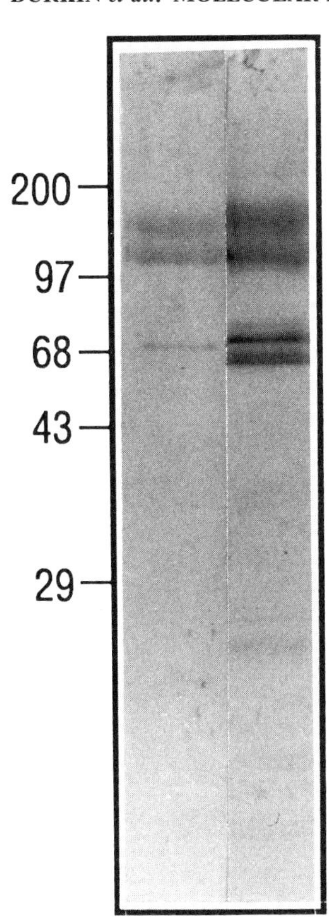

FIGURE 3. Western blot of partially purified exchanger using polyclonal antibodies to the canine cardiac exchanger (right lane) or the bovine retinal rod exchanger (left lane). Material after density gradient centrifugation (FIG. 2, lane D) was run on an SDS polyacrylamide gel, transferred to Immobilon and incubated with a 1 : 500 dilution of the antibodies after blocking with 5% bovine serum albumin. Detection of bound antibodies utilized akaline phosphatase-conjugated anti-rabbit IgG antibodies. Molecular weight markers (in kDa) are indicated on the left. We thank Dr. K. Philipson and Dr. N. Cook for the antibodies used.

NH_2-TERMINAL SEQUENCE

The NH_2-terminal amino acid sequences of both the 120- and 160-kDa bands were determined[4] and found to be identical.[c] The sequence begins with the following residues: ETEMEG. This is identical to a sequence beginning at residue 33 of the deduced sequence obtained from the cDNA clone of the canine cardiac exchanger.[3] The first 32 residues of the canine exchanger have the attributes of a cleavable signal sequence,[13,14] including a stretch of hydrophobic amino acids (LLAVVALLF in the cDNA sequence) and the presence of basic amino acids in the region NH_2-terminal to the hydrophobic region. The NH_2-terminal sequence of the protein begins immediately after an amino acid with a small side chain (A_{32} in the cDNA sequence), which corresponds to a predicted cleavage site for a signal sequence, as pointed out by Nicoll *et al.*[3]

A signal sequence is found at the NH_2-terminus of secreted proteins in both eucaryotic and procaryotic cells; it is an important recognition element for the machinery

[c] We thank Dr. Yu-Ching Pan and Jeffrey Hulmes, Department of Protein Biochemistry, Roche Research Center, for carrying out the protein sequencing.

involved in translocating secreted proteins across biomembranes during their synthesis on the ribosome (see Refs. 14-16). After translocation across the membrane, the signal sequence is cleaved by a specific peptidase, releasing the newly synthesized protein. Certain membrane proteins also have signal sequences, although many others do not. To the best of our knowledge, the cardiac Na^+-Ca^{2+} exchanger is the only membrane transport carrier for which a signal sequence has been described. For other carrier proteins, the NH_2-terminus is thought to lie on the cytoplasmic side of the membrane and the first transmembrane segment appears to function as an internal, *uncleaved* signal sequence. For membrane proteins without a signal sequence, the characteristics of the flanking regions on either side of the first transmembrane segment determine the topological orientation of the protein within the membrane.[17-21]

What is the function of the signal sequence for the cardiac Na^+-Ca^{2+} exchanger? This question cannot be answered with certainty, but the amino terminus of the exchanger is highly polar (6 of the first 13 residues are glutamates), and a signal sequence may be necessary to insure the efficient translocation of this portion of the protein to the exterior.[4] This raises the question as to whether the extracellular location of the NH_2-terminal domain might be functionally important. Because of its high polarity, the NH_2-terminal segment would probably be well removed from the membrane interior and therefore would not be expected to participate in transport *per se*. However, it is possible that it might be involved in interactions with other membrane proteins that would place the exchanger at sites on the membrane where its efficiency in regulating intracellular Ca^{2+} levels would be topographically advantageous. It will be interesting to see if the apparent external location and high polarity of the amino-terminal region are features of Na^+-Ca^{2+} exchangers from other tissues as their sequences become available.

FUNCTIONAL STUDIES WITH PARTIALLY PURIFIED EXCHANGER

As mentioned previously, the initial steps in the purification scheme (through the density gradient step; FIG. 1) can be utilized for the rapid preparation of quite large amounts of partially purified exchange carrier in a highly active and concentrated form (FIG. 2D). These preparations would appear to be suitable for examining the pre-steady-state kinetics of Na^+-Ca^{2+} exchange, that is, attempting to observe a single turnover event. This should be possible in these preparations, since, with an approximate turnover number of 1,000 sec^{-1} (see Cheon & Reeves[5]), the amount of active exchanger estimated from the V_{max} (1-2 μmol/mg protein·sec) would be 1-2 nmol/mg protein, an easily detectable quantity. Assuming that the exchanger is a 108-kDa protein (from the cDNA sequence), a site density of 1-2 nmol/mg protein would be equivalent to 11-22% of the protein in these partially purified preparations being active exchanger, a figure which seems reasonable in view of the SDS gel shown in FIGURE 2D.

A simplified diagram for a consecutive (or "ping-pong") reaction mechanism for Na^+-Ca^{2+} exchange is depicted in FIGURE 4 and will be our guide for the design of experiments along this line. In this scheme, the translocation of Ca^{2+} and Na^+ across the membrane occurs during separate steps in the reaction mechanism. Ping-pong mechanisms are to be distinguished from "simultaneous" (also called "sequential") models in which both ions are translocated at the same time. Much of the early kinetic data in various experimental systems favored the simultaneous model,[22,23] but the more recent kinetic studies of Khananshvili,[24] Hilgemann *et al.*,[25] Kimura,[26] and Niggli and

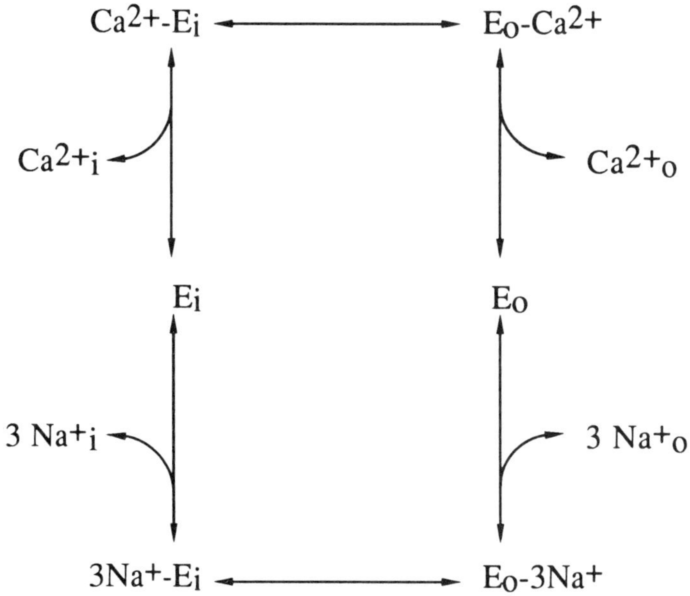

FIGURE 4. Diagram of the consecutive (ping-pong) model for the Na^+-Ca^{2+} exchange reaction. The subscripts *o* and *i* refer to the external and internal sides of the membrane; E indicates the exchanger.

Lederer[27] have each provided data that favor the consecutive reaction mechanism. Although a number of kinetic tests have been formulated for distinguishing between the two types of mechanisms,[29] kinetic data alone do not always provide an unequivocal resolution of this issue. It is clear, however, that the two types of reaction schemes can be sharply differentiated in terms of their transient, pre-steady-state behavior, that is, in the consecutive model, the exchanger can translocate Ca^{2+} in the complete absence of Na^+ by a half-turnover event (FIG. 4), whereas translocation of Ca^{2+} always requires Na^+ in the simultaneous model.

HALF-TURNOVER OF THE EXCHANGER

The procedure we followed in attempting to detect a half-turnover is depicted in FIGURE 5. Proteoliposomes were reconstituted in 200-300 mM NaCl containing 10 mM EDTA and then treated with 8 μM monensin, an ionophore that catalyzes Na^+-H^+ and K^+-H^+ exchanges. The proteoliposomes (250 μl) were centrifuged twice through 2.5-ml columns of Sephadex G-50 (equilibrated with 160 mM KCl containing 20 μM EGTA) to remove the external NaCl and EDTA. Because of the presence of monensin, the internal Na^+ will exchange for external K^+ during passage of the proteoliposomes through the columns. The proteoliposomes were then mixed with four times their volume of 160 mM KCl (pH 7.0) containing 35 μM EGTA and 30 μM $^{45}Ca^{2+}$ (final $[Ca^{2+}] \cong 1.2$ μM) and assayed for Ca^{2+} uptake. As a control, the uptake reaction was also carried out in 160 mM NaCl under the same conditions; at this low $[Ca^{2+}]$, the high concentration of Na^+ should block any Ca^{2+} uptake by the exchanger.

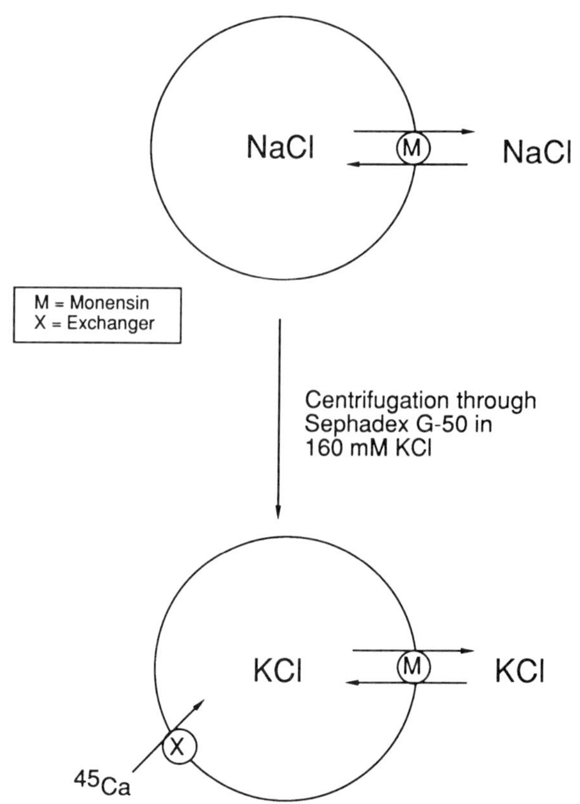

FIGURE 5. Outline of protocol followed for measuring half-turnover of the Na^{+}-Ca^{2+} exchanger (see results in FIG. 7). See text for details.

These experiments were conducted at 4°C to minimize passive Ca^{2+} entry into the proteoliposomes; control studies showed that the Na^{+}-Ca^{2+} exchanger was still active at this temperature.

The results, shown in FIGURE 6, indicate that Ca^{2+} uptake was the same in KCl and in NaCl, that is, there was no detectable exchanger-mediated Ca^{2+} uptake by the proteoliposomes. The low levels of Ca^{2+} uptake that were observed in this experiment were identical when the experiment was repeated using a reconstituted preparation of *unpurified* exchanger, in which the content of exchange carriers is 10- to 20-fold lower. This background Ca^{2+} accumulation is probably due to binding to the phospholipids and/or passive entry of Ca^{2+} into leaky proteoliposomes. Note that Ca^{2+} entry due to a half-turnover of the exchanger, expected to be 1-2 nmol/mg protein, would have been easily detectable in this experiment.

While this result is unexpected in terms of the consecutive model (FIG. 4), it is possible that in the absence of internal Na^{+}, all the Ca^{2+}-binding sites on the exchange carrier are sequestered within the vesicle interior and are therefore unavailable to mediate Ca^{2+} uptake by a half-turnover. To test this possibility, the experiment depicted in FIGURE 7 was undertaken. Monensin-treated proteoliposomes were centrifuged through Sephadex G-50 in KCl as described above and then NaCl was added to a final

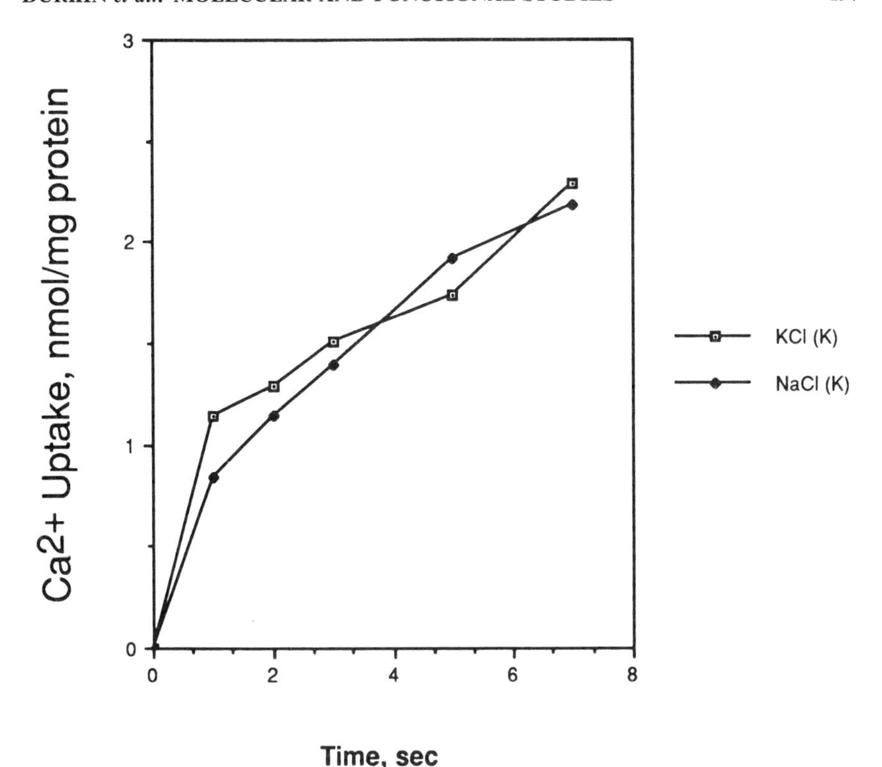

FIGURE 6. Ca^{2+} uptake at 4°C by K^+-loaded proteoliposomes in 160 mM KCl (open squares) or 160 mM NaCl (closed diamonds) containing 1.2 μM free Ca^{2+}.

concentration of 8 mM. After an equilibration period of 5 min, the proteoliposomes were diluted 10-fold into either (a) 8 mM NaCl-152 mM KCl or (b) 160 mM KCl, and ^{45}Ca was added after the delay times indicated in FIGURE 7. When the vesicles are diluted into 160 mM KCl, the monensin will allow the internal Na^+ to exchange rapidly for external K^+ to achieve a final concentration of 0.8 mM, a concentration that is very inefficient in activating Na^+-Ca^{2+} exchange.

It should be noted that these experiments, unlike those described above, were carried out at 25°C so that the internal Na^+ would be rapidly removed by the monensin; control studies showed that 1 sec was a sufficient time to remove nearly all the internal Na^+ under these conditions. The rationale behind the experiment is that the initial presence of internal Na^+ should permit the translocation of a substantial fraction of the exchange carrier binding sites to the exterior where they would be available to interact with external Ca^{2+}. Based on the fact that the exchanger does not mediate unidirectional ion fluxes in the absence of an exchange partner, the "externalized" carriers would not be expected to translocate back to the vesicle interior in the absence of external Na^+ or Ca^{2+} and should remain available to mediate a half-turnover of Ca^{2+} uptake for a considerable period of time.

The results of this experiment are shown in FIGURE 8, where the data represent the initial rates (1 sec) of Ca^{2+} uptake after various intervals between the dilution step and $^{45}Ca^{2+}$ addition (cf. FIG. 7). Also shown is a control experiment that was carried

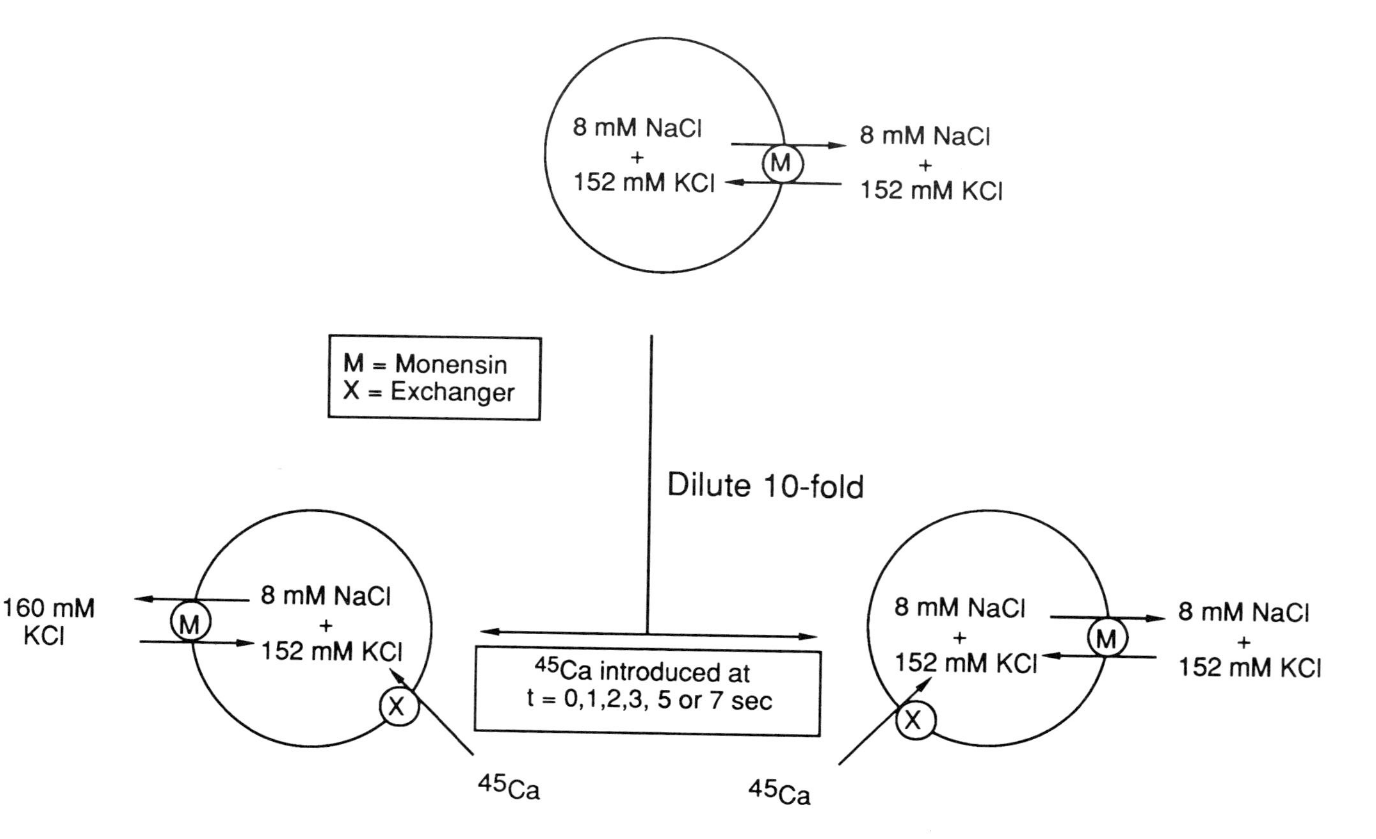

FIGURE 7. Outline of protocol for measuring decay of Ca^{2+} uptake after removal of internal Na^+. See text for details.

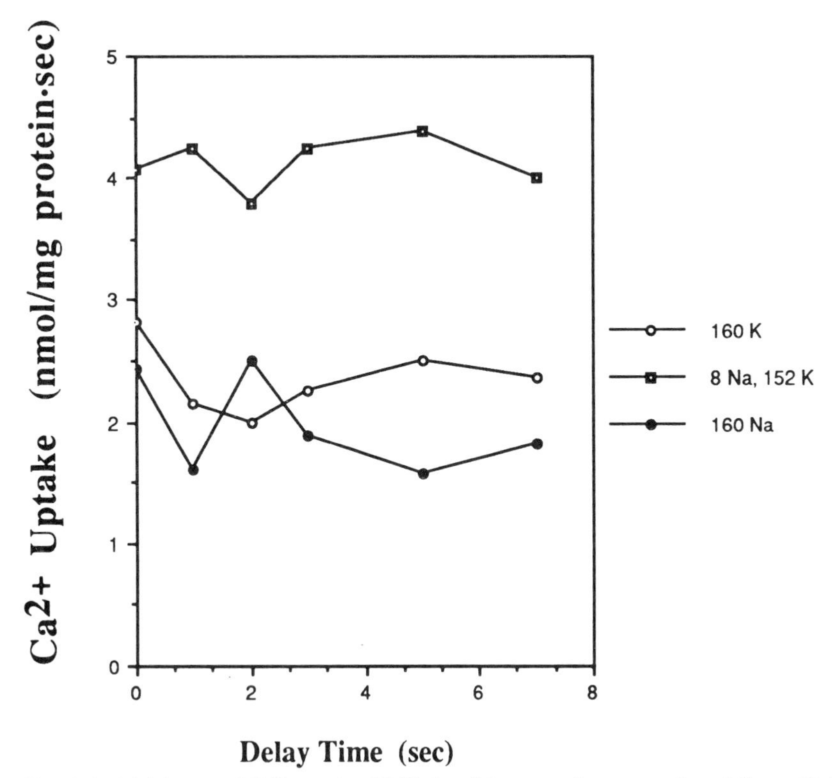

FIGURE 8. Initial rates of Ca^{2+} uptake (25 °C; 1 sec) by proteoliposomes of partially purified exchanger following the protocol outlined in FIGURE 7. The vesicles were diluted into either 160 mM KCl (open circles), 160 mM NaCl (closed circles), or 8 mM NaCl/152 mM KCl (closed squares) containing 1.7 μM free Ca^{2+}.

out in 160 mM NaCl to block any Ca^{2+} uptake by the exchanger. The results show that Ca^{2+} uptake by the vesicles diluted into 160 mM KCl is only slightly above that of the NaCl control at each time point assayed (including t = 0, in which $^{45}Ca^{2+}$ was present initially in the KCl dilution medium). This low level of Ca^{2+} uptake over background is equal to that which would be seen if Na^+-Ca^{2+} exchange were measured in monensin-treated proteoliposomes equilibrated and assayed in 0.8 mM NaCl, the final $[Na^+]$ in these experiments after the dilution step (data not shown). In contrast, the initial rate of Ca^{2+} uptake measured in the 8 mM NaCl medium is well above background and shows that 8 mM internal Na^+, when present throughout the experiment, is high enough to activate exchange activity to a readily detectable level.

These results are difficult to reconcile with the consecutive model of Na^+-Ca^{2+} exchange (FIG. 4). The rapid rundown of exchange activity as Na^+ is removed from the vesicle interior (FIG. 8) indicates that the exchanger retains no "memory" of the previous presence of internal Na^+. Ping-pong mechanisms, by definition, retain a memory of the previous presence of a substrate through the formation of a relatively stable intermediate complex in the reaction mechanism, that is, externalized carriers in the case of Na^+-Ca^{2+} exchange. The results in FIGURES 6 and 8 suggest that either no such intermediate exists, or its amount is too small to be detectable in these studies.

However, if a half-turnover of the exchanger were undetectably small (< 0.3 nmol/mg protein), then the amount of active exchanger in the partially purified preparations would have to be $< 3\%$ of the total protein and this would not be consistent with either the amounts of the exchange protein seen in SDS gels of these preparations (FIG. 2D) or the excellent preservation of total activity up to this point in the purification procedure. The results provide the most direct evidence to date for the simultaneous mechanism for Na^{+}-Ca^{2+} exchange. Obviously, more experimental work must be done to reconcile the present results with the wealth of kinetic data[24-27] that seems to favor the consecutive model.

REFERENCES

1. HILGEMANN, D. W. 1986. Extracellular calcium transients and action potential configuration changes related to post-stimulatory potentiation in rabbit atrium. J. Gen. Physiol. **87:** 675-706.
2. BERS, D. M. & J. H. B. BRIDGE. 1989. Relaxation of rabbit ventricular muscle by Na-Ca exchange and sarcoplasmic reticulum calcium pump. Ryanodine and voltage sensitivity. Circ. Res. **65:** 334-342.
3. NICOLL, D. A., S. LONGONI & K. D. PHILIPSON. 1990. Molecular cloning and functional expression of the cardiac sarcolemmal Na^{+}-Ca^{2+} exchanger. Science **250:** 562-565.
4. DURKIN, J. T., D. C. AHRENS, Y.-C. PAN & J. P. REEVES. 1991. Purification and amino-terminal sequence of the bovine cardiac sodium-calcium exchanger. Evidence for the presence of a signal sequence. Arch. Biochem. Biophys. In press.
5. CHEON, J. & J. P. REEVES. 1988. Site density of the sodium-calcium exchange carrier in reconstituted vesicles from bovine cardiac sarcolemma. J. Biol. Chem. **263:** 2309-2315.
6. PHILIPSON, K. C., A. A. MCDONOUGH, J. S. FRANK & R. WARD. 1987. Enrichment of Na^{+}-Ca^{2+} exchange in cardiac sarcolemmal vesicles by alkaline extraction. Biochim. Biophys. Acta **945:** 298-306.
7. MIYAMOTO, H. & E. RACKER. 1980. Solubilization and partial purification of the Ca^{2+}/Na^{+} antiporter from the plasma membrane of bovine heart. J. Biol. Chem. **255:** 2656-2658.
8. PHILIPSON, K. D., S. LONGONI & R. WARD. 1988. Purification of the cardiac Na^{+}/Ca^{2+} exchange protein. Biochim. Biophys. Acta **945:** 298-306.
9. REEVES, J. P., C. A. BAILEY & C. HALE. 1986. Redox modification of sodium-calcium exchange activity in cardiac sarcolemmal vesicles. J. Biol. Chem. **261:** 4948-4955.
10. COOK, N. J. & U. B. KAUPP. 1988. Solubilization, purification and reconstitution of the sodium-calcium exchanger from bovine retinal rod outer segments. J. Biol. Chem. **263:** 11382-11388.
11. SCHNETKAMP, P. P. M., D. K. BASU & R. T. SZERENSCEI. 1989. Na^{+}-Ca^{2+} exchange in bovine rod outer segments requires and transports K^{+}. Am. J. Physiol. **257:** C153-C157.
12. CERVETTO, L., L. LAGNADO, R. J. PERRY, D. W. ROBINSON & P. A. MCNAUGHTON. 1989. Extrusion of calcium from rod outer segments is driven by both sodium and potassium gradients. Nature **337:** 740-743.
13. VON HEIJNE, G. 1983. Patterns of amino acids near signal-sequence cleavage sites. Eur. J. Biochem. **133:** 17-21.
14. GIERASCH, L. M. 1989. Signal sequences. Biochemistry **28:** 923-930.
15. BLOBEL, G. 1980. Intracellular protein topogenesis. Proc. Natl. Acad. Sci. USA **77:** 1496-1500.
16. RAPOPORT, T. A. 1990. Protein transport across the ER membrane. Trends Biochem. Sci. **15:** 355-358.
17. VON HEIJNE, G. & Y. GAVEL. 1988. Topogenic signals in integral membrane proteins. Eur. J. Biochem. **174:** 671-678.
18. HARTMANN, E., T. A. RAPOPORT & H. F. LODISH. 1989. Predicting the orientation of eukaryotic membrane-spanning proteins. Proc. Natl. Acad. Sci. USA **86:** 5786-5790.

19. BELTZER, J. P., K. FIEDLER, C. FUHRER, I. GEFFEN, C. HANDSCHIN, H. P. WESSELS & M. SPIESS. 1991. Charged residues are major determinants of the transmembrane orientation of a signal-anchor sequence. J. Biol. Chem. **266:** 973-978.
20. WESSELS, H. P. & M. SPIESS. 1988. Insertion of a multispanning protein occurs sequentially and requires only one signal sequence. Cell **55:** 61-70.
21. PARKS, G. D. & R. A. LAMB. 1991. Topology of eukaryotic type II membrane proteins: Importance of N-terminal positively charged residues flanking the hydrophobic domain. Cell **64:** 777-787.
22. REEVES, J. P. 1985. The sarcolemmal sodium-calcium exchange system. Curr. Top. Membr. Transp. **25:** 77-127.
23. HILGEMANN, D. W. 1989. Numerical probes of sodium-calcium exchange. *In* Sodium-Calcium Exchange. T. J. A. Allen, D. Noble & H. Reuter, Eds.: 126-152. Oxford University Press. Oxford, England.
24. KHANANSHIVILI, D. 1990. Distinction between the two basic mechanisms of cation transport in the cardiac Na^+-Ca^{2+} exchange system. Biochemistry **29:** 2437-2442.
25. HILGEMANN, D., A. COLLINS, G. A. NAGEL & D. P. CASH. 1991. Ann. N. Y. Acad. Sci. This volume.
26. LI, J. & J. KIMURA. 1990. Translocation mechanism of Na-Ca exchange in single cardiac cells of guinea pig. J. Gen. Physiol. **96:** 777-788.
27. NIGGLI, E. & W. J. LEDERER. 1991. Molecular operations of the sodium-calcium exchanger revealed by conformation currents. Nature **349:** 621-624.
28. LÄUGER, P. 1987. Voltage dependence of sodium-calcium exchange: Predictions from kinetic models. J. Membr. Biol. **99:** 1-12.

Characterization of Monoclonal Antibodies Cross-reacting with Myocardial and Retinal Sodium-Calcium Exchange Proteins

H. PORZIG[a]

Department of Pharmacology
University of Bern
CH 3010 Bern, Switzerland

INTRODUCTION

In many excitable tissues a transport protein that catalyzes the transmembrane exchange of Ca and Na plays a vital part in maintaining a low intracellular Ca concentration against a large inwardly directed electrochemical gradient for this ion.[1] Mammalian myocardial membranes and retinal rod outer segments have been identified as comparatively rich sources for this protein. However, a comparison of the two cation exchange systems has revealed a number of significant differences in their biochemical and functional properties. Thus, the two proteins differ in their apparent molecular weight, in their behavior during chromatographic separation, as well as in their transport kinetics and stoichiometries.[2-4] These differences clearly suggested that the exchange function in heart and in retinal cells is catalyzed by two distinct proteins. This conclusion was further supported by the apparent lack of a marked cross-reactivity among antibodies directed against either one of the two proteins.[5] Up until now, no specific high-affinity ligands are available for these proteins that would allow a detailed pharmacological dissection of their properties. Therefore, we have now used monoclonal antibodies (mAb) directed against the retinal exchanger as specific labels to explore the cardiac exchanger on the basis of shared epitopes.

METHODS

The retinal Na-Ca exchange protein was prepared from bovine retinas according to the method of Cook and Kaupp.[3] Polyacrylamid gel electrophoresis (PAGE) of the final preparation showed the exchanger as a major high-molecular-weight (HMW) protein band in the 215-kDa range as well as several minor bands with apparent molecular weights below 100 kDa. After a PAGE run in 6% gels, the proteins were transblotted onto nitrocellulose sheets. The HMW band was located by Indian ink staining, excised, and transformed into a fine powder using a Duall micro tissue grinder (Kontes, Vineland, NJ). A suspension of this powder was then used as antigen for the immunization of BALB/c mice.[6] The preparation of monoclonal antibodies by fusion

[a]Address for correspondence: Dr. H. Porzig, Pharmakologisches Institut der Universitaet, Friedbuehlstr. 49, CH 3010 Bern/Switzerland.

of spleen lymphocytes with NSO plasmocytoma cells followed essentially the methods given by Westerwoudt.[7] Hybridomas secreting mAbs that recognized the retinal HMW band were cloned by limiting dilution.

Myocardial membranes were prepared from equine hearts combining methods given by Slaughter *et al.*[8] and by Kuwayama and Kanazawa.[9] Partial purification of the Na-Ca exchanger from detergent-solubilized (CHAPS, DesoxyBIGCHAP) total membrane proteins was reached by successive ion-exchange chromatography on Fractogel EMD-TMAE 650 (Merck, Darmstadt, FRG), gel exclusion chromatography on Superose 6 (Pharmacia, Dübendorf, Switzerland), and reactive dye chromatography on Fractogel TSK AF-Red (Merck). The exchange activity of each preparation was monitored by reconstitution of proteins into Na-loaded asolectin vesicles and testing for Na-dependent ^{45}Ca uptake. After each purification step, myocardial membrane proteins were subjected to PAGE separation and electrophoretic blotting onto nitrocellulose strips. These blots were used to screen for cross-reactivity of mAbs directed against the retinal HMW protein band.

RESULTS AND DISCUSSION

Characterization of mAbs against the Retinal Exchanger

The Na-Ca exchange protein from bovine retina has been identified by Cook and Kaupp[3] as a HMW protein with an electrophoretic mobility corresponding to an apparent molecular weight of $\approx$215 kDa. Following their protocol, we also found that the exchange activity was enriched in parallel with a HMW protein after successive ion exchange and reactive dye chromatography. The stability of this protein in solution is known to be markedly enhanced in the presence of high concentrations of KCl (100-700 mM) and $CaCl_2$ (5-10 mM). Therefore, as a test for the specificity of the HMW protein, we repeated the same purification protocol in the presence of NaCl as the only cation. Furthermore, we added 0.5 mM EGTA to reduce the free Ca ion concentration to below 0.1 μM. Under these conditions we observed pronounced changes in the behavior of the Na-Ca exchanger during purification and reconstitution. During ion-exchange or gel-exclusion chromatography, the specific exchange activity, measured after reconstitution in lipid vesicles, no longer eluted in a defined peak. Rather, the activity appeared as a long tail trailing behind the major protein peaks and extending well beyond the total column volume in the case of gel filtration. These observations point to significant (hydrophobic?) interactions of the exchanger with the column materials. The total exchange activity that could be recovered exceeded the initial activity in native membrane vesicles by up to 300%. PAGE separation of the active fractions showed no HMW band. Instead, several new bands appeared in the 30-70-kDa range, none of which was characteristically enriched.

We concluded from these results that in the absence of K and Ca the HMW exchanger protein fragments into smaller subunits that still contain the exchange function but have lost some endogenous inhibitory control. Consequently, we decided to use the HMW protein band as antigen for mAb production.

Following established techniques, we isolated a total of 46 hybridomas secreting mAbs that reacted with retinal membrane proteins. Among these, eight immunolabeled a HMW protein band on Western blots of a partially purified preparation of the retinal

Na-Ca exchange protein. FIGURE 1 shows typical examples. The differences in the labeling patterns suggest that the antibodies interact with different epitopes.

To assess the specificity of these immunoreactive antibodies, we tested their capacity to immunoprecipitate the exchange activity from a mixture of detergent-solubilized total retinal rod membrane proteins. The antibodies were first bound to protein G agarose. The solubilized membrane proteins were equilibrated with the protein G-mAb complex and then reconstituted into lipid vesicles. FIGURE 2 shows the result obtained with mAb P8G10, which caused maximally a 60-70% reduction in vesicular transport activity. Two other mAbs reduced the activity by about 30%, while the remaining five were without effect in this assay. It is well known that mAbs failing to immunoprecipitate the target protein are not necessarily unspecific. In our case it is possible that these mAbs are directed against epitopes that are not exposed in the detergent-solubilized functional protein but are available after denaturation and Western blotting.

Preparation of Partially Purified Myocardial Na-Ca Exchanger

A possible immunological cross-reactivity of mAbs directed against the retinal exchanger with the corresponding myocardial protein was assessed in preparations that contained either unpurified sarcolemmal proteins or were partially enriched in exchange activity. In an attempt to avoid possible epitope distortion due to variable proteolytic degradation, we developed a purification scheme for the cardiac exchanger similar to the one used for the retinal protein. It included successive ion-exchange chromatogra-

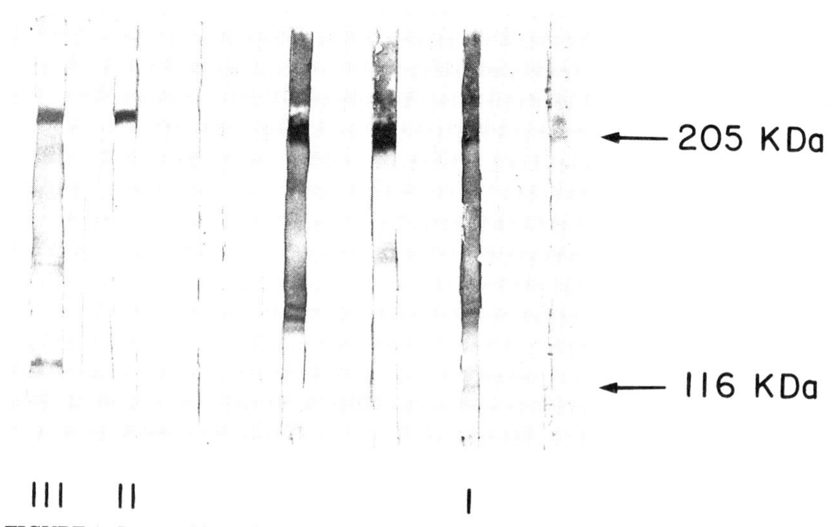

FIGURE 1. Immunoblots of partially purified Na-Ca exchange protein from bovine retinal rod outer segments (ROS). CHAPS-solubilized proteins, after ion-exchange and reactive dye AF-Red chromatography, were separated by PAGE on a 6% gel and transblotted onto nitrocellulose. Blots were stained with the indirect immunoperoxidase method after reaction with five different monoclonal antibodies (mAb), directed against the retinal 215-kDa exchanger protein. the lanes labeled I, II, and III indicate those mAbs that cross-reacted with myocardial membrane proteins. The empty lane gives the result of a transblot treated with a mAb not reacting with transblotted ROS membrane proteins.

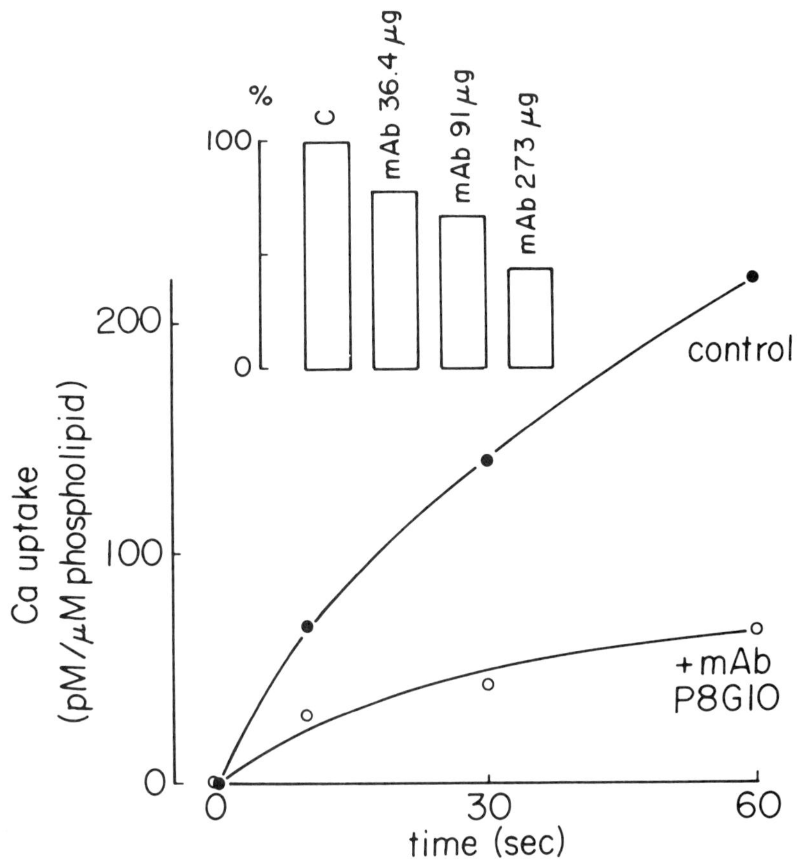

FIGURE 2. Na-dependent Ca transport after reconstitution of the retinal exchanger protein in lipid vesicles. CHAPS-solubilized membrane proteins supplemented with 20 mg/ml asolectin were reconstituted by dialysis in sodium phosphate buffer in the absence (control) or presence of mAb P8G10 bound to protein G agarose. After 48 hr, the Na-loaded lipid vesicles were sedimented and used to measure Na-dependent ^{45}Ca uptake in a standard filtration assay. The inset gives the dependency of the exchange rate on the mAb concentration during vesicle reconstitution.

phy, gel filtration, and a modified version of reactive dye chromatography in the presence of KCl (60-350 mM), NaCl (60-350 mM), and $CaCl_2$ (5-10 mM). The best preparations resulted in an about 200-fold purification of the exchange activity after reconstitution into lipid vesicles. In most experiments, no protein larger than ~150 kDa showed up in partially purified preparations even though gel exclusion chromatography suggested an apparent molecular weight for the exchanger close to 270 kDa. The active fraction after the second purification step usually contained two bands at ~130 and at 55-60 kDa that had no match in inactive fractions. However, among the bands remaining after the last step, none appeared larger than 100 kDa and none of the unmatched bands was consistently enriched.

Cross-reacting Antibodies

Among the 8 mAbs that immunolabeled a HMW retinal protein, three cross-reacted with nonpurified myocardial membrane proteins on Western blots (FIG. 3). The labeling patterns were significantly different, suggesting that the three mAbs recognized different epitopes on sarcolemmal proteins. The labeling of multiple bands could indicate epitope sharing among nonrelated proteins or it could reflect binding to differently sized fragments of a large protein all containing the same epitope. At least the mAb labeled III (Q13F10) on FIGURE 3 seems to interact also with proteins not related to the exchanger. The strong reaction with a protein at ~55 kDa was also seen in preparations without Na-Ca exchange activity. Each of the three antibodies labeled a 200-kDa protein in this preparation, but such a HMW protein was never observed in any of the partially purified preparations. In the immunoblot after ion-exchange fractionation (FIG. 4), all mAbs labeled a protein band in the 120- to 130-kDa range. An additional band close to 100 kDa was recognized by two of the three mAbs. The pattern of labeled proteins remained rather similar on immunoblots of the active fraction after both ion exchange and gel exclusion chromatography. After the final separation on reactive dye AF-Red all antibodies immunolabeled a single band with an apparent molecular weight close to 60 kDa. These results are compatible with the assumption that the three mAbs all recognize the same sarcolemmal protein that is gradually degraded in the course of purification. A similar interpretation has been given for analogous results of earlier

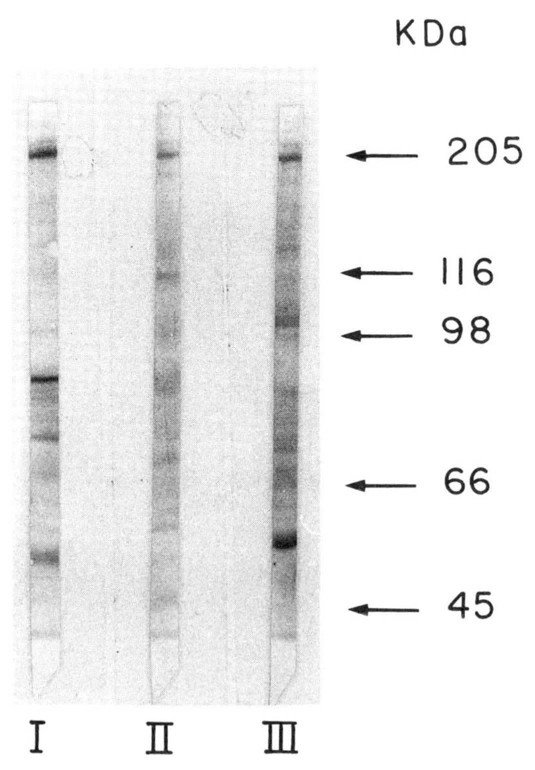

FIGURE 3. Immunoblot of nonpurified equine myocardial membrane proteins. CHAPS-solubilized cardiac membrane vesicles (300 μg protein/lane) were separated by PAGE on a 6% gel and transblotted onto nitrocellulose. Blots were reacted with three different mAbs directed against the retinal 215-kDa exchanger protein (I = mAb P15D4, II = mAb P15E1, III = mAb Q13F10). The blots were stained with the indirect immunoperoxidase method.

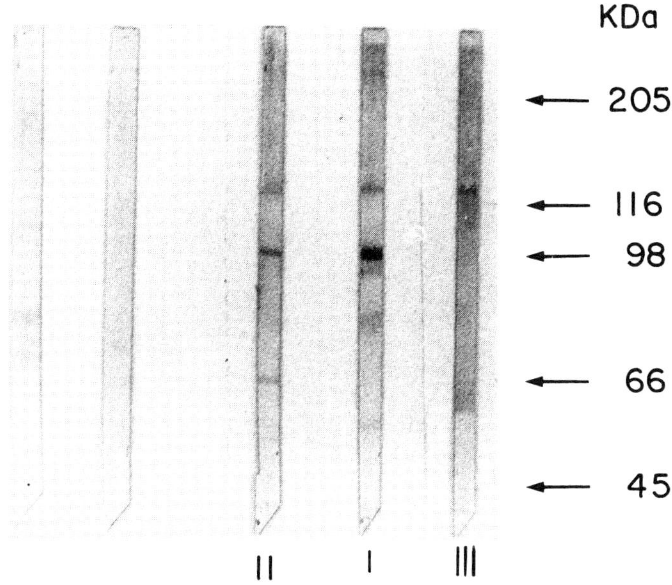

FIGURE 4. Immunoblots of equine myocardial membrane proteins after ion-exchange chromatography. Same method as described in FIGURE 3, except that the proteins were solubilized in desoxyBIGCHAP. The lanes labeled I-III contained the exchange activity and were reacted with the same three mAbs that were identified in the legend to FIGURE 3. The two empty lanes on the left give the result of immunoblots with proteins eluting with an ionic strength below 0.16 and containing no exchange activity. These lanes were reacted with mAb II and III.

attempts at purifying the exchanger. Yet, the apparent molecular weights of the proteins suspected to represent "fragments" of the native exchanger are somewhat variable.[2,9] Although the mAb-labeled proteins comigrate with Na-Ca exchange activity, there is no way to determine unequivocally their identity with the exchanger or its subunits by this technique.

Epitope Characterization

To get some additional topographical information on the location of the epitopes for the cross-reactive mAbs, we studied their interaction with increasing dilutions of intact sarcolemmal membrane vesicles in an ELISA filtration plate assay. Under these conditions, one of the three antibodies (P15E1) gave a positive response down to 3 μg membrane proteins/filter while the other two were negative with less than 25 μg membrane protein/filter. Possibly, this particular antibody interacts with an epitope located on the intra- or extracellular surface of myocardial membranes. Selecting among the possible candidate peptide sequences suggested by the model of Nicoll *et al.*[10] for the cloned exchanger protein, we tested a synthetic peptide containing the 12

Ser–Ser–Leu–Glu–Ala–Tyr–Cys(Acm)–His–Ile–Lys–Gly–Phe

FIGURE 5. Amino acid sequence of the synthetic peptide S-12-F. This sequence is identical to the published[11] COOH-terminal sequence of the cloned myocardial exchanger.

carboxy-terminal amino acids (FIG. 5) Unexpectedly, if titrated against constant antibody concentrations (100 μg/ml) in an ELISA assay, high concentrations of this peptide (200 μg/ml) were recognized by all three mAbs.

As indicated in FIGURE 5, the cysteine residue present in the carboxy-terminal sequence was protected in the synthetic peptide by an acetaminomethyl group to avoid dimerization during lyophilization. After removal of this protective group, the threshold peptide concentration detected by mAb P15E1 (labeled II in FIGS. 1, 3, and 4) was reduced 100-fold. For the two other antibodies the detection limit decreased by a factor of 10. These observations suggest that the three antibodies all recognize cysteine-containing epitopes. The carboxy-terminal sequence may form part of the epitope structure.

Further support for this conclusion came from experiments designed to evaluate the effect of SH-group reducing agents during chromatographic purification of the exchanger protein. Typically, the presence or absence of dithiothreitol (DTT) during the preparation procedure had little effect on the transport activity of the reconstituted cardiac exchanger. This is different from what has been reported for the retinal exchanger.[11] However, most Western blots of partially purified protein fractions that were prepared and PAGE-separated in the absence of DTT could not be immunolabeled with the antibody P15E1. The same was true for mAb P15D4 (labeled "I" in FIGS. 1, 3, 4). The third antibody still labeled the unspecific 55-kDa protein mentioned above while other bands were no longer visible.

CONCLUSION

Our results suggest that the retinal and the myocardial Na-Ca exchange proteins share epitopes that can be recognized by monoclonal antibodies. One such epitope could be traced to a cysteine-containing sequence that, in the cardiac exchanger, is part of the carboxy-terminal sequence. However, in spite of an extensive search, only a few cross-reactive mAbs were detected. None of these antibodies had an obvious functional effect on the transport function of the protein, and none was found to immunoprecipitate the myocardial target protein. The antibodies can be used to trace the labile myocardial exchanger during chromatographic purification. The results indicate that the labeled protein is progressively decreasing in size, while the preparation was enriched in Na-dependent Ca transport activity. The ion exchange function may be localized in a small fragment of the native exchanger. A similar conclusion may be drawn for the retinal exchanger, which is activated in spite of rapid fragmentation in the absence of its "substrate" cations K and Ca.

REFERENCES

1. REEVES, J. P. 1990. Sodium-calcium exchange. *In* Intracellular Calcium Regulation. F. Bonner, Ed.: 305-347. Wiley-Liss. New York.

2. PHILIPSON, K. D., S. LONGONI & R. WARD. 1988. Purification of the cardiac Na^{+}-Ca^{2+} exchange protein. Biochim. Biophys. Acta **945:** 298-306.
3. COOK, N. J. & U. B. KAUPP. 1988. Solubilization, purification, and reconstitution of the sodium-calcium exchanger from bovine retinal rod outer segments. J. Biol. Chem. **263:** 11382-11388.
4. LAGNADO, L. & P. A. MCNAUGHTON. 1990. Electrogenic properties of the Na : Ca exchange. J. Membr. Biol. **113:** 177-191.
5. VEMURI, R., M. E. HABERLAND, D. FONG & K. D. PHILIPSON. 1990. Identification of the cardiac sarcolemmal Na^{+}-Ca^{2+}exchanger using monoclonal antibodies. J. Membr. Biol. **118:** 279-283.
6. DIANO, M., A. LE BIRIC & M. HIRN. 1987. A method for the production of highly specific polyclonal antibodies. Anal. Biochem. **166:** 224-229.
7. WESTERWOUDT, R. J. 1985. Improved fusion methods. IV. Technical aspects. J. Immunol. Methods **77:** 181-196.
8. SLAUGHTER, R. S., J. L. SUTKO & J. P. REEVES. 1983. Equilibrium calcium-calcium exchange in cardiac sarcolemmal vesicles. J. Biol. Chem. **258:** 3183-3190.
9. KUWAYAMA, H. & T. KANAZAWA. 1982. Purification of cardiac sarcolemmal vesicles: High sodium pump content and ATP-dependent, calmodulin-activated calcium uptake. J. Biochem. (Tokyo) **91:** 1419-1426.
10. SOLDATI, L., S. LONGONI & E. CARAFOLI. 1985. Solubilization and reconstitution of the Na^{+}/Ca^{2+} exchanger of cardiac sarcolemma. Properties of the reconstituted system and tentative identification of the protein(s) responsible for the exchange activity. J. Biol. Chem. **260:** 13321-13327.
11. NICOLL, D. A., S. LONGONI & K. D. PHILIPSON. 1990. Molecular cloning and functional expression of the cardiac sarcolemmal Na^{+}-Ca^{2+} exchanger. Science **250:** 562-565.
12. NICOLL, D. A. & M. L. APPLEBURY. 1989. Purification of the bovine rod outer segment Na^{+}/Ca^{2+} exchanger. J. Biol. Chem. **264:** 16207-16213.

Molecular and Mechanistic Heterogeneity of the Na^+-Ca^{2+} Exchanger[a]

HANNAH RAHAMIMOFF, DEBBIE DAHAN, IAN FURMAN, RIVKA SPANIER, AND MICHELA TESSARI

Department of Biochemistry
Hebrew University-Hadassah Medical School
Jerusalem, 91010 Israel

INTRODUCTION

The Na^+-Ca^{2+} exchanger is one of the major Ca^{2+}-regulating proteins in excitable and many nonexcitable cells.[1] It transports Ca^{2+} across the membrane in a bidirectional fashion in response to a driving Na^+ gradient. The extent and direction of the Ca^{2+} flux depends on the respective $[Na^+]$ and $[Ca^{2+}]$ gradients across the membrane and the membrane potential.[2,3] Its abundancy is low in most membranes,[4,5] with the possible exception of rod outer segments.[6] No pharmacological ligand is known to date that binds to the Na^+-Ca^{2+} exchanger in a specific manner and modulates its activity.[7] Consequently, its purification, molecular identification, and elucidation of its mechanism of action poses considerable difficulties. Recent experimental evidence suggests that the Na^+-Ca^{2+} exchange function might be carried out by a heterogeneous family of proteins differing in molecular and mechanistic properties. We would like to summarize some of the molecular and mechanistic data supporting this notion.

MECHANISTIC HETEROGENEITY IN Na^+-Ca^{2+} EXCHANGE

Early experiments in the squid giant axon[8] suggested that, in order to maintain the intracellular $[Ca^{2+}]$ at or below 10^{-6} M, a stoichiometry of 3-5 Na^+ ions exchanged for each Ca^{2+} would be acceptable. On theoretical grounds, Mullins[9] favored a stoichiometry of four. Different studies with intact heart, myocytes, or cardiac sarcolemmal vesicles[10-12] resulted in stoichiometries from 3 Na^+/1 Ca^{2+} to 6 Na^+/1 Ca. In rod outer segments, a stoichiometry of 4 Na^+/1 Ca^{2+} + 1 K^+ was proposed, K^+ ion being an obligatory component of the system.[13,14]

In synaptic plasma membranes (SPMs), kinetic and steady-state measurements[15] indicated that the stoichiometry was 4-5 Na^+/1 Ca^{2+}. Examining the possible role of K^+ in rat brain synaptic plasma membrane vesicles, we have found that, although Na^+-Ca^{2+} exchange could proceed in the absence of K^+, it was consistently stimulated

[a] The work from the authors' laboratory has been supported in part by a research grant to H.R. from the U.S.-Israel Binational Science Foundation.

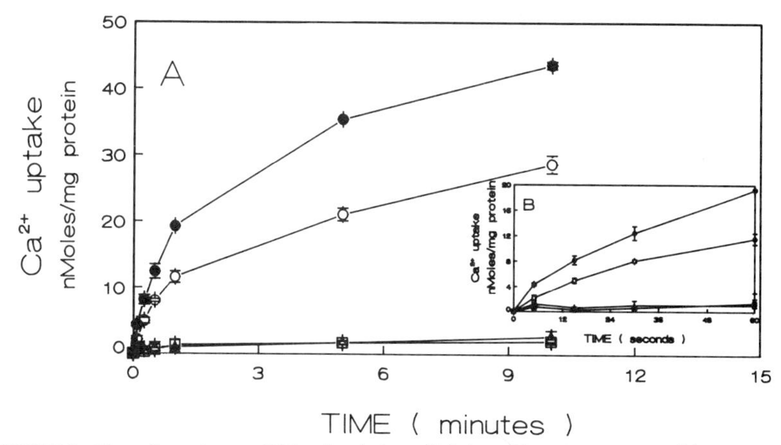

FIGURE 1. Time dependence of K^+-stimulation of Na^+ gradient-dependent Ca^{2+} uptake: SPM vesicles were preloaded with 0.19 M NaPi and either 0.01 M choline Cl or 0.005 M KCl and 0.005 M choline Cl. The K^+-free vesicles were diluted into a K^+-free external medium: 0.2 M choline Cl (○) or 0.19 M NaCl and 0.01 M choline Cl (□). The vesicles that contained 0.005 M KCl were diluted into an external medium that contained 0.195 M choline Cl and 0.005 M KCl (●) or 0.19 M NaCl, 0.005 M KCl, and 0.005 M choline Cl (△). Three μl of vesicles (about 15 μg protein) was diluted into 250 μl medium. All media also contained 0.01 M Tris HCl pH 7.5, 50 μM $^{45}CaCl_2$ and 20 μM valinomycin. **A.** The entire time range. **B.** The initial time range.

by inclusion of K^+ ions in the reaction medium.[16] The stimulation was noticeable already at 0.5 mM K^+, and it saturated with respect to $[K^+]$ at 2 mM.[16]

FIGURE 1 shows an experiment where the effect of K^+ on Na^+ gradient-dependent Ca^{2+} influx into rat brain SPM vesicles was studied. Equimolar K^+ (15 mM) was added both inside the NaPi-preloaded SPM vesicles and into the iso-osmotic choline-chloride-containing extravesicular medium. Valinomycin was added to prevent the build-up of membrane potential due to the electrogenic exchange, although in its absence, similar stimulation of the Na^+ gradient-dependent Ca^{2+} influx was obtained.[16] Na^+ gradient-dependent Ca^{2+} influx in the absence of K^+ is also shown. The insert (FIG. 1B) contains the expanded time scale of the initial rate. It can be seen that a 1.7-fold stimulation of Na^+ gradient-dependent Ca^{2+} influx is obtained in this experiment. The range of stimulation of Na^+ gradient-dependent Ca^{2+} influx by equimolar K^+ within the vesicles and in their extravesicular medium was 1.3-2.5-fold. Unlike the rod outer segment Na^+-Ca^{2+} exchanger[13,14] in rat brain SPMs, Na^+ gradient-dependent Ca^{2+} influx could also be obtained in the complete absence of K^+.

K^+ can stimulate Na^+-Ca^{2+} exchange in several ways: It can shunt the build-up of negative inside membrane potential resulting from the electrogenic Na^+ gradient-dependent Ca^{2+} influx. Distribution of equimolar K^+ across the vesicle's membrane can clamp the membrane potential at different values and thus modulate Na^+-Ca^{2+} exchange. K^+ can modulate the protein without being cotransported, and it can be cotransported with Ca^{2+} in a Na^+ gradient-dependent manner. To distinguish between these possibilities, several experiments were performed.

In the experiment shown in TABLE 1, K^+ was added only to the extravesicular medium and FCCP (H^+ ionophore) was added to dissipate any build-up of membrane potential. In this experiment, SPMs were reconstituted into a phospholipid-enriched preparation[15–18] where, presumably due to the high phospholipid to protein ratio (30 : 1), the native SPM membrane proteins are separated among a large number of liposomes.[19]

TABLE 1. The Effect of K^+ and FCCP on Na^+-Dependent Ca^{2+} Uptake in Reconstituted SPMs

$[Medium]^{out}$	Na^+-Dependent Ca^{2+} Uptake (nmol/mg protein · 10 min)	
	No FCCP	With FCCP
Choline Cl	52.6	72.86
Choline Cl + 10 mM K^+	68.39	102.57
Stimulation by K^+	1.3	1.4

NOTE: Reconstituted SPM vesicles were loaded during reconstitution with 0.19 M Na-phosphate buffer, pH 7.5, and 0.01 M choline Cl. Three μl of these vesicles (about 1.5 μg protein and 60 μg calf brain phospholipids) were diluted into 250 μl external isoosomotic medium composed of either choline Cl or choline Cl and 0.01 M KCl as specified, 50 μM $^{45}CaCl_2$ (0.1 μCi/100 μl) and either 5 μM FCCP dissolved in ethanol or an equivalent amount of ethanol (0.5 μl/ml) when no FCCP was added. The Na^+ gradient-independent component of the Ca^{2+} taken up by the vesicles was determined by diluting the same vesicles into 0.19 M NaCl-containing solution with either 0.01 M choline Cl or 0.01 M KCl. $^{45}Ca^{2+}$, FCCP, or ethanol were included as well. All external solutions were buffered with 0.01 M Tris HCl, pH 7.5. The Ca^{2+} associated with the reconstituted vesicles in the absence of a Na^+ gradient, was less than 5% of the values obtained in the presence of a Na^+ gradient, and these were subtracted. Each time point is an average of a triplicate determination.

Consequently, the probability that the Na^+-Ca^{2+} exchanger and an unrelated K^+ transporter or channel reside in the same vesicle is very small. It can be seen, that FCCP stimulates Na^+ gradient-dependent Ca^{2+} influx to about 1.4-fold in the absence of added K^+. Addition of K^+ to the extravesicular medium in the absence of FCCP stimulates Na^+ gradient-dependent Ca^{2+} influx 1.3-fold as compared to the choline chloride-containing medium. Addition of FCCP to the K^+-containing external medium results in a higher stimulation than FCCP alone, indicating that the role of K^+ in Na^+-Ca^{2+} exchange in rat brain SPMs does not involve modulation of membrane potential alone.

Indirect evidence of K^+ cotransport with Ca^{2+} can be obtained by measuring the amount of $[^3H]TPP^+$ taken up by the reconstituted SPM vesicles in the presence of K^+ and in its absence. FIGURE 2 shows such an experiment. It can be seen that the amount of $[^3H]TPP^+$ taken up by the reconstituted SPM vesicles when K^+ is included in the reaction mixture (closed symbols) is less than the amount of $[^3H]TPP^+$ taken up in the course of the Na^+ gradient-dependent Ca^{2+} influx in the absence of K^+ (open symbols). The Na^+ gradient-dependent Ca^{2+} uptake, however, measured in a parallel experiment that contained unlabeled TPP^+ and $^{45}Ca^{2+}$ (FIG. 2B) indicated that higher amounts of Ca^{2+} were transported in the presence of K^+ than in its absence. These findings indicate that the negative inside membrane potential that develops in response to activation of Na^+ gradient-dependent Ca^{2+} influx is smaller in the presence of K^+ than in its absence. If the overall coupling ratio between Na^+ and Ca^{2+} in the presence and in the absence of K^+ remains the same, this finding suggests that K^+ ions are cotransported with Ca^{2+} in a Na^+ gradient-dependent manner.

To test directly whether K^+ ions are cotransported in a Na^+ gradient-dependent manner with Ca^{2+}, we have used $^{86}Rb^+$. Rb^+ could substitute for K^+ (as tested in experiments similar to those shown in FIG. 1 and TABLE 1) and stimulate Na^+-Ca^{2+}

exchange. The experiment described in FIGURE 3 demonstrates that $^{86}Rb^+$ cotransport with Ca^{2+} can be initiated by activation of Na^+-Ca^{2+} exchange. Reconstituted NaPi-containing SPM vesicles were pre-equilibrated with $^{86}Rb^+$ for 16 hours in the absence of a Na^+ gradient (see legend to FIG. 3). Dilution of these vesicles into a choline chloride-containing external medium that contained $^{86}Rb^+$ of identical concentration to that used for pre-equilibration in the presence of unlabeled Ca^{2+} resulted in Na^+ gradient-dependent $^{86}Rb^+$ uptake. In parallel experiments, Na^+ gradient-dependent $^{45}Ca^{2+}$ uptake was determined in the presence of unlabeled Rb^+. The $^{45}Ca^{2+}$ and $^{86}Rb^+$ uptake in the absence of a Na^+ gradient or the $^{86}Rb^+$ uptake in the absence of Ca^{2+} is also shown. Since in all the experiments that were done, smaller amounts of Rb^+

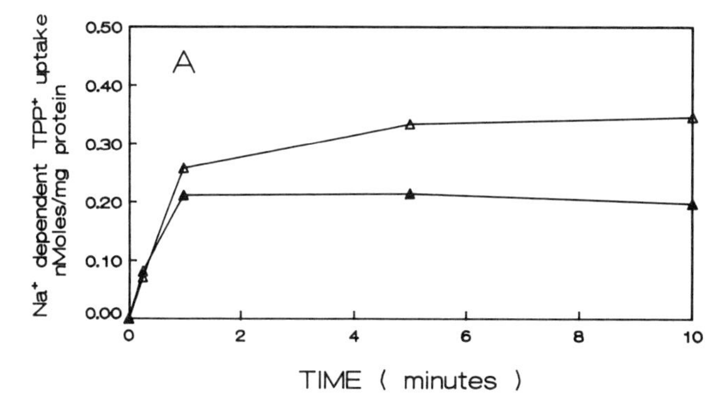

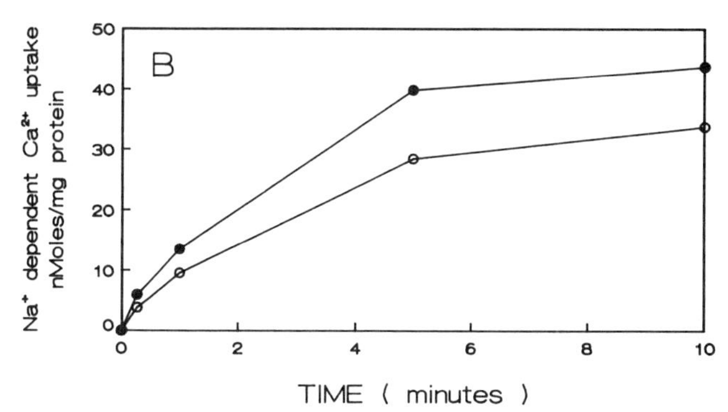

FIGURE 2. K^+ modulation of Na^+ gradient-dependent [^{3}H]TPP^+ and $^{45}Ca^{2+}$ uptake in reconstituted SPM vesicles: 3 μl of SPM vesicles (about 1.5 μg protein and 60 μg calf brain phospholipids) were preloaded with either 0.195 M NaPi (pH 7.5), 0.005 M choline Cl or 0.195 M NaPi, 0.005 M KCl. The K^+-containing vesicles were diluted into 250 μl of 0.195 M choline Cl, 0.005 M KCl or 0.195 M NaCl, 0.005 M KCl (▲ ; ●). The vesicles that did not contain K^+ were diluted into either 0.2 M choline Cl or 0.195 M NaCl, 0.005 M choline Cl (△; ○). In addition:

A: 0.0466 μM [^{3}H]TPP^+ (0.25 μCi/250 μl) and 50 μM unlabeled $CaCl_2$ were added. **B:** 50 μM $^{45}CaCl_2$ (0.25 μCi/250 μl) and 0.0466 unlabeled TPP^+ were added.

The Na^+ gradient-dependent component of [^{3}H]TPP^+ uptake in the absence of K^+ (△) or in its presence (▲) is shown in FIGURE 2A and the Na^+-gradient component of $^{45}Ca^{2+}$ uptake in the absence of K^+ (○) or in its presence (●) is shown in FIGURE 2B.

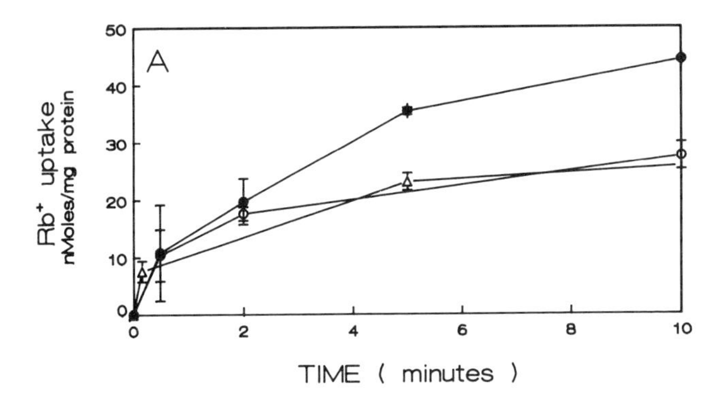

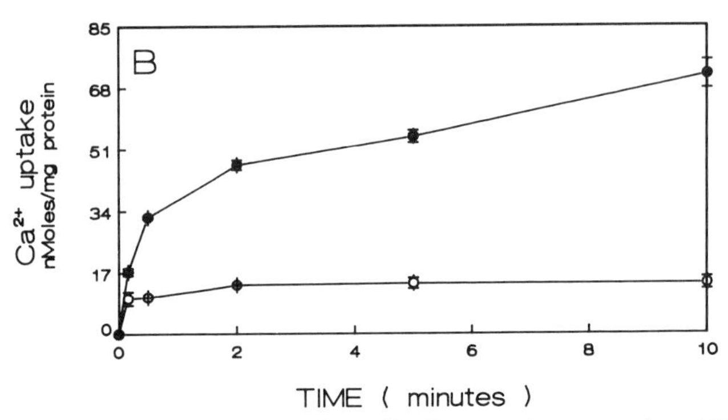

FIGURE 3. Na^+ gradient-dependent $^{86}Rb^+$ and $^{45}Ca^{2+}$ uptake into reconstituted SPM vesicles.

A: ^{86}Rb uptake: SPM vesicles were reconstituted as described.[15,18] They were preloaded during reconstitution with 0.2 M NaPi buffer, pH 7.4. The reconstituted vesicles were pre-equilibrated for 16 hours at 37°C with a fivefold volume excess of 0.2 M NaCl, 0.01 M Tris HCl (pH 7.4), and 0.5 mM $^{86}RbCl$ (0.5 μCi/250 μl). Five μl of the $^{86}Rb^+$-containing vesicles were diluted into 250 μl of either 0.2 M choline Cl, 0.01 M Tris HCl (pH 7.4), 50 μM $^{40}CaCl_2$, and 0.5 mM $^{86}Rb^+$ (●) or 0.2 M NaCl, 0.01 M Tris HCl (pH 7.4), 50 μM $^{40}CaCl_2$, and 0.5 mM $^{86}Rb^+$ (○) or 0.2 M choline Cl, 0.01 M Tris HCl (pH 7.4), 0.5 mM $^{86}Rb^+$, and 0.1 mM EGTA (△). $^{86}Rb^+$ uptake was determined.[16]

B: ^{45}Ca uptake: The same reconstituted vesicles were preincubated in an identical fashion to that described in FIGURE 3A, except that unlabeled RbCl was used. Five μl of Rb^+-containing vesicles were diluted into identical media to those used in FIGURE 3A except that $^{45}Ca^{2+}$ (0.2 μCi/250 μl) was used and the Rb^+ was unlabeled. $^{45}Ca^{2+}$ uptake in the presence of Na^+ gradient (●) or in the absence of Na^+ gradient (○).

were transported in conjunction with Na^+ gradient-dependent Ca^{2+} uptake than the corresponding amounts of Ca^{2+}, and since the exchange was only partially dependent on K^+ it is suggested that in rat brain multiple molecular forms (K^+-sensitive and K^+-insensitive) of Na^+-Ca^{2+} exchangers co-exist. In three different SPMs tested, the ratio of Ca^{2+}/Rb^+ transported in a Na^+ gradient-dependent manner was between 5 and 8, indicating that the SPM preparations we tested contained about 10-20% transporters that cotransport K^+.

MOLECULAR AND IMMUNOLOGICAL HETEROGENEITY OF THE Na^+-Ca^{2+} EXCHANGER

In the past five years, different protein entities were identified as the tentative Na^+-Ca^{2+} exchanger. Using different protein purification techniques in conjunction with immunological studies in cardiac sarcolemma led to identification of proteins of 33 kDa, 70 kDa 120 kDa, 140 kDa, and 160 kDa as potential candidates.[20–22]

Recently, the canine cardiac Na^+-Ca^{2+} exchanger has been successfully cloned and its primary structure was elucidated.[23] The protein was found to contain 970 amino acids of 108 kDa molecular mass prior to glycosylation. A polyclonal antibody prepared against a 14 amino acid long synthetic peptide corresponding to a highly hydrophilic segment within the cytoplasmic domain of the canine Na^+-Ca^{2+} exchanger[23] bound to protein bands of 70 kDa, 120 kDa, and 160 kDa on Western blots of the denatured canine cardiac sarcolemmal membranes. In rod outer segments, a protein of 220 kDa molecular mass was identified as the Na^+-Ca^{2+} exchanger.[6]

In rat brain synaptic membranes (SPMs) we have identified a protein of 70 kDa molecular mass as the most likely candidate to contain the catalytic part of Na^+-Ca^{2+} exchanger.[17] A polyclonal antibody prepared against this protein immunoprecipitated Na^+-Ca^{2+} exchange activity from solubilized synaptic membranes.[17] The anti-70-kDa-protein antibody prepared against the rat brain Na^+-Ca^{2+} exchanger bound also to a 70-kDa protein in bovine cardiac sarcolemma,[18] suggesting that some immunological cross-reactivity existed between these two membranes. Moreover, a polyclonal antibody prepared against the 220-kDa rod outer segment Na^+-Ca^{2+} exchanger binds to proteins of 70-kDa molecular mass in rat brain synaptic membranes (antibody provided by Dr. N. Cook, experiment carried out by Ian Furman in our laboratory), indicating that the 220-kDa protein from rod outer segments shares some common epitopes with a 70-kDa protein in rat brain SPMs. The anti-220-kDa-protein antibody prepared against the rod outer segment Na^+-Ca^{2+} exchanger cross-reacted with proteins of 70 and 120 kDa in cannine cardiac sarcolemma.[27]

Expression of Na^+-Ca^{2+} exchange activity in *Xenopus* oocytes injected with chick heart[24] and rabbit heart[25] mRNA indicated that a preparation enriched in mRNA sedimenting at 25 S coded for the first and between 3-5 kb for the later activity. Injection of oocytes with mRNA obtained from 1-day-old rat brains indicated[26] that Na^+ gradient-dependent Ca^{2+} uptake activity could be obtained either when total mRNA was injected or when a fraction enriched in 14-18 S mRNA was injected into oocytes. These results are shown in TABLE 2.

In order to identify the protein(s) synthesized in oocytes in response to rat brain mRNA injection, we have prepared oocyte membranes, separated them by SDS-PAGE, and tested the binding of the polyclonal antibody that was prepared against the purified rat brain 70-kDa protein on immunoblots.[18] The antibody bound specifically to proteins of about 70 kDa in membranes prepared from oocytes injected with either total polyadenylated mRNA or a fraction enriched in 14-18 S mRNA. A similar experiment is shown in FIGURE 4, except that the size-fractionated mRNAs were added to an mRNA-free reticulocyte lysate, the proteins synthesized were separated by SDS-PAGE, and binding of the anti-70-kDa antibody[18] was tested on immunoblots (FIG. 4A). It can be seen that when the reticulocyte lysate is supplemented with total polyadenylated mRNA (lane B) or 14-18 S mRNA (lane E) a protein of about 70 kDa that is recognized by the antibody against the rat brain Na^+-Ca^{2+} exchanger is formed. Trace amounts of protein are recognized by the antibody also in lane F, where 9-13 S mRNA was added to the reticulocyte lysate. Lane A shows an immunoblot of native rat

TABLE 2. Expression of Na^+-Ca^{2+} Exchange Activity in *Xenopus* Oocytes

mRNA Injected	Na^+-Dependent Ca^{2+} Uptake (pmoles/oocyte · 10 min) (SD)
Total poly A	1.05 (0.4)
14-18S	4.6 (1.8)
H_2O	Not detected

NOTE: Total polyadenylated mRNA from 1-day-old rat brains was size fractionated on a linear 5-25% (wt/vol) sucrose gradient also containing 0.5% (wt/vol) SDS. Individual fractions were collected and the contents of each three fractions were pooled, precipitated by alcohol, suspended in H_2O at a concentration of 1 ng/nl, and 50 ng were injected into oocytes. Na^+ gradient-dependent Ca^{2+} uptake was determined as described in Furman and Rahamimoff.[26] The Na^+ gradient-independent component of the $^{45}Ca^{2+}$ taken up by the oocytes was subtracted. Injection of any other mRNA fraction beside the 14-18 S mRNA did not result in Na^+ gradient-dependent $^{45}Ca^{2+}$ uptake activity in the oocytes. The transport assays were done each time with groups of 8-10 oocytes and three different mRNA preparations were tested.

brain SPMs for comparison. FIGURE 3B shows an autoradiogram of the corresponding [^{35}S]methionine-labeled protein profile.

TORPEDO ELECTRIC ORGAN Na^+-Ca^{2+} EXCHANGER

Nerve terminals isolated from the *Torpedo* electric organ contain a high specific activity Na^+-Ca^{2+} exchanger.[28] Its maximal reaction velocity is about two orders of magnitude higher then that of the rat brain Na^+-Ca^{2+} exchanger, and its magnitude is comparable to the cardiac sarcolemmal Na^+-Ca^{2+} exchanger.[29] This Na^+-Ca^{2+} exchanger differs from rat brain SPM and sarcolemmal Na^+-Ca^{2+} exchanger in at least two respects: its affinity to Na^+ and its temperature dependence.

Studying its kinetic properties revealed[28] that the dependence of the initial rate of the Na^+ gradient-dependent Ca^{2+} influx on internal [Na^+] exhibits a sigmoidal curve that reaches half-maximal reaction rate ($K_{0.5}$) at 170 mM [Na^+]. In comparison, rat brain[30] and bovine heart[31] Na^+-Ca^{2+} exchangers reach the half-maximal reaction rate at about 20 mM Na^+. High $K_{0.5}$ (110-300 mM) to Na^+ has also been detected under ATP depletion in the squid giant axon. Unlike the dialyzed squid giant axon, however, in *Torpedo* synaptic plasma membrane vesicles, the affinity to Na^+ could not be increased by addition of ATPγS.[28]

The second difference between the *Torpedo* electric organ and rat brain SPM Na^+-Ca^{2+} exchangers is in their temperature-dependence profile. Na^+ gradient-dependent Ca^{2+} uptake in *Torpedo* SPMs reaches its maximal initial rate between 15 and 20°C,[28] whereas in rat brain SPMs the maximal initial rate is obtained between 30 and 40°C.[28] The same temperature dependence is obtained also when steady-state rates are determined. These are shown in FIGURE 5A-C. This difference in temperature dependence could be a result of the difference in the fatty acids[28] esterified to rat brain and *Torpedo* phospholipids, which presumably determine the overall membrane fluidity.[28] The different temperature dependence could also result from differences in the protein itself. To distinguish between these two possibilities, *Torpedo* and rat brain SPMs were each reconstituted into a large excess of both brain and *Torpedo* phospholipids. The temperature dependence of the Na^+ gradient-dependent Ca^{2+} influx has been determined. FIGURE 6A-D show these experiments. It can be seen that each transporter retains its native temperature dependence in an independent fashion from the bulk lipid composition of the membrane into which they are reconstituted. It cannot be ruled out that during solubilizaton the proteins retained some of their native tightly bound lipids. These, however, did not contribute to the overall membrane fluidity.[28]

SUMMARY

1. Studying the effect of K^+ on Na^+-Ca^{2+} exchange in rat brain SPMs revealed that a consistent stimulation was obtained. This stimulation persisted also when FCCP was included in the K^+-containing reaction mixture to minimize the effect of membrane potential on the electrogenic process.

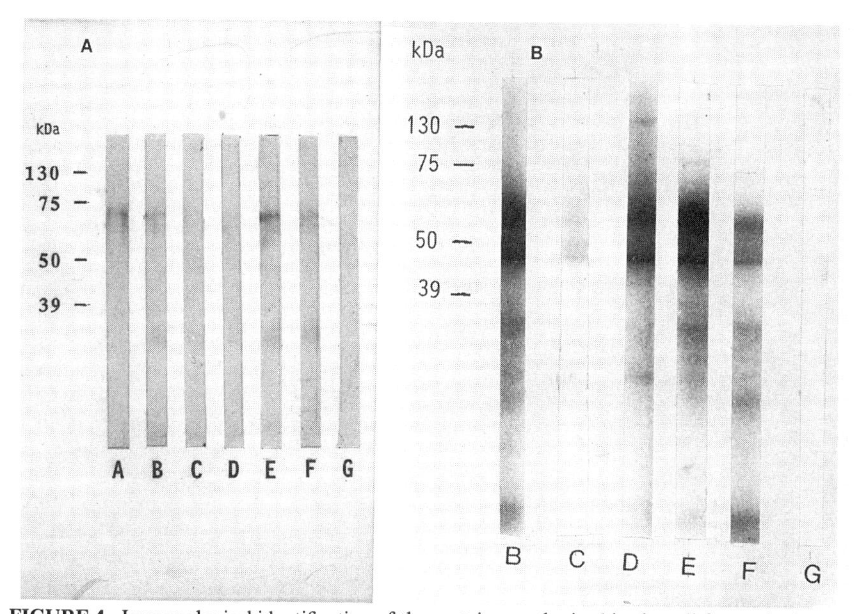

FIGURE 4. Immunological identification of the proteins synthesized in the cell-free reticulocyte lysate with different mRNA fractions. **A:** Protein synthesis in the reticulocyte lysate was carried out essentially as described in the manufacturer's kit (NEN-Dupont, U.S.A.). Total lysate proteins (40 μg), were separated by SDS-containing polyacrylamide gel (10%) electrophoresis. Western blots were treated with a 1 : 5000 dilution of the anti-70-kDa protein antibody.[18] Prior to usage, the antibody was pretreated with acetone-precipitated proteins from rabbit erythrocytes to reduce unspecific background. Antigen-antibody complexes were detected with a secondary alkaline phosphatase-conjugated anti-rabbit antibody (Promega). The molecular weight markers are the Bio-Rad pre-stained molecular weight markers for immunoblotting. **Lane A:** Native rat brain synaptic plasma membrane proteins. **Lanes B-E** present the immunoblot of the proteins synthesized in the reticulocyte lysate supplied with 1 μg of the following mRNAs/25 μl lysate. **B:** total polyadenylated mRNA. **C:** 24-28 S; **D:** 19-23 S; **E:** 14-18 S; **F:** 9-13 S; **G:** lysate proteins without added mRNA.

B: Autoradiography of the same SDS-containing polyacrylamide gel as shown in FIGURE 4A. [^{35}S]methionine-labeled proteins were detected after two days of exposure of the film and treatment by Amplify (Amersham, England).

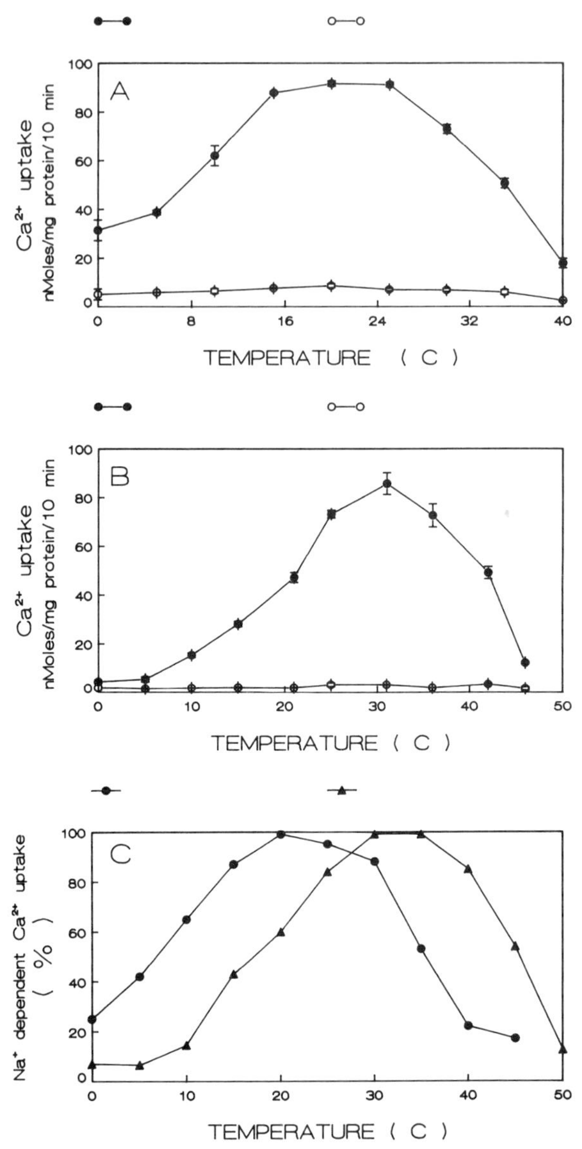

FIGURE 5. Temperature dependence of Na^+ gradient-dependent Ca^{2+} influx of native *Torpedo* and rat brain SPMs. **A.** Three μl of *Torpedo* SPM vesicles (about 10 μg protein) preloaded with 0.4 M NaCl, 0.01 M Tris HCl (pH 7) were diluted into 0.4 M KCl (●) or 0.4 M NaCl (○) also containing 0.01 M Tris Hcl (pH 7.4) and 50 μM $^{45}CaCl_2$. The averaged data presented in this figure (the bar represents SD) were obtained from three different experiments using the same SPM preparation. **B.** The experiment was identical to A except that rat brain SPMs were used and the osmolarity of internal NaCl and external KCl-containing media was 0.2 M. **C.** The data in this figure were compiled from three different preparations each, of *Torpedo* (●) and rat brain (▲) SPMs. To permit comparison, the highest value of Na^+ gradient-dependent Ca^{2+} uptake in each individual experiment was taken as 100%, and all other data points were normalized accordingly.

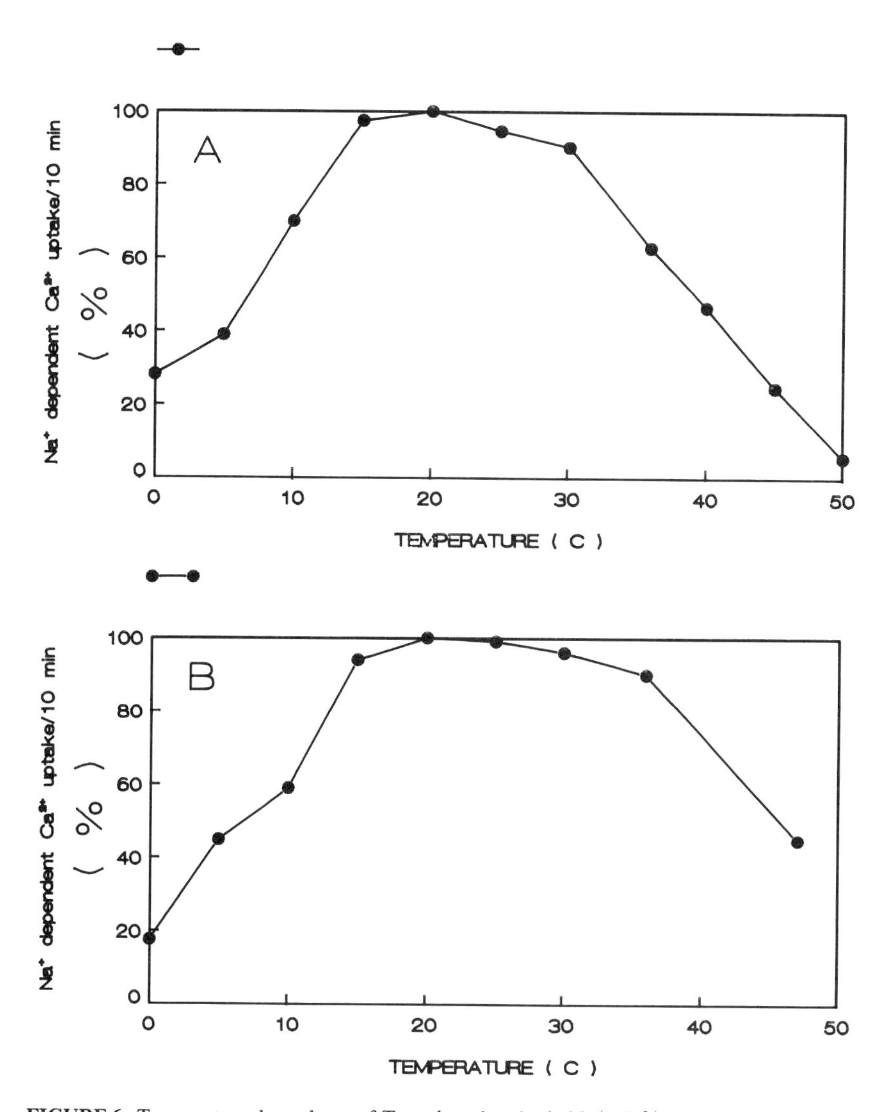

FIGURE 6. Temperature dependence of *Torpedo* and rat brain Na^+-Ca^{2+} exchangers reconstituted into *Torpedo* and brain phospholipids. *Torpedo* or rat brain SPMs were reconstituted at a 1 : 30 (protein : phospholipid) weight ratio. Three μl of reconstituted vesicles (1-2 μg protein) were diluted into 250 μl of external medium composed of either: 0.4 M KCl (when *Torpedo* SPMs were used) or 0.2 M KCl (when rat brain SPMs were used), buffered with 0.01 M Tris HCl (pH 7.4) and 50 μM $^{45}CaCl_2$ (0.1 μCi), or, respectively, 0.4 M and 0.2 M NaCl, 0.01 M Tris HCl (pH 7.4) and 50 μM $^{45}CaCl_2$ (0.1 μCi). The normalized net Na^+ gradient component of Ca^{2+} influx is presented. The highest Na^+ gradient-dependent Ca^{2+} uptake value in each individual experiment was taken as 100%, and all other values were normalized accordingly. The amount of Na^+ gradient-independent component of $^{45}Ca^{2+}$ associated with the vesicles was very low (less than 5% of the total amount of $^{45}Ca^{2+}$ taken up by the vesicles). **A:** *Torpedo* SPMs reconstituted into TPLs. **B:** *Torpedo* SPMs reconstituted into BPLs. **C:** Rat brain SPMs reconstituted into TPLs. **D:** Rat brain SPMs reconstituted into BPLs. (FIG. 6 C and D are shown on the following page.)

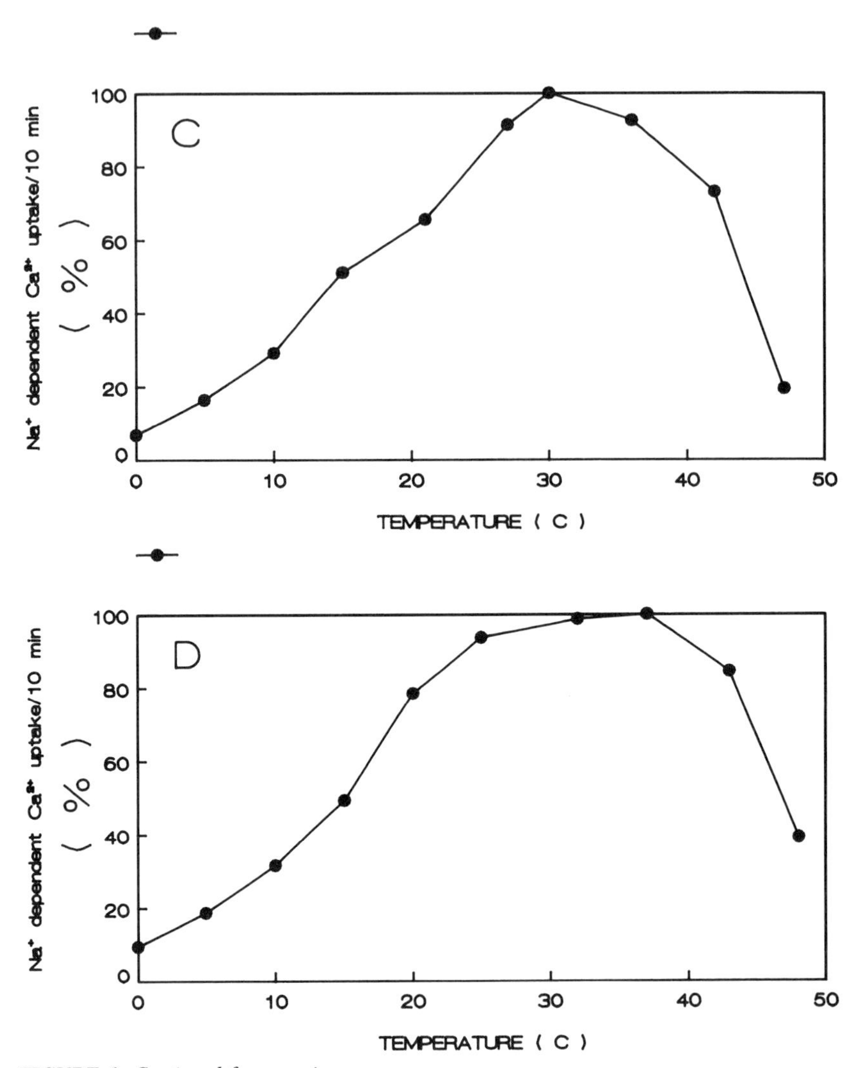

FIGURE 6. *Continued from previous page.*

2. Using $^{86}Rb^{+}$ as a K^{+} analogue revealed that it was cotransported with Ca^{2+} in a Na^{+} gradient-dependent manner. The ratio between the amount of Ca^{2+}/Rb^{+} transported in rat brain SPMs in a Na^{+} gradient-dependent manner suggests that not all the Na^{+}-Ca^{2+} exchangers in that preparation cotransport Rb^{+} (K^{+}) with Ca^{2+}. This is supported also by the finding that Na^{+} gradient-dependent Ca^{2+} influx can proceed in rat brain SPMs in the complete absence of K^{+} although to a lesser extent.

3. Protein purification studies and immunological characterization indicate that a 70-kDa protein is consistently detected in rat brain SPMs. Immunological characterization of the proteins expressed in the 14-18 S mRNA-injected *Xenopus* oocyte in conjunction with Na^{+} gradient dependent Ca^{2+} uptake activity or in the same mRNA-

fortified reticulocyte lysate suggest that proteins of about 70 kDa are specifically synthesized.

4. *Torpedo* electric organ Na^+-Ca^{2+} exchanger differs at least in two respects from the rat brain Na^+-Ca^{2+} exchanger: It has a low affinity to Na^+ ($K_{0.5}$ = 170 mM), and it reaches maximal activity between 15-20°C. Reconstitution studies suggest that the temperature difference might reflect a difference in the proteins themselves rather then a difference in membrane fluidity due to a difference in the membrane lipid composition.

REFERENCES

1. ALLEN, T. J. A., D. NOBLE & H. REUTER, Eds. 1989. Sodium-Calcium Exchange. Oxford University Press. Oxford, England.
2. LAGNADO, L. & P. A. MCNAUGHTON. 1990. J. Membr. Biol. **113:** 177-191.
3. RAHAMIMOFF, H. 1990. Curr. Top. Cell. Reg. **31:** 241-271.
4. CHEON, J. & J. P. REEVES. 1988. J. Biol. Chem. **263:** 2309-2315.
5. RAHAMIMOFF, H. 1989. *In* Sodium-Calcium Exchange. T. J. A. Allen, D. Noble & M. Reuter, Eds. Oxford University Press. Oxford, England.
6. COOK, N. J. & U. B. KAUPP. 1988. J. Biol. Chem. **263:** 11382-11388.
7. KACZOROWSKI, G. J., R. S. SLAUGHTER, V. F. KING & M. L. GARCIA. 1989. Biochim. Biophys. Acta **988:** 287-302.
8. BAKER, P. F. 1972. Prog. Biophys. Mol. Biol. **24:** 177-223.
9. MULLINS, L. J. 1981. Ion Transport in Heart. Raven Press. New York.
10. HORACKOVA, M. & G. VASSORT. 1979. J. Gen. Physiol. **73:** 402-424.
11. REEVES, J. P. & C. C. HALE. 1984. J. Biol. Chem. **259:** 7733-7739.
12. KIMURA, J., A. NOMA & H. IRISAWA. 1986. Nature **319:** 596-597.
13. CERVETTO, L., L. LAGNADO, R. J. PERRY, D. W. ROBINSON & P. A. MCNAUGHTON. 1989. Nature **337:** 740-743.
14. SCHNETKAMP, P. P. M., D. K. BASU & R. T. SZERENCSEI. 1989. Am. J. Physiol. **257:** C153-C157.
15. BARZILAI, A. & H. RAHAMIMOFF. 1987. Biochemistry **26:** 6113-6118.
16. DAHAN, D., R. SPANIER & H. RAHAMIMOFF. 1991. J. Biol. Chem. **266:** 2067-2075.
17. BARZILAI, A., R. SPANIER & H. RAHAMIMOFF. 1984. Proc. Natl. Acad. Sci. USA **81:** 6521-6525.
18. BARZILAI, A., R. SPANIER & H. RAHAMIMOFF. 1987. J. Biol. Chem. **262:** 10315-10320.
19. HERMONI-LEVINE, M. & H. RAHAMIMOFF. 1990. Biochemistry **29:** 4940-4950.
20. SOLDATI, L., S. LONGONI & E. CARAFOLI. 1985. J. Biol. Chem. **260:** 13321-13327.
21. LONGONI, S. & E. CARAFOLI. 1987. Biochem. Biophys. Res. Commun. **145:** 1059-1063.
22. PHILIPSON, K. D., S. LONGONI & R. WARD. 1988. Biochim. Biophys. Acta **945:** 298-306.
23. NICOLL, D. A., S. LONGONI & K. D. PHILIPSON. 1990. Science **250:** 562-565.
24. SIGEL, E., H. BAUR & H. REUTER. 1988. J. Biol. Chem. **263:** 14614-14616.
25. LONGONI, S., M. J. COADY, T. IKEDA & K. D. PHILIPSON. 1988. Am. J. Physiol. **255:** C870-C873.
26. FURMAN, I. & H. RAHAMIMOFF. 1990. Brain Res. **532:** 41-46.
27. VEMURI, R., M. E. HABERLAND, D. FONG & K. D. PHILIPSON. 1990. J. Membr. Biol. **118:** 279-283.
28. TESSARI, M. & H. RAHAMIMOFF. 1991. Biochim. Biophys. Acta (in press).
29. REEVES, J. P. & K. D. PHILIPSON. 1989. *In* Na-Ca Exchange. T. J. A. Allen *et al.*, Eds.: 27-53. Oxford University Press. Oxford, England.
30. BARZILAI, A. 1987. Ph.D. Thesis. Approved by the Senate of the Hebrew University. Jerusalem, Israel.
31. REEVES, J. P. & J. L. SUTKO. 1983. J. Biol. Chem. **258:** 3178-3182.
32. BLAUSTEIN, M. P. 1977. Biophys. J. **20:** 79-111.

Distinctive Properties of the Purified Na-Ca Exchanger from Rod Outer Segments [a]

GEORGE P. LESER,[b] DEBORA A. NICOLL,[c] AND MEREDITHE L. APPLEBURY

Visual Sciences Center
University of Chicago
Chicago, Illinois 60637

INTRODUCTION

Na-Ca exchangers are important components in the cellular control of Ca^{2+} levels in a variety of tissues. Members of a family of exchangers have been identified in photoreceptors, heart, brain, axons, and other tissues.[1-5] Although these proteins share similar functions, recent studies indicate that distinct differences exist for tissue-specific subtypes. Most of the exchangers are believed to have a stoichiometry of 3Na : 1Ca.[2] However, in some instances other ions play a stimulatory role, and, in at least the case of the rod outer segment (ROS) exchanger, K ions are specifically required to achieve transport.[6,7] There are differences in the rate of exchange as well. The ROS exchanger has been reported to transport approximately 60-120 Ca ions per second per exchanger, while the exchanger from heart has been described as having the capacity to move 1000 ions per second per molecule.[8-10] The diversity of this family is underscored by the molecular weights that have been assigned to the Na-Ca exchangers from different tissues: 220 kDa in ROS, 120 kDa in heart, and 70 kDa in brain.[4,8,9,11,12]

In photoreceptor cells, Ca^{2+} plays an important role in the visual transduction pathway. In the dark, the intracellular Ca^{2+} levels are determined by the balance of influx through a cGMP-gated ion channel and efflux through the Na-Ca exchanger. Upon exposure to light, a cascade of events ensues (for review see Chabre & Applebury[13]): Activated rhodopsin catalyzes the exchange of GTP for GDP bound to the alpha subunit of transducin. This event frees the alpha from the beta-gamma subunits of transducin, which can then activate the cGMP-phosphodiesterase by removing an inhibitory subunit. These events cause a rapid drop in the cellular cGMP concentration. This reduction in the cGMP level results in the closure of the cGMP-dependent ion channels' subsequent hyperpolarization of the membrane, and a halt in the influx of Ca^{2+} ions. The exchanger continues to extrude Ca^{2+} ions lowering the internal Ca^{2+} concentration within the cell. Restoration of the photoreceptor to the "dark" state occurs when the activation of this cascade is turned off and the concentra-

[a] This work was funded in part by National Institutes of Health Grant EY04801 (MLA) and National Research Service Award 1-F32-EY06261-01 (GPL).

[b] Address for correspondence: Visual Sciences Center, University of Chicago, 939 E. 57th Street, Chicago, IL 60637.

[c] Current address: Dept. of Physiology, School of Medicine, University of California, Los Angeles, Los Angeles, CA 90024.

tion of cGMP is returned to dark levels by the turning on of guanylate cyclase. The latter enzyme is indirectly activated by the decrease in cellular Ca^{2+} concentration.[14,15] Therefore, the rise in cGMP levels that results in the reopening of the ion channel is due to the lowering of the Ca^{2+} concentration by the Na-Ca exchanger. Adaptation, the partial recovery of light-sensitive response by a photoreceptor in the presence of a sustained light stimulus, is thought to occur by resetting the balance between the activation and recovery phases of the visual pathway, involving both cGMP and Ca^{2+} levels[16] (for review see Fain & Matthews[17]).

To better understand the role of the Na-Ca exchanger in controlling intracellular Ca^{2+} concentrations in photoreceptor cells, we have undertaken studies of the purified Na-Ca exchanger, thus making possible the examination of its properties in the absence of extraneous exchanger or channel activity. We used specific polyclonal antibodies to characterize the ROS exchanger and examine possible shared domains with Na-Ca exchangers from brain and heart. Here we describe the activity of the purified ROS exchanger in a proteoliposome assay system. This system allowed an examination of the ion requirements of the purified exchanger. By studying ^{42}K uptake, we were able to demonstrate that K ions are not only required for activity but are transported as well.

MATERIALS AND METHODS

Materials

Fresh bovine eyes were obtained from local slaughterhouses. Trasylol (aprotinin) was obtained from Mobay Chemical Corp. (New York). Leupeptin, CHAPS, and pepstatin were purchased from Boehringer Mannheim (Indianapolis, IN). Sucrose was purchased from ICN Biochemicals (Cleveland, OH), ultrapure glycerol from International Biotechnologies, Inc. (New Haven, CT), polyvinylidene difluoride (PVDF) membrane was obtained from Millipore Corp. (Bedford, MA), and the Centricon 30 microconcentrators were obtained from Amicon (Danvers, MA). The protein assay kit was purchased from Bio-Rad, Inc. (Richmond, CA). The rabbit anti-chicken secondary antibody was obtained from Jackson Immunoresearch, Inc. (West Grove, PA). All other chemicals were purchased from Sigma (St. Louis, MO). $^{45}CaCl_2$ was obtained from Amersham (Arlington Heights, IL). Both ^{42}KCl and $^{86}RbCl$ were obtained from Dupont NEN (Wilmington, DE).

ROS Preparation and Exchanger Purification

Rod outer segments (ROS) were prepared from bovine eyes in the light and the Na-Ca exchanger was purified as described[9] with minor modifications. Briefly, washed ROS membranes were solubilized at 0.5 mg/ml in solubilization buffer (225 mM NaCl, 20 mM bis-Tris, pH 7.0, 7.5 mM CHAPS, 0.1% soy asolectin) for 1 hr at 4°C with gentle agitation. Unsolubilized material was removed by centrifugation in a Beckman Ti-70 rotor in a Beckman L7 ultracentrifuge at 50,000 rpm for 30 min. The solubilized ROS was loaded upon a 1 $\times$ 6.5 cm DEAE-Sepharose column that had been pre-equilibrated with solubilization buffer. Solubilized ROS membranes from 100 bovine retina were loaded, in a volume of 150-200 ml at a flow rate of 0.1 ml/min. The column was then washed with solubilization buffer, 1.5-2 times the volume loaded, at a flow rate of 0.2 ml/min. Bound protein was eluted with a linear gradient of 225-500 mM

NaCl in solubilization buffer. Fractions containing the exchanger were concentrated using a Centriprep 30 microconcentrator device according to the manufacturer's directions. Concentrated exchanger was made to 20% glycerol, aliquoted, and stored at −80°C. Canine cardiac sarcolemmal vesicles and rat brain synaptic plasma membranes were the kind gifts of Drs. K. Philipson and H. Rahamimoff, respectively.

Protein concentration was determined by either the Bio-Rad protein assay, or, in the presence of high lipid concentrations, by the method of Kaplan and Pederson.[18] In both cases bovine serum albumin was used as the protein standard.

Assay of Na-Ca Exchanger Activity

The purified Na-Ca exchanger was reconstituted into proteoliposomes following the protocol of Miyamoto and Racker.[19] Typically, 20-30 μg/ml exchanger was added to a volume of lipid solution (50 mM cholate, 20 mM Mes Tris, pH 7.4, 160 mM NaCl, 2.4% soy asolectin). Vesicles were formed by dilution with five times the volume of buffer (160 mM NaCl, 20 mM Mes-Tris, pH 7.4) and pelleted by centrifugation in a TL-100.3 rotor in a Beckman TLA-100 tabletop ultracentrifuge at 90,000 rpm for 40 min. Pelleted vesicles were washed once in dilution buffer and pelleted as before. Finally, vesicles were resuspended in a small volume of dilution buffer, usually 0.4 times the volume of the starting lipid solution, and immediately used for assays.

Exchanger activity was measured by diluting vesicles 1 : 10 into reaction buffer (160 mM KCl, 20 mM Mes-Tris, pH 7.4, 40 μM $CaCl_2$). To assess ion flux, 100-μl aliquots were removed at the indicated times and placed on a 1-ml column of Dowex 50 × 8-400 prepared by equilibration with 160 mM sucrose contained 25 mg/ml bovine serum albumin. The vesicles were collected from the column using a stream of compressed air. Calcium not sequestered within vesicles was bound by the ion-exchange resin. The vesicles were washed through the column with two successive 1-ml volumes of sucrose/BSA solution. Sequestered Ca was determined by liquid scintillation counting in a Beckman LS5000 CE liquid scintillation counter. Background was determined by measuring the amount of Ca bound to vesicles in the absence of K and with no Na gradient, that is, proteoliposomes were diluted into 160 mM NaCl, 20 mM Mes-Tris, pH 7.4, 40 μM $CaCl_2$, 0.2 μCi $^{45}CaCl_2$. The results presented have the background subtracted from them unless otherwise indicated.

Transport of K was assessed by diluting vesicles loaded with 160 mM NaCl into reaction buffer containing 20 mM KCl, 20 mM mes-Tris, 40 μM $CaCl_2$ and 25 μCi ^{42}KCl. The exchange of Rb was measured as above except the reaction buffer contained 20 mM RbCl instead of KCl and $^{86}RbCl$ instead of ^{42}KCl. Background was determined in both cases by diluting the vesicles into the identical reaction buffer minus the $CaCl_2$.

Preparation of Antibodies and Blotting

To prepare material for immunization, the DEAE-enriched exchanger was isolated using preparative polyacrylamide gel electrophoresis. The gels were stained with Coomassie blue and the protein band identified as the exchanger was cut out. The excised gel fragments were destained with 50% methanol, frozen in liquid nitrogen,and finally ground to a powder with a mortar and pestle. The ground gel was resuspended in incomplete Freund's adjuvant. Two white leghorn chickens approximately 22 weeks old were used for immunization; we adapted the technique of Stuart *et al.*[20] Approximately 20 μg protein were injected subcutaneously under the wing per immunization. The chickens were immunized a total of four times at 2-3-week intervals. Published reports suggest high titers of antibodies that extend for several months may be achieved within 2-4 weeks following injection.[21] Pre-immune eggs were collected from each chicken. Eggs were continuously collected following the first injection. They were labeled with a marking pen and stored at 4°C.

Antibodies were purified from the yolks of the collected eggs. The following protocol is adapted from several published reports.[21,22] The eggs were cracked open and the egg whites discarded. The yolks were carefully washed with phosphate-buffered saline (PBS). Two egg yolks were combined and processed together, their combined volume was approximately 40 ml. One volume of PBS was added to the combined egg yolks. The yolks were disrupted in the PBS and one volume of 10.5% polyethylene glycol (PEG) (ave. mol. wt. 8000) was added. The mixture was stirred at room temperature for 20 minutes, then spun in a J-14 rotor in a Beckman Model J-21C preparative centrifuge at 8,250 rpm for 20 min. The supernatant was filtered through two thicknesses of cheesecloth and saved. One-half the total yolk volume, that is, 20 ml, of 42% PEG was added to the supernatant and stirred at room temperature for 30 min. The mixture was pelleted by centrifugation as before. The supernatant was discarded while the pellet was carefully resuspended in 10-20 ml of PBS. After the pellet was thoroughly resuspended, the volume was brought up to 2.5 times the original yolk volume with PBS. An equal volume of saturated ammonium sulfate, freshly adjusted to pH 7.0 with NH_4OH, was added with stirring. The precipitate was pelleted in the J-14 rotor at 8,250 rpm for 30 min. The supernatant was carefully aspirated, and the pellet was resuspended in PBS containing 0.1 mM EDTA and 0.02% NaN_3, then dialyzed against several liters of the same buffer. Finally the antibody was made to 25% with glycerol, aliquoted, and stored at −80°C.

Antibodies testing positive on spot blots of total ROS proteins were used for further analysis. Approximately 2 μg ROS membrane proteins and 4 μg each of brain synaptic membrane proteins and cardiac sarcolemma polypeptides were separated on SDS-polyacrylamide gels[23] and subsequently electrophoretically transferred onto PVDF membrane using a semi-dry blotting apparatus according to the manufacturer's directions. The reaction of blots was carried out as described[24] except 3% nonfat milk was used to block remaining reactive sites on the membrane and antibody binding was detected using an alkaline phosphatase coupled rabbit anti-chicken secondary antibody. 5-Bromo-4-chloro-3-indolyl phosphate (BCIP) and nitro blue tetrazolium (NBT) were used as substrates. Total proteins bound to the membrane were visualized by staining with colloidal gold as detailed elsewhere.[25]

Polyclonal antibodies specific for the cardiac Na-Ca exchanger were kindly provided by Dr. K. Philipson.

RESULTS

Our objective for the following studies was to obtain a nearly pure preparation of the ROS Na-Ca exchanger that could be examined in the absence of other ion-transporting proteins. The retinal rod Na-Ca exchanger can be purified from solubilized ROS membranes by a combination of ion-exchange and affinity chromatography.[8,9] A slight modification of our laboratory's published purification procedure[9] entails extended washing of the ion-exchange column; this procedure yields a 200-300-fold enrichment

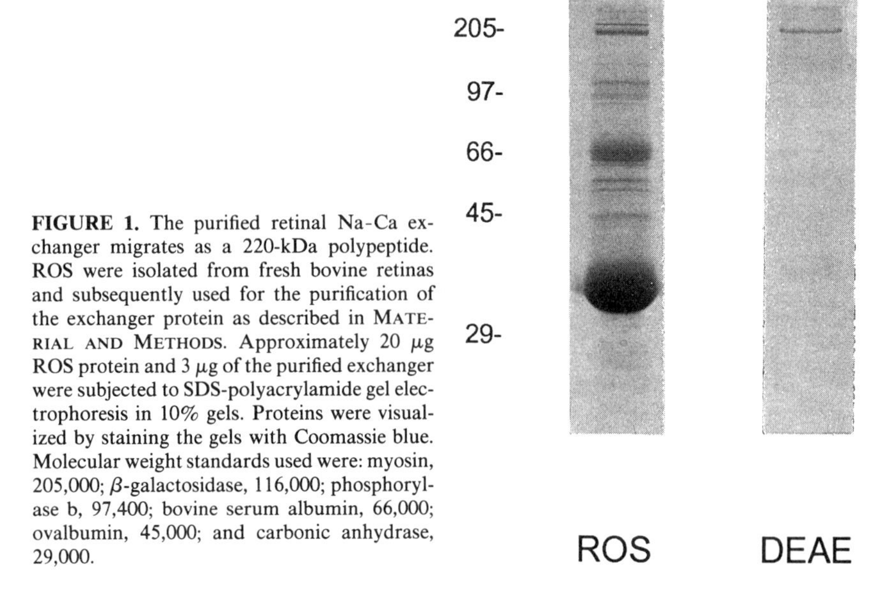

FIGURE 1. The purified retinal Na-Ca exchanger migrates as a 220-kDa polypeptide. ROS were isolated from fresh bovine retinas and subsequently used for the purification of the exchanger protein as described in MATERIAL AND METHODS. Approximately 20 μg ROS protein and 3 μg of the purified exchanger were subjected to SDS-polyacrylamide gel electrophoresis in 10% gels. Proteins were visualized by staining the gels with Coomassie blue. Molecular weight standards used were: myosin, 205,000; β-galactosidase, 116,000; phosphorylase b, 97,400; bovine serum albumin, 66,000; ovalbumin, 45,000; and carbonic anhydrase, 29,000.

over the exchanger activity in washed ROS membranes (FIG. 1). The exchanger migrates as a single band of approximately 220 kDa under reducing conditions in SDS polyacrylamide gels. Gold-stained blots reveal a small amount of opsin contamination as well as minor bands at 60 and 110 kDa. Coomassie staining suggests the 220-kDa band constitutes greater than 98% of the protein in the fraction. This enriched exchanger preparation, substantially free of other ROS membrane proteins was used in subsequent flux assay experiments.

Distinct Size and Antigenicity of the ROS Exchanger

Specific polyclonal antibodies were prepared to examine purity, size, and cross-reactivity of the Na-Ca exchanger. We formed antibodies against gel-purified bovine retinal exchanger by injecting chickens with polyacrylamide gel-purified protein, collecting eggs, and fractionating the yolks to isolate IgY immunoglobulins. Initial attempts at reacting Western blots of ROS with these antibodies occasionally showed two bands at 68 and 55 kDa. The appearance of these bands was more frequent if the samples, in dithiothreitol-containing sample buffer, were heated before electrophoresis. If β-mercaptoethanol was used in place of dithiothreitol, protein samples consistently showed these two bands on Western blots (FIG. 2a, left lane). Moreover, these bands were visible regardless of the nature of the protein blotted; both ROS proteins and molecular weight markers reacted equally well. The appearance of these 68- and 55-kDa bands could be eliminated in samples containing dithiothreitol by incubating the sample with iodoacetamide before electrophoresis. The presence of iodoacetamide in β-mercaptoethanol-containing buffer eliminated or decreased their appearance (FIG. 2a, right lane). Hence, in subsequent experiments where protein samples were blotted, iodoacetamide was used routinely with dithiothreitol-containing sample buffers.

In initial studies, multiple lower molecular weight polypeptides in addition to the 220-kDa protein were frequently detected on Western blots using exchanger-specific polyclonal antisera. There were two polypeptides at 96 and 80 kDa in addition to several minor bands (FIG. 2b, left lane). These bands were rarely detectable by staining with Coomassie but could be visualized by silver staining. It was necessary to determine whether these bands were minor components with similar epitopes or were degradation products of the 220-kDa protein. Furthermore, the addition of small amounts of proteinase K to the 220-kDa protein yielded fragments of the same molecular weights as those shown in FIGURE 2b. Therefore, it appears that the ROS Na-Ca exchanger is very susceptible to proteolytic cleavage. Adjusting the concentration of a variety of protease inhibitors in all stages of the purification procedure along with maintaining the exchanger at 4°C markedly reduced the presence of these polypeptide fragments (FIG. 2b, right lane).

We examined the degree of antibody cross-reactivity between photoreceptor, brain, and cardiac Na-Ca exchangers. Total proteins from bovine rod outer segments, rat brain synaptic plasma membranes, and canine cardiac sarcolemmal vesicles were electrophoretically transferred onto PVDF membrane. Total proteins were stained on duplicate blots with colloidal gold (FIG. 3). Blots were reacted with antibody specific for bovine ROS exchanger or antibody specific for the 160-, 120-, and 70-kDa polypeptides associated with the cardiac exchanger (FIG. 2). The retinal-specific immunoglobulin showed no cross-reactivity with the proteins from synaptic or sarcolemmal membranes. Similarly antibodies known to react with the cardiac exchanger[12] did not bind to ROS

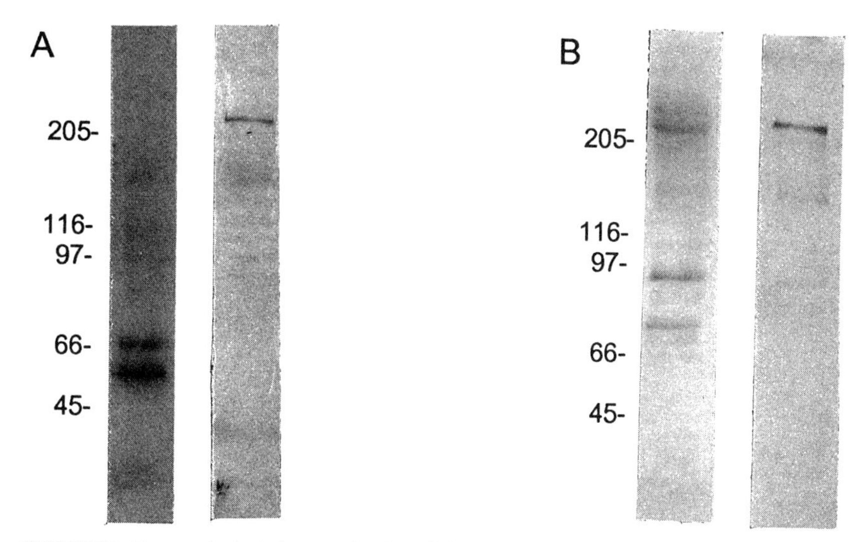

FIGURE 2. Immunological characterization of the exchanger. Total proteins from bovine ROS (2 μg) (**A**) or DEAE-enriched exchanger (0.5 μg) (**B**) were separated on 7.5% SDS-polyacrylamide gels. Proteins were electrophoretically transferred onto PVDF membrane. Blots were reacted with polyclonal antibody specific for Na-Ca exchanger of retina. The blots in A show multiple lower molecular weight fragments that frequently copurify with the exchanger (left lane) and the exchanger purified in the presence of protease inhibitors (right lane). The proteins in B were prepared in sample buffer containing β-mercaptoethanol (left lane) or the same sample buffer plus iodoacetamide (right lane). Molecular weight standards used were as in FIGURE 1.

or synaptic membrane proteins. Negative immunological results should be approached cautiously, and the existence of shared antigenic determinants among some or all of the exchangers that are not recognized by these polyclonal antibodies cannot be ruled out. However, these results suggest that the exchangers from ROS, heart, and brain each possess significant antigenic determinants that are unique.

K Ions Are Transported by the Purified ROS Exchanger

Studies of the exchanger in intact ROS or whole photoreceptors have demonstrated that transport activity depends specifically on the presence of K ions. In light of this, we examined the ion requirements of the purified retinal exchanger in a proteoliposome assay system. Purified exchanger protein was reconstituted into soy asolectin vesicles containing NaCl (160 mM). These vesicles were diluted into reaction buffer containing [Ca] varying from 10 nM to 200 μM and containing 0.2 μCi of ^{45}Ca. Five seconds after dilution into the reaction buffer, the Ca sequestered within the vesicles was determined. Velocity curves obtained under conditions of varying Ca, Na, or K concentrations in the presence of trace amounts of $^{45}CaCl_2$ indicate the exchanger has a V_{max} of approximately 3800 nmol Ca·min·mg protein (data not shown). Maximal exchanger activity required concentrations of 20 μM Ca, 130 mM Na, and 20 mM K. Lineweaver-Burk analysis of this data gave K_m values of 14 μM for Ca, 85 mM for Na, and 1 mM for K (data not shown). The Hill plots of the log of activity versus the log of the ion concentrations indicate a stoichiometry of 4Na : 1Ca,1K, although this does not prove whether K ions are actually transported or are required as a stimulatory cofactor.

To examine whether K is transported by the purified exchanger, proteoliposomes loaded with Na (160 mM), were diluted into reaction buffer containing 20 mM KCl and trace ^{42}K. The balance of the osmolarity was made up with choline Cl. Nonspecific binding was determined by diluting vesicles into reaction buffer containing K as before without Ca present. ^{42}K is specifically sequestered by vesicles containing the purified ROS exchanger. FIGURE 4 shows the time course of K transport. This result is more easily confirmed by studies with ^{86}Rb which has a half-life much longer than ^{42}K. Both K and Rb transport are dependent upon the presence of Ca. A time course of ion sequestration into liposomes shows that the nmol of ^{42}K or ^{86}Rb accumulated/mg of protein was roughly equivalent to that of ^{45}Ca (FIG. 4). Thus, the requirement of the ROS Na-Ca exchanger reflects the role of K as an ion cotransported with Ca in exchange for Na.

DISCUSSION

In photoreceptor cells, Ca acts as an important messenger in restoring and/or resetting light sensitivity following a light stimulus. The levels of intracellular Ca are regulated by the Na-Ca exchanger of ROS in conjunction with the cGMP-gated cation channel. Given the essential role of the exchanger, we have sought to identify its functional properties in the absence of other ROS components that mediate ion flux. Studies of the purified exchanger have enabled us to examine to what extent the purified ROS exchanger shares common traits with other members of the family of transporters and to examine features that distinguish its role in the photoreceptor outer segment.

We prepared polyclonal antibodies to the purified exchanger to better characterize its molecular properties and to examine potential cross-reactivity with other exchangers. The initial characterization of the antibody showed strong apparent reaction

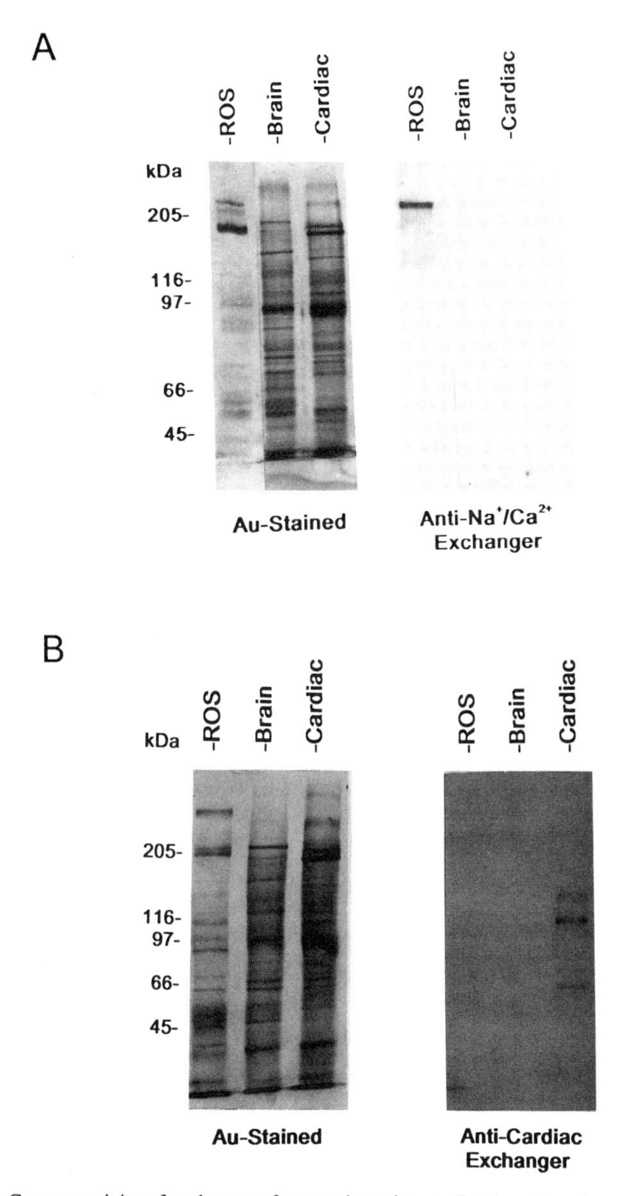

FIGURE 3. Cross-reactivity of exchangers from various tissues. Total proteins from bovine ROS (2 μg), rat brain synaptic plasma membranes (4 μg), and canine heart sarcolemmal vesicles (4 μg) were separated on 7.5% SDS-polyacrylamide gels. Proteins then were blotted electrophoretically onto PVDF membrane. Blots were reacted with polyclonal antibodies specific for either the ROS exchanger (**A**) or the cardiac exchanger (**B**). Blots were reacted and antibody binding was detected as described in MATERIALS AND METHODS. Total proteins bound to the PVDF membrane were visualized by staining duplicate blots with colloidal gold. Molecular weight standards used were as in FIGURE 1.

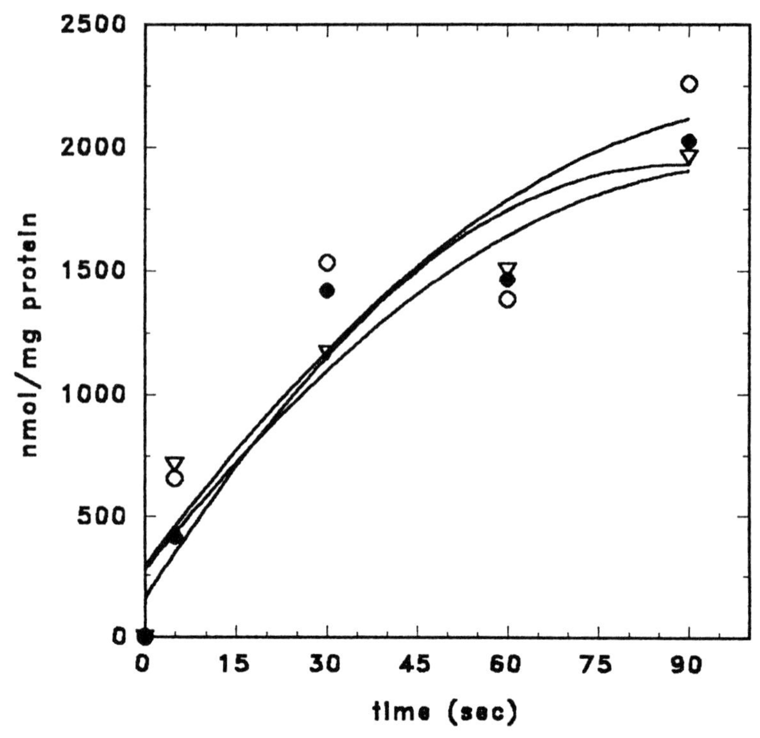

FIGURE 4. The retinal exchanger transports K^+ and Rb^+ ions at the same level as Ca^{2+}. The ability of the retinal exchanger to physically transport K^+ and Rb^+ ions was tested by forming vesicles with the purified Na-Ca exchanger containing 160 mM NaCl and 20 mM Mes Tris, pH 7.4. These proteoliposomes were diluted into reaction buffer containing 140 mM choline-Cl, 20 mM Mes Tris, pH 7.4, 100 μM $CaCl_2$, and either 20 mM KCl with trace amounts of ^{42}KCl (open circles) or 20 mM RBCl with ^{86}RbCl (closed circles). A control curve showing the time course of ^{45}Ca uptake by vesicles diluted into 160 mM KCl, 20 mM Mes Tris, pH 7.4, 100 μM $CaCl_2$ with trace amounts of $^{45}CaCl_2$ (open triangles) is shown for comparison. At 5, 30, 60, and 90 sec following dilution of the vesicles into reaction buffer, aliquots were removed, passed through an ion-exchange column, and the amount of sequestered isotope was counted. Background was determined by diluting vesicles into the same reaction buffer minus the $CaCl_2$. Background was subtracted from the data shown.

with bands at 68 and 55 kDa on Western blots. Since the antibody was prepared against a 220-kDa polypeptide excised from a polyacrylamide gel, we investigated the possibility that the observed staining was artifactual. The appearance of these bands was independent of the presence of ROS proteins; a lane of molecular weight markers inadvertently included in a blot showed strong staining (FIG. 2a). Artifactual bands detected by Coomassie or silver staining migrating at between 68-70 and 54-55 kDa in SDS-polyacrylamide gels have been previously documented.[26-28] Their appearance has been attributed to the use of β-mercaptoethanol or dithiothreitol in sample buffers and could be eliminated by the use of iodoacetamide.[27] Our samples showed a weak and variable presence of these bands when stained with Coomassie or silver, but the polyclonal antibodies reacted with these bands strongly on blots. Pretreating samples with iodoacetamide, however, abolishes the cross-reaction artifact. This reactivity is a peculiar property of these antibodies and may be attributable to the use of preparative gels to purify the immunogen or could even be inherent in the structure of a fraction of the immunoglobulins specific for this exchanger.

The antibodies made to the 220-kDa exchanger protein also detected polypeptides of 80 and 96 kDa. Their presence was variable and Coomassie or silver staining suggested they were minor components. Their cross-reactivity with the antibody suggested they might be proteolytic fragments of the 220-kDa protein or minor copurifying proteins sharing common epitopes. We were able to eliminate these fragments by adjusting the use of protease inhibitors throughout the preparation and thus modestly improve the state of intactness of the purified exchanger. In our hands, the exchanger appears particularly susceptible to proteolysis, as has been found for other exchangers. The cardiac Na-Ca exchanger has been shown to yield multiple fragments upon purification thought to be due to proteolytic cleavage.[12,29]

We made polyclonal antibodies that might detect both common and distinct epitopes of the exchanger protein. The ROS Na-Ca exchanger is a polypeptide that migrates as a 220-kDa band in SDS-polyacrylamide gels.[8,9] Molecular weights of other members of the Na-Ca family are varied, reported to be 120-kDa in heart[12,29] and 70 kDa in brain.[4,11] The antiserum specific for the 220-kDa ROS exchanger fails to react with any proteins in cardiac sarcolemma or brain synaptic plasma membrane preparations, although it is extremely sensitive to minor amounts of the ROS exchanger. Likewise, cardiac exchanger specific antibody fails to detect the 220-kDa protein or any polypeptides in the brain preparation. The differing sizes of members of this functional family coupled with the apparent lack of cross-reactivity between exchangers from three different tissues with specific polyclonal antisera clearly suggest that each exchanger has distinct molecular features.

A distinguishing feature of the Na-Ca exchanger is its specific requirement for K.[6,9] It was of interest whether the purified exchanger itself was sufficient for transport, or whether it was coupled to another protein for K transport. In testing for ion specificities, we noted that Rb would substitute for K. We observed ^{42}K or ^{86}Rb uptake in proteoliposomes reconstituted with purified exchanger, confirming that these ions are transported and thus eliminating any influences of other channels or exchangers. Our results show that the Na-Ca exchanger of rod outer segments transports K ions at a similar rate and equimolar amount compared to Ca. Analysis of the velocities of ion fluxes obtained by varying the Ca, Na, or K ions establish a stoichiometry of 4Na : 1Ca,1K. The assessment of the purified exchanger is consistent with recent electrophysiological findings studying intact ROS.[6]

The role K plays in the functioning of other exchangers is not completely clear. In the case of the cardiac exchanger, K does not appear to be a required cofactor nor is it transported, resulting in a 3Na : 1Ca stoichiometry (for reviews see Philipson[2] and Reeves[3]). The exchanger of the squid axon is stimulated in the presence of K but does not transport it.[30] K ions exert a mild stimulatory effect on the exchanger of brain synaptic plasma membranes, and K is apparently transported at a level about 20% that of Ca.[31] Whether this means that K is transported by the brain exchangers at a low rate or that there is a K-transporting subset of exchangers is not clear.[31] Thus the role of K underscores one of the key differences in this family. As the primary structures of these molecules are elucidated through molecular cloning, it will be of interest to examine whether the ability to transport K arises from a subtle change in structure or has been achieved by addition of an extra domain for the ROS Na-Ca exchanger.

SUMMARY

The Na-Ca exchanger of rod outer segments plays an important role in the regulation of Ca levels in photoreceptor cells. While this transporter shares functional properties with other Na-Ca exchangers, it has several unique features. The purified ROS exchanger migrates as a single band at 220 kDa in SDS-polyacrylamide gels, indicating that the unit size of its polypeptide is larger than other known Na-Ca exchangers (and most transporters). A specific antiserum to the ROS exchanger does not bind to the Na-Ca exchangers found in sarcolemmal vesicles or brain synaptic plasma membranes. Similarly, polyclonal antiserum specific for the cardiac exchanger does not react with ROS or brain proteins. The ROS exchanger requires K for transport activity. By incorporating the purified exchanger into proteoliposomes and measuring the sequestration of ^{42}K, the actual transport of K is demonstrated. A stoichiometry of 4Na : 1Ca,1K for the exchanger of ROS has been measured.

ACKNOWLEDGMENTS

The authors would like to thank Dr. J. Tanaka for her helpful discussions during the course of this work. The cardiac sarcolemmal proteins and the cardiac exchanger specific polyclonal antisera were the generous gift of Dr. K. Philipson. The fraction of brain synaptic plasma membrane proteins was the kind gift of Dr. H. Rahamimoff.

REFERENCES

1. YAU, K.-W. & K. NAKATANI. 1985. Nature **309:** 352-354.
2. PHILIPSON, K. D. 1985. Annu. Rev. Physiol. **47:** 561-571.
3. REEVES, J. P. 1985. Curr. Topics Membr. Transp. **25:** 77-127.
4. BARZILAI, A., R. SPANIER & H. RAHAMIMOFF. 1984. Proc. Natl. Acad. Sci. USA **81:** 6521-6525.
5. DIPOLO, R. & L. BEAUGE. 1990. J. Gen. Physiol. **95:** 819-835.
6. CERVETTO, L., L. LAGNADO, R. J. PERRY, D. W. ROBINSON & P. A. MCNAUGHTON. 1989. Nature **337:** 740-743.
7. SCHNETKAMP, P. P. M., R. T. SZERENCSEI & D. K. BASU. 1991. J. Biol. Chem. **266:** 198-206.
8. COOK, N. J. & U. B. KAUPP. 1988. J. Biol. Chem. **263:** 11382-11388.
9. NICOLL, D. & M. L. APPLEBURY. 1989. J. Biol. Chem. **264:** 16207-16213.
10. CHEON, J. & J. P. REEVES. 1988. J. Biol. Chem. **263:** 2309-2315.
11. BARZILAI, R., R. SPANIER & H. RAHAMIMOFF. 1987. J. Biol. Chem. **262:** 10315-10320.
12. VEMURI, R., M. E. HABERLAND, D. FONG & K. D. PHILIPSON. 1990. J. Membr. Biol. **118:** 279-283.
13. CHABRE, M. & M. L. APPLEBURY. 1986. *In* The Molecular Mechanisms of Photoreception. H. Stieve, Ed.: 51-66. Springer-Verlag. New York.
14. RAY, S., A. DIZHOOR, S. KUMA, K. WALSH, J. HURLEY & L. STRYER. 1990. J. Cell Biol. **111:** 2444-2452.
15. LAMBRECHT, H.-G. & K.-W. KOCH. 1991. EMBO J. **10:** 793-798.
16. MATTHEWS, H. R., R. L. W. MURPHY, G. L. FAIN & T. L. LAMB. 1988. Nature **334:** 67-69.
17. FAIN, G. L. & H. R. MATTHEWS. 1990. TINS **13:** 378-384.
18. KAPLAN, R. S. & P. L. PEDERSON. 1985. Anal. Biochem. **150:** 97-104.
19. MIYAMOTO, H. & E. RACKER. 1980. J. Biol. Chem. **255:** 2656-2658.
20. STUART, C. A., R. A. PIETRZYK, R. W. FURLANETTO & A. GREEN. 1988. Anal. Biochem. **173:** 142-150.

21. POLSON, A., B. VON WECHMAR & M. H. V. VAN REGENMORTEL. 1980. Immunol. Commun. **9:** 475-493.
22. JENSENIUS, J. C., I. ANDERSEN, J. HAU, M. CRONE & C. KOCH. 1981. J. Immunol. Methods **46:** 63-68.
23. LAEMMLI, U. 1970. Nature **227:** 680-685.
24. LESER, G. P. & T. E. MARTIN. 1987. J. Cell Biol. **105:** 2083-2094.
25. MOEREMANS, M., G. DANEELS & J. DEMEY. 1985. Anal. Biochem. **145:** 315-321.
26. TASHEVA, B. & G. DESSEV. 1983. Anal. Biochem. **129:** 98-102.
27. HASHIMOTO, F., T. HORIGOME, M. KANBAYASHI, K. YOSHIDA & H. SUGANO. 1983. Anal. Biochem. **129:** 192-199.
28. MARSHALL, T. & K. M. WILLIAMS. 1984. Anal. Biochem. **139:** 502-505.
29. PHILIPSON, K. D., S. LONGONI & R. WARD. 1988. Biochim. Biophys. Acta. **945:** 298-306.
30. CONDRESCU, M., H. ROJAS, A. GERARDI, R. DIPOLO & L. BEAUGE. 1990. Biochim. Biophys. Acta **1024:** 198-202.
31. DAHAN, D., R. SPANIER & H. RAHAMIMOFF. 1991. J. Biol. Chem. **266:** 2067-2075.

Biochemical and Molecular Characterization of the Sodium-Calcium Exchanger from Bovine Rod Photoreceptors [a]

ANITA ACHILLES,[b] UTE FRIEDEL,
WINFRIED HAASE, HELMUT REILÄNDER, AND
NEIL J. COOK

Max-Planck-Institut für Biophysik
Abteilung für Molekulare Membranbiologie
D-6000 Frankfurt am Main 71,
Federal Republic of Germany

INTRODUCTION

Vertebrate rod photoreceptors hyperpolarize in response to light by means of the closure of cGMP-gated channels, present in the plasma membrane of the outer segment, which mediate an inwardly directed cation flow referred to as the dark current.[1-3] Although these channels are most permeable to monovalent cations, they also exhibit an appreciable permeability to the divalent cations Ca^{2+} and Mg^{2+}.[4,5] Indeed, calcium ions entering the cell via the cGMP-gated channel constitute 10-15% of the dark current.[5]

The sodium-calcium exchanger of rod photoreceptors is responsible for the extrusion of calcium ions entering the outer segment cytosol through the cGMP-gated channel under conditions of darkness.[6-8] The driving force for this calcium extrusion process is an inwardly directed electrochemical Na^+ gradient together with an outwardly directed electrochemical K^+ gradient.[9,10] The stoichiometry of the exchange process under physiological conditions is now accepted to be 4 Na_o^+ : 1 Ca_i^{2+} + 1 K_i^+ and is therefore electrogenic.[11,12]

PURIFICATION AND RECONSTITUTION OF THE SODIUM-CALCIUM EXCHANGER FROM BOVINE ROD PHOTORECEPTORS

A major obstacle in the purification of membrane transport or receptor proteins is the requirement for a method of detection of the protein in question after solubilization

[a] This work was supported by the Deutsche Forschungsgemeinschaft (SFB 169 Projekt C4) and a Leibniz-Programm Grant to Prof. Hartmut Michel.

[b] Address for correspondence: Max-Planck-Institut für Biophysik, Abteilung für Molekulare Membranbiologie, Heinrich-Hoffmann Str. 7, D-6000 Frankfurt am Main 71, Federal Republic of Germany.

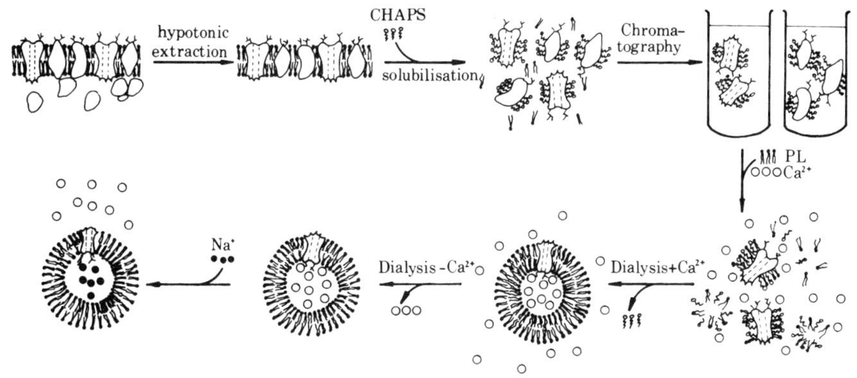

FIGURE 1. Solubilization, purification, and reconstitution of the sodium-calcium exchanger from bovine rod outer segments. Rod outer segment membranes are stripped of peripheral proteins and then solubilized using the zwitterionic detergent CHAPS. After purification by conventional column chromatography, phospholipids and calcium are added to the purified exchanger. Liposomes are formed by detergent elimination through dialysis in the presence of calcium. A transmembrane calcium gradient is established by a further dialysis step against calcium-free buffer. The addition of Na^+ to liposomes containing the purified exchanger results in the release of entrapped Ca^{2+}, thereby constituting a functional assay for the exchanger protein. Alternatively, it is possible to reconstitute the sodium-calcium exchanger in the presence of intraliposomal Na^+, and to subsequently monitor exchange activity by measuring Ca^{2+} uptake (see FIG. 3).

from its native membrane. For this reason, most purification procedures for membrane proteins have utilized the binding of radiolabeled ligands (agonists or antagonists) as an assay for the target protein during its purification by chromatographic separation after solubilization. Unfortunately, in the case of sodium-calcium exchangers, no ligands exist that bind to the exchanger protein with sufficient affinity and specificity to permit such a method of detection.

During previous work aimed at purifying and characterizing the cGMP-gated channel from bovine rod photoreceptors,[13,14] we developed a functional assay for the channel protein based on its reconstitution into Ca^{2+}-containing liposomes after solubilization. Entrapped calcium could be released by the addition of cGMP and could be conveniently monitored using the calcium-sensitive dye Arsenazo III. When applying the reconstitution procedure to solubilized rod outer segment membrane extracts, we noticed that entrapped calcium could be released not only by cGMP, but also by Na^+.[15] The addition of other alkali cations at the same concentration did not induce calcium release, therefore it was concluded that the sodium-calcium exchanger was responsible and not some unspecific release mechanism due to, say, the rise in ionic strength in the extraliposomal medium. The reconstitution procedure was thus found to be applicable and quantitative for the sodium-calcium exchanger of bovine rod outer segments and was used as an assay while purifying the protein to essential homogeneity by conventional column chromatography.[15] A summary of the reconstitution procedure is shown in FIGURE 1.

The purification procedure for the sodium-calcium exchanger from bovine rod outer segment membranes is summarized in FIGURE 2. The procedure was found to be rapid and efficient; tens of micrograms of purified exchanger protein could be purified within one day. An essential part of the purification procedure was the inclusion of phospholipids and calcium in all buffers used in order to stabilize the exchanger protein during chromatography. The purified exchanger protein was found to have a molecular

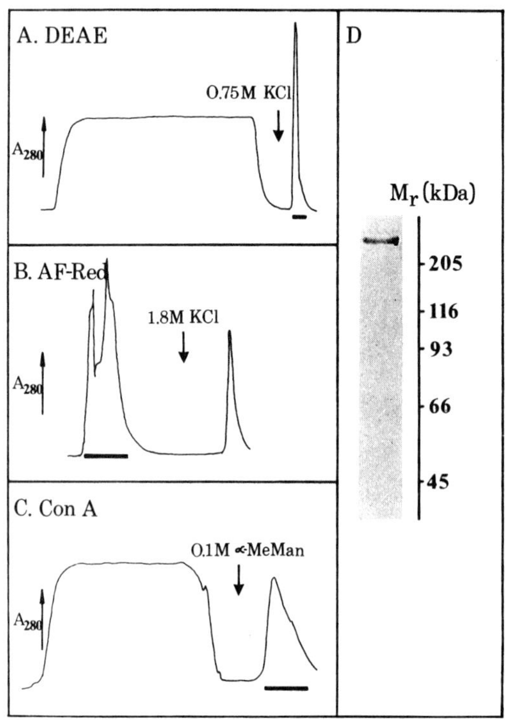

FIGURE 2. Chromatographic purification of the sodium-calcium exchanger from bovine rod photoreceptors.[15] **A:** Membranes are solubilized in the presence of 0.15 M KCl and applied to a DEAE column. After unbound proteins are washed from the column, the exchanger is eluted by 0.75 M KCl. **B:** The exchanger is then applied to an AF Red affinity column in the presence of 0.75 M KCl and is found in the unbound protein fraction. **C:** The active fraction is then applied to a Con A column, and the purified exchanger is eluted by 0.1 M α-methylmannoside. In A-C, all column fractions containing Na-Ca exchange activity are underlined. **D:** SDS-polyacrylamide gel electrophoresis of the purified rod photoreceptor sodium-calcium exchanger.

weight of $\approx$220 kDa as determined by SDS gel electrophoresis. This result was confirmed independently by Nicoll and Applebury,[16] who reported that the exchanger has a molecular weight of $\approx$215 kDa. The exchanger was found to be heavily glycosylated,[17] and to be distinct from the so-called "rim protein," a structural glycoprotein of similar molecular weight also found in rod outer segments.[18]

FUNCTIONAL PROPERTIES OF THE PURIFIED AND RECONSTITUTED SODIUM-CALCIUM EXCHANGER

The availability of the rod outer segment sodium-calcium exchanger in its purified and functional form permits the direct investigation of the functional properties of this membrane transport protein in the complete absence of other transport processes. Therefore, all transport properties observed are intrinsic to the sodium-calcium exchanger and cannot be attributed to other transport systems. An additional advantage of the reconstitution procedure is that it is possible to impose completely defined ionic conditions on the system, thereby eliminating the possibility of ionic contamination which might complicate the interpretation of results obtained *in situ.*

During the course of our work on the purified exchanger, we investigated two modes of Ca^{2+} transport using the metallochromic dye Arsenazo III. These are summarized in FIGURE 3. The first mode (FIG. 3, top) consists of loading the liposomes, into which the exchanger is reconstituted, with calcium; calcium release can then be induced by adding sodium to the extraliposomal medium. As can be seen in FIGURE 4, of all the alkali cations tested only Na^+ was capable of releasing Ca^{2+}, thereby indicating that the Na^+-binding site(s) of the exchanger is highly specific for this cation. Such calcium release experiments also allowed the determination of the K_m and cooperativity of the Na^+ dependence of the purified exchanger.[15,19] We were also able to demonstrate that transliposomal K^+ gradients were capable of influencing Na^+-induced Ca^{2+} release in a way that is consistent with the thesis that K^+ is cotransported by the exchanger.[19]

A second way in which the Ca^{2+}-transport properties of the purified exchanger were investigated is shown in FIGURE 3 (bottom). In these experiments, liposomes into which the exchanger was reconstituted were produced in the presence of 150 mM Na^+ (inside and outside). These liposomes were then added to a cuvette containing calcium and Arsenazo III, thereby diluting the extraliposomal Na^+ concentration and creating a transmembrane sodium gradient, which in turn induces calcium uptake. As shown

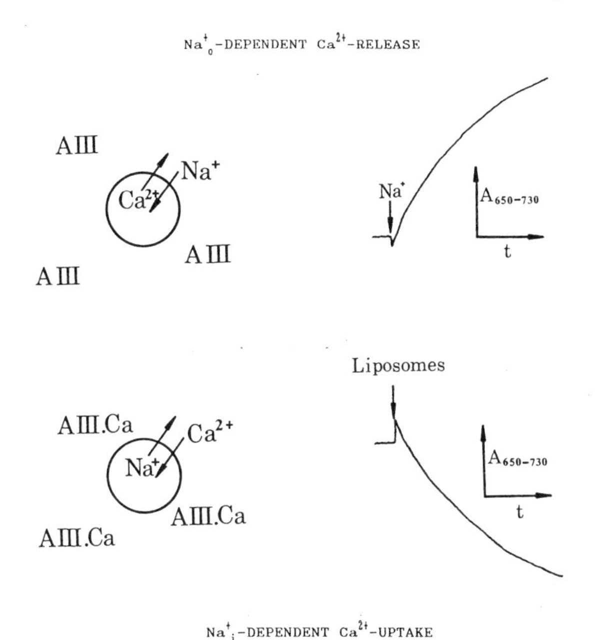

FIGURE 3. Spectroscopic determination of sodium-calcium exchange activity after reconstitution using the calcium-sensitive metallochromic dye Arsenazo III. **Above.** Na_o^+-dependent Ca^{2+} release. Liposomes are formed with entrapped calcium and then suspended in a cuvette containing Arsenazo III (AIII). Addition of Na^+ causes the release of Ca^{2+}, formation of the AIII·Ca complex, and an increase in absorbance at 650 nm. **Below.** Na_i^+-dependent Ca^{2+} uptake. Liposomes are prepared in the presence of symmetrical sodium (150 mM) and then added to a cuvette containing the AIII·Ca complex. Extraliposomal sodium is thereby diluted, leading to the uptake of Ca^{2+}, a decrease in the concentration of the AIII·Ca complex and a decrease in absorbance at 650 nm.

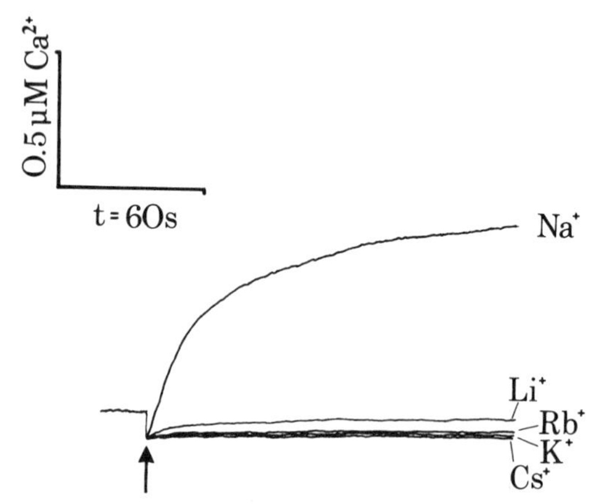

FIGURE 4. Alkali cation dependence of Ca^{2+}release by the purified and reconstituted sodium-calcium exchanger. Measurements were carried out in the presence of 50 mM KCl and 100 mM choline chloride (concentrations both inside and outside). At the time point indicated by the arrow, an aliquot of a concentrated stock solution of the alkali cation under investigation was injected into the cuvette to give a final concentration of 50 mM.

in FIGURE 5, calcium uptake was greatly facilitated by the presence of extraliposomal K^+. We also observed a residual uptake in the complete absence of potassium[19]; however, this is probably not of physiological importance since K^+ is never likely to be absent on the side of the membrane away from which calcium is transported. We also determined the selectivity of this K^+-binding site to be $K^+ \geq Rb^+ > Cs^+ >> Li^+$.[19] The K_m for the K^+ dependence of sodium-calcium exchange was determined to be about 1 mM and was found to follow Michaelis-Menten kinetics, thereby implicating 1 K^+ ion in the transport of 1 Ca^{2+} ion.[19] The results on the functional properties of the purified and reconstituted rod photoreceptor sodium-calcium exchanger are summarized in TABLE 1, and compared with the functional properties of the exchanger *in situ.*

LOCALIZATION AND DENSITY OF THE SODIUM-CALCIUM EXCHANGER

The availability of the rod outer segment exchanger in its purified form permits the generation of specific antibodies, and the subsequent immunochemical localization of this protein in the bovine retina and within the rod outer segment. As shown in FIGURE 6, anti-exchanger antibodies intensely and specifically label the outer segment layer of bovine retina. Other cell layers only show residual, nonspecific labeling. Interestingly, cone cells are not labeled by the anti-exchanger polyclonal serum, even though there is ample evidence to demonstrate that a sodium-calcium exchanger does indeed exist in these cells.[20] We were also unable to demonstrate any significant and specific immunoreactivity of the polyclonal serum in other tissues (heart, brain, liver, kidney), thereby suggesting that this type of sodium-calcium exchanger is highly tissue specific.

Within the rod outer segment, immunoreactivity was found to be localized exclusively to the plasma membrane.[17,18] This result was confirmed by Western blotting of purified outer segment disk and plasma membrane extracts.[17] Furthermore, after

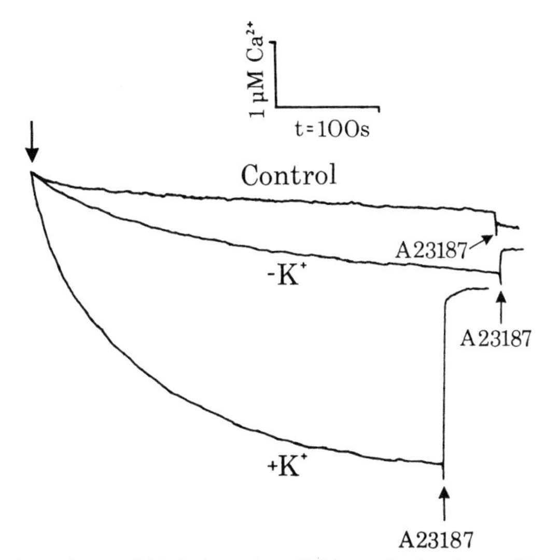

FIGURE 5. K^+ dependence of Na_o^+-dependent Ca^{2+} uptake by the purified and reconstituted sodium-calcium exchanger. Sodium-loaded liposomes were injected into the cuvette in the presence or absence of potassium. Uptaken Ca^{2+} could be rereleased by addition of the calcium ionophore A23187. In the control experiment, liposomes were injected into the cuvette in the presence of 150 mM sodium (the same concentration as that inside the liposomes). Reproduced from Friedel *et al.*[19] with the permission of the editors of *Biochimica Biophysica Acta.*

TABLE 1. Comparison of the Functional Properties of the Purified and Reconstituted Sodium-Calcium Exchanger with Those of the Rod Photoreceptor Exchanger *in Situ*

	Reconstituted[a]	*In Situ*
K_m Na^+ (mM)	26.1	30-93[b-d]
Cooperativity Na^+ (n)	3.01	2-2.3[b-d]
K_m K^+ (mM)	1.0	1.2-1.5[e]
K_m Ca^{2+} (μM)	n.d.	5[f]
Turnover number (Ca^{2+} ions sec^{-1} $exchanger^{-1}$)	115	250-750[g]

NOTE: n.d., not determined.
[a] Friedel *et al.*[19]
[b] Schnetkamp.[7]
[c] Schnetkamp & Bownds.[22]
[d] Lagnado *et al.*[23]
[e] Schnetkamp *et al.*[12]
[f] Barrios *et al.*[24]
[g] Calculated from Schnetkamp & Bownds[22] as described in Friedel *et al.*[19]

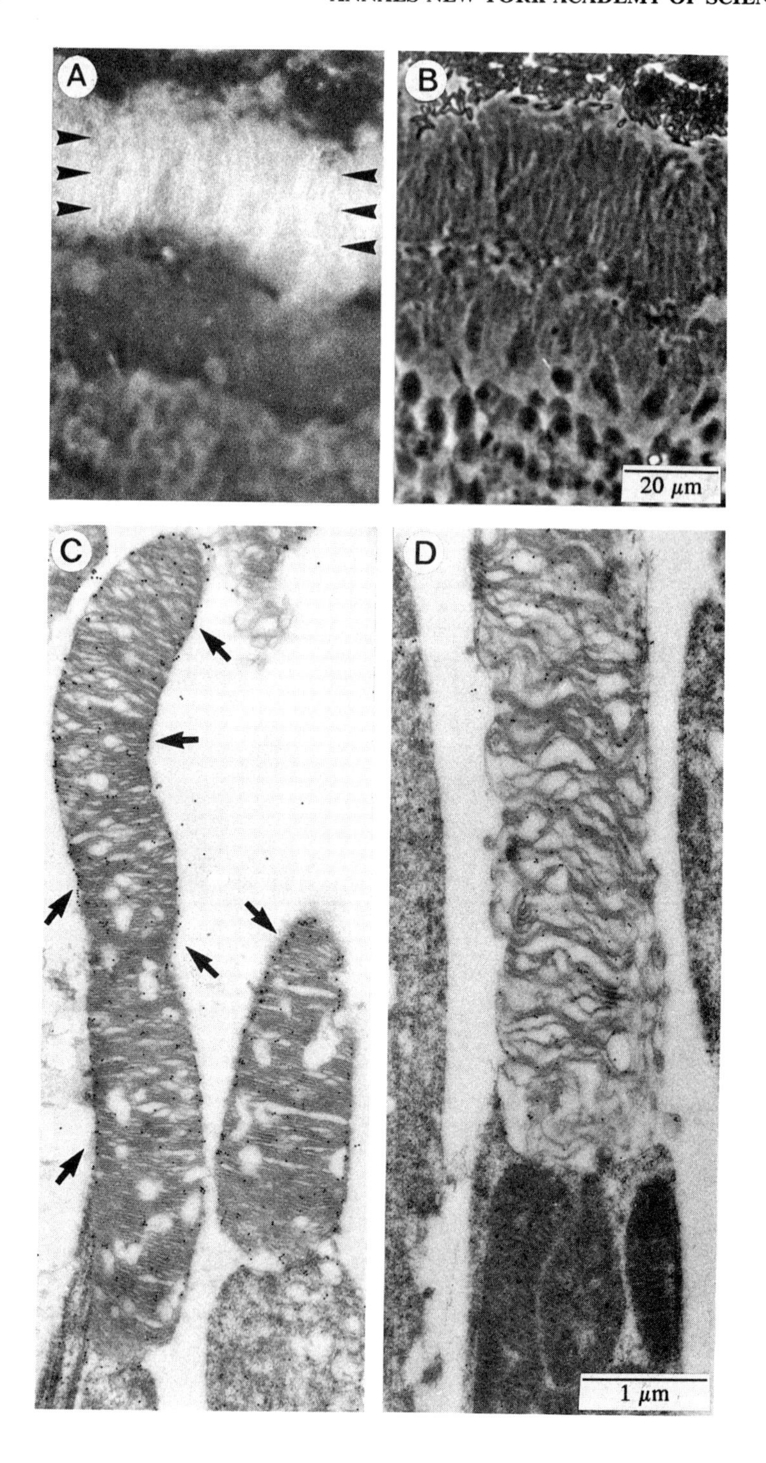
A
B
20 μm
C
D
1 μm

solubilization and reconstitution, only plasma membrane extracts and not disks showed sodium-calcium exchange activity.[17] Through biochemical means, we were able to calculate the density of the sodium-calcium exchanger in the rod outer segment to lie in the range 200-450 μm^{-2}.[17]

MOLECULAR PROPERTIES OF THE ROD PHOTORECEPTOR SODIUM-CALCIUM EXCHANGER

In order to determine the primary structure of the rod outer segment sodium-calcium exchanger, we screened a lambda gt11 cDNA expression library from bovine retina with the previously described polyclonal serum.[18] Thirty-three positive clones were obtained, each of which was further probed with the monoclonal antibody 1B3.[17] This procedure gave rise to two further positive clones, which were then subjected to sequence analysis after cloning the respective phage DNA-inserts into the pBluescript vector. Sequence determination revealed that the two double-positive clones were identical.

In a further approach, we subjected the purified sodium-calcium exchanger polypeptide to cyanogen bromide cleavage and succeeded in purifying four peptides that were suitable for amino acid sequence analysis by the Edman degradation procedure. On the basis of one of the peptide sequences, a degenerated oligonucleotide probe was synthesized and used to screen a lambda gt10 cDNA library from bovine retina. We obtained several positive clones which were subjected to DNA sequence analysis. We were able to obtain approximately 1600 base pairs corresponding to the COOH-terminal end of the sodium-calcium exchanger by sequencing these clones. The DNA sequence obtained from the antibody-screened clones could be localized within this sequence. In order to obtain the complete NH_2-terminal region of the exchanger, we isolated mRNA from bovine retina and constructed a lambda ZAP cDNA library by priming with an internal oligonucleotide. From this library the complete sequence of the exchanger could be established.

A Kyte-Doolittle hydrophobicity plot of the complete sequence is shown in FIGURE 7. The protein has a hydrophobic region at the NH_2-terminal end, which possibly constitutes a signal sequence. This is followed by a long, hydrophilic region that contains several potential *N*-glycosylation sites, followed by a cluster of at least three, but possibly as many as five, potential transmembrane segments. These are followed by a large hydrophilic loop that was concluded to be cytosolic because it contains the sequence cloned using the monoclonal antibody 1B3, which is known to bind on the cytosolic side of the membrane.[17] This loop is then followed by a cluster of six potential transmembrane segments and a very short hydrophilic COOH terminus.

◆ **FIGURE 6.** Immunocytochemical localization of the sodium-calcium exchanger in bovine retina with a polyclonal antibody directed against the purified sodium-calcium exchanger. As shown by fluorescence microscopy, the antibody intensively binds to the photoreceptor outer segment layer of bovine retina (**A**, arrowheads). **B** is the corresponding phase-contrast image of the bovine retina cryosection depicted in A. By postembedding immunogold labeling, specific antibody binding was only detected along the plasma membrane of the rod outer segment (**C**, arrows). Only background immunoreactivity is observed over the disk membranes of rod photoreceptors (**C**) and over both plasma and disk membranes of cone photoreceptors (**D**). The inner segments of both photoreceptor types show no specific immunoreactivity. Background immunoreactivity was verified using preimmune serum (not shown). Magnifications: A and B, 656×; C and D, 17,000×.

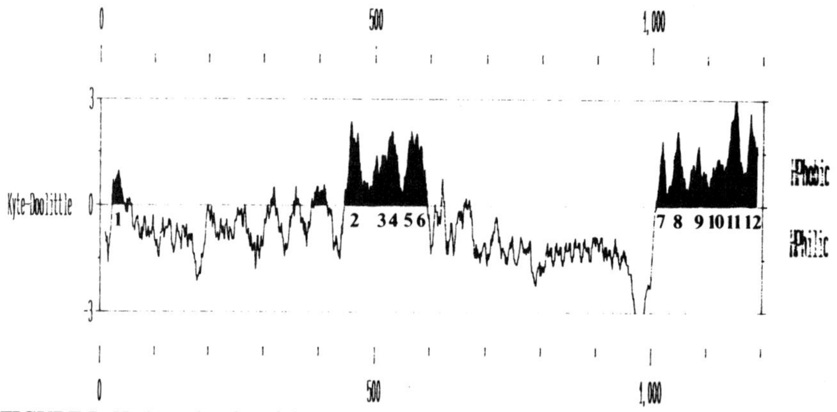

FIGURE 7. Hydropathy plot of the primary sequence of the rod outer segment sodium-calcium exchanger. The plot was determined by the method of Kyte and Doolittle[25] with a window of 20 amino acids. Hydrophobicity is represented on the ordinate by positive values and hydrophilicity by negative values. Potential transmembrane segments are labeled 1-12.

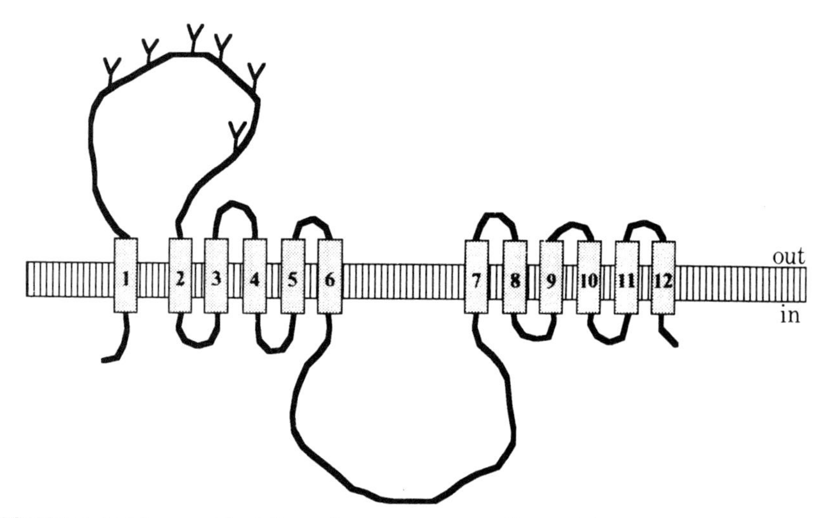

FIGURE 8. Folding model of the rod outer segment sodium-calcium exchanger across the plasma membrane; **out,** outside; **in,** inside (i.e., the cytosolic side).

A topological model for the folding of the sodium-calcium exchanger within the rod outer segment plasma membrane is shown in FIGURE 8. Interestingly, although there is no apparent sequence homology with the recently sequenced cardiac sodium-calcium exchanger,[21] the general topological distribution within the membrane is remarkably similar. The lack of sequence homology is, however, not completely surprising given the fact that the rod photoreceptor exchanger is now known to transport K^+, which the cardiac exchanger clearly does not. Further work is necessary before a complete understanding of how the rod sodium-calcium exchanger executes its functions at the molecular level can be attained.

ACKNOWLEDGMENTS

We are grateful to the laboratories of Robert S. Molday (Vancouver, Canada) and U. Benjamin Kaupp (Osnabrück & Jülich, F.R.G.) in collaboration with which the above work was undertaken. We are also grateful to Hartmut Michel (Frankfurt, F.R.G.) for support and encouragement.

REFERENCES

1. STRYER, L. 1986. Cyclic GMP cascade of vision. Annu. Rev. Neurosci. **9:** 87-119.
2. FESENKO, E. E., S. S. KOLESNIKOV & A. L. LYUBARSKY. 1985. Induction by cyclic GMP of cationic conductance in plasma membrane of retinal rod outer segment. Nature **313:** 310-313.
3. MCNAUGHTON, P. A. 1990. Light response of vertebrate photoreceptors. Physiol. Rev. **70:** 847-883.
4. HODGKIN, A. L., P. A. MCNAUGHTON, B. J. NUNN & K.-W. YAU. 1985. Effect of ions on retinal rods from *Bufo marinus.* J. Physiol. (London) **350:** 649-680.
5. YAU, K.-W. & K. NAKATANI. 1984. Cation selectivity of light-sensitive conductance in retinal rods. Nature **309:** 352-354.
6. YAU, K.-W. & K. NAKATANI. 1984. Electrogenic Na-Ca exchange in retinal rod outer segments. Nature **311:** 661-663.
7. SCHNETKAMP, P. P. M. 1986. Sodium-calcium exchange in the outer segments of bovine rod photoreceptors. J. Physiol. (London) **373:** 25-45.
8. HODGKIN, A. L., P. A. MCNAUGHTON & B. J. NUNN. 1987. Measurement of sodium-calcium exchange in salamander rods. J. Physiol. (London) **391:** 347-370.
9. SCHNETKAMP, P. P. M. 1989. Na-Ca or Na-Ca-K-exchange in rod photoreceptors. Prog. Biophys. Molec. Biol. **54:** 1-29.
10. LAGNADO, L. & P. A. MCNAUGHTON. 1990. Electrogenic properties of Na : Ca exchange. J. Membr. Biol. **113:** 177-191.
11. CERVETTO, L., L. LAGNADO, R. J. PERRY, D. W. ROBINSON & P. A. MCNAUGHTON. 1989. Extrusion of calcium from rod outer segments is driven by both sodium and potassium gradients. Nature **337:** 740-743.
12. SCHNETKAMP, P. P. M., D. K. BASU & R. T. SZERENCSEI. 1989. Na-Ca exchanger in the outer segments of bovine rod photoreceptors requires and transports potassium. Am. J. Physiol. **257:** C152-C157.
13. COOK, N. J., C. ZEILINGER, K.-W. KOCH & U. B. KAUPP. 1986. Solubilization and functional reconstitution of the cGMP-dependent cation channel from bovine rod outer segments. J. Biol. Chem. **261:** 17033-17039.
14. COOK, N. J., W. HANKE & U. B. KAUPP. 1987. Identification, purification, and functional reconstitution of the cyclic GMP-dependent channel from rod photoreceptors. Proc. Natl. Acad. Sci. USA **84:** 585-589.
15. COOK, N. J. & U. B. KAUPP. 1988. Solubilization, purification and functional reconstitution of the sodium-calcium exchanger from bovine rod outer segments. J. Biol. Chem. **263:** 11382-11388.
16. NICOLL, D. A. & M. L. APPLEBURY. 1989. Purification of the bovine rod outer segment Na^+/Ca^{2+} exchanger. J. Biol. Chem. **264:** 16207-16213.
17. REID, D. M., U. FRIEDEL, R. S. MOLDAY & N. J. COOK. 1990. Identification of the sodium-calcium exchanger as the major ricin-binding protein of bovine rod outer segments and its localization to the plasma membrane. Biochemistry **29:** 1601-1607.
18. HAASE, W., W. FRIESE, R. D. GORDON, H. MÜLLER & N. J. COOK. 1990. Immunological characterization and localization of the Na^+/Ca^{2+}-exchanger in bovine retina. J. Neurosci. **10:** 1486-1494.
19. FRIEDEL, U., G. WOLBRING, P. WOHLFART & N. J. COOK. 1991. The sodium-calcium exchanger of bovine rod photoreceptors: K^+-dependence of the purified and reconstituted protein. Biochim. Biophys. Acta **1061:** 247-252.

20. NAKATANI, K. & K.-W. YAU. 1989. Sodium-dependent calcium extrusion and sensitivity regulation in retinal cones of the salamander. J. Physiol. (London) **409:** 525-548.
21. NICOLL, D. A., S. LONGONI & K. D. PHILIPSON. 1990. Molecular cloning and functional expression of the cardiac sarcolemmal Na^{+}-Ca^{2+}-exchanger. Science **250:** 562-565.
22. SCHNETKAMP, P. P. M. & M. D. BOWNDS. 1987. Sodium and cGMP-induced Ca^{2+} fluxes in frog rod photoreceptors. J. Gen. Physiol. **89:** 481-500.
23. LAGNADO, L., L. CERVETTO & P. A. MCNAUGHTON. 1988. Ion transport by the Na : Ca exchanger in isolated rod outer segments. Proc. Natl. Acad. Sci. USA **85:** 4548-4552.
24. BARRIOS, B., D. A. NICOLL & K. D. PHILIPSON. 1990. A comparison of the properties of the Na/Ca exchanger from sarcolemma and rod outer segments. Biophys. J. **57:** 180a.
25. KYTE, J. & R. F. DOOLITTLE. 1982. A simple method for displaying the hydropathic nature of a protein. J. Mol. Biol. **157:** 105-132.

Effect of Polyclonal Antibodies on the Cardiac Sodium-Calcium Exchanger[a]

ANTHONY AMBESI, ELDWIN L. VANALSTYNE,[b]
ERVIN E. BAGWELL, AND
GEORGE E. LINDENMAYER[c]

Departments of Pharmacology and Medicine
Medical University of South Carolina
Charleston, South Carolina 29425

We have previously reported the purification of the cardiac sodium-calcium exchanger from canine myocardium and the generation of polyclonal antibodies against the purified exchanger.[1] The polyclonal antibodies immunoprecipitated 97% of the sodium-calcium exchange activity from detergent-solubilized sarcolemma and reacted with prominent proteins of 75, 120, and 140 kDa (reducing conditions) on immunoblots of the purified exchanger. Only one major protein, centered at 140 kDa, was detected under nonreducing conditions. Subsequently, antibodies against the 75-, 120-, and 140-kDa proteins were antigen purified and found to immunoprecipitate 92, 91, and 83% of the exchange activity, respectively. Furthermore, the antigen-purified antibodies exhibited cross-reactivity with each of the other two prominent proteins. These data are consistent with those reported by Philipson *et al.*,[2] with the exception of minor differences in the molecular weights reported, and suggest that the three proteins are immunologically related and that all are related to the sodium-calcium exchanger.

The purpose of the present study was to determine the effect of these antibodies on sodium-calcium exchange activity manifested by sarcolemmal vesicles (70% sealed R/O; 0-12% sealed I/O; remainder leaky) from canine ventricle. The vesicles were exposed to increasing concentrations of affinity-purified IgG from preimmune or immune serum. Antibodies from the immune serum stimulated exchange activity 3.5-fold in a dose-dependent manner with half-maximal stimulation at 0.5 μg IgG/μg sarcolemmal protein (FIG. 1). Conversely, IgG from preimmune serum had little or no effect between 0.01 and 10 μg IgG/μg. A separate experiment was carried out to further test whether the stimulation observed with IgG from the immune serum (anti-NCX) was specific. Four IgG fractions were tested: (1) from the rabbit that was subsequently immunized against the sodium-calcium exchanger (preimmune); (2) anti-NCX; (3) from a rabbit immunized against total sarcolemmal proteins (anti-SL); and (4) from a rabbit immunized against a prominent sarcolemmal protein of 82 kDa enriched by a

[a]This work was supported in part by National Institutes of Health Grants HL29566 and HL42042.

[b]Eldwin L. VanAlstyne's current address: Lilly Research Labs, MC931, 98c/4, Lilly Corporate Center, Indianapolis, IN 46285.

[c]Address for correspondence: Dr. George E. Lindenmayer, Department of Pharmacology, Medical University of South Carolina, 171 Ashley Avenue, Charleston, SC 29425.

protease treatment[4] (anti-82). Exposure to 3 μg IgG/μg sarcolemmal protein showed that the preimmune and the anti-82 fractions caused an increase in sodium-calcium exchange activity of 50 and 46%, respectively, whereas the anti-NCX increased activity by 184% (*inset,* FIG. 1). Interestingly, the anti-SL caused nearly as much of an increase (145%) as the anti-NCX but neither the anti-SL nor anti-82 sera were able to immunoprecipitate sodium-calcium exchange activity from solubilized sarcolemma. Eadie-Hofstee plots suggested that the stimulation by anti-NCX and anti-SL was due to a small increase in V_{max} and a larger decrease in the $K_{0.5}$ for outside calcium (FIG. 2).

It seems clear that these IgG fractions are capable of causing some nonspecific stimulation of sodium-calcium exchange activity (results with preimmune and anti-82 fractions). Questions remain as to the explanation of the larger stimulation seen with the anti-NCX and anti-SL fractions. Several possibilities are: (1) stimulation by both the anti-NCX and anti-SL reflect nonspecific interactions; (2) stimulation may reflect interaction of the anti-NCX with the exchanger protein and of the anti-SL with a closely associated protein in the sarcolemmal membrane; or (3) the effects reflect interaction of the anti-NCX and a subfraction of the anti-SL with the exchanger protein

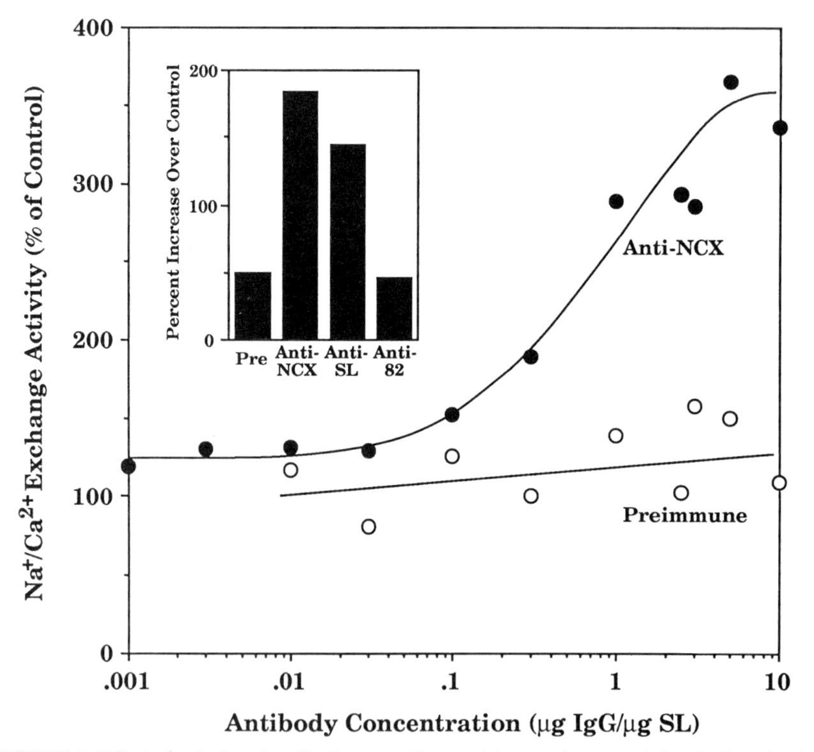

FIGURE 1. Effect of polyclonal antibodies on sodium-calcium exchange activity. Sodium-loaded (160 mM NaCl, 10 mM Mops-Tris, pH 7.4) sarcolemmal vesicles were exposed for one hour at 37° to varying concentrations of affinity-purified IgG from preimmune or immune (anti-NCX) serum and assayed for exchange activity[3] (n = 1-4). *Inset*: Sarcolemmal vesicles were exposed to affinity-purified IgG (3 μg IgG/μg sarcolemmal protein) from preimmune serum, anti-NCX, anti-SL, or anti-82 and then assayed for exchange activity (n = 4-5). Sodium-calcium exchange assays were carried out using 40 μM outside $^{45}Ca/CaCl_2$.

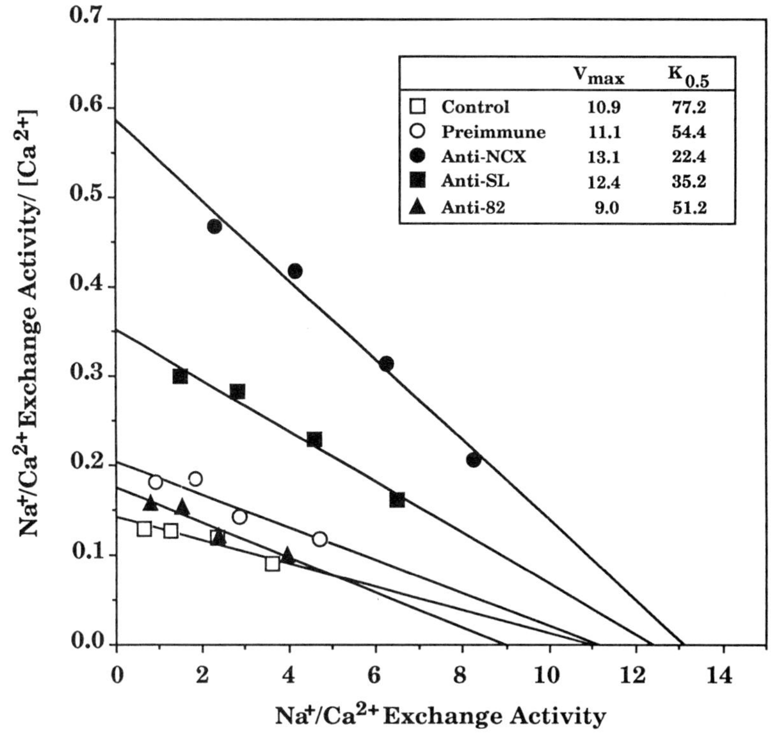

FIGURE 2. Effect of polyclonal antibodies on sodium-calcium exchange activity versus outside calcium concentration. Sodium-loaded sarcolemmal vesicles were exposed to affinity-purified IgG at 3 μg IgG/μg sarcolemmal protein for one hour at 37°C. Sodium-calcium exchange activity was then assessed at varying outside calcium concentrations (5-40 μM). Results (n = 2) are plotted as ν/[calcium] versus ν. Least-squares analysis was used to determine values (*inset*) for V_{max} (nmol·mg·sec) and $K_{0.5}$ (μM).

because inability of the anti-SL to immunoprecipitate the exchanger in one form (detergent-solubilized) does not necessarily preclude specific interactions of a subset of antibodies in the IgG fraction with the exchanger in another form (sarcolemmal membrane).

REFERENCES

1. Ambesi, A., E. E. Bagwell & G. E. Lindenmayer. 1991. Biophys. J. **59:** 138a.
2. Philipson, K. D., S. Longoni & R. Ward. 1988. Biochim. Biophys. Acta **945:** 298-306.
3. Reeves, J. P. & J. L. Sutko. 1983. J. Biol. Chem. **258:** 3178-3182.
4. Hale, C. C., R. S. Slaughter, D. C. Ahrens & J. P. Reeves. 1984. Proc. Natl. Acad. Sci. USA **81:** 6569-6573.

At Least Three Functional Isoforms of the Cardiac Na^+-Ca^{2+} Exchange Exist

I. DRUBAIX,[a] N. KASSIS, AND L. G. LELIÈVRE

Laboratoire de Pharmacologie des Transports Ioniques Membranaires
Hall de Biotechnologies
Université Paris 7
75251 Paris Cedex 05, France

The regulatory mechanisms of Ca^{2+} movements involved in the excitation-contraction coupling in cardiac muscles vary during ontogenesis. Myocardial perinatal development is characterized by ultrastructural changes,[1,2] a progressive decline in duration of the ventricular action potential plateau,[3] a decrease in time to peak tension,[4] and an incomplete functional maturation of the sarcoplasmic reticulum (SR). The Ca^{2+}-induced Ca^{2+} release from the SR appears at birth or at the latest two days postpartum.[4] Convergent data from cat, dog, rat, and rabbit hearts show that the neonatal heart is more dependent on transsarcolemmal Ca^{2+} influx and efflux than the adult heart. The predominant roles of the sarcolemma-bound Ca^{2+} transports, such as the Na^+-Ca^{2+} exchange, in neonatal hearts would counterbalance either the absence of SR structures or its incomplete maturation at the neonatal stage.[5,6] The SR calcium uptake (μmol per g of muscle) in the newborn is 60% of the adult value.[7]

The question arises: What could be the properties of the Na^+-Ca^{2+} exchange(s) in newborn rat heart inasmuch as intracellular Ca^{2+} is higher (400 nM) than in adult[8] (130 nM)?

The Na^+-Ca^{2+} exchange has been studied in cardiac sarcolemma (SL) isolated from newborn (6 hours after birth), adult Wistar, and adult SHR rats.

The dose-response curves of Na^+-Ca^{2+} exchange activity to free Ca^{2+} concentrations are very different in adult and newborn SL preparations. For adult hearts, the curves are monophasic. The apparent affinities for Ca^{2+} transport are 1 μM and 10 μM in Wistar and SHR, respectively. In newborn heart, the dose-response curve spans over four logarithmic units suggesting the existence of two types of saturable and independent Ca^{2+} carriers: apparent affinities of 50 nM and 10 μM and proportional contributions of 73% and 27%, respectively (TABLE 1).

We have tested whether the switch in sensitivity to Ca^{2+} could result from the expression of different isoforms of the Na^+-Ca^{2+} exchange. According to Western blots in newborn cardiac SL, two polypeptides have been identified under reduced conditions. Their apparent molecular masses (n = 15) were 119,000 ± 2,000 daltons (so called isoform I_1) and 126,000 ± 2,000 (isoform I_3). The same polyclonal antibodies (raised against purified canine cardiac Na^+-Ca^{2+} exchange preparations, a generous gift of Prof. K. D. Philipson) react, in adult Wistar cardiac preparations, with a single molecular form (I_2) exhibiting an apparent molecular weight of 123,000 ± 3,000

[a] Recipient of a fellowship from le Ministère de la Recherche et de la Technologie.

TABLE 1. Characteristics and Distribution of the Three Rat Cardiac Isoforms Detected

	In vitro[a] Properties		Distribution		
Isoforms	Sensitivity to Ca^{2+} (M)[b]	Molecular Mass (kDa)[c]	Wistar Adult	Wistar Newborn	SHR Adult
I_1	10^{-5}	119	NO	YES	YES
I_2	10^{-6}	123	YES	NO	NO
I_3	5.10^{-8}	126	NO	YES	NO

[a] Cardiac SL vesicles were isolated by the same procedure[10] from adult and newborn (6 hours after birth) rat hearts. The modifications were no lysis and a longer period of homogenization (2 $\times$ 5 sec). All the vesicles were found to be of inside-out orientation.

[b] Na^+-Ca^{2+} exchange was measured as the Na^+-dependent $^{45}Ca^{2+}$ uptake, at a well-defined potential (85 mV) as previously described.[11] The sensitivity to free Ca^{2+} was evaluated by the amount of $^{45}Ca^{2+}$ leading to half-maximal transport capacity.

[c] Samples from adult, newborn, and adult SHR hearts were electrophoresed as described by Laemmli[12] and immunoblotted with polyclonal antiserum (Western blot) according to the method described by Towbin *et al.*[13]

daltons (n = 11). In adult SHR heart (n = 6), we found a single band of 119,000 ± 3,000 daltons (I_1; see TABLE 1).

A new question arises: Inasmuch as in adult rat heart the contribution of the Na^+-Ca^{2+} exchanger I_2 is 20-23%[9] in Ca^{2+} movements, in the absence of a functional SR, how can I_1 and I_3 regulate the Ca^{2+} entry and efflux in newborn heart?

REFERENCES

1. LEGATO, M. J. 1979. Circ. Res. **44:** 250-262.
2. OLIVETTI, G., P. ANVERSA & A. V. LOUD. 1980. Circ. Res. **46:** 503-512.
3. LANGER, G. A., A. J. BRADY, S. T. TAN & S. D. SERENA. 1975. Circ. Res. **36:** 744-752.
4. FABIATO, A. & F. FABIATO. 1978. Ann. N. Y. Acad. Sci. **307:** 491-522.
5. MAHONY, L. & L. R. JONES. 1986. J. Biol. Chem. **261:** 15257-15265.
6. NAYLER, W. G. & E. FASSOLD. 1977. Cardiovasc. Res. **11:** 213-237.
7. NAKANISHI, T., M. SEGUCHI & A. TAKAO. 1988. Experientia **44:** 936-944.
8. BORZAK, S., R. A. KELLY, B. K. KRAMER, Y. MATOBA, J. D. MARSH & M. REERS. 1990. Am. J. Physiol. **259:** H973-H978.
9. BERS, D. M., W. LEDERER & J. R. BERLIN. 1990. Am. J. Physiol. **258:** C944-C954.
10. MANSIER, P., D. CHARLEMAGNE, B. ROSSI, M. PRETESEILLE, B. SWYNGHEDAUW & L. G. LELIEVRE. 1983. J Biol. Chem. **258:** 6628-6635.
11. HANF, R., I. DRUBAIX, F. MAROTTE & L. G. LELIEVRE. 1988. FEBS Lett. **236:** 145-149.
12. LAEMMLI, U. K. 1970. Nature **227:** 680-685.
13. TOWBIN, H., T. STAEHELIN & J. GORDON. 1979. Proc. Natl. Acad. Sci. USA **76:** 4350-4354.

Some Molecular Properties of the Synaptic Plasma Membrane Na^+-Ca^{2+} Exchanger [a]

M. L. MICHAELIS,[b] J. L. WALSH,
C. JAYAWICKREME, S. SCHUELER, AND
M. HURLBERT

Department of Pharmacology/Toxicology
University of Kansas
Lawrence, Kansas 66045

The identification of the brain Na^+-Ca^{2+} exchanger protein has not yet been resolved. We have developed a procedure that involves the use of sequential gel filtration, ion exchange, and immobilized metal affinity chromatography in the purification of exchanger activity associated with two protein bands on SDS-PAGE.[1] The two protein bands have estimated molecular sizes of 36 and 50 kDa, considerably smaller than the proteins isolated from sarcolemmal (70, 120, and 160 kDa)[2] and rod outer segment (ROS) membranes (215-220 kDa).[3,4] Polyclonal antibodies (Abs) were raised against the electroeluted 36-kDa protein, and these Abs cross-reacted with the electroeluted 50-kDa protein on ELISAs, suggesting that the two proteins are closely related or that one may be an abnormally migrating form of the other. The anti-36-kDa serum immunoextracted >95% of the Na^+-Ca^{2+} exchange activity in solubilized synaptic membranes while preimmune serum extracted none.[1] These results indicate that the 36-kDa band is a critical component of the brain Na^+-Ca^{2+} exchanger protein. The addition of many protease inhibitors did not lead to the isolation of higher molecular weight proteins. This does not totally rule out the possibility that these bands represent proteolytic fragments from a larger protein. The exchanger activity present in brain membranes differs significantly from the sarcolemmal exchanger in terms of sensitivity to treatment with proteases. Philipson and colleagues reported that treatment of sarcolemmal membranes with a variety of proteases led to marked activation (200%) of antiporter activity.[5] Similar treatment of synaptic membranes, on the other hand, never stimulated exchanger activity and consistently led to inhibition of activity at higher protease concentrations (FIG. 1). These results may indicate that brain proteases have already activated the exchanger in the brain membranes, preventing further stimulation by exogenous proteases and leading to purification of active proteolytic fragments that associate with each other during the reconstitution step. It is also possible, however, that the 36-kDa brain protein represents a specific isoform of the exchanger that is expressed in high levels in the brain.

[a] The work described in this report was supported by Public Health Service Grant #AA04732 and Army Research Office Grant #DAAL 03-88-K0017.

[b] Address for correspondence: M. L Michaelis, Ph.D., Department of Pharmacology/Toxicology, University of Kansas, Malott Hall Room 5064, Lawrence, KS 66045.

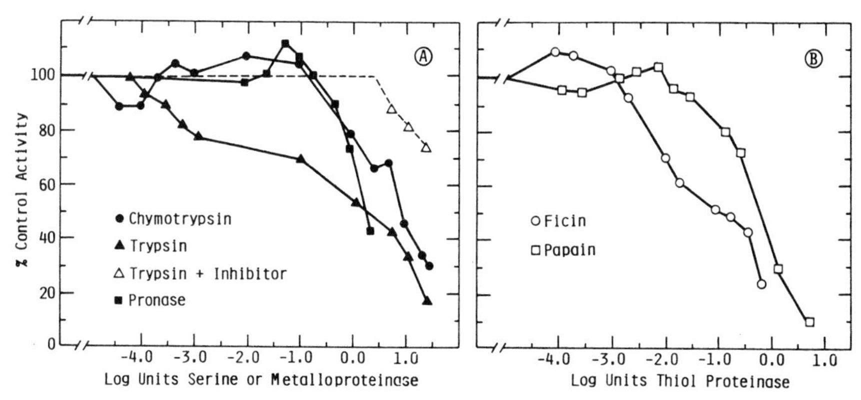

FIGURE 1. Effects of proteinase treatment on Na^+-Ca^{2+} exchange activity in synaptic plasma membrane vesicles. The synaptic membranes (~50 μg protein) were preincubated with the indicated amounts of serine proteinase or metalloproteinase (**A**) or thiol proteinase (**B**), and the Na^+-dependent Ca^{2+} influx was determined in the presence of 10 μM $CaCl_2$ for 8 sec at 35°C. Each point represents the mean of 8-12 determinations from three experiments, and the SEMs were less than 10% of the mean values. The mean Na^+-dependent Ca^{2+} transport activity for all the control samples in these experiments was 3.75 ± 0.15 nmol Ca^{2+}·mg protein·min.

TABLE 1. Effects of Cholestrol Incorporation into Proteoliposomes on Na^+-Ca^{2+} Exchange Activity of Partially Purified Brain Antiporter

Lipid Composition	Specific Activity (nmol Ca^{2+}/mg protein)[a]
Asolectin 1.0	68 ± 4[b]
PC : PS : PE 3 : 5 : 2	44 ± 4
Asolectin : Cholesterol 4 : 1	10 ± 1.1
PC : PS : PE : Cholesterol 3 : 4 : 2 : 1	24 ± 3

[a] Activity determined at 10 μM Ca^{2+} for 60 sec at 30°C.
[b] SEM for 4-8 determinations in two experiments.

In addition to the failure of protease treatment to activate the brain protein, the exchanger activity in brain membranes differs from that of sarcolemmal and ROS membranes in other respects. For example, the ROS exchanger, even in its isolated form, requires K^+ for Ca^{2+} transport activity,[4] whereas the brain protein, whether in the membranes or isolated and reconstituted, exhibits no requirement for K^+.[1] Another example of a difference between the sarcolemmal and brain exchangers involves the effects of cholesterol addition to the liposome mixture. The sarcolemmal exchange activity is strongly enhanced by the addition of cholesterol to the lipids used for the reconstitution.[6] However, cholesterol markedly decreased the activity of the partially purified brain exchanger relative to that observed in either asolectin alone or a defined mixture of PC, PS,and PE (TABLE 1). Thus it appears that some properties of the brain protein are quite different from those of the sarcolemmal and ROS exchangers, and the molecular entity responsible for brain exchange activity may differ as well. Western blot analysis of synaptic plasma membranes using the serum that Philipson and colleagues prepared against the partially purified sarcolemmal exchanger revealed immunoreactivity against a number of synaptic membrane proteins ranging from about 66 kDa down to about 30 kDa, and, in addition, there was a weak reaction with the electroeluted 36-kDa brain protein. The serum showed little or no reaction with brain proteins at molecular sizes of 100 kDa or greater. Despite the significant progress that has recently been made in characterizing the exchanger protein in sarcolemmal and ROS membranes, final resolution of the brain exchanger identity must await the cloning of the exchanger protein(s) from brain tissue.

REFERENCES

1. MICHAELIS, M. L., E. W. NUNLEY, C. JAYAWICKREME, M. HURLBERT, S. SCHUELER & C. GUILLY. 1991. Purification of a synaptic membrane Na^+/Ca^{2+} antiporter and immunoextraction with antibodies to a 36-kDa protein. J. Neurochem. In press.
2. PHILIPSON, K. D., S. LONGONI & R. WARD. 1988. Purification of the cardiac Na^+-Ca^{2+} exchange protein. Biochim. Biophys. Acta **945:** 298-306.
3. COOK, N. J. & U. B. KAUPP. 1988. Solubilization, purification, and reconstitution of the sodium-calcium exchanger from bovine retinal rod outer segments. J. Biol. Chem. **263:** 417-426.
4. NICOLL, D. A. & M. L. APPLEBURY. 1989. Purification of the bovine rod outer segment Na^+/Ca^{2+} exchanger. J. Biol. Chem. **264:** 16207-16213.
5. PHILIPSON, K. D. & A. Y. NISHIMOTO. 1982. Stimulation of Na^+-Ca^{2+} exchange in cardiac sarcolemmal vesicles by proteinase pretreatment. Am. J. Physiol. **243:** C191-C195.
6. VEMURI, R. & K. D. PHILIPSON. 1987. Phospholipid composition modulates the Na^+-Ca^{2+} exchange activity of cardiac sarcolemma in reconstituted vesicles. Biochim. Biophys. Acta **937:** 258-268.

Physiological Roles of Sodium-Calcium Exchange

Introduction to Part IV

The widespread distribution of the plasmalemmal Na-Ca exchanger, its relatively low affinity for Ca^{2+} but large capacity or maximal rate of transport, and the fact that it usually operates in parallel with the plasmalemmal ATP-driven Ca^{2+} pump, raise interesting questions about the physiological roles of the exchanger. The following group of articles focuses on key aspects of the exchanger's function in vertebrate and invertebrate photoreceptors and in neurons and astroglia.

As a result of the 4 Na^+ : 1 Ca^{2+} + 1 K^+ exchanger coupling in vertebrate photoreceptors, the Na^+ and K^+ electrochemical gradients provide sufficient energy to maintain the low "resting" $[Ca^{2+}]_i$ despite the relatively low resting potentials and high resting $[Na^+]_i$. Moreover, by modulating $[Ca^{2+}]_i$ during activation, the exchanger plays a critical role in light adaptation.

In many types of cells, the ATP-driven Ca^{2+} pump in the plasmalemma plays a dominant role in regulating "resting" $[Ca^{2+}]_i$, but this level may be "fine-tuned" by the Na-Ca exchanger. Perhaps much more important, however, is the exchanger's indirect influence on the amount of Ca^{2+} stored in intracellular organelles (especially the endoplasmic reticulum, (ER). Consequently, as discussed in this section, the Na-Ca exchanger may help to regulate the amount of Ca^{2+} release from the ER during cell activation and, thus, may modulate cell responsiveness.

The large capacity of the exchanger has raised some new ideas, including: (i) the possibility that exchanger-mediated Ca^{2+} efflux at mammalian nerve terminals might be sufficiently rapid to play a significant role in swift termination of neurotransmitter release, and (ii) the suggestion that neuronal resistance to insults (e.g., excitotoxins, which promote Ca^{2+} entry and Ca^{2+}-dependent cell injury and death) is directly related to the capacity of the cell's Na-Ca exchanger to extrude Ca^{2+}.

In sum, these observations illustrate increasing recognition of the vital importance of the Na-Ca exchanger in the function of cells in the nervous system, including neurons, astrocytes, and very specialized cells such as photoreceptors.

Physiological Roles of the Sodium-Calcium Exchanger in Nerve and Muscle[a]

MORDECAI P. BLAUSTEIN,[b,c]
WILLIAM F. GOLDMAN,[b] GIOVANNI FONTANA,[b]
BRUCE K. KRUEGER,[b] ELIGIO M. SANTIAGO,[b]
THOMAS D. STEELE,[b,d] DANIEL N. WEISS,[e] AND
PAUL J. YAROWSKY[f]

[b]*Departments of Physiology,* [e]*Medicine (Cardiology Division), and*
[f]*Pharamacology & Experimental Therapeutics*
University of Maryland School of Medicine
Baltimore, Maryland 21201

INTRODUCTION

Sodium-calcium exchange has been well studied in cardiac muscle and in photoreceptors; the physiological roles of the exchanger in these tissues are widely recognized and appreciated (see other articles in this volume). A prominent Na-Ca exchanger has also been identified in both vertebrate and invertebrate neurons, but the physiological role(s) of the exchanger in neurons is (are) poorly understood. In other types of cells, such as vascular smooth muscle (VSM) cells, the physiological significance of an Na-Ca exchanger, and even its presence, have been questioned. This uncertainty has arisen because even large changes in the trans-plasmalemmal Na^+ electrochemical gradient, $\Delta\mu_{Na}$, often do not markedly alter the resting cytosolic free-Ca^{2+} concentration, $[Ca^{2+}]_{c(rest)}$, or, in VSM, "resting" tension.

Here we review the evidence that there is a prominent, physiologically important Na-Ca exchanger in the plasmalemma of mammalian neurons as well as in astrocytes and VSM cells, and in barnacle "skeletal" muscle fibers. The exchangers in these cells clearly modulate $[Ca^{2+}]_{c(rest)}$, even though this parameter is mainly under the control of the ATP-driven Ca^{2+} pumps in the plasmalemma and in the endoplasmic reticulum (ER) or, in muscle, the sarcoplasmic reticulum (SR). A very different situation prevails during cell activation, however, because even small effects of the exchanger on $[Ca^{2+}]_{c(rest)}$ are reflected by relatively large changes in the amount of Ca^{2+} stored in the ER or SR. Thus, a key role of the exchanger in many types of cells appears to be

[a]Supported by National Institutes of Health Grants NS-16106 and HL-45215 to MPB, HL-43091 to WFG, NS-16285 to BKK, and National Science Foundation Grant BNS 8711829 to PJY. GF was funded, in part, by a stipend from the Ministero della Universita e della Ricerca Scientifica e Tecnologica (M.U.R.S.T.), Rome, Italy.

[c]Address for correspondence: Mordecai P. Blaustein, M.D., Department of Physiology, University of Maryland School of Medicine, 655 West Baltimore Street, Baltimore, MD 21201.

[d]Present address: Department of Neurology, Johns Hopkins University, Francis Scott Key Medical Center, Baltimore, MD 21224.

modulation of the amount of stored Ca^{2+}. During cell activation, this is translated into a correspondingly altered amount of inositol-trisphosphate-induced Ca^{2+} release or Ca^{2+}-induced Ca^{2+} release from the internal stores. Furthermore, activation of many types of cells, including myocytes, neurons, and secretory cells, is associated with a substantial transient elevation of $[Ca^{2+}]_c$ due to Ca^{2+} entry as well as Ca^{2+} release from the ER or SR. In these cells, the exchanger, because of its large capacity (or maximal flux rate), may also play a major role in the rapid extrusion of Ca^{2+} during recovery from activation. These two key roles of the Na-Ca exchanger, regulation of ER/SR Ca^{2+} stores and, thus, cell responsiveness, and rapid extrusion of Ca^{2+} following cell activation and the elevation of $[Ca^{2+}]_c$, are the focus of this report.

METHODS

Barnacle Muscle

Unidirectional ^{45}Ca and ^{22}Na influxes and effluxes were measured in single, internally perfused, giant barnacle (*Balanus nubilus*) muscle cells. The Na-Ca exchange-mediated fluxes were identified as either the external Na^+ (Na_o-) activated Ca^{2+} efflux, or cytosolic Na^+ (Na_c-) activated Ca^{2+} influx (both components of "Ca^{2+} entry mode" exchange), or the external Ca^{2+} (Ca_o-) activated Na^+ efflux (a component of "Ca^{2+} exit mode" exchange). Details of the methods are published.[1]

In some experiments, ^{45}Ca influx and net ionic contents were measured in intact muscle fibers; the cation content of barnacle hemolymph was also measured. Total ion concentrations were determined by atomic absorption (Ca and Mg) or flame emission (Na and K) spectroscopy; extracellular cations were removed by washing the muscle fibers in three 10-min changes of ice-cold isotonic (960 mOsM) sucrose solution containing 2 mM $LaCl_3$ and buffered to pH 7.0 with 6 mM Tris-HCl.

Vascular Smooth Muscle

Contraction Experiments

Isometric tension was measured in rings of rat aorta and a small (second-order) branch of the rat mesenteric artery. The rings were placed in a small-volume (0.75 ml) organ bath and were continally superfused with standard physiological salt solution (Na-PSS; HEPES- and/or bicarbonate-buffered, as noted in RESULTS). In some instances, as described in RESULTS, the ionic composition of the medium was altered (some or all of the Na^+ was replaced isosmotically, or Ca^{2+} was removed), and drugs or vasoconstrictor agents were added to the superfusion medium. Details of these methods are published.[2,3]

Calcium Imaging Experiments

The apparent distribution of intracellular ionized Ca^{2+}, $[Ca^{2+}]_{App}$, was determined in cultured A7r5 cells (a stable cell line derived from fetal rat aorta) using Fura-2 and digital imaging methods. The cells were obtained from American Type Culture Collection (Bethesda, MD) and were cultured on 25-mm plastic coverslips. The cells were loaded with Fura-2 in the membrane-permeable acetoxymethyl ester (Fura-2/

AM) form. A Nikon Diaphot inverted microscope equipped with fluorescence objectives (for epifluorescence measurements) was employed for the imaging studies. The coverslips served as the floor of the experimental chamber which was mounted on the microscope stage. The cells were continuously superfused with modified Krebs solution; vasoconstrictors and other agents were applied by adding them to the inflowing superfusion medium. Details of the imaging methods are published.[4]

Neurons and Glia

Rat Brain Synaptosomes

Synaptosomes were prepared from the forebrains of female Sprague-Dawley rats (150-180 gm) as described.[5] ^{45}Ca influxes and effluxes were measured following brief (1-6 sec) incubations. To determine the Na_c-dependent Ca influx, ^{45}Ca uptake by Li-loaded synaptosomes was subtracted from the ^{45}Ca uptake by Na-loaded synaptosomes ($= \Delta Na_c$). Uptake was measured in low-Na^+ PSS (all NaCl replaced by *N*-methylglucamine = NMG-PSS) or in standard Na-PSS; the difference corresponds to the Ca^{2+} influx activated by removal of external Na^+ ($= \Delta Na_o$). For Ca^{2+} efflux experiments, synaptosomes were loaded with ^{45}Ca by incubating them in Na-PSS for 4 min or in depolarizing medium containing 100 mM K^+ for 9 sec, then washing away the extracellular isotope with Ca-free NMG-PSS, and measuring the efflux into Ca-free Na-PSS or Ca-free NMG-PSS for 1-4 sec; the difference corresponds to the Na_o-dependent Ca^{2+} efflux. Additional information is provided in the RESULTS section; further details of the influx and efflux methods have been published.[6,7]

Cultured Rat Striatal Neurons

$[Ca^{2+}]_{App}$ was determined in primary cultured striatal neurons using the same Fura-2 and digital imaging methods as were employed for the A7r5 cells. The neurons were cultured from fetal rat corpus striatum for 7-10 days using standard methods.[8]

Astrocytes

$[Ca^{2+}]_{App}$ was also determined in primary cultured brain astrocytes using the same Fura-2 and digital imaging methods as were employed for the A7r5 cells. The astrocytes were cultured from neonatal (postnatal day 1) rat cerebral cortex for 7-10 days, as described.[9]

RESULTS AND DISCUSSION

Barnacle Muscle

The giant barnacle muscle fiber is a convenient preparation in which to study ion-transport processes because the large size of the individual cells (1-1.5 mm diameter × 20-25 mm length) permits intracellular perfusion as well as intracellular tracer injection. Tracer flux studies[1,10-12] as well as contraction experiments[13] demonstrate that barnacle muscle fiber sarcolemma contains a Na-Ca exchanger with similar prop-

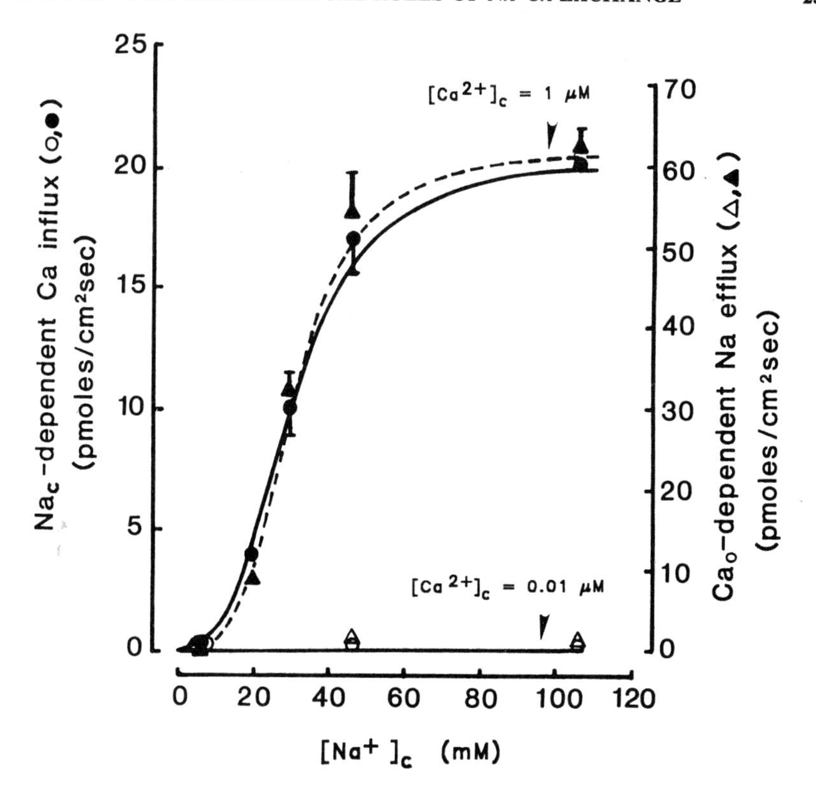

FIGURE 1. Effects of $[Na^+]_i$ (cytosolic Na^+ concentration) on the Ca_o-dependent Na^+ efflux (Δ, ▲ ; right-hand ordinate scale) and the Na_i-dependent Ca^{2+} influx (○ , ●; left-hand ordinate scale) in internally perfused barnacle muscle cells. Data from 27 cells are summarized in the figure; data for at least two different $[Ca^{2+}]_c$ and/or $[Na^+]_c$ were obtained in each cell. Each symbol represents the mean of at least three flux measurements; the bars indicate ± SE for each of the data points where the errors extend beyond the symbols. The external solution in all experiments was (Na-free) Tris SW (seawater with the Na^+ replaced by isotonic Tris, buffered to pH 7.2 with HCl) containing 0.1 mM ouabain. $[Ca^{2+}]_c$ was either 0.01 μM (open symbols) or 1.0 μM (solid symbols). Note that the ordinate scale for the Ca^{2+} influx is expanded threefold, relative to the scale for Na^+ efflux. The solid line fits the Hill equation with a Hill coefficient of 3.0, a K_{Na} (half-maximal $[Na^+]_c$ for activation) of 30 mM, and maximal fluxes of 20.5 pmol/cm² × sec for Na_i-dependent Ca^{2+} influx and 61.5 pmol/cm² × sec for Ca_o-dependent Na^+ efflux. The discontinuous line is the best fit to the data and has a Hill coefficient of 3.7 ± 0.4 and calculated maximal Na^+ efflux of 62.4 ± 1.8 pmol/cm² × sec and K_{Na} = 30.0 mM. All the data in this figure were obtained with V_M between −33 and −43 mV; most of the data were obtained with V_M = −37 ± 3 mV. Reproduced from Rasgado-Flores *et al.*[1] with permission from the *Journal of General Physiology.*

erties to those of the squid axon and mammalian cardiac Na-Ca exchangers. The kinetic properties of the barnacle muscle Na-Ca exchanger are described elsewhere.[1,14] Here we will focus on the physiological role of the barnacle muscle exchanger, although, with this goal in mind, we do need to consider some of its kinetic features.

FIGURE 1 shows data on the Na_c dependence of Ca_c-activated (and Ca_o-dependent) Na^+ efflux and Ca^{2+} influx in internally perfused barnacle muscle fibers. These data provide information about the maximal flux rate ($J_{Na/Ca(max)}$), or "capacity," of the Na-Ca exchanger operating in the Ca^{2+}influx mode (Na^+ exit/Ca^{2+} entry), as well as its Na^+ : Ca^{2+} coupling ratio (or stoichiometry) and its activation by cytosolic Na^+ and Ca^{2+}. As discussed in detail elsewhere,[1] we can draw several conclusions about the barnacle muscle Na-Ca exchanger from these data: (i) The ratio of the Ca_o-dependent Na^+ efflux to the Na_c-dependent Ca^{2+} influx is 3 Na^+ : 1 Ca^{2+} over the entire range of $[Na^+]_c$, and *both* fluxes require cytosolic Ca^{2+} for activation. The exchange is also activated by K^+, but does not absolutely require K^+ (see Rasgado-Flores *et al.*[14]). Thus, the coupling ratio of the barnacle muscle exchanger is 3 Na^+ : 1 Ca^{2+} and is identical to that of the mammalian cardiac myocyte exchanger,[15] but different from that of the retinal rod exchanger (4 Na^+ : 1 Ca^{2+} + 1 K^+; (see Cervetto *et al.*[16]). (ii) Ca^{2+} entry via the exchanger is activated by cytosolic Na^+ concentrations, $[Na^+]_c$, within the normal dynamic physiological range because the apparent $[Na^+]_c$ for half-maximal activation of the exchanger is about 30 mM (FIG. 1). The normal resting $[Na^+]_c$ is about 20-24 mM (TABLE 1; see also Brinley[17]), a concentration at which the exchanger may be activated by as much as 15-20%, depending upon $[Ca^{2+}]_c$. (iii) Ca^{2+} entry via the exchanger is also activated by $[Ca^{2+}]_c$ within the normal dynamic physiological range because the apparent $[Ca^{2+}]_c$ for half-maximal activation of the exchanger is about 0.5 μM.[1] $[Ca^{2+}]_{c(rest)}$ is about 0.1 μM,[18,19] and $[Ca^{2+}]_c$ can rise to about 5-10 μM during activation.[18,20] (iv) $J_{Na/Ca(max)}$ is about 25 pmol $Ca^{2+}/cm^2 \times$ sec (FIG. 1, see also Rasgado-Flores *et al.*[1]). In contrast, the *total* resting Ca^{2+} influx and Ca^{2+} efflux are both only about 2 pmol $Ca^{2+}/cm^2 \times$ sec, and less than half of these fluxes can be attributed to Na-Ca exchange (FIGS. 2 and 3, see also Russell & Blaustein[10] and Ashley *et al.*[11]). Thus, under normal resting conditions the exchanger has a very large reserve capacity because it operates at less than 1/25th of its maximal rate.

From the data in TABLE 1, with $[Na^+]_c$ = 22 mM and $[Ca^{2+}]_{c(rest)}$ = 0.1 μM,[18,19] the apparent reversal potential ($E_{Na/Ca}$) for a Na-Ca exchanger with a coupling ratio of 3 Na^+ : 1 Ca^{2+} is $E_{Na/Ca} = 3E_{Na} - 2E_{Ca} = -58$ mV. Thus, at the normal resting potential (V_M) of about -68 mV,[17] the electrochemical driving force on the exchanger, $V_M - E_{Na/Ca}$, may only be about -10 mV; the negative sign indicates that the exchanger mediates net Ca^{2+} extrusion under these circumstances. Other Ca^{2+} extrusion systems, probably dominated by an ATP-driven Ca^{2+} pump, account for more than half of the resting Ca^{2+} efflux, and may therefore exert primary control over $[Ca^{2+}]_{c(rest)}$. The Na-Ca exchanger may then, depending upon $[Na^+]_c$ and V_M, modulate $[Ca^{2+}]_{c(rest)}$ around the "set point" determined by the ATP-driven Ca^{2+} pump.

These observations and considerations raise the question: If the Na-Ca exchanger operates at only a small fraction of its maximal capacity in resting barnacle muscle cells, what is its main physiological role in these cells? During activation, when the

TABLE 1. Cationic Content of Barnacle Hemolymph and Muscle

Tissue	Cation (mmoles/liter of hemolymph or fiber water)			
	Na	K	Ca	Mg
Hemolymph	489 ± 22 (4)[a]	10.7 ± 0.4 (4)	10.8 ± 0.3 (4)	31.9 ± 1.4 (4)
Muscle Fibers	23.9 ± 1.4 (5)[a,b]	180 ± 3.1 (5)	2.1 ± 0.2 (10)	21.7 ± 1.8 (5)

[a] Data are means ± SE; numbers of barnacles are given in parentheses.
[b] Cationic content was determined on 9 ± 3 (SD) muscle fibers from each barnacle.

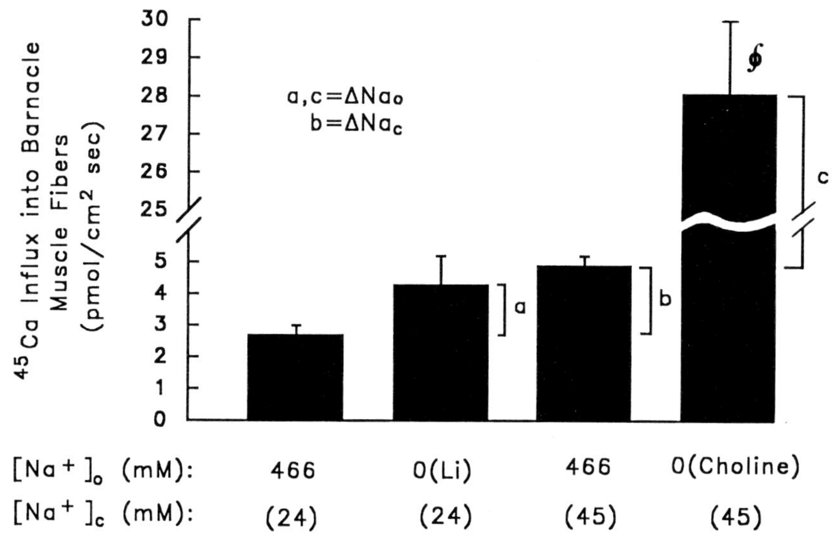

FIGURE 2. ^{45}Ca influx into giant barnacle muscle fibers. Uptake of ^{45}Ca was measured in intact, fresh muscle fibers for 10 min at 15°C, followed by a 56-min wash in Ca-free isotonic (840 mM) sucrose + 2 mM $LaCl_3$ to displace and wash out extracellular ^{45}Ca. Data are calculated in terms of pmol/cm^2 × sec based on a surface area estimate of 1 cm^2/25 mg wet wt. The data are means ± SE of determinations on 5-10 individual muscle fibers, all from a single barnacle. Where indicated, external Na^+ was replaced by equimolar Li^+ or choline during the 10-min incubation with ^{45}Ca; some fibers were preloaded with Na^+ by a 210-min preincubation in K-free medium containing 1 mM ouabain and only 2 mM $CaCl_2$. The latter treatment increased $[Na^+]_c$ from the normal level of 24 mM (TABLE 1) to about 45 mM (determined in separate experiments). ΔNa_o = the component of Ca^{2+} influx activated by removal of external Na^+; ΔNa_c = the component of Ca^{2+} influx that was dependent upon the increase in $[Na^+]_c$. §All of these Na^+-loaded muscle fibers contracted when external Na^+ was removed. Previously unpublished data of Santiago and Blaustein.

cells are depolarized, the driving force on the exchanger ($V_M - E_{Na/Ca}$) will reverse sign so that Ca^{2+} entry is favored; however, initially there will be little exchanger-mediated flux because of the low $[Ca^{2+}]_c$ (see FIG. 1). But, as $[Ca^{2+}]_c$ rises because of Ca^{2+} entry via voltage-gated channels and Ca^{2+} release from the SR, the Ca^{2+} regulatory and Ca^{2+} transport sites on the cytoplasmic face of the exchanger will begin to saturate. This will also make $E_{Na/Ca}$ more positive, but the depolarization will tend to keep the driving force small. Then, when the membrane repolarizes during recovery, the sudden, large increase in the driving force will increase extrusion of Ca^{2+} via the exchanger and thereby help to lower $[Ca^{2+}]_c$ very rapidly. In practice, however, the situation may be more complex because barnacle muscle contracts in a graded fashion and may remain partially contracted for a long period of time; the exchanger may then help to maintain $[Ca^{2+}]_c$ above the contraction threshold during prolonged depolarization and tonic contractions.

The data in FIGURES 2 and 4 also hint at another role for the Na-Ca exchanger: modulation of intracellular Ca^{2+} stores. In these experiments, the muscle fibers with a normal $[Na^+]_c$ exhibited an increase in Ca^{2+}influx and net gain of Ca^{2+} when external Na^+ was removed, because Ca^{2+} efflux decreases under these circumstances (FIG. 3; see also Russell & Blaustein[10]); but the cells did not contract.[13] The implication is that the entering Ca^{2+} was sequestered and buffered so as to keep $[Ca^{2+}]_c$ below the

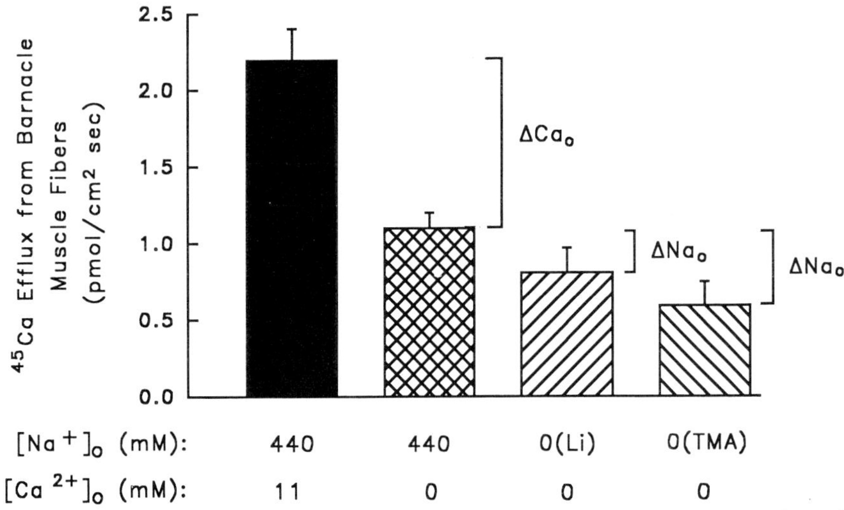

FIGURE 3. ^{45}Ca efflux from giant barnacle muscle fibers calculated from the data in Table II of Russell and Blaustein.[10] Values for Ca^{2+} efflux in pmol/cm^2 $\times$ sec at 19°C from the ^{45}Ca-injected muscle fibers are based on: (i) the assumption that the injected ^{45}Ca mixed completely with the intracellular ^{40}Ca (2.1 mmoles/liter fiber water; see TABLE 1) to give a uniform specific activity, and (ii) experiments and calculations which indicated that (a) fiber surface area was 1 cm^2/25 mg wet wt and (b) there was 0.69 liter fiber water/kg wet wt. Where noted, Ca^{2+} efflux was determined in Ca-free media (Ca^{2+} replaced by equimolar Mg^{2+}) and in Ca-free, Na-free medium with Na^+ replaced by isotonic Li^+ or tetramethylammonium (TMA) ions. The external Ca-dependent (ΔCa_o) and external Na-dependent (ΔNa_o) components of the Ca^{2+}efflux are indicated by the brackets.

contraction threshold of about 0.8 μM.[18] Only when the muscle fibers were first loaded with Na^+ (by prolonged incubation in ouabain-containing K-free medium to inhibit the Na^+ pump), did we observe a large net Ca^{2+} gain (FIGS. 2 and 4) and contraction[13] upon removal of extracellular Na^+. These data indicate that modest changes in $\Delta\bar{\mu}_{Na}$ may alter $[Ca^{2+}]_{c(rest)}$ only slightly; when there is a net gain of Ca^{2+}, almost all of this Ca^{2+} will be sequestered—presumably in the SR. Thus, another function of the Na-Ca exchanger in barnacle muscle may be to modulate the amount of Ca^{2+} stored in the SR. Indeed, data from several other tissues, discussed below, are consistent with this concept.

Vascular Smooth Muscle

Evidence that a component of Ca^{2+} influx in VSM is Na_c-dependent and is promoted by removal of external Na^+, as well as evidence that Ca^{2+} efflux is, in part, Na_o-dependent, led to the view that VSM has a prominent Na-Ca exchanger.[21] Nevertheless, as in barnacle muscle, even large reductions of the Na^+ gradient induced by removal of external Na^+ or by incubation with cardiotonic steroid do not always induce contractions in VSM.[2] This has fostered controversy about the physiological role of Na-Ca exchange in VSM.[21,22] Yet, there can be little doubt that VSM possesses a relatively large-capacity Na-Ca exchanger: Evoked contractions (FIG. 5, and see Bova *et al.*[3]) and Ca^{2+} gain[23] are substantially augmented after Na^+ loading. Furthermore, relaxation

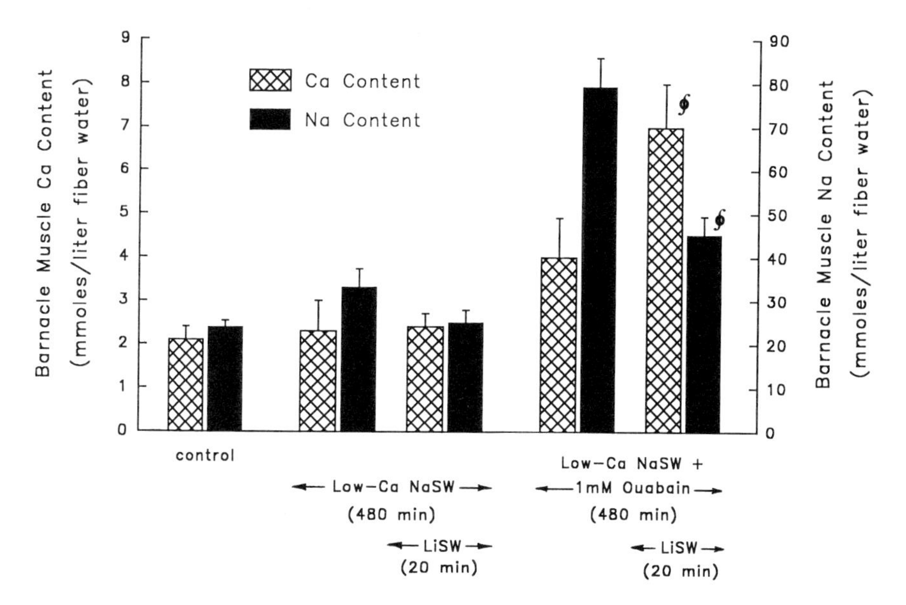

FIGURE 4. Effects of internal and external sodium on net calcium accumulation in giant barnacle muscle fibers. Standard NaSW contained (in part) 466 mM NaCl, 20 mM $CaCl_2$ and 12 mM $MgCl_2$; Low-Ca NaSW contained (in part) 466 mM NaCl, 2 mM $CaCl_2$ and 30 mM $MgCl_2$; LiSW contained (in part) 466 mM LiCl, 20 mM $CaCl_2$ and 12 mM $MgCl_2$. All incubations were at 18°C. See METHODS and FIGURE 2 legend for details about washout of extracellular ions and calculations. §All of these Na^+-loaded muscle fibers contracted when external Na^+ was removed. Data are means ± SE of values from 6-12 individual fibers, all from a single barnacle. Unpublished data of Santiago and Blaustein.

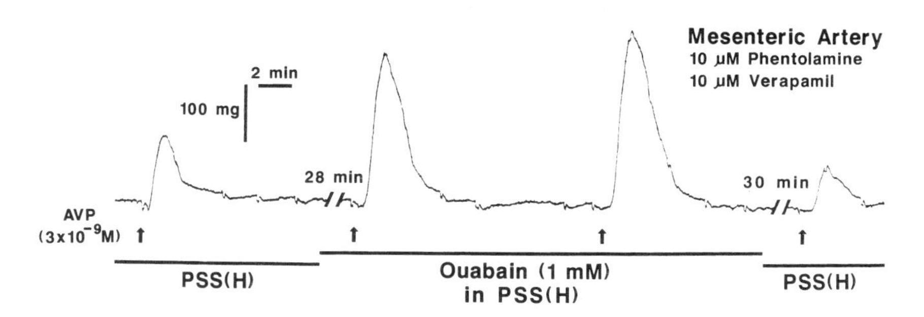

FIGURE 5. Effect of 1 mM ouabain on the contractile responses of a ring of rat mesenteric artery to brief exposures (at the arrows) to 3 nM arginine-vasopressin (AVP) in the presence of 10 μM phentolamine (to block α-receptors) and 10 μM verapamil (to block voltage-gated Ca^{2+} channels). The tissue was exposed to ouabain for 30 min before the second exposure to AVP. Tissue wet wt = 76 μg; original resting tension = 320 mg; Temp. = 37°C. Reproduced from Bova *et al.*[3] with permission from the *American Journal of Physiology*.

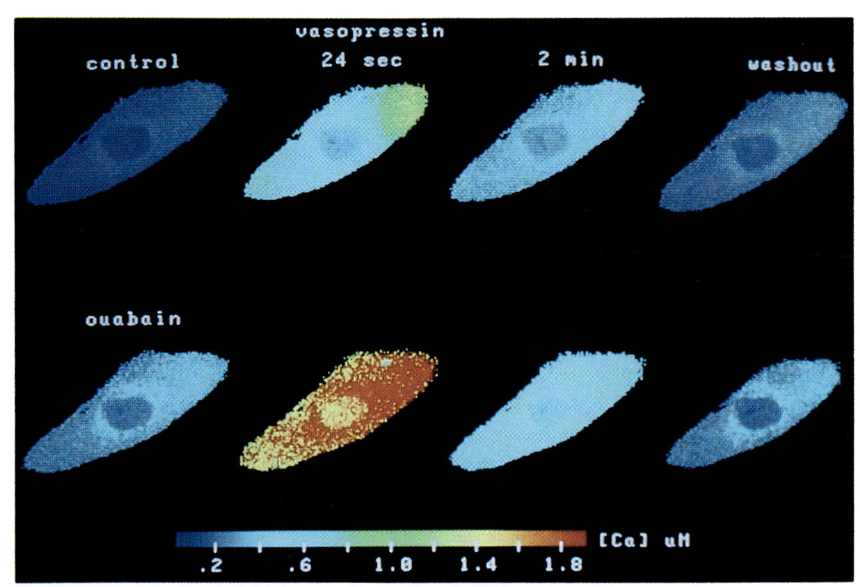

FIGURE 6. Calculated Ca^{2+} images illustrating the effect of ouabain on $[Ca^{2+}]_{App}$ in a Fura-2 loaded A7r5 cell before stimulation (left-hand images), and during exposure to 1 nM AVP. *Upper row*: response to AVP under control conditions (incubation in HEPES-bicarbonate-buffered Na-PSS). *Lower row*: response to the same dose of AVP after a 15-min exposure to 3 mM ouabain. The second column of images (from the left) shows the peak responses to AVP in the absence (upper image) and presence (lower image) of ouabain. "Washout" indicates Ca^{2+} images obtained after a 5-min washout of AVP. Color scale shows $[Ca^{2+}]_{App}$ in μM. Data reproduced from Bova *et al.*[3] with permission from the *American Journal of Physiology.*

following evoked contractions,[2] and reduction of $[Ca^{2+}]_c$ following evoked Ca^{2+} transients,[3] are markedly slowed when external Na^+ is removed, especially under conditions where Ca^{2+} sequestration in the SR is limited. Thus, as in barnacle muscle (above) and mammalian cardiac muscle (see articles in Part V of this volume), one important role of the Na-Ca exchanger in VSM may be to extrude Ca^{2+} rapidly following cell activation and the transient rise in $[Ca^{2+}]_c$. Unlike cardiac muscle, however, VSM does not usually exhibit regular, large cyclic rises and falls in $[Ca^{2+}]_c$, so the maximal capacity of the VSM exchanger may not need to be as large as that of cardiac muscle.

Further clues about the role of the Na-Ca exchanger in VSM may be obtained from direct examination of $[Ca^{2+}]_{App}$ in cultured cells from a VSM cell line (A7r5), using digital imaging methods. When these cells are loaded with the membrane permeable Fura-2/AM, the dye enters not only the cytosol, but intracellular organelles as well, before being hydrolyzed to the Ca-sensitive Fura-2 free acid.[4,24] Therefore, the dye reports information about sequestered Ca^{2+} as well as cytosolic free Ca^{2+}, especially when $[Ca^{2+}]_c$ is low. As seen in FIGURE 6, incubation of A7r5 cells with 3 mM ouabain for 15 min induced only a modest rise in $[Ca^{2+}]_{App}$ at the cell periphery, where the data reflect primarily a change in $[Ca^{2+}]_c$ because there are few organelles in this region. There is, however, a substantial increase in $[Ca^{2+}]_{App}$ in the perinuclear region of the cytoplasm where many of the intracellular organelles including SR (ER) and mitochondria are located (compare upper right and lower left images).

Additional evidence that the store of Ca^{2+} in the SR was increased comes from the observation that the arginine-vasopressin (AVP-) evoked transient rise in $[Ca^{2+}]_c$ (the

"Ca^{2+} transient") was greatly augmented after exposure to ouabain (FIG. 6). The rate of rise and peak amplitude of the Ca^{2+} transients are much greater than we ever observed with depolarization (maximum $[Ca^{2+}]_{App}$ = 500-600 nM), and therefore cannot be attributed to Ca^{2+} entry via voltage-gated Ca^{2+} channels.

More direct evidence that Na^+ gradient reduction increases the Ca^{2+} store in the SR comes from the data in FIGURE 7 (and see Ashida & Blaustein[2]). Here we employed the caffeine-evoked increase in VSM tension as a measure of the amount of Ca^{2+} stored in the SR. The upper record shows the contractions of a ring of rat aorta evoked in response to repeated, brief applications of 10 mM caffeine before, during, and after a 75-min exposure to 1 mM ouabain. Ouabain elevated baseline (unstimulated or "resting") tension and markedly augmented the evoked tension; the latter indicates that the SR Ca^{2+} stores were greatly increased. To determine whether these effects were due to Ca^{2+} entry via voltage-gated Ca^{2+} channels as a consequence of depolarization (i.e., as a result of electrogenic Na^+ pump inhibition and/or the decline in the K^+ equilibrium potential because of net K^+ loss), the experiment was repeated in the presence of 10 μM verapamil to block these channels. Verapamil blocked virtually all of the ouabain-induced rise in baseline tension (FIG. 7, lower record), but did not diminish ouabain's effect on caffeine-evoked contractions. The latter effect depended upon external Ca^{2+} as well as Na^+, and was abolished by ryanodine (Weiss & Blaustein, unpublished data) which selectively depletes the SR Ca^{2+} stores.[25] However, the external Na^+ was required only for cell Na^+ loading (i.e., to raise $[Na^+]_c$); reduction of $[Na^+]_o$ to 15 or 30 mM also augmented caffeine-evoked contractions in the rat aorta without altering baseline tension.[2] Augmentation of caffeine-evoked responses by ouabain and low $[Na^+]_o$ has also been observed in human resistance arteries.[26]

In sum, these studies indicate that reduction of the Na^+ gradient across the VSM sarcolemma may not only modulate $[Ca^{2+}]_c$, but also, more importantly, may regulate the amount of Ca^{2+} stored in the SR. For example, consider a (hypothetical) cell in which Na^+ pump inhibition reduces the Na^+ concentration gradient, $[Na^+]_o/[Na^+]_c$, from 145/14.5 mM to 145/16 mM with V_M and $[Ca^{2+}]_o$ maintained at -60 mV and 1.0 mM, respectively. At equilibrium, the Na-Ca exchanger with a 3 : 1 coupling ratio should, thoretically, raise $[Ca^{2+}]_{c(rest)}$ from 90 to 134 nM. Normally, however, $[Ca^{2+}]_{c(rest)}$ will be higher than the calculated equilibrium value (90 nM) because of Ca^{2+} "leak" and because exchanger turnover in resting VSM is only a small fraction of $J_{Na/Ca(max)}$ (see above). Furthermore, following partial Na^+ pump inhibition, the rise in $[Ca^{2+}]_{c(rest)}$ will be attenuated because of Ca^{2+} extrusion via the sarcolemmal ATP-driven Ca^{2+} pump. At the same time, the SR ATP-driven Ca^{2+} pump will maintain a Ca^{2+} concentration gradient between the SR and the cytosol, $[Ca^{2+}]_{SR}/[Ca^{2+}]_c$, of about 5×10^4.[27,28] Thus, even if SR volume is only 3-5% of total cell volume,[28] the SR Ca^{2+} pump will amplify the rise in $[Ca^{2+}]_{c(rest)}$ by a factor of about $0.03\text{-}0.05 \times 5 \times 10^4$ (= $1.5\text{-}2.5 \times 10^3$). In other words, total cell calcium will increase by about 15-25 μmoles per liter of cell water for every 10 nM rise in $[Ca^{2+}]_{c(rest)}$; virtually all of this additional calcium will be stored in the SR. Then, because release of Ca^{2+} from the SR is proportional to the amount stored,[29] evoked responses that involve SR Ca^{2+} release should be significantly amplified as a result of partial Na^+ pump inhibition and a small rise in $[Ca^{2+}]_c$.

Neurons and Glia

Na-Ca exchange was initially identified in squid giant axons[30,31] (as well as cardiac muscle) and was subsequently observed in a variety of vertebrate neuronal preparations including brain slices and synaptosomes.[32] Under resting conditions, when $[Ca^{2+}]_c$ is

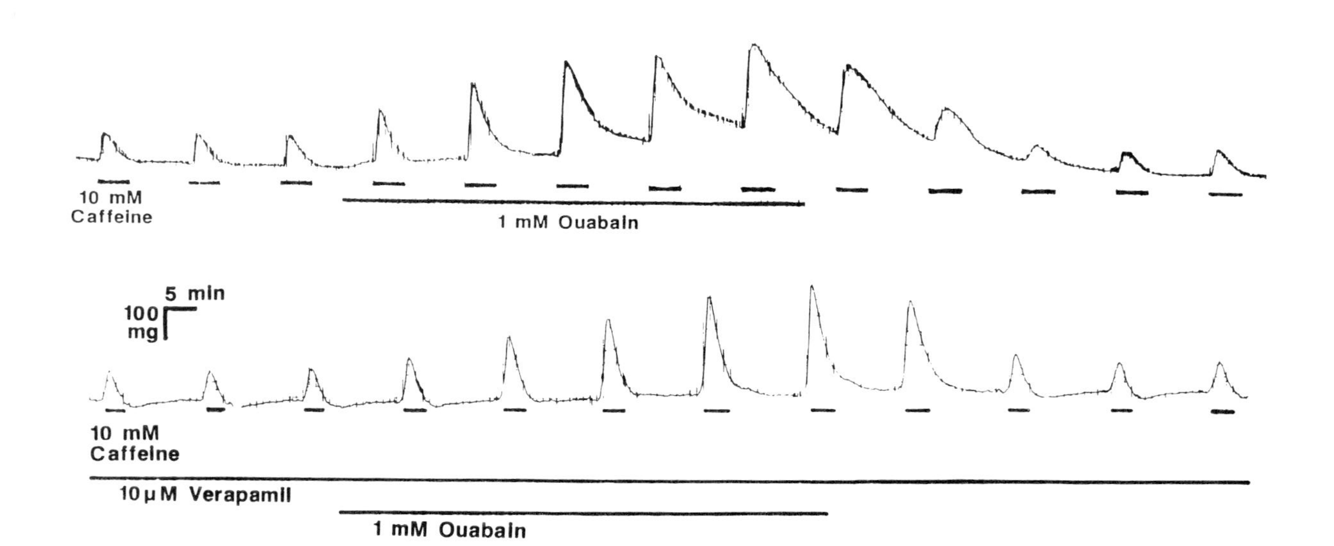

FIGURE 7. Effect of 1 mM ouabain on the contractile responses of rings of rat aorta to brief exposures to 10 mM caffeine (indicated by the short bars immediately below the tension records). Ouabain was added to the HEPES/bicarbonate-buffered Na-PSS for the periods indicated. In the second experiment (lower recording), 10 μM verapamil was included in the Na-PSS to block voltage-gated Ca^{2+} channels. Original resting tension = 500 mg; Temp. = 37°C. Unpublished data of Weiss and Blaustein.

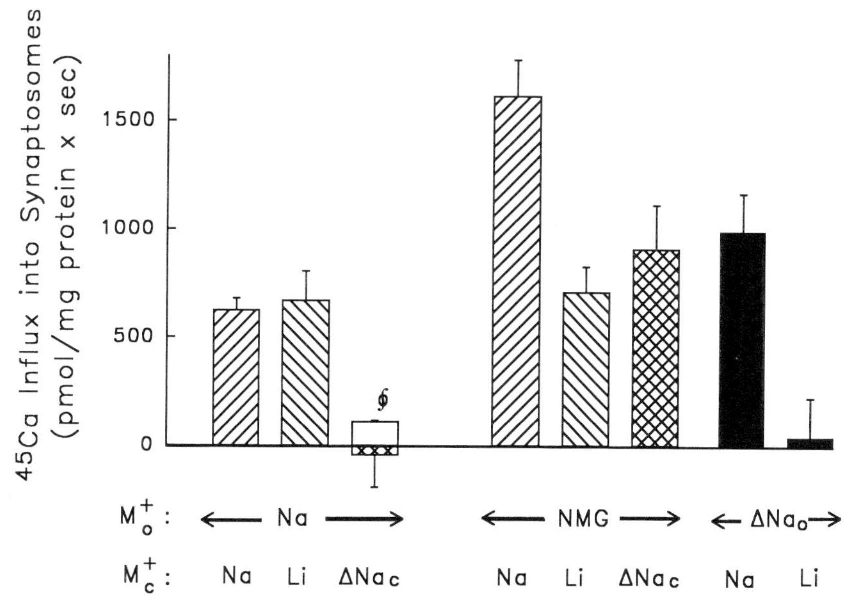

FIGURE 8. ^{45}Ca influx into rat brain synaptosomes. Synaptosomes were preincubated for 15 min at 37°C in Na-PSS (M_c^+ = Na) or in Ca-free Li-PSS to deplete internal Na^+ (M_c^+ = Li), and then incubated for 1 sec in Na-PSS ($M_o^+ = Na^+$) or NMG-PSS (M_o^+ = NMG) containing 1.2 mM $CaCl_2$ labeled with ^{45}Ca. The values shown include the "total" ^{45}Ca uptake as well as the internal Na-dependent component (ΔNa_c) and the component activated by removal of external Na^+ (ΔNa_o). §The Na_c-dependent Ca^{2+} uptake from Na-PSS was difficult to measure during a 1-sec incubation because of the small fluxes and relatively large errors. The open bar represents averaged data from this and one other experiment in which this component of Ca^{2+} uptake was more accurately measured at both 3 and 6 sec. All the original data (diagonally striped bars) include a "background (time-independent) ^{45}Ca uptake" of about 550-600 p/mgsec. Each value is the mean of five determinations ± SE. Unpublished data of Fontana and Blaustein.

about 100 nM, the squid axon Na-Ca exchanger operates at a small fraction of its maximal rate, and $[Ca^{2+}]_{c(rest)}$ is controlled primarily (but not completely) by the ATP-driven Ca^{2+} pump in the plasmalemma.[33] The physiological role of the Na-Ca exchanger in neurons has not yet been resolved.

Nerve Terminals

In rat brain synaptosomes, too, under normal resting conditions (low $[Ca^{2+}]_c$), and normal $[Na^+]_o$ and $[Na^+]_c$, the Ca^{2+} fluxes mediated by the exchanger are small fractions of the respective maximal fluxes mediated by the exchanger. The "resting" Na_c-dependent Ca^{2+} influx is about 100 pmol/mg protein × sec (p/mgsec; FIG. 8, open bar). In terminals with a small (physiological) ^{45}Ca load (430 pmol/mg protein), the Na_o-dependent Ca^{2+} efflux is about 80 p/mgsec (Fig. 2 in Sanchez-Armass & Blaustein[7]).

The Ca^{2+} influx into Na-containing synaptosomes that is activated by removing external Na^+ (FIG. 8, ΔNa_o) is about 1,000 p/mgsec. Virtually all of this increment is

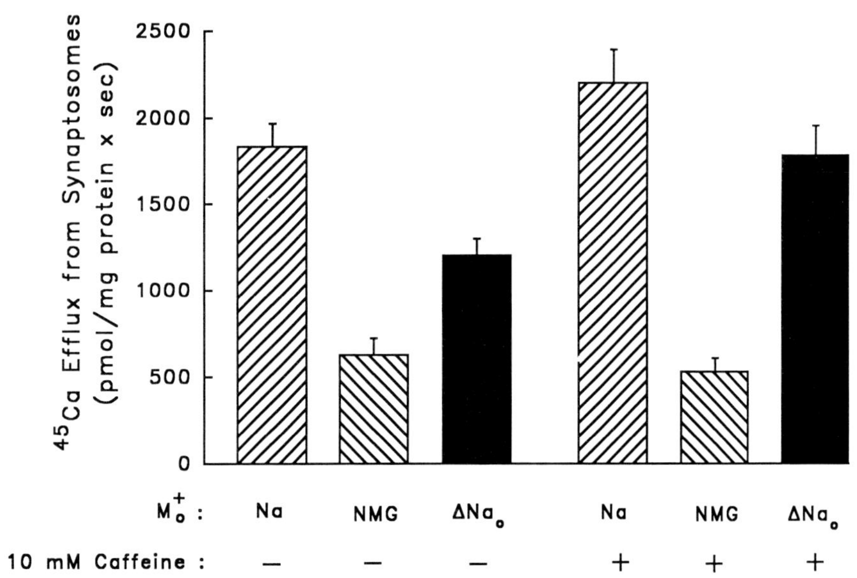

FIGURE 9. ^{45}Ca efflux from rat brain synaptosomes. Synaptosomes were loaded with ^{45}Ca by incubating them in 100 mM K-PSS (100 mM K^+ and only 50 mM Na^+, plus 1.2 mM $CaCl_2$ labeled with ^{45}Ca) to give a load of 6660 ± 304 pmol/mg protein. Ca^{2+} efflux into Ca-free Na-PSS (M_o^+ = Na) or Ca-free NMG-PSS (M_o^+ = NMG), either without (−) or with (+) 10 mM caffeine, was measured at 1 sec and 37°C. The caffeine was used to release Ca^{2+} from the ER in order to raise $[Ca^{2+}]_c$ to saturate the internal Ca^{2+}-binding sites on the Na-Ca exchanger. The data are means ± SE from eight experiments, including five with caffeine, with four replicate determinations in each experiment. The Na_o-dependent Ca^{2+} effluxes (ΔNa_o) are indicated by the solid bars. Unpublished data of Fontana and Blaustein.

Na_c-dependent (FIG. 8, ΔNa_c); it is abolished if the terminals are depleted of cytosolic Na^+ by preincubation in Li-PSS before measuring the Ca^{2+} influx. At saturating $[Ca^{2+}]_o$, ΔNa_o in Na-loaded synaptosomes (i.e., $J_{Na/Ca(max)}$) is 1427 ± 111 (n = 5) p/mgsec (not shown). The maximal Na_o-dependent component of the Ca^{2+} efflux is about 1,800 p/mgsec when the terminals are given a large load of Ca^{2+} and then treated with caffeine to release Ca^{2+} from internal stores and maximally elevate $[Ca^{2+}]_c$ (FIG. 9). Then, taking the synaptosome internal volume as 3.5 μl/mg protein,[34] the exchanger is capable of transporting about 500 micromoles of Ca^{2+} per liter of synaptosomes per sec, or 500 nmol/L × msec.[g]

What is the physiological role of this prevalent exchanger at nerve terminals? One possibility is that it plays a role in the rapid decline of $[Ca^{2+}]_c$ at the end of the action

[g] Another way to view these data is that, with an exchanger turnover rate of about 1,000 sec^{-1}[35] and a synaptosome surface area of 3×10^{10} μm^2/mg protein,[33] the calculated density of Na-Ca exchangers is about 40/μm^2 if all the terminals are functional. But, if only about one-third of the protein is associated with functional nerve terminals,[36] a better estimate of the density of exchangers in synaptosome plasma membrane (assuming uniform distribution) is about 120/μm^2, half the density in cardiac muscle sarcolemma (about 250/μm^2).[37] This calculated high density of exchangers at nerve terminals is consistent with our recent immunocytochemical localization of the exchangers at presynaptic nerve terminals in neuromuscular preparations.[38]

potential, and thus may be a major contributor to the termination of transmitter release. When terminals are depolarized and $[Ca^{2+}]_c$ rises (perhaps to levels exceeding 10 μM just beneath the plasmalemma, especially in the active zone[39]), the Ca^{2+}-binding sites at the internal face of the plasmalemma (both the regulatory sites and the transporting sites) will be saturated with Ca^{2+}. However, the driving force on the exchanger will be small and Ca^{2+} extrusion will be slow because of the depolarization. Then, when the terminals repolarize, the large increase in driving force on the exchanger will suddenly favor Ca^{2+} extrusion, and the exchanger could, within a couple of exchanger cycles (i.e., in a couple of milliseconds), lower subplasmalemmal $[Ca^{2+}]_c$ and thereby contribute to the rapid decline of transmitter release.[40] Indeed, this voltage-dependent Ca^{2+} extrusion might account for the apparent voltage sensitivity of transmitter release.[41]

The exchanger contributes to the potentiation of transmitter release following tetanic stimulation (post-tetanic potentiation, PTP).[42,43] For example, during a tetanus, $[Na^+]_c$ may rise progressively. This may slow the extrusion of Ca^{2+} and increase the amount of Ca^{2+} stored in intraterminal organelles such as the ER. As a consequence, the Ca^{2+} buffers should begin to saturate, and depolarization-evoked release of Ca^{2+} from the ER should increase. The net effect will be increased $[Ca^{2+}]_{c(rest)}$ and augmented depolarization-evoked Ca^{2+} transients which could account for, respectively, the enhanced spontaneous transmitter release and augmented evoked transmitter release.[42,43]

Cultured Neurons

Calcium has many other functions in neurons[44] in addition to its central role in neurotransmitter release: For example, Ca^{2+} helps to control excitability[45] and to regulate growth cone behavior.[46] Na-Ca exchange has also been implicated in processes related to protecting neurons from the Ca^{2+} overload associated with cell injury and cell death.[47]

To study the influence of the Na^+ gradient on $[Ca^{2+}]_{App}$ in resting and activated cultured rat striatal neurons, we employed Fura-2 with digital imaging methods. Calculated Ca^{2+} images from representative experiments are shown in FIGURES 10 (upper row) and 11. These data are, in many ways, comparable to those obtained in VSM cells (above). Under control conditions resting neurons (upper middle image in FIG. 10; upper left image in FIG. 11) had a low $[Ca^{2+}]_{App}$ (100-200 nM); however, the Ca^{2+} was not distributed uniformly and appeared to be highest in the perinuclear regions of the cell bodies (indicated by the brightest area in the upper left [fluorescent] image in FIG. 10). When neurons were activated by 0.1 mM glutamate (FIG. 11) or 100 mM K^+ (not shown), Ca^{2+} rose transiently and asynchronously in the processes and cell bodies; in most portions of the cell bodies, $[Ca^{2+}]_{App}$ reached a peak of about 0.6-1 μM (FIG. 11, left-hand middle image). After washout of the glutamate (lower left-hand image), the neurons were exposed to 1 mM ouabain for 15 min. $[Ca^{2+}]_{App}$ rose slightly, but nonuniformly (FIGS. 10 and 11, upper right-hand images); the most marked increases were seen in the perinuclear regions of the cell bodies, and in a few "hot spots" along the major processes (FIG. 10 illustrates this with an expanded color scale). Then, when the neurons were again activated with 0.1 mM glutamate, the Ca^{2+} transients were greatly augmented (FIG. 11, right-hand middle image). Similar results (not shown) were obtained when neurons were activated by depolarizing them with K^+-rich media, and when the Na^+ gradient was reduced by removing extracellular Na^+. From these data it is not possible to decide whether most of the Ca^{2+} for the Ca^{2+} transients (under both control and reduced Na^+ gradient conditions) came from the extracellular medium or from intracellular stores (perhaps in the ER). However, estimates based on the time course of the rising phase of the transients suggest that the major source of

Ca^{2+} was the intracellular stores—perhaps those located in the perinuclear regions of the cells that appeared to increase their Ca^{2+} content during exposure to ouabain. If these preliminary observations can be verified, it would imply that the Na-Ca exchanger in neurons, as in VSM cells, helps to modulate the size of the ER Ca^{2+} stores, and thereby plays a critical role in the control of neuronal responsiveness to activating agents.

Astrocytes

Relatively little is known about Ca^{2+} metabolism in astrocytes, although these cells account for a large fraction of brain volume and may help to govern the extracellular ionic environment around brain neurons.[48] Digital imaging experiments on cultured cortical astrocytes yielded results comparable to those obtained with the striatal neurons. Representative data are shown in FIGURES 10, 12, and 13. The Ca^{2+} images in FIGURE 10 illustrate the apparent distribution of Ca^{2+} in unstimulated astrocytes before (lower center) and after (lower right) 15 min of exposure to 1 mM ouabain. As in VSM and striatal neurons, reduction of the Na^+ gradient raised $[Ca^{2+}]_{App}$ only slightly in unstimulated astrocytes; the largest changes were observed in the perinuclear regions (the bright areas in the fluorescent image at the lower left), where most of the

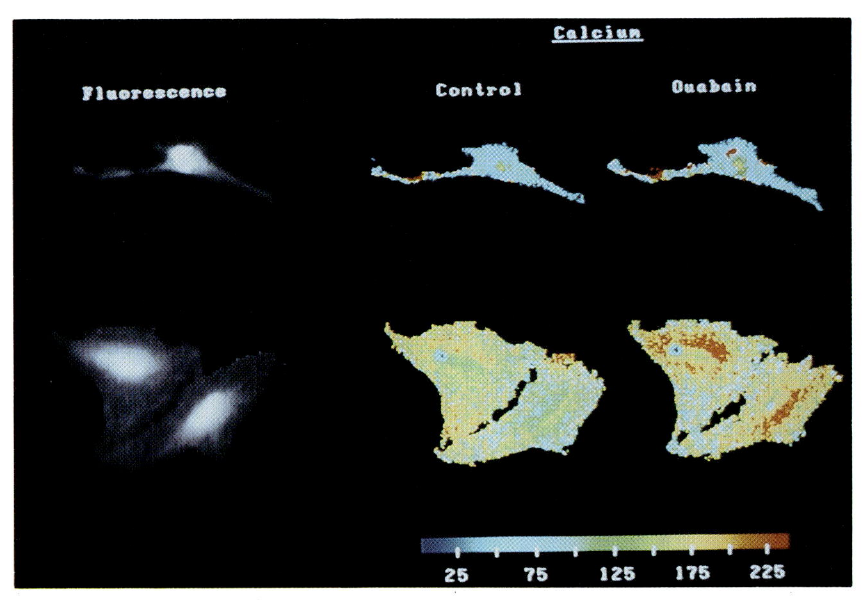

FIGURE 10. Fluorescent (black and white) and calcium (color) images of a cultured striatal neuron (upper row) and two cultured cortical astrocytes (bottom row) loaded with Fura-2. The black and white images are the fluorescent emission (510-nm) images of cells excited with 360-nm light. The large bright areas near the cell centers correspond to the nuclear regions of these cells. The middle column (labeled "control") shows the calculated Ca^{2+} images recorded during exposure to control medium (HEPES/bicarbonate buffered Na-PSS). The right-hand images were calculated after the cells were exposed to Na-PSS containing 1 mM ouabain for 15 min. Note the expanded color scale compared to the ones used in FIGURES 6 and 11: color scale shows $[Ca^{2+}]_{App}$ in nM. Unpublished experiments of Goldman, Krueger, Steele, Yarowsky, and Blaustein.

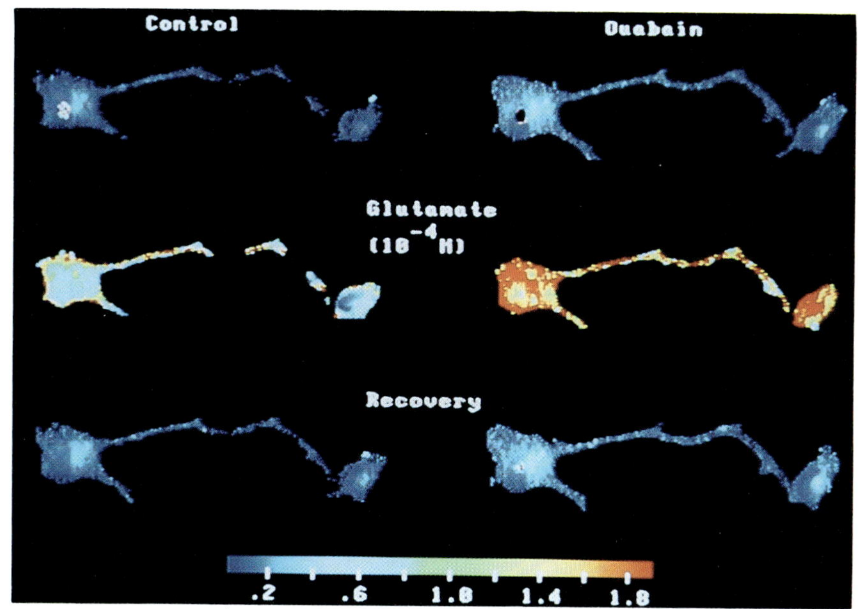

FIGURE 11. Calcium images illustrating the effect of ouabain on $[Ca^{2+}]_{App}$ in two Fura-2 loaded cultured striatal neurons before stimulation (upper images), during a 90-sec exposure to 100 μM glutamate in Na-PSS (middle row), and after a 5-min washout of glutamate (bottom). The neurons were incubated with 1 mM ouabain for 15 min after the left-hand images were obtained. The middle images were obtained at the peaks of the glutamate-evoked rises in $[Ca^{2+}]_{App}$, before (left) and during (right) exposure to ouabain. Color scale shows $[Ca^{2+}]_{App}$ in μM. Unpublished data of Goldman, Steele, and Blaustein.

intracellular organelles are located. FIGURE 12 shows the mean cytoplasmic $[Ca^{2+}]_{App}$ graphed as a function of time during the application of 0.1 mM glutamate in standard Na-PSS (●), in NMG-PSS (▼), and in Na-PSS containing 1 mM ouabain (∇). Note that the Ca^{2+} transients were markedly augmented when the Na^+ gradient was reduced—either by reducing $[Na^+]_o$ (▼) or increasing $[Na^+]_c$ (∇), even though resting $[Ca^{2+}]_{App}$ (at "0" time) was only minimally increased. Similar results were obtained when the astrocytes were stimulated with K^+-rich medium (FIG. 13). These effects depended upon extracellular Ca^{2+} and were attenuated by 10 mM caffeine (not shown). The latter observation suggests that much of the Ca^{2+} for the Ca^{2+} transient, especially under reduced Na^+ gradient conditions, is derived from intracellular stores—most likely from the ER. Clearly, the responsiveness of these cells, too, is regulated by the amount of Ca^{2+} stored in the ER and thus, indirectly, by the Na-Ca exchanger.

CONCLUSIONS

A coherent view of cell Ca^{2+} regulation emerges from the examination of Na^+ gradient-dependent Ca^{2+} transport in these varied types of cells. Barnacle muscle and mammalian VSM, neurons, and astroglia all have prominent Na-Ca exchangers in their plasma membranes. In all of these cells, the Na-Ca exchanger likely operates in parallel with a low-capacity, plasmalemmal ATP-driven Ca^{2+} pump. The latter appears

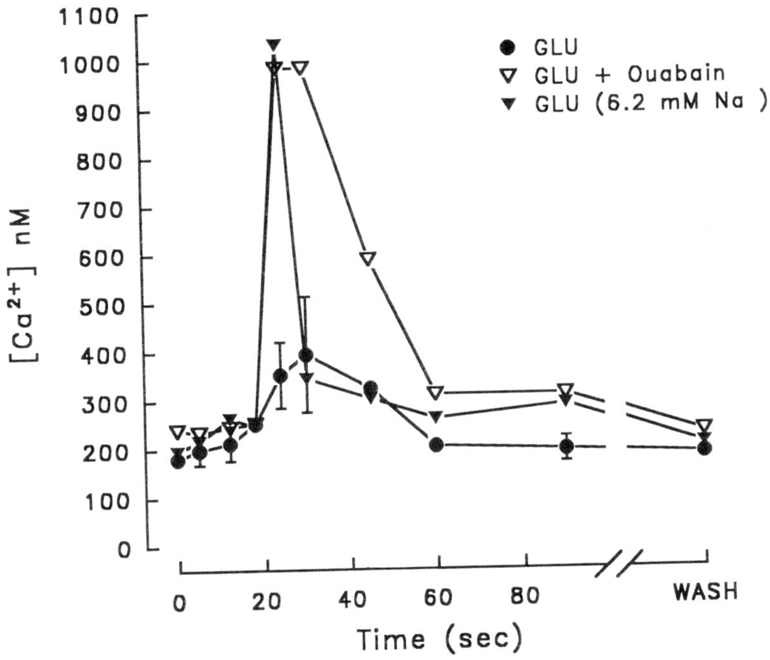

FIGURE 12. Time course of the effects of glutamate on $[Ca^{2+}]_{App}$ in Fura-2 loaded cortical astrocytes under control conditions (●) and in the presence of ouabain (▽) and low-Na^+ medium (NMG-PSS) (▼). The average resting $[Ca^{2+}]_{App}$ in the control cells was 181 ± 5 nM. Low-Na^+ increased the level only slightly (to 196 nM), whereas 1 mM ouabain increased the level to 240 nM. Both ouabain and low-Na^+ media (in different cell preparations) greatly augmented the responses to 100 μM glutamate. Unpublished data of Goldman, Krueger, Yarowsky, and Blaustein.

to play a dominant role in controlling $[Ca^{2+}]_c$ in these cells under resting (nonstimulated) conditions when $[Ca^{2+}]_c$ is low. Nevertheless, a low rate of exchanger-mediated Ca^{2+} flux may help to modulate $[Ca^{2+}]_{c(rest)}$ in these cells when $\Delta\bar{\mu}_{Na}$ is altered.

More important than the direct modulation of $[Ca^{2+}]_{c(rest)}$, however, is the indirect regulation of the intracellular Ca^{2+} stores in the ER or SR, because the Ca^{2+} concentration gradient between the intra-organellar space in the ER or SR and the cytosol may be about 5×10^4.[27,28] In this way, a small change in $[Ca^{2+}]_c$ may be converted into a relatively large change in ER or SR Ca^{2+}: The ATP-driven Ca^{2+} pump in the ER or SR membrane serves as an amplifier. Therefore, since most of the second messenger Ca^{2+} in these cells may come from the ER or SR, the amplitude and duration of the Ca^{2+} transient and the cell response to that transient (e.g., muscle contraction or neurotransmitter secretion), likely depends upon the amount of Ca^{2+} in the ER or SR store and thus, indirectly, upon the Na^+ gradient across the plasma membrane. The implication is that the responsiveness of these cells to a variety of activating agents is regulated, to a large extent, by $\Delta\bar{\mu}_{Na}$ via the plasma membrane Na-Ca exchanger. Indeed, this amplification system may play a key role in mediating the action of the recently discovered endogenous ouabain-like compound.[49]

Another critical role of the exchanger is the rapid extrusion of Ca^{2+} following cell activation. The marked dependence of Ca^{2+} extrusion on extracellular Na^+ when

$[Ca^{2+}]_c$ is high (during cell activation) in neurons and astrocytes, as well as in barnacle muscle and VSM, suggests that the plasmalemmal ATP-driven Ca^{2+} pumps in all of these cells have a much lower capacity to transport Ca^{2+} than do the Na-Ca exchangers in these cells. When $[Ca^{2+}]_c$ is high during the Ca^{2+} transient, the exchanger's Ca^{2+} transport and Ca^{2+} regulatory sites at the inner face of the plasmalemma become saturated with Ca^{2+}; then, when V_M is large and negative, the large driving force on the exchanger ($V_M - E_{Na/Ca}$) will favor the extrusion of Ca^{2+}. In nerve terminals this rapid extrusion of Ca^{2+} may play an important role in the removal of Ca^{2+} from the transmitter release sites, and may thereby help to terminate release. In addition, the exchanger may play a role in the facilitation and post-tetanic potentiation of transmitter release. In barnacle muscle and in VSM, the exchanger may play a role in the tonic elevation of $[Ca^{2+}]_c$ and, thus, in the control of tonic tension.

This view of Na-Ca exchanger function appears to reconcile a widely recognized dilemma: namely, how to account for the evidence that many types of cells possess large capacity Na-Ca exchangers, but that these exchangers exert only limited influence on $[Ca^{2+}]_{c(rest)}$. It is now apparent that the Na-Ca exchangers in these cells serve mainly as regulators of SR (or ER) Ca^{2+} and, thus, of cell responsiveness and as a mechanism for the rapid extrusion of Ca^{2+} following cell activity.

ACKNOWLEDGMENTS

We thank R. F. Rogowski for assistance with preparation of the figures.

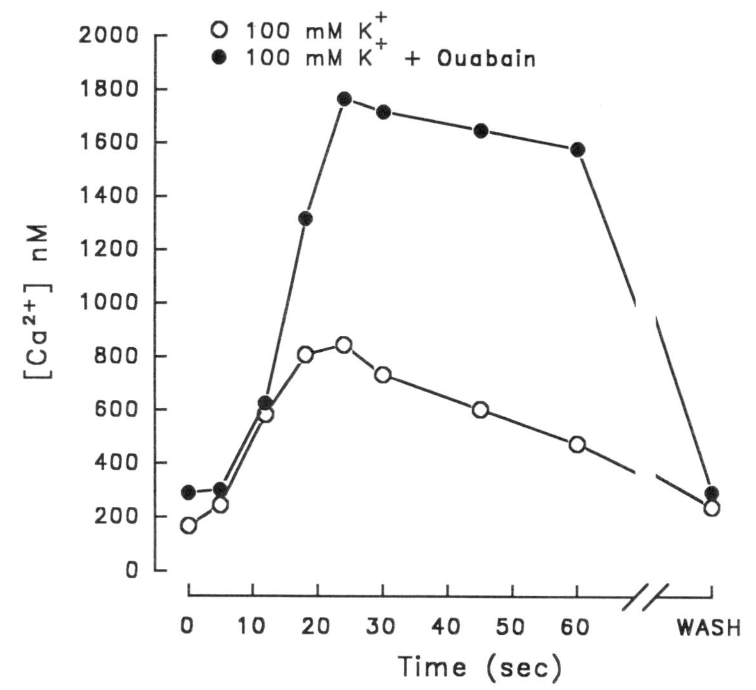

FIGURE 13. Time course of the effects of 100 mM K^+ on $[Ca^{2+}]_{App}$ in a group of Fura-2 loaded cortical astrocytes under control conditions (○) and in the presence of ouabain (●). The average resting $[Ca^{2+}]_{App}$ in the control cells was 167 ± 15 nM. Ouabain greatly amplified and prolonged the response to 100 mM K^+. Unpublished data of Goldman, Krueger, Yarowsky, and Blaustein.

REFERENCES

1. RASGADO-FLORES, H., E. M. SANTIAGO & M. P. BLAUSTEIN. 1989. Kinetics and stoichiometry of coupled Na efflux and Ca influx (Na/Ca exchange) in barnacle muscle cells. J. Gen. Physiol. **93:** 1219-1241.
2. ASHIDA, T. & M. P. BLAUSTEIN. 1987. Regulation of cell calcium and contractility in mammalian arterial smooth muscle: The role of sodium-calcium exchange. J. Physiol. **392:** 617-635.
3. BOVA, S., W. F. GOLDMAN, X.-J. YUAN & M. P. BLAUSTEIN. 1990. Influence of Na^+ gradient on Ca^{2+} transients and contraction in vascular smooth muscle. Am. J. Physiol. **259:** H409-H423.
4. GOLDMAN, W. F., S. BOVA & M. P. BLAUSTEIN. 1990. Measurement of intracellular Ca^{2+} in cultured arterial smooth muscle cells using Fura-2 and digital imaging microscopy. Cell Calcium **11:** 221-231.
5. RAITERI, M., G. BONANNO, M. MARCHI & G. MAURA. 1984. Is there a functional linkage between neurotransmitter uptake mechanisms and presynaptic receptors? J. Pharmacol. Exp. Ther. **231:** 671-677.
6. NACHSHEN, D. A. & M. P. BLAUSTEIN. 1989. Some properties of potassium-stimulated calcium influx in presynaptic nerve endings. J. Gen. Physiol. **76:** 709-728.
7. SANCHEZ-ARMASS, S. & M. P. BLAUSTEIN. 1987. Role of Na/Ca exchange in the regulation of intracellular Ca^{2+} in nerve terminals. Am J. Physiol. (Cell Physiol.) **252:** C595-C603.
8. WALICKE, P. A. 1988. Basic and acidic fibroblast growth factors have trophic effects on neurons from multiple CNS regions. J. Neurosci. **8:** 2618-2627.
9. YAROWSKY, P. J. & B. K. KRUEGER. 1989. Development of saxitoxin-sensitive and insensitive sodium channels in cultured neonatal rat astrocytes. J. Neurosci. **9:** 1055-1061.
10. RUSSELL, J. M. & M. P. BLAUSTEIN. 1974. Calcium efflux from barnacle muscle fibers: Dependence on external cations. J. Gen. Physiol. **63:** 144-167.
11. ASHLEY, C. C., P. C. CALDWELL & A. G. LOWE. 1972. The efflux of calcium from single crab and barnacle muscle fibres. J. Physiol. (London) **223:** 733-755.
12. ASHLEY, C. C., J. C. ELLORY & K. HAINAUT. 1974. Calcium movements in single crustacean muscle fibres. J. Physiol. (London) **242:** 255-272.
13. BLAUSTEIN, M. P. 1976. Sodium-calcium exchange and the regulation of cell calcium in muscle fibers. The Physiologist **19:** 525-540.
14. RASGADO-FLORES, H., J. DE SANTIAGO & R. ESPINOSA-TANGUMA. 1991. Stoichiometry and regulation of the Na-Ca exchanger in barnacle muscle cells. Ann. N.Y. Acad. Sci. This volume.
15. YASUI, K. & J. KIMURA. 1990. Is potassium co-transported by the cardiac Na-Ca exchange? Pflugers Arch. **415:** 513-515.
16. CERVETTO, L., L. LAGNADO, R. J. PERRY, D. W. ROBINSON & P. A. MCNAUGHTON. 1989. Extrusion of calcium from rod outer segments is driven by both sodium and potassium gradients. Nature **337:** 740-743.
17. BRINLEY, F. J., JR. 1968. Sodium and potassium fluxes in isolated barnacle muscle fibers. J. Gen. Physiol. **51:** 445-477.
18. HAGIWARA, S. & S. NAKAJIMA. 1966. Effects of the intracellular Ca ion concentration upon the excitability of the muscle fiber membrane of a barnacle. J. Gen. Physiol. **49:** 807-818.
19. ASHLEY, C. C. 1970. An estimate of calcium concentration changes during the contraction of single muscle fibres. J. Physiol. (London) **210:** 133-134P.
20. ASHLEY, C. C., & D. G. MOISESCU. 1977. Effect of changing the composition of the bathing solutions upon the isometric tension-pCa relationship in bundles of crustacean myofibrils. J. Physiol. (London) **270:** 627-652.
21. REUTER, H., M. P. BLAUSTEIN & G. HAUSLER. 1973. Na/Ca exchange and tension development in arterial smooth muscle. Phil. Trans. R. Soc. (London) **B265:** 87-94.
22. MULVANY, M. J. 1985. Changes in sodium pump activity and vascular contraction. J. Hypertens. **3:** 429-436.
23. SOMLYO, A. P., R. BRODERICK & A. V. SOMLYO. 1986. Calcium and sodium in vascular smooth muscle. Ann. N.Y. Acad. Sci. **488:** 228-239.

24. GOLDMAN, W. F. 1991. Spatial and temporal resolution of serotonin-induced changes in intracellular calcium in a cultured arterial smooth muscle cell line. Blood Vessels **28:** 252-261.
25. ASHIDA, T., J. SCHAEFFER, W. F. GOLDMAN, J. B. WADE & M. P. BLAUSTEIN. 1988. Role of sarcoplasmic reticulum in arterial contraction: Comparison of ryanodine's effect in a conduit and a muscular artery. Circ. Res. **62:** 854-863.
26. WOOLFSON, R. G., P. J. HILTON & L. POSTON. 1990. The effects of ouabain and low sodium on the contractility of human subcutaneous resistance vessels. Hypertension **15:** 583-590.
27. TANFORD, C. 1981. Equilibrium state of ATP-driven ion pumps in relation to physiological ion concentration gradients. J. Gen. Physiol. **77:** 223-229.
28. LEIJTEN, P. A. A. & C. VAN BREEMEN. 1984. The effects of caffeine on the noradrenaline-sensitive calcium store in rabbit aorta. J. Physiol. (London) **357:** 327-339.
29. HASHIMOTO, T., N. HIRATA, T. ITOH, Y. KANAMURA & H. KURIYAMA. 1986. Inositol 1,4,5,-triphosphate activates pharmacomechanical coupling in smooth muscle of the rabbit mesenteric artery. J. Physiol. (London) **370:** 605-618.
30. BAKER, P. F., M. P. BLAUSTEIN, A. L. HODGKIN & R. A. STEINHARDT. 1969. The influence of calcium ions on sodium efflux in squid axons. J. Physiol. (London) **200:** 431-458.
31. BLAUSTEIN, M. P. & A. L. HODGKIN. 1969. The effect of cyanide on the efflux of calcium from squid axons. J. Physiol. (London) **200:** 497-527.
32. BLAUSTEIN, M. P. 1974. The interrelationship between sodium and calcium fluxes across cell membranes. Rev. Physiol. Biochem. Pharmacol. **70:** 33-82.
33. DI POLO, R. & L. BEAUGÉ. 1979. Physiological role of ATP-driven calcium pump in squid axons. Nature **278:** 271-273.
34. BLAUSTEIN, M. P. 1975. Effects of potassium, veratridine and scorpion venom on calcium accumulation and transmitter release by nerve terminals *in vitro.* J. Physiol. (London) **247:** 617-644.
35. CHEON, J. & J. P. REEVES. 1988. Site density of the sodium-calcium exchange carrier in reconstituted vesicles from bovine cardiac sarcolemma. J. Biol. Chem. **262:** 2309-2315.
36. FRIED, R. C. & M. P. BLAUSTEIN. 1978. Retrieval and recycling of synaptic vesicle membrane in pinched-off nerve terminals (synaptosomes). J. Cell Biol. **78:** 685-700.
37. NIGGLI, E. & W. J. LEDERER. 1991. Molecular operations of the sodium-calcium exchanger revealed by conformation currents. Nature **349:** 621-624.
38. LUTHER, P. W., R. K. YIP, R. J. BLOCH, A. AMBESI, G. E. LINDENMAYER & M. P. BLAUSTEIN. 1991. Localization of Na/Ca exchangers in neuromuscular preparations by immunofluorescence microscopy. Soc. Neurosci. Abstr. In press.
39. SMITH, S. J. & G. J. AUGUSTINE. 1988. Calcium ions, active zones and synaptic transmitter release. Trends Neurosci. **11:** 458-464.
40. KATZ, B. & R. MILEDI. 1968. The role of calcium in neuromuscular facilitation. J. Physiol. (London) **195:** 481-492.
41. PARNAS, I. & H. PARNAS. 1988. The Ca-voltage hypothesis for neurotransmitter release. Biophys. Chem. **29:** 85-93.
42. ATWOOD, H. L., M. P. CHARLTON & C. S. THOMPSON. 1983. Neuromuscular transmission in crustaceans is enhanced by a sodium ionophore, monensin, and by prolonged stimulation. J. Physiol. (London) **335:** 179-195.
43. ZUCKER, R. S., K. R. DELANEY, R. MULKEY & D. W. TANK. 1991. Presynaptic calcium in transmitter release and posttetanic potentiation. Ann. N.Y. Acad. Sci. **635:** In press.
44. BLAUSTEIN, M. P. 1988. Calcium and synaptic function. Handbk. Exp. Pharmacol. **83:** 275-304.
45. AGHAJANIAN, G. K., C. P. VANDERMAELEN & R. ANDRADE. 1983. Intracellular studies on the role of calcium in regulating the activity and reactivity of locus caoeruleus neurons *in vivo.* Brain Res. **273:** 237-243.
46. KATER, S. B., & L. R. MILLS. 1991. Regulation of growth cone behavior by calcium. J. Neurosci. **11:** 891-899.

47. MILLS, L. R. 1991. Neuron-specific and state-specific differences in calcium regulation: Their role in the development of neuronal architecture. Ann. N.Y. Acad. Sci. This volume.
48. KIMELBERG, H. K. & M. D. NORENBERG. 1989. Astrocytes. Sci. Am. April: 66-76.
49. HAMLYN, J. M., M. P. BLAUSTEIN, S. BOVA, D. W. DUCHARME, D. W. HARRIS, F. MANDEL, W. R. MATHEWS & J. H. LUDENS. 1991. Identification and characterization of a ouabain-like compound from human plasma. Proc. Natl. Acad. Sci. USA **81:** 6259-6263.

Sodium-Calcium Exchange and Phototransduction in Retinal Photoreceptors [a]

K.-W. YAU, K. NAKATANI, AND T. TAMURA[b]

Howard Hughes Medical Institute
and
Department of Neuroscience
Johns Hopkins University School of Medicine
Baltimore, Maryland 21205

INTRODUCTION

Phototransduction is the process by which light triggers an electrical response in retinal rod and cone receptors. Rapid progress has been made over the past several years in the understanding of this process. Furthermore, it is now known that the Na^+-Ca^{2+} exchange carrier participates in an important negative feedback control on this process. We briefly summarize this feedback here.

OUTLINE OF PHOTOTRANSDUCTION

Retinal rod and cone receptors respond to light with a graded membrane hyperpolarization. This is relayed to second-order neurons in the retina through a modulation of neurotransmitter release at the synaptic terminals of the receptor cells. This transmitter release is high in darkness and is reduced in the light by the hyperpolarization.[1]

The way in which the hyperpolarizing response to light is generated is as follows. In darkness, a cation conductance on the plasma membrane of the receptor's outer segment is kept open by the cyclic nucleotide cGMP (guanosine 3′ : 5′-cyclic monophosphate), letting cations into the cell.[2] This "dark" inward current depolarizes the cell and maintains the steady release of synaptic transmitter described above. Light activates the following reaction cascade: light → photoisomerization of visual pigment → G protein activation → cGMP phosphodiesterase stimulation → cGMP hydrolysis.[2-4] As a result, the free cGMP level falls in the outer segment, causing the cGMP-gated conductance to close and producing the membrane hyperpolarization as the light response. This phototransduction process is the same for both rods and cones, with

[a] The experiments described here were supported by National Institutes of Health Grant EY 06837.

[b] Present address: Department of Ophthalmology, Kanazawa University School of Medicine, 13-1 Takara-machi, Kanazawa, Ishikawa 920, Japan.

only quantitative differences between them. The main features of this process are summarized in FIGURE 1.

In dark steady state, there is a circulation of Ca^{2+} at the rod outer segment, consisting of an influx through the cGMP-gated conductance and an equal efflux through a Na^+-Ca^{2+} exchange carrier[5,6] (see FIG. 1). This carrier in photoreceptor cells involves K^+ in addition to Na^+ and Ca^{2+}.[7] Biochemical experiments have also indicated that Ca^{2+} has a suppressive effect on the intracellular cGMP level,[8-10] at least part of which is apparently through a negative modulation on guanylate cyclase, the cGMP-synthesizing enzyme.[11-16] From these pieces of information, it can be inferred that Ca^{2+} mediates a negative feedback control on the free cGMP concentration in the outer segment, and hence the cGMP-gated conductance. Thus, any sudden increase in cGMP concentration will lead to more channels being open, resulting in a larger Ca^{2+} influx, which in turn tends to restore the original cGMP concentration; the same feedback would act on a sudden decrease in cGMP concentration. In darkness, this negative feedback should reduce background electrical noise by dampening any fluctuations in the basal cGMP metabolism. Because only about 1% of the cGMP-gated channels on the outer segment membrane are normally open in darkness (this being due to the low concentration of free cGMP compared to the $K_{1/2}$ of activation for the channel),[17,18] the feedback also prevents any large accidental increase in cGMP, and hence a surge in cation influx, that can be detrimental to the cell. In the light, the same feedback should operate. In this case, the decrease in internal free Ca^{2+} concentration[5,6,19,20] resulting from the reduction in Ca^{2+} influx and the continuing operation of the exchange carrier leads to an increase in cGMP synthesis, which counteracts the

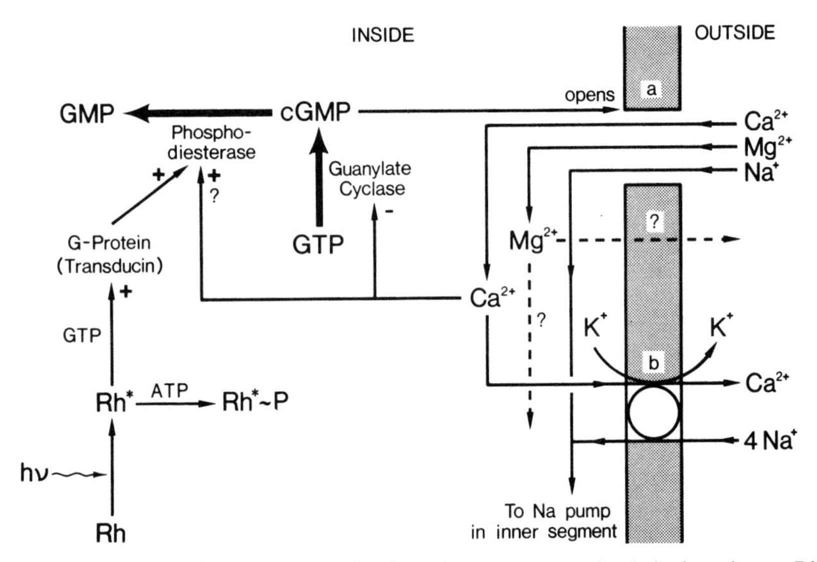

FIGURE 1. Scheme of visual transduction in rod outer segment. Symbols: hν, photon; Rh, rhodopsin molecule; Rh*, photoisomerized rhodopsin molecule; Rh* ~ P, phosphorylated form of Rh*; a, light-sensitive conductance; b, Na^+-Ca^{2+} exchange; +, stimulation; −, inhibition. The nature of the enhancement by Ca^{2+} of the light-activated cGMP phosphodiesterase activity is still not well understood at present. The pathway for Mg^{2+} out of the outer segment is also unknown. (Modified from Yau *et al.*[29] Reprinted by permission from Gustav Fischer Verlag.)

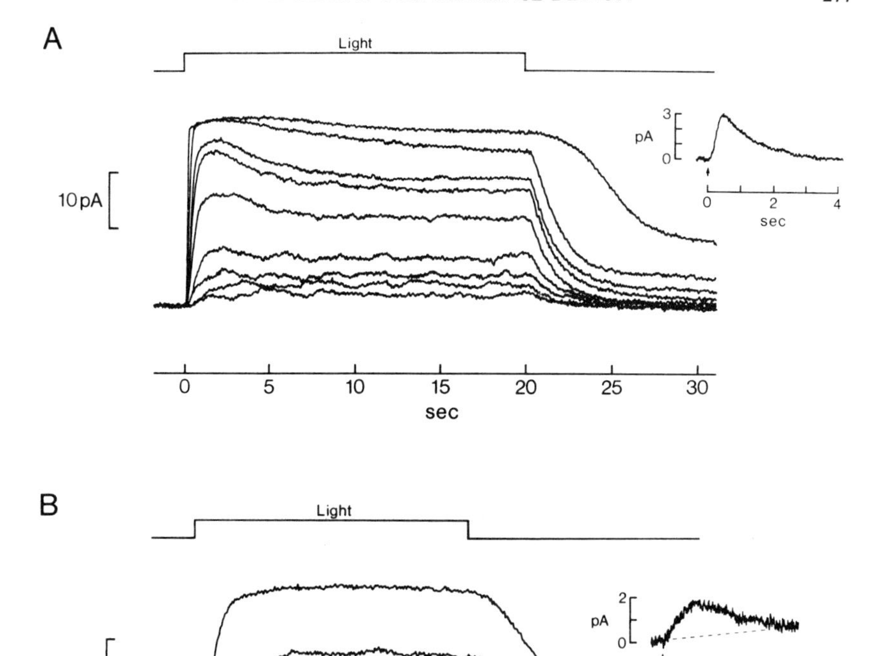

FIGURE 2. Response-intensity families recorded from two salamander rods. **A,** normal Ringer's solution; **B,** test solution with the Ca^{2+} feedback removed (see text). Single trials in all traces. Insets: averaged responses of the cells to dim flashes in control and test conditions. (From Nakatani & Yau.[25] Reprinted by permission from *Nature.*)

increase in cGMP hydrolysis triggered by light. As such, it should constitute a potential mechanism for the phenomenon of background light adaptation in photoreceptors.[5]

Ca^{2+} FEEDBACK AND BACKGROUND LIGHT ADAPTATION

To find out how much of background light adaptation in photoreceptors is due to the Ca^{2+} feedback described above, the feedback can be removed, and the effect on light adaptation examined. This can be achieved by removing external Ca^{2+} so as to eliminate the Ca^{2+} influx and simultaneously replacing all external Na^+ with a cation such as guanidinium, which is able to carry dark current through the cGMP-gated conductance but is unable to drive the Na^+-Ca^{2+} exchange carrier to extrude Ca^{2+}.[6] With both Ca^{2+} influx and efflux removed, the closure of the conductance should no longer lead to a decline in internal Ca^{2+}, and so the negative feedback should disappear. FIGURE 2 shows the experimental results obtained from salamander rods.[21] In control conditions, a recorded salamander rod shows prominent adaptation to light, as indicated by the rapid relaxation (i.e. decline) of the cell's response to a step of light (FIG. 2A). Also, if the normalized response amplitude is plotted against light step intensity, it is found that this relation is steep at the rising phase of the response, but becomes progressively shallower at the transient peak and the response plateau.[21] Thus, light adaptation develops progressively with time. In the test condition, however, no relaxation of the response to a step of light is observed (FIG. 2B). Furthermore, it is found that the relation between response amplitude and light intensity is identical at the early rising phase of the response as at the plateau level, indicating the absence of light adaptation.[21] It can therefore be concluded that the Ca^{2+} feedback is the primary mechanism underlying background light adaptation in rods. Matthews *et al.*[22] have made the same observations. A similar conclusion can also be arrived at for cones.[21,23]

Ca^{2+} FEEDBACK AND ABSOLUTE LIGHT SENSITIVITY OF CONES

The ability of the Ca^{2+} feedback to regulate photoreceptor sensitivity in background light as described above raises the question of whether the low sensitivity of cones (about 1/50-1/100th of rods[24]) is due to a more powerful feedback in these cells. Since the cone outer segment has a much higher surface-to-volume ratio, internal Ca^{2+} might indeed be expected to be pumped down more rapidly by the Na^+-Ca^{2+} exchange carrier during illumination, bringing about a faster feedback.[25]

The rate of decrease in internal free Ca^{2+} due to pumping by the exchange can be studied by loading the outer segment of a rod or cone cell with Ca^{2+} in darkness using a $0Na^+$ external solution (so as to eliminate Ca^{2+} efflux via the exchange), and then reactivating the exchange in the light by restoring external Na^+. The rate at which internal Ca^{2+} is pumped down can then be monitored by measuring the membrane current associated with the exchange activity, which is electrogenic.[26] FIGURE 3A shows three experiments on salamander cones.[25] In all three cases the decline time course of the current is roughly exponential, with a time constant of around 100 msec. For comparison, FIGURE 3B shows an identical experiment on a salamander rod[25]; the decline time constant in this case is about 400 msec, or four times slower. The difference may even be higher if the decline time constant observed in cones is limited by the membrane time constant of the cell.

To see whether the faster decrease in internal Ca^{2+} in the light constitutes the mechanism for the low light sensitivity of cones, the Ca^{2+} feedback can again be removed using a $0Na^+$-$0Ca^{2+}$ external solution as described above and the light sensitivity tested. Rather surprisingly, it is found[25] that removal of the feedback increases cone sensitivity only by several fold (FIG. 4, top), versus the 50-100-fold expected if the feedback were the primary mechanism underlying the low sensitivity. Indeed, rod sensitivity also increases by several fold under a similar treatment[25] (FIG. 4, bottom). Thus, the Ca^{2+} feedback has little to do with the difference in light sensitivity between rods and cones. The faster feedback in cones appears to serve simply to match the phototransduction kinetics in cones, which are also faster than in rods (compare the two traces represented by open circles in FIG. 4, top and bottom panels), so that the two processes can function in unison.

Regarding the difference in absolute sensitivity between rods and cones, one simple explanation[25] at present is that the active intermediates in the cGMP cascade (i.e., photoexcited pigment, active G-protein, and active cGMP phosphodiesterase) all have shorter (say, 1/4th for salamander) lifetimes in cones. This difference not only can

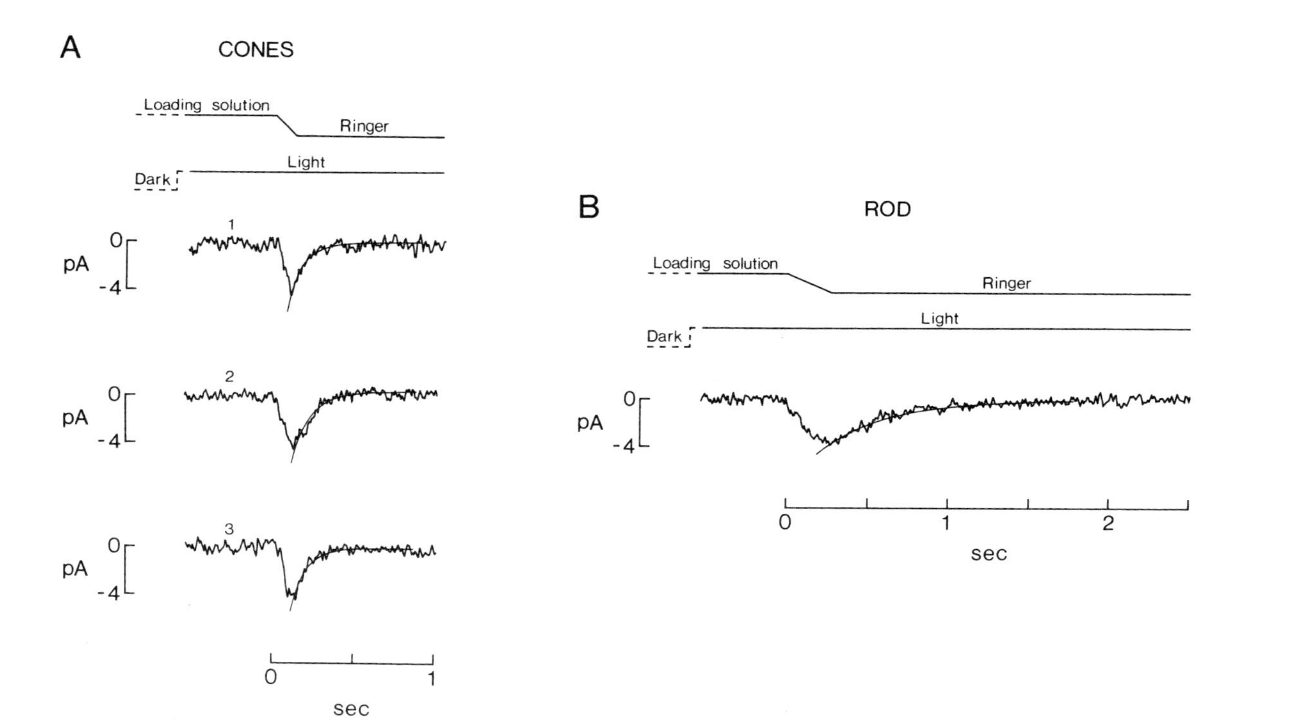

FIGURE 3. Comparison between the rates of decline of the Na^+-Ca^{2+} exchange current in salamander cone and rod cells. **A,** three different cones. In each case a small Ca^{2+} load was effected in darkness by briefly replacing external Na^+ with Li^+ (see text). The smooth curves indicate exponential declines, with $\tau = 90$ msec (top), 100 msec (middle), and 90 msec (bottom), respectively. Single trials. **B,** similar experiment on a salamander rod. Loading solution contained choline to replace Na^+. Smooth curve is an exponential decline, with $\tau = 400$ msec. Average of three trials. (From Nakatani & Yau.[25] Reprinted by permission from the *Journal of Physiology*.)

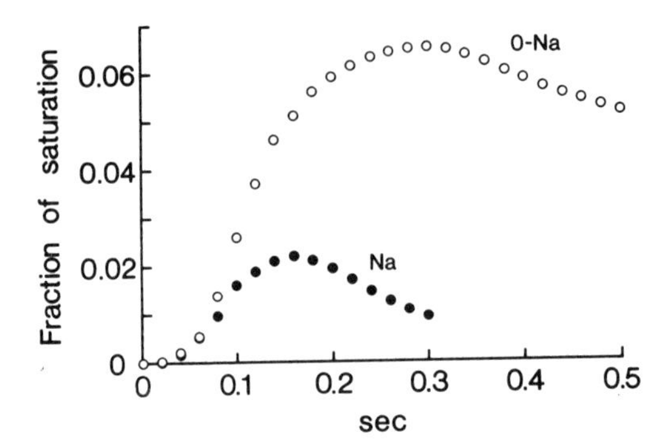

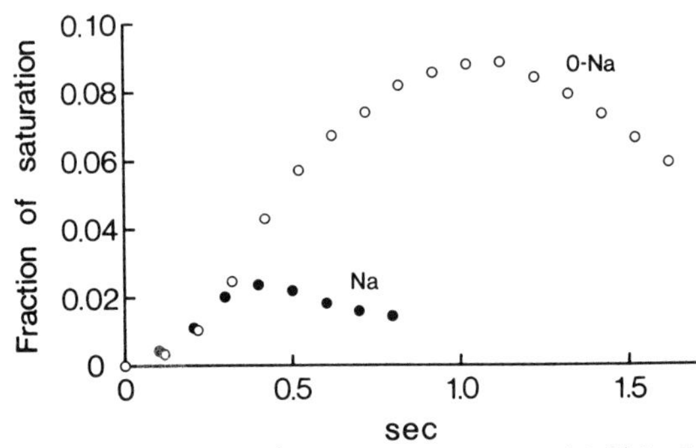

FIGURE 4. Dim flash sensitivity in the presence and the absence of Ca^{2+} feedback. *Top,* salamander cone. *Bottom,* salamander rod. Filled circles, flash response in Ringer's solution. Open circles, response to identical flash in test solution with feedback removed. Flash intensity about 40 times brighter for cone than for rod. (From Nakatani & Yau.[25] Reprinted by permission from the *Journal of Physiology.*)

explain the faster kinetics of the cone response, but since the total gain of the cascade (hence light sensitivity) depends on the lifetimes of the active intermediates in a multiplicative manner, it can also explain a sensitivity difference of 50-100-fold between rods and cones.

QUANTITATIVE RELATION BETWEEN CA^{2+} FEEDBACK AND BACKGROUND ADAPTATION IN PRIMATE RODS

It was described earlier that when the Ca^{2+} feedback is removed, background light adaptation essentially disappears. A more difficult question to ask is whether background adaptation can be *quantitatively* explained by the known properties of the feedback.

We have recently studied this question in primate rods.[27] As pointed out above, the Ca^{2+} feedback consists of a Ca^{2+} influx, a Ca^{2+} efflux, and a negative modulation of the guanylate cyclase enzyme by Ca^{2+}. The detailed properties of each of these components have to be specified in order to make predictions. We first characterized the Ca^{2+} efflux by loading a single primate rod with Ca^{2+} (using a $0Na^+$ external solution) and

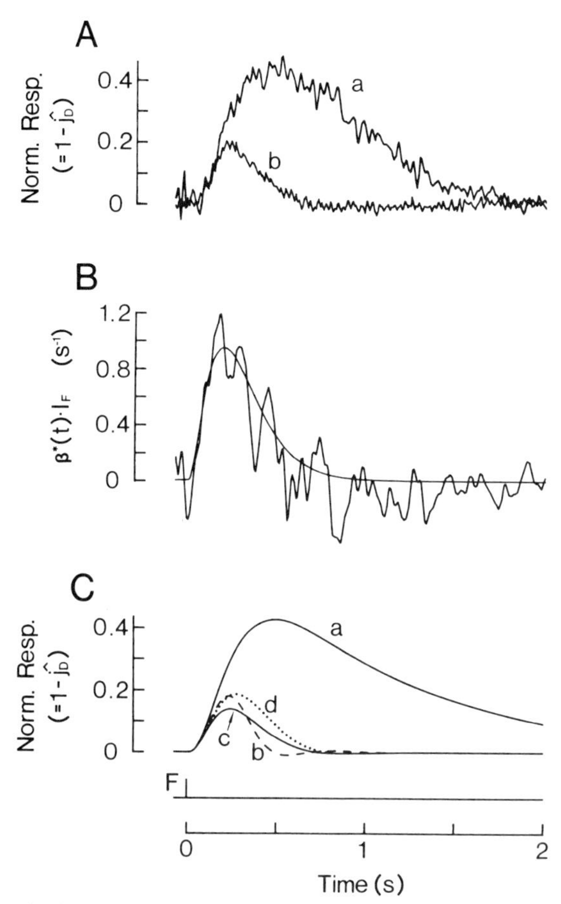

FIGURE 5. Quantitative prediction of dim flash response of primate rod from the Ca^{2+} feedback. **A,** trace *a* shows the response to a dim flash in a $0Na^+$-$0Ca^{2+}$ solution, and trace *b* shows the response to an identical flash in control Ringer's solution. Both responses are normalized against the dark currents in the respective conditions. Flash (F) was delivered at time zero. External Na^+ was replaced by guanidinium. Traces *a* and *b* represent averages of 4 and 8 flash trials, respectively; 36-38°C. **B,** time course of the flash-triggered phosphodiesterase activity extracted from trace *a* in A (see text). Smooth curve is drawn from the convolution of exponential time constants 0.1, 0.1, and 0.1 sec. **C,** flash responses computed from phosphodiesterase time course shown in B. Trace *a* is computed in the absence of Ca^{2+} feedback; traces *b* to *d* are all computed with the feedback imposed, but they differ in respect to the detailed assumption about internal Ca^{2+} buffering. (From Tamura *et al.*[27] Reprinted by permission from the *Journal of General Physiology*.)

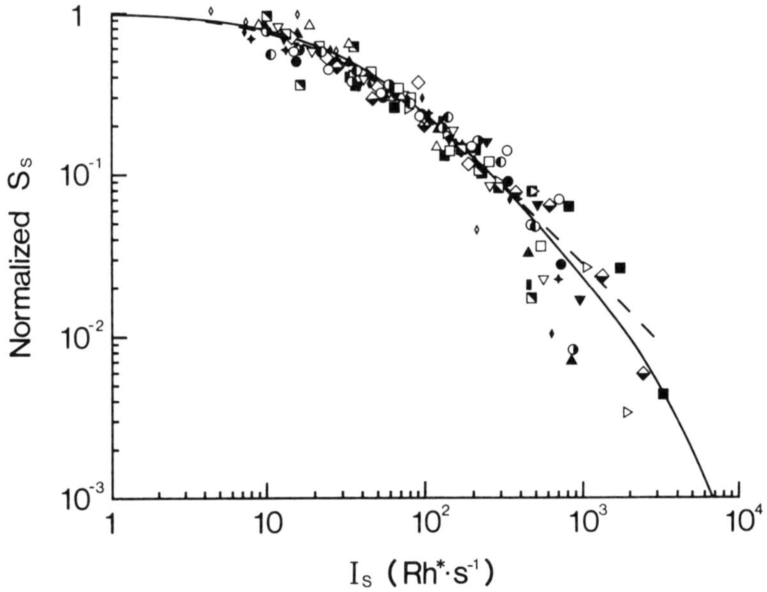

FIGURE 6. Reduction in step sensitivity (S_s) of primate rods by steady light. *Continuous curve,* theoretical prediction from Ca^{2+} feedback model (see text). *Symbols,* actual measurements from cells; the data from each cell has been shifted horizontally so that the experimental relation overlaps with the continuous curve at $S_s = 0.5$. Dashed curve, Weber-Fechner relation. (From Tamura *et al.*[27] Reprinted by permission from the *Journal of General Physiology.*)

observing the time course of decline of the Na^+-Ca^{2+} exchange current. As for the Ca^{2+} influx, the activation of the cGMP-gated conductance through which this occurs has already been characterized by recording either from an excised patch of rod outer segment membrane[28–30] or from a truncated rod outer segment.[17] We assume that the same activation characteristics apply to the channel in primate rods. The relation between guanylate cyclase activity and Ca^{2+} concentration has also been examined by Koch and Stryer[14] on bovine rods. This relation is very steep, showing approximately a fourth-power dependence on Ca^{2+} concentration; half-inhibition of the cyclase occurs at about 0.1 μM Ca^{2+}. Having completely specified the Ca^{2+} feedback, the response of a primate rod to a dim flash can be generated if both the dark metabolic flux rate for cGMP and the time course of light-activated phosphodiesterase activity are known. The basal metabolic flux rate for cGMP in darkness can be measured by again rapidly removing external Na^+ in darkness to stop the Na^+-Ca^{2+} exchange, a strategy first used by Hodgkin and Nunn[13] on amphibian rods. This manipulation causes a rapid increase in internal Ca^{2+}, which in turn reduces the dark guanylate cyclase activity to near zero. As a result, the free cGMP level should decline exponentially, with a time constant reflecting the dark phosphodiesterase activity. We indeed detected such an exponential decline using the dark current as an indicator of the cGMP concentration.[27] Also, in the simplified experimental condition of having the Ca^{2+} feedback removed, the observed time course of a cell's response to a dim flash should simply be the convolution of the dark cGMP flux rate and the light-activated phosphodiesterase time course. Thus, by mathematically deconvolving this light response with the basal metabolic flux rate in darkness, the time course of the light-activated phosphodiesterase activity can be derived. By applying back the equations describing the Ca^{2+} feedback to this time course of light-activated phosphodiesterase activity, the physiological response to a dim flash can be predicted. We found this predicted response to match the experimentally measured response quite well (FIG. 5). Good agreement between prediction and measurement was likewise found for responses to steps of light. At the same time, the calculations reproduced the adaptational behavior shown by primate rods to background light (FIG. 6). Thus, it appears that the Ca^{2+} feedback can indeed predict the phenomenon of background adaptation reasonably well. A similar approach has been taken by Forti *et al.*[31] and Sneyd and Tranchina[32] to study the light responses of amphibian rods and cones, though the work described here adheres more closely to experimentally derived parameters.

CONCLUSION

The Na^{+}-Ca^{2+} exchange carrier plays an important role in retinal phototransduction by participating in a negative feedback control on this process. In darkness, this feedback stabilizes the dark current and reduces background noise that might otherwise interfere with the detection of dim light. In the light, this feedback underlies the phenomenon of background light adaptation. Mechanistically, the Ca^{2+} feedback seems to involve a negative modulation by Ca^{2+} on the guanylate cyclase enzyme. In addition, there now appears to be also an enhancement of the light-activated cGMP phosphodiesterase activity by Ca^{2+}, though the nature of this enhancement and its relative importance are still unclear at present.[33-36]

REFERENCES

1. DOWLING, J. E. 1987. The Retina: An Approachable Part of the Brain. Belknap/Harvard. Cambridge, MA.
2. YAU, K.-W. & D. A. BAYLOR. 1989. Cyclic GMP-activated conductance of retinal photoreceptor cells. Ann. Rev. Neurosci. **12:** 289-327.
3. PUGH, E. N., JR. & W. H. COBBS. 1986. Visual transduction in vertebrate rods and cones: A tale of two transmitters, calcium and cyclic GMP. Vision Res. **26:** 1613-1643.
4. STRYER, L. 1986. Cyclic GMP cascade of vision. Ann. Rev. Neurosci. **9:** 87-119.
5. YAU, K.-W. & K. NAKATANI. 1985. Light-induced reduction of cytoplasmic free calcium in retinal rod outer segment. Nature **313:** 579-582.
6. NAKATANI, K. & K.-W. Yau. 1988. Calcium and magnesium fluxes across the plasma membrane of the toad rod outer segment. J. Physiol. (London) **395:** 695-729.
7. CERVETTO, L., L. LAGNADO, R. J. PERRY, D. W. ROBINSON & P. A. MCNAUGHTON. 1989. Extrusion of calcium from rod outer segments is driven by both sodium and potassium gradients. Nature **337:** 740-743.
8. COHEN, A. I., I. A. HALL & J. A. FERRENDELLI. 1978. Calcium and cyclic nucleotide regulation in incubated mouse retina. J. Gen. Physiol. **71:** 595-612.
9. KILBRIDE, P. 1980. Calcium effects on frog retinal cyclic guanosine 3' : 5'-monophosphate levels and their light-initiated rate of decay. J. Gen. Physiol. **75:** 457-465.
10. WOODRUFF, M. L. & G. L. FAIN. 1982. Ca^{2+}-dependent changes in cyclic GMP levels are not correlated with opening and closing of the light-dependent permeability of toad photoreceptors. J. Gen. Physiol. **80:** 537-555.
11. LOLLY, R. N. & E. RACZ. 1982. Calcium modulation of cyclic GMP synthesis in rat visual cells. Vision Res. **22:** 1481-1486.
12. PEPE, I. M., I. PANFOLI & C. CUGNOLI. 1986. Guanylate-cyclase in rod outer segments of the toad retina. FEBS Lett. **203:** 73-76.

13. HODGKIN, A. L. & B. J. NUNN. 1988. Control of light-sensitive current in salamander rods. J. Physiol. (London) **403:** 439-472.
14. KOCH, K.-W. & L. STRYER. 1988. Highly co-operative feedback control of retinal rod guanylate cyclase by calcium ions. Nature **334:** 64-66.
15. KAWAMURA, S. & M. MURAKAMI. 1989. Regulation of cGMP levels by guanylate cyclase in truncated frog rod outer segments. J. Gen. Physiol. **94:** 649-668.
16. DIZHOOR, A. M. *et al.* 1991. Recoverin: A calcium-sensitive activator of retinal rod guanylate cyclase. Science **251:** 915-918.
17. YAU, K.-W. & K. NAKATANI. 1985. Light-suppressible, cyclic GMP-sensitive conductance in the plasma membrane of a truncated rod outer segment. Nature **317:** 252-255.
18. NAKATANI, K. & K.-W. YAU. 1988. Guanosine 3′ : 5′-cyclic monophosphate-activated conductance studied in a truncated rod outer segment of the toad. J. Physiol. (London) **395:** 731-753.
19. MCNAUGHTON, P. A., L. CERVETTO & B. J. NUNN. 1986. Measurement of the intracellular free calcium concentration in salamander rods. Nature **322:** 261-263.
20. RATTO, G. M., R. PAYNE, W. G. OWEN & R. Y. TSIEN. 1988. The concentration of cytosolic free calcium in vertebrate rod outer segments measured with Fura-2. J. Neurosci. **8:** 3240-3246.
21. NAKATANI, K. & K.-W. YAU. 1988. Calcium and light adaptation in retinal rods and cones. Nature **334:** 69-71.
22. MATTHEWS, H. R., R. L. W. MURPHY, G. L. FAIN & T. D. LAMB. 1988. Photoreceptor light adaptation is mediated by cytoplasmic calcium concentration. Nature **334:** 67-69.
23. MATTHEWS, H. R., G. L. FAIN, R. L. W. MURPHY & T. D. LAMB. 1990. Light adaptation in cone photoreceptors of the salamander: A role for cytoplasmic calcium. J. Physiol. (London) **420:** 447-469.
24. SCHNAPF, J. L. & R. N. MCBURNEY. 1980. Light-induced changes in membrane current in cone outer segments of tiger salamander and turtle. Nature **287:** 239-241.
25. NAKATANI, K. & K.-W. YAU. 1989. Sodium-dependent calcium extrusion and sensitivity regulation in retinal cones of the salamander. J. Physiol. (London) **409:** 525-548.
26. YAU, K.-W. & K. NAKATANI. 1984, Electrogenic Na-Ca exchange in retinal rod outer segment. Nature **311:** 661-663.
27. TAMURA, T., K. NAKATANI & K.-W. YAU. 1991. Calcium feedback and sensitivity regulation in primate rods. J. Gen. Physiol. **98:** 95-130.
28. FESENKO, E. E., S. S. KOLESNIKOV & A. L. LYUBARSKY. 1985. Induction by cyclic GMP of cationic conductance in plasma membrane of retinal rod outer segment. Nature **313:** 310-313.
29. YAU, K.-W., L. W. HAYNES & K. NAKATANI. 1986. Roles of calcium and cyclic GMP in visual transduction. *In* Membrane Control of Cellular Activity. H. Ch. Lüttgau, Ed.: 343-366. Gustav Fischer. Stuttgart, Germany.
30. LÜHRING, H. & U. B. KAUPP. 1989. Study of the mammalian cGMP-gated channels in excised membrane patches. Biophys. J. **55:** 377a.
31. FORTI, S., A. MENINI, G. RISPOLI & V. TORRE. 1989. Kinetics of phototransduction in retinal rods of the newt *Triturus cristatus.* J. Physiol. (London) **419:** 265-295.
32. SNEYD, J. & D. TRANCHINA. 1989. Phototransduction in cones: An inverse problem in enzyme kinetics. Bull. Math. Biol. **51:** 749-784.
33. ROBINSON, P. R., S. KAWAMURA, B. ABRAMSON & M. D. BOWNDS. 1980. Control of the cyclic GMP phosphodiesterase of frog photoreceptor membranes. J. Gen. Physiol. **76:** 631-645.
34. KAWAMURA, S. & M. D. BOWNDS. 1981. Light adaptation of the cyclic GMP phosphodiesterase of frog photoreceptor membranes mediated by ATP and calcium ions. J. Gen. Physiol. **77:** 571-591.
35. BARKDOLL III, A. E., E. N. PUGH, JR. & A. SITARAMAYYA. 1989. Calcium dependence on the activation and inactivation kinetics of the light-activated phosphodiesterase of retinal rods. J. Gen. Physiol. **93:** 1091-1108.
36. KAWAMURA, S. & M. MURAKAMI. 1991. Calcium-dependent regulation of cyclic GMP phosphodiesterase by a protein from frog retinal rods. Nature **349;** 420-423.

Sodium-Calcium Exchange in Invertebrate Photoreceptors[a]

PETER M. O'DAY

Institute of Neuroscience
and
Department of Biology
University of Oregon
Eugene, Oregon 97403

INTRODUCTION

The importance of Ca^{2+} in visual function has been recognized for many years. In photoreceptors from vertebrates and invertebrates, Ca^{2+} participates in visual excitation and adaptation. As we learn more about visual physiology, the proposed functional roles of Ca^{2+} seem to increase in number and complexity. The importance of Ca^{2+} in visual function underscores the importance of the processes by which intracellular Ca^{2+} is regulated. As in many other systems, Na^+-Ca^{2+} exchange is a vitally important mechanism for Ca^{2+} regulation in photoreceptors.

In this article, I will discuss the roles of Ca^{2+}, the regulation of intracellular Ca^{2+}, the evidence for Na^+-Ca^{2+} exchange, and the physiological importance of Na^+-Ca^{2+} exchange in invertebrate photoreceptor physiology.

ROLES OF Ca^{2+} IN INVERTEBRATE PHOTORECEPTOR FUNCTION

Photoreceptors are the cells in the visual system that transduce light stimuli into neural information, a process called phototransduction. Photoreceptors contain light-sensitive pigment that resides in specialized, highly invaginated membranous structures called rhabdomeres. When a pigment molecule absorbs light, a cascade of intracellular biochemical reactions begins; the result is electrical excitation that spreads over the plasma membrane to the photoreceptor synapse. The processes that give rise to excitation have been extensively studied; however, the overall intracellular phototransduction scheme remains unresolved. Nonetheless, it has been well established that intracellular Ca^{2+} is involved in phototransduction in several observable ways. In this section, I will discuss the roles of Ca^{2+} in transduction and some models that have been proposed to explain the involvement of Ca^{2+} in transduction. I will consider primarily the photoreceptors from the ventral eye of the horseshoe crab, *Limulus polyphemus,* a widely used model preparation for photoreceptor physiology, and the photoreceptors in the retina of the fruitfly *Drosophila melanogaster,* an excellent preparation for genetic and molecular biological studies.

[a]This work was supported by National Science Foundation Grant BNS-84024463 and grants from the Oregon affiliate of the American Heart Association and the Medical Research Foundation of Oregon.

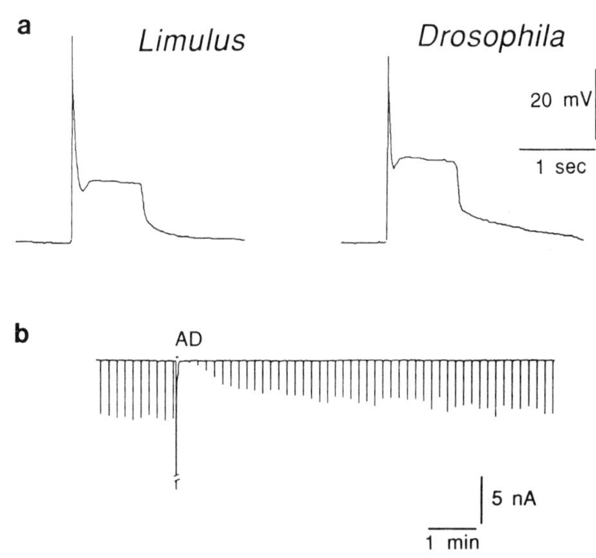

FIGURE 1. Excitation and adaptation in invertebrate photoreceptors. **a.** Depolarizing receptor potentials were recorded with single intracellular microelectrodes from *Limulus* and *Drosophila* photoreceptors. Moderate intensity illumination induces depolarizing electrical responses with transient and plateau components and a repolarization to baseline after termination of stimulus. **b.** Light adaptation and dark adaptation in *Limulus* photoreceptors. Under two microelectrode voltage clamps, response to brief test flashes appear as transient downward deflections (inward currents). At the bar marked AD, a bright adapting light was presented to the cell. *Light adaptation:* After the AD light, test-flash response amplitudes were vastly diminished, reflecting massive desensitization. *Dark adaptation:* As time progresses after light adaptation, the test-flash response amplitudes recover toward levels observed before light adaptation.

Invertebrate photoreceptors generally respond to illumination with a depolarizing receptor potential[1,2] (FIG. 1). In *Limulus* ventral photoreceptors, light also causes release of Ca^{2+} from intracellular compartments[3,4] (called subrhabdomeral cisternae, or SRC). The resulting elevation of $[Ca_i^{2+}]$ causes a reduction of the sensitivity of the photoreceptors to subsequent illumination.[5] This process, called light adaptation, can be detected as a reduction in the electrical response amplitude. Light adaptation can be inhibited by blocking the light-induced elevation of $[Ca_i^{2+}]$ by injecting Ca^{2+}-buffers.[6] Light adaptation can be mimicked by injection of Ca^{2+} intracellularly.[7] Thus, Ca^{2+} is said to be an intracellular messenger of adaptation.[8,9]

In addition to its role in adaptation, Ca^{2+} appears to play a direct role in excitation. Intracellular Ca^{2+} is *required* for excitation in low concentrations,[10] and exogenous introduction of Ca^{2+} in the dark can be sufficient to *cause* excitation.[11] Further, the waveshape of the electrical light response is strongly dependent on intracellular Ca^{2+}: The rising edge of the response rises faster as a result of light-induced Ca^{2+} release,[12] and in low $[Ca_i^{2+}]$ conditions, light responses are slow, but response amplitude is large.[1]

At least part of the Ca^{2+} dependence of phototransduction may arise from the role that the phosphoinositide pathway plays in excitation. It has been established over the past few years that phosphoinositide metabolism is a fundamental component of phototransduction in both *Limulus* ventral photoreceptors and *Drosophila* retina.[13-15] One end product of phosphoinositide metabolism, IP_3 (inositol 1,4,5-trisphosphate, a soluble second messenger), causes Ca^{2+} release from *Limulus* intracellular stores.[16] Further, exogenous introduction of IP_3 mimics excitation.[13–15] It appears that light-induced, IP_3-mediated Ca^{2+} release mediates light adaptation and that Ca^{2+} release is involved in feedback inhibition of further Ca^{2+} release by IP_3.[17] This feedback may play a major role in light adaptation by suppressing the light-dependent activation of the phosphoinositide pathway.

Ca^{2+} involvement through the phosphoinositide pathway, however, is insufficient to explain excitation. The electrical light response results from the opening of plasma membrane channels that are gated by an intracellular second messenger.[18] Recently, in *Limulus* photoreceptors, it was demonstrated that light-activated channels were opened by cyclic-GMP[19] (the same ligand that gates the light-sensitive channels in vertebrate photoreceptors[20]) but not by Ca^{2+}. This observation demonstrates that cyclic-GMP can also be involved directly in transduction, implying that both cyclic-GMP metabolism and phosphoinositide metabolism are components of the intracellular biochemistry of transduction.

The observation, however, raises the important question how phosphoinositide metabolism is coupled to the opening of the membrane channels that underlie excitation. One possibility is that cyclic-GMP-gated channels are a different class of channels than the putative channels activated by the phosphoinositide pathway. However, it has been proposed that the phosphoinositide and cyclic-GMP metabolic pathways in *Limulus* photoreceptors interact *via* Ca^{2+} to yield excitation.[21] At the basis of this proposal is the observation that inhibitors of cyclic-GMP regulatory enzymes exhibit Ca^{2+} dependence. Cyclic-GMP levels are governed by synthesis (with the enzyme guanylate cyclase) and degradation (with the enzyme cyclic-GMP phosphodiesterase). The physiological effects of cyclic-GMP phosphodiesterase inhibition are Ca^{2+} dependent, and cyclic-GMP phosphodiesterase activity is inhibited by light.[21–23] It has been suggested[21] that there is a dark resting flux of cyclic-GMP synthesis and degradation and that this flux is depressed by intracellular Ca^{2+}. There is also evidence of a Ca^{2+}-dependent inhibition of guanylate cyclase.[24]

It has further been proposed that the phosphoinositide pathway is involved as one of at least two parallel pathways in *Limulus* and in *Drosophila* photoreceptors.[12,21] If this is the case, it is additionally possible that other parallel pathways are Ca^{2+} sensitive as well. Other invertebrate photoreceptors may operate in somewhat different ways; in excised plasma membrane patches from photoreceptors of the scallop, both Ca^{2+} and cyclic-GMP activated membrane currents,[25] suggesting that another Ca^{2+}-dependent mechanism can contribute to excitation.

Excitation is triggered by light, and excitation must be "shut off" for repolarization of the membrane to occur. Normally, shut-off spontaneously occurs after termination of the light stimulus. However, in several invertebrate photoreceptors, shut-off can be delayed, resulting in a prolonged, depolarizing afterpotential (PDA).[26–28] In *Limulus* photoreceptors, shut-off can be delayed by prolonged exposure to low extracellular Ca^{2+}, suggesting that the processes underlying shut-off are Ca^{2+} sensitive.[21]

In addition to its roles in the excitation and modulation of light-activated channels, Ca^{2+} affects voltage-gated channels.[29] In *Limulus* ventral photoreceptors, a voltage-gated (delayed rectifier) K^+ channel is modulated by Ca^{2+} released during excitation.[30,31] *Balanus* and *Lima* photoreceptors possess Ca^{2+}-activated K^+ conductances that would be enhanced by a light-induced Ca^{2+} rise.[32,33] Further, in *Limulus* and *Lima,* there are voltage-gated Ca^{2+} channels that contribute to the light response and to light adaptation;[34,32] a large light-induced Ca^{2+} rise can in principle dramatically change the driving force for these plasma membrane Ca^{2+} channels.

Since intracellular Ca^{2+} plays so many roles in transduction, its proper spatial and temporal regulation must be absolutely essential to support the complicated biochemical pathways that underlie normal photoreceptor function. So it is important to examine the factors that influence intracellular Ca^{2+} in photoreceptors.

REGULATION OF INTRACELLULAR Ca^{2+} IN INVERTEBRATE PHOTORECEPTORS

In *Limulus* ventral photoreceptors, several factors can influence intracellular $[Ca^{2+}]$, including (i) light-activated release from intracellular stores[3,4]; (ii) voltage-activated Ca^{2+} channels in the plasma membrane[29]; (iii) intracellular Ca^{2+} buffers[35]; (iv) Na^+-Ca^{2+} exchange in the plasma membrane[35]; and (v) Ca^{2+} uptake into intracellular stores.[36] In addition, in *Balanus* photoreceptors, Ca^{2+} can enter the cell through the light-activated plasma membrane channels.[37] Analogous to other neurons, there are probably other factors that influence intracellular $[Ca_i^{2+}]$ in invertebrate photoreceptors, such as (i) Ca^{2+}-binding proteins; (ii) active uptake into intracellular organelles; and (iii) active extrusion across the plasma membrane.

In *Limulus*, the rate of Na^+-Ca^{2+} exchange seems to be strongly dependent on $[Ca_i^{2+}]$, and at low Ca_i^{2+} levels the primary means of Ca^{2+} extrusion is not Na^+-Ca^{2+} exchange.[35] It seems likely that Ca^{2+} extrusion in *Limulus* is similar to that in other neurons, relying on bulk removal of Ca^{2+} by Na^+-Ca^{2+} exchange at high $[Ca^{2+}]$, and relying on low volume Ca^{2+} removal by a high-affinity Ca^{2+} pump when $[Ca_i^{2+}]$ is low.

EVIDENCE FOR Na^+-Ca^{2+} EXCHANGE IN INVERTEBRATE PHOTORECEPTORS

Na^+-Ca^{2+} exchange was proposed for *Limulus* photoreceptors to explain the fact that intracellular injection of Na^+ caused a reduction in the sensitivity to light,[7] mimicking light adaptation and mimicking the effects of raising $[Ca_i^{2+}]$.[38] This desensitization was dependent on extracellular Ca^{2+} levels, being absent in very low Ca^{2+} saline. Na^+-Ca^{2+} exchange was also proposed[39] for retinas of honeybee drones to explain experiments in which intracellular injection of Na^+ mimicked the effects of raising $[Ca_i^{2+}]$. Na^+-dependent elevations of extracellular $[Ca^{2+}]$ and $[K^+]$ were induced by prolonged intense illumination of the bee retina.[40] Electrogenic Na^+-Ca^{2+} exchange activated by light was also invoked to explain a transient depolarization following the light response in fly photoreceptors induced by the mitochondrial Ca^{2+} uptake inhibitor, ruthenium red; the effect is dependent upon extracellular $[Ca^{2+}]$ and $[Na^+]$.[41]

In *Limulus* photoreceptors, Ca_o^{2+}-dependent desensitization was observed[42] following removal of extracellular Na^+. A similar result has been reported for *Balanus* photoreceptors.[43] Further, an elevation of intracellular Ca^{2+} was measured in *Limulus* during procedures designed to reduce the transmembrane electrochemical gradient for Na^+, removal of extracellular Na^+, and an increase in intracellular Na^+, either by blocking the Na^+-K^+ pump or by direct injection of Na^+.[35,44] The elevation of $[Ca^{2+}{}_i]$ was not observed in very low Ca^{2+} saline. These results suggested that Na^+-Ca^{2+} exchange could be reversed (i.e., Ca^{2+} entry coupled with Na^+ efflux). Restoration of extracellular Na^+ following desensitization and $[Ca_i^{2+}]$ elevation caused a recovery of sensitivity and a rapid decline in $[Ca_i^{2+}]$.[35] These observations suggested that forward exchange is responsible for maintaining low $[Ca_i^{2+}]$. The voltage dependence of the elevation of $[Ca_i^{2+}]$ induced by bright and prolonged illumination of *Limulus* photoreceptors is also consistent with a voltage-dependent and reversible Na^+-Ca^{2+} exchange process.[45] Direct measurement of membrane currents under voltage-clamp suggested

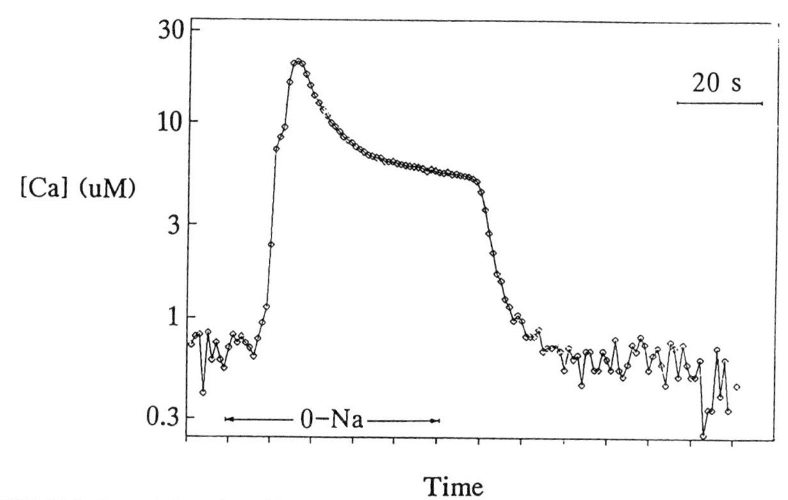

FIGURE 2. Reversible Na^+-Ca^{2+} exchange. Reduction of extracellular Na^+ caused a rapid rise in $[Ca_i^{2+}]$. Restoration of Na_o^+ quickly reduced $[Ca_i^{2+}]$. The ordinate is the estimate of $[Ca_i^{2+}]$ obtained from measurements using aequorin. Arrows indicate extracellular Na^+ removal. Reprinted with permission from the *Journal of General Physiology.*

that Na^+-Ca^{2+} exchange is electrogenic in *Limulus.*[35] Inward currents resulted from raising extracellular $[Na^+]$ and outward currents resulted from raising extracellular $[Ca^{2+}]$ in saline initially containing reduced concentrations of Na^+ and Ca^{2+}.

In examining Ca^{2+} regulation by Na^+-Ca^{2+} exchange in *Limulus* photoreceptors, Na^+ removal experiments (FIG. 2) were particularly instructive.[35] When extracellular Na^+ was removed, the $[Ca_i^{2+}]$ rose to a peak in two phases; $[Ca_i^{2+}]$ then fell with at least two time constants and approached a plateau. Subsequent restoration of extracellular Na^+ resulted in a fast decline in $[Ca_i^{2+}]$. Assuming the $[Ca_i^{2+}]$ rise comes only from reverse Na^+-Ca^{2+} exchange and Ca^{2+} leakage into the cell and that Ca^{2+} pump activity is negligible, then a maximum rate of reverse exchange can be estimated to be approximately one $\log_{10}$ unit per 2 seconds or 0.6 pmoles cm^{-2} sec^{-1} at $[Ca_i^{2+}] = 5$ μM. Then, assuming that the $[Ca_i^{2+}]$ decline induced by restoration of extracellular Na^+ is due solely to forward exchange, the maximum rate of forward exchange would be approximately one $\log_{10}$ unit per 9 seconds or 0.5 pmoles cm^{-2} sec^{-1} at $[Ca_i^{2+}] =$ 5 μM.

The question arises as to why the $[Ca_i^{2+}]$ rise induced by Na^+ removal in FIGURE 2 is transient. Ultimately, the rate of Ca^{2+} entry by reverse exchange must decline as $[Ca_i^{2+}]$ is raised and intracellular Na^+ is lowered, and the $[Ca_i^{2+}]$ rise must stop when the rate of Ca^{2+} removal overtakes that of Ca^{2+} elevation. After that point. $[Ca_i^{2+}]$ declines. If it is assumed that Ca^{2+} removal is due solely to active Ca^{2+} ATPase (Ca^{2+} pump), the minimum rate of Ca^{2+}-pump activity in this situation would be approximately one $\log_{10}$ unit per 22 seconds, or 0.04 pmoles cm^{-2} sec^{-1} at $[Ca_i^{2+}] =$ 5 μM.

If some assumptions are made about the buffering capacity of the cell, the total calcium in the cell, and the Ca^{2+}-pump rates, it is possible to work backward from the time-dependent changes in $[Ca_i^{2+}]$ (from FIG. 2) to calculate values for the exchange currents, the time dependence of intracellular $[Na^+]$, and E_x, the reversal potential of

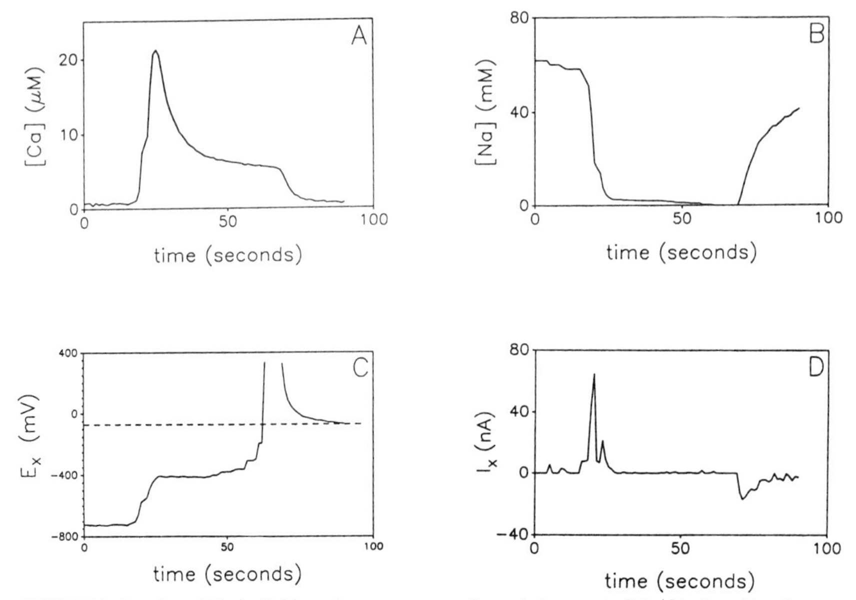

FIGURE 3. Predicted Na^+-Ca^{2+} exchange currents, I_x, and changes in $[Na_i^+]$ induced by changes in extracellular $[Na^+]$. **(A)** Measured $[Ca_i^{2+}]$ versus time; **(B)** predicted $[Na_i^+]$ versus time; **(C)** predicted I_x versus time; **(D)** predicted E_x versus time. $t = 0$ is the time at which extracellular Na^+ was removed. For the analysis, it was assumed that the Ca^{2+} pump rate was maximal and that the fraction of free to bound Ca^{2+} was constant at all $[Ca_i^{2+}]$.

Na^+-Ca^{2+} exchange (FIG. 3). These values suggest that $[Na_i^+]$ may fall quite dramatically soon after extracellular Na^+ is removed; but, as $[Na_i^+]$ falls and $[Ca_i^{2+}]$ rises, the driving force for reverse exchange falls, reducing the Ca^{2+} influx. This, along with a maximal Ca^{2+}-pump activity, can account for the transient nature of the $[Ca_i^{2+}]$ rise induced by Na^+.

CONTRIBUTIONS OF Na^+-Ca^{2+} EXCHANGE TO INVERTEBRATE PHOTORECEPTOR PHYSIOLOGY

Resting Sensitivity

Because a rise in $[Ca_i^{2+}]$ normally triggers light adaptation, maintenance of low intracellular Ca^{2+} levels must be required for maintenance of resting dark sensitivity. Thus, it seems reasonable to hypothesize that Na^+-Ca^{2+} exchange is important in maintaining resting sensitivity because of its role in lowering $[Ca_i^{2+}]$. This was examined in *Limulus* photoreceptors in which blocking Na^+-Ca^{2+} exchange by removing extracellular Na^+ reduces the resting sensitivity in a Ca^{2+}-dependent fashion.[46] The reduction of resting sensitivity depends strongly on the adaptational state of the photoreceptor. In very dark-adapted cells, removal of extracellular Na^+ does not affect on the sensitivity; but in less dark-adapted conditions there is a dramatic desensitization (FIG.

4). These results are consistent with the observations[35] that Na^+-Ca^+ exchange rates are low at low levels of intracellular Ca^{2+}.

These observations suggest that Na^+-Ca^+ exchange is important in reducing [Ca_i^{2+}] when it has been elevated by illumination; however, in low [Ca_i^{2+}] (dark-adapted) conditions, Na^+-Ca^+ exchange may not contribute significantly to maintenance of low [Ca_i^{2+}] and high sensitivity.

Excitation

Na^+-Ca^+ exchange seems to influence the waveshapes of individual light responses of invertebrate photoreceptors.[46,41] Very dark-adapted *Limulus* photoreceptors bathed in Na^+-free saline show no loss of sensitivity compared with cells bathed in control saline. However, the receptor potentials evoked by bright flashes in these photoreceptors differed from those measured in control saline, containing a Ca^{2+}-dependent persistent depolarizing phase that was maintained for many seconds after the termination of light.[46] The Na^+- and Ca^{2+}-dependent depolarizing phase may be due at least in part to Na^+-Ca^+ exchange activity. In the fly retina, a transient afterhyperpolarization (TA) is seen after treatment with ruthenium red.[41] The Na^+ and Ca^{2+} dependencies of the TA suggest that it arises from the electrogenic nature of Na^+-Ca^+ exchange.[40] This suggests that Na^+-Ca^+ exchange currents may contribute to the receptor potentials in normally operating photoreceptors. The effects of manipulating Na^+-Ca^{2+} exchange parameters on the receptor potential waveshapes in invertebrate photoreceptors appear rather late in the response time. Thus, it may be that Ca^{2+} plays an important role in shaping the electrical light response at later times. Due to variations in the response waveshapes, it has been difficult to examine experimentally whether

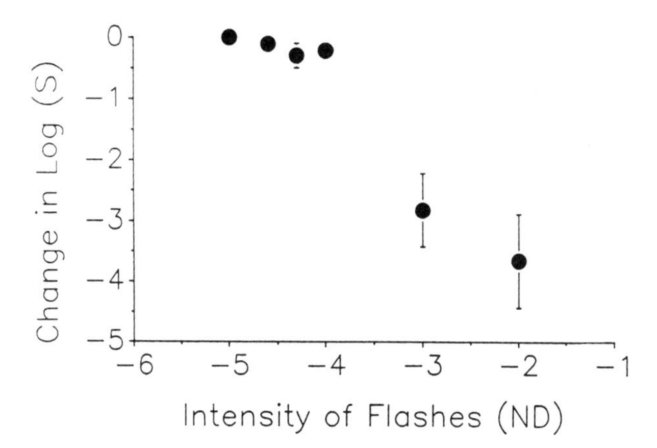

FIGURE 4. Na^+-Ca^+ exchange contributes to maintenance of resting sensitivity when the photoreceptors are moderately dark adapted. The effect on resting sensitivity of inhibiting Na^+-Ca^+ exchange is substantial in conditions of low resting sensitivity, but not in very dark-adapted cells. The ordinate indicates the difference in sensitivity measured in reduced extracellular Na^+ and that measured in normal extracellular Na^+ conditions. Test flashes were used to maintain particular levels of resting sensitivity; the abscissa indicates the test flash intensity, thus increasing I represents less dark-adapted conditions. Reprinted from O'Day *et al.*[46] with permission from the *Journal of General Physiology.*

Na^+-Ca^{2+} exchange might affect the response waveshapes at earlier times; however, a theoretical examination of the question is possible (see below).

Light Adaptation

Because light adaptation in *Limulus* photoreceptors is triggered by elevation of intracellular [Ca^{2+}], rapid Ca^{2+} removal processes that act sufficiently quickly to counteract the [Ca_i^{2+}] rise can diminish the adapting effects of light. Removal of extracellular Na^+ greatly enhances light adaptation in previously dark-adapted cells,[46] and this effect is strongly dependent on the intensity of the adapting light (FIG. 5a). Light adaptation also exhibits strong Ca^{2+}-dependence, being less effective when extracellular Ca^{2+} is lowered (FIG. 5b). The overall results suggest that Na^+-Ca^+ exchange normally mitigates the desensitizing effects of light. This would presumably occur as Ca^{2+} removal by forward exchange limits the elevation of [Ca_i^{2+}] caused by the light-induced Ca^{2+} release that produces light adaptation.

Dark Adaptation

Dark adaptation is the recovery of sensitivity following desensitization by light. Because intracellular Ca^{2+} triggers light adaptation, dark adaptation requires a reduction in [Ca_i^{2+}]. Further, since [Ca_i^{2+}] can rise to levels at which Na^+-Ca^+ exchange is very active,[35] one might expect Na^+-Ca^+ exchange would be important in dark adaptation when the reduction of [Ca_i^{2+}] is rate limiting for dark adaptation. FIGURE 6 illustrates that in *Limulus* photoreceptors reduction of extracellular Na^+ slows the rate of dark adaptation from mild light adaptation.[46] This figure also illustrates that when the extent of light adaptation is severe, recovery does not go to completion. These observations are consistent with the idea that Na^+-Ca^+ exchange is important in the normal course of dark adaptation. They also suggest that the reduction of [Ca_i^{2+}] is rate limiting for dark adaptation at least under some experimental conditions.

Theoretical Considerations

The effects of Na^+-Ca^{2+} exchange on photoreceptor physiology can be considered in a theoretical framework based on the formulations created for and applied to other systems.[47–49] It is possible to use the theoretical voltage dependence of the Na^+-Ca^{2+} exchange current to estimate whether and to what extent the exchange might affect the photoresponse at early times. Similarly, the question can be addressed whether, under normal physiological conditions, reverse exchange might occur and contribute to the response waveshape and/or to the light-induced elevation of [Ca_i^{2+}]. The exchange current has been modeled in several ways; one simplification from the formulation of Mullins[47] is an asymmetric hyperbolic sine function of the energy gradient. A modification of this formulation[48] to consider significant changes in intracellular Ca^{2+} concentration is given in Equation (1):

$$I_x(t) = \frac{\mathrm{Exp}(V_m(t))\,F/2RT)\,\{([Na_i](t))^3[Ca_o]\} - \mathrm{Exp}(-V_m(t)\,F/2RT)\,\{[Na_o]^3\,[Ca_i](t)\}}{(1 + d_{NaCa}\{[Na_o]^3\,[Ca_i](t) + ([Na_i](t))^3\,[Ca_o]\})/K_{NaCa}} \tag{1}$$

a

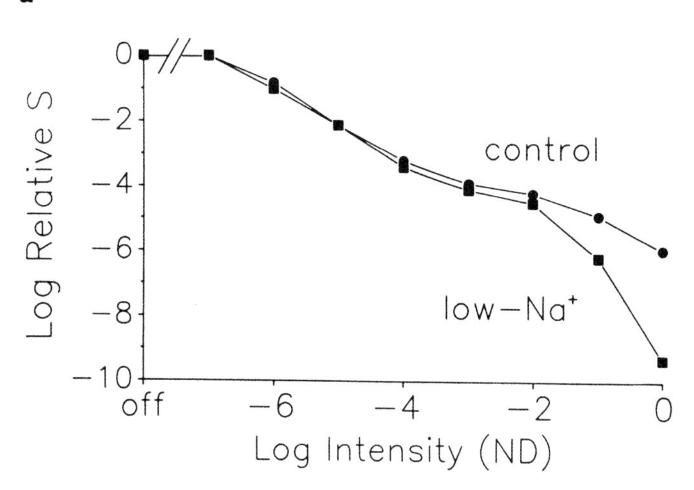

b

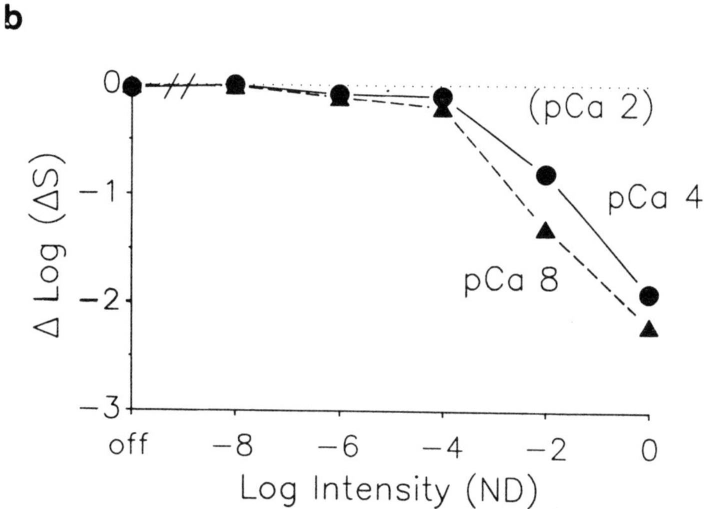

FIGURE 5. Na^+-Ca^+ exchange mitigates effects of adaptation to bright illumination. In (a) and (b), the ordinates indicate the intensity of the adapting light. a. Reduction of extracellular [Na^+] enhances desensitization induced by bright lights but not that induced by dim lights. The abscissa is the relative sensitivity of the cell measured after the adapting illumination; circles are data from control, and squares are data taken in low-Na_o^+ saline. **b.** Reduction of extracellular [Ca^{2+}] diminishes desensitization induced by bright lights but not that induced by dim lights. The ordinance is the difference between the sensitivity measured in normal extracellular Ca^{2+} and that measured in the indicated extracellular Ca^{2+}; circles are data from pCa4 (0.1 mM Ca^{2+}) saline, and squares are from pCa8 (0.01 μM Ca^{2+}) saline. Reprinted from O'Day *et al.*[46] with permission from the *Journal of General Physiology.*

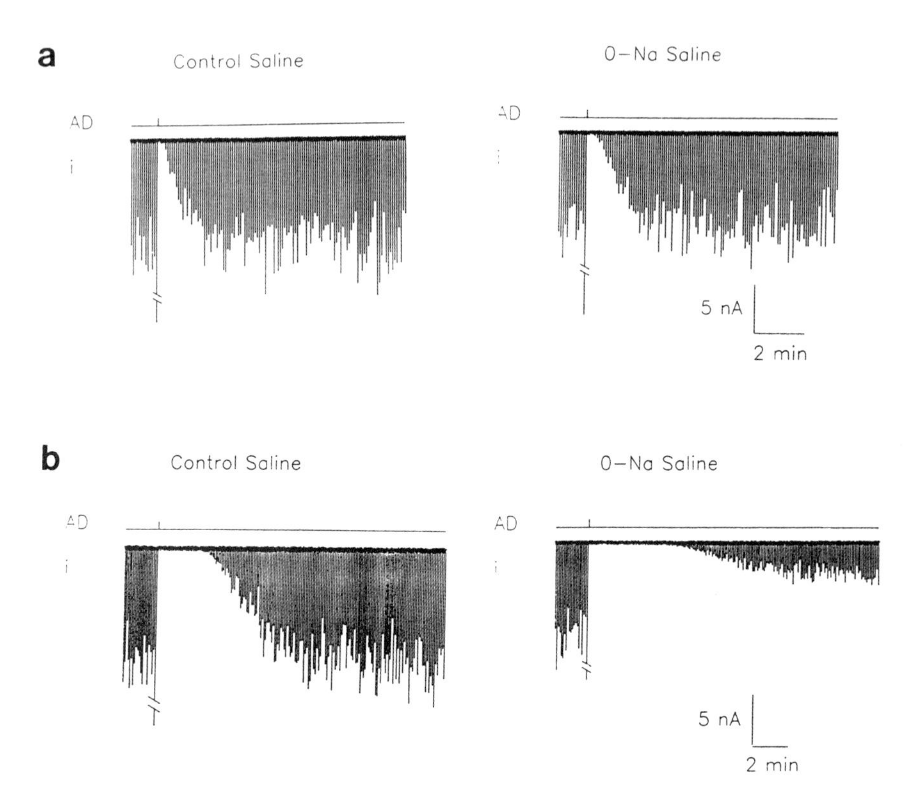

FIGURE 6. Na^+-Ca^+ exchange contributes to dark adaptation. Dark adaptation is enhanced by Na^+ restoration. Dark-adapted photoreceptors, voltage-clamped to −70 mV, were stimulated with dim test flashes, then light adapted with a bright flash (indicated at AD) in low Na^+ saline. The recovery of test flash responses occurred in: *left traces,* normal-Na^+ saline; *right traces,* low-Na^+ saline. **a.** Adaptation to moderate intensity lights. **b.** Adaptation to bright lights. Recovery was slowed in low-Na^+ saline. After adaptation to bright illumination, recovery was incomplete. Reprinted from O'Day *et al.*[46] with permission from the *Journal of General Physiology.*

where k_{NaCa} is a scaling factor, and d_{NaCa} is a correction factor intended to furnish a more accurate representation of the dependence of I_x on $[Ca_i^{2+}]$. The changes in ion concentrations that *Limulus* photoreceptors undergo during normal visual processing may exceed the limits within which Equation (1) strictly applies; however, its application may offer a rough evaluation of the effects of I_x on the photoresponse.

A family of membrane voltage versus I_x curves generated from Equation (1) is plotted in FIGURE 7 for a hypothetical cell under three different adaptational states. This figure illustrates that, in principle, the Na^+-Ca^{2+} exchange current should be much greater during light adaptation than during dark adaptation. This is consistent with the results that suggest that Na^+-Ca^{2+} exchange bears less significance in sensitive dark-adapted (low $[Ca_i^{2+}]$) conditions than in less sensitive light-adapted (high $[Ca_i^{2+}]$) conditions. It is also consistent with results of FIGURE 5 that suggest that the effects of Na^+-Ca^+ exchange on the phenomenology of LA should be greater at greater

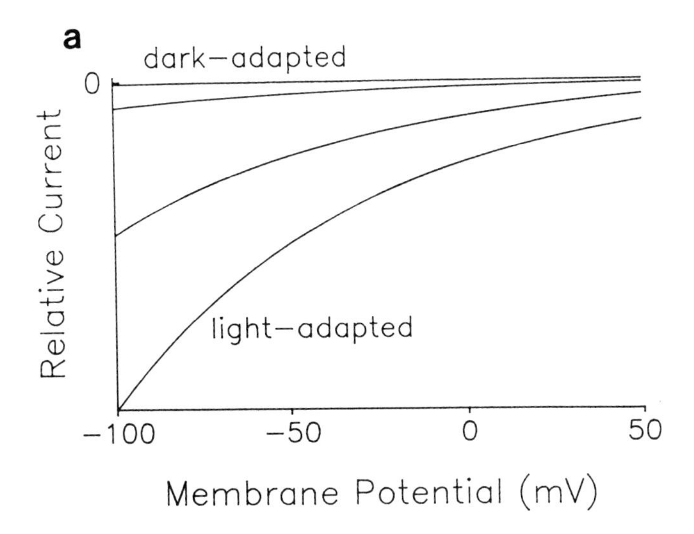

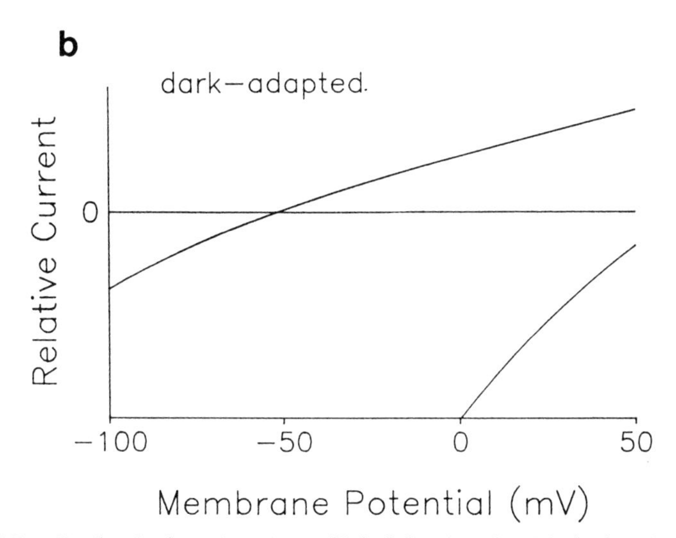

FIGURE 7. **a.** Predicted voltage dependence of I_x in light-adapted and dark-adapted conditions. Traces are from Equation (1) with the following parameters: $[Na_o^+]$ = 425 mM; $[Ca_o^{2+}]$ = 10 mM. In the top trace (dark adapted), $[Na_i^+]$ = 12 mM; $[Ca_i^{2+}]$ = 10 μM. In the second trace from the top, $[Na_i^+]$ = 10 mM; $[Ca_i^{2+}]$ = 1 μM. In the third trace from the top, $[Na_i^+]$ = 6 mM; $[Ca_i^{2+}]$ = 100 nM. And in the bottom trace, $[Na_i^+]$ = 6 mM; $[Ca_i^{2+}]$ = 10 nM. **b.** To illustrate the expected reversal potential of exchange in dark-adapted conditions, the calculations in **a** are replotted with the ordinate expanded, illustrating a reversal potential of about −50 mV.

degrees of light adaptation. Moreover, it is consistent with the results of FIGURE 6 that suggest that the rate of dark adaptation is enhanced by Na^+-Ca^+ exchange.

This formulation is also in agreement with the analysis above (FIG. 3) that suggests that following removal of Na_o^+ the transient nature of the $[Ca_i^{2+}]$ rise is due to the dramatic reduction in Na^+-Ca^+ exchange driving force by the rise in $[Ca_i^{2+}]$ and the decline in $[Na_i^+]$. This treatment is not consistent, however, with the observation that sensitivity does not fall during Na_o^+ reduction (reverse exchange) in very dark-adapted cells, unless $[Na_i^+]$ is assumed to fall to very low levels during dark adaptation under the experimental conditions. There may be threshold levels of $[Na_i^+]$ and/or $[Ca_i^{2+}]$ below which reverse exchanges cannot operate that are not expressed in this formulation.

The equation for the equilibrium potential for Na^+-Ca^{2+} exchange,

$$E_x = 3\ E_{Na} - 2\ E_{Ca} \tag{2}$$

follows from setting $I_x = 0$ in Equation (1). In FIGURE 7 above, it would appear that under dark-adapted conditions the reversal potential of the exchange currents, E_x, is below that of the light-induced currents (shown by arrow), suggesting that a fast, depolarizing light response might yield an outward exchange current and Ca^{2+} entry. The dynamic changes in $[Na_i^+]$ and $[Ca_i^{2+}]$ during the light response, however, should alter E_x so that for each point in time E_x, $[Na_i^+]$, and $[Ca_i^{2+}]$ must be recalculated to determine $I_x(t)$.

To examine whether reverse exchange might result during normal excitation, initial conditions of $[Na_i^+]$, $[Ca_i^{2+}]$, and V_m were estimated, and Equation (1) was used reiteratively to yield time-dependent estimates of I_x, $[Na_i^+]$, $[Ca_i^{2+}]$, and V_m. Comparison of the results with a similar analysis that assumes $I_x(t) = 0$ (i.e., absence of exchange) suggests that the waveshape of a typical response to a bright illuminating flash is affected somewhat by forward exchange, and the contribution of forward exchange to the light-induced change in $[Ca_i^{2+}](t)$ can be significant. Under some dark-adapted conditions, the exchange current may be briefly outward (reverse exchange), but the actual Ca^{2+} entry by reverse exchange seems unlikely to contribute significantly to a light-induced elevation of $[Ca_i^{2+}]$.

CONCLUSION

Na^+-Ca^{2+} exchange appears to be a major factor in the regulation of $[Ca_i^{2+}]$ in active invertebrate photoreceptor cells. Evidence suggests that it operates principally when $[Ca_i^{2+}]$ is elevated (i.e., when the photoreceptors are light adapted) to lower $[Ca_i^{2+}]$. In darkness, when $[Ca_i^{2+}]$ is lowered, the rate of exchange is greatly reduced, operating minimally in very dark-adapted conditions. Thus, Na^+-Ca^{2+} exchange seems to play a large role in dark adaptation and the maintenance of photoreceptor sensitivity only at elevated levels of $[Ca_i^{2+}]$. Light adaptation appears to be opposed by forward exchange at bright adapting intensities. In addition, the waveshapes of the receptor potentials appear to be affected by Na^+-Ca^{2+} exchange activity. The exchange process appears to be similar to that in other neurons as a high-capacity, low-affinity Ca^{2+}-extrusion process that operates in elevated $[Ca_i^{2+}]$ conditions. But the cells probably rely on a high-capacity Ca^{2+} pump to extrude Ca^{2+} when $[Ca_i^{2+}]$ is low.

ACKNOWLEDGMENTS

I very gratefully acknowledge the help of Dr. Edwin C. Johnson, Dr. Mark P. Gray-Keller, Matthew Lonergan, and Dr. Judith Eisen.

REFERENCES

1. MILLECCHIA, R. & A. MAURO. 1969. The ventral photoreceptor cells of *Limulus*. III. A voltage-clamp study. J. Gen. Physiol. **54:** 331-351.
2. ALAWI, A. A. & W. L. PAK. 1971. On-transient of insect electroretinogram: Its cellular origin. Science **172:** 1055-1057.
3. BROWN, J. E. & J. R. BLINKS. 1974. Changes in intracellular free calcium concentration during illumination of invertebrate photoreceptors. J. Gen. Physiol. **64:** 643-665.
4. PAYNE, R., B. WALZ, S. LEVY & A. FEIN. 1988. The localization of calcium release by inositol trisphosphate in *Limulus* photoreceptors and its control by negative feedback. Phil. Trans. R. Soc. London B **320:** 359-379.
5. LISMAN, J. E. & J. E. BROWN. 1975. Light-induced changes of sensitivity in *Limulus* ventral photoreceptors. J. Gen. Physiol. **66:** 473-488.
6. LISMAN, J. E. & J. E. BROWN. 1975. Effects of intracellular injection of calcium buffers on light-adaptation in *Limulus* ventral photoreceptors. J. Gen. Physiol. **66:** 473-488.
7. LISMAN, J. E. & J. E. BROWN. 1972. The effects of intracellular iontophoretic injection of calcium and sodium ions on the light response of *Limulus* ventral photoreceptors. J. Gen. Physiol. **59:** 701-719.
8. FAIN, G. L. & J. E. LISMAN. 1981. Membrane conductances photoreceptors. Prog. Biophys. Mol. Biol. **37:** 91-147.
9. BROWN, J. 1977. Calcium ion, a putative intracellular messenger for light-adaptation in *Limulus* ventral photoreceptors. Biophys. Struct. Mech. **3:** 141-143.
10. BOLSOVER, S. R. & J. E. BROWN. 1985. Calcium ion, an intracellular messenger of light-adaptation, also participates in excitation of *Limulus* photoreceptors. J. Physiol. **364:** 381-393.
11. PAYNE, R., D. W. CORSON & A. FEIN. 1986. Pressure injection of calcium both excites and adapts *Limulus* ventral photoreceptors. J. Gen. Physiol. **88:** 107-126.
12. PAYNE, R. & A. FEIN. 1986. The initial response of *Limulus* ventral photoreceptors to bright flashes. J. Gen. Physiol. **87:** 243-269.
13. BROWN, J. E., L. J. RUBIN, A. J. GHALAYINI, A. L. TARVER, R. F. IRVINE, M. J. BERRIDGE & R. E. ANDERSON. 1984. Myo-inositol polyphosphate may be a messenger for visual excitation in *Limulus* ventral photoreceptors. Nature **311:** 160-162.
14. FEIN, A., R. PAYNE, D. W. CORSON, M. J. BERRIDGE & R. F. IRVINE. 1984. Phororeceptor excitation and adaptation by inositol 1,4,5-trisphosphate. Nature **311:** 157-160.
15. DEVARY, O., O. HEICHAL, A. BLUMENFELD, A. CASSEL, A. BARASH, T. RUBINSTEIN, B. MINKE & Z. SELINGER. 1987. Coupling of photoexcited rhodopsin to phosphoinositide hydrolysis in fly photoreceptors. Proc. Natl. Acad. Sci. USA **84:** 6939-6943.
16. PAYNE, R., D. W. CORSON, A. FEIN & M. J. BERRIDGE. 1986. Excitation and adaptation of *Limulus* ventral photoreceptors by inositol 1,4,5 trisphosphate result from a rise in intracellular calcium. J. Gen. Physiol. **88:** 127-142.
17. PAYNE, R., T. M. FLORES & A. FEIN. 1990. Feedback inhibition by calcium limits the release of calcium by inositol trisphosphate in *Limulus* ventral photoreceptors. Neuron **4:** 547-555.
18. BACIGALUPO, J. & J. E. LISMAN. 1983. Single-channel currents activated by light in *Limulus* ventral photoreceptors. Nature **304:** 268-270.
19. BACIGALUPO, J., E. C. JOHNSON, C. VERGARA & J. E. LISMAN. 1991. Cyclic GMP opens light-dependent channels in excised patches of *Limulus* ventral photoreceptors. Biophys. J. **59:** 530a.
20. FESENKO, E. E., S. S. KOLESNIKOV & A. LYUBARSKY. Induction by cyclic GMP of cationic conductance in plasma membrane of retinal rod outer segment. Nature **313:** 310-313.

21. O'DAY, P. M., E. C. JOHNSON & M. BAUMGARD. 1991. Effects of lithium, calcium, and PDE-inhibitors on excitation in *Limulus* photoreceptors. Biophys. J. **59:** 540a.
22. FADDIS, M. & J. E. BROWN. 1988. Effects of drugs presumed to change intracellular cGMP on voltage-clamp current in *Limulus* ventral photoreceptors. Invest. Ophthalmol. Vis. Sci. **29:** 350.
23. INOUE, M. & J. E. BROWN. 1988. Cyclic GMP phosphodiesterase in *Limulus* ventral eye. Invest. Ophthalmol. Vis. Sci. **29:** 218.
24. BOLSOVER, S. R. & J. E. BROWN. 1982. Injection of guanosine and adenosine nucleotides into *Limulus* ventral photoreceptor cells. J. Physiol. **322:** 325-342.
25. NASI, E. & M. GOMEZ. 1991. Light-activated channels in scallop photoreceptors: Recordings from cell-attached and perfused excised patches. Biophys. J. **59:** 540a.
26. NOLTE, J. J., J. E. BROWN & T. G. SMITH. 1968. A hyperpolarizing component of the receptor potential in the median ocellus of *Limulus.* Science **162:** 677-679.
27. COSENS, D. & D. BRISCOE. 1972. A switch phenomenon in the compound eye of the white-eyed mutant of *Drosophila melanogaster.* J. Insect Physiol. **18:** 627-632.
28. HOCHSTEIN, S., B. MINKE & P. HILLMAN. 1973. Antagonistic components of the late receptor potential in the barnacle photoreceptor arising from different stages of the pigment process. J. Gen. Physiol. **62:** 105-128.
29. LISMAN, J. E., G. L. FAIN & P. M. O'DAY. 1982. Voltage-dependent conductances in *Limulus* ventral photoreceptors. J. Gen. Physiol. **79:** 187-209.
30. LEONARD, R. J. & J. E. LISMAN. 1981. Light modulates voltage-dependent potassium channels in *Limulus* ventral photoreceptors. Science **212:** 1273-1275.
31. CHINN, K. & J. E. LISMAN. 1984. Calcium mediates the light-induced decrease in maintained K^+ current in *Limulus* ventral photoreceptors. J. Gen. Physiol. **84:** 447-462.
32. NASI, E. 1991. Whole-cell clamp of dissociated photoreceptors from the eye of *Lima scabra.* J. Gen. Physiol. **97:** 35-54.
33. BOLSOVER, S. R. 1981. Calcium-dependent potassium current in barnacle photoreceptor. J. Gen. Physiol. **78:** 617-636.
34. O'DAY, P. M., J. E. LISMAN & M. GOLDRING. 1982. Functional significance of voltage-dependent conductances in *Limulus* ventral photoreceptors. J. Gen. Physiol. **79:** 211-232.
35. O'DAY, P. M. & M. P. GRAY-KELLER. 1989. Evidence for electrogenic Na^+-Ca^{2+}-exchange in *Limulus* ventral photoreceptors. J. Gen. Physiol. **93:** 473-392.
36. WALZ, B. & A. FEIN. 1983. Evidence for calcium-sequestering smooth ER in *Limulus* ventral photoreceptors. Invest. Opthalmol. Vis. Sci. **24**(suppl): 281.
37. BROWN, H. M., S. HAGIWARA, H. KOIKE & R. M. MEECH. 1970. Membrane properties of a barnacle photoreceptor examined by the voltage clamp technique. J. Physiol. **208:** 385-413.
38. BROWN, J. E. & J. E. LISMAN. 1975. Intracellular calcium modulates sensitivity and time scale in *Limulus* ventral photoreceptors. Nature **258:** 252-254.
39. BADER, C. R., F. BAUMANN & J. BERTRAND. 1976. The role of intracellular calcium and sodium in light-adaptation in the retina of the honeybee drone (*Apis mellofera*). J. Gen. Physiol. **67:** 475-491.
40. MINKE, B. & M. TSACOPOULOS. 1986. Light-induced sodium-dependent accumulation of calcium and potassium in the extracellular space of bee retina. Vis. Res. **26:** 679-690.
41. ARMON, E. & B. MINKE. 1983. Light activated electrogenic Na^+-Ca^{2+} exchange in fly photoreceptors: Modulation by Na^+/K^+ pump activity. Biophys. Struct. Mech. **9:** 349-357.
42. BROWN, J. E. & M. I. MOTE. 1974. Ionic dependence of reversal voltage of the light response in *Limulus* ventral photoreceptors. J. Gen. Physiol. **63:** 337-350.
43. WALOGA, G., J. E. BROWN & L. H. PINTO. 1975. Detection of changes in $[Ca^{2+}{}_i]$ from *Limulus* ventral photoreceptors using Arsenazo III. Biol. Bull. **149:** 449-450.
44. SMOLLEY, J. & H. M. BROWN. 1991. Adaptational changes in *Balanus* photoreceptors associated with alterations in Na^+-Ca^{2+} exchange. Biophys. J. **59:** 530a.
45. DECKERT, A. & H. STIEVE. 1991. Electrogenic Na^+-Ca^{2+} exchanger, the link between intracellular and extracellular calcium in the *Limulus* ventral photoreceptor. J. Physiol. **433:** 467-482.

46. O'DAY, P. M., M. P. GRAY-KELLER & M. LONERGAN. 1991. Physiological roles of Na^{+}-Ca^{2+}-exchange in *Limulus* ventral photoreceptors. J. Gen. Physiol. **97:** 369-391.
47. MULLINS. L. 1977. A mechanism of Na^{+}-Ca^{2+}-exchange. J. Gen. Physiol. **70:** 681-696.
48. CAMPBELL, D. L., W. R. GILES, K. ROBINSON & E. F. SHIBATA. 1988. Studies of the sodium-calcium exchanger in bull-frog atrial myocytes. J. Physiol. **403:** 317-340.
49. DIFRANCESCO, D. & D. NOBLE. 1985. A model of cardiac electrical activity incorporating ionic pumps and concentration changes. Phil. Trans. R. Soc. London B **307:** 353-398.

Sodium-Calcium Exchange in Nerve Terminals

Influence on Internal Ca^{2+} and Neurosecretion[a]

A. P. CARVALHO,[b] C. BANDEIRA-DUARTE,[b]
I. L. FERREIRA,[b] O. P. COUTINHO,[c] AND
C. M. CARVALHO[b]

[b]*Centro de Biologia Celular*
Departamento de Zoologia
Universidade de Coimbra
3049 Coimbra, Portugal

[c]*Departamento de Biologia*
Universidade do Minho
Braga, Portugal

INTRODUCTION

The Na^{+}-Ca^{2+} exchanger of the synaptic plasma membrane can mediate Ca^{2+} fluxes both ways. Normally, it transports Ca^{2+} out of the cells in exchange for extracellular Na^{+}, but if the transmembrane Na^{+} electrochemical gradient is reduced by decreasing the Na^{+} gradient or the membrane potential, the Na^{+}-Ca^{2+} exchanger operates to transport Ca^{2+} into the cells or isolated nerve terminals.[1-6] We have used this property of the Na^{+}-Ca^{2+} exchanger to raise the intrasynaptosomal Ca^{2+} concentration, $[Ca^{2+}]_i$, to study the role of this ion in neurosecretion.

An increase in $[Ca^{2+}]_i$ plays a crucial role in neurotransmitter release, but the Ca^{2+} requirement for neurosecretion is usually shown in experiments in which Ca^{2+} presumably enters the nerve terminal by voltage-dependent Ca^{2+} channels that open in response to depolarization.[7-9] When the membrane is depolarized two alterations occur, that is, membrane depolarization and Ca^{2+} entry, both of which have been shown to be closely associated with neurosecretion, although it is usually considered that depolarization is needed in neurosecretion merely to trigger the entry of Ca^{2+} through the Ca^{2+} channels.[7-10]

Clearly, it is of interest to test the Ca^{2+} hypothesis under conditions in which only $[Ca^{2+}]_i$ increases in the nerve terminal in the absence of depolarization. Calcium can be injected into cells ionophoretically; or, more recently, photosensitive Ca^{2+} chelators, the so-called "caged Ca^{2+} compounds," have been used to release pulses of Ca^{2+} in the cells to induce secretion, and such experiments have been carried out on synapses in culture[11] and on the neuromuscular junction of the small crayfish[12] with contradictory results. At least in the latter preparation the results were interpreted to indicate that the membrane potential plays a role in neurosecretion in addition to its classical role of opening Ca^{2+} channels to allow Ca^{2+} to enter.[12]

[a]This work was financed by grants from INIC and JNICT (Portuguese Research Councils).

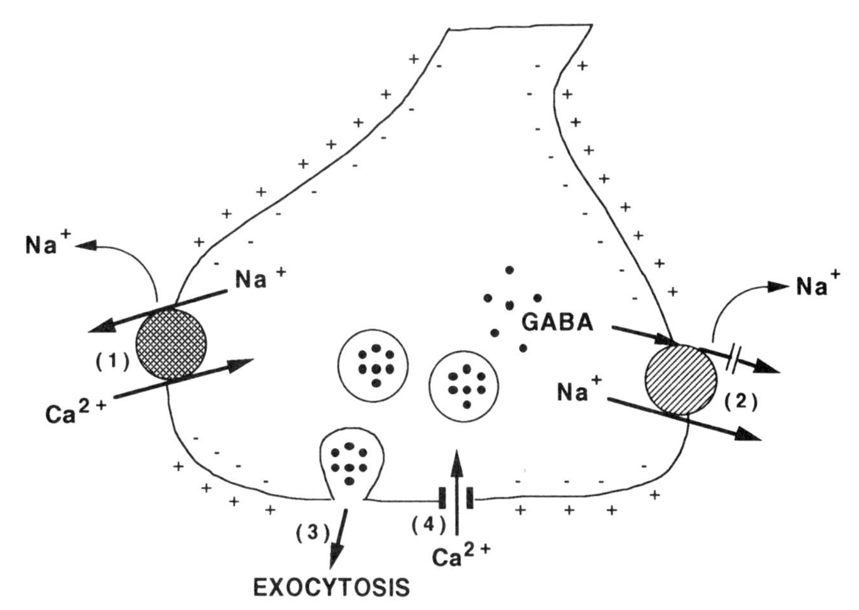

FIGURE 1. Removal of external Na^+ causes influx of Ca^{2+} by Na^+-Ca^{2+} exchange (**1**) into the nerve terminal (synaptosomes) and blocks release of GABA by the GABA carrier (**2**). Thus, in the absence of external Na^+, GABA is released only by exocytosis (**3**). These experimental conditions were used to load the synaptosomes with Ca^{2+} by Na^+-Ca^{2+} exchange. The entry of Ca^{2+} may also occur through the Ca^{2+} channels (**4**).

We have used the reversal of the Na^+-Ca^{2+} exchanger to raise the $[Ca^{2+}]_i$ of synaptosomes, in the absence of depolarization, beyond the level required for neurosecretion to test whether Ca^{2+} alone is sufficient to induce exocytosis of previously accumulated [^{3}H]GABA by rat brain synaptosomes. This permitted testing, subsequently, the role of membrane depolarization at different $[Ca^{2+}]_i$ values.

In FIGURE 1, we summarize several points of importance about the system we used; removal of Na^+ from the external media of the synaptosomes reverses the Na^+-Ca^{2+} exchanger[1,2,15,16] and blocks the release of [^{3}H]GABA through reversal of the GABA carrier.[13-15] Therefore, by removing Na^+ from the external medium of the synaptosomes, previously loaded with Na^+, we can load the synaptosomes with Ca^{2+} by Na^+-Ca^{2+} exchange and be certain that under these conditions the release of [^{3}H]GABA occurs only by exocytosis, that is, by a Ca^{2+}-dependent mechanism, since external Na^+ is required for [^{3}H]GABA release through the carrier.[13-16] Thus, this experimental set-up provides an ideal system to test the Ca^{2+} requirement of exocytosis and to study the role played by depolarization in this mechanism of neurotransmitter release. Our results show that it is possible to raise the $[Ca^{2+}]_i$ of synaptosomes by Na^+-Ca^{2+} exchange, without depolarization, above the value required for exocytosis without inducing release, but release will occur if the membrane is subsequently depolarized without further increasing the $[Ca^{2+}]_i$. Membrane depolarization alone, in the absence of Ca^{2+}, does not induce [^{3}H]GABA release under the conditions of our experiments in which the GABA carrier is blocked. Therefore, it appears that membrane depolarization is essential for Ca^{2+}-dependent release of [^{3}H]GABA from synaptosomes and that depolarization alone is insufficient for exocytosis.

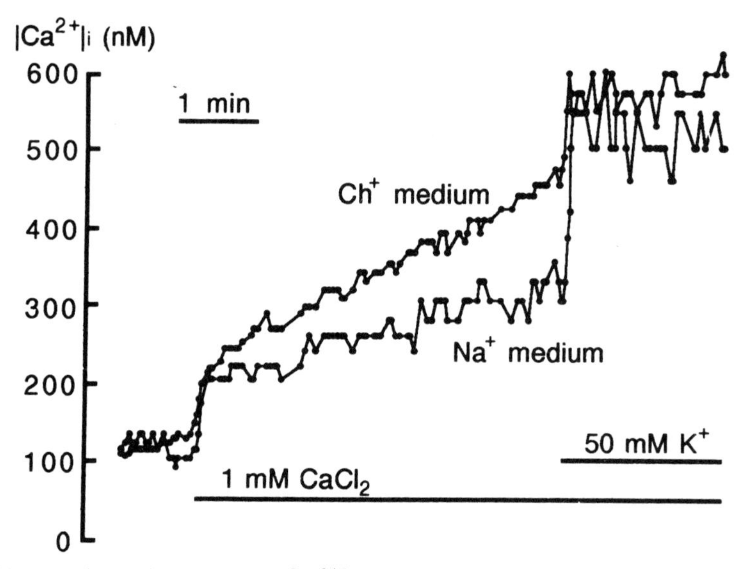

FIGURE 2. Changes in synaptosomal $[Ca^{2+}]_i$, as measured by Indo-1 fluorescence. Synaptosomes were loaded with Indo-1 in Na^+ medium without Ca^{2+}, as described previously,[21] and were transferred to incubation media rich in either Na^+ or choline, containing (in mM): 1 $MgCl_2$, 10 glucose, 10 HEPES-Tris, pH 7.4, in addition to 133 NaCl or choline chloride (Ch^+), as indicated in the traces. After 2 min, 1 mM $CaCl_2$ was added, and $[Ca^{2+}]_i$ changes were followed for 4 more minutes, after which K^+ (50 mM) was added and the additional changes in $[Ca^{2+}]_i$ were recorded. Representative result of similar experiments performed in four different preparations.

CALCIUM LOADING OF SYNAPTOSOMES BY Na^+-Ca^{2+} EXCHANGE AND BY K^+ DEPOLARIZATION

Changes in intrasynaptosomal $[Ca^{2+}]_i$ in Na^+-loaded synaptosomes were measured by Indo-1 fluorescence in Na^+ or choline media. FIGURE 2 shows that in Na^+ medium the $[Ca^{2+}]_i$ rises from about 100 nM to 300 nM within 3 min, whereas in a choline medium the $[Ca^{2+}]_i$ invariably reaches values of 400-500 nM before K^+ depolarization, and the subsequent addition of 50 mM K^+ raises the $[Ca^{2+}]_i$ to about 600 nM in the case of both Na^+ and Ch^+ media. Thus, more Ca^{2+} enters due to depolarization in a Na^+ medium than in a Ch^+ medium in which a larger amount of Ca^{2+} had entered by Na^+-Ca^{2+} exchange. It appears, therefore, that the amount of Ca^{2+} entering by the voltage-dependent Ca^{2+} channels is influenced by the level of previous Ca^{2+} loading by Na^+-Ca^{2+} exchange.

To test the idea that the $[Ca^{2+}]_i$ accumulated by Na^+-Ca^{2+} exchange influences the Ca^{2+} influx due to depolarization, we allowed the synaptosomes to take up Ca^{2+} by Na^+-Ca^{2+} exchange in a Ch^+ medium containing 1.0 mM Ca^{2+} during either 30 sec or 3 min, and then added 50 mM KCl. This allowed us to test, under identical experimental conditions, the role of two levels of Ca^{2+} accumulated by Na^+-Ca^{2+} exchange on the voltage-dependent entry of Ca^{2+}. The results depicted in FIGURE 3 show that at the end of 30 sec the $[Ca^{2+}]_i$ was about 450 nM as compared to about 600 nM at the end of 3 min, and that K^+ depolarization raised the $[Ca^{2+}]$ by about 400 nM and 200 nM, respectively, confirming that the higher the $[Ca^{2+}]_i$ the lower the influx of Ca^{2+} due to K^+ depolarization. These results suggest inactivation of the Ca^{2+}

channels by Ca^{2+} entering by means of Na^+-Ca^{2+} exchange. Similar inactivation of the Ca^{2+} channels was also observed due to predepolarization,[16] as has also been reported previously.[17,18]

A good correlation was obtained between $[Ca^{2+}]_i$ before depolarization and K^+-stimulated $[Ca^{2+}]_i$ rise, but the decreased Ca^{2+} influx at the higher $[Ca^{2+}]_i$ could partly be due to a decreased electrochemical driving force for Ca^{2+} entry after Ca^{2+} loading, increased rate of Ca^{2+} efflux, or limited capacity to take up Ca^{2+}.[17] In FIGURE 4, we summarize the results showing inactivation by $[Ca^{2+}]_i$ of the Ca^{2+} influx due to depolarization obtained both in a choline and Na^+ media. The increase in $[Ca^{2+}]_i$ due to K^+ depolarization in Na^+ medium is always higher than that observed in choline medium, probably because of the lower $[Ca^{2+}]_i$ observed in Na^+ than in Ch^+ just before depolarization. It should be noted that the experiments reported in FIGURES 3 and 4 gave initial $[Ca^{2+}]_i$ values higher by about 100 nM than those reported in FIGURES 2 and 5 due to different conditions for loading the synaptosomes with Indo-1. We have found that high levels of loading of the synaptosomes with Indo-1 give lower $[Ca^{2+}]_i$.

We have found earlier that the Na^+-Ca^{2+} exchanger studied in synaptic plasma membranes has a K_D of about 18 μM Ca^{2+},[5] which is similar to that reported by Gill *et al.*,[27] but recently we have obtained values as low as 0.2 μM. Therefore, it would be expected that the Na^+-Ca^{2+} exchanger would be efficient in transporting Ca^{2+} in the μM range. This was confirmed in experiments in which it was shown that when synaptosomes were placed for 3 min in a medium containing 11 μM Ca^{2+}, the $[Ca^{2+}]_i$ rose from about 100 nM to about 350 nM, which is only slightly lower than the value attained when the synaptosomes were placed in 1.0 mM Ca^{2+} (FIG. 5). Subsequent K^+ depolarization raised the $[Ca^{2+}]_i$ by the same amount (by about 100 nM) whether the

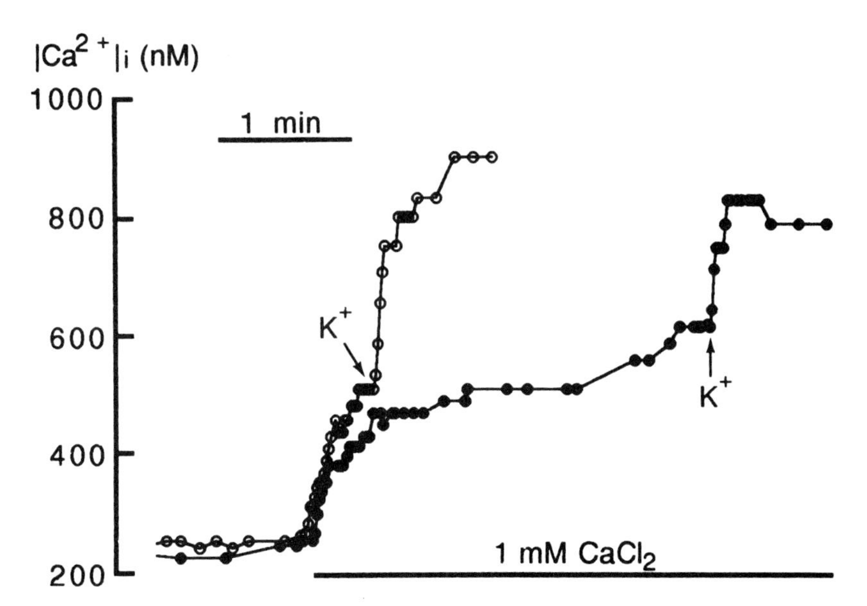

FIGURE 3. K^+-stimulated $[Ca^{2+}]_i$ changes determined after loading the synaptosomes with Ca^{2+} by Na^+-Ca^{2+} exchange in choline medium containing 1 mM $CaCl_2$ during either 30 sec or 3 min. At the arrows, 30 sec or 3 min after adding Ca^{2+} to the choline medium, the synaptosomes were depolarized with 50 mM KCl and the $[Ca^{2+}]_i$ was measured by Indo-1 fluorescence.[21]

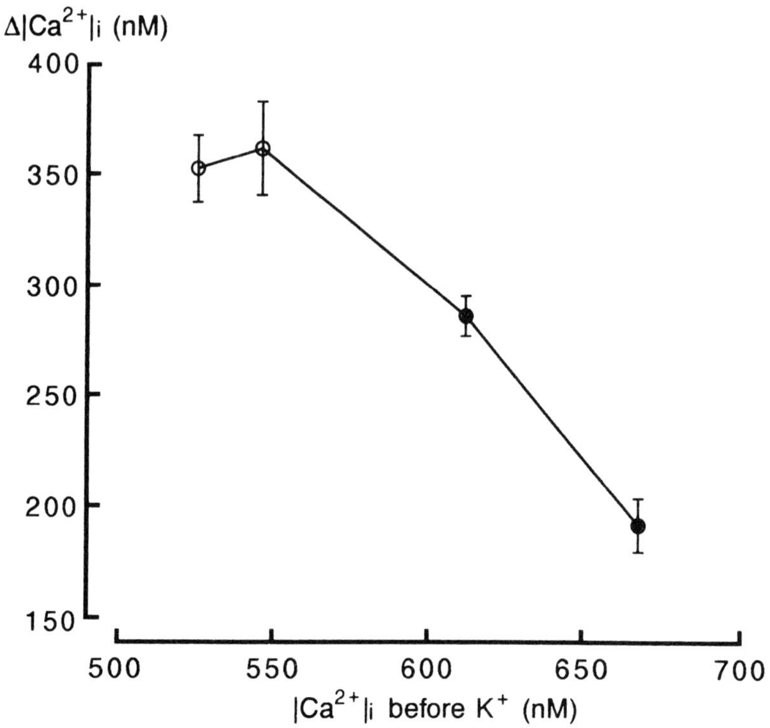

FIGURE 4. Changes in $[Ca^{2+}]_i$ due to K^+ depolarization as a function of $[Ca^{2+}]_i$ attained by Na^+-Ca^{2+} exchange before K^+ depolarization. The data are means ± SE of 3-5 experiments performed in different preparations, as illustrated in FIGURE 3. Maximal $[Ca^{2+}]_i$ changes due to K^+ depolarization were calculated and are represented as a function of the level of $[Ca^{2+}]_i$ before K^+ addition. Empty or closed circles refer to data obtained in Na^+ or Ch^+ media, respectively. (NOTE: These experiments (FIGS. 3 and 4) were performed in synaptosomal preparations that had a lower level of loading with Indo-1, which gave higher $[Ca^{2+}]_i$ than those shown in FIGURES 2, 5, and 8).

synaptosomes were in 11 μM or 1.0 mM Ca^{2+}. It should be noted that the level of Ca^{2+} accumulated by Na^+-Ca^{2+} exchange in 1.0 mM Ca^{2+}, during 3 min, is inhibitory (FIGS. 3 and 4), so that the fact that the $[Ca^{2+}]_i$ increased by the same extent at the two external Ca^{2+} concentrations is a coincidence resulting from this inhibition and from the lower Ca^{2+} gradient out → in at 11 μM external Ca^{2+}.

In FIGURE 6, it is shown that the Ca^{2+} loaded by Na^+-Ca^{2+} exchange in a choline medium can be released by a pulse of Na^+ added to a medium superfusing the synaptosomes after Ca^{2+} loading. Similar results have been reported using a different technique to measure Ca^{2+} efflux through the exchanger.[26] This indicates that we are in fact dealing with a fraction of Ca^{2+} that is handled reversibly by the exchanger across the membrane.

RELATIONSHIP BETWEEN CA^{2+} UPTAKE AND $[^3H]$GABA RELEASE

It can be estimated that a large part of the Ca^{2+} taken up by Na^+-Ca^{2+} exchange is buffered since only modest increases in $[Ca^{2+}]_i$ were observed for relatively large uptakes of ^{45}Ca. Thus, we observed that the uptake of about 10 nmol Ca^{2+}/mg protein produced an increase in $[Ca^{2+}]_i$ of only about 250 nM. If no Ca^{2+} buffering occurred, we would expect a total increase, $(\Delta[Ca^{2+}]_T)$, in $[Ca^{2+}]_i$ of 2.5 mM, assuming that the internal volume of synaptosomes is 4 μl/mg protein,[19] that is, the buffering capacity $(\beta = \Delta[Ca^{2+}]_T/\Delta[Ca^{2+}]_i)$ can be calculated to be about 10,000. This value is slightly lower than that reported earlier for similar studies.[2] When calculations are done for the buffering capacity of the Ca^{2+} entering by the voltage-dependent Ca^{2+} channels, we obtain a similar value suggesting that the Ca^{2+} entering by the two mechanisms is handled similarly by the buffering systems of the synaptosomes.

It became of interest to test whether raising the $[Ca^{2+}]_i$ by Na^+-Ca^{2+} exchange to levels attained by depolarization, and which normally are associated with neurosecretion, was sufficient to produce release of [^{3}H]GABA. In FIGURE 7A, it is shown that maximal loading of the synaptosomes with $^{45}Ca^{2+}$ by Na^+-Ca^{2+} exchange in choline media containing increasing concentrations of Ca^{2+} from 10^{-7} M to 10^{-3} M did not causes release of [^{3}H]GABA. Under the conditions of these experiments the $[Ca^{2+}]_i$ rose from about 150 nM to about 450 nM when 1.0 mM Ca^{2+} was present in the external medium, and 11 nmol Ca^{2+}/mg protein were taken up (FIG. 7A). The K_D value for Ca^{2+} calculated from the data presented in FIGURE 7 is 0.35 μM, which is close to the value of 0.2 μM obtained by us in earlier experiments carried out with synaptic plasma membranes, but it is lower than that reported for experiments carried out with synaptosomes.[20]

In FIGURE 7B we see that K^+ depolarization after ^{45}Ca uptake by Na^+-Ca^{2+} exchange in a choline medium containing Ca^{2+} concentrations above 10^{-6}M, increased the ^{45}Ca uptake from about 11 nmol Ca/mg protein to 13 nmol Ca^{2+}/mg protein, and

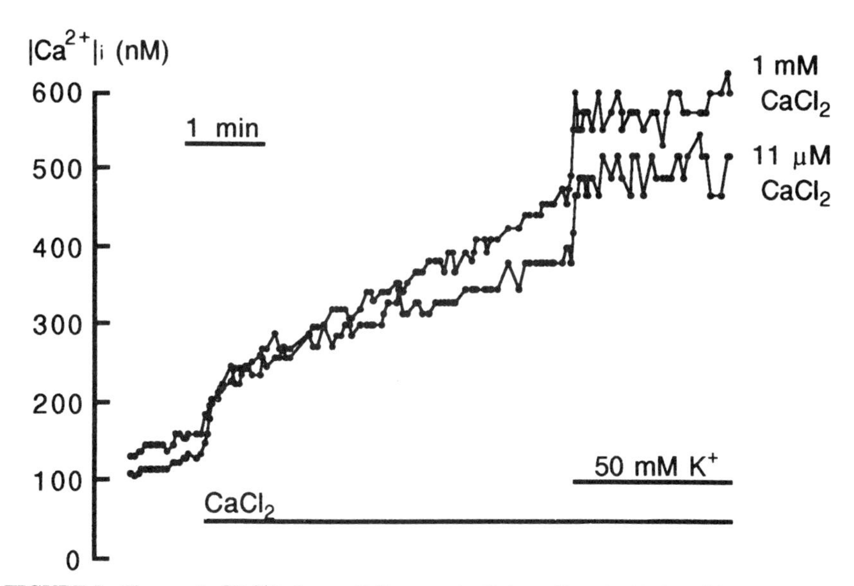

FIGURE 5. Changes in $[Ca^{2+}]_i$ due to Ca^{2+} entry in Ch^+ medium by Na^+-Ca^{2+} exchange or by K^+ depolarization, for two different external free $[Ca^{2+}]$ (1 mM and 11 μM) controlled with EGTA. Experimental details are described in the legend of FIGURE 2. Representative of 3-4 experiments performed in different preparations.

this increase in ^{45}Ca uptake was accompanied by [^{3}H]GABA release. One might argue that this additional Ca^{2+} is essential for Ca^{2+}-dependent release to occur. However, we were able to devise experimental conditions that permitted raising the $[Ca^{2+}]_i$ by Na^+-Ca^{2+} exchange in choline medium to values higher than those observed under conditions of depolarization which produce release of [^{3}H]GABA. The results of these experiments showed that no release was ever observed when the $[Ca^{2+}]_i$ was raised only by Na^+-Ca^{2+} exchange, without depolarization (FIG. 8).

As observed in FIGURE 7A, about 10^{-5}M external Ca^{2+} already gave maximal ^{45}Ca uptake by Na^+-Ca^{2+} exchange in a choline medium, and this uptake of Ca^{2+} gave a $[Ca^{2+}]_i$ of about 350 nM, whereas in equivalent experiments carried out in a Na^+ medium, the $[Ca^{2+}]_i$ did not rise above 200 nM (FIG. 8). Neither of these conditions induced exocytosis of [^{3}H]GABA from the synaptosomes. However, subsequent depo-

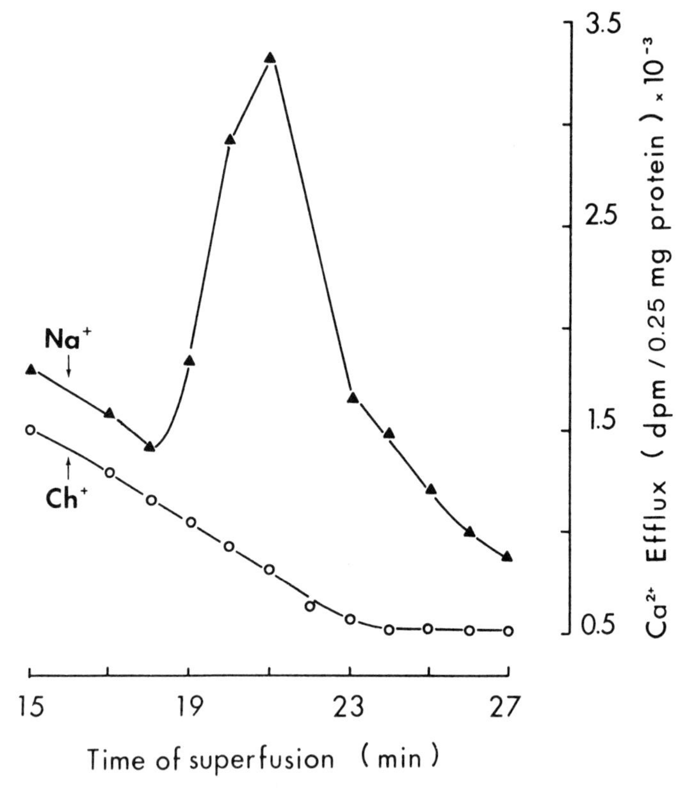

FIGURE 6. Na^+-induced Ca^{2+} efflux from synaptosomes preloaded with $^{45}Ca^{2+}$, as determined by superfusion. Synaptosomes (0.25 mg protein) were collected on the filter of a superfusion system[16] and were washed with Na^+ medium. A pulse of 1 min of K^+-depolarization medium (50 mM KCl + 83 mM choline chloride) containing 1 mM $^{45}CaCl_2$ was applied to load synaptosomes with $^{45}CaCl_2$ in conditions similar to those used to induce [^{3}H]GABA release. The synaptosomes were subsequently perfused with Ch^+ medium for 15 min to remove external $^{45}Ca^{2+}$, after which Na^+-induced Ca^{2+} release was initiated by changing to a Na^+-rich medium, as indicated by the upper arrow. In a parallel chamber of the superfusion system, another sample of synaptosomes continued to be perfused with Ch^+ medium and no Ca^{2+} efflux occurred (lower curve).

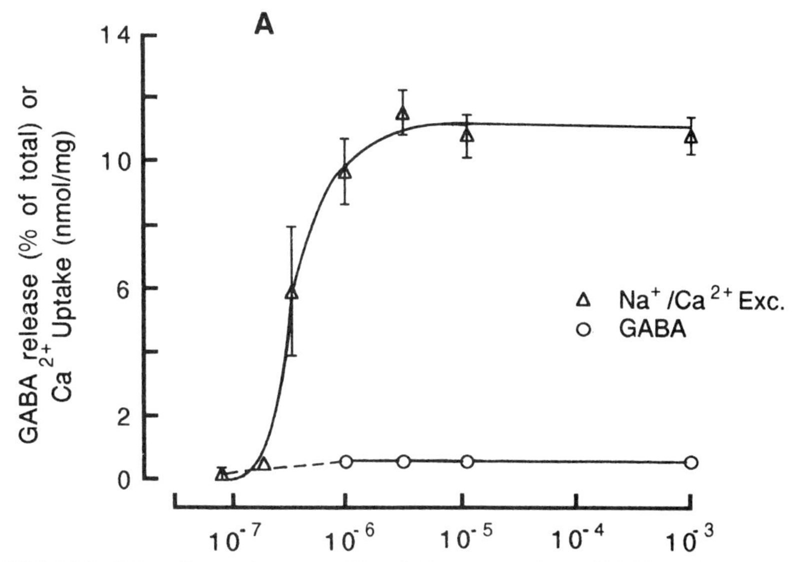

FIGURE 7A. Effect of increasing external free-Ca^{2+} concentration on $^{45}Ca^{2+}$ uptake and GABA release in synaptosomes incubated in choline chloride media. $^{45}Ca^{2+}$ uptake due to Na^+-Ca^{2+} exchange was determined in Na^+-free media containing (in mM): 133 choline chloride, 1 $MgCl_2$, 10 glucose, 10 HEPES-Tris, pH 7.4, 1 EGTA, and increasing $CaCl_2$ concentrations to give free $[Ca^{2+}]$ as indicated in the abscissa. The reaction was initiated by diluting synaptosomes 20-fold in the uptake media and was stopped after 4 min of incubation at 30°C, as described previously.[21] The GABA release experiments were performed as described previously,[16] after loading the synaptosomes with 0.5 μM [^{3}H]GABA for 15 min, at 30°C. The [^{3}H]GABA released over the basal, in ChCl media containing increasing $[Ca^{2+}]$, similar to those used for $^{45}Ca^{2+}$ uptake, was determined as percentage of total GABA for periods of 4 min of perfusion. The data are presented as means ± SE for 4-5 experiments performed in different preparations.

larization with 50 mM K^+ produced [^{3}H]GABA release that was Ca^{2+} dependent, although the rise in $[Ca^{2+}]_i$ due to K^+ depolarization in the Na^+ medium never rose to the value obtained by Na^+-Ca^{2+} exchange alone in choline without depolarization, which does not cause [^{3}H]GABA release (FIG. 8). Therefore, it seems that the rise in $[Ca^{2+}]_i$ by Na^+-Ca^{2+} exchange to the levels attained by K^+ depolarization is not sufficient to induce Ca^{2+}-dependent release of [^{3}H]GABA.

DISCUSSION AND CONCLUSIONS

The above observations are at variance with the calcium hypothesis of neurotransmitter release in which it is proposed that release is triggered by membrane depolarization whose function is to open Ca^{2+} channels to allow the $[Ca^{2+}]_i$ to rise sufficiently to induce exocytosis.[7-9] On the other hand, our observations, while confirming the requirement for a rise in $[Ca^{2+}]_i$ for exocytosis, suggest that membrane depolarization is also required without being necessarily linked to the influx of Ca^{2+}. In fact, we have observed previously[15] that K^+ depolarization of synaptosomes can occur without Ca^{2+} influx due to depolarization provided that the Na^+-Ca^{2+} exchange occurs first in a

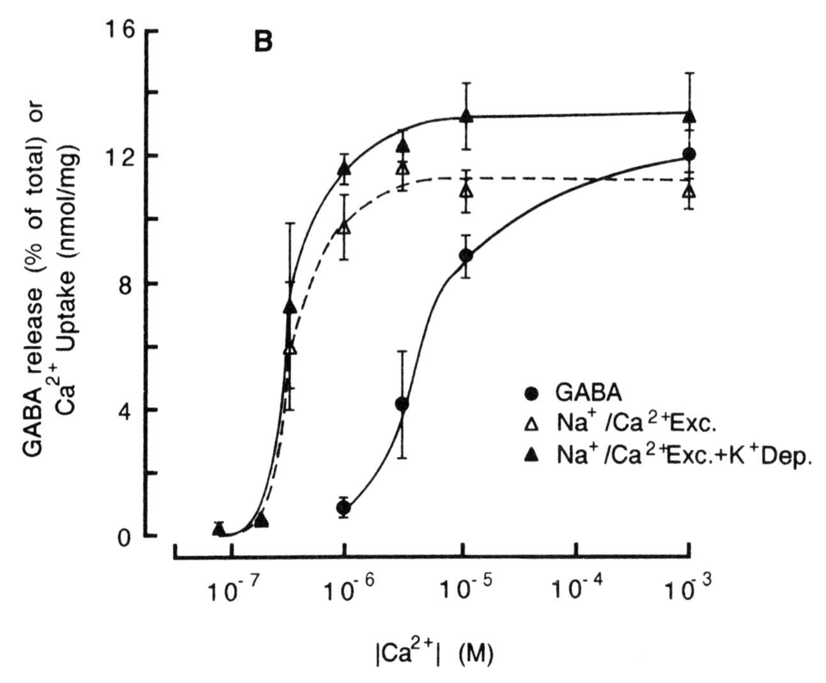

FIGURE 7B. Relationship between $^{45}Ca^{2+}$ uptake due to Na^+-Ca^{2+} exchange plus K^+ depolarization, and [^{3}H]GABA release, at increasing free external [Ca^{2+}]. The experimental conditions were similar to those described in Part A, except that, after allowing Ca^{2+} uptake due to Na^+-Ca^{2+} exchange for a period of 4 min, K^+ depolarization (by adding 50 mM KCl) was induced and both $^{45}Ca^{2+}$ uptake (▲) (1 min after K^+ addition) and [^{3}H]GABA release (●) (for the next 4 min of perfusion) were determined. The curve showing Na^+-Ca^{2+} exchange alone (△) is also shown for comparison. As shown in Part A, Na^+-Ca^{2+} exchange alone does not induce [^{3}H]GABA release. The results are means ± SE for 5-10 experiments in different preparations.

choline medium containing 5 mM K^+. These conditions permit maximal loading of the synaptosomes with Ca^{2+}, but do not induce [^{3}H]GABA release. Subsequent K^+ depolarization induces release without further Ca^{2+} entry.[15]

It might be argued that the Ca^{2+} entering by Na^+-Ca^{2+} exchange does not contribute as efficiently to the triggering of [^{3}H]GABA release as the Ca^{2+} entering through the Ca^{2+} channels. Recently, McMahon and Nicholls[22] concluded that glutamate exocytosis is evoked by localized Ca^{2+} entering through voltage-dependent Ca^{2+} channels and that nonlocalized Ca^{2+} entry with ionomycin is inefficient. However, our results show that Ca^{2+} that enters by Na^+-Ca^{2+} exchange during a preincubation of the synaptosomes in a choline medium containing 1.0 mM Ca^{2+} facilitates [^{3}H]GABA release when the synaptosomes are subsequently K^+ depolarized in media containing concentrations of Ca^{2+} between 10^{-7}M and 10^{-5}M (results not shown). It should also be noted that when the synaptosomes were Ca^{2+} loaded by Na^+-Ca^{2+} exchange before K^+ depolarization, we never observed [^{3}H]GABA release unless external Ca^{2+} was present during depolarization. This suggests that Ca^{2+} either acts externally or the small additional Ca^{2+} influx through the Ca^{2+} channels is essential for [^{3}H]GABA release, probably because it enters near the active zones.[8]

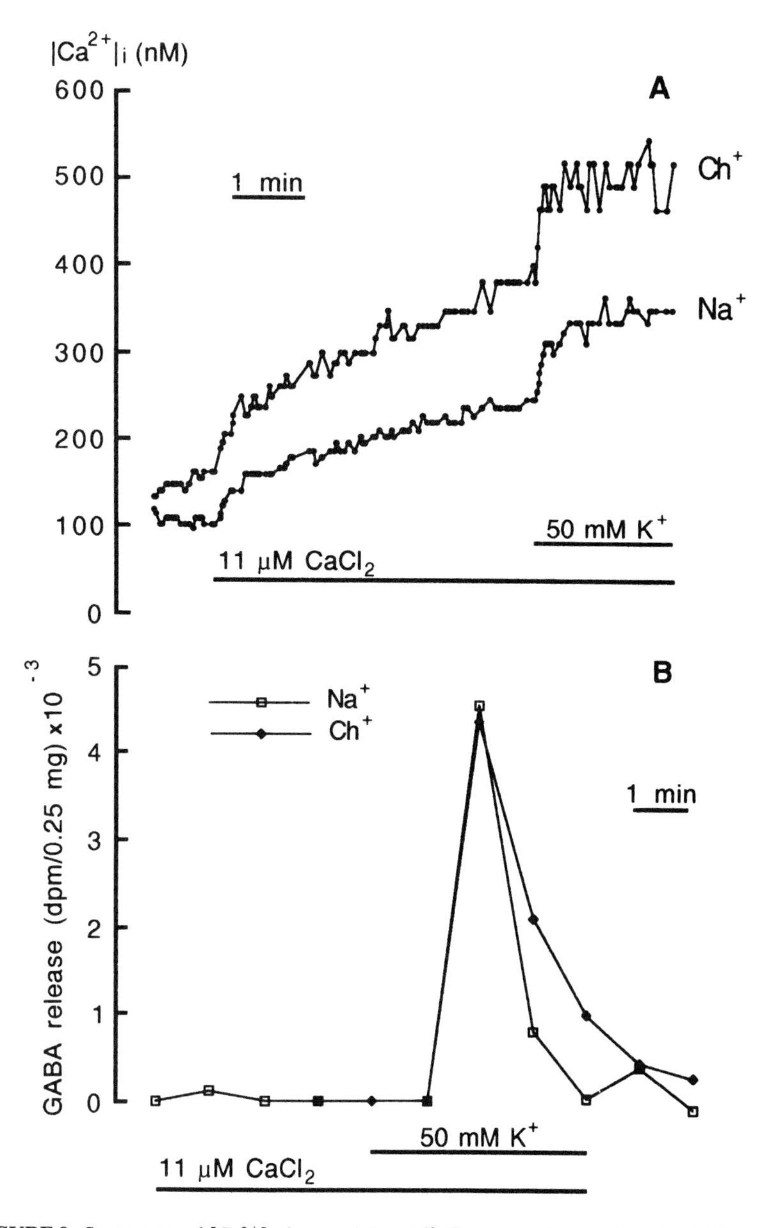

FIGURE 8. Synaptosomal $[Ca^{2+}]_i$ changes (**A**) and $[^3H]$GABA release (**B**) due to Ca^{2+} entry by Na^+-Ca^{2+} exchange and K^+ depolarization in Ch^+ or Na^+ media, at low external free $[Ca^{2+}]$ (11 μM, controlled with 1.0 mM EGTA). Experimental details are similar to those described for FIGURES 2 and 7. Individual representative traces for one of four independent experiments are shown. The Ca^{2+}-independent release of $[^3H]$GABA was subtracted when Na^+ was present.

The Na^+-Ca^{2+} exchange mechanism was used in this study essentially as a technique to load synaptosomes with Ca^{2+} without depolarizing the membrane. The intrasynaptosomal [Ca^{2+}] was increased by reversal of the Na^+-Ca^{2+} exchange, and the changes in $[Ca^{2+}]_i$ were related to the Ca^{2+}-dependent release of [^{3}H]GABA to see whether Ca^{2+} entering through the Na^+-Ca^{2+} exchanger can trigger exocytosis. The $[Ca^{2+}]_i$ could be raised by Na^+-Ca^{2+} exchange to values that triggered exocytosis provided that depolarization also took place, but we never observed [^{3}H]GABA release due to raising the $[Ca^{2+}]_i$ by Na^+-Ca^{2+} exchange to the same values without depolarization. It appears that raising the $[Ca^{2+}]_i$ alone is not sufficient for exocytosis to occur, and our results show that depolarization probably plays a role in the mechanism of release of [^{3}H]GABA in addition to opening the Ca^{2+} channels that allow Ca^{2+} in. However, it has been possible to induce release in several types of electropermeabilized cells by directly changing the [Ca^{2+}] in contact with the release mechanism.[28]

In experiments using a similar approach to load the synaptosomes with Ca^{2+} in Na^+-free media, Arias and Tapia[23] reported some Ca^{2+}-dependent release of acetylcholine and GABA when they changed the superfusion medium from a Na^+-rich to a sucrose medium. We find that, in the case of GABA, release under these conditions is mostly through the GABA carrier due to the presence of some Na^+ on the outside immediately after shifting to the sucrose solution and the favorable Na^+ electrochemical gradient to drive GABA out.

It should be pointed out that recently[24] it was reported that catecholamines can be released from primary cultures of chromaffin cells exposed to low external Na^+ (5 mM), and that this effect could be inhibited by removing Ca^{2+} from the external medium. These results would suggest that in this system a mere increase in Ca^{2+} influx by Na^+-Ca^{2+} exchange is sufficient to support excytosis without depolarization. This could also be of physiological significance because it shows that if the electrochemical gradient of Na^+ is sufficiently reversed, exocytosis linked to Ca^{2+} entering by Na^+-Ca^{2+} exchange may occur in these cells in accordance with the Ca^{2+} hypothesis of neurotransmitter release. Taglialatela *et al.*[25] have also related selective inhibition of Na^+-Ca^{2+} exchange in synaptosomes by DMB (2′,4′-dimethylbenzamil amiloride) to inhibition of dopamine release, which would suggest a role for Na^+-Ca^{2+} exchange in dopamine release. However, at least in the case of [^{3}H]GABA release, depolarization, in additional to the level of $[Ca^{2+}]_i$, is essential for release to occur. Thus, inhibitors of Na^+-Ca^{2+} exchange are also expected to modulate the release of GABA through their effect on the $[Ca^{2+}]_i$, but this modulation would also require membrane depolarization, which is essential for release.

REFERENCES

1. Blaustein, M. P. & M. T. Nelson. 1982. Sodium-calcium exchange: Its role in the regulation of cell calcium. *In* Membrane Transport of Calcium. Ernesto Carafoli, Ed.: 217-236. Academic Press. New York.
2. Nachshen, D. A., S. Sanchez-Armass & A. M. Weinstein. 1986. The regulation of cytosolic calcium in rat brain synaptosomes by sodium-dependent calcium efflux. J. Physiol. **381:** 17-28.
3. Nachshen, D. A. & S. Kongsamut. 1989. "Slow" K^+-stimulated Ca^{2+} influx is mediated by Na^+-Ca^{2+} exchange: A pharmacological study. Biochim. Biophys. Acta **979:** 305-310.
4. DiPolo, R. & L. Beaugé. 1988. Ca^{2+} transport in nerve fibers. Biochim. Biophys. Acta **947:** 549-569.
5. Carvalho, A. P., O. P. Coutinho, V. M. C. Madeira & C. A. M. Carvalho. 1984. Calcium transport in synaptosomes and synaptic plasma membrane vesicles. *In* Biomem-

branes. Dynamics and Biology. R. M. Burton & F. C. Guerra, Eds.: 291-316. Plenum Press. New York.
6. COUTINHO, O. P., C. A. M. CARVALHO & A. P. CARVALHO. 1984. Calcium uptake related to K^+-depolarization and Na^+-Ca^{2+} exchange in sheep brain synaptosomes. Brain Res. **290:** 261-271.
7. KATZ, B. & R. MILEDI. 1967. The timing of calcium action during neuromuscular transmission. J. Physiol. (London) **189:** 535-544.
8. SMITH, S. U. & G. J. AUGUSTINE. 1988. Calcium ions, active zones and synaptic transmitter release. Trends Neurosci. **11:** 458-464.
9. ZUCKER, R. S. & L. LANDÒ. 1986. Mechanism of transmitter release: Voltage hypothesis and calcium hypothesis. Science **231:** 574-579.
10. DRAPEAU, P. & M. P. BLAUSTEIN. 1983. Initial release of [^{3}H]dopamine from rat striatal synaptosomes: Correlation with calcium entry. J. Neurosci. **3:** 703-713.
11. ZUCKER, R. S. & P. G. HAYDEN. 1988. Membrane potential has no direct role in evoking neurotransmitter release. Nature **335:** 360-362.
12. HOCHNER, B., H. PARNAS & I. PARNAS. 1989. Membrane depolarization evokes neurotransmitter release in the absence of calcium entry. Nature **342:** 433-435.
13. NELSON, M. T. & M. P. BLAUSTEIN. 1982. GABA efflux from synaptosomes: Effects of membrane potential, and external GABA and cations. J. Membr. Biol. **69:** 213-223.
14. CARVALHO, C. A. M., S. V. SANTOS & A. P. CARVALHO. 1986. Aminobutyric acid release from synaptosomes as influenced by Ca^{2+} and Ca^{2+} channel blockers. Eur. J. Pharmacol. **131:** 1-12.
15. SANTOS, M. S., P. P. GONÇALVES & A. P. CARVALHO. 1991. Release of [^{3}H]aminobutyric acid from synaptosomes: Effect of external cations and of ouabain. Brain Res. **547:** 135-141.
16. CARVALHO, C. A. M., C. BANDEIRA-DUARTE, I. L. FERREIRA & A. P. CARVALHO. 1991. Regulation of carrier-mediated and exocytotic release of ^{3}H-GABA in rat brain synaptosomes. Neurochem. Res. **16:** 763-772.
17. NACHSHEN, D. A. 1985. The early time course of potassium-stimulated calcium uptake in synaptic nerve terminals isolated from rat brain. J. Physiol. **361:** 251-268.
18. SUSKIW, J. B., M. E. O'LEARY, M. M. MURAWSKY & T. WANG. 1986. Presynaptic calcium channels in rat cortical synaptosomes: Fast kinetics of phasic calcium influx, channel inactivation, and relationship to nitrendipine receptors. J. Neurosci. **6:** 1349-1357.
19. BLAUSTEIN, M. P. & C. J. OSBORN. 1975. The influence of sodium on calcium fluxes in pinched off presynaptic nerve terminals *in vitro.* J. Physiol. **247:** 657-686.
20. MCGRAW, F. CATHERINE, D. A. NACHSHEN & M. P. BLAUSTEIN. 1982. Calcium movement and regulation in presynaptic nerve terminals. Calcium Cell Funct. **11:** 81-110.
21. BANDEIRA-DUARTE, C., C. A. M. CARVALHO, E. J. CRAGOE, JR. & A. P. CARVALHO. 1990. Influence of isolation media on synaptosomal properties: Intracellular pH, pCa, and Ca^{2+} uptake. Neurochem. Res. **15:** 313-320.
22. MCMAHON, H. T. & D. G. NICHOLLS. 1991. Transmitter glutamate release from isolated nerve terminals: Evidence for biphasic release and triggering by localized Ca^{2+}. J. Neurochem. **56:** 86-94.
23. ARIAS, C. & R. TAPIA. 1986. Differential calcium dependence of γ-aminobutyric acid and acetylcholine release in mouse brain synaptosomes. J. Neurochem. **47:** 396-404.
24. TÖRÖK, T. L. & D. A. POWIS. 1990. Catecholamine release from bovine chromaffin cells: The role of sodium-calcium exchange in ouabain-evoked release. Exp. Physiol. **75:** 573-586.
25. TAGLIALATELA, M., L. M. T. CANZONIERO, E. J. CRAGOE, JR., G. DI RENZO & L. ANNUNZIATO. 1990. Na^+-Ca^{2+} exchange activity in central nerve endings. II. Relationship between pharmacological blockade by amiloride analogues and dopamine release from tuberoinfundibular hypothalamic neurons. Mol. Parmacol. **38:** 393-400.
26. SANCHEZ-ARMASS & M. P. BLAUSTEIN. 1987. Role of sodium-calcium exchange in regulation of intracellular calcium in nerve terminals. Am. J. Physiol. **252:** C595-C603.
27. GIL, D., E. F. GROLLMAN & L. D. KOHN. 1981. Calcium transport mechanisms in membrane vesicles from guinea pig brain synaptosomes. J. Biol. Chem. **256:** 184-192.
28. KNIGHT, D. E., H. VON GRAFENSTEIN & C. M. ATHAYDE. 1989. Calcium-dependent and calcium-independent exocytosis. Trends Neurosci. **12:** 451-458.

Neuron-Specific and State-Specific Differences in Calcium Regulation

Their Role in the Development of Neuronal Architecture[a]

LINDA R. MILLS

Playfair Neuroscience Unit and
Department of Physiology
University of Toronto and the Toronto Hospital
Toronto, Ontario, Canada, M5T 2S8

INTRODUCTION

Intracellular calcium is a key regulatory molecule in the nervous system. In addition to its role in functional neuronal circuitry (e.g., the regulation of excitability[1] and the release of neurotransmitter[2,3]) intracellular calcium plays a central role in the regulation of neurite outgrowth, specific growth cone behaviors,[4-12] and even the selective pruning of existing neural architecture.[13] Additionally, inappropriate levels of intracellular calcium can initiate pathophysiological changes that result in degeneration and cell death.[14-18]

Given this multiplicity of roles, it is hardly surprising that neurons possess a complex system for regulating intracellular calcium. Calcium regulation is achieved through the integration of multiple components that control different aspects of calcium influx, efflux, and sequestration. These include voltage-dependent plasma membrane calcium channels,[19] and ATP-dependent calcium pump,[20] the Na^{+}-Ca^{2+} exchanger,[21] and a buffering system that includes both sequestration organelles, for example, the endoplasmic reticulum,[22] mitochondria,[23] calcisomes,[24] and binding proteins.[25] Together these mechanisms maintain calcium concentration at normal intracellular levels (i.e., about 10-7M), despite a 10,000-fold concentration gradient across the cell membrane. Additionally, they determine the temporal and spatial characteristics of calcium signals (i.e., those changes in intracellular calcium that occur in response to extrinsic and intrinsic stimuli).

Although there is an extensive literature on neuronal calcium regulation[20,26,27] and the relative importance of different regulatory mechanisms under various conditions, rather less attention has been paid to the idea that calcium homeostasis may vary significantly among individual neurons. Similarly, while much discussion has focused on the potential role of a disruption in calcium homeostasis in aging and chronic neurodegenerative diseases,[15,28-30] the idea that changes in calcium homeostasis play an important role during the course of normal development has been largely ignored (but see Kater *et al.*[31]).

[a]This work was supported by National Institutes of Health Grants NS24683 and NS15350 to S. B. Kater and by the Toronto Hospital.

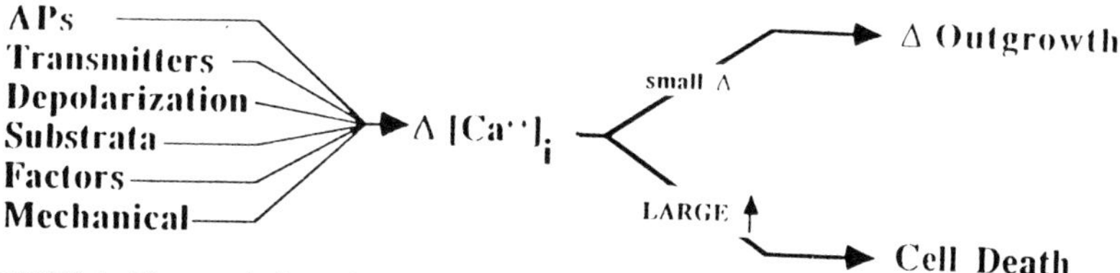

FIGURE 1. The regulation of growth cone behavior. A wide variety of stimuli ranging from action potentials to mechanical stimulation can alter intracellular calcium levels. Because the growth cone behavior is highly sensitive to changes in calcium levels, these stimuli acting through calcium can alter growth cone behavior. Small changes in intracellular calcium can alter growth cone behavior and neurite outgrowth rates; large changes can result in degeneration and cell death.

This paper presents the view that calcium homeostasis is an inherently plastic process. It will describe work that demonstrates calcium regulation can vary in different neurons, and can even change in the same neuron. Additionally, it will identify one mechanism, the sodium-calcium exchanger, as a site at which changes in calcium regulation can occur. Finally, it will consider the implications of a dynamically variable calcium regulatory system within the context of the development and modification of neuronal morphology.

INTRACELLULAR CALCIUM AND THE REGULATION OF GROWTH CONE BEHAVIOR

The neuronal growth cone is the primary structure responsible for the development, repair, and modification of neuronal circuitry. A wide variety of stimuli that regulate growth cone behavior, and thus ultimately neuronal morphology (e.g., neurotransmitters, growth factors, substrata, electrical activity), are now known to do so by altering intracellular calcium levels[9,32] (see FIG. 1). Our current working model[9,12] views growth cone behavior, and ultimately neuronal morphology, as a function of specific levels of intracellular calcium. At very low levels of calcium, outgrowth is inhibited, at moderate levels, outgrowth is permitted, and within an optimal range, even enhanced. At calcium levels above the optimal range, neurite outgrowth ceases, and at higher levels still neurite pruning, and ultimately cell death, occur. It is important to emphasize that the absolute values of high and low calcium are quite neuron-specific, that is, a level that is high for one neuron may in fact be within the optimal outgrowth range of another. Implicit in this model is the recognition that the effects of a stimulus that changes intracellular calcium will depend upon both the initial rest calcium levels and the magnitude of change. Additionally, it is clear that calcium regulatory mechanisms play a critical role; depending upon their efficacy, a given calcium signal may be modulated or even negated.

NEURON-SPECIFIC AND STATE-SPECIFIC DIFFERENCES IN CALCIUM HOMEOSTASIS

All neurons display a considerable capacity for calcium regulation.[5,33-36] The finding that different types of neuron have unique calcium "set points" or resting levels[37] also suggests that neurons differ in their capacity to regulate intracellular calcium.

This question was addressed directly using the calcium ionophore A23187 to deliver a calcium challenge to fura-2 loaded *Helisoma* neurons *in vitro*. *Helisoma* neurons have similar affinities for ionophore[18]; consequently its use, while clearly nonphysiological, avoids complications associated with receptor-dependent calcium signals. This approach permitted us not only to examine calcium regulation in different identified neurons but also to examine the consequences of such differences within the context of normal neuronal behavior. In addition since *Helisoma* neurons in culture spontaneously undergo a transformation from an actively growing state to a nongrowing immotile stable state,[38] comparisons could be made between neurons in different stages of growth.

Exposure to A23187 (0.4 μM) revealed distinctive neuron-specific differences in calcium regulatory capacity (FIG. 2A). In growing buccal neurons B5 ionophore evokes a rapid twofold rise in intracellular calcium. This increase is transient; despite the continuous presence of ionophore, calcium levels subsequently decline, and final calcium levels are not significantly above rest levels. In B19s, A23187 also initially produces a twofold rise in calcium. However, in marked contrast to B5s, calcium levels in B19s continue to rise and final levels are 15-20-fold higher than pretreatment values. Striking differences in calcium responses are also observed between growing and nongrowing cells. Doses of A23187 that produce a massive and sustained rise in calcium in growing cells evoke only a transient rise in their nongrowing (stable-state) counterparts (e.g., FIG. 2B). Interestingly some neuron-specific differences remain, for example, doses that caused a sustained rise in stable-state B19s produced only a transient rise in stable-state B5s.[18]

Parallel morphological studies indicate that these differences in calcium regulatory capacity have profound consequences on the generation and degeneration of neuronal architecture. A striking example of one such an effect, neuron-selective cell death, is shown in FIGURE 3. In this case a B5 (i.e., "good" calcium regulator), and a B19, (i.e., a "poor" calcium regulator) were plated as a pair. In the B19, exposure to A23187 rapidly produces progressive degenerative changes (e.g., beading of neurites and filopodia) and, ultimately, cell death. In contrast, the B5 not only survives but shows no degenerative changes. Similarly, stable-state (nongrowing) neurons, which can also be generally classified as "good" calcium regulators compared to growing neurons, show an increased resistance to calcium-induced degeneration and cell death. Concentrations of A23187 that are toxic for growing neurons have no effect on their nongrowing counterparts.

The morphological effects of ionophore such as the ionophore-induced rise in calcium[18] are dose dependent. Low doses of ionophore (which produce a small but sustained rise in calcium in growing B19s but only a transient rise in B5s) produce neuron-selective inhibition of neurite outgrowth. Actively growing B19s cease elongation while outgrowth of B5s is unaffected. Similarly moderate doses of ionophore, (which produce correspondingly larger and sustained rises in calcium in B19s) consistently produce neuron-specific degeneration of growth cones and neurite retraction in B19s. B5s in the same culture are unaffected by exposure to A23187 and continue to grow.[18] These more subtle localized effects are of particular interest since they may well mirror changes that occur during neuronal remodeling.

DIFFERENCES IN CALCIUM REGULATION: THE ROLE OF THE NA$^+$-CA^{2+} EXCHANGER

A number of candidate mechanisms could be responsible for the observed differences in the ability of neurons to handle a calcium load. One of these is a sodium-calcium exchange system, a high-capacity, low-affinity mechanism that is known to be

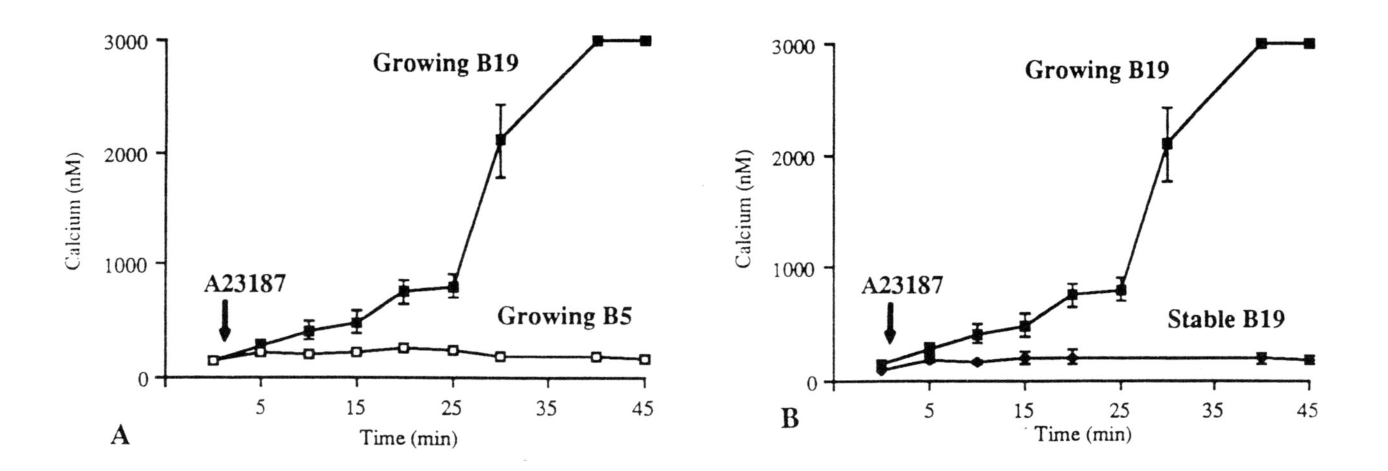

FIGURE 2. Neuron-specific and state-specific differences in calcium regulation. In A and B, intracellular calcium concentrations were measured in fura-2 loaded cells before and after addition of A23187. Arrows indicate the points at which A23187 was added. **A**: Neuron-specific changes in the response to calcium ionophore. Addition of A23187 causes a transient rise in growing B5s but a massive and sustained rise in calcium in B19s. Note: Error bars for B5s are not detectable at this scale. **B**: State-specific differences in the response to ionophore. In nongrowing, stable-state B19s, the ionophore-evoked rise is transient but in growing B19s the rise is sustained.

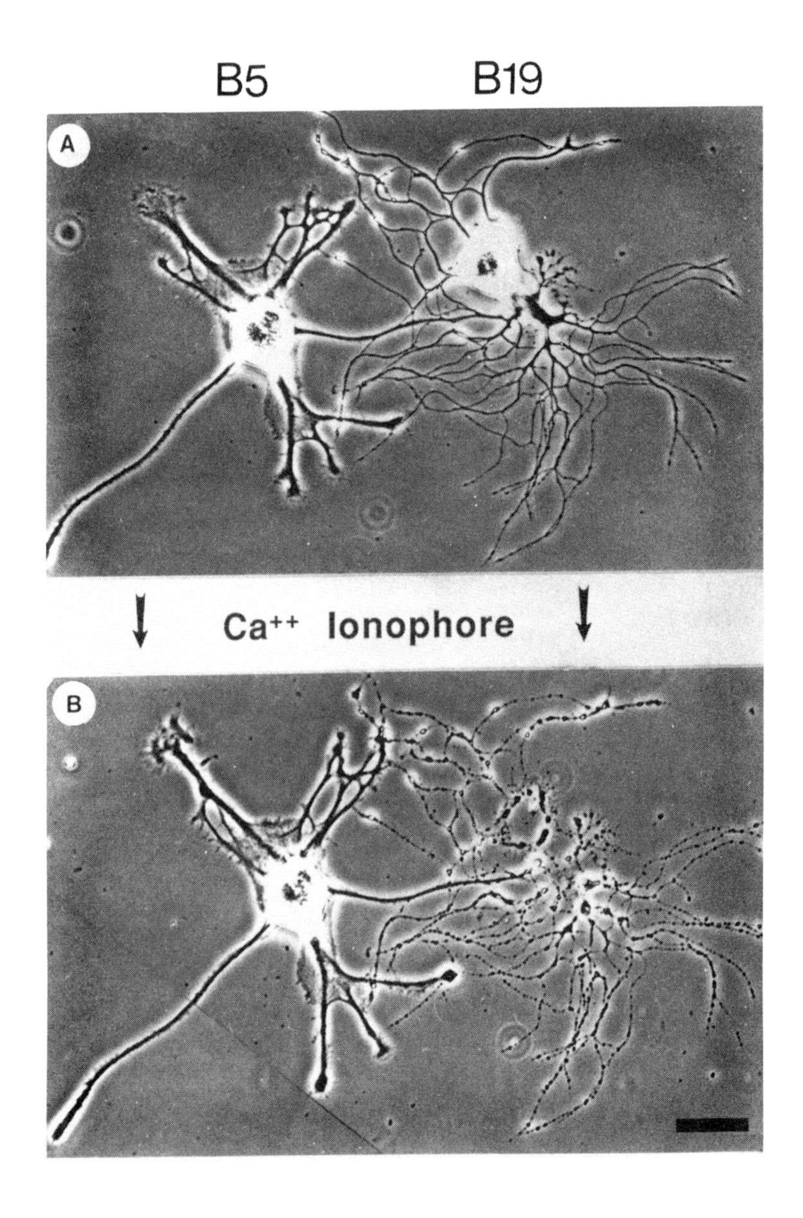

FIGURE 3. Ionophore-induced selective neuronal cell death. **A**: Neurons B5 and B19 immediately before treatment with A23187. **B**: The same cells 45 minutes after addition of A23187 (0.4 μM). The B5 remains intact, while B19 has completely disintegrated. Calibration: 75 microns.

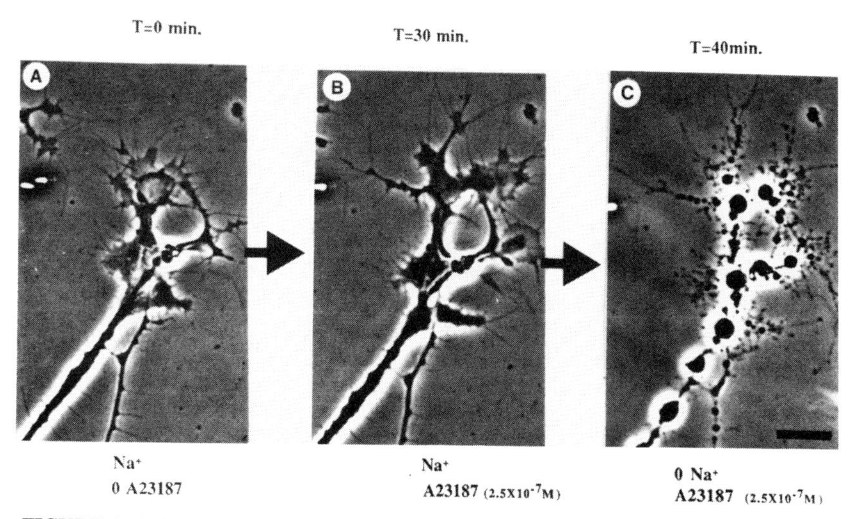

FIGURE 4. Calcium regulation and cell viability are reduced in zero extracellular sodium. **A:** B5 growth cones immediately before the application of A23187 in normal medium (54 mM sodium). **B:** The same growth cones after a 30-min exposure to ionophore. Typically, this dose produced no degeneration of growing B5s. Immediately after this time the (normal) medium was replaced with medium containing no sodium but the same concentration of A2318. Although some residual A23187 likely remained from the initial treatment, and thus contributed to the overall concentration of ionophore, the final concentration was assumed to be close to 2.5×10^{-7} M. **C:** Ten minutes after the medium replacement in B, extensive degeneration of growth cones has occurred. Control cells where the replacement medium contained A23187 in normal medium showed no such changes. Calibration: 20 microns.

activated in response to an increase in intracellular calcium.[39] This mechanism, which is known to be present in a wide variety of neurons including at least one species of mollusc,[40] is dependent upon the presence of extracellular sodium.[39,41]

In the absence of extracellular sodium, neuron-specific differences in calcium regulatory capacity are significantly diminished. For example, in the absence of external sodium "good" regulators lose their ability to compensate for a calcium load and also fail to survive exposure to normally nontoxic doses of ionophore (see FIGS. 4 and 5A,B). Elimination of external sodium also reduces the ability of stable-state neurons to reduce a calcium load and survive exposure to ionophore (FIGS. 5C,D). Calcium regulation and cell survival is also compromised in "poor" calcium regulators, although to a lesser extent.[18] These results indicate that differences in a sodium-dependent mechanism, likely the sodium-dependent calcium exchanger, can determine whether a neuron initiates outgrowth, ceases elongation, or even degenerates in response to a calcium challenge. However, the fact that in the absence of sodium all neurons are not equal (i.e., stable-state cells still compensate better for a calcium load than growing cells) argues that other calcium regulatory mechanisms also play a role in the observed differences in calcium regulation.

THE IMPLICATIONS OF DIFFERENCES IN CALCIUM HOMEOSTASIS

The emerging picture of calcium regulation suggests that it is a labile characteristic. Indeed, the observed variation in calcium resting levels between different neurons

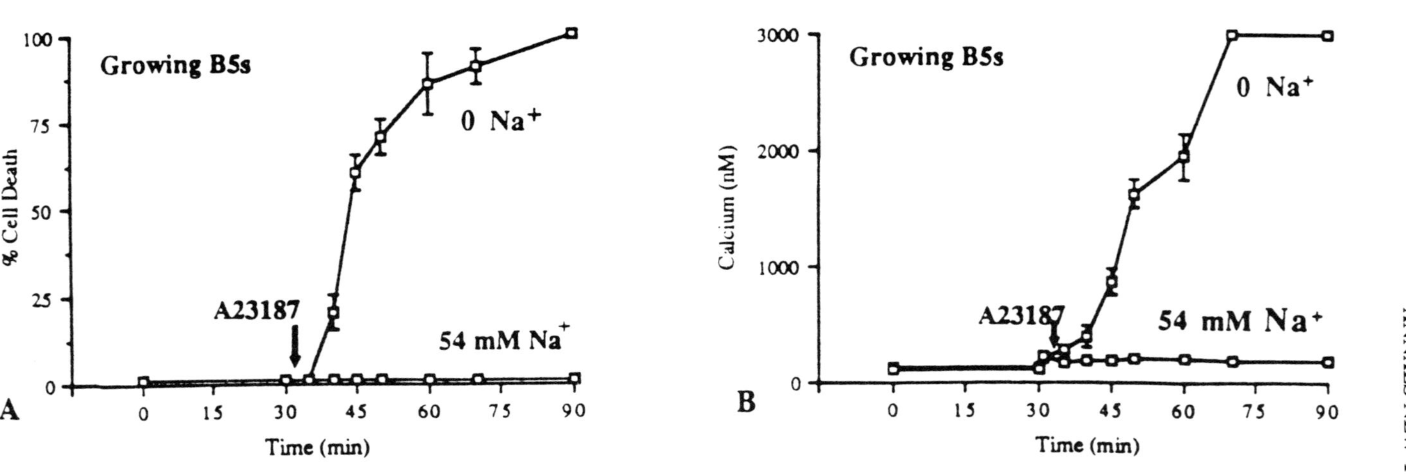

FIGURE 5. Neuron-specific and state-specific differences in calcium homeostasis are diminished in the absence of external sodium. At t = 0 experimental cells and the control cells were in normal medium. Immediately after this point the experimental cells were switched to sodium-free medium. Control cells were similarly treated with normal medium. Thirty minutes later (see arrow), A23187 was added to the cultures.

A: Cell viability of growing B5s is reduced after exposure to A23187 in the absence of sodium. In sodium-free medium all growing B5s ($n = 15$) died within 60 min of exposure to 0.4 μM A23187. This dose produced no cell death of controls ($n = 10$) in normal medium. Sodium substitution alone did not produce any detectable morphological change. **B**: Calcium regulation of growing B5s is compromised in sodium-free medium. In sodium-free medium the addition of A23187 resulted in a massive and sustained increase in calcium. In control medium (54 mM sodium) the same dose of A23187 produced only a small and transient increase in calcium. Final calcium levels in low-sodium medium were approximately 3000 nM compared to about 150 nM in control cells. Sodium substitution caused no detectable effects on rest calcium level: values at t = 0 and t = 30 minutes were not significantly different ($p < 0.005$).

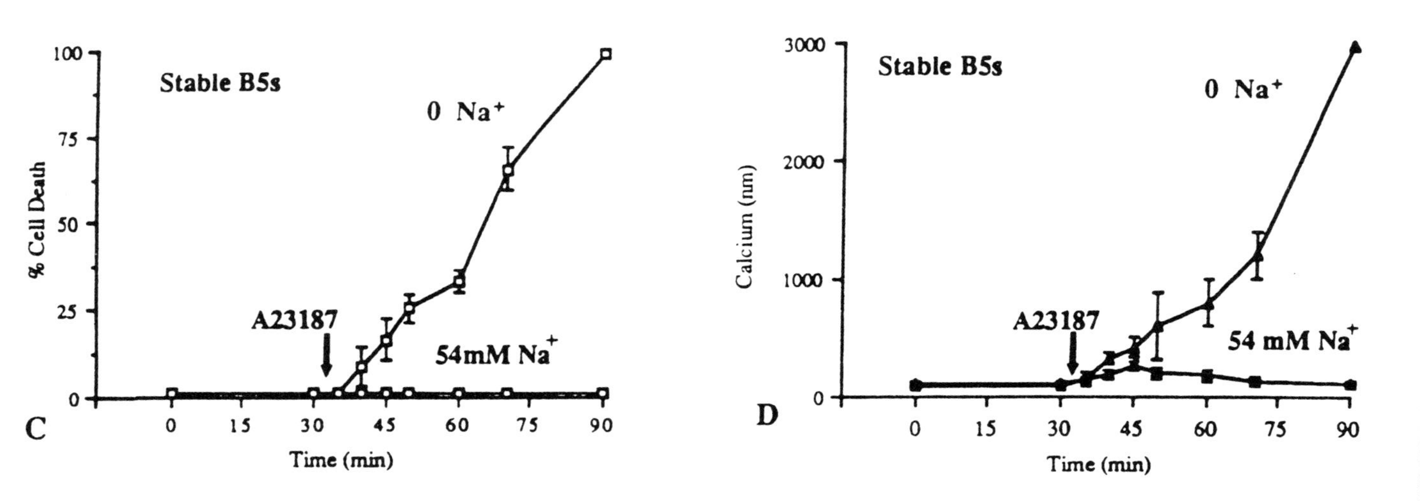

FIGURE 5 (*continued*). **C**: Cell viability of stable state B5s following exposure to A23187 (0.7 μM) is reduced in the absence of sodium. In sodium-free medium all B5s (n = 15) were killed following exposure to A23187. In control medium cell no cells died (n = 10). **D**: Calcium regulation of stable-state B5s is severely compromised in sodium-free medium. Intracellular calcium levels in stable-state B5s increased dramatically after exposure to A23187 in sodium-free medium. In control medium the same dose of A23187 produced only a small and transient increase in calcium. Resting calcium levels in normal medium (t = 0) were not significantly different in zero sodium (t = 30 min, $p < 0.005$). Although the ability of stable-state cells to survive was severely compromised in the absence of sodium, they remained quantitatively more resistant to the effects of A23187 than growing cells, requiring twice the concentrations of ionophore to produce the same effects.

suggests that each class of neuron may have a unique calcium-regulating "personality" or profile. Furthermore, this profile can change during different growth states and can be modified by exogenous agents. For example, the addition of fibroblast growth factor can alter calcium homeostasis in cultured hippocampal neurons.[42] Similarly, preliminary studies indicate that the addition of nerve growth factor modifies calcium regulation in PC12 cells. Work on mouse dorsal root ganglion neurons indicates that electrical stimulation can also change calcium regulation.[43] In this case action potentials inhibit outgrowth through a rise in intracellular calcium. However, after chronic stimulation there is a long-term modification in some as yet unidentified component of calcium homeostasis; continued stimulation no longer causes a large increase in calcium, and outgrowth resumes.

The examples discussed above are particularly notable because they occur in intact cells within the context of what can be considered to be normal development. Numerous other examples using cell-free systems demonstrate that individual components of calcium homeostasis can be modified.[44,45] Cells in abnormal developmental states, for example, transformed cells, also appear to be "better" able to regulate intracellular calcium. Mouse neuroblastoma hybrid cells survive doses of ionophore that are toxic for primary hippocampal neurons.[46] Similar differences have also been reported between neoplastic and normal thymocytes.[47] Finally, although their significance has yet to be established, it has long been recognized that changes in calcium regulation occur on at least one end of the normal developmental spectrum, that is, in aging.

What Are the Implications of a Dynamically Variable Calcium Regulatory System for Neuronal Development?

First, depending upon the degree to which a given neuron can compensate for a calcium load, a given stimulus could produce very different results in two different neurons. Second, even in a single neuron the same stimulus could produce radically different effects at different stages of development. Indeed, changes in calcium regulatory capacity may plausibly be regarded as one of the primary events that underlie neuronal plasticity.

For example, changes in calcium homeostasis could be the basis of the neuronal cell death that occurs in many areas of the developing nervous system.[48] They could also play a role in the conversion of active growth cones to mature synaptic terminals: Contact between growth cone and target cell could signal a change in one or more of the components of calcium regulation that ultimately reduces intracellular calcium to levels below the range permissible for outgrowth, thus transforming the formerly motile growth cone to an immotile, potentially synaptic, ending. Such a possibility is consistent with previous studies demonstrating that calcium levels in stable-state growth cones are typically lower than in active-growth cones.[5] Plausibly mature terminals require an increased capacity for calcium regulation to maintain endings in a stable, nongrowing, state in the face of the large calcium influxes associated with synaptic transmission.

Changes in calcium homeostasis could also be the basis for the remodeling of neuronal circuitry, that is, changes in calcium regulatory mechanisms could reset intracellular calcium levels in stable-state growth cones in such a way as to permit them to reinitiate outgrowth. This renewal of outgrowth could result in both the loss of existing connections, and the formation of new ones. Similarly, alterations in calcium regulation may be important in repair processes.

Changes calcium homeostasis within specific neuronal populations could also underlie the selective cell death that characterizes aging and chronic neurodegenerative

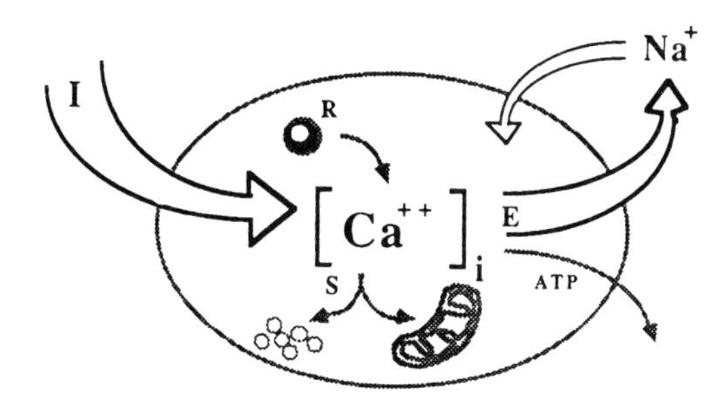

Ca^{2+} REGULATION (NEURON X) ≠ Ca^{2+} REGULATION (NEURON Y)

FIGURE 6. Schematic view of calcium homeostatic mechanisms. The regulation of intracellular calcium involves multiple components; influx (I) occurs via ion channels, and release (R) from intracellular stores. These stores also sequester calcium (S). Efflux occurs via an ATPase-dependent mechanism (ATP) and the sodium-calcium exchanger. A large influx of calcium (I) activates both sequestration and efflux (E) mechanisms, in particular the sodium-calcium exchanger (Na^+), which plays a major role in rapidly reducing calcium to resting levels. Differences in the sodium-calcium exchanger or other regulatory mechanisms could produce differences in calcium regulation illustrated by the equation below the figure where X equals an individual neuron and Y another neuron, or the same neuron in a different growth state.

diseases.[15,28,30] Equally plausibly, selective cell death could be the consequence of innate differences in the efficacy of one or more of the calcium-extrusion mechanisms.

A primary locus at which calcium regulation can be modified appears to be the sodium-calcium exchanger (see FIG. 6). As such it also plays a potentially key role in neuroplasticity. Since it becomes the dominant calcium-extrusion mechanism only at relatively high calcium concentrations,[20,39,41] differences in the efficacy of the exchanger may be important only during those times when calcium changes rapidly (e.g., in response to a calcium signal).[1,20] However, the exchanger plays only a minor role in the maintenance of resting calcium levels. Consequently, neuron-specific and state-specific differences in resting calcium levels must be due to differential contributions by other calcium regulatory mechanisms, for example, the ATPase-dependent pump or sequestration organelles.

DISCUSSION AND CONCLUSIONS

This paper has focused on the potential role of calcium homeostasis in the development of the nervous system, in particular how changes in calcium regulation can alter neuronal architecture. Clearly, alterations in calcium regulation will also have profound effects on neuronal function, both because of the obvious structural-functional relationship and because of the major role played by intercellular calcium in information processing. Understanding the extent to which even a minor change in calcium regulation may alter the nervous system is formidable task—one that is not yet possible. Not only will it require knowledge of the relative contributions of multiple calcium regulatory mechanisms (and the conditions under which those change), but also the ability to determine how these mechanisms are integrated. In addition we must be able to identify those various aspects of neuronal structure and function that are regulated by intracellular calcium. Nevertheless, the existing evidence does suggest that the calcium regulatory system is one of the primary mechanisms that confers flexibility upon the nervous system, providing both stability and the opportunity for change.

ACKNOWLEDGMENTS

I wish to thank Dr. S. B. Kater for support and many insightful discussions.

REFERENCES

1. MILLER, R. J. 1988. Calcium signalling in neurons. TINS **11:** 415-419.
2. KATZ, B. 1969. The Release of Neural Transmitter Substances. Liverpool University Press. Liverpool, England.
3. SMITH, S. J. & G. J. AUGUSTINE. 1988. Calcium ions, active zones and synaptic transmitter release. TINS **11:** 458-464.
4. GUNDERSON, R. W. & J. N. BARRET. 1980. Characterization of the turning response of dorsal root neurites towards nerve growth factor. J. Cell. Biol. **87:** 546-554.
5. COHAN, C. S., J. A. CONNOR & S. B. KATER. 1987. Electrically and chemically mediated increases in intracellular calcium in neuronal growth cones. J. Neurosci. **7:** 3588-3599.
6. HAYDON, P. G., D. P. MCCOBB & S. B. KATER. 1987. The regulation of neurite outgrowth, growth cone motility, and electrical synaptogenesis by serotonin. J. Neurobiol. **18:** 197-215.
7. MATTSON, M. & S. B. KATER. 1987. Calcium regulation of neurite elongation and growth cone motility. J. Neurosci. Res. **7:** 4034-4043.
8. MCCOBB, D. P. & S. B. KATER. 1988. Dopamine and serotonin inhibition of neurite elongation of different identified neurons. J. Neurosci. Res. **19:** 19-26.
9. KATER, S. B., M. MATTSON, C. COHAN & J. CONNOR. 1988. Calcium regulation of the neuronal growth cone. TINS **11:** 315-321.
10. LANKFORD, K. & P. LETOURNEAU. 1991. Roles of actin filaments and three second messenger systems in short term regulation of chick dorsal root ganglion neurite outgrowth. Cell Motil. Cytoskel. **20:** 7-29.
11. KATER, S. B., L. R. MILLS & P. B. GUTHRIE. 1989. Intracellular calcium and the control of neuronal growth and form. *In* Trophic Factors and the Nervous System. N. Neff, Ed. Raven Press. New York.
12. KATER, S. B. & L. R. MILLS. 1991. Regulation of growth cone behavior by calcium. J. Neurosci. **11:** 891-899.
13. MATTSON, M. P. & S. B. KATER. 1989. Excitatory and inhibitory neurotransmitters in the generation and degeneration of hippocampal neuroarchitecture. Brain Res. **478:** 337-348.
14. SCHANNE, F. A. X., A. B. KANE, E. E. YOUNG & J. L. FARBER. 1979. Calcium dependence of toxic cell death. Science **206:** 700-702.
15. GIBSON, G. E. & C. PETERSON. 1987. Calcium and the aging nervous system. Neurobiol. Aging **8:** 329-343.
16. BONDY, S. C. & H. KOMULAINEN. 1988. Intracellular calcium as an index of neurotoxic damage. Toxicology **49:** 35-49.
17. CHOI, D. W. 1988. Glutamate neurotoxicity and diseases of the nervous system. Neuron. **1:** 623-634.
18. MILLS, L. R. & S. B. KATER. 1990. Neuron-specific and state-specific differences in calcium homeostasis regulate the generation and degeneration of neuronal architecture. Neuron **2:** 149-163.
19. MEYER, F. B. 1988. Calcium, neuronal hyperexcitability and ischemic injury. Brain Res. Rev. **14:** 227-243.
20. CARAFOLI, E. 1987. Intracellular calcium homeostasis. Ann. Rev. Biochem. **56:** 395-433.
21. BLAUSTEIN, M. P. 1984. *In* Electrogenic Transport: Fundamental Principals and Physiological Implications. M. P. Blaustein & M. Leiberman, Eds.: 129-147. Raven Press. New York.

22. MELDOLOSI, J., P. VOLPE & T. POZZAN. 1988. The intracellular distribution of calcium. TINS **11:** 449-452.
23. RASSMUSSEN, H. & P. Q. BARRETT. 1984. Calcium messenger system: An integrated view. Physiol. Rev. **64:** 938-984.
24. VOLPE, P., K-H. KRAUSE, S. HASHIMOTO, F. ZORZATO, T. POZZAN, J. MELDOLESI & D. P. LEW. 1988. Calcisomes, a cytoplasmic organelle: The inositol 1,4,5-trisphosphate-sensitive Ca^{2+} store of nonmuscle cells? Proc. Natl. Acad. Sci. USA **85:** 1091-1095.
25. WOOD, J. G., R. W. WALLACE & W. Y. CHEUNG. 1980. *In* Calcium and Cell Function. W. Y. Cheung, Ed.: 291-202. Academic Press. New York.
26. MCBURNEY, R. N. & I. R. NEERING. 1987. Neuronal calcium homeostasis. TINS **10:** 164-169.
27. MICHAELIS, M. 1989. Ca^{2+} handling systems and neuronal aging. *In* Calcium, Membranes, and Alzheimer's Disease. Z. S. Khachaturian, C. Cotman & J. W. Pettegrew, Eds. Ann. N.Y. Acad. Sci. **658:** 89-94.
28. KHACHATURIAN, Z. 1984. Towards theories of brain aging. *In* Handbook of Studies on Psychiatry and Old Age. D. S. Kay & G. W. Burrows, Eds.: 7-30. Elselvier Amsterdam.
29. KATER, S. B., M. P. MATTSON & P. B. GUTHRIE. 1990. Calcium-induced neuronal degeneration: A normal growth cone regulating signal gone awry (?). *In* Calcium, Membranes, Aging and Alzheimer's Disease. C. W. Khatchaturian, C. Cottman & J. W. Pettigrew, Eds. Ann. N.Y. Acad. Sci. **568:** 252-261.
30. SETO-OSHIMA, A., P. C. EMSON, E. LAWSON, C. Q. MONTJOY & L. H. CARRASCO. 1988. Loss of matrix calcium-binding protein-containing neurons in Huntingdon's disease. Lancet **1:** 1252-1255.
31. KATER, S. B., P. B. GUTHRIE & L. R. MILLS. 1990. Integration by the neuronal growth cone: A continuum from neuroplasticity to neuropathology. Prog. Brain Res.: 1-31.
32. MILLS, L. R. & S. B. KATER. 1989. Integration of environmental and intracellular signals: The calcium hypothesis for the control of neuronal growth cones. *In* Assembly of the Nervous System. L. Landmesser, Ed. Alan R. Liss. New York.
33. TANK, D. W., S. SUGIMORI, J. A. CONNOR & R. R. LLINAS. 1988. Spatially resolved calcium dynamics of mammalian Purkinje cells in cerebellar slice. Science **242:** 773-776.
34. THAYER, S. A., T. M. PERNEY & R. J. MILLER. 1988. Regulation of calcium homeostasis in sensory neurons by bradykinin. J. Neurosci. **8:** 4089-4097.
35. LIPSCOMBE, D., D. V. MADISON, M. POENIE, H. REUTER, R. TSIEN, R. W. TSIEN & R. Y. TSIEN. 1988. Imaging of cytosolic Ca^{2+} transients arising from Ca^{2+} stores and Ca^{2+} channels in sympathetic neurons. Neuron **1:** 355-365.
36. REHDER, V., J. R. JENSEN, P. DOU & S. B. KATER. 1991. A comparison of calcium homeostasis in isolated and attached growth cones of the snail *Helisoma.* J. Neurobiol. **22:** 499-511.
37. GUTHRIE, P. B., M. MATTSON, L. R. MILLS & S. B. KATER. 1988. Calcium homeostasis in molluscan and mammalian neurons: Neuron-selective set-point of calcium rest concentrations. Soc. Neurosci. Abstr. **14:** 582.
38. HAYDON, P. G., C. S. COHAN, D. P. MCCOBB, P. MILLER & S. B. KATER. 1985. Neuron-specific growth cone properties as seen in identified neurons of *Helisoma.* J. Neurosci. Res. **13:** 135-147.
39. BLAUSTEIN, M. P. 1977. Effects of internal and external cations and of ATP on sodium-calcium and calcium-calcium exchange in squid axons. Biophys. J. **20:** 79-110.
40. LEVY, S. & D. TILLOTSON. 1988. Effects of Na^+ and Ca^{2+} gradients on intracellular free-Ca^{2+} in voltage-clamped *Aplysia* neurons. Brain Res. **474:** 333-347.
41. BAKER, P. F. 1986. The sodium-calcium exchange system. *In* Calcium and the Cell. Ciba Foundation Symposium **12:** 73-97. Wiley. Chichester, England.
42. MATTSON, M., M. MURRAIN, P. G. GUTHRIE & S. B. KATER. 1989. Fibroblast growth factor and glutamate: Opposing roles in the generation of neuronal architecture. J. Neurosci. **9:** 3728-3738.
43. FIELDS, R. D., P. B. GUTHRIE & S. B. KATER. 1990. Calcium homeostatic capacity is regulated by patterned activity in the growth cones of mouse DRG neurons. Soc. Neurosci. Abstr. **16:** 457.

44. MICHAELIS, M., E. K. MICHAELIS, E. W. NUMLEY & N. GALTON. 1987. Effects of chronic alcohol administration on synaptic membrane Na^+-Ca^{2+} exchange activity. Brain Res. **414:** 239-244.
45. JENSEN, J., G. LYNCH & M. BAUDRY. 1989. Allosteric activation of brain mitochondrial Ca^{2+} uptake by spermine and by Ca^{2+}: Developmental changes. J. Neurochem. **53:** 1173-1181.
46. MATTSON, M., P. B. GUTHRIE & S. B. KATER. 1989. A role for Na^+-dependent calcium extrusion in protection against neuronal excitotoxicity. FASEB J. **3:** 3728-3738.
47. CALVIELLO, G., D. BOSSI & A. CITTADINI. 1988. Further observations on the effect of calcium ionophores on ascites tumor cells. Arch. Biochem. Biophys. **259:** 38-45.
48. OPPENHEIM, R. W. 1985. Naturally occurring cell death during neural development. TINS **11:** 487-493.

Na^+-Ca^{2+} Exchange Activity Is Increased in Alzheimer's Disease Brain Tissues

ROBERT A. COLVIN, JONATHAN W. BENNETT, AND SHARON L. COLVIN

Department of Zoological and Biomedical Sciences
Ohio University College of Osteopathic Medicine
Athens, Ohio 45701

The cognitive decline associated with aging and the human neurodegenerative diseases (e.g., Alzheimer's disease, Parkinson's disease, and Huntington's disease) is most certainly caused by neuronal dysfunction and loss of synapses. The "calcium hypothesis of brain aging"[1] proposes that changes in the cellular mechanisms that act to modulate the concentration of free intracellular calcium ($[Ca_i]$) within the neuron contribute to the causative factors leading to neuronal dysfunction and degeneration. A similar hypothesis has also been proposed to explain the neurodegenerative processes occurring in Alzheimer's disease (AD). Unfortunately, very little is known about the effects of human aging and AD on neuronal Ca^{2+} homeostatic processes. Important cellular mechanisms for maintaining low intracellular Ca^{2+} within the neuron include the Na^+-Ca^{2+} exchanger. In the present study we tested the hypothesis that differences exist in the Na^+-Ca^{2+} exchange activity measured in cerebral plasma membrane vesicles purified from human post-mortem brain tissues of normal, AD, and non-Alzheimer's origin dementia (NAD).

RESULTS AND DISCUSSION

The major finding of this study was that Na^+-Ca^{2+} exchange activity measured in cerebral plasma membrane vesicles derived from AD post-mortem brain tissues was significantly greater than that measured in vesicles from normal or NAD brain tissues. This conclusion was supported by a kinetic analysis of the effect of increasing extravesicular Ca^{2+} on the initial velocity of Na^+-Ca^{2+} exchange in each of the three tissue types. A summary of the kinetic data obtained for Na^+-Ca^{2+} exchange in all three tissue samples studied is shown in TABLE 1. The mean values obtained for the K_m of Na^+-Ca^{2+} exchange were not significantly different when compared among the three groups (normal, AD, and NAD). On the other hand, the mean V_{max} of Na^+-Ca^{2+} exchange was significantly increased for AD tissue ($p < 0.05$) when compared with either normal or NAD tissue.

What is the possible role of Na^+-Ca^{2+} exchange in the pathogenesis of nerve cell degeneration in AD? The recent study of Mattson *et al.*[2] has clearly shown cell-specific differences in calcium homeostatic systems and that cells with superior Na^+-Ca^{2+} exchange capacity are more likely to survive an insult of increased $[Ca_i]$. It is attractive to speculate that the increased Na^+-Ca^{2+} exchange activity observed in the plasma membranes of neurons from AD brain reflects the properties of surviving neurons with increased capacity for Na^+-Ca^{2+} exchange. The neurons may have survived the

TABLE 1. Summary of the Kinetic Constants Obtained for Na^+-Ca^{2+} Exchange

Origin	n	K_m (μM)[a]	V_{max} (nmol/mg·min)[a]
Normal	6	57.9 ± 28.1	4.45 ± 0.58
AD	6	71.2 ± 29.3	6.68 ± 1.58[b]
NAD	4	66.1 ± 3.3	3.76 ± 1.50

NOTE: Cerebral plasma membrane vesicles were loaded with Na^+ by preincubation in 132 mM NaCl, 5 mM KCl, 1.3 mM $MgCl_2$, 10 mM glucose, 10 mM Hepes (pH 7.4) for 15 minutes at 37°C. The membrane suspension was then diluted to 0.5 ml by the same buffer (A) or buffer B in which choline-Cl was substituted for NaCl. Each buffer contained various concentrations of $CaCO_3$ (0.01 to 0.3 mM) with the addition of $^{45}CaCl_2$ (0.01 mCi/ml). Reactions were stopped by the addition of 200 μM $LaCl_3$ and placing the tubes on ice. Calcium content was determined by rapid filtration of the vesicles on glass fiber filters. Initial velocities were estimated as the change in Ca^{2+} content in buffer B minus that measured in buffer A after 30 seconds.

[a] Values are means ± standard deviation.

[b] Significantly different from either normal or NAD brain ($p < 0.05$).

neurodegenerative process of AD by virtue of their increased Na^+-Ca^{2+} exchange activity. If this were true, it is probable that increased [Ca_i] played a key role in the death of the nonsurvivors as is suggested by the scheme of FIGURE 1, which proposes that overproduction of β-A4 and amyloid deposition are early events in the pathogenesis of AD as Selkoe has suggested.[3] Overproduction of β-A4 is suggested to result from deranged post-translational processing of amyloid precursor protein (APP) caused by aging and/or genetic factors. No link has yet been discovered between amyloid deposition, senile plaque formation, and derangements in cell calcium metabolism. However, β-A4 can enhance excitatory amino acid neurotoxicity in cultured mouse cortical neurons.[4] The scheme of FIGURE 1 proposes that a combination of genetic factors, aging, ischemic insult, and amyloid deposition would contribute to increased sensitivity of certain brain regions and neuron types to increases in excitatory amino acid input,

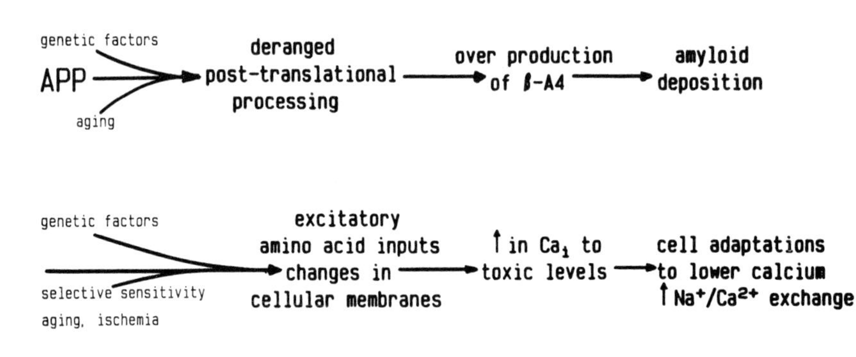

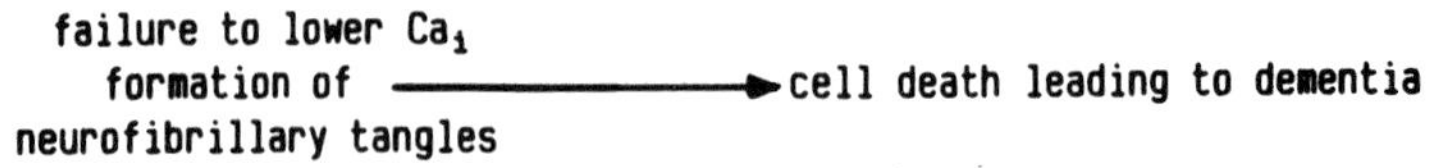

FIGURE 1. Speculative scheme to illustrate the role of increased [Ca_i] in the pathogenesis of Alzheimer's disease. See text for further details.

ultimately producing toxic levels of [Ca_i]. The failure to respond adequately to increases in [Ca_i] could lead to abnormal phosphorylation of tau, neurofibrillary tangle formation, and neuronal death. In support of this, it has been found that the antigenic changes in neuronal cytoarchitecture seen in cultured hippocampal cells after treatment with glutamate and Ca^{2+} influx are similar to that seen in the neurofibrillary tangles of AD.[5] Confirmation of the hypothesis that a loss of Ca^{2+} homeostasis is the final common pathway to neuronal degeneration in AD must await more detailed knowledge of the neurotoxic properties of β-A4 and the temporal sequence of events in the pathogenesis of AD.

REFERENCES

1. KHACHATURIAN, Z. S. 1989. Ann. N.Y. Acad. Sci. **568:** 1-4.
2. MATTSON, M. P., P. B. GUTHRIE & S. B. KATER. 1989. FASEB J. **3:** 2519-2526.
3. SELKOE, D. J. 1989. N. Engl. J. Med. **320:** 1484-1487.
4. KOH J.-Y., L. L. YANG & C. W. COTMAN. 1990. Brain Res. **533:** 315-320.
5. MATTSON, M. P. 1990. Neuron **4:** 105-117.

Reverse Operation of the Na^+-Ca^{2+} Exchanger Mediates Ca^{2+} Influx during Anoxia in Mammalian CNS White Matter[a]

PETER K. STYS,[b] STEPHEN G. WAXMAN, AND BRUCE R. RANSOM

Department of Neurology
Yale University School of Medicine
New Haven, Connecticut 06510
and
Neuroscience Research Center (127A)
Veterans Administration Hospital
West Haven, Connecticut 06516

Central white matter (WM) tracts in the mammalian central nervous system (CNS), such as subcortical pathways and spinal cord tracts that are critical to the functional integrity of the CNS, suffer irreversible injury after anoxia/ischemia. Anoxia-induced cell death appears to be caused by sustained increases in intracellular Ca^{2+}. In gray matter, Ca^{2+} influx into the cytoplasm during anoxia is thought to occur via NMDA-receptor-gated channels.[1,2] The mechanisms of anoxic injury in CNS WM are less well understood; specifically, the critical step by which extracellular Ca^{2+} enters the cytoplasmic compartment is not known. We have studied this question using the *in vitro* rat optic nerve, a representative CNS WM tract. Our results indicate that a large part of the damaging Ca^{2+} influx that occurs during anoxia in WM is mediated by the Na^+-Ca^{2+} exchanger, forced to operate in the reverse mode due to Na^+ influx via voltage-gated Na^+ channels.

METHODS AND RESULTS

The methods have been described in detail elsewhere.[3,4] Briefly, optic nerves were dissected from adult Long Evans rats and perfused in an interface brain slice chamber with artificial cerebrospinal fluid (CSF) containing, in mM: NaCl 126, KCl 3.0, $MgSO_4$ 2.0, $NaHCO_3$ 26, NaH_2PO_4 1.25, $CaCl_2$ 2.0, dextrose 10 (pH 7.45). Temperature was maintained at 37°C. The tissue was aerated with a 95% O_2/5% CO_2 gas mixture. The functional integrity of the nerves was monitored electrophysiologically as the area

[a] This work was supported by grants from the National Institute of Neurological Disorders & Stroke, the American Paralysis Association, and by the Medical Research Service, Veterans Administration. P.K.S. was supported by a fellowship from the Blinded Veterans Association and by a Centennial Fellowship from the Canadian Medical Research Council.

[b] Address for correspondence: Dr. Peter K. Stys, Yale University School of Medicine, Department of Neurology, 710 LCI, 333 Cedar Street, New Haven, CT 06510.

under the compound action potential (CAP) using suction electrodes.[4] Optic nerves were subjected to a standard 60-min period of anoxia by switching to 95% N_2/5% CO_2. The degree of functional recovery after anoxic insult, measured at its maximum typically 60 min after re-oxygenation, was calculated as the ratio of the CAP area postanoxia to the control CAP area.

When subjected to a standard 60-min anoxic insult, CAP area recovers to 32.0 ± 9.4% of control (FIG. 1). We have previously shown that extracellular Ca^{2+} is critical to the production of irreversible anoxic injury in WM,[3] and that Ca^{2+} does not enter through voltage-gated Ca^{2+} channels[5] or via NMDA-gated channels.[6] Blocking Na^+ channels with tetrodotoxin during anoxia markedly improved functional recovery of WM (81.5 ± 11%, SD, CAP area recovery; $p < 0.0001$). Similarly, inhibitors of Na^+-Ca^{2+} exchange also significantly improved recovery in a dose-dependent manner (FIG. 1); optimal protection was seen with bepridil at a concentration of 50 μM (69.0 ± 14%, SD; $n = 7$; $p < 0.0001$), and benzamil at 500 μM (71.0 ± 15%, SD; $n = 10$; $p < 0.0001$). Addition of tetrodotoxin (TTX) to bepridil or benzamil further enhanced recovery by 5-10% (see FIG. 1).

Perfusing the tissue with zero-Na^+ solution (choline substitution) beginning 20 min before anoxia was also very protective (88.0 ± 5% CAP area recovery; $p < 0.0001$, FIG. 2). However, introducing zero-Na^+ solution after the start of anoxia (e.g., at +40 min, FIG. 2), resulted in increased injury (13.7 ± 6% CAP area recovery; $p < 0.0001$).

DISCUSSION

In CNS WM, extracellular Ca^{2+} is required for irreversible anoxic injury.[3] Our results indicate that Na^+ influx through voltage-gated Na^+ channels is also important to the development of WM anoxic injury. Presumably, under conditions of energy failure as seen in anoxia or ischemia, this Na^+ influx results in a rise in intracellular $[Na^+]$ ($[Na^+]_i$). This rise in $[Na^+]_i$ would then cause the Na^+-Ca^{2+} exchanger to operate in reverse, raising $[Ca^{2+}]_i$ to injurious levels. Consistent with this proposal are the observations that either Na^+ channel blockers or Na^+-Ca^{2+} exchanger blockers are protective. The Ca^{2+} channel-blocking properties of bepridil or benzamil are unlikely to play a role since we have previously shown that Ca^{2+} channels do not mediate WM anoxic injury.[5]

The results of the zero-Na^+ substitution experiments also support involvement of the exchanger (FIG. 2). Perfusing the nerves with zero-Na^+ solution prior to anoxia will lower $[Na^+]_i$ from its resting level of about 20-25 mM.[7] This will tend to inhibit reverse exchange as this mode requires a finite $[Na^+]_i$ (≈25-30 mM for half-maximal activation in squid giant axon.[8,9] Delaying the introduction of zero-Na^+ until 20 or 40 min into the anoxic period allows $[Na^+]_i$ to increase above its normal level, as ionic gradients collapse; introducing zero-Na^+ at this point results in a large *reverse* Na^+ gradient (i.e., $[Na^+]_i > [Na^+]_o$) which would tend to drive reverse exchange even more strongly than during perfusion with solution containing normal (153 mM) Na^+, where the Na^+ gradient could, at worst, collapse completely ($[Na^+]_i = [Na^+]_o$) but would never be reversed. We hypothesize that under such conditions, even more Ca^{2+} is admitted through reverse exchange, manifested by increased cellular injury.

Experimental work to date indicates that cytoplasmic Ca^{2+} loading is a final common pathway for cellular injury.[10–12] In WM, the mechanism of anoxia-induced Ca^{2+} influx appears primarily to involve reverse operation of the Na^+-Ca^{2+} exchanger. This pathophysiological situation suggests protective strategies. Drugs that limit Na^+ influx should be protective; preliminary studies from our laboratory with local anesthetics

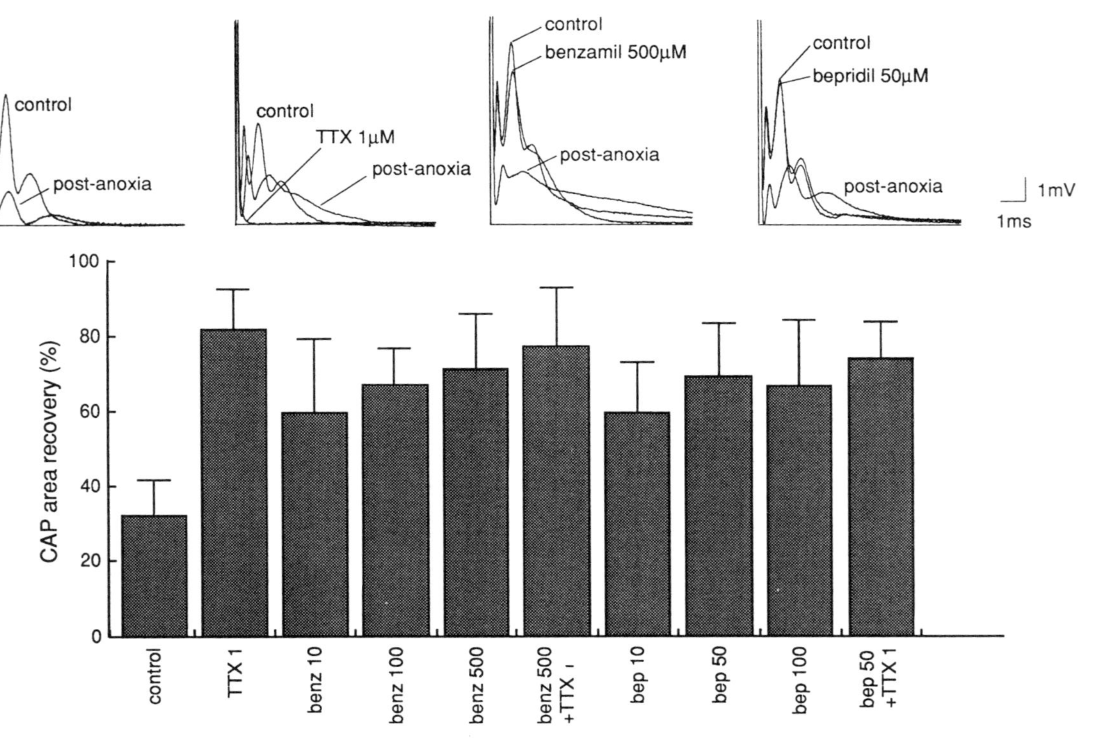

FIGURE 1. Effects of blockers of Na^+ channels and Na^+-Ca^{2+} exchange on recovery of optic nerve function after 60 min of anoxia. In normal CSF, CAP area recovered to 32.0 ± 9.4% of control. Addition of TTX (1 μM), applied 20 min before anoxia and continued during the anoxic period, significantly enhanced CAP area recovery to 81.5 ± 11%. Blockers of Na^+-Ca^{2+} exchange, bepridil and benzamil, were applied in a similar fashion, and also improved recovery in a dose-dependent manner. Addition of TTX to either bepridil or benzamil further enhanced recovery by an additional 5-10%. Top panels show recordings of CAPs in normal CSF, in the presence of drug, and 1-3 hr after the anoxic period in normal CSF.

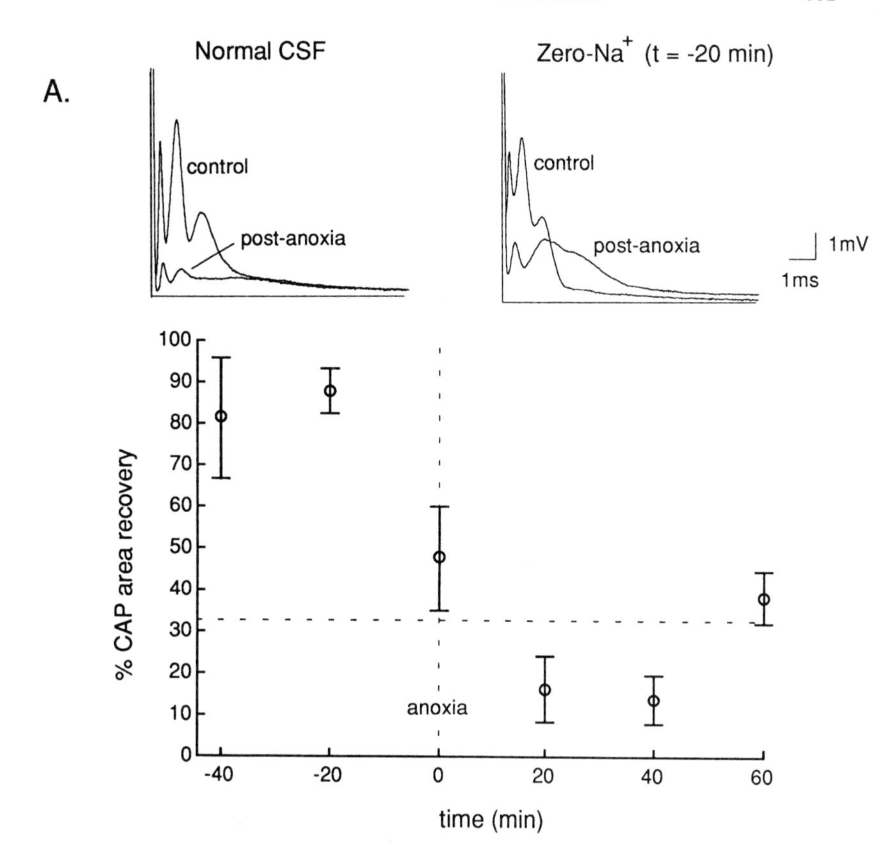

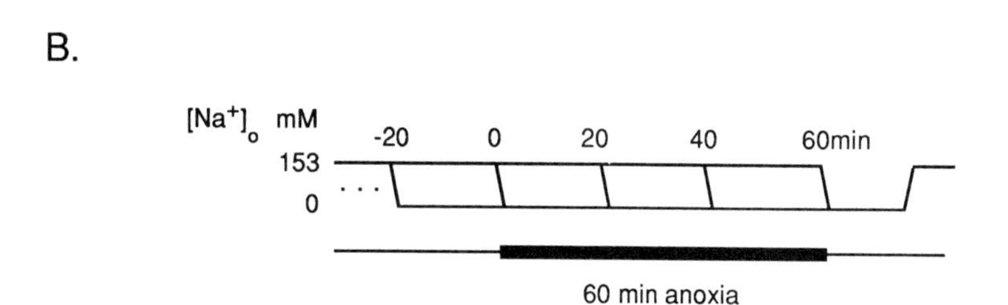

FIGURE 2. Effects of zero-Na^+ solution (choline substitution) on recovery of CAP area after 60 min of anoxia. **A:** the x-axis of the graph represents the time after anoxia when zero-Na^+ CSF was introduced. When zero-Na^+ CSF was applied 20 min before anoxia ($t = -20$ min), recovery of CAP area was markedly enhanced to 88.0 ± 5% vs. 32.0 ± 9.4% in normal (153 mM) Na^+ (horizontal dashed line). As the introduction of zero-Na^+ CSF was delayed with respect to the start of anoxia, progressively less recovery was seen. Introducing zero-Na^+ CSF at 20 or 40 min after anoxia resulted in more injury than was seen with normal [Na^+] maintained throughout (horizontal dashed line). Top panels show representative traces of control and post-anoxic CAPs in normal CSF, and with zero-Na^+ CSF introduced 20 min before anoxia. **B:** diagram showing timing of solutions.

support this expectation. Compounds that selectively block reverse Na^{+}-Ca^{2+} exchange, leaving intact the forward Ca^{2+} efflux mode, should be particularly valuable; increasing knowledge about the molecular character of the exchanger protein may facilitate the design of such agents.

REFERENCES

1. MacDermott, A. B., M. L. Mayer, G. L. Westbrook, S. J. Smith & J. L. Barker. 1986. NMDA-receptor activation increases cytoplasmic calcium concentration in cultured spinal cord neurones. Nature **321:** 519-522.
2. Rothman, S. M. & J. W. Olney. 1986. Glutamate and the pathophysiology of hypoxic-ischemic brain damage. Ann. Neurol. **19:** 105-111.
3. Stys, P. K., B. R. Ransom, S. G. Waxman & P. K. Davis. 1990. Role of extracellular calcium in anoxic injury of mammalian central white matter. Proc. Natl. Acad. Sci. USA **87(11):** 4212-4216.
4. Stys, P. K., B. R. Ransom & S. G. Waxman. 1991. Compound action potential of nerve recorded by suction electrode: A theoretical and experimental analysis. Brain Res. **546:** 18-32.
5. Stys, P. K., B. R. Ransom & S. G. Waxman. 1990. Effects of polyvalent cations and dihydropyridine calcium channel blockers on recovery of CNS white matter from anoxia. Neurosci. Lett. **115:** 293-299.
6. Ransom, B. R., S. G. Waxman & P. K. Davis. 1990. Anoxic injury of CNS white matter: Protective effect of ketamine. Neurology (Minneapolis) **40:** 1399-1403.
7. Erecińska, M. & I. A. Silver. 1991. Intra- and extracellular ion concentrations and transmembrane potentials in neurons and glia in rat brain. J. Physiol. In press.
8. Requena, J., L. J. Mullins, J. Whittembury & F. J. J. Brinley. 1986. Dependence of ionized and total Ca in squid axons on Nao-free or high-Ko conditions. J. Gen. Physiol. **87**(1): 143-159.
9. Requena, J., J. Whittembury & L. J. Mullins. 1989. Calcium entry in squid axons during voltage clamp pulses. Cell Calcium **10(6):** 413-423.
10. Schanne, F. A., A. B. Kane, E. E. Young & J. L. Farber. 1979. Calcium-dependence of toxic cell death: A final common pathway. Science **206:** 700-702.
11. Choi, D. W. 1988. Calcium-mediated neurotoxicity: Relationship to specific channel types and role in ischemic damage. Trends Neurosci. **11:** 465-469.
12. Siesjö, B. K. 1986. Calcium and ischemic brain damage. Eur. Neurol. **25:** 45-56.

Cardiac Physiology

Introduction to Part V

Exchange current measurements and optical techniques for measuring intracellular Ca^{2+} concentrations in cardiac myocytes were the most prominent experimental approaches described in the articles in this section. The results obtained from these studies leave little doubt that the Na-Ca exchange system is the predominant Ca^{2+} efflux mechanism in cardiac cells. A role for the exchanger in mediating Ca^{2+} entry has also been suggested, although this would require the presence of a restricted space that would permit the transient generation of a cytoplasmic Na^+ gradient. Na-Ca exchange currents appear to be an experimentally useful index of subsarcolemmal Ca^{2+} concentrations; indeed, under certain conditions, changes in subsarcolemmal $[Ca^{2+}]$ appear to be distinguishable from changes in the bulk cytoplasmic $[Ca^{2+}]$. Thus, measurements of exchange activity may provide experimental access to the difficult problem of cytoplasmic compartmentation. This issue is likely to become increasingly important in future research on cardiac myocyte physiology.

The Role of Sodium-Calcium Exchange during the Cardiac Action Potential[a]

D. NOBLE,[b] S. J. NOBLE, AND G. C. L. BETT

Department of Physiology
University of Oxford
Oxford, England

Y. E. EARM, W. K. HO, AND I. K. SO

Department of Physiology
Seoul National University
Seoul, Republic of Korea

INTRODUCTION

The idea that sodium-calcium exchange might generate a current during cardiac electrical activity was first raised as a possible mechanism for the pathological situation when calcium-overloaded tissue generates the transient inward current responsible for ectopic beating. This was one of the mechanisms proposed by Kass, Lederer, Tsien, and Weingart in 1978.[1] Mullins[2] (see also Mullins[3]) had already suggested on theoretical grounds that a variety of currents flowing across the cardiac cell membrane might be carried by sodium-calcium exchange. With regard to the transient inward current, the evidence that it is carried mostly (though not entirely) by sodium-calcium exchange is now very strong in ventricular cells,[4] and it has proved possible to reproduce this mechanism in computer models of cardiac electrical activity.[5] The purpose of the present paper is to review the evidence, experimental and theoretical, for a major role of the sodium-calcium exchange current during normal cardiac rhythm.

Brown *et al.*[6] and Arlock and Noble[7] recorded slow components of inward current following the calcium current that were attributed to sodium-calcium exchange activated by the intracellular calcium transient, while Mitchell *et al.*[8,9] and Schouten and ter Keurs[10] showed that the exchange current could play a major role in maintaining the late low plateau in rat ventricular cells. Several years previously Simurda *et al.*[11] showed that a component of inward current during weak depolarizations is directly proportional to contractile force, suggesting that it was calcium activated.

[a]The work on single atrial cells in the Seoul laboratory was supported by a research grant from the Korea Research Foundation and by a basic medical research grant from Seoul National University College of Medicine. The ventricular work in the Oxford laboratory described in this paper is supported by the British Heart Foundation and the Medical Research Council. The computer program HEART, from which some of the figures were computed is available from OXSOFT Ltd., 49 Old Road, Oxford, OX3 7JZ, U.K.

[b]Address for correspondence: University Laboratory of Physiology, University of Oxford, Parks Road, Oxford OX1 3PT, U.K.

Until 1987 it has to be admitted that these explanations were always subject to the criticism that the currents concerned were measured in highly complex situations with many other current components also present. Of course, if the role of sodium-calcium exchange during normal cellular activity is to be assessed, such experiments are necessary, and we will describe more of them later in this paper. But they need to be backed up by experiments done under conditions where the exchange current is isolated from other components. This was first achieved by Kimura, Miyamae, and Noma in 1987.[12] The results were very relevant indeed, not only because they showed the electrogenic nature of the sodium-calcium exchange in cardiac cells (including demonstrating that its most likely stoichiometry was 3 : 1) but also because the current-voltage relations obtained were very similar to those required theoretically in modeling work [13-17] to explain the results obtained in normal beating cardiac tissue.

It is no longer necessary, therefore, to prove the hypothesis that sodium-calcium exchange contributes a significant electric current. The question now is rather a functional one: How large can this current be and can it be sufficient to be primarily responsible for maintaining calcium balance in the heart by generating the major component of calcium efflux? This is the hypothesis that led to the experimental and modeling work done by Hilgemann and Noble.[15]

In this paper, we will describe experiments done on atrial and ventricular cells that attempt to refine this hypothesis. We will raise the question of how the contribution of sodium-calcium exchange to the action potential varies with the region of the heart from which the cells come, with the ionic concentration levels (in particular of internal ion concentrations) and with inotropic state.

ATRIAL CELLS

The atrial cells of most species display a characteristic, almost triangular, action potential waveform, very distinct from the "square" waveform of most ventricular cells. The exception is the rat ventricle, which also has the "atrial" waveform. Not surprisingly, the ionic mechanisms differ substantially according to waveform and the rat ventricular action potential closely resembles the atrium in its ionic mechanisms.

We start with the atrial waveform because its two phases of repolarization reflect an almost complete separation between the contributions of the calcium current, i_{Ca}, and the sodium-calcium exchange current, i_{NaCa}. The calcium current flows during the depolarization and is rapidly deactivated once the transient outward current produces the initial fast repolarization to the beginning of the low plateau at about -40 mV. The exchange current flows during, and maintains, this late plateau. This functional separation leads to a very useful experimental protocol for studying i_{NaCa}.[17] First the calcium current, and consequent internal release of calcium, is activated with a brief depolarization for 2-5 msecs that mimics the rapid phase of the action potential. The membrane potential is then stepped back to various potentials within the voltage range of the late slow plateau. The slow inward current recorded then reflects almost entirely current activated by internal calcium, including i_{NaCa}.

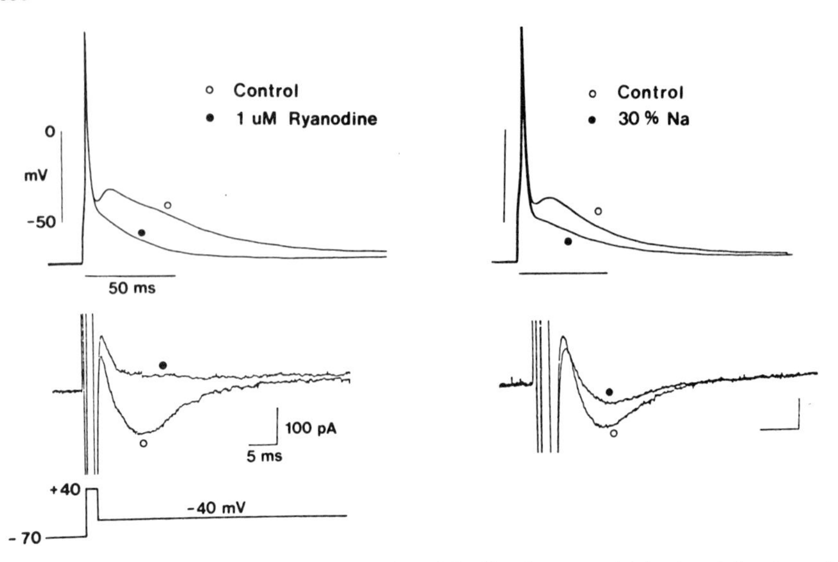

FIGURE 1. *Left:* effect of ryanodine on rabbit atrial cell action potential (top) and slow inward current during late plateau (bottom). Note that abolishing calcium release with ryanodine abolishes the inward current and the late plateau collapses. *Right:* Effect of reducing external sodium to 30% normal concentration on action potential (top) and inward current (bottom).[17]

FIGURE 1 shows experimental results obtained using this protocol. In the left-hand records, internal calcium release was blocked by using ryanodine. On the right, the sodium-calcium exchange was reduced by substituting external sodium with lithium ions. In both cases, the late plateau collapses and the slow inward current is almost abolished. This is very similar to the behavior of the rat ventricle.[8,11]

The evidence that the slow inward current recorded during the late plateau is calcium-activated is now very strong. FIGURE 2 shows the effect of infusing a high (1 mM) concentration of the rapid calcium chelator BAPTA. The initial concentration was low (20 μM) and allowed a large inward tail current to be recorded. As 1 mM BAPTA is infused from the patch electrode, the inward current is gradually suppressed.

Evidence that the current is carried by the sodium-calcium exchange mechanism comes from measurements of its voltage dependence. FIGURE 3 shows the voltage dependence of the current separated by a variety of methods employed by Earm, Ho, and So.[17] In each case the current-voltage relation shows the nearly exponential dependence on voltage that is characteristic of the inward mode of the sodium-calcium exchange.

FIGURE 4 shows the reconstruction of the rabbit atrial action potential in the single-cell model[18] developed from the Hilgemann-Noble[15] model. The main panel shows the computed ionic currents during the action potential, showing i_{NaCa} as the main depolarizing current during the late plateau.

What is the functional significance of this very substantial contribution of the Na-Ca exchange to the cardiac action potential? This was the question that initially led to the development of the Hilgemann-Noble model. The aim of the model was to reproduce the net surface membrane calcium fluxes determined by measuring fast extracellular calcium transients.[19] In this the model was entirely successful. The quantitative significance of this result is that the sodium-calcium exchange activity required

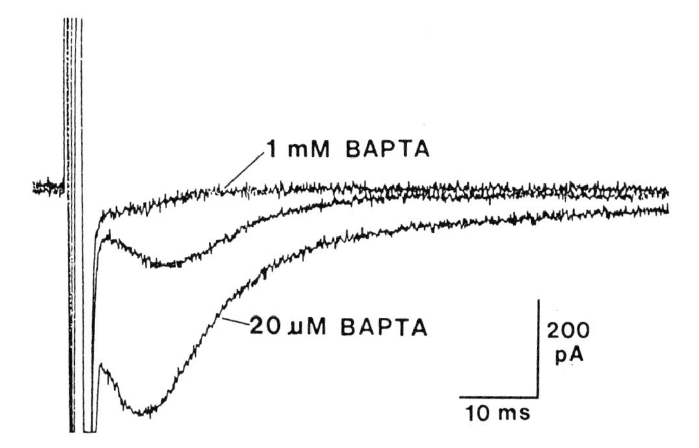

FIGURE 2. Effect of intracellular perfusion with BAPTA. The slow inward current during the late plateau of the rabbit atrial cell was obtained using the same protocol as in FIGURE 1 with 20 μM BAPTA in the patch electrode. When 1 mM BAPTA is perfused into the cell, the slow current first diminishes and is then almost abolished.

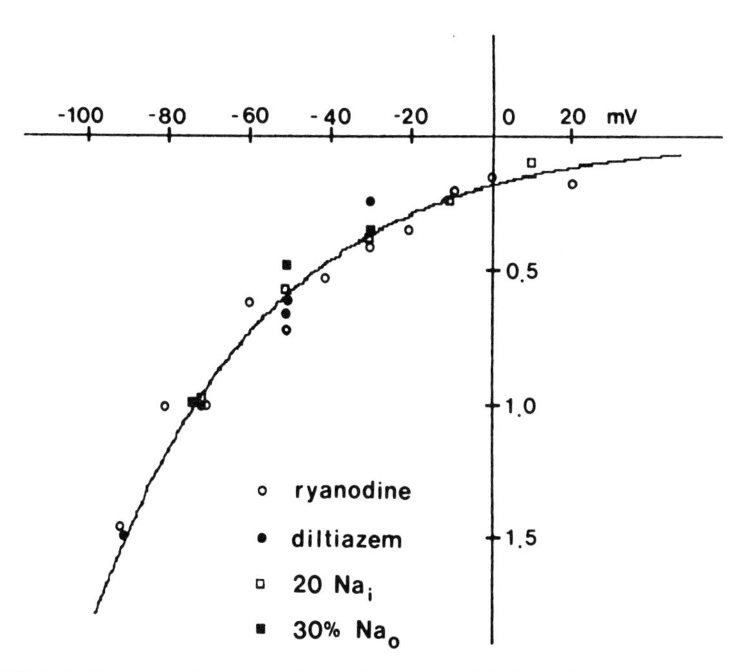

FIGURE 3. Fully activated current-voltage relationship of the inward current in rabbit atrial cell. The difference currents measured before and after treatment of the cells with ryanodine, diltiazem, 20 mM Na_i^+ or 30% Na_o^+ were plotted against the membrane potential. The magnitude of the difference current was normalized and expressed as a fraction of the difference current magnitude at the holding potential (usually −70 mV). The relationship is nearly an exponential curve characteristic of that of the inward mode of the sodium-calcium exchanger.[17]

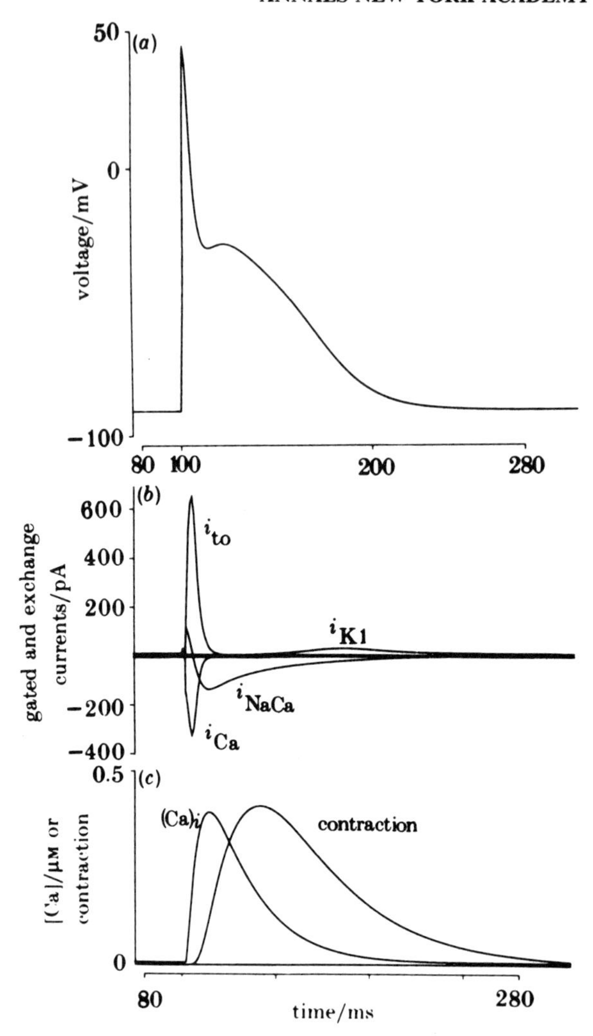

FIGURE 4. The Earm-Noble model of the single rabbit atrial cell, based on the multicellular model of Hilgemann and Noble.[15] *Top:* computed action potential. *Middle:* computed currents. *Bottom:* Computed $[Ca]_i$ and contraction.[18]

to maintain the late plateau is also entirely sufficient to enable calcium balance to be achieved by pumping out a quantity of calcium equivalent to that entering the cell via the voltage-gated calcium channels.

This is, presumably, the main functional significance of sodium-calcium exchange during normal beating of the heart, and, in a sense, the influence of the sodium-calcium exchange current on the action potential can be viewed as a necessary side effect of the fact that the exchange is electrogenic.

Nevertheless, this electrogenic effect does itself also have important functional consequences. Thus, any factor that leads to an enhancement of the inward sodium-calcium exchange current will automatically prolong and enhance the late plateau. One such factor is inotropic state. This can best be illustrated by using a simple physiological intervention that enhances contraction: the rested-state contraction.

FIGURE 5 shows experimental and computed results obtained by initiating action potentials and contractions following a long rest period. The contraction, and intracellular calcium transient, is then greatly enhanced (as a result of continuous calcium loading of the SR during the rest period), and it can be seen that the late plateau is then also greatly enhanced. As the cell returns to the normal steady state during repetitive beating following the rested-state contraction, the late plateau returns to its normal amplitude and duration.

Not all interventions that enhance the calcium transient (and the contraction) lead to this result, however. Thus, a very different result is obtained when the contraction

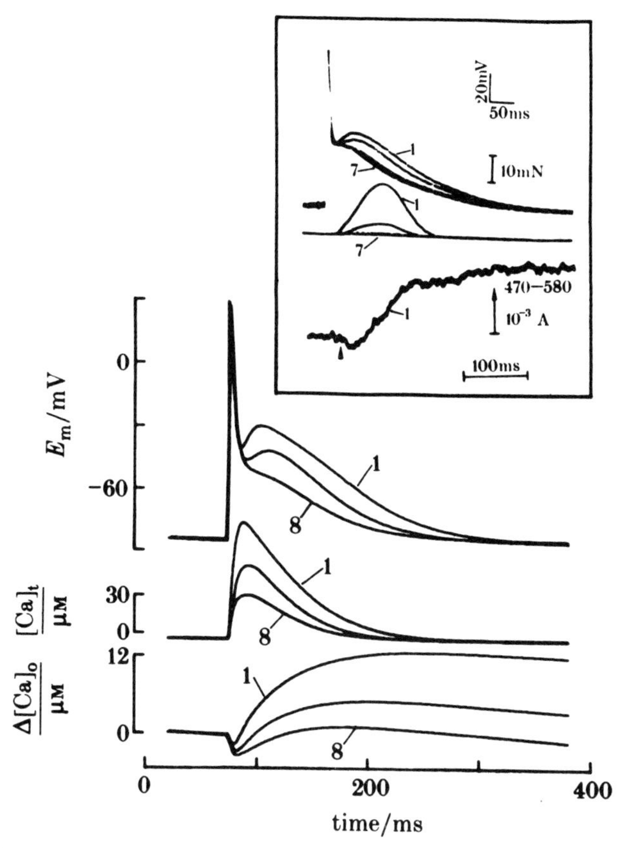

FIGURE 5. Experimental and computed reconstruction of post-rest stimulation at low frequency in rabbit atrium. In the inset (the experimental records) the bottom record (1) shows the extracellular calcium transient obtained during a post-rest potential contraction. The top two records of the inset show the developed tension and the action potentials during post-rest stimulation at 0.2 Hz. As the contractions decline the late low plateau shortens. The bottom records show the computed results using the Hilgemann-Noble model. Here also the plateau repolarizes more quickly as the calcium transient falls progressively with successive beats.[18]

is increased by inhibiting the sodium-potassium exchange pump, as would occur during the action of cardiac glycosides.

FIGURE 6 shows the results obtained in the single atrial cell model when the sodium-potassium exchange pump is reduced to 10% of its normal level of activity. The initial effect (top left) is a small prolongation of the action potential attributable to reduction of the outward sodium pump current, i_p. As in experiments with cardiac glycosides,[20] this effect is very transient and over a period of minutes the action potential shortens (top right). This change is attributable to a rise in internal sodium (bottom left), which reduces the sodium gradient that drives the exchange process. Since the stoichiometry of the exchange is 3 : 1, this effect of sodium gradient is very steep. Even a few mM change in internal sodium can effectively abolish the contribution of i_{NaCa} (see also FIG. 12), so that the plateau collapses. Note that this effect occurs despite the strong positive inotropic effect of pump inhibition (which is also successfully reproduced by the model—bottom left). The influence of the reduced sodium gradient must therefore outweigh the opposite effect of an increased calcium transient. The explanation for this is that the exchange current is simply proportional to internal calcium and so increases linearly with calcium (at least in some experimental conditions), whereas the dependence on sodium is nonlinear.[12] Together with the high stoichiometry (which means that changes in $[Na]_i$ have a very large effect on the equilibrium potential for the exchange process), this ensures that the relatively small increase in internal sodium concentration more than outweighs even a relatively large increase in the intracellular calcium transient.

This massive reduction in action potential duration also occurs experimentally and has already been attributed to the secondary consequences of sodium pump inhibition on the sodium gradient and hence on the sodium-calcium exchange current.[20]

VENTRICULAR CELLS

Some ventricular cells, for example, those of the adult rat heart, show action potential waveforms very similar to that of the atrium, and not surprisingly a similar analysis can be made. Indeed, the contribution of the sodium-calcium exchange current to this type of action potential was first established in work on rat ventricular cells.

In most species, however, including man, guinea-pig, rabbit, cat, and dog, the ventricular action potential shows a characteristic high ("square") plateau. The cell membrane repolarizes very slowly until it reaches negative potentials, when the voltage-dependent reactivation of i_{K1} rapidly re-establishes the resting potential. In this kind of action potential, the plateau occurs at potentials where the inactivation of calcium current occurs relatively slowly, and it is then not possible to separate the sodium-calcium exchange component of current in the neat way in which this has been done in atrial cells. This gives rise to two major problems, neither of which has yet been solved completely satisfactorily. The first is experimental: How to achieve the separation? The second is theoretical: As we will show later in this paper, the high plateau makes the time course, amplitude, and even the direction of the sodium-calcium exchange current extremely sensitive to small variations in either membrane potential, internal sodium, or the amplitude of the internal calcium transient.

First, then, how can the separation be achieved experimentally? The method used in the Oxford laboratory[21] is based on the ability to separate the exchange current at very negative potentials (negative to -40 mV), combined with extrapolating the results to work out what the exchange current does at positive potentials. The principle is illustrated in FIGURE 7, which shows results obtained on guinea-pig ventricular cells.

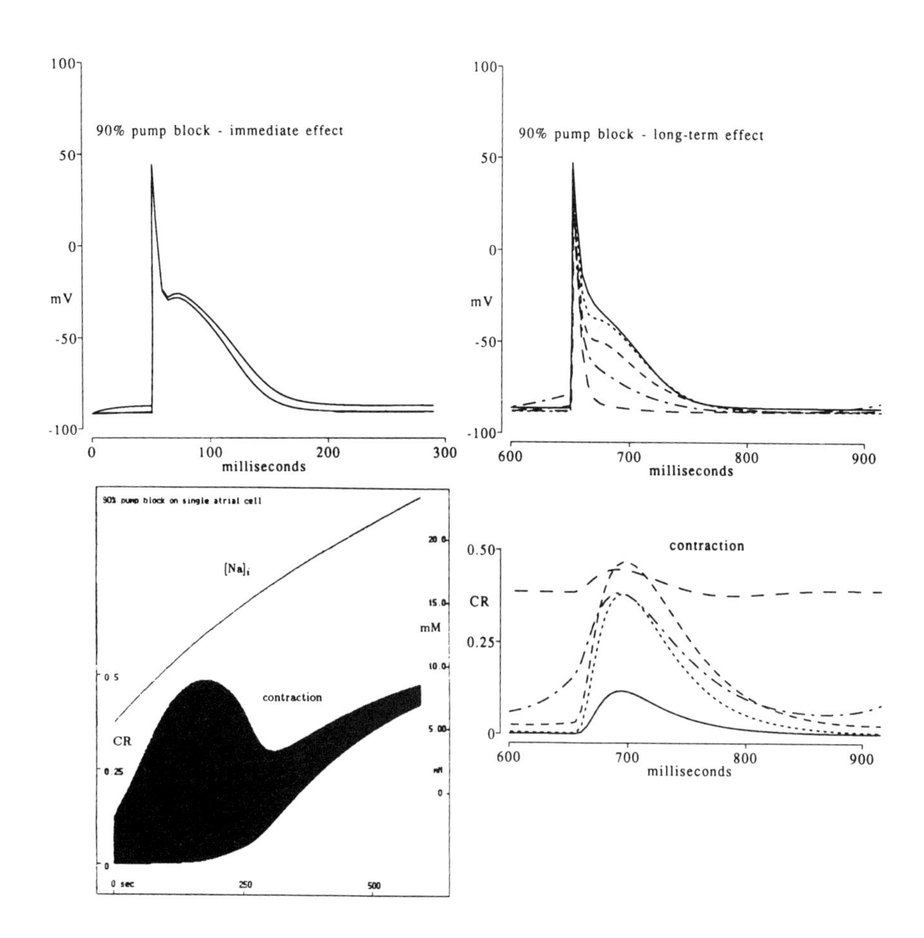

FIGURE 6. Reconstruction of the actions of cardiac glycosides using the Earm-Noble[16] model of the single rabbit atrial cell. *Top left:* immediate effect of 90% sodium pump block is a small depolarization of the resting potential and a moderate lengthening of repolarization. *Bottom left:* computed internal sodium and contractions (fraction of cross-bridge formations, CR) over a time scale of 600 seconds following 90% sodium pump block. Note the massive positive inotropic effect, which turns into a negative inotropic effect as the resting contraction starts to appear. *Top right:* action potentials computed at the beginning (0 seconds—continuous line), 100 seconds (dotted line), 200 seconds (dashed line), 250 seconds (dot-dashed line), and 500 seconds (long dashed line). *Bottom right:* corresponding computed contractions. Note that at 250 seconds there is oscillation of contraction (and hence of internal calcium).[5]

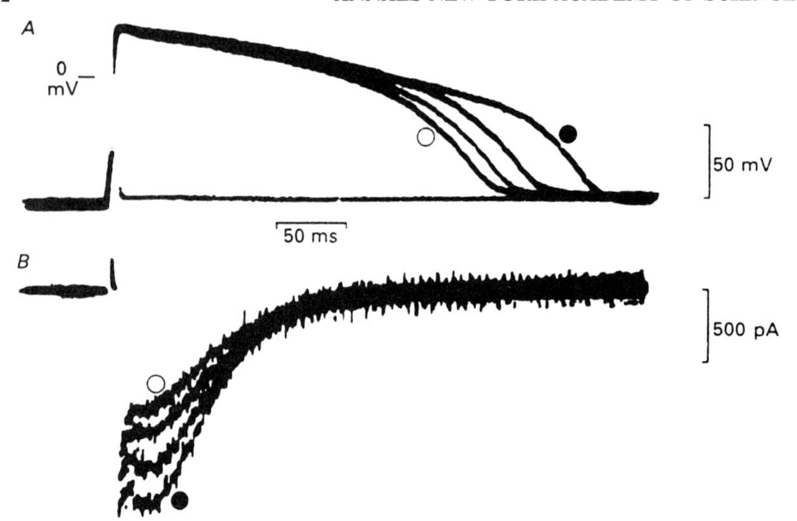

FIGURE 7. Recording of sodium-calcium exchange current tails in guinea-pig ventricular cell by imposing a repolarizing clamp to the resting potential very early (10 msec) during the action potential plateau. This experiment shows that the tail currents evoked at this early time show a rounded time course reaching a peak inward current after about 30 msec before decaying over a period of 200 msec. The experiment also shows the effect of isoprenaline on the action potential and on the exchange current tail. Isoprenaline increased the duration of the action potential and increased the amplitude of the tail current.

The method is to initiate action potentials at regular intervals (in this case every 10 seconds), and during some of the action potentials (usually alternate ones) to interrupt the plateau by clamping back to the resting potential (about −90 mV). Provided that the intracellular calcium transient is not buffered, this protocol produces a slowly decaying inward tail current, similar to that obtained by clamping back to negative holding potentials after activating the calcium current.[21] The amplitude of this tail current varies in the way expected of a process whose activation time course reflects that of the intracellular calcium transient. It increases on increasing the rate of stimulation (reflecting the well-known staircase effect—see Egan *et al.*[22] (FIG. 3). It is reduced by interrupting every action potential (which prevents loading of the SR—see Egan *et al.*,[22] FIG. 2), and it is increased by increasing SR loading, for example by applying adrenaline. This was the manipulation used in FIGURE 7.

FIGURE 7 also illustrates an important feature of the tail current obtained in this way, which is that very early interruptions (within the first 20-50 msec of the action potential plateau) produce a current that first shows a small increase to its inward peak before decaying towards zero within around 200 msec. The atrial exchange current also shows this phenomenon (see FIGS. 1 and 2). Its interpretation will be discussed later in this paper.

FIGURE 8 shows that if later interruptions are used, the inward tail current then simply decays exponentially. It is also clear from this experiment that the envelope of the tail current amplitudes obtained by interrupting at progressively later times during the plateau resembles that of the intracellular calcium transient. This is what would be expected if the current is largely sodium-calcium exchange current responding rapidly (within a msec or two) to the variations in internal calcium. Moreover, from the work of Kimura, Miyamae, and Noma,[12] we know what the relation between these two variables should be. The exchange current is roughly linearly related to internal calcium (at least the relation is roughly linear above 100 nM $[Ca]_i$) and, for a guinea-pig ventricular cell, the constant of proportionality is about 1 nA of current recorded at −80 mV per μM of calcium. We can therefore convert the experimental current tail envelope of FIGURE 8 into a presumed calcium transient. This has been done in FIGURE 9. The result is a peak calcium transient a little above 1 μM, which is similar to that recorded experimentally with fura-2.

Knowing the variation of internal calcium, the current recorded at various times at one potential, the external calcium and sodium concentrations, and the known current-voltage relation for the exchange current, only one variable is missing to enable the exchange current to be calculated at any time and at any voltage. This is the level of internal sodium. In the calculation done in FIGURE 9, we assumed that this was 7 mM. This assumption enabled calculation of the reversal potential of the sodium-calcium exchanger to be made at each time, and of the net current it should carry during the plateau of the action potential. The result shows that in this case the current would be expected to be inward throughout the plateau after 50 msec. It should therefore be transporting calcium out of the cell during this time, and contributing a small current that would help to maintain the plateau itself. Egan *et al.*[22] calculated that the net

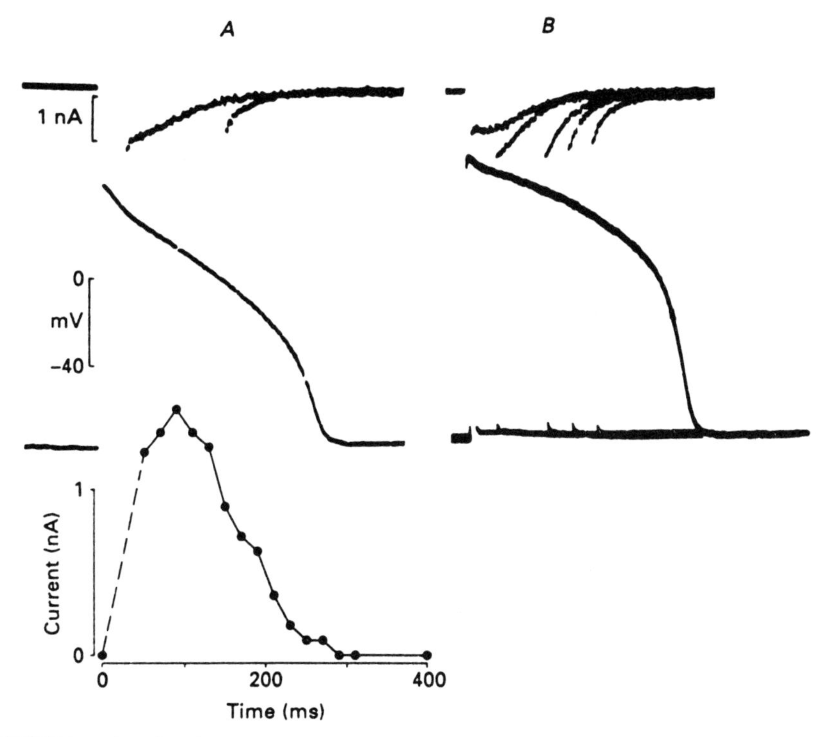

FIGURE 8. Results of two experiments on isolated guinea-pig ventricular cells in which clamp interruptions of the action potential were applied at various times following the onset of repolarization. **A:** Two examples of the tail currents (top), the uninterrupted action potential (middle) and a graph of the tail current amplitude envelope (bottom). **B:** The results of interruptions at five intervals in a different cell.[22]

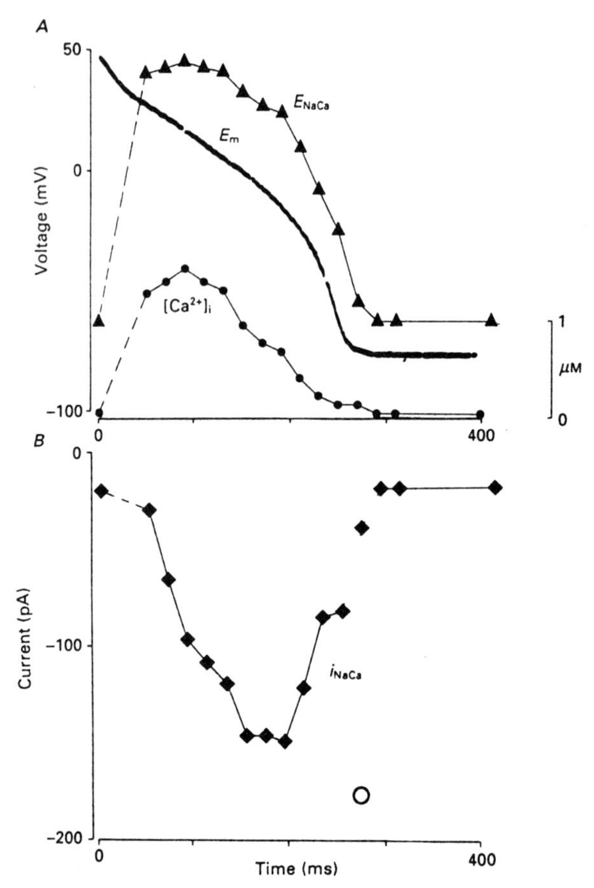

FIGURE 9. This figure is constructed from that of FIGURE 8 using the assumptions described in the text.[21] The filled circles show the tail current amplitude converted to internal calcium concentrations using a conversion factor of 1 μM/nA (see Kimura *et al.*[12]). The filled triangles show the computed values for the reversal potential for Na-Ca exchange using these $[Ca]_i$ values and an assumed $[Na]_i$ of 7 mM. The filled squares show the computed exchange current during the plateau and repolarization using the DiFrancesco-Noble equation for i_{NaCa}. We are grateful to Dr. George Hart for pointing out to us that one of the points on this graph was misplotted. This is the point corresponding to the last tail current significantly different from zero. To check the matter, we have recalculated the data throughout the action potential. The open circle indicates where the point should be and how easily it was misplotted: It is 10 times larger than the "expected" point! It could, therefore, be taken as experimental evidence for the narrow second inward peak computed in the model shown in FIGURE 11. We would not, however, like to press the point too strongly: The experimental data at this very late time during repolarization depends on tail currents that are only a little larger than zero. The reason why such a small data point can nevertheless give a relatively large inward current is that the voltage at this time is repolarizing very rapidly and the exchange current is very strongly voltage-dependent in precisely this range of potential. Clearly, this region of the data should be investigated more carefully in future experiments to determine whether a second interval peak really does exist. Ryder, Bryant, and Hart[27] have used the same method in normal and hypertrophied guinea-pig ventricular cells and their data (Hart, personal communication) is consistent with the existence of a second inward peak.

extrusion of calcium so produced could be similar to the net entry of calcium via the calcium channels. It is therefore plausible that, as in rabbit atrium, the exchange process is the main route by which calcium leaves the cell during the repolarization phase.

These results have been simulated recently using a model of the guinea-pig ventricular cell constructed in the Oxford laboratory. The results are shown in FIGURE 10. The model was developed from the single atrial cell model described earlier in this paper. The major differences are:

(1) The inward rectifier current, i_{K1}, plays the major role both in maintaining the resting potential and in the late phase of repolarization. The equations for this current (see APPENDIX) were developed from the original DiFrancesco-Noble[13] equations by varying the steepness of the voltage dependence of activation, and the K_m for potassium activation, until the computed current fits that obtained experimentally in guinea-pig ventricular cells by Sakmann and Trube.[23]

(2) The transient outward current was greatly reduced. This current is not completely absent in the guinea-pig ventricular cell, but it is smaller than in the atrium.

(3) Inactivation of the calcium current was slowed compared to the atrial kinetics.

(4) The parameters determining the release and uptake of calcium by the SR stores were adjusted to give an intracellular calcium transient similar to that recorded experimentally in ventricular cells using fura-2 and Indo-1.

The computations were performed using version 3.3 of the OXSOFT HEART program. The complete input file is reproduced in the APPENDIX.

It can be seen that the stimulation of the exchange current tails in FIGURE 10 is very successful indeed. Early interruptions generate a tail current that rises slowly to a peak before decaying, while interruptions later than 50 msec activate simple exponential current decays. The overall envelope of the current peaks resembles that shown experimentally in FIGURE 8. The right-hand part of FIGURE 10 shows the change in time course as the time of interruption is prolonged in more detail on an expanded time scale. This computation also shows that, over this period of time, the envelope of the peak exchange current tails would be a very poor measure of the rising phase of the calcium transient. The envelope is a much better indicator of the falling phase. (The rising phase is also only poorly measured by calcium indicators, like fura-2 and Indo-1.[24] At high concentrations of these indicators the falling phase is also seriously distorted and becomes much longer than the envelope of the sodium-calcium exchange tails).

We will use this model later in this paper to investigate some of the features that might control the flow of the exchange current during the action potential. First, however, it is worth explaining how the model succeeds so well in reproducing the experimental tail currents. The main reason lies in the fact that early interruptions occur at a time when the calcium-dependent calcium release process is still occurring. This means that, although hyperpolarization immediately causes a large efflux of calcium to occur through the exchanger, release from SR continues to cause $[Ca]_i$ to increase. This gives rise to the rounded shape of the early exchange current tails. The implication, of course, is that the SR calcium release is at least partly regenerative. In the atrial model from which the present model was developed, this is the case.[18] The exchange current tails recorded experimentally show a steeper dependence on voltage than does the calcium current,[17] an experimental feature that the atrial model successfully reproduces (Earm & Noble,[18] FIG. 3).

We will now use the ventricular cell model to explore some of the factors that control the exchange current during the action potential.

FIGURE 11 shows the computed exchange current during the action potential. The result shows three peaks, including an early peak of *outward* current, which would correspond to calcium entering the cell via the sodium-calcium exchange. In the atrial model this entry is in fact sufficient to activate calcium release at very strong

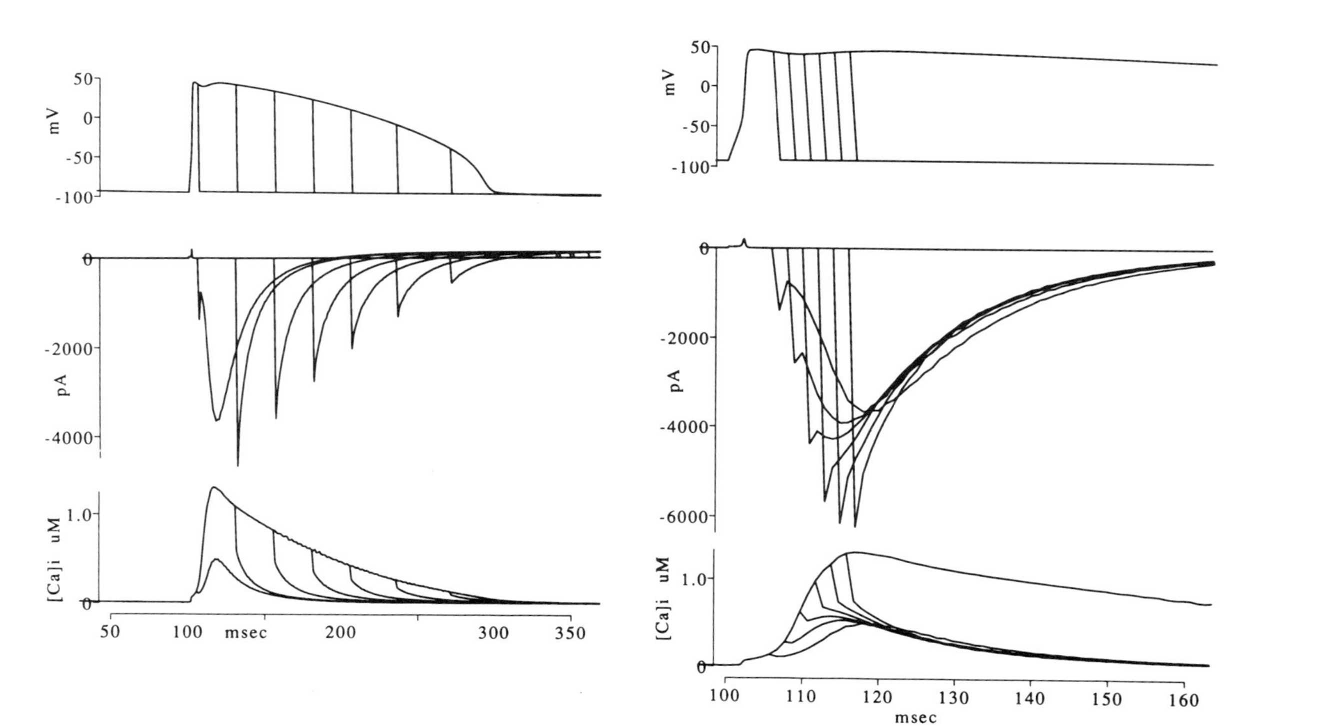

FIGURE 10. Reconstruction of results of the type shown in FIGURE 8 using a computer model of the guinea-pig ventricular action potential. The left-hand traces show the results of interrupting the action potential at various times (*top traces*) to produce a family of inward current tails similar to those observed experimentally (*middle traces*). The bottom traces show the computed internal calcium transients. The right-hand traces show interruptions during the rising phase of the calcium transient plotted on an expanded time scale to show how the current tails change from a rounded to an exponential time course.

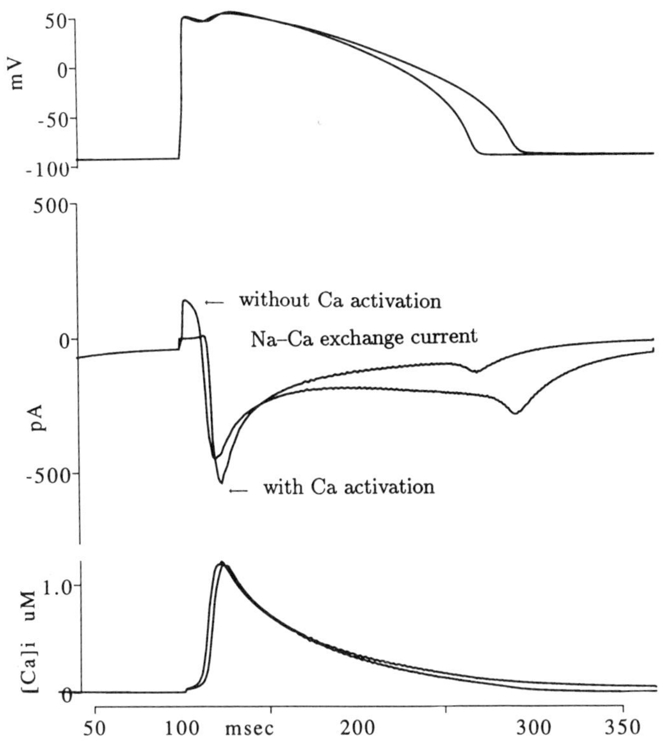

FIGURE 11. Computation of the time course of the exchange current during the normal model action potential. The computation was first done with the original DiFrancesco-Noble[13] formulation of the equations for sodium-calcium exchange. It was then repeated with calcium activation included as described in the text. To achieve this the following parameters were set immediately after the upstroke of the action potential: KMNACA = 0.0002, KMCA = 0.005, KNACA = 0.001. KMNACA sets the K_m for calcium activation of the exchange (200 nM). The adjustments to the other parameters compared to their normal values (see APPENDIX) was done in order to keep the overall amplitudes of the calcium transient and the inward peak of the exchange current roughly constant.

depolarizations when very little calcium would enter through the calcium channels (see Earm & Noble,[18] FIG. 4). This possibility has also been shown experimentally by Hume *et al.*[25] How certain, though, can we be that it really does occur?

The rather large amplitude of the outward peak of current computed in FIGURE 11 depends on assuming that the outward mode of the exchange can be activated at very strong depolarization even when internal calcium is very low (i.e., at the resting level before the SR release is activated). It is by no means certain that this is the case. On the contrary, experimental results show that there is a calcium regulatory site that ensures that at very low intracellular calcium concentrates the exchange process becomes inactivated. We have therefore explored the effect of such a calcium-dependent activation of the exchange mechanism by including a calcium-binding reaction in the equations for i_{NaCa}. The current calculated using the original DiFrancesco-Noble[13] equation was modified by multiplying it by the factor $([Ca]_i/([Ca]_i + K_{mNaCa})^3$ This corresponds to the steady state of an activation binding reaction with three calciums binding to the activation site. For the computation in FIGURE 11 we chose to set K_{mNaCa} to 200 nM. The result is very striking: The outward peak of exchange current is completely removed. This arises because, at this time, calcium is still too low to activate the exchange, even when it is carrying calcium into the cell.

The point of this calculation is not, however, to claim that an outward peak of current (and hence calcium entry that could trigger release) does *not* occur, but rather to emphasize that the amplitude of this process must depend on the activation characteristics of the exchange at low internal calcium. It is therefore of great importance functionally to determine the activation kinetics by low calcium as accurately as possible.

The second peak (the first inward peak) of the exchange current occurs fairly early during the plateau. This peak is in fact earlier than in the exchange current computed from our experimental results (compare with FIG. 10). This arises from the fact that the plateau in the modeled action potential repolarizes more slowly than in the experimental response. The difference in potential is not large, which emphasizes how sensitive the system is when E_{NaCa} is so close to the plateau potential. Finally, as already noted in the case of the experimental result, there can also be a third peak (a second inward peak) during the final phase of repolarization. This possibility was also noted by Hilgemann.[26] It arises from the fact that rapid repolarization activated the exchange current very strongly. If the calcium transient has not fully returned to its resting value during the final phase of repolarization, a second peak is possible. Again, though, it should be noted that its amplitude would be strongly dependent on the characteristics of the calcium-activation process at low calcium. When calcium activation is included, the late peak becomes very small. Again, this emphasizes the functional importance of knowing these characteristics more accurately.

In FIGURE 12 we have investigated the expected dependence on internal sodium concentration. The computations were done using the calcium-activated exchange current equations at internal sodium concentrations of 3, 4, 5, 6, 7, and 8 mM. We chose this range because it covers the range of concentrations measured experimentally using sodium ion-sensitive electrodes.

The first conclusion to be drawn from the results is that the amplitude, time course, and even the direction of the exchange current are all very strongly dependent on internal sodium. As expected, when the sodium gradient is very high (i.e., internal sodium very low), the exchange current is strongly inward and would be very effective in pumping calcium out of the cell. When the sodium gradient is reduced (internal sodium increased) the calcium efflux is reduced and can even reverse. The majority of calcium efflux through the exchange would then occur late in the action potential. Note that the late inward peak *increases* as the early inward peak is reduced or reversed.

For the purpose of calculating the exchange current during the plateau from experimental results of the kind we have obtained, these results are extremely disappointing because we do not know the absolute level of internal sodium with the accuracy required. Most recent estimates of internal sodium give lower values than originally thought (4-5 mM, rather than 7-10 mM, would now be thought reasonable), but the range is still large. A 2-3 mM difference in $[Na]_i$ has too large an effect on the computed current to be able to say with any certainty what its amplitude and time course really should be.

From a functional point of view, by contrast, the results are very exciting. It is now very well established that, by controlling internal calcium via the sodium-calcium exchanger, the level of internal sodium is one of the main determinants of the SR calcium loading and hence of inotropic state. These effects would, over time, counter the effects computed in FIGURE 12. Thus, when internal sodium is increased, although the efflux through the exchange during each beat would initially be reduced or even

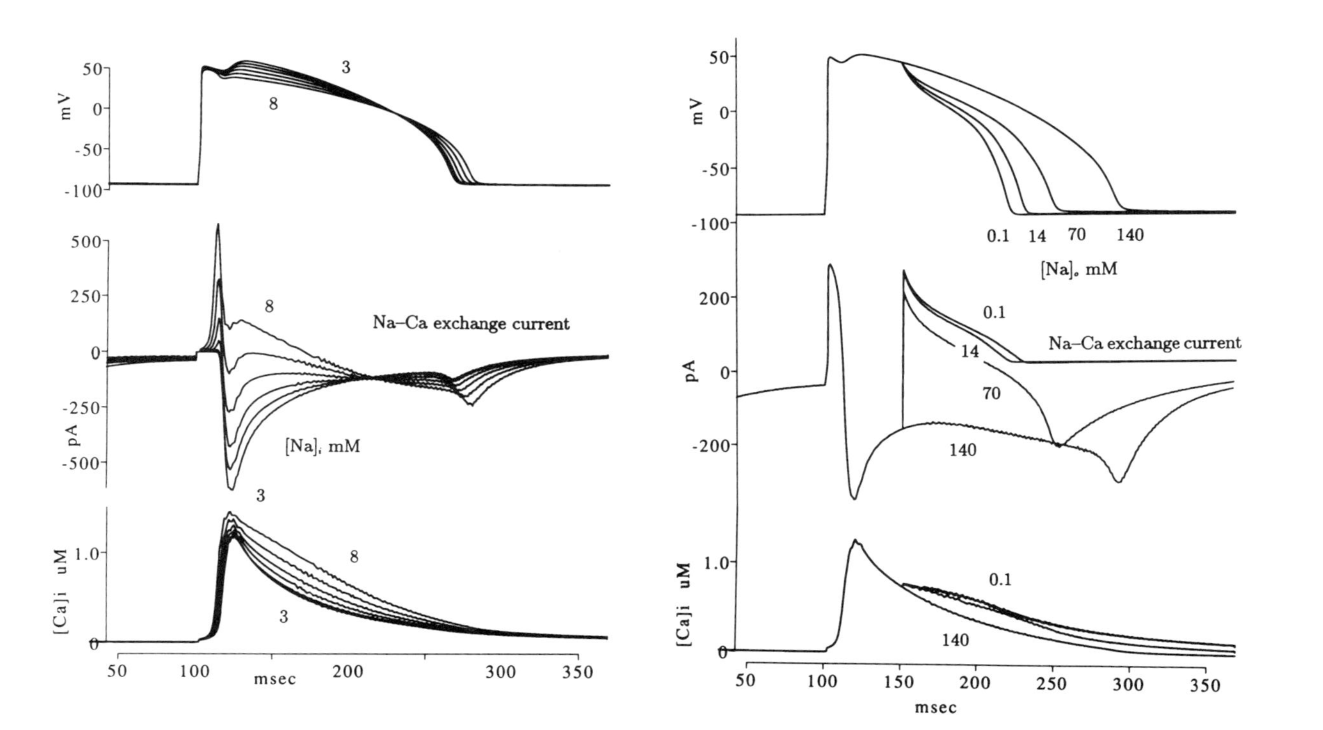

FIGURE 12. *Left:* effect of varying internal sodium on exchange current (*middle traces*), action potentials (*top traces*), and calcium transient (*bottom traces*). During each run internal sodium was changed immediately after the onset of the action potential to obtain the *immediate* effect of internal sodium. As explained in the text, this would not be the long-term steady-state effect. *Right:* Similar computations done by varying external sodium. External sodium was suddenly reduced after 50 msec to 70, 14, or 0.1 mM.

reversed as computed here, the *later* result of this disturbance of calcium balance would be that much less calcium would leave during each beat than enters. This is the dynamic reason why stored calcium then accumulates to give the positive inotropic effect. But, as the calcium transient grows with increased storage of calcium, the exchange current will once again shift in the inward direction. We have not yet computed these effects in the ventricular model, but they certainly occur in the atrial model[15] from which it was developed and should therefore also be present in the new model.

This is, of course, an exquisitely sensitive control mechanism. The exchange current amplitude and direction is strongly sodium-sensitive over precisely the range of sodium concentrations that have a strong inotropic effect. If it is correct to assume in ventricle, as it is in the atrium, that the sodium-calcium exchange is the main mechanism by which calcium is extruded from the cell on a beat-to-beat basis, then the necessary feedback control to bring this process into a steady state in a wide variety of inotropic conditions clearly exists as an inherent property of the system. It is quite possible that no other control process is actually required. It will be an exciting challenge in the future to work this out in detail in the ventricle (for example, during staircase effects, during sodium pump changes, and during the actions of inotropic agents) along the lines already established in the atrium.[15] The problem, though, will be to obtain experimental data accurate enough to validate the modeling.

Finally, the right-hand traces in FIGURE 12 show the effects of sudden variations in the extracellular sodium concentration. We have calculated the results of reducing $[Na]_o$ from 140 mM to 70, 14, and 0.1 mM 50 msec after the onset of the action potential. Even though the net current carried by the exchange is fairly small (less than 200 pA at this time) the effects on the speed of repolarization are very substantial indeed. This is attributable, first, to the fact that the net membrane conductance during the action potential plateau is very low, so that fairly small currents can have a large effect on the rate of repolarization. Second, it can be seen that large variations in external sodium cannot only remove the inward current but also strongly reverse the current flow. The net change in the exchange current can therefore be large, and it can change from being a current that helps to maintain the plateau to one that helps to terminate it. Changes similar to those computed here have also been observed experimentally in isolated guinea-pig ventricular cells subjected to rapid changes in extracellular sodium (J. Bridge, J. Smolley & K. Spitzer, personal communication).

It has, of course, been known for a very long time (see, for example, Weidmann's classic monograph[28]) that low external sodium has a large effect on the action potential duration. This would formerly have been attributed to residual activation during the plateau of sodium channels or to sodium ions flowing through the calcium channel. In fact the sodium channel blocker, tetrodotoxin, has very little influence on the ventricular action potential plateau,[25] and sodium enters via calcium channels only at very low external calcium. It is likely, therefore, that the most important mechanism of the external sodium dependence of the plateau duration is sodium-calcium exchange.

REFERENCES

1. KASS, R. S., W. J. LEDERER, R. W. TSIEN & R. WEINGART. 1978. Role of calcium ions in transient inward current and after contractions induced by strophanthidin in cardiac Purkinje fibres. J. Physiol. **281:** 187-208.
2. MULLINS, L. J. 1977. A mechanism for Na/Ca transport. J. Gen. Physiol. **70:** 681-695.
3. MULLINS, L. J. 1981. Ion Transport in the Heart. Raven Press. New York.
4. FEDIDA, D., D. NOBLE, A. C. RANKIN & A. J. SPINDLER. 1987. The transient inward current, i_{TI}, and related contraction in guinea-pig ventricular myocytes. J. Physiol. **392:** 523-542.

5. NOBLE, D. 1991. Ionic mechanisms determining the timing of ventricular repolarization: Significance for cardiac arrhythmias. Ann. N. Y. Acad. Sci. In Press.
6. BROWN, H. F., J. KIMURA, D. NOBLE, S. J. NOBLE & A. I. TAUPIGNON. 1984. Mechanisms underlying the slow inward current, i_{si}, in the rabbit sino-atrial node investigated by voltage clamp and computer simulation. Proc. R. Soc. London **B 222:** 305-328.
7. ARLOCK, P. & D. NOBLE. 1985. Two components of "second inward current" in ferret papillary muscle. J. Physiol. **369:** 88P.
8. MITCHELL, M. R., T. POWELL, D. A. TERRAR & V. W. TWIST. 1984. The effects of ryanodine, EGTA and low-sodium on action potentials in rat and guinea-pig ventricular myocytes: Evidence for two inward currents during the plateau. Br. J. Pharmacol. **81:** 543-550.
9. MITHCELL, M. R., T. POWELL, D. A. TERRAR & V. W. TWIST. 1987. Calcium-activated inward current and contraction in rat and guinea-pig ventricular myocytes. J. Physiol. **391:** 545-560.
10. SCHOUTEN, V. J. A. & H. E. D. J. TER KEURS. 1985. The slow repolarization phase of the action potential in rat heart. J. Physiol. **360:** 13-25.
11. SIMURDA, J., M. SIMURDOVA, P. BRAVENY & J. SUMBERA. 1981. Activity-dependent changes of slow inward current in ventricular heart muscle. Pflugers Arch. **391:** 277-283.
12. KIMURA, J., S. MIYAMAE & A. NOMA. 1987. Identification of sodium-calcium exchange current in single ventricular cells of guinea-pig. J. Physiol. **384:** 199-222.
13. DIFRANCESCO, D. & D. NOBLE. 1985. A model of cardiac electrical activity incorporating ionic pumps and concentration changes. Phil. Trans. R. Soc. London **B 307:** 353-398.
14. NOBLE, D. 1986. Sodium-calcium exchange and its role in generating electric current. *In* Excitation and Regulation of Contraction. R. Nathan, Ed.: 171-199. Academic Press. New York.
15. HILGEMANN, D. W. & D. NOBLE. 1987. Excitation-contraction coupling and extracellular calcium transients in rabbit atrium: Reconstruction of basic cellular mechanisms. Proc. R. Soc. London **B 230:** 163-205.
16. HILGEMANN, D. W. 1988. Numerical approximations of sodium-calcium exchange. Prog. Biophys. Mol. Biol. **51:** 1-45.
17. EARM, Y. E., W. K. HO & I. S. SO. 1990. Inward current generated by sodium-calcium exchange during the action potential in single atrial cells of the rabbit. Proc. R. Soc. London **B 240:** 61-81.
18. EARM, Y. E. & D. NOBLE. 1990. A model of the single atrial cell: Relation between calcium current and calcium release. Proc. R. Soc. London **240:** 83-96.
19. HILGEMANN, D. W. 1986. Extracellular calcium transients at single excitations in rabbit atrium. J. Gen. Physiol. **87:** 707-735.
20. LEVI, A. 1989. The effect of strophanthidin on action potential and contraction in isolated ventricular myocytes from the guinea-pig. J. Physiol. **416:** 44P.
21. FEDIDA, D., Y. SHIMONI, D. NOBLE & A. J. SPINDLER. 1987. Inward currents related to contraction in guinea-pig ventricular myocytes. J. Physiol. **385:** 565-589.
22. EGAN, T. M., D. NOBLE, S. J. NOBLE, T. POWELL, A. J. SPINDLER & V. W. TWIST. 1989. Sodium-calcium exchange during the action potential in guinea-pig ventricular cells. J. Physiol. **411:** 639-661.
23. SAKMANN, B. & G. TRUBE. 1984. Conductance properties of single inwardly rectifying potassium channels in ventricular cells from guinea-pig heart. J. Physiol. **347:** 641-657.
24. NOBLE, D. & T. POWELL. 1990. The attenuation and slowing of calcium signals in cardiac muscle by fluorescent indicators. J. Physiol. **425:** 54P.
25. LEBLANC, N. & J. R. HUME. 1990. Sodium-current induced release of calcium from cardiac sarcoplasmic reticulum. Science **248:** 372-375.
26. HILGEMANN, D. W. 1990. "Best Estimates" of physiological Na/Ca exchange function: Calcium conservation and the cardiac electrical cycle. *In* Cardiac Electrophysiology: From Cell to Bedside. D. P. Zipes & J. Jalife, Eds.: 51-61. W. B. Saunders. New York.
27. RYDER, K. O., S. M. BRYANT & G. HART. 1990. Calcium and calcium-activated currents in hypertrophied left ventricular myocytes isolated from guinea-pig. J. Physiol. **430:** 71P.
28. WEIDMANN, S. 1956. Elektrophysiologie der Herzmuskelfaser. Huber. Bern.

APPENDIX

We include here the data input file that generates the guinea-pig ventricular cell model used in this paper. This input file is valid for HEART version 3.3 and later versions. Version 3.2 can easily be edited to become valid because only two new variables are defined in version 3.3. STEEPK1 determines the steepness of the i_{K1} rectifier. Its default value is 2 (corresponding to the factor multiplying F/RT in Eq. 13 of the DiFrancesco-Noble[13] model). KMNACA is the K_m for calcium activation of the sodium-calcium exchange. In FIGURES 11 and 12 a value of 0.0002 (i.e. 200 nM) was used. The default value of KMNACA is zero. The model is developed from the single atrial cell model described by Earm and Noble,[18] which is activated in this input file by the command PREP : YATRIUM. This model was in turn developed from the Hilgemann-Noble[15] and DiFrancesco-Noble[13] models. Parameters not set in this input file retain their values in the original model defining the single atrial cell.

We have given the units of each parameter where relevant in a bracketed comment on each line (HEART 3.4 permits these comments to be included in the data file itself for future reference). In the equations used by the program, currents are in nA (but are plotted here in pA), conductances in μSiemens, voltages in mV, time in seconds (but plotted here in msec), concentrations in mM, and permeabilities (e.g. PCA) are the constant field permeabilities (cm$\cdot$sec^{-1}) *multiplied by the Faraday constant, F* (this gives a more manageable parameter than the permeability itself since it eliminates a row of at least five zeros which are readily subject to copying errors).

Experimental version of guinea-pig ventricular cell model.

This input data file works with HEART version 3.3 and later versions.

$ **** PREP : YATRIUM ******

IPULSESIZE = −6 (depolarizing stimulus of 6 nA)
ON = 0.1 (applied at 100 msec)
OFF = 0.102 (of duration 2 msec)
REP = 1 (repeated every second)
NAI = 5 (internal sodium 5 mM)
CAMODE = 1 (use standard DiFrancesco-Noble i_{Ca} equations)
SPEEDF = 0.5 (i_{Ca} inactivation 50% of standard rate)
BUFFAST = 1 (select full Hilgemann calcium-buffering equations)
CTROP = 0.05 (concentration of c-troponin 50 μM)
ALPHA[12] = 0.4 (SR calcium uptake rate constant)
BETA[12] = 0.03 (SR calcium back rate constant)
ALPHA[26] = 0 (remove voltage dependence of Ca release: Ca dependent only)
TOMODE = 5 (Hilgemann-Noble equations for i_{to})
GTO = 0.005 (g_{to} set to 5 nS)
Q = 0 (i_{to} activation starts at zero)
R = 1 (i_{to} inactivation starts at 1)
GNA = 2.5 (g_{Na} set to 2.5 μSiemens)
PCA = 0.25 (Ca channel permeability [constant field units] multiplied by F)
IKM = 1 (maximum current carried by i_K channels 1 nA)
KMODE = 0 (use standard DiFrancesco-Noble i_K equations)
X = 0 (i_K activation starts at zero)
GBNA = 0.0006 (background sodium conductance 0.6 nS)
GBCA = 0.00025 (background calcium conductance 0.25 nS)

GBK = 0.0006 (background K conductance 0.6 nS)
GFK = 0 (K component of i_f set to zero: no i_f in ventricle)
GFNA = 0 (Na component of i_f set to zero)
PUMP = 0.7 (maximum Na pump current 0.7 nA)
KNACA = 0.0005 (scaling factor for sodium-calcium exchange current)
DNACA = 0 (denominator factor for sodium-calcium exchange current)
GK1 = 1 (maximum i_{K1} conductance 1 nS)
STEEPK1 = 2.1 (steepness factor for i_{K1} rectification)
SHIFTK1 = 11 (shift voltage dependence of i_{K1} rectifer 11 mV)
KMK1 = 14 (activation constant for K activation of i_{K1} = 14 mM)
E = −89 (initial voltage −89 mV)
LENGTH = 0.08 (cell length 80 μm)
DX = 1.5 (cell radius 10 × 1.5 = 15 μm)
CAPACITANCE = 0.0002 (cell capacitance 200 pF)
TIMESCALE = 1000 (convert time for plotting from seconds to milliseconds)
DT = 0.001 (step length 1 msec)
TABT = 0.001 (tabulation interval 1 msec)
ISCALE = 1000 (convert output currents for plotting to pA)
OUT = 4 (tabulate standard ionic currents in data output files)
TEND = 0.4 (terminate computation at 400 msec)
$$$$$$$$ (row of dollar symbols terminates input)

Spatial Properties of Ca^{2+} Transients in Cardiac Myocytes Studied by Simultaneous Measurement of Na^{+}-Ca^{2+} Exchange Current and Indo-1 Fluorescence[a]

LUTZ POTT AND PETER LIPP

Department of Cell Physiology
Ruhr-University Bochum
W-4630 Bochum, Federal Republic of Germany

GEERT CALLEWAERT AND EDWARD CARMELIET

Laboratory of Physiology
Department of Medicine
KU Leuven, Campus Gasthuisberg
B-3000 Leuven, Belgium

INTRODUCTION

Contraction of a cardiac cell is controlled by a phasic rise of intracellular free Ca^{2+} concentration (Ca^{2+} transient) which is the result of Ca^{2+} entry via sarcolemmal Ca^{2+} channels and Ca^{2+} release from the sarcoplasmic reticulum (SR, for review see Refs. 1-3). The role of transmembrane Ca^{2+} current is to serve as a signal, triggering release from the SR (Ca^{2+}-induced Ca^{2+} release, CICR[4]). In addition, Ca^{2+} entry must be involved in refilling of releasable Ca^{2+} stores.[5] The concept of CICR has been developed by means of measurements of contraction and Ca^{2+}-sensitive aequorin luminescence in skinned cardiac cells.[6] By means of recent techniques for measuring intracellular Ca^{2+} concentrations using fluorescent dyes,[7,8] strong evidence has been provided in support of CICR as the mechanism generating the Ca^{2+} transient also in mammalian cardiac myocytes with intact sarcolemma.[9,10] Using an intracellular Ca^{2+}-buffering condition that causes slowing of the Ca^{2+} transient, in atrial myocytes the latter can be separated into two components. By means of simultaneous measurement of the current carried by electrogenic Na^{+}-Ca^{2+} exchange as a signal indicating subsarcolemmal [Ca^{2+}] and Indo-1 fluorescence, which provides information on global cytosolic [Ca^{2+}], it can be shown that these components represent different compartments of the sarcoplasmic reticulum. Our experimental approach permits us to study spatial aspects of Ca^{2+} signaling in cardiac cells.

[a]Supported by the Deutsche Forschungsgemeinschaft and the National Fund for Scientific Research (Belgium).

METHODS

Cell Isolation and Culture

The method of enzymatic isolation of atrial myocytes from hearts of adult guinea pigs has been described in detail previously,[11,12] with the exception that for UV-microfluorometry the cells were not cultured in plastic tissue culture dishes. A drop of cell suspension in medium (HEPES-buffered M199, Gibco) was plated on a thin coverslip. The coverslips were stored in multiwell culture chambers in an incubator at 37°C, 5% CO_2, and 90% humidity. Cells were used from day 1 to day 7 after plating. The cells were spherical, with a diameter ranging from about 15 to 20 μm. Their membrane capacitance ranged from around 10 to 20 pF. Loading of these cells with Indo-1 via the patch-clamp pipette ($R \leq 2$ MΩ) was complete within less than 5 min.[13]

Current Measurement

A coverslip with the myocytes attached was mounted in a sandwich chamber that was placed on the stage of an inverted microscope. The chamber was continuously perfused at 1-2 ml/min with a solution of the following composition (mM): NaCl 140; $CaCl_2$ 2.0; CsCl 2.0; KCl 2.0; $MgCl_2$ 1.0; HEPES/NaOH 10.0, pH 7.4; TTX (1 mg/ml) was included in some experiments. Temperature was 21 to 23°C. The solution for filling the patch-clamp pipettes (internal solution) contained, if not otherwise stated, (mM): Cs_3-citrate 60; CsCl 10; NaCl 5-10; $MgCl_2$ 1.0; HEPES 20; MgATP 4; K_5Indo-1 0.1 (or 0.1-0.2 mM EGTA, respectively, if only membrane current was measured). Solutions were adjusted to pH 7.4 with CsOH. Membrane currents were recorded by means of standard patch-clamp technique (whole-cell mode[14]), using a List EPC-7 amplifier. Membrane currents and fluorescence signals were filtered with an 8-pole bessel filter (f_c = 1 KHz), and stored for later analysis on a computer equipped with an A/D-board (data-translation DT 2821) at sample intervals between 0.1 and 1 msec.

Fluorescence Measurement

The setup was based on a Nikon TMD inverted microscope equipped for epifluorescence. The fluorescence probe was excited at 360-nm wavelength with a 75 W xenon arc lamp. Fluorescence was measured from a circular field slightly larger than the cell under study. Fluorescence emitted from the Indo-1-loaded cell was split by a dichroic mirror centered at 450 nm and detected by two parallel photomultiplier tubes (Hamamatsu type R928) at 405 and 485 nm.

Calibration of $[Ca^{2+}]_i$

Background fluorescence due to cell autofluorescence and the Indo-1-containing pipette was measured in the cell-attached mode before breaking the membrane under the tip of the pipette by a suction pulse. The ratio of emitted fluorescence at 405 nm/485 nm (R), after subtraction of background fluorescence at both wavelengths was then used to calculate $[Ca^{2+}]$ according to the equation[7]

$$[Ca^{2+}]_i = K_d \cdot \beta \, (R - R_{min})/(R_{max} - R)$$

wherein R_{min} and R_{max} are the fluorescence ratios obtained in the absence of Ca^{2+} and at saturating $[Ca^{2+}]$, respectively. β is the ratio of the 485-nm signal in the absence and at saturating $[Ca^{2+}]$. The limiting ratios were determined *in vivo* under metabolic inhibition in the presence of the bromo-derivative of the ionophore A23187.[10,15] For K_d, the dissociation constant, a value of 213 nM was assumed.[16] A detailed description of the calibration procedure can be found in Callewaert *et al.*[13]

RESULTS

In mammalian cardiac myocytes Ca^{2+} release from the sarcoplasmic reticulum causes an inward current that, under various experimental conditions, has been demonstrated to be carried by Na^+-Ca^{2+} exchange.[12,17-20] Under the Ca^{2+}-buffering condition described above, this inward current, which provides a signal indicating Ca^{2+} release from the sarcoplasmic reticulum, is slowed and delayed with regard to the triggering Ca^{2+} current. This permits direct separation and analysis of the release signal, particularly with regard to its rising phase, without contamination by I_{Ca}.

FIGURE 1 illustrates an example of a frequent observation made under these experimental conditions. The cell was stepped at 0.3 sec^{-1} from -50 mV holding potential to $+5$ mV (step duration 50 msec) in order to elicit I_{Ca}. I_{Ca} flow served as a trigger for CICR. In that experiment each depolarization was followed by a Ca^{2+}-release-dependent inward current of constant shape and amplitude (trace a) for a period of time of about 3 min. Upon shortening the clamp pulses to 25 msec, release signals became variable. Trace b represents a failure, that is, no release was elicited; d to f were selected because these traces display release signals consisting of two components (c_1, c_2), the first one of which (c_1) can be seen in isolation in trace c. A separation of the release-dependent exchange current into two distinct components could be verified in most of the myocytes loaded with citrate-based solution by trying different pulse protocols. FIGURE 1B illustrates that c_2 was strongly dependent on c_1; it was only seen if c_1 exceeded a certain amplitude. The time interval between I_{Ca} activation and peak c_2 was shorter, the bigger the amplitude of c_1. The amplitude of c_2, on the other hand, was constant throughout.

Two components could also be identified in inward current fluctuations caused by spontaneous Ca^{2+} release. FIGURE 2 displays a recording of membrane current from a myocyte that was loaded with a solution containing 25 mM citrate and 200 μM EGTA as exogenous Ca^{2+} chelators. At -50 mV holding potential ($[Na^+]_i$ = 10 mM), the cell is continuously loaded with Ca ions via the exchanger, resulting in spontaneous unsynchronized release from the SR. These events are irregular in frequency and amplitude, and are typical for a Ca^{2+}-overload condition (e.g., Kass & Tsien[21] and Berlin *et al.*[22], see January & Fozzard[23] for review). After a train of depolarizing voltage steps (for details see legend) the spontaneous release signals first became synchronized. This was followed by a transient dissociation into two components, before the initial pattern of irregular transient inward currents reappeared.

As two-component signals have been found frequently, whereas more than two components were never detected, it can be assumed that an intrinsic property of Ca^{2+} release is involved, rather than stochastic fluctuations of $[Ca^{2+}]_i$ due to focal release events. In order to obtain further information on the nature of the two components of the I_{NaCa} transient, we measured intracellular Ca^{2+} concentration by means of the fluorescent indicator Indo-1 under otherwise identical conditions, except that the high-

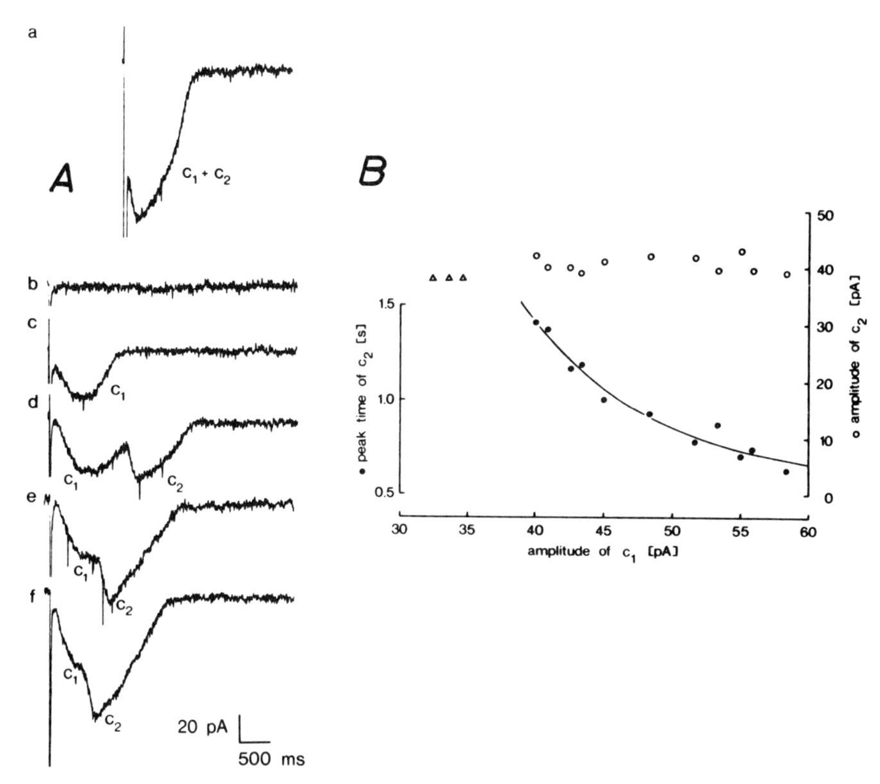

FIGURE 1. Separation of two components of inward current caused by Ca^{2+} release. **A.** Trace a: A Ca^{2+}-release signal triggered by depolarizing the cell from -50 to $+5$ mV at 0.1 sec^{-1} with pulses 50 msec in duration. (The resulting Ca^{2+} current is off scale.) This trace is representative for a period of 3 min. Thereafter, duration of the voltage step was reduced to 25 msec, resulting in variable current traces upon termination of the clamp steps. Trace b lacks a release signal; c through d have been selected because they show two-component release signals with various degrees of overlap. **B.** Plot of time-to-peak of second component (closed symbols) and amplitude of second component (open symbols) against amplitude of first component. The three triangles symbolize first components that were not followed by a second component (e.g., trace c in A). (From Budde *et al.*[41] with permission from *Pflügers Archiv.*)

affinity chelator EGTA in the pipette solution was replaced by 100 μM Indo-1 (see METHODS). In combination with simultaneous measurement of I_{NaCa}, two release signals are obtained. As the changes of Indo-1 fluorescence are recorded from the entire cell, the resulting Ca^{2+} signal represents "global" or "mean" $[Ca^{2+}]_i$. This signal does not provide any information on putative spatial gradients or inhomogeneities of $[Ca^{2+}]_i$, respectively. The magnitude of I_{NaCa}, on the other hand, is only determined by the concentration of Ca^{2+} seen by the sarcolemmal exchanger molecules, that is, this current provides a signal that is determined by subsarcolemmal $[Ca^{2+}]$. The magnitude of I_{NaCa} has been demonstrated to be linearly related to $[Ca^{2+}]_i$.[24] Such a linear relation is in line with the established stoichiometry (3 Na^+ : 1 Ca^{2+}) of cardiac Na^+-Ca^{2+} exchange, provided that $[Ca^{2+}]_i$ is far from saturation of the internal Ca^{2+}-binding site. A linear dependence of current on $[Ca^{2+}]_i$ is confirmed by the result illustrated in

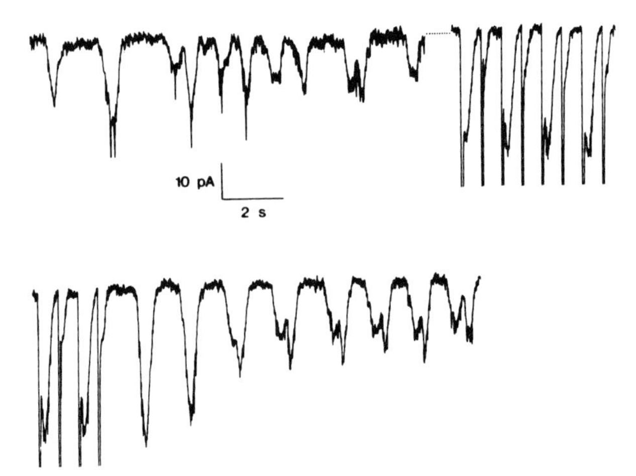

FIGURE 2. Synchronization of spontaneous transient inward currents unmasks two components. Spontaneous current events were recorded at −50-mV holding potential (top left trace). Repetitive depolarizations at 1.3 sec^{-1} (50 msec; −50 to +5 mV) resulted in two types of release signals. (Peak I_{Ca} is off scale). After cessation of repetitive depolarizations (bottom trace) synchronized spontaneous inward currents were recorded transiently. (From Budde *et al.*[41] with permission from *Pflügers Archiv.*)

FIGURE 3. FIGURE 3A (left) shows a pair of release signals triggered by activation of I_{Ca}. No symptoms of two components can be detected in these traces. Release obviously started during I_{Ca} flow. A plot of change of current against $[Ca^{2+}]_i$ between peak $[Ca^{2+}]_i$ and baseline yields an almost perfectly straight line over a range of concentrations 150 nM $\leq [Ca^{2+}]_i \leq 1$ μM (**B**). Both the Ca^{2+} transient as well as the inward I_{NaCa} are almost entirely due to Ca^{2+} release. If I_{Ca} failed to trigger release, as shown in FIGURE 3A (right), a change of $[Ca^{2+}]_i$ was recorded that was smaller by an order of magnitude. A discrete inward exchange current is lacking under that condition. Signals caused by spontaneous Ca^{2+} release from the SR had similar properties. In FIGURE 4A, a spontaneous Ca^{2+} transient and the corresponding change of I_{NaCa} have been traced. A plot of the current change against $[Ca^{2+}]_i$ again yields a straight line between 200 nM and 1 μM Ca^{2+}. Below 200 nM, that is, toward the very end of the transients, a steeper decay of the current was observed in this cell. This was found occasionally and might reflect the catalytic Ca^{2+}-binding site of the exchanger.[25] With regard to the decay phase of the Ca^{2+} transient, a linear relation could be confirmed in all recordings analyzed under this aspect for both triggered and spontaneous transients. This supports the view that during the decay phase, that is, late after release, subsarcolemmal $[Ca^{2+}]$ seen by the exchanger and global $[Ca^{2+}]$ detected by the fluorescence measurement are identical.

Results obtained from a cell that during trains of step depolarizations responded with variable release signals, sometimes with two clearly identifiable components, are illustrated in FIGURE 5. The three sets of data show (a) a rapidly rising transient without direct evidence for two components, (b) a transient that clearly showed two components, and (c) a transient that obviously represents the isolated first component. The dense traces represent current, whereas the thin lines are the Ca^{2+} signals. The latter were inverted and scaled by a common factor that was determined by adjusting the amplitude of the Ca^{2+} trace in (a) to match the point marked by the arrow. This point represents decay to 50% of peak I_{NaCa}. In accordance with the linear dependence of I_{NaCa} on $[Ca^{2+}]_i$ demonstrated in the previous figures, there is an almost perfect match of the traces below this point. At higher $[Ca^{2+}]_i$ ($\geq$ 1 μM) a deviation in terms of a flattening of the I_{NaCa} versus $[Ca^{2+}]_i$ relation is observed, which is likely to reflect saturation of the exchanger. Immediately after termination of the triggering I_{Ca},

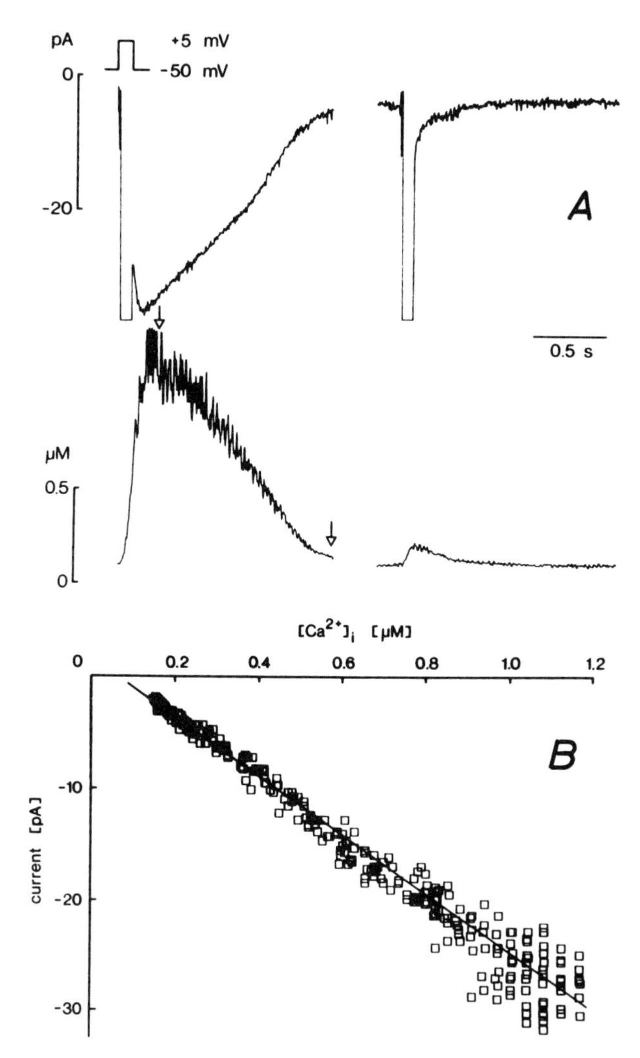

FIGURE 3. Linear dependence of release-dependent inward current on $[Ca^{2+}]_i$. **A.** Membrane current (top) and $[Ca^{2+}]_i$ (bottom) upon activation of I_{Ca} by depolarization from -50 to $+5$ mV. Amplitude of I_{Ca} was constant (240 pA, off scale) throughout a train of 20 pulses (0.2 sec^{-1}). Currents and Ca^{2+} signals alternated between the two representative responses illustrated. **B.** Plot of change of membrane current against $[Ca^{2+}]_i$ for the period of time marked by the arrows in A. One point per 10 msec has been plotted. The straight line was calculated by linear regression for 200 nM $\leq$ $[Ca^{2+}]_i$ $\leq$ 1.0 μM. (From Lipp *et al.*[42] with permission from *FEBS Letters.*)

development of inward current preceded $[Ca^{2+}]_i$, resulting in an earlier peak of the former. This positive deviation of current from $[Ca^{2+}]_i$ is a property of the first component of release. In (b) both components can be identified because of a distinct incision in the current trace. The two traces significantly deviate during the first component, whereas from the peak of the second component to the end of the traces they perfectly coincide. For the isolated first component (c) both traces deviate from each other throughout. It should be noted that these deviations occur in a range of $[Ca^{2+}]_i$ in which during the decay of the signals in (a), a complete match of the traces is observed. This safely excludes these deviations to be caused by methodological artifacts, such as erroneous calibration of the fluorescence signals.

The first component is not a signal caused by Ca^{2+} entry via I_{Ca}. As shown in FIGURE 3, Ca^{2+} entry does not result in an identifiable exchange current, although there may be some contamination of the late deactivation phase of I_{Ca} by I_{NaCa}. The small Ca^{2+} transient due to Ca^{2+} entry starts to decay immediately upon termination of I_{Ca}. Furthermore, two components can also be identified in spontaneous transient inward current caused by cyclic Ca^{2+} release from the SR at constant membrane potential. An example is illustrated in FIGURE 6. Like in the previous figure, current traces and inverted $[Ca^{2+}]_i$ traces have been superimposed. For scaling, the peaks of signals in (a) have been matched. The transients recorded in that experiment were faster than in the above figures, since the pipette solution contained only 30 mM citrate. Signal pairs (a) and (b) clearly show two components in the rising phases of the current transients, whereas the Ca^{2+} transients rose monotonically. Even more dramatically than in FIGURE 5, the current preceded the Ca^{2+} transient. During their decay both signals were fairly correlated. The isolated first component is characterized by a complete dissociation of $[Ca^{2+}]_i$ and I_{NaCa}. The latter reached 80% of its peak level before any change of $[Ca^{2+}]_i$ could be detected. It was a consistent finding that the current signal rose faster than the Ca^{2+} signal. This was particularly pronounced if two compo-

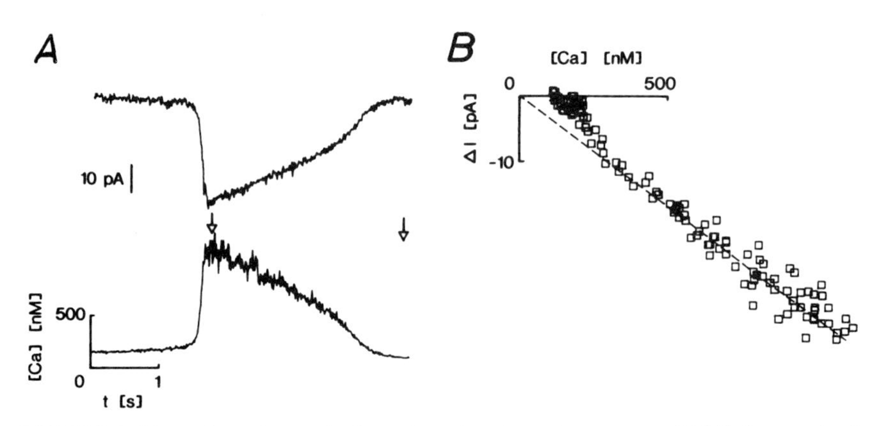

FIGURE 4. Linear dependence of release-dependent inward current on $[Ca^{2+}]_i$ for a pair of signals reflecting spontaneous Ca^{2+} release. At -50 mV holding potential ($[Na^+]_i = 10$ mM). Spontaneous release-dependent I_{NaCa} transients occurred at a regular frequency ($\approx 0.18\ sec^{-1}$) for a period of time of about 6 min. Fluorescence of Indo-1 could be recorded by manually opening the shutter for illumination at appropriate times. **A.** Top trace: current; bottom trace: $[Ca^{2+}]_i$. **B.** Plot of current against $[Ca^{2+}]_i$ for the period between the arrows in A. The straight line was drawn by eye.

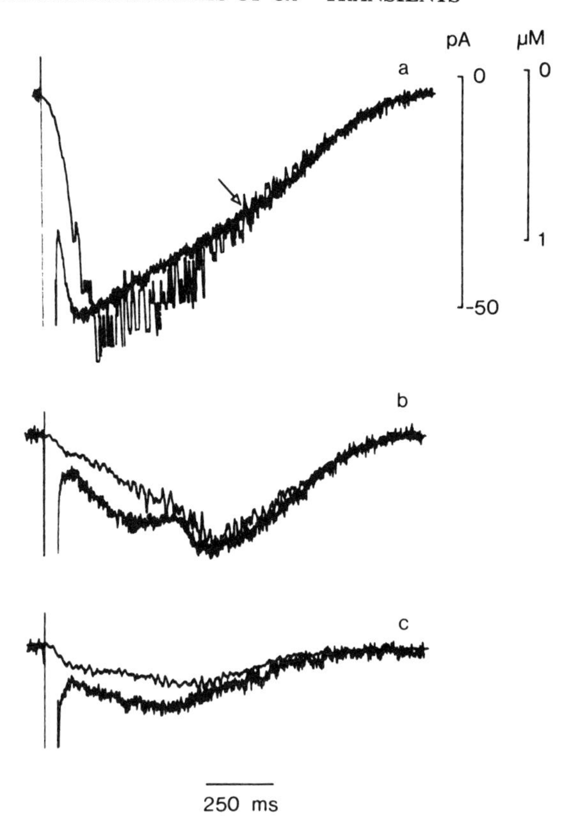

FIGURE 5. Lack of correlation of I_{NaCa} and $[Ca^{2+}]_i$ during first component of Ca^{2+} release. Samples of pairs of signals recorded during a train of step depolarizations from −50 to +5 mV (50 msec; 0.15 sec^{-1}; I_{Ca} off scale). I_{NaCa} (thick lines) and $[Ca^{2+}]_i$ (thin lines) have been superimposed; that is, in contrast to FIGURES 3 and 4, a rise in $[Ca^{2+}]_i$ is indicated by a downward deflection. **(a)** Pair of traces without direct evidence for two components (compare FIG. 1). The Ca^{2+} trace has been scaled to match the point indicated by the arrow on the current trace. This corresponds to 50% of peak inward I_{NaCa}, or 0.95 μM Ca^{2+}. Identical scaling has been applied to signal pairs b and c. **(b)** Current signal displays two distinct components. **(c)** Isolated first component.

nents were seen in the current signal, or a first component was seen in isolation. A deviation of the two signals in the opposite direction was never observed.

DISCUSSION

In the present study it could be demonstrated that in atrial myocytes inward current caused by Ca^{2+} release from the sarcoplasmic reticulum and carried by Na^+-Ca^{2+} exchange is linearly related to $[Ca^{2+}]_i$ during the decay phase of the Ca^{2+} transient, that is, late after release. During the rising phase of the release signal(s) a deviation from the linear relation is frequently observed. This deviation is always in the same direction: inward I_{NaCa} is larger than the linear relation would predict. The identity of

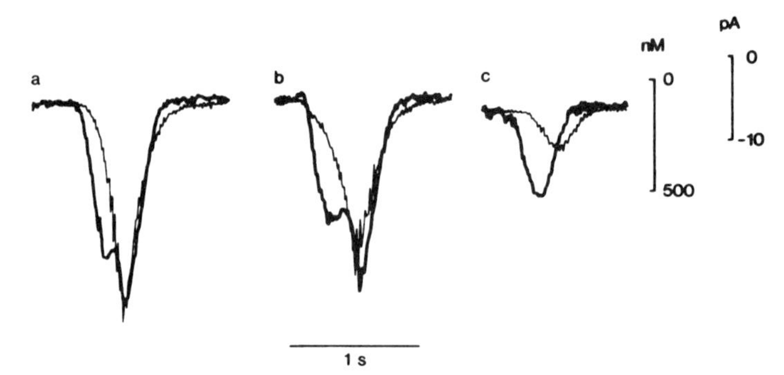

FIGURE 6. Dissociation of I_{NaCa} and $[Ca^{2+}]_i$ during spontaneous Ca^{2+} transients. I_{NaCa} (thick traces) and $[Ca^{2+}]_i$ (thin traces) have been superimposed as in FIGURE 5. $[Ca^{2+}]_i$ traces were scaled by matching peak $[Ca^{2+}]_i$ and peak I_{NaCa} of (a). Pairs of signals labeled (**a**) and (**b**) represent different degrees of overlap of two components; (**c**) represents isolated first component.

release-dependent inward current with I_{NaCa} has been proven under similar experimental conditions.[12,17,26] This is further supported by the linear dependence on $[Ca^{2+}]_i$, which has been found previously also in ventricular myocytes, both in the absence[24] and in the presence[27] of a functional sarcoplasmic reticulum (see also Miura & Kimura[25]). Such a linear function is in line with the stoichiometry,[28] which for the cardiac exchanger has been determined as 3 Na^+ : 1 Ca^{2+},[29–31] provided the range of $[Ca^{2+}]_i$ is far from saturation of the internal Ca^{2+}-binding site. Assuming simple saturation kinetics, a graph that would be hardly distinguishable from a straight line is obtained for concentrations of a ligand of up to about 0.1 to 0.2 times the apparent K_d. As we reproducibly found a fairly linear behavior up to around 1 μM, this would require an apparent K_d (in the presence of 5 or 10 mM Na^+) on the order of 5 to 10 μM. This is in line with measurements on sarcolemmal vesicles[32] (for review see Reeves & Philipson[33] and Carafoli[34]). It contradicts, however, recent measurements on intact myocytes.[25] These authors reported a K_d of 0.6 μM. $[Ca^{2+}]_i$ was varied in a range between 50 and 500 nM by perfusion of a patch-clamp pipette. Evidence is growing that in a comparatively large cell this method does not permit perfect control of the internal, or at least subsarcolemmal, ionic composition (e.g., Luk & Carmeliet[35]).

The current transients in the majority of myocytes can be split into two but never more than two discrete components. Whenever such a dissociation occurred, the positive deviation from linearity of the dependence of I_{NaCa} on $[Ca^{2+}]_i$ turned out to result from the first component only. It is conceivable that during the first component apart from the exchanger an additional current-generating system is activated. In that case the question arises, Why does this mechanism remain silent during the second component? As Ca^{2+} release from the SR is involved in generating both components, these phenomena are more likely to reflect a property of the release mechanism in atrial myocytes. There is hardly any doubt that in mammalian cardiac muscle Ca^{2+} release from the SR is triggered by transmembrane Ca^{2+} entry, or the resulting rise in $[Ca^{2+}]_i$, respectively (Ca^{2+}-induced Ca^{2+} release[6,9,10]). Ultrastructural studies have shown that in cardiac cells a significant fraction of the SR can be found in close association with the sarcolemma.[36] It is evident that a rise in $[Ca^{2+}]_i$ due to Ca^{2+} entry, either via I_{Ca} or reverse-mode exchanger, will first be detected by a subsarcolemmal layer of SR. Release from a subsarcolemmal store, on the other hand, will result in a temporary spatial gradient of $[Ca^{2+}]$. The concentration will be higher near the membrane than in the center of the cell. As a consequence the Ca^{2+} concentration "seen" by the exchanger will be higher than the "mean" or integral Ca^{2+} concentration detected by the optical measurement. A schematic representation of this hypothesis is shown in FIGURE 7. In *A*, a selected pair of spontaneous release signals has been traced with two clearly identifiable components in the current recording. In the resting condition (1) Ca^{2+} entering the cell via reverse-mode Na^{+}-Ca^{2+} exchange is taken up by the SR, preferentially by the subsarcolemmal compartment. When this compartment becomes saturated, further Ca^{2+} entry results in a rise in free [Ca^{2+}] in the space between the sarcolemma and the superficial SR, initiating CICR (2). This gives rise to a Ca^{2+} wave that spreads centrally and triggers Ca^{2+} release from deeper stores (3). The latter assumption, namely that release from central stores is in fact triggered via release from the subsarcolemmal compartment, is supported by the finding that the second component (its peak time) depends on the amplitude of the first one, and that a minimal amplitude of the first component is necessary to induce the second one (FIG. 1). The above consideration is not limited to spontaneous release events but also applies to triggered release with Ca^{2+} current as the major entry pathway. This scheme does not necessarily imply that the two compartments of the SR are different entities that are not connected. On the other hand, the existence of one continuous network of SR, which is usually suggested on the basis of morphological data (e.g., Sommer & Jennings[37]), hitherto has not been proven. The discontinuous two-step behavior of Ca^{2+} release that was consistently found suggests at least some structural and functional separation of the two compartments.

Spatial inhomogeneities of [Ca^{2+}]$_i$ have been described to occur in cardiac ventricular cells under various conditions of mild and severe Ca^{2+} overload.[22,38–40] These appear as waves or spots, respectively, of increased [Ca^{2+}]$_i$. Spontaneous release of Ca^{2+} from the SR has been demonstrated as the primary mechanism. For several reasons we suggest that these phenomena are different from the kind of spatial gradients of [Ca^{2+}]$_i$ described here: (i) the cells used in the present investigation usually are spherical with a diameter of around 15 μm. This approximately corresponds to the extension of the

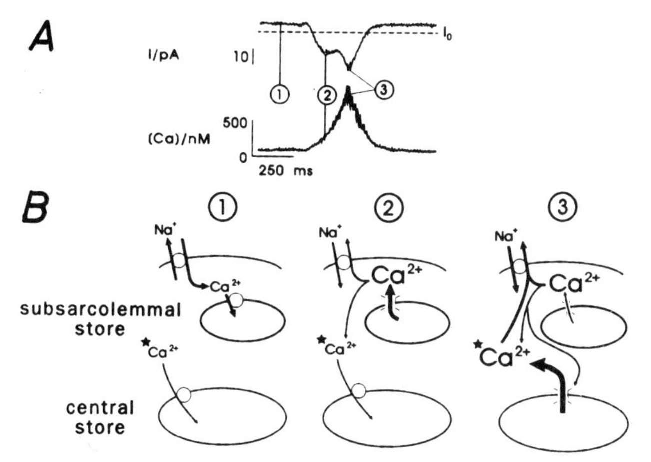

FIGURE 7. Scheme of two-component Ca^{2+} release. **A.** Sample recording of a pair of spontaneous release signals (top: current; bottom: [Ca^{2+}]$_i$). **B.** Ca^{2+} movements during the times indicated by the numbers. Changes in [Ca^{2+}] are indicated by the size of the symbols; movements of Ca^{2+} by the thickness of the arrows. The cytosolic fraction of Ca^{2+} that primarily contributes to the fluorescence signal is marked by the asterisks. For further details see text.

waves seen in ventricular cells (e.g., Takamatsu & Wier[39]); (ii) as for the dissociation of I_{NaCa} transients, the same observations made in spherical cells could be confirmed in freshly isolated, elongated atrial myocytes that have a length of about 100-150 μm. If the spreading of a Ca^{2+} signal along a cell contributed to the phenomena described in the present study, these two extremes in cell geometry should have a profound influence on the properties of the signals.

REFERENCES

1. REITER, M. 1988. Calcium mobilization and cardiac inotropic mechanisms. Pharmacol. Rev. **40:** 189-217.
2. MORAD, M. & L. CLEEMANN. 1987. Role of Ca^{2+} channel in development of tension in heart muscle. J. Mol. Cell. Cardiol. **19:** 527-553.
3. WIER, W. G. 1990. Cytoplasmic [Ca^{2+}] in mammalian ventricle: Dynamic control by cellular processes. Annu. Rev. Physiol. **52:** 467-485.
4. FABIATO, A. 1983. Calcium-induced release of calcium from the cardiac sarcoplasmic reticulum. Am. J. Physiol. **245:** C1-C14.
5. FABIATO, A. 1985. Simulated calcium current can both cause calcium loading and trigger calcium release from the sarcoplasmic reticulum of a skinned canine cardiac Purkinje cell. J. Gen. Physiol. **85:** 291-320.
6. FABIATO, A. 1985. Time and calcium dependence of activation and inactivation of calcium-induced release of calcium from the sarcoplasmic reticulum of a skinned canine cardiac Purkinje cell. J. Gen. Physiol. **85:** 247-289.
7. GRYNKIEWICZ, G., M. POENIE & R. Y. TSIEN. 1985. A new generation of Ca^{2+} indicators with greatly improved fluorescence properties. J. Biol. Chem. **260:** 3440-3450.
8. TSIEN, R. Y. 1989. Fluorescence probes of cell signalling. Ann. Rev. Neurosci. **12:** 227-253.
9. NÄBAUER, M., G. CALLEWAERT, L. CLEEMANN & M. MORAD. 1989. Regulation of calcium release is gated by calcium current, not gating charge, in cardiac myocytes. Science **244:** 800-803.
10. BEUCKELMANN, D. J. & W. G. WIER. 1988. Mechanism of release of calcium from sarcoplasmic reticulum of guinea-pig cardiac cells. J. Physiol. (London) **405:** 233-255.
11. BECHEM, M., L. POTT & H. RENNEBAUM. 1983. Atrial muscle cells from hearts of adult guinea-pigs in culture: A new preparation for cardiac cellular electrophysiology. Eur. J. Cell Biol. **31:** 366-369.
12. LIPP, P. & L. POTT. 1988. Transient inward current in guinea-pig atrial myocytes reflects a change of sodium-calcium exchange current. J. Physiol. (London) **397:** 601-630.
13. CALLEWAERT, G., P. LIPP, L. POTT & E.CARMELIET. 1991. High-resolution measurement and calibration of Indo-1 fluorescence in guinea-pig atrial myocytes under voltage-clamp. Cell Calcium **72:** 629-277.
14. HAMILL, O. P., A. MARTY, E. NEHER, B. SAKMANN & F. J. SIGWORTH. 1981. Improved patch-clamp techniques for high-resolution current recording from cells and cell-free membrane patches. Pflügers Arch. **391:** 85-100.
15. LI, Q., R. A. ALTSCHULD & B. T. STOKES. 1987. Quantitation of intracellular free calcium in single adult cardiomyocytes by fura-2 fluorescence microscopy: Calibration of fura-2 ratios. Biochem. Biophys. Res. Commun. **147:** 120-126.
16. BENHAM, C. D. 1989. Voltage-gated and agonist-mediated rises in intracellular Ca^{2+} in rat clonal pituitary cells (GH_3) held under voltage clamp. J. Physiol. (London) **415:** 143-158.
17. MECHMANN, S. & L. POTT. 1986. Identification of Na-Ca exchange current in single cardiac myocytes. Nature **319:** 597-599.
18. EGAN, T. M., D. NOBLE, S. J. NOBLE, T. POWELL, A. J. SPINDLER & V. W. TWIST. 1989. Sodium-calcium exchange during the action potential in guinea-pig ventricular cells. J. Physiol. (London) **411:** 639-661.
19. CALLEWAERT, G., L. CLEEMANN & M. MORAD. 1989. Caffeine-induced Ca^{2+} release activates Ca^{2+} extrusion via Na^+-Ca^{2+} exchanger in cardiac myocytes. Am. J. Physiol. **257:** C147-C152.

20. EARM, Y. E., W. K. HO & I. S. SO. 1990. Inward current generated by Na-Ca exchange during the action potential in single atrial cells of the rabbit. Proc. R. Soc. London B **240:** 61-81.
21. KASS, R. S. & R. W. TSIEN. 1982. Fluctuations in membrane current driven by intracellular calcium in cardiac Purkinje fibers. Biophys. J. **38:** 259-269.
22. BERLIN, J. R., M. B. CANNELL & W. J. LEDERER. 1989. Cellular origins of the transient inward current in cardiac myocytes. Role of fluctuations and waves of elevated intracellular calcium. Circ. Res. **65:** 115-126.
23. JANUARY, C. T. & H. A. FOZZARD. 1988. Delayed afterdepolarizations in heart muscle: Mechanisms and relevance. Pharmacol. Rev. **40:** 219-227.
24. BEUCKELMANN, D. J. & W. G. WIER. 1989. Sodium-calcium exchange in guinea-pig cardiac cells: Exchange current and changes in intracellular Ca^{2+}. J. Physiol. (London) **414:** 499-520.
25. MIURA, Y. & J. KIMURA. 1989. Sodium- calcium exchange current. Dependence on internal Ca and Na and competitive binding of external Na and Ca. J. Gen. Physiol. **93:** 1129-1145.
26. LIPP, P. & L. POTT. 1988. Voltage dependence of sodium-calcium exchange current in guinea-pig atrial myocytes determined by means of an inhibitor. J. Physiol. (London) **403:** 355-366.
27. BEUCKELMANN, D. J., L. POTT & W. G. WIER. 1989. Transient inward current and intracellular calcium concentration in guinea-pig ventricular myocytes. J. Physiol. (London) **314:** 110P.
28. DIFRANCESCO, D. & D. NOBLE. 1985. A model of cardiac electrical activity incorporating ionic pumps and concentration changes. Philos. Trans. R. Soc. London B **307:** 353-398.
29. REEVES, J. P. & C. C. HALE. 1984. The stoichiometry of the cardiac sodium-calcium exchange system. J. Biol. Chem. **259:** 7733-7739.
30. EHARA, T., S. MATSUOKA & A. NOMA. 1989. Measurement of reversal potential of Na^{+}-Ca^{2+} exchange current in single guinea-pig ventricular cells. J. Physiol. (London) **410:** 227-249.
31. CRESPO, L. M., C. J. GRANTHAM & M. B. CANNELL. 1990. Kinetics, stoichiometry and role of the Na-Ca exchange mechanism in isolated cardiac myocytes. Nature **345:** 618-621.
32. PHILIPSON, K. D. & A. Y. NISHIMOTO. 1982. Na^{+}-Ca^{2+} exchange in inside-out cardiac sarcolemmal vesicles. J. Biol. Chem. **257:** 5111-5117.
33. REEVES, J. P. & K. D. PHILIPSON. 1989. Sodium-calcium exchange activity in plasma membrane vesicles. *In* Sodium-Calcium Exchange. T. J. A. Allen, D. Noble & H. Reuter, Eds.: 27-53. Oxford University Press. Oxford, England.
34. CARAFOLI, E. 1987. Intracellular calcium homeostasis. Annu. Rev. Biochem. **56:** 395-433.
35. LUK, H.-N. & E. CARMELIET. 1990. Na^{+}-activated K^{+} current in cardiac cells: Rectification, open probability, block and role in digitalis toxicity. Pflügers Arch. **416:** 766-768.
36. JOHNSON, E. A. & J. R. SOMMER. 1967. A strand of cardiac muscle: Its ultrastructure and the electrophysiologic implications of its geometry. J. Cell. Biol. **33:** 103-129.
37. SOMMER, J. R. & R. B. JENNINGS. 1986. Ultrastructure of cardiac muscle. *In* The Heart and Cardiovascular System. Scientific Foundations. H. A. Fozzard, E. Haber, R. B. Jennings, A. M. Katz & H. E. Morgan, Eds.: 61-100. Raven Press. New York.
38. WIER, W. G., M. B. CANNELL, J. R. BERLIN, E. MARBAN & W. J. LEDERER. 1987. Cellular and subcellular heterogeneity of $[Ca^{2+}]_i$ in single heart cells revealed by fura-2. Science **235:** 325-328.
39. TAKAMATSU, T. & W. G. WIER. 1990. Calcium waves in mammalian heart: Quantification of origin, magnitude, waveform, and velocity. FASEB J. **4:** 1519-1525.
40. ISHIDE, N., T. URAYAMA, K.-I. INOUE, T. KOMARU & T. TAKISHIMA. 1990. Propagation and collision characteristics of calcium waves in rat myocytes. Am. J. Physiol. **259:** H940-H950.
41. BUDDE, T., P. LIPP & L. POTT. 1991. Measurement of Ca^{2+}-release-dependent inward current reveals two distinct components of Ca^{2+}-release from sarcoplasmic reticulum in guinea-pig atrial myocytes. Pflügers Arch. **417:** 638-644.
42. LIPP, P., L. POTT, G. CALLEWAERT & E. CARMELIET. 1990. Simultaneous recording of Indo-1 fluorescence and Na^{+}/Ca^{2+} exchange current reveals two components of Ca^{2+} release from sarcoplasmic reticulum of cardiac atrial myocytes. FEBS Lett. **275:** 181-184.

Sodium-Calcium Exchange in Intact Cardiac Cells

Exchange Currents and Intracellular Calcium Transients

W. GIL WIER[a]

Department of Physiology
University of Maryland School of Medicine
Baltimore, Maryland 21201

INTRODUCTION

Of all the ions inside cardiac cells, Ca^{2+} undergoes the largest change in concentration during each cardiac cycle.[1] Thus, of the quantities known to affect Na-Ca exchange (internal and external Na^+, Ca^{2+}, K^+, ATP, membrane potential, and others) the two that are known unequivocally to change significantly during each cardiac cycle are $[Ca^{2+}]_i$ and membrane potential. Thus, Ca^{2+} fluxes through the exchanger during the time course of the cardiac action potential will be dependent on the voltage-dependence of exchange processes, the thermodynamic driving force[2] for Na-Ca exchange, and on occupancy of the exchanger by Ca^{2+}. Recently, experiments[3,4] have been performed in my laboratory to provide information on the relationship between $[Ca^{2+}]_i$ and exchanger current (I_{NaCa}) at a given membrane voltage, and on the relationship between I_{NaCa} and membrane voltage at a given $[Ca^{2+}]_i$. Membrane currents and changes in intracellular calcium-ion concentration ($[Ca^{2+}]_i$) have been recorded that can be attributed to the operation of an electrogenic, voltage-dependent sodium-calcium (Na-Ca) exchanger in mammalian heart cells. Single guinea-pig ventricular myocytes under voltage clamp have been perfused internally with the fluorescent Ca^{2+} indicator, fura-2, and changes in $[Ca^{2+}]_i$ and membrane current that resulted from Na-Ca exchange were isolated through the use of various organic channel blockers (verapamil, tetrodotoxin [TTX]), impermeant ions (Cs^+, Na^{2+}), and inhibitors of sarcoplasmic reticulum (ryanodine). Na-Ca exchange current has been identified putatively as the Ni^{2+}-sensitive current in the presence of these inhibitors. (The reversal potential of the Ni^{2+}-sensitive ramp current was observed to vary linearly with $\ln([Ca^{2+}]_i)$. The results indicate that Ca^{2+} fluxes through the exchanger during the cardiac action potential can be understood quantitatively by considering mainly the binding of Ca^{2+} to the exchanger during the $[Ca^{2+}]_i$ transient, and the effects of membrane voltage on the exchanger.

METHODS

In general, the experimental methods employed in the studies reviewed here have been described in detail elsewhere.[4–6] Guinea-pig ventricular myocytes were isolated

[a] Address for correspondence: Dr. W. Gil Wier, Dept. of Physiology, University of Maryland at Baltimore, 660 West Redwood St., Baltimore, MD 21201.

using a modification of the procedure described by Mitra and Morad[7] with retrograde perfusion of the isolated heart with a collagenase-protease solution. Cells were loaded over 5-15 minutes by internal perfusion via the microelectrode pipette (1.8-3 MOhm resistance) with 0.070 mM fura-2 (molecular probes), 135 mM CsCl, 2 mM $MgCl_2$, 10 to 20 mM NaCl, and 10 mM Hepes (cesium salt), pH 7.2 (with CsOH).

In some of the studies,[4] the superfusing modified saline solution contained: 2.0 mM $CaCl_2$; 135 mM NaCl; 10 mM CsCl to block K^+-channels; 1 mM $MgCl_2$; 10 mM dextrose; 10 mM Hepes (sodium salt); 10 μM verapamil to block Ca^{2+} channels and 10 μM ryanodine were added unless more physiological conditions were desired; pH was 7.3 with addition of NaOH. Thirty μM tetrodotoxin (TTX) was added except when a holding potential of -40 mV was employed.

To calibrate fluorescence signals,[4,5] (fura-2) R_{min} and R_{max} (the fluorescence ratio 380-360 nm in 0 mM $Ca_2{}^+$and in a saturating concentration of Ca^{2+}) were determined intracellularly by metabolic inhibition and subsequent extracellular exposure to 0 or 2 mM Ca^{2+} in the presence of the Ca^{2+}ionophore Bromo A 23187, 10 μM. By analysis of thin solutions in the microscope, the dissociation constant for the reaction Ca^{2+} + fura-2 $\rightarrow$ Ca^{2+} : fura-2 was 200 nM.

RESULTS

First, I will present recordings of $[Ca^{2+}]_i$ transients and membrane currents under physiological conditions in intact guinea-pig cardiac cells. Even under these circumstances, it is possible to identify putatively changes in $[Ca^{2+}]_i$ and current that might be attributable to Na-Ca exchange. Farther on in this manuscript, I will review the experiments in intact cardiac cells in which other, highly nonphysiological experimental conditions have been used to isolate more conclusively the changes in $[Ca^{2+}]_i$ and currents that are attributable to Na-Ca exchange.

FIGURE 1, from work by Beuckelmann and Wier[8] shows a cell superfused with a solution containing TTX (30 μM), but no verapamil and no ryanodine, using a holding potential of -68 mV. The voltage dependence of the amplitudes of the $[Ca^{2+}]_i$ transients elicited by depolarization, measured as the average of the $[Ca^{2+}]_i$ during the period 50 to 100 msec after the step is bell-shaped, with a maximum at 10 mV. At more positive and more negative potentials, transients elicited by depolarization were smaller. However, under all conditions in which a fast $[Ca^{2+}]_i$ transient could be elicited by depolarization, a "tail-transient" could also be observed on repolarization from potentials of 50 mV and higher. The tail transient increased as the test pulse voltage increased over the range of 50 to 80 mV. These transients, first recorded by Barcenas-Ruiz and Wier[9] are now known[8] to be triggered by "tails" of Ca^{2+} currents during the time after repolarization during which the driving force for Ca^{2+} entry is increased and the Ca^{2+} channel deactivates. They are now regarded as an indication for Ca^{2+}-induced release of Ca^{2+} in heart. Both the fast-rising transients on depolarization and tail transients are abolished by ryanodine,[8-10] thus showing that they are dependent on Ca^{2+} release from the sarcoplasmic reticulum (SR). Depolarization to very positive membrane potentials (+90 mV in FIG. 1) elicits a slow increase in $[Ca^{2+}]_i$ during the depolarizing pulse. Even when release of Ca^{2+} from the SR and Ca^{2+} entry via L-type Ca^{2+} channels were blocked, the slow increase in $[Ca^{2+}]_i$ at +70 and +90 mV (see FIG. 1) could still be recorded. Furthermore, the peak amplitude of these slow $[Ca^{2+}]_i$ transients increased monotonically with membrane voltage, a pattern that is distinctly different from the bell-shaped dependence on membrane voltage of $[Ca^{2+}]_i$ transients attributed to release of Ca^{2+} from the SR.[8-11] This suggests that the slow changes

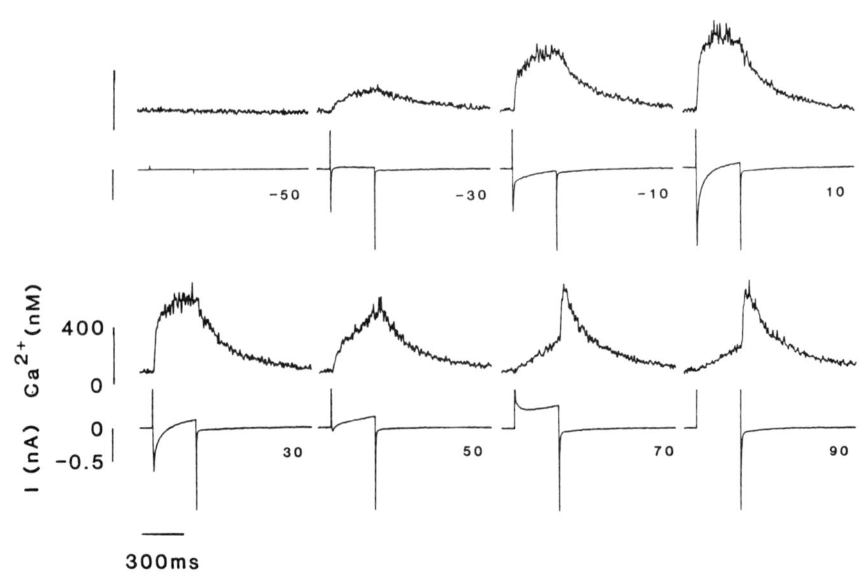

FIGURE 1. $[Ca^{2+}]_i$ transients and membrane currents in the presence of Na^+. As explained in the text, the slow increases in $[Ca^{2+}]_i$ upon depolarization to very positive membrane potentials are attributed to Ca^{2+} influx via the Na-Ca exchanger. Rapid increases in $[Ca^{2+}]_i$ upon depolarization or upon repolarization from positive membrane potentials are attributable to Ca^{2+} released from the SR. The figure presents original current records and calculated $[Ca^{2+}]_i$ during depolarizing pulses from a holding potential of −68 mV to various levels as indicated. Six conditioning pulses from a holding potential of −80 mV (amplitude 80 mV; duration, 300 msec; frequency, 1 Hz) were followed by a test pulse to various potentials, as indicated. TTX was present at a concentration of 30 μM. (Taken from Beuckelmann & Wier[8]; used with permission.)

in Ca^{2+} may be attributable to the operation of an electrogenic Na-Ca exchange. Identification of a current arising from Na-Ca exchange would be difficult under these conditions, but the slowly declining inward current occurring after repolarization could, in theory, be the result of Ca^{2+} efflux via the exchanger.

In the absence of a highly specific inhibitor of Na-Ca exchange[12] (which was not available at the time of these experiments), the best way to remove components of $[Ca^{2+}]_i$ transients and membrane currents arising from Na-Ca exchange is to remove Na^+ completely. This is accomplished[6] by perfusing the cells, both internally and externally, with solutions free of Na^+. In this case, Ca^{2+}transients can still be recorded, although there are certain distinct differences between $[Ca^{2+}]_i$ transients recorded under such conditions (FIGS. 2 and 3) and those recorded in the presence of Na^+. Upon depolarization to moderate membrane potentials, the Ca^{2+} transient rises rapidly, as in the presence of Na^+. The subsequent decline of $[Ca^{2+}]_i$ is substantially slower, however. In the data of FIGURE 2, the half-time of decline after repolarization is about 450 msec or three times longer than under physiological conditions (FIG. 1). The amplitudes of the $[Ca^{2+}]_i$ transients are larger than normal, probably reflecting the fact that the SR has become more heavily loaded with Ca^{2+}, in the absence of operation of one of the cell's mechanisms for extruding Ca^{2+}. Most striking however, is the total absence of a slow rise in $[Ca^{2+}]_i$ during a depolarizing pulse to positive membrane potentials, as shown in FIGURE 3. In this case,[6] the cell membrane was depolarized to the reversal potential of the Ca^{2+} current (+60 mV) for 300 msec, an experiment that would, in the presence of Na^{+}, elicit a slow rise in $[Ca^{2+}]_i$ during the pulse, and a pronounced "tail transient" on repolarization. It can be seen that the tail transient was prominent, but that the slow rise in $[Ca^{2+}]_i$ was absent. These results support the interpretation given above and elsewhere[1] that rapid increases in $[Ca^{2+}]_i$ upon depolarization and repolarization (from positive membrane potentials) arise from Ca^{2+} released from the SR, and the slow increases in $[Ca^{2+}]_i$ during maintained depolarization to positive potentials may be due to Na-Ca exchange.

As mentioned above, isolating the current[13–17] and $[Ca^{2+}]_i$ transients[3,9] that are generated by Na^{+}-Ca^{2+} exchange has proved difficult, without unusual experimental conditions. In this section, such experiments are described. In theory, the Na-Ca exchange current could be obtained by blocking it selectively, by activating it selectively, or by blocking all other currents, so that only Na-Ca exchange current remained. The approach used by Beuckelmann and Wier[4] was to first block as many other currents as possible and then to block the Na-Ca exchange current selectively. Nickel (Ni^{2+}) has been reported to block Na-Ca exchange relatively selectively.[15] Any action of Ni^{2+} to block Ca^{2+} channels and Na^{+} channels in the experiments of Beuckelmann and

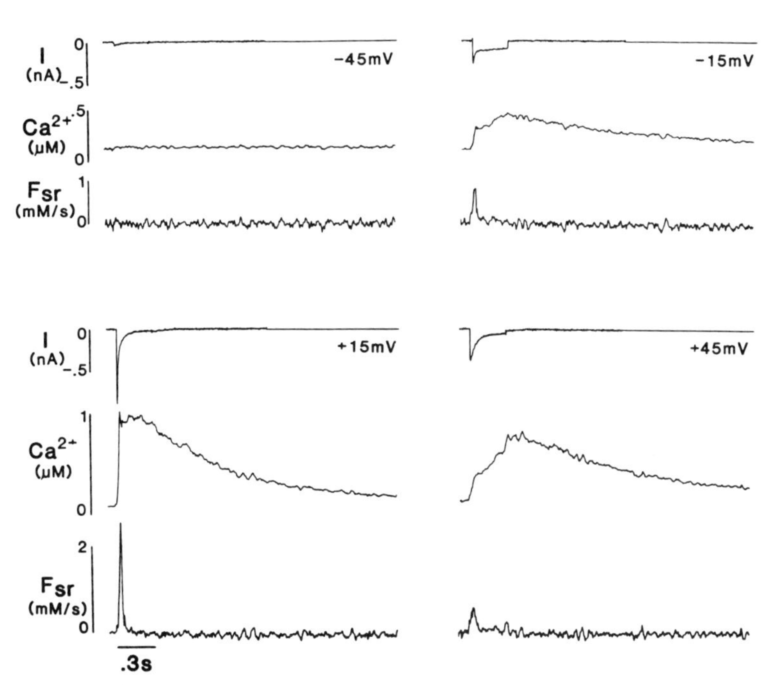

FIGURE 2. $[Ca^{2+}]_i$-transients in the total absence of Na^{+}. Na^{+} was excluded from the external solution and from the solution perfusing the inside of the cell. Ca^{2+} transients decline much more slowly than normal, but components of the $[Ca^{2+}]_i$ transients attributable to release of Ca^{2+} from the SR are still present. Depolarizing pulses to −45 mV, −15 mV, +15 mV, and +45 mV. Holding voltage, −80 mV; I, verapamil-sensitive current; Ca^{2+}, $[Ca^{2+}]_i$ transient; $F_{s.r.}$, net flux of Ca^{2+} from the SR. (Taken from Sipido & Wier[6] which can be consulted for details. Used with permission.)

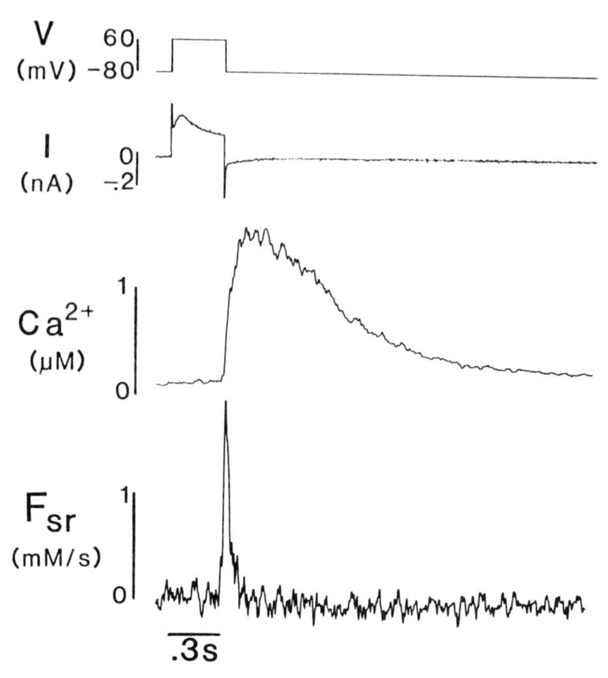

FIGURE 3. In the total absence of Na^+, the slow increase in Ca^{2+} upon depolarization to positive membrane potentials is absent, but the tail transient, a component of the $[Ca^{2+}]_i$ transients attributable to release of Ca^{2+} from SR remains. I; total whole cell current; the tail of Ca^{2+} current on repolarization could not be measured. $F_{s.r.}$ indicates net flux of Ca^{2+} from the SR as defined in Sipido & Wier.[6]

Wier[4] was not important because those channels were already blocked or inactivated. FIGURE 4A shows changes in $[Ca^{2+}]_i$, and membrane current elicited by 250-msec voltage-clamp depolarizations in the presence of Ca^{2+}, K^+-channel blockade and inhibition of SR Ca^{2+} release. $[Ca^{2+}]_i$ increases slowly throughout the depolarizing clamp pulse, while membrane current is flat or declines. Upon repolarization $[Ca^{2+}]_i$ declines and an inward current develops instantaneously. After the addition of 5 mM Ni^{2+}(FIG. 4B) to the superfusing solution the outward current during depolarizing pulses was reduced, and time-dependent changes of inward current flowing after repolarization were abolished. Changes in $[Ca^{2+}]_i$ were greatly attenuated. To obtain the "Ni^{2+}-sensitive current," the current recorded in the presence of Ni^{2+} was subtracted from that under control conditions. This "Ni^{2+}-sensitive current" is illustrated in FIGURE 4C. Beuckelmann and Wier[4] also showed that Ni^{2+}-sensitive current and $[Ca^{2+}]_i$ transients increase monotonically with voltage up to +160 mV, a pattern that is distinctly different from at least one model of a sequential step model of ion transfer.[17] An important issue in all experiments that are concerned with Na^+-Ca^{2+} exchange is whether the current attributed to Na-Ca exchange is indeed due to Na^+-Ca^{2+}exchange or if it is carried by Ca^{2+}-activated nonspecific cation channels.[18,19] Such Ca^{2+}-activated channel currents are expected to be outward at voltages positive to the reversal potential and to increase as $[Ca^{2+}]_i$ rises. FIGURE 4 shows clearly that the current that is associated with the increase in $[Ca^{2+}]_i$ does not increase but remains flat or decreases as $[Ca^{2+}]_i$ rises. More direct evidence was obtained by determination of reversal potentials and their dependence on $[Ca^{2+}]_i$, as the predictions for currents carried by nonspecific Ca^{2+}-activated cation channels and Na^+-Ca^{2+} exchange currents have distinct differences. The reversal potential is expected to become more *positive* as $[Ca^{2+}]_i$ rises. The reversal potential of a current through nonspecific cation channels will remain *unchanged,* if it does not conduct Ca^{2+}, or, if it also carries Ca^{2+} ions, it will become more *negative* as $[Ca^{2+}]_i$ increases. To measure reversal potentials, Beuckelmann and Wier[4] produced ramped repolarizations preceded by conditioning depolarizations of various durations to increase $[Ca^{2+}]_i$ to different levels. During the ramp, an inward current developed that was more inward at higher $[Ca^{2+}]_i$. Upon termination of the ramp an outwardly directed capacity current was followed by a declining inward current that again was larger at higher $[Ca^{2+}]_i$. The reversal potential of the current increased linearly with $\log([Ca^{2+}]_i)$. It is clear, therefore, that the current sensitive to Ni^{2+} in the presence of internal and external Cs^+, verapamil, and ryanodine, has many of the properties expected of the Na-Ca exchange current, I_{NaCa}.

The relationship between the Ni^{2+}-sensitive current (I_{NaCa}) and $[Ca^{2+}]_i$ at a constant membrane voltage was also investigated. As reported previously,[3,4,20,21] this relationship is approximately linear over the range of $[Ca^{2+}]_i$ that can be examined in intact cells. This can be interpreted as evidence that the K_D of the internal binding site for Ca^{2+} is higher than the $[Ca^{2+}]_i$ that was attained in those experiments. The K_D for this site in the absence of Na^+ is 0.6 μM[14], but in the presence of Na^+ it will be somewhat higher.[22] Determination of the relationship between I_{NaCa} and $[Ca^{2+}]_i$ is important for understanding the molecular mechanism of Na-Ca exchange, as well as for understanding the role of Na-Ca exchange in cardiac cell function.

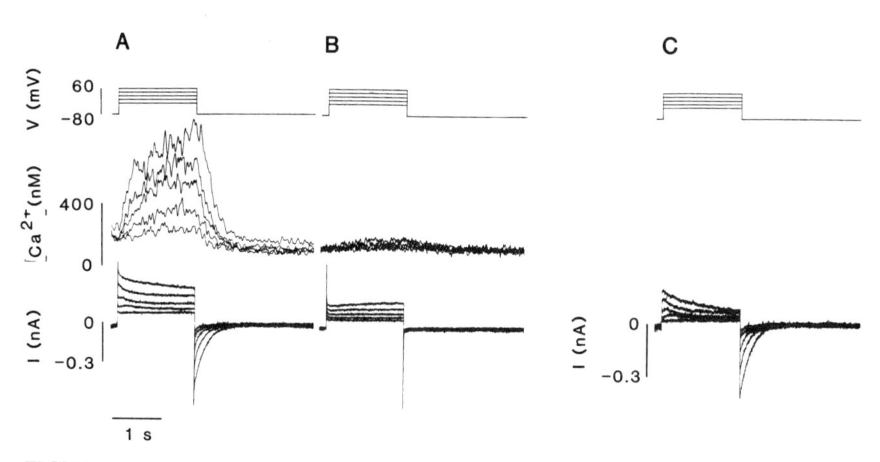

FIGURE 4. Experimental isolation of changes in $[Ca^{2+}]_i$ and membrane currents attributable to Na-Ca exchange. $[Ca^{2+}]_i$ changes and membrane currents during 250-msec depolarizing clamp pulses from a holding potential of −80 mV to 0-+100 mV in the presence of Na^+-, K^+-, and Ca^{2+}-channel blockade (**A**). Changes in $[Ca^{2+}]_i$ and time-dependent changes in current were greatly diminished after addition of 5 mM Ni^+ (**B**). Current traces in (**C**) were calculated by subtracting current records obtained during exposure to Ni^+ from those recorded previously under control conditions (from Beuckmann & Wier[4]). Current traces in (C) represent the putative Na-Ca exchange current. $[Ca^{2+}]_i$ transients are attributable to Na-Ca exchange, since all other sources of Ca^{2+} are blocked.

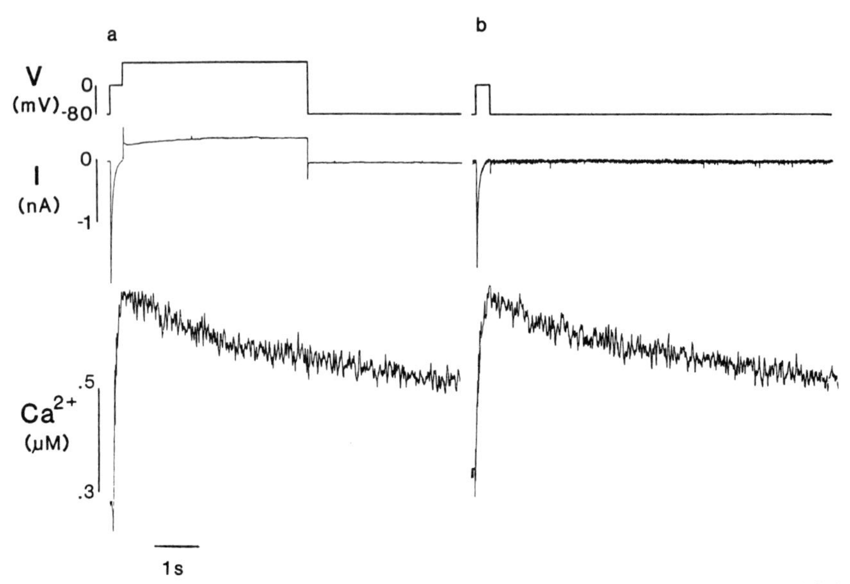

FIGURE 5. Decline of $[Ca^{2+}]_i$ in Na^+-free solution in the presence of caffeine (10 mM) is extremely slow and voltage-independent. External $[Ca^{2+}]_o$ was 1 mM. Upper trace, voltage-clamp protocol; middle trace, whole-cell current (mainly carried by Ca^{2+}); lower trace, $[Ca^{2+}]_i$ transient. The rate of decline of $[Ca^{2+}]_i$ is the same at membrane voltages of −80 mV (**B**) and +60 mV (**A**). The increase in $[Ca^{2+}]_i$ during the short depolarizing pulse (300 msec) is presumed to be due to influx of Ca^{2+} via the sarcolemmal Ca^{2+} channel (see text of Sipido & Wier[6] from which figure is taken. Used with permission.)

In the presence of caffeine and the absence of Na^+, $[Ca^{2+}]_i$ increase during a depolarizing pulse is due entirely to influx of Ca^{2+} through the sarcolemmal L-type Ca^{2+} channel (I_{Ca}). When I_{Ca} is turned off in this case, either by repolarization or by depolarization to very positive potentials, the decline of $[Ca^{2+}]_i$ is extremely slow (FIG. 5). In the experiments of Sipido and Wier,[6] the mean half-time ($T_{1/2}$) of the decline in $[Ca^{2+}]_i$ was 14.0s + 2.4 (mean + SEM, $n = 5$). The $T_{1/2}$ of decline of a $[Ca^{2+}]_i$ transient under normal conditions is about 0.15 sec.[8] It can also be seen that, in contrast to the control condition,[4,23] the rate of decline is not voltage-dependent, which confirms the absence of appreciable contributions by a voltage-dependent Ca^{2+} transporter, such as Na-Ca exchange, or an electrodiffusive leak of Ca^{2+} across the sarcolemma (FIG. 5A). Sipido and Wier[6] concluded that, on the time scale of a single Ca^{2+} transient, only the SR and Na-Ca exchange are capable of removing Ca^{2+} from the cytoplasm at a significant rate.

In conclusion, the data indicate that in guinea-pig cardiac centricular cells, Ca^{2+} entry via Na-Ca exchange can be significant at very positive membrane potentials, particularly when intracellular Na^+ is elevated.[3,4] Under such circumstances the thermodynamic gradient favors exchange of internal Na^+ for external Ca^{2+} and net influx of Ca^{2+} via the exchanger will result. When $[Ca^{2+}]_i$ is elevated and the membrane potential is repolarized (after voltage-clamp pulse or during repolarization of the action potential), Ca^{2+} efflux will be favored. The evidence indicates that efflux of Ca^{2+} by this mechanism contributes to the decline of the $[Ca^{2+}]_i$ transient under normal circumstances, but that the sarcoplasmic reticulum is normally much more effective at removing Ca^{2+} from the cytoplasm. Since the sarcolemmal Ca^{2+}-ATPase appears to remove very little Ca^{2+} on the time scale of a single Ca^{2+}transient,[1,6] it must be concluded that the Na-Ca exchanger removes nearly the exact amount of Ca^{2+} from the cell that enters via the Ca^{2+} current on each beat. The contribution of the Na-Ca exchange current to total membrane current under physiological conditions is very difficult to determine experimentally and remains to be done in detail.

REFERENCES

1. WIER, W. G. 1990. Cytoplasmic $[Ca^{2+}]$ in mammalian ventricle: Dynamic control by cellular processes. Ann. Rev. Physiol. **52:** 467-485.
2. MULLINS, L. J. 1979. The generation of electric currents in cardiac fibers by Na/Ca exchange. Am. J. Physiol. **236:** C103-110.
3. BARCENAS-RUIZ, L., D. J. BEUCKELMANN & W. G. WIER. 1987. Sodium-calcium exchange in heart: Currents and changes in $[Ca^{2+}]_i$. Science **238:** 1720-1722.
4. BEUCKELMANN, D. J. & W. G. WIER. 1989. Sodium-calcium exchange in guinea-pig cardiac cells: Exchange current and changes in intracellular Ca^{2+}. J. Physiol. (London) **414:** 499-520.
5. WIER, W. G., M. B. CANNELL, J. R. BERLIN, E. MARBAN & W. J. LEDERER. 1987. Cellular and subcellular heterogeneity of $[Ca^{2+}]_i$ in single heart cells revealed by fura-2. Science **235:** 325-328.
6. SIPIDO, K. R. & W. G. WIER. 1991. Flux of Ca^{2+} across the sarcoplasmic reticulum of guinea-pig cardiac cells during excitation-contraction coupling. J. Physiol. (London) **435:** 605-630.
7. MITRA, R. & M. MORAD. 1981. A uniform enzymatic method for dissociation of myocytes from hearts and stomachs of vertebrates. Pflueger's Archiv. Eur. J. Physiol. **391:** 85-100.
8. BEUCKELMANN, D. J. & W. G. WIER. 1988. Mechanism of release of calcium from sarcoplasmic reticulum of guinea-pig cardiac cells. J. Physiol. (London) **405:** 233-255.
9. BARCENAS-RUIZ, L. & W. G. WIER. 1987. Voltage-dependence of intracellular $[Ca^{2+}]_i$ transients in guinea pig ventricular myocytes. Circ. Res. **61:** 148-154.
10. CALLEWAERT, G., L. CLEEMANN & M. MORAD. 1988. Epinephrine enhances Ca^{2+}-current regulated Ca^{2+}release and Ca^{2+}-reuptake in rat ventricular myocytes. Proc. Natl. Acad. Sci. USA **85:** 2009-2013.
11. CANNELL, M. B., J. R. BERLIN & W. J. LEDERER. 1987. Effects of membrane potential changes on the calcium transient in single rat cardiac muscle cells. Science **238:** 1419-1423.
12. LI, Z., D. A. NICOLL, A. COLLINS, D. W. HILGEMANN, A. G. FILOTEO, J. T. PENNISTON, J. N. WEISS, J. M. TOMICH & K. D. PHILLIPSON. 1991. Identification of a peptide inhibitor of the cardiac sarcolemmal Na-Ca exchanger. J. Biol. Chem. **286(2):** 1014-1021.
13. HUME, J. R. & A. UEHARA. 1986. "Creep currents" in single frog atrial cells may be generated by electrogenic Na/Ca exchange. J. Gen. Physiol. **87:** 857-884.
14. KIMURA, J. & Y. MIURA. 1988. Effects of intracellular sodium and calcium on sodium-calcium exchange current. J. Mol. Cell. Cardiol. (Supp. IV) **20:** S.19.
15. KIMURA, J., S. MIYAMAE & A. NOMA. 1987. Identification of sodium-calcium exchange current in single ventricular cells of guinea-pig. J. Physiol. (London) **384:** 199-222.
16. LIPP, P. & L. POTT. 1988. Voltage dependence of sodium-calcium exchange current in guinea-pig atrial myocytes determined by means of an inhibitor. J. Physiol. **403:** 355-366.
17. EISNER, D. A. & W. J. LEDERER. 1985. Na-Ca exchange: Stoichiometry and electrogenicity. Am. J. Physiol. **248:** C189-202.
18. COLQUHOUN, D., E. NEHER, H. REUTER & C. F. STEVENS. 1981. Inward current channels activated by intracellular Ca in cultured cardiac cells. Nature **294:** 752-754.
19. EHARA, T., A. NOMA & K. ONO. 1988. Calcium-activated non-selective cation channel in ventricular cells isolated from adult guinea-pig hearts. J. Physiol. (London) **403:** 117-133.
20. BERLIN, J. R., J. R. HUME & W. J. LEDERER. 1988. $[Ca^{2+}]_i$-activated 'creep' currents in guinea-pig ventricular myocytes. J. Physiol. (London) **407:** 128p.

21. LIPP, P., L. POTT, G. CALLEWAERT & E. CARMELIET. 1990. Simultaneous recording of indo-1 fluorescence and Na/Ca exchange current reveals two components of Ca^{2+} release from sarcoplasmic reticulum of cardiac atrial myocytes. FEBS Lett. **275(1,2):** 181-184.
22. CERVETTO, L., L. LAGNADO & P. A. MCNAUGHTON. 1989. The effects of internal Na on the activation of the Na : Ca exchange in isolated salamander rods. Proceedings of the Physiological Society, 22-23 July, Cambridge Meeting, Abstract C.78.
23. BRIDGE, J. H. B., K. W. SPITZER & P. R. ERSHLER. 1988. Relaxation of isolated ventricular cardiomyocytes by a voltage-dependent process. Science **241:** 823-825.

Species Differences and the Role of Sodium-Calcium Exchange in Cardiac Muscle Relaxation[a]

DONALD M. BERS

Division of Biomedical Sciences
University of California
Riverside, California 92521-0121

INTRODUCTION

Contraction of cardiac muscle is activated by an increase of intracellular [Ca] ($[Ca]_i$). Some of this activating Ca enters the cell via sarcolemmal Ca channels, and some may also enter via Na-Ca exchange, but the latter seems only to be significant when intracellular Na activity is elevated (thereby biasing the Na-Ca exchange in favor of Ca influx).[1,2] Ca entering the cell can also trigger Ca release from the sarcoplasmic reticulum (SR) via the Ca-induced Ca-release process.[3] While Ca current is generally considered to be the trigger in Ca-induced Ca release, there is evidence to support the possibility that Ca entry via Na-Ca exchange can also induce SR Ca release[2,4] (see also Levesque *et al.,* this volume). Thus, a combination of Ca influx and SR Ca release activates cardiac myofilaments during a normal twitch contraction (with the latter normally contributing more from a quantitative perspective[5]).

For relaxation to occur, Ca must be effectively removed from the myofilaments. At least three transport processes can compete for cytoplasmic Ca during cardiac relaxation: (1) the SR Ca-ATPase pump, (2) the sarcolemmal Ca-ATPase pump, and (3) sarcolemmal Na-Ca exchange. We examined the competition among these systems by selective inhibition of these transport pathways during relaxation.[6-9]

NA-CA EXCHANGE CONTRIBUTES TO CARDIAC RELAXATION

In some of the experiments described below, we have used the approach of rapidly cooling and rewarming cardiac muscles or myocytes. In mammalian cardiac muscle, rapid cooling from 30°C to 0-1°C, leads to the release of apparently all of the SR Ca, while simultaneously inhibiting other Ca transport systems.[10-12] The release Ca then activates a contracture that develops slowly at 1°C (e.g., see FIGS. 3, 5). The amplitude of the rapid cooling contracture (RCC) can be used as an index of the relative amount of Ca available for release from the SR at the moment of cooling. Rapid rewarming to 30°C increases myofilament Ca sensitivity, resulting in a rapid increase in force (or rewarming spike).[13] Rewarming also reactivates the processes involved in reducing $[Ca]_i$ (which had been inhibited by the cold). We took advantage of the relatively stable elevation of $[Ca]_i$ during an RCC to change the extracellular solution before rewarming

[a] This work was supported in part by a Research Career Development Award and a grant from the National Institutes of Health (HL-30077).

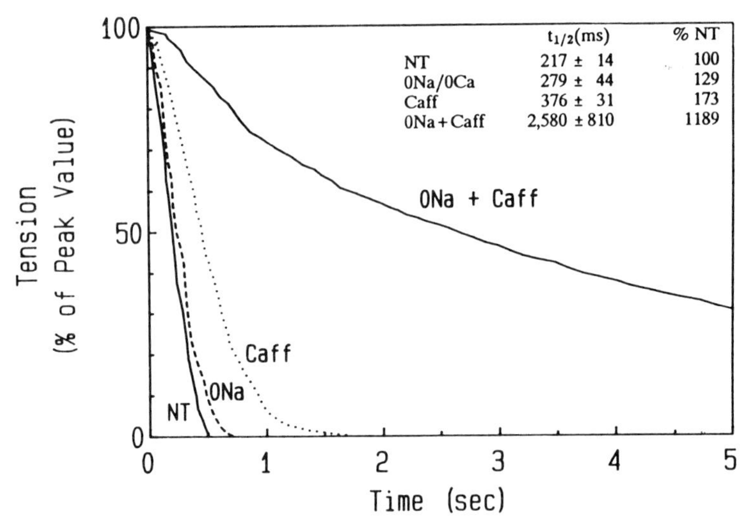

FIGURE 1. Time-course of relaxation in rabbit ventricular muscle after RCCs with inhibition of either Na-Ca exchange, the SR Ca pump, or both. During the RCC at ~ 1°C, the superfusion solution was changed from a normal Tyrode's (NT) to either a Na-free, Ca-free (0 Na), a NT with 10 mM caffeine (Caff), or a Na-free, Ca-free solution with 10 mM caffeine (0 Na + Caff). Relaxation half-times and percents of the NT value for pooled results are shown in the inset (from Bers & Bridge,[6] with permission of the American Heart Association).

from RCCs of similar amplitudes.[6] In this way, we could compare the relaxation phases of RCCs in a single rabbit ventricular muscle under different experimental conditions (see FIG. 1). When Na-Ca exchange is prevented by removal of extracellular Na and Ca (0 Na), relaxation is slowed by about 30%. When SR Ca accumulation is prevented by inclusion of 10 mM caffeine, relaxation is slowed by about 70%. When both systems are inhibited (0 Na + Caff), relaxation is dramatically slowed (i.e., by more than 1000%) and is often incomplete even after 20 sec. This result indicates that the SR Ca pump and to a lesser extent sarcolemmal Na-Ca exchange can produce cardiac relaxation. It also shows that other systems (such as the sarcolemmal Ca ATPase or mitochondrial Ca uptake) are too slow to account for cardiac relaxation. Indeed, the contribution of the sarcolemmal Ca-ATPase pump and mitochondrial Ca flux to normal relaxation in the heart would be expected to be very small.

We also examined the effect of membrane potential on relaxation that was attributable either to Na-Ca exchange (caffeine in FIG. 2), or to SR Ca accumulation (0 Na in FIG. 2). As can be seen, when relaxation depended upon Na-Ca exchange, the relaxation was voltage dependent, in a manner that would be expected for an electrogenic (3 Na : 1 Ca) Na-Ca exchange. In contrast, when relaxation depended upon SR Ca accumulation, there was no apparent voltage dependence. Data from Crespo *et al.*[14] is also shown in FIGURE 2 (open symbols). They separated the declining phase of intracellular Ca transients into a fast, E_m-insensitive component (τ_1) and a slow, E_m-sensitive component (τ_2). These components may reflect the SR and Na-Ca exchange components, respectively, as discussed above.

We also used another strategy to quantitatively assess the competition between Na-Ca exchange and the SR Ca pump in cardiac myocytes. FIGURES 3A and B show paired RCCs in an isolated rabbit ventricular myocyte.[8] If all the Ca released at the first RCC were reaccumulated by the SR, a second RCC should be as large as the first. This is indeed the case when Na-Ca exchange is prevented by Na-free, Ca-free solution (FIG. 3B). However, when Na-Ca exchange can occur (FIG. 3A), the second RCC is about 25% smaller than the first (23 ± 3%, $n = 14$). This indicates that Ca responsible for about 25% of the contractile force is extruded by Na-Ca exchange, while 75% is reaccumulated by the SR.

FIGURES 3C and D show a similar experiment in a guinea pig ventricular myocyte loaded with the intracellular Ca indicator, Indo-1.[7] When Na-Ca exchange is prevented by Na-free, Ca-free solution, three successive RCCs result in similar $[Ca]_i$ transients, without appreciable decrement from one to the next (FIG. 3D). When Na-Ca exchange is allowed (FIG. 3C), the successive $[Ca]_i$ transients are reduced by 36 and 51% (compared to their preceding RCCs). This result is consistent with Na-Ca exchange being responsible for removal of about 30 to 50% of the activator Ca in guinea pig ventricular myocytes at resting membrane potential.

FIGURE 4 shows another experiment in which we sought to evaluate the relative contributions of Na-Ca exchange in the SR Ca pump to relaxation. In this case we studied rat ventricular myocytes under voltage-clamp conditions, with intracellular Indo-1.[9] With short depolarizing voltage-clamp pulses (top) from −50 to +15 mV, the absence of Na_o slowed the decline in $[Ca]_i$ by about 23%. In this case, Ca entry via Na-Ca exchange was prevented by dialyzing the cell with very low $[Na]_i$. At the end of the long pulses (1 second, lower panels), when the cell was repolarized to −50 mV, the $[Ca]_i$ decline was also slowed by about 20% in Na-free solution. However, during the 1-second pulse at +15 mV, extracellular Na had little effect on the rate of $[Ca]_i$ decline. This is consistent with the finding that relaxation via Na-Ca exchange was slow at positive membrane potentials (see FIG. 2). This positive E_m would bias the competition between the SR Ca-ATPase pump and the Na-Ca exchanger in favor of

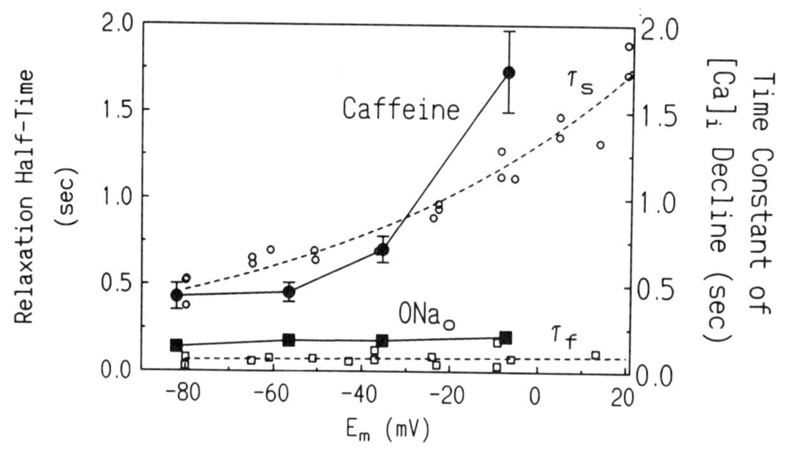

FIGURE 2. Voltage dependence of relaxation via Na-Ca exchange or the SR Ca pump. Na-Ca exchange is prevented by Na-free, Ca-free solution (0 Na_o, ■) and when SR uptake is prevented by 10 mM caffeine (Caffeine, ●). Membrane potential was altered by changing $[K]_o$ during the RCC and rewarming-induced relaxation. Nifedipine (10 μM) was included to block Ca current, which could occur at the depolarized potentials.[6] The open symbols are taken from a voltage-clamp study of Crespo *et al.*[14] where the decline of the $[Ca]_i$ transient at each E_m was resolved into two kinetic components, one with a fast time constant (τ_1) and one with a slower time constant (τ_2). The broken lines are curve fits to the open symbols.[14]

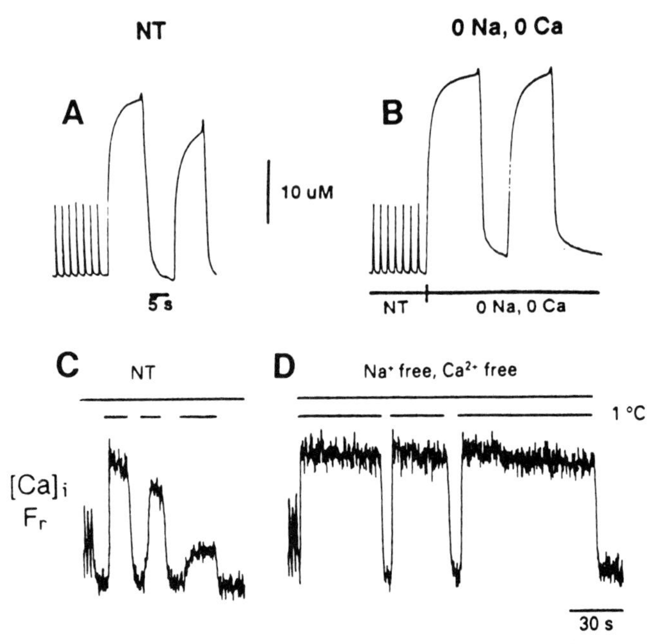

FIGURE 3. Sequential rapid cooling contractures (RCCs) used to assess the resequestration of Ca by the SR. (**A-B**) Cell shortening in a rabbit ventricular myocyte during a train of electrically evoked twitches and paired RCCs. The RCCs are induced in either normal Tyrode's (**A**) or when Na-Ca exchange is prevented from extruding Ca during rewarming by using a Na-free, Ca-free solution (**B**). (**C-D**) A similar protocol in a guinea-pig ventricular myocyte except that $[Ca]_i$ transients were measured during the RCCs (using Indo-1) and three successive RCCs were used instead of two (from Hryshko *et al.*[8] (top) and Bers *et al.*[7] (bottom), with permission).

the SR. Thus extracellular Na would have less influence on the decline of the $[Ca]_i$ transient.

This reiterates that Na-Ca exchange may be responsible for the extrusion of a significant fraction of activating Ca during relaxation in cardiac muscle (e.g., ~20-30%). If 20-30% of Ca required for activation is extruded from the cell via Na-Ca exchange during relaxation, it may be expected that during steady state a similar quantity of Ca should enter the cell during the activation phase. That is, over a single cardiac cycle at steady state, Ca influx must equal the Ca efflux. It is probable that the dominant pathway for Ca influx during the early part of the contraction is via sarcolemmal Ca channels. Indeed, the integral of Ca entry via Ca current could be sufficient to account for about 15-25% of the activation of the myofilaments.[5,15,16] In addition, an elegant experiment by Bridge *et al.*[17] showed a stoichiometric relationship between Ca entry via Ca current and Ca extrusion via Na-Ca exchange.

DIASTOLIC CA EFFLUX IS PRIMARILY DUE TO NA-CA EXCHANGE

During rest, rabbit ventricular muscle gradually becomes Ca depleted. This process is called rest decay and has been attributed to a gradual loss of SR Ca, assessed by post-rest contractions or RCCs[7,12,18,19] (see also FIGS. 5A-C). This is presumed to be due to a finite rate of SR Ca leak at rest and a subsequent extrusion of Ca from the cell. FIGURES 5A-C show that RCCs become progressively smaller as the duration of rest is increased from 2 sec to 2 min. If Na-Ca exchange is prevented by conducting these rest periods in Na-free, Ca-free solution, the rest decay is almost completely abolished for up to 5 min (see FIGS. 5D-F and 6). If the rate of SR Ca leak is increased (e.g. by ryanodine), the rate of rest decay can be dramatically increased (see FIG. 6). However, even the rest decay in the presence of ryanodine can be slowed by decreasing the transsarcolemmal [Na] gradient.[20] These results indicate that the process of rest decay and gradual cellular Ca depletion depends on Na-Ca exchange and not on other efflux mechanisms. That is, in intact cardiac cells, Ca efflux during diastole (as well as during relaxation), is mainly attributed to Na-Ca exchange with the sarcolemmal Ca pump making only a very small contribution. Indeed, one might have expected the sarcolemmal Ca pump to make a larger than normal contribution in the absence of extracellular Ca (i.e., when the gradient for the Ca pump is relatively small).

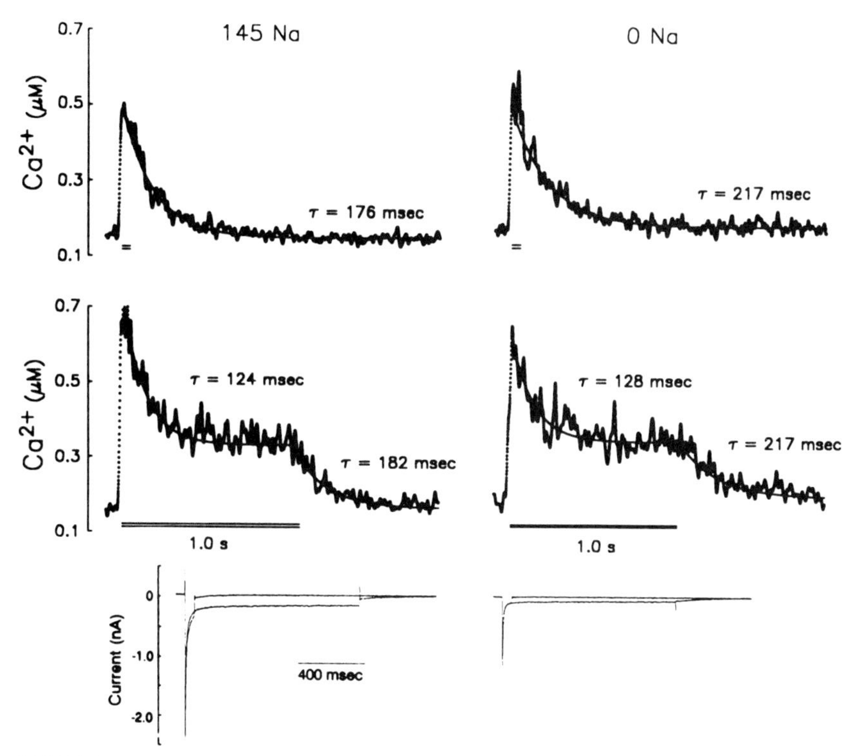

FIGURE 4. The effect of Na_o on the decline of $[Ca]_i$ in a voltage-clamped rat ventricular myocyte. Depolarizing pulses from −50 mV to +15 mV were used with durations of 50 msec (top) or 1 sec (bottom). The dialyzing patch-clamp pipette contained ≤0.5 mM Na to prevent Ca influx via Na-Ca exchange and 70 μM Indo-1 to allow assessment of $[Ca]_i$. During the long pulses and after repolarization, the $[Ca]_i$ was fit with a monoexponential decline (curves superimposed and time constants, τ indicated). During the long depolarizations there may be a small residual Ca influx via Ca channels (see current records at bottom). (From Bers *et al.*[9] with permission.)

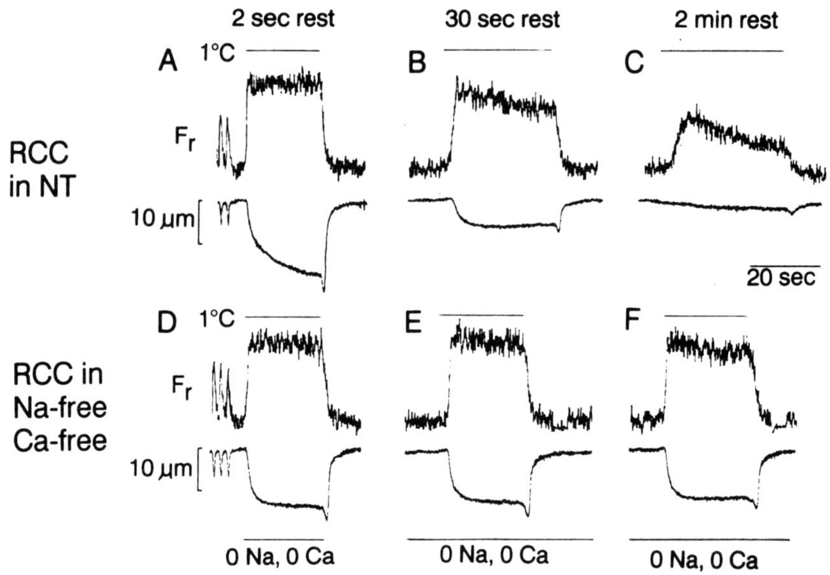

FIGURE 5. RCCs induced after various rest intervals in an isolated guinea-pig ventricular myocyte. Shortening (in µm) and $[Ca]_i$ (F_r) are shown when the rest (2 sec-2 min) and RCCs were in normal Tyrode's (NT, **A-C**) or in Na-free and Ca-free solution with 500 µM EGTA (**D-F**). The last few stimulated twitches (at 0.5 Hz) are shown in **A** and **D** in addition to the RCCs shown in each panel. The horizontal bar indicates the time during which the superfusate was at 1°C (from Bers *et al.*[7] with permission).

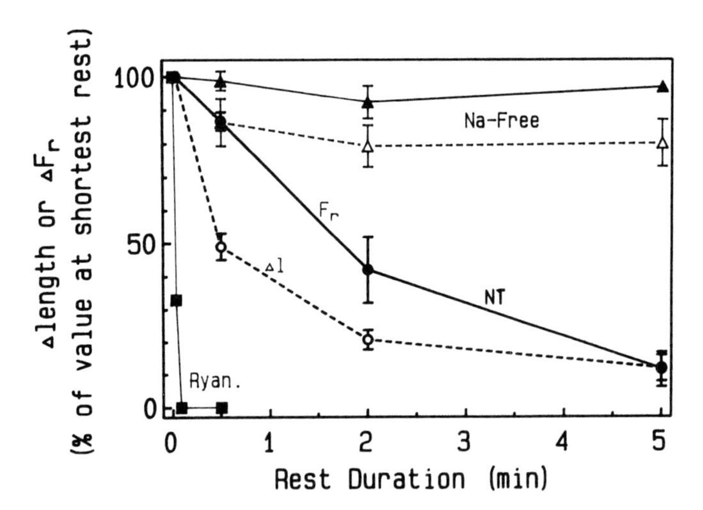

FIGURE 6. The effect of rest duration on the amplitude of myocyte shortening (Δl) or $[Ca]_i$ transients (F_r) during RCCs in guinea-pig ventricular myocytes. Results are from 15 experiments like that shown in FIGURE 5. The more rapid decline of the shortening NT curve can be attributed to different manner in which the fluorescence signal and the myofilaments depend on [Ca]. Ryanodine was also equilibrated with cells (■) and did not prevent RCCs, but greatly accelerated the rate of rest decay (from Bers *et al.*[7] with permission).

DIFFERENCES IN CA FLUXES IN RAT VERSUS RABBIT VENTRICULAR MUSCLE

In contrast to the process of rest decay discussed above for rabbit ventricle, rest potentiation, and a negative force-frequency relationship are observed in rat ventricle. That is, in rat ventricle the first contraction after a rest period of up to ~5 min is larger than the steady-state twitch contraction (i.e. rest potentiation) and increased frequency leads to decreased force (negative staircase). After the potentiated post-rest contraction, the subsequent contractions become gradually smaller as steady state is re-achieved. Rat ventricular SR (and cells) also appear to gain Ca during rest and lose Ca during stimulation after rest based on results with rapid cooling contractures and caffeine-induced contractures.[19,21,22]

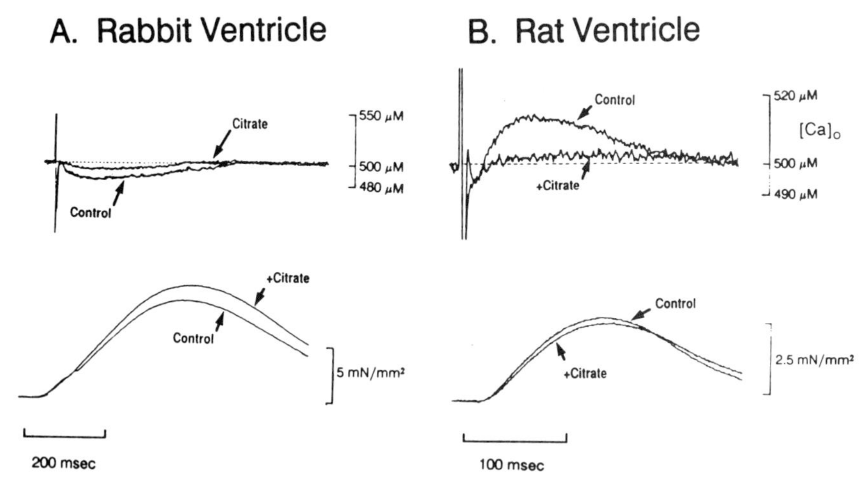

FIGURE 7. Transsarcolemmal Ca fluxes assessed with extracellular double-barreled, Ca-selective microelectrodes during individual contractions in rabbit (**A**) and rat (**B**) ventricular muscle (0.5 Hz, 30°C). The traces show $[Ca]_o$ (top) and tension (bottom) in the absence and presence of 10 mM citrate (which limits Ca_o depletion by buffering $[Ca]_o$, although we now know it also inhibits Ca current $[I_{Ca}]$[29]). The bath $[Ca]_o$ = 0.5 mM and is indicated by the dotted line. (A is from Shattock & Bers[24] with permission.)

These results indicate a potentially major fundamental difference in transsarcolemmal Ca fluxes in rat versus rabbit ventricle during both rest and contraction. However, this conclusion relies on indirect estimates of SR Ca content to infer transsarcolemmal Ca fluxes. To examine these Ca fluxes more directly, we used extracellular Ca microelectrodes in rat and rabbit ventricular muscle during the cardiac cycle (see FIG. 7). During the twitch in rabbit ventricle net Ca influx occurs as evidenced by the transient depletion of $[Ca]_o$. This is consistent with Ca entering the cell during the action potential and being extruded from the cell during relaxation and rest. This is also consistent with our intuitive expectation that Ca enters during contraction and is extruded afterwards.

Rat ventricle presents a striking contrast to this observation. FIGURE 7B shows that in rat ventricle there is a net Ca efflux occurring during the twitch. The net efflux of Ca during the contraction of rat ventricular muscle was sensitive to caffeine[24] (which also contrasts with the $[Ca]_o$ depletion in rabbit ventricle[23]). This, coupled with the fact that Na-Ca exchange is the main mechanism for Ca extrusion from the cell, led us to

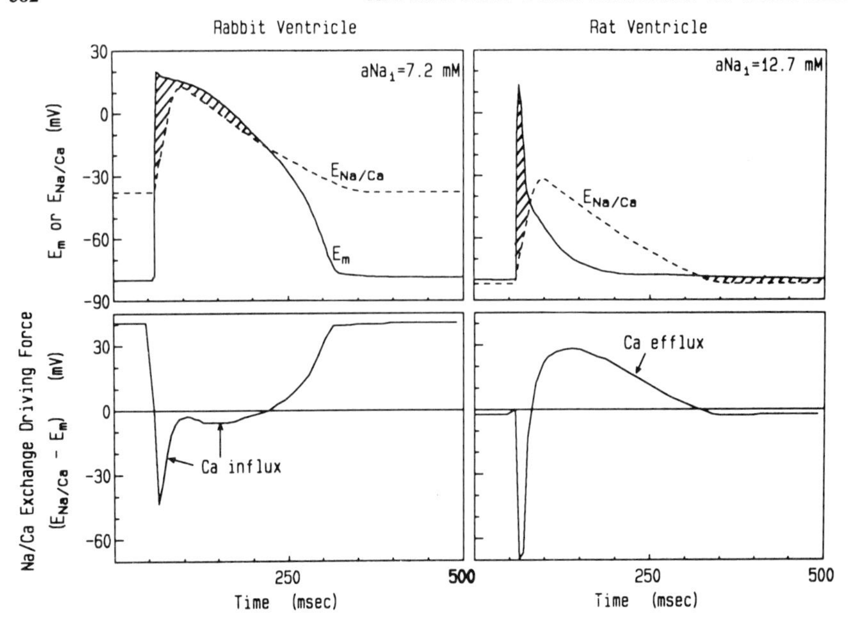

FIGURE 8. Action potentials (E_m) and estimated changes in the reversal potential of the Na-Ca exchange ($E_{Na/Ca}$) in rabbit and rat ventricle during the cardiac cycle (*top*). The electrochemical driving force for Ca entry via Na-Ca exchange ($E_{Na/Ca}$-E_m) is also shown (*bottom*). We assumed a stoichiometry of 3Na : 1Ca for the Na-Ca exchanger (such that $E_{Na/Ca} = 3\ E_{Na}$-$2E_{Ca}$) a Na_i values measured in these preparations.[24] For simplicity, the Ca transient accompanying the contraction has been assumed to be the same for both species (i.e., resting $[Ca]_i = 150$ nM, rising to a peak of 1 μM, 40 msec after the start of the action potential). Top panels were redrawn after Shattock and Bers[24] with permission from the *American Journal of Physiology.*

conclude that the large Ca efflux recorded in rat was due to Ca released from the SR being extruded via Na-Ca exchange.[24] In this same comparative study we also measured resting intracellular Na activity (aNa_i) and found that aNa_i was significantly higher in rat (12.7 mM) than in rabbit ventricle (7.2 mM).

FIGURE 8 shows a simple thermodynamic scheme that helps to illustrate how Ca flux via Na-Ca exchange may be expected to vary during the cardiac cycle in rat versus rabbit ventricular muscle. The upper panels show the membrane potential (E_m) and the predicted reversal potential for Na-Ca exchange ($E_{Na/Ca}$) during the action potential. $E_{Na/Ca}$ increases during the contraction as a consequence of the transient rise in $[Ca]_i$ (according to the equation, $E_{Na/Ca} = 3E_{Na} - 2E_{Ca}$). The lower panels show how the driving force for Ca entry via Na-Ca exchange ($E_{Na/Ca} - E_m$) changes over this same time. In rabbit ventricle Ca extrusion via Na-Ca exchange is thermodynamically favored at rest since E_m is positive to $E_{Na/Ca}$. During the action potential E_m exceeds $E_{Na/Ca}$ (even though $[Ca]_i$ is elevated) such that there is a modest driving force favoring Ca entry during the action potential. This is consistent with the transient Ca_o depletions observed during rabbit ventricular contractions (FIG. 7A) and gradual Ca loss during rest (rest decay). The actual rate of Ca extrusion, of course, depends on both the thermodynamic driving force and also kinetic factors (e.g., the fraction saturation and turnover rate of the exchanger). Thus, in rabbit ventricle during rest the Ca efflux may be kinetically limited by the low diastolic $[Ca]_i$ despite the large thermodynamic driving force favoring extrusion.

The resting aNa_i measured in rat ventricle is such that the $E_{Na/Ca}$ would be near the resting membrane potential (FIG. 8, lower right). This would be particularly important after a train of stimuli where aNa_i would be even higher, such that $E_{Na/Ca}$ would be negative to E_m and net Ca uptake would be favored at rest (at least until aNa_i recovers). This may provide a simple explanation for the rest-dependent increase in SR Ca and rest potentiation in rat ventricle. The action potential in rat ventricle is also very short compared to rabbit ventricle and lacks a plateau phase (see FIG. 8). Thus, when $[Ca]_i$ is high in rat ventricle there is a large driving force favoring Ca extrusion via Na-Ca exchange. The short action potential duration in rat ventricle thus biases the competition between the SR Ca pump and the Na-Ca exchanger toward the latter and net Ca efflux occurs during the contraction in rat ventricle (compare FIGS. 7 and 8).

The large Ca efflux during the potentiated post-rest contraction in rat ventricle reduces the SR Ca content available for release at the next contraction. Thus, there is a progressive decrease in the contraction amplitude, or a "negative staircase" as the SR becomes somewhat Ca-depleted as the steady state is approached. This framework provides a clear explanation for the well-known negative staircase and force-frequency relationship in rat ventricle. Spurgeon *et al.*[25] used voltage-clamped rat ventricular myocytes to show that if the duration of depolarization is prolonged, the negative staircase of contractions is converted to a positive staircase (as is seen in rabbit ventricular muscle). Similarly, Isenberg and Wendt-Gallitelli[26] demonstrated that in guinea-pig ventricle, increasing the duration of a voltage-clamp pulse leads to a positive "staircase," while reduction of pulse duration leads to a negative staircase. The prolongation of depolarization can be expected to increase intracellular (and SR) Ca loading by limiting the extrusion of Ca via Na-Ca exchange (or enhancing net influx), but also by allowing continued Ca entry via sarcolemmal Ca channels.

Manipulation of the transsarcolemmal [Na] and [Ca] gradients can also alter transmembrane Ca movements via Na-Ca exchange. For example, acetylstrophanthidin (which inhibits the sarcolemmal Na-pump and raises aNa_i) can change rest decay in rabbit ventricle into rest potentiation and cause Ca efflux to occur during the contraction much as observed in rat ventricle.[20,24,27,28] Rest potentiation in rat ventricle can also be converted to rest decay (as seen in rabbit ventricle) by reduction of $[Ca]_o$.[20] Thus, the direction and net movements of Ca via Na-Ca exchange can be strongly influenced by the transmembrane concentration gradients for Na and Ca as well as E_m.

In rabbit ventricle ryanodine greatly accelerates rest decay of RCCs in rabbit and guinea-pig ventricle such that post-rest contractions are abolished after a very few seconds of rest ($t_{1/2} \sim 1$ sec, see FIG. 6 and Refs. 7, 12, 20). In rat ventricle, ryanodine produces a stronger suppression of steady-state contractions but does not abolish RCCs even after a 5-min rest.[20] The foregoing discussion provides a clear explanation for this observation as well. The lower transsarcolemmal [Na] gradient in the rat ventricle may limit Ca extrusion via Na-Ca exchange such that the SR may still be able to retain some Ca, even with the ryanodine-induced SR Ca leak. In conclusion, the higher aNa_i and rapid action potential repolarization in rat compared to rabbit ventricle can explain many of the functional differences observed between these two species.

SUMMARY

During normal relaxation in rabbit, guinea-pig, and rat ventricular muscle, the Na-Ca exchange system competes with the SR Ca pump, with the former being responsible for about 20-30% of the Ca removal from the cytoplasm. Ca extrusion via Na-Ca exchange is E_m-sensitive, whereas Ca uptake by the SR is not. Neither the sarcolemmal Ca-ATPase pump nor mitochondrial Ca uptake appear to contribute significantly to the decline of $[Ca]_i$ during relaxation. Furthermore, the diastolic efflux of Ca from cardiac muscle cells appears to be primarily attributable to Na-Ca exchange and not the sarcolemmal Ca-ATPase pump. In rabbit ventricle Ca entry via Na-Ca exchange is favored thermodynamically during much of a normal twitch contraction and Ca extrusion occurs primarily between beats. In rat ventricle Ca efflux via Na-Ca exchange occurs during the contraction and net Ca influx may occur between beats. This fundamental difference in Ca fluxes during the cardiac cycle in rat versus rabbit ventricle may be a simple consequence of the shorter action potential duration and higher aNa_i in rat ventricle (due to the effects of E_m and [Na] and [Ca] gradients on Na-Ca exchange).

ACKNOWLEDGMENTS

The results presented here are based on recent studies done in collaboration with Drs. J. R. Berlin, J. H. B. Bridge, L. V. Hryshko, W. J. Lederer, M. J. Shattock and K. W. Spitzer.

REFERENCES

1. BARCENAS-RUIZ, L., D. J. BEUCKELMANN & W. G. WIER. 1987. Sodium-calcium exchange in heart: Membrane currents and changes in $[Ca^{2+}]_i$. Science **238:** 1720-1722.
2. BERS, D. M., D. M. CHRISTENSEN & T. X. NGUYEN. 1988. Can Ca entry via Na-Ca exchange directly activate cardiac muscle contraction? J. Mol. Cell. Cardiol. **20:** 405-414.
3. FABIATO, A. 1985. Time and calcium dependence of activation and inactivation of calcium-induced release of calcium from the sarcoplasmic reticulum of a skinned canine cardiac Purkinje cell. J. Gen. Physiol. **85:** 247-290.
4. LEBLANC, N. & J. R. HUME. 1990. Sodium current-induced release of calcium from cardiac sarcoplasmic reticulum. Science **248:** 372-376.
5. BERS, D. M. 1991. Excitation-Contraction Coupling and Cardiac Contractile Force. Kluwer Academic Press. Dordrecht, the Netherlands. 258 pp.
6. BERS, D. M. & J. H. B. BRIDGE. 1989. Relaxation of rabbit ventricular muscle by Na-Ca exchange and sarcoplasmic reticulum Ca-pump: Ryanodine and voltage sensitivity. Circ. Res. **65:** 334-342.
7. BERS, D. M., J. H. B. BRIDGE & K. W. SPITZER. 1989. Intracellular Ca transients during rapid cooling contractures in guinea-pig ventricular myocytes. J. Physiol. **417:** 537-553.
8. HRYSHKO, L. V., V. M. STIFFEL & D. M. BERS. 1989. Rapid cooling contractures as an index of SR Ca content in rabbit ventricular myocyte. Am. J. Physiol. **257:** H1369-1377.
9. BERS, D. M., W. J. LEDERER & J. R. BERLIN. 1990. Intracellular Ca transients in rat cardiac myocytes: Role of Na/Ca exchange in excitation-contraction coupling. Am. J. Physiol. **258:** C944-C954.
10. KURIHARA, S. & T. SAKAI. 1985. Effects of rapid cooling on mechanical and electrical responses in ventricular muscle of guinea pig. J. Physiol. **361:** 361-378.
11. BRIDGE, J. H. B. 1986. Relationships between the sarcoplasmic reticulum and transarcolemmal Ca transport revealed by rapidly cooling rabbit ventricular muscle. J. Gen. Physiol. **88:** 437-473.
12. BERS, D. M., J. H. B. BRIDGE & K. T. MACLEOD. 1987. The mechanism of ryanodine action in cardiac muscle assessed with Ca selective microelectrodes and rapid cooling contractures. Can. J. Physiol. Pharmacol. **65:** 610-618.

13. HARRISON, S. M. & D. M. BERS. 1989. The influence of temperature on the calcium sensitivity of the myofilaments of skinned ventricular muscle from the rabbit. J. Gen. Physiol. **93:** 411-427.
14. CRESPO, L. M., C. J. GRANTHAM & M. B. CANNELL. 1990. Kinetics, stoichiometry and role of the Na-Ca exchange mechanism in isolated cardiac myocytes. Nature **345:** 618-621.
15. ISENBERG, G. 1982. Ca entry and contraction as studied in isolated bovine ventricular myocytes. Z. Naturforsch. Teil C **37:** 502-512.
16. FABIATO, A. 1983. Calcium-induced release of calcium from the cardiac sarcoplasmic reticulum. Am. J. Physiol. **245:** C1-C14.
17. BRIDGE, J. H. B., J. R. SMOLLEY & K. W. SPITZER. 1990. Isolation of the sodium-calcium exchange current underlying sodium-dependent relaxation in heart muscle. Science **248:** 376-378.
18. ALLEN, D. G., B. R. JEWELL & E. H. WOOD. 1976. Studies of the contractility of mammalian myocardium at low rates of stimulation. J. Physiol. **254:** 1-17.
19. BERS, D. M. 1989. SR Ca loading in cardiac muscle preparations based on rapid cooling contractures. Am. J. Physiol. **256:** C109-C120.
20. BERS, D. M. & D. M. CHRISTENSEN. 1990. Functional interconversion of rest decay and ryanodine effects in rabbit or rat ventricle depends on Na/Ca exchange. J. Mol. Cell. Cardiol. **22:** 715-723.
21. LEWARTOWSKI, B. & K. ZDANOWSKI. 1990. Net Ca^{2+} influx and sarcoplasmic reticulum Ca^{2+} uptake in resting single myocytes of the rat heart: Comparison with guinea-pig. J. Mol. Cell Cardiol. **22:** 1221-1229.
22. BANIJAMALI, H. S., W. D. GAO & H. E. D. J. TER KEURS. 1990. Induction of calcium leak from the sarcoplasmic reticulum of rat cardiac trebeculae by ryanodine. Circulation **82:** III-215.
23. BERS, D. M. 1985. Ca influx and SR Ca release in cardiac muscle activation during postrest recovery. Am. J. Physiol. **248:** H366-H381.
24. SHATTOCK, M. J. & D. M. BERS. 1989. Rat vs. rabbit ventricle: Ca flux and intracellular Na assessed by ion-selective microelectrodes. Am. J. Physiol. **256:** C813-C822.
25. SPURGEON, H. A., G. ISENBERG, A. TALO, M. D. STERN, M. C. CAPOGROSSI & E. G. LAKATTA. 1988. Negative staircase in cytosolic Ca^{2+} in rat myocytes is modulated by depolarization duration. Biophys. J. **53:** 601a.
26. ISENBERG, G. & M. F. WENDT-GALLITELLI. 1989. Cellular mechanisms of excitation contraction coupling. *In* Isolated Adult Cardiomyocytes. H. M. Piper & G. Isenberg, Eds. Vol. II: 213-248. CRC Press. Boca Raton, FL.
27. SUTKO, J. L., D. M. BERS & J. P. REEVES. 1986. Postrest inotropy in rabbit ventricle: Na^{+}-Ca^{2+} exchange determines sarcoplasmic reticulum Ca^{2+} content. Am. J. Physiol. **250:** H654-H661.
28. BERS, D. M. 1987. Mechanisms contributing to the cardiac inotropic effect of Na-pump inhibition and reduction of extracellular Na. J. Gen. Physiol. **90:** 479-504.
29. BERS, D. M., L. V. HRYSHKO, S. M. HARRISON & D. DAWSON. 1991. Citrate decreases contraction and Ca current in cardiac muscle independent of its buffering action. Am. J. Physiol. **260:** C900-C909.

Role of Reverse-Mode Na^+-Ca^{2+} Exchange in Excitation-Contraction Coupling in the Heart[a]

PAUL C. LEVESQUE,[b] NORMAND LEBLANC,[c] AND JOSEPH R. HUME[b]

[b]*Department of Physiology*
University of Nevada School of Medicine
Reno, Nevada 89557-0046
and
[c]*Department of Physiology*
University of Manitoba
Winnipeg, Canada R2H 2A6

INTRODUCTION

The electrical and contractile properties of heart muscle cells are critically dependent upon changes in $[Ca^{2+}]_i$. Depolarization causes Ca^{2+} entry through sarcolemmal voltage-sensitive calcium channels and perhaps via Na^+-Ca^{2+} exchange.[1] The elevated $[Ca^{2+}]_i$ triggers release of additional Ca^{2+} from the SR,[2] resulting in cell contraction. Ca^{2+} must then be removed from the contractile apparatus, via uptake into the SR or extrusion across the sarcolemma, for relaxation to occur. The sarcolemmal Na^+-Ca^{2+} exchange mechanism contributes to the regulation of $[Ca^{2+}]_i$ in many cell types, including heart.[3] An interesting feature of the Na^+-Ca^{2+} exchanger is that it can move Ca^{2+} inward or outward, depending on the thermodynamic driving force and electrochemical gradient for Na^+.[4] Thus, the Na^+-Ca^{2+} exchanger could potentially play a role in both contraction and relaxation of cardiac cells. At present, the extent to which Na^+-Ca^{2+} exchange contributes to Ca^{2+} fluxes that occur during the various phases of the cardiac cycle is unclear.

There is widespread agreement that Na^+-Ca^{2+} exchange is an important mechanism for Ca^{2+} extrusion and cell relaxation under physiological conditions. Evidence suggesting the importance of the exchanger in mediating extrusion of Ca^{2+} comes from studies in which trans-sarcolemmal Na^+ and Ca^{2+} gradients were altered while extracellular[5,6] and intracellular[7,8] $[Ca^{2+}]_i$ transients as well as isolated exchange currents[9-12] were measured during activation of cell contraction. Although there is a growing consensus regarding the role of the exchanger in Ca^{2+} extrusion, there is still uncertainty over whether Na^+-Ca^{2+} exchange promotes sarcolemmal Ca^{2+} entry during action potentials and thus contributes to EC coupling under physiological conditions. Sarcolemmal voltage-sensitive Ca^{2+} channels are considered to be the primary source of trigger Ca^{2+} for SR Ca^{2+} release during action potentials.[13] Theoretically, Ca^{2+} influx via Na^+-Ca^{2+} exchange could occur during the initial phase of action potentials when the membrane is depolarized and the potential transiently

[a]This work was supported by National Institutes of Health Grant HL 30143. J.R.H. was supported by an American Heart Association Established Investigator Award.

becomes positive to the reversal potential for the exchanger.[6] Thus, outward exchange currents (Ca^{2+} influx) would develop simultaneously with the upstroke of the action potential, and inward currents (Ca^{2+} efflux) would occur later during action potential repolarization.[1] Consistent with this notion, $[Ca^{2+}]_i$ transients attributable to Ca^{2+} influx via Na^+-Ca^{2+} exchange (SR Ca^{2+} release and sarcolemmal Ca^{2+} channels were blocked by ryanodine and verapamil, respectively) have been observed in guinea pig ventricular myocytes depolarized to potentials similar to those achieved during action potentials.[14]

A fundamental feature of the Na^+-Ca^{2+} exchanger is that an elevation of $[Na^+]_i$ will promote the exchange of intracellular Na^+ for extracellular Ca^{2+}. Thus, manipulations that elevate $[Na^+]_i$ could augment the ability of the exchanger to elevate $[Ca^{2+}]_i$ and to cause contraction. Even minimal increases in $[Na^+]_i$ profoundly affect contractions in cardiac Purkinje fibers.[15] Bers *et al.*[16] reported that in the absence of Ca^{2+} influx through Ca^{2+} channels, Ca^{2+} entry via Na^+-Ca^{2+} exchange caused contraction of cardiac muscle after treatment with a cardiac glycoside (to inhibit the Na^+-K^+ pump) but not beforehand, suggesting that an elevation of $[Na^+]_i$ is required for the exchanger to contribute significantly to contraction. In these experiments, however, loss of contraction in the presence of a Ca^{2+} channel antagonist and absence of a cardiac glycoside may be due to reduced loading of the SR Ca^{2+} store rather than a reduced source of trigger Ca^{2+}. More recently, Bers *et al.*[7] utilized Indo-1 to measure Ca^{2+} transients in isolated rat ventricular myocytes under voltage-clamp control and found that Ca^{2+} influx via Na^+-Ca^{2+} exchange was unable to contribute to SR Ca^{2+} release. However, the myocytes were dialyzed using patch pipettes with internal solutions that contained only 0.5 mM Na^+, which is well below physiological levels of intracellular Na^+.[17] The results of a similar study were consistent in that Na^+-Ca^{2+} exchange did not significantly contribute to $[Ca^{2+}]_i$ in cardiac cells dialyzed with low Na^+.[18] These studies suggest that $[Na^+]_i$ is a critical determinant of the ability of Na^+-Ca^{2+} exchange to contribute to the $[Ca^{2+}]_i$ transient in cardiac myocytes.

Crespo *et al.*[8] found that exchanger-mediated increases in $[Ca^{2+}]_i$ occurred in single guinea pig ventricular myocytes, in the absence of functional SR, when patch pipettes contained physiological $[Na^+]_i$. Despite observing exchanger-mediated Ca^{2+} transients, Ca^{2+} entry through the exchanger was considered too slow to markedly affect the rise of Ca^{2+} transients observed with action potentials. Philipson and Ward[19] have also estimated from sarcolemmal vesicle experiments that Ca^{2+} entry via Na^+-Ca^{2+} exchange under normal cellular conditions would be insignificant. These studies, however, fail to consider the possibility that the Na^+ that enters the cell during activation of Na^+ current (I_{Na}) may support, at least in part, influx of Ca^{2+} via Na^+-Ca^{2+} exchange. For instance, $[Ca^{2+}]_i$ transients that accompany action potentials are larger in amplitude and occur more rapidly than those that accompany peak Ca^{2+} currents elicited in the absence of I_{Na} under physiological conditions in the same cell.[20] This suggests that the I_{Na} may contribute to the $[Ca^{2+}]_i$ transient-accompanying action potentials. Leblanc and Hume[21] found that I_{Na} can release Ca^{2+} from cardiac SR in the absence of Ca^{2+} influx through Ca^{2+} channels. These TTX-sensitive $[Ca^{2+}]_i$ transients were dependent upon extracellular Ca^{2+} and ryanodine-sensitive Ca^{2+} stores, suggesting that perhaps the Na^+-Ca^{2+} exchanger was involved in mediating influx of trigger Ca^{2+} for SR release. Thus, when I_{Na} is not suppressed, Na^+-Ca^{2+} exchange may be involved in fast events such as SR Ca^{2+} release.

The present study was designed to investigate further the contribution of I_{Na} to EC coupling in heart. Also, we tested more directly whether Na^+-Ca^{2+} exchange mediates the I_{Na}-induced release of Ca^{2+} from SR. Replacing extracellular Na^+ with equimolar lithium, which carries current through Na^+ channels but does not readily substitute for Na^+ on the Na^+-Ca^{2+} exchanger, abolished I_{Na}-induced Ca^{2+} transients. Lithium also inhibited nisoldipine-insensitive Ca^{2+} transients elicited by action potentials. Thus, I_{Na}- and action potential-induced, nisoldipine-insensitive Ca^{2+} transients occur only under conditions that allow for normal functioning of the sarcolemmal Na^+-Ca^{2+} exchanger. The results support the hypothesis that in response to a I_{Na}-mediated transient increase in $[Na^+]_i$, Na^+-Ca^{2+} exchange promotes Ca^{2+} entry into cardiac cells and triggers SR Ca^{2+} release during physiologic action potentials.

METHODS

Single guinea pig ventricular myocytes were enzymatically isolated using a variation of a method described previously.[22] Briefly, the heart was rapidly removed from a guinea pig (200-300 g), and the coronary arteries were cleared of blood by perfusing the aorta with a Krebs-Henseleit buffer (KHB) solution containing (mM): 120, NaCl; 4.8, KCl; 1.5, $CaCl_2$; 2.2, $MgSO_4$; 1.2, NaH_2PO_4; 25, $NaHCO_3$; and 10, glucose, equilibrated with 95% O_2-5% CO_2 at 37°C. The aorta was then cannulated and the heart was attached to a perfusion apparatus and gravity fed at a constant pressure with normal KHB. The perfusate was then switched to a nominally Ca^{2+}-free KHB for 5 min, after which time collagenase (Sigma Type 1) was added to the Ca^{2+}-free KHB to give a final concentration of about 0.2 mg/ml. Digestion was allowed to continue for 45 min. The heart was then removed from the perfusion apparatus and the right ventricle was dissected, cut into small pieces and incubated for 20-30 min in normal KHB containing collagenase (0.8 mg/ml) at 37°C. The tissue was then washed free of collagenase and reintroduced into normal KHB. Aliquots of disaggregated cells were harvested by gentle trituration for electrophysiolgical studies.

Membrane currents and potentials were recorded using the whole-cell variant of the patch-clamp technique.[23] Microelectrodes with tip resistances of 1-2 MΩ allowed for control of the intracellular milieu by dialysis of the cell interior with the pipette solution. Pipette resistances were compensated for electronically. For current-clamp experiments (FIGS. 1 and 5), the patch electrode was filled with (mM): 110, K^+ gluconate; 30, KCl; 10, NaCl; 5, Hepes; 5, ATP (4, diK^+; 1, Mg^{2+}); and 0.05 Indo-1 (pH 7.2 with KOH). The external solution contained (mM): 140, NaCl; 5.4, KCl; 2.5, $CaCl_2$; 0.5, $MgCl_2$; 5, Hepes; 11, glucose; ± 0.005 nisoldipine (pH 7.4 with NaCl). For voltage-clamp experiments (FIGS. 2, 3, and 4) K^+ channels were inhibited with cesium and tetra ethylammonium chloride (TEA). The internal solution contained (mM): 105, Cs-aspartate; 20, CsCl; 20, TEA; 10, NaCl; 5, Hepes; 5, ATP (4, diK^+; 1, Mg^{2+}); and 0.05 Indo-1 (pH 7.2 with CsOH). The external solution contained (mM): 130, NaCl; 10, CsCl; 0.5, MgCl; 2.5, $CaCl_2$; 5, Hepes; 11, glucose; ± 0.005 nisoldipine (pH 7.4 with NaCl). All experiments were carried out at room temperature. The data were digitized at a sampling rate of 1 kHz during current-clamp experiments. Two clock speeds were used during voltage-clamp experiments; the first 50 msec were digitized at 20 kHz and the last 950 msec at 1 kHz. The ability to observe I_{Na}-induced Ca^{2+} release from SR in the absence of Ca^{2+} entry through Ca^{2+} channels was facilitated by using a conditioning protocol designed to maintain the SR Ca^{2+} load.[20] The SR Ca^{2+} store was loaded 5-10 sec before the test pulse by a series of voltage-clamp steps (5 at 0.5 Hz) to either 0 mV or +60 mV (when nisoldipine was present) from a holding potential of −80 mV. In the presence of nisoldipine, the conditioning pulses to positive potentials allow Ca^{2+} entry via reverse-mode Na^+-Ca^{2+} exchange to load SR Ca^{2+} stores. The Ca^{2+} indicator Indo-1 (pentapotassium salt, 50 μM) was used to measure $[Ca^{2+}]_i$ transients in isolated myocytes. The method used has been described in detail previously.[21]

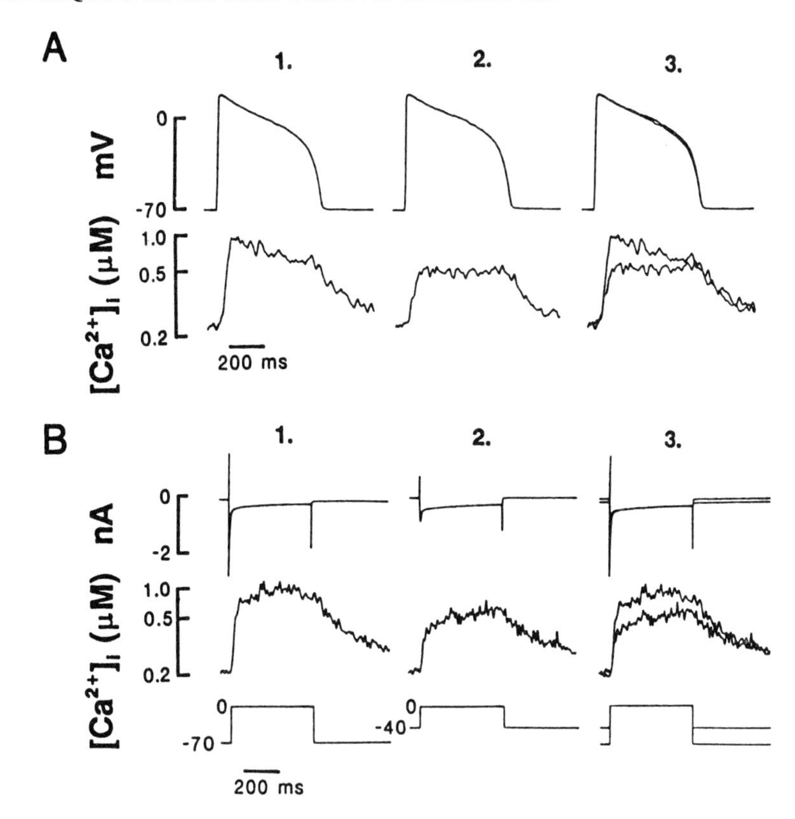

FIGURE 1. A. Effects of TTX on the action potential and $[Ca^{2+}]_i$ transient. The action potential recorded in control conditions (A1) was preceded by a train of five action potentials evoked at a frequency of 0.1 Hz to load SR Ca^{2+} stores. During exposure to TTX (5 μM), action potentials were elicited every 30 sec. The action potential and corresponding Ca^{2+} signals shown in A2 were measured after 3 min in TTX (steady-state Ca^{2+} transient). Action potentials and $[Ca^{2+}]_i$ transients from A1 and A2 are superimposed in A3. **B.** Effects of two different holding potentials on membrane current and amplitude of the Ca^{2+} signal evoked at 0 mV. Different cells were used for A and B. Recordings in B1 and B2 were obtained following a train of five conditioning test pulses to 0 mV and 500-msec duration from a holding potential of −70 mV. For A and B, the cells were dialyzed with normal internal and external solutions containing K^+ and the ratio, 400 nm/500nm, of the collected light signals was calibrated as described.[21] Resting $[Ca^{2+}]_i$ under control conditions was 186 ± 14.7 nM (mean ± SEM, $n = 20$). The peak Ca^{2+} transient during the action potential was 1374 ± 360 nM (mean ± SEM, $n = 15$). Similar results were obtained in five other cells. (Taken from Leblanc & Hume[21]; used with permission from *Science.* Copyright 1990 by the AAAS.)

RESULTS

Effect of I_{Na} on $[Ca^{2+}]_i$ Transients

The role of I_{Na} in EC coupling in heart has not been systematically studied previously. This prompted us to examine the effect of I_{Na} on $[Ca^{2+}]_i$ transients in isolated guinea pig ventricular myocytes dialyzed with Indo-1. Action potentials elicited under current-clamp conditions were accompanied by a rapid, transient increase in $[Ca^{2+}]_i$ (FIG. 1A). $[Ca^{2+}]_i$ returned toward previous levels during repolarization of the action potential. The effects of TTX on the action-potential $[Ca^{2+}]_i$ transient were tested to determine whether Na^+ influx through sarcolemmal Na^+ channels might contribute to EC coupling during the action potential. Exposure of the cells to TTX reduced the size of the $[Ca^{2+}]_i$ transient without significantly altering the action-potential plateau or duration (FIG. 1, A2 and A3). The effect of TTX on the $[Ca^{2+}]_i$ transient was not likely due to decreased Ca^{2+} influx through Ca^{2+} channels since the action potential was unaltered from control.

We then conducted separate experiments in voltage-clamped myocytes to directly investigate the contribution of I_{Na} to $[Ca^{2+}]_i$ transients. Holding potentials were varied so that the contribution of different components of inward current to the $[Ca^{2+}]_i$ transients could be determined (FIG. 1B). The inward current elicited by a 500-msec test pulse to 0 mV from a holding potential of −70 mV is shown in FIGURE 1, B1. This current results from the activation of TTX-sensitive Na^+ channels and dihydropyridine-sensitive Ca^{2+} channels. The test pulse induced a rapid rise in $[Ca^{2+}]_i$, which peaked at 1 μM before returning to resting levels. Next, a 500-msec test pulse from − 40 mV to 0 mV was applied and the corresponding inward current and $[Ca^{2+}]_i$ transient were measured. Na^+ channels were inactivated at this holding potential and the inward current reflects the activation primarily of only L-type Ca^{2+} channels (FIG. 1, B2). The $[Ca^{2+}]_i$ transient elicited by Ca^{2+} influx through Ca^{2+} channels was significantly smaller in amplitude than those elicited by the test pulse from −70 mV (FIG. 1, B3). These data suggest that Na^+ currents, whether occurring during action potentials or elicited by voltage-clamp depolarizations, contribute, in some way, to $[Ca^{2+}]_i$ transients in cardiac myocytes. The findings are consistent with those of a previous study which showed that the $[Ca^{2+}]_i$ transients elicited by action potentials are larger in amplitude than those resulting from maximal activation of sarcolemmal voltage-sensitive Ca^{2+} channels.[20]

Since I_{Na} appears to influence the magnitude of action potential $[Ca^{2+}]_i$ transients, we were curious as to whether I_{Na} alone, in the absence of Ca^{2+} influx through sarcolemmal Ca^{2+} channels, could elicit Ca^{2+} transients. Voltage-clamp depolarizations were applied during constant exposure of Indo-1-loaded myocytes to 5 μM nisoldipine to block dihydropyridine-sensitive Ca^{2+} channels (FIG. 2, A1). Fast, large amplitude inward currents were elicited by 500-msec test pulses to −40 mV from a holding potential of −80 mV in the absence of Ca^{2+} currents. The inward current induced a rapid increase in $[Ca^{2+}]_i$, which peaked and then declined during the test pulse. Both the inward current and $[Ca^{2+}]_i$ transient were blocked after exposure to 30 μM TTX for 1 min (FIG. 2, A2). These results demonstrate that $[Ca^{2+}]_i$ transients can be elicited by activation of Na^+ influx through TTX-sensitive Na^+ channels in the absence of Ca^{2+} entry through voltage-sensitive Ca^{2+} channels.

Although the TTX-sensitive Ca^{2+} transients were presumed to be due to Na^+ entry through Na^+ channels, alternative mechanisms could underlie the I_{Na}-induced transients. Incomplete block of Ca^{2+} channels or loss of voltage control during attempts to clamp Na^+ currents could give rise to the $[Ca^{2+}]_i$ transients. Therefore, we measured I_{Na}-induced $[Ca^{2+}]_i$ transients in voltage-clamped myocytes under conditions that optimized Ca^{2+} channel blockade and voltage control of I_{Na} (FIG. 2,B). Complete block of L-type Ca^{2+} channels was attained by dialyzing the cell internally and externally with the Ca^{2+} channel antagonist D-600 (10 μM). Voltage control was improved by partially inactivating Na^+ currents by using a holding potential of −60 mV. Fast inward currents of smaller amplitude were elicited under these conditions (FIG. 2, B1, and B3). These inward currents were due to activation of Na^+ channels since they were inhibited by 30 μM TTX (FIG. 2, B2). The absence of inward current during exposure to TTX also shows that D-600 completely blocked Ca^{2+} influx.

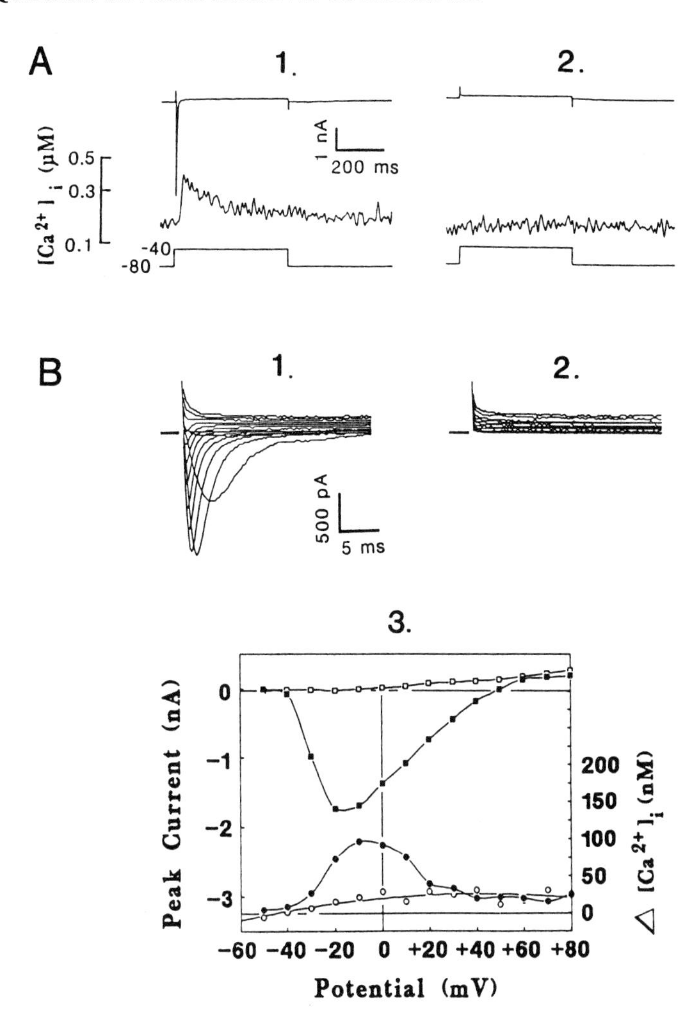

FIGURE 2. Ca^{2+} transients in cesium-loaded cells in the absence of Ca^{2+} current. The data in A and B are from different cells. **A.** The cell was continuously superfused with 5 μM nisoldipine; 500-msec step depolarizations to −50 mV from a holding potential of −80 mV were applied to the cell every 30 sec. In the absence of TTX (A1), this protocol elicited Na^+ currents and $[Ca^{2+}]_i$ transients. The traces shown in A2 were obtained after 1 min of exposure to 30 μM TTX. Similar observations were made in nine other cells. No conditioning protocol was used. **B.** The cell was internally dialyzed and externally perfused with 10 μM D600. Currents were elicited in the absence (B1) and presence (B2) of 30 μM TTX. Na^+ channels were partly inactivated by holding the membrane potential at −60 mV. Five-hundred-msec pulses from −50 to +80 mV were applied in 10-mV increments at 0.2 Hz. In B3, peak inward currents (left ordinate; squares) and resulting changes in peak $[Ca^{2+}]_i$ from resting level (right ordinate; circles) were plotted as a function of pulse potential. Control, filled symbols; TTX (3 min), empty symbols. Similar observations were made in four other cells. Resting $[Ca^{2+}]_i$ = 158 nM. A conditioning protocol was used as described in METHODS. Resting $[Ca^{2+}]_i$ in nisoldipine-treated cells was 175 ± 14 nM (mean ± SEM, n = 9). (Taken from Leblanc & Hume[21]; used with permission from *Science.* Copyright 1990 by the AAAS.)

The voltage dependence of Na^+ currents and corresponding $[Ca^{2+}]_i$ transients in the absence and presence of TTX are shown in FIGURE 2, B3. The membrane potentials at which the Na^+ currents activated, peaked, and reversed indicate that reasonably good voltage control was maintained. $[Ca^{2+}]_i$-transient amplitudes showed similar voltage dependence. As expected, TTX abolished both the inward currents and $[Ca^{2+}]_i$ transients. These results provide further evidence that the $[Ca^{2+}]_i$ transients measured under these conditions are due to Na^+ influx through Na^+ channels and are not derived from Ca^{2+} influx caused by activation of unblocked Ca^{2+} channels or by the loss of voltage control. In earlier studies, we found that SR Ca^{2+} stores provide the Ca^{2+} for these I_{Na}-induced $[Ca^{2+}]_i$ transients since the transients are completely eliminated in cells pretreated with ryanodine.[21]

Na^+-Ca^{2+} Exchange Mediates Ca^{2+} Influx and SR Ca^{2+} Release

The experiments with ryanodine strongly implicate SR Ca^{2+} release but do not suggest the mechanism by which activation of I_{Na} and SR release are coupled. SR Ca^{2+} release is probably not induced by voltage *per se,*[24] because the same voltage steps in the presence of TTX do not elicit $[Ca^{2+}]_i$ transients. Since Na^+ influx appears necessary for inducing $[Ca^{2+}]_i$ transients, two other mechanisms that could explain the coupling are that (a) Na^+ ions may directly activate SR Ca^{2+} release or (b) release may be triggered by Ca^{2+} ions subsequent to Ca^{2+} influx via reverse-mode Na^+-Ca^{2+} exchange induced by a transient increase in $[Na^+]_i$. To distinguish between these mechanisms, we tested whether I_{Na}-induced transients were dependent on extracellular Ca^{2+} ($[Ca^{2+}]_o$). Nisoldipine-insensitive, I_{Na}-induced transients were elicited by voltage-clamp steps to -40 mV from a holding potential of -80 mV in 2.5 mM extracellular Ca^{2+} (FIG. 3a). Replacement of external Ca^{2+} with equimolar Mg^{2+} had no effect on I_{Na} elicited by the same test pulse but did abolish the $[Ca^{2+}]_i$ transient (FIG. 3b). In the absence of external Ca^{2+}, the SR still contained a releasable Ca^{2+} store because exposure of the cells to caffeine caused an increase in $[Ca^{2+}]_i$ (FIG. 3c). The requirement for $[Ca^{2+}]_o$ rules out the possibility that Na^+ ions directly trigger SR Ca^{2+} release. Also, the possibility that SR release is voltage-dependent or tightly coupled to Na^+ current activation is unlikely. Thus, release of SR Ca^{2+} induced by I_{Na} most likely involves a Ca^{2+}-induced release mechanism. The trigger Ca^{2+} for SR release may enter the cell via reverse-mode Na^+-Ca^{2+} exchange.

We next sought to test more directly whether Na^+-Ca^{2+} exchange mediates the I_{Na}-induced release of Ca^{2+} from SR. Following a conditioning protocol[21] to load SR Ca^{2+}, I_{Na} and associated Ca^{2+} transient were recorded during a test pulse to -50 mV from a holding potential of -80 mV in control external solutions containing Ca^{2+} (2.5 mM) and nisoldipine (5 μM) (FIG. 4A). The conditioning protocol maintains constant availability of SR Ca^{2+} stores.[20] A second conditioning protocol was applied to the same cell and the external solution was changed to one in which Na^+ was replaced by equimolar Li^+. Li^+ carries current through the Na^+ channel but does not readily replace Na^+ for transport on the Na^+-Ca^{2+} exchanger. An identical test pulse activated Li^+ currents but failed to elicit a $[Ca^{2+}]_i$ transient (FIG. 4B). This finding supports the hypothesis that Na^+-Ca^{2+} exchange mediates I_{Na}-induced Ca^{2+} release from cardiac SR. In this cell, basal $[Ca^{2+}]_i$ eventually increased until the cell contracted irreversibly due to inhibition of the exchanger by Li^+.

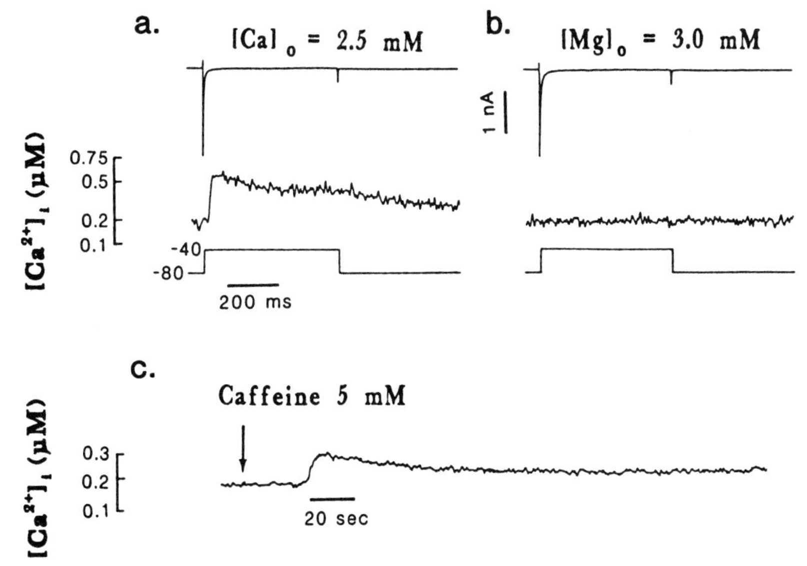

FIGURE 3. Dependence of I_{Na}-induced $[Ca^{2+}]_i$ transients on extracellular Ca^{2+}. Nisoldipine (5 μM) was used to block Ca^{2+} currents. Following a conditioning protocol to load SR Ca^{2+} stores (methods), Na^+ current and associated $[Ca^{2+}]_i$ transient were recorded during a test pulse to −40 mV from a holding level of −80 mV in the presence of normal $[Ca^{2+}]_o$ (**a**). The perfusate was then switched to a medium in which $[Ca^{2+}]_o$ was proportionately replaced by Mg^{2+} to maintain the total divalent cation concentration constant. During perfusion with this solution, no voltage-clamp pulses were applied to the cell. The traces shown in (**b**) were obtained after a 5-min incubation in this nominally Ca^{2+}-free solution. Still in Ca^{2+}-free solution, the cell was subsequently exposed to 5 mM caffeine (**c**). The membrane potential was clamped at −80 mV during application of the compound. Similar results were observed in four other cells. (Taken from Leblanc & Hume[21]; used with permission from *Science*. Copyright 1990 by the AAAS.)

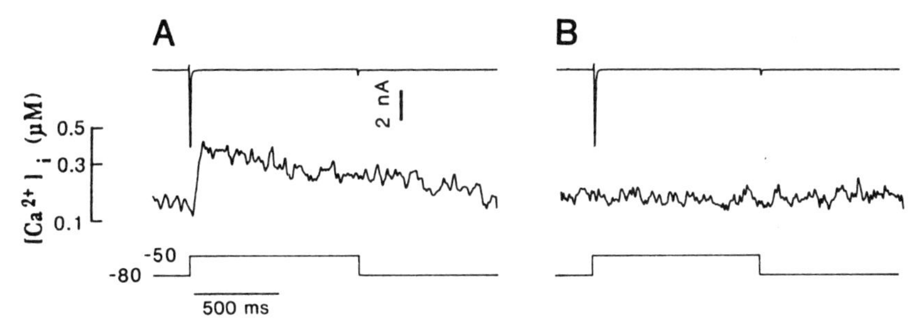

FIGURE 4. Effects of Li^+ substitution on Na^+ current-induced $[Ca^{2+}]_i$ transients. (**A**) After a SR Ca^{2+}-loading protocol, I_{Na} and associated $[Ca^{2+}]_i$ transient were recorded during a test pulse to −50 mV from a holding potential of −80 mV in the presence of 5 μM nisoldipine. (**B**) Another SR-conditioning protocol was applied to the cell and the external solution was changed to one in which Na^+ was replaced by equimolar Li^+. An identical test pulse was applied to the cell 5 min after exposure to Li^+. Similar results were obtained in four other cells. (Taken from Hume *et al.*[26]; used with permission from *Science*. Copyright 1991 by the AAAS.)

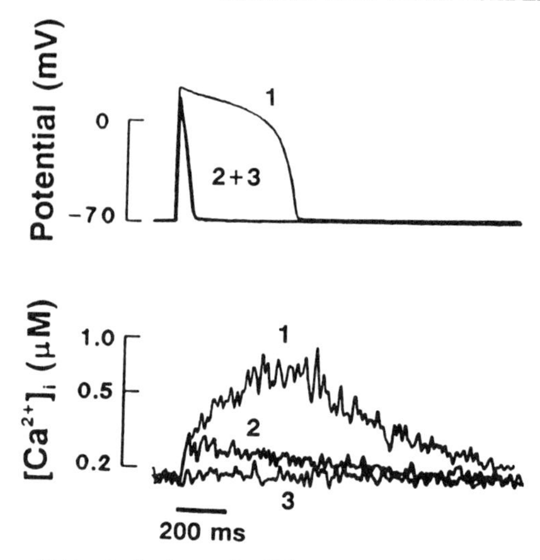

FIGURE 5. Effects of Li^+ substitution on $[Ca^{2+}]_i$ transients elicited by action potentials. Action potentials (*top traces*) and associated $[Ca^{2+}]_i$ transients (*bottom traces*) were recorded simultaneously under current-clamp conditions. After recording control action potentials and transients (**1**), the external solution was switched to one containing 5 μM nisoldipine (**2**) and then eventually to one containing nisoldipine and equimolar Li^+ substituted for Na^+ (**3**). Steady-state action potentials are shown for each condition. Similar results were obtained in three other cells.

Physiological Role of Na^+-Ca^{2+} Exchange in SR Ca^{2+} Release

The ability of Na^+-Ca^{2+} exchange to mediate SR Ca^{2+} release during action potentials was tested to determine whether the exchanger has a role under more physiologic conditions. Action potentials and accompanying $[Ca^{2+}]_i$ transients were recorded simultaneously under current-clamp conditions (FIG. 5,1). After recording a control action potential and transient, external solutions were switched to ones containing 5 μM nisoldipine. As expected, block of voltage-sensitive Ca^{2+} channels by nisoldipine resulted in a markedly shortened, spike-like action potential. Also, the peak amplitude of the accompanying $[Ca^{2+}]_i$ transient was significantly diminished (FIG. 5,2). When the external solution was switched to one containing nisoldipine and equimolar Li^+ substituted for Na^+, the action potential-induced $[Ca^{2+}]_i$ transient was abolished (FIG. 5,3). Thus, under these conditions, Na^+-Ca^{2+} exchange mediated SR Ca^{2+} release during the action potential. This result indicates that Na^+-Ca^{2+} exchange may play a role in EC coupling under physiologic conditions. The demonstration of the I_{Na}-induced SR release mechanism in current-clamp experiments argues against the possibility that I_{Na}-induced $[Ca^{2+}]_i$ transients measured in voltage-clamp experiments may have been artifactual, perhaps due to loss of voltage control during attempts to clamp Na^+ currents.

DISCUSSION

The contribution of I_{Na} to EC coupling was investigated in voltage- and current-clamped, isolated cardiac myocytes dialyzed with Indo-1. In the absence of Ca^{2+} entry through voltage-sensitive Ca^{2+} channels, membrane depolarization elicits release of Ca^{2+} from ryanodine-sensitive internal stores. This process is dependent on Na^+ entry through TTX-sensitive Na^+ channels and on extracellular Ca^{2+}, suggesting that Ca^{2+} entry into the cell via reverse-mode Na^+-Ca^{2+} exchange triggers SR Ca^{2+} release. The hypothesis that Na^+-Ca^{2+} exchange mediates SR Ca^{2+} release is further supported by experiments that show that equimolar replacement of external Na^+ by Li^+ abolishes nisoldipine-insensitive $[Ca^{2+}]_i$ transients induced by I_{Na} and action potentials.

The results provide strong evidence in support of the participation of the Na^+-Ca^{2+} exchanger in the coupling between activation of Na^+ channels and SR Ca^{2+} release. Johnson and Lemieux[25] have suggested that Ca^{2+} influx through voltage-dependent, TTX-sensitive Na^+ channels, rather than Na^+-Ca^{2+} exchange, provides the trigger for SR Ca^{2+} release. We feel that this alternative explanation of our results is rather unlikely.[26] First, Ca^{2+} flux through cardiac Na^+ channels, even under conditions favoring Ca^{2+} entry (20 mM external Ca^{2+}), is extremely small.[27] Second, our results show that no nisoldipine-insensitive $[Ca^{2+}]_i$ transients are elicited by activation of Na^+ channels when equimolar Li^+ replaces Na^+ in the extracellular solution. This finding provides direct evidence for the involvement of Na^+-Ca^{2+} exchange because the I_{Na}-induced SR Ca^{2+} release is abolished when the exchanger is inhibited.

The mechanism proposed for I_{Na}-induced release of SR Ca^{2+} assumes that activation of TTX-sensitive Na^+ channels results in an increase in $[Na^+]_i$ of sufficient magnitude to promote reserve-mode Na^+-Ca^{2+} exchange. The transient increase in $[Na^+]_i$ on the inner surface of the sarcolemma would shift the reversal potential of the exchanger toward more negative potentials. This shift would induce a transient influx of Ca^{2+} through the exchanger that would trigger SR Ca^{2+} release. A significant elevation of $[Na^+]_i$ may be necessary for activation of reverse-mode Na^+-Ca^{2+} exchange.[28] Lederer *et al.*[29] proposed that the ability of I_{Na} to elevate $[Na^+]_i$ to levels capable of activating the exchanger is due to the existence of an incomplete barrier to diffusion located just beneath the sarcolemma. Na^+ entering the cell via Na^+ channels would temporarily increase $[Na^+]_i$ within the restricted space and subsequently activate reverse-mode Na^+-Ca^{2+} exchange. This hypothesis is appealing in that it may explain how I_{Na} elevates $[Na^+]_i$ to levels capable of activating the exchanger, but whether such a barrier to diffusion actually exists is unknown.

Perhaps, even without a diffusion barrier, activation of I_{Na} may initially elevate $[Na^+]_i$ close to the sarcolemmal membrane to a greater extent and much more rapidly than it does deeper within the cell. This short-lived spatial inhomogeneity in $[Na^+]_i$ could be important in activating reverse-mode Na^+-Ca^{2+} exchange. Mazzanti and DeFelice[30] reported that the reversal potential of Na^+ channels is significantly less than the predicted Na^+ equilibrium potential in spontaneously beating chick ventricle cells, suggesting that $[Na^+]_i$ near the sarcolemmal membrane is independent of $[Na^+]_i$ in the remainder of the cell. The calculated $[Na^+]_i$ near the mouths of Na^+ channels was nearly 40 mM in beating cells. This concentration is much higher than cytosolic $[Na^+]_i$ (13 mM) in this type of cell.[31] Whether the difference in $[Na^+]_i$ is simply due to the rapid influx of Na^+ before diffusion or a subsarcolemmal diffusion barrier is not clear.

Two principal sources of Ca^{2+} that may provide the activating Ca^{2+} for Ca^{2+}-induced Ca^{2+} release from SR are the sarcolemmal Ca^{2+} channel and the Na^+-Ca^{2+} exchanger. Although voltage-sensitive Ca^{2+} channels are thought to provide the major source of Ca^{2+} for SR release,[32] we show in this study that the exchanger may contribute to EC coupling in heart under physiological conditions. The results suggest that trigger Ca^{2+} may also enter the cell via reverse-mode Na^+-Ca^{2+} exchange, subsequent to a transient rise in $[Na^+]_i$ caused by I_{Na}. Rate-dependent changes in cardiac inotropy and in the inotropic effects of cardiac glycosides and antiarrhythmic drugs that interact with Na^+ channels may, at least in part, involve the I_{Na}-induced Ca^{2+} release mechanism.

SUMMARY

A mechanism capable of eliciting SR Ca^{2+} release independent of Ca^{2+} entry through voltage-gated Ca^{2+} channels was investigated using whole-cell voltage- and current-clamped guinea pig ventricular myocytes dialyzed with the Ca^{2+} indicator, Indo-1. Depolarization-induced Na^+ influx through TTX-sensitive Na^+ channels caused a rapid, transient increase in intracellular Ca^{2+} concentration ($[Ca^{2+}]_i$). The I_{Na}-induced $[Ca^{2+}]_i$ transients (a) occur after blocking voltage-sensitive sarcolemmal Ca^{2+} channels with nisoldipine or D-600, (b) are inhibited by ryanodine, and (c) are dependent upon extracellular Ca^{2+}. These results indicate that the I_{Na}-induced $[Ca^{2+}]_i$ transients arise from SR Ca^{2+} release triggered by Ca^{2+} entering the myocyte, after a transient rise in $[Na^+]_i$, via a pathway distinct from sarcolemmal Ca^{2+} channels. One such pathway for Ca^{2+} entry into cardiac cells is reverse-mode Na^+-Ca^{2+} exchange. Depolarization-induced Na^+ influx failed to elicit Ca^{2+} transients when extracellular Na^+ was replaced with equimolar lithium, which carries current through Na^+ channels but does not readily substitute for Na^+ on the exchanger. This result provides direct evidence that Ca^{2+} entry via reverse-mode Na^+-Ca^{2+} exchange mediates the I_{Na}-induced SR Ca^{2+} release. Lithium also inhibited nisoldipine-insensitive $[Ca^{2+}]_i$ transients elicited by action potentials indicating that I_{Na} and Na^+-Ca^{2+} exchange may play a role in EC coupling under physiological conditions. Taken together, the results suggest that depolarization-induced Na^+ influx through Na^+ channels can trigger SR Ca^{2+} release in cardiac myocytes by activating Ca^{2+} influx via reverse-mode Na^+-Ca^{2+} exchange. The I_{Na}-induced release of Ca^{2+} from SR may partially account for the positive inotropic effects of cardiac glycosides and the negative inotropic effects of antiarrhythmic drugs that block Na^+ channels.

REFERENCES

1. MULLINS, L. J. 1979. Am. J. Physiol. **236:** C103-C110.
2. FABIATO, A. 1985. J. Gen. Physiol. **85:** 291-320.
3. ALLEN, T. J. A., D. NOBLE & H. REUTER. 1989. Sodium-Calcium Exchange. Oxford University Press. Oxford, England.
4. MULLINS, L. J. 1981. Ion Transport in Heart. Raven Press. New York.
5. HILGEMANN, D. W. 1986. J. Gen. Physiol. **87:** 675-706.
6. BERS, D. M. 1987. J. Gen. Physiol. **90:** 479-504.
7. BERS, D. M., W. J. LEDERER & J. R. BERLIN. 1990. Am. J. Physiol. **258:** C944-C954.
8. CRESPO, L. M., C. J. GRANTHAM & M. B. CANNELL. 1990. Nature **345:** 618-621.
9. HUME, J. R. & A. UEHARA. 1986. J. Gen. Physiol. **87:** 857-884.
10. KIMURA, J., A. NOMA & H. IRISAWA. 1986. Nature **319:** 596-597.
11. BRIDGE, J. H. B., J. R. SMOLLEY & K. W. SPITZER. 1990. Science **248:** 376-378.
12. EARM, Y. E., W. K. HO & I. S. SO. 1990. Proc. R. Soc. London **240**(1297): 61-81.
13. NABAUER, M., G. CALLEWAERT, L. CLEEMAN & M. MORAD. 1989. Science **244:** 800-803.
14. BARCENAS-RUIZ, L., D. J. BEUKELMANN & W. G. WIER. 1987. Science **238:** 1720-1722.
15. EISNER, D. A., W. J. LEDERER & R. D. VAUGHAN-JONES. 1984. J. Physiol. (London) **355:** 251-266.
16. BERS, D. M., D. M. CHRISTENSEN & T. X. NGUYEN. 1988. J. Mol. Cell. Cardiol. **20:** 405-414.
17. WALKER, J. L. 1986. *In* The Heart and Cardiovascular System. H. A. Fozzard, Ed.: 561-572. Raven Press. New York.
18. CANNELL, M. B., J. R. BERLIN & W. J. LEDERER. 1987. Science **238:** 1419-1423.
19. PHILIPSON K. D. & R. WARD. 1986. J. Mol. Cell. Cardiol. **18:** 943-951.
20. BEUCKELMANN, D. J. & W. G. WIER. 1988. J. Physiol. (London) **405:** 233-255.
21. LEBLANC, N. & J. R. HUME. 1990. Science **248:** 372-376.

22. HUME, J. R. & A. UEHARA. 1985. J. Physiol. (London) **368:** 525-544.
23. HAMILL, O. P., A. MARTY, E. NEHER, B. SACKMANN & F. J. SIGWORTH. 1981. Pflügers Arch. **391:** 85-100.
24. NIGGLI, E. & W. J. LEDERER. 1990. Science **250:** 565-568.
25. JOHNSON, E. A. & R. D. LEMIEUX. 1991. Science **251:** 1370.
26. HUME, J. R., P. C. LEVESQUE & N. LEBLANC. 1991. Science **251:** 1370-1371.
27. NILIUS, B. 1988. J. Physiol. (London) **399:** 537-558.
28. MIURA, Y. & J. KIMURA. 1989. J. Gen. Physiol. **93:** 1129-1145.
29. LEDERER, W. J., E. NIGGLI & R. W. HADLEY. 1990. Science **248:** 283.
30. MAZZANTI, M. & L. J. DEFELICE. 1987. Biophys. J. **52:** 95-100.
31. FOZZARD, H. A. & S. S. SHEU. 1980. J. Physiol. (London) **306:** 579-586.
32. CLEEMAN, L. & M. MORAD. 1991. J. Physiol. (London) **432:** 283-312.

Receptor-Mediated Inotropic Effect in Heart

Role of Sodium-Calcium Exchange[a]

SHEY-SHING SHEU,[b] VIRENDRA K. SHARMA,
MICHAEL KORTH,[c] ADRIANA MOSCUCCI,[d]
ROBERT T. DIRKSEN, AND MEI-JIE JOU

Department of Pharmacology
University of Rochester
School of Medicine and Dentistry
Rochester, New York 14642

INTRODUCTION

In mammalian heart, the precise mechanisms responsible for the dynamic changes in cytosolic Ca^{2+} concentration ($[Ca^{2+}]_i$) during the excitation-contraction-relaxation cycle are still not fully understood. The general consensus is that during the early portion of the cardiac action potential, the influx of Ca^{2+} through voltage-gated L-type Ca^{2+} channels triggers the release of Ca^{2+} from the sarcoplasmic reticulum (SR). As a result, $[Ca^{2+}]_i$ rises and the muscle contracts. During the repolarization phase of the action potential, the ATP-driven Ca^{2+} pump in the SR, with its high affinity and fast kinetics, sequesters Ca^{2+} into the SR and, thereby, causes muscle relaxation. Recently, this conventional view of Ca^{2+} mobilization during the cardiac cycle has been modified with the addition of the involvement of an electrogenic Na^+-Ca^{2+} exchange. The Na^+-Ca^{2+} exchange has been suggested to either deliver or remove cytosolic Ca^{2+} during the cardiac cycle depending on the membrane potential and Na^+ and Ca^{2+} chemical gradients. Therefore, Na^+-Ca^{2+} exchange regulates cardiac contraction on a beat-to-beat basis.[1]

The close relationship between $[Ca^{2+}]_i$ level and muscle contraction is subject to neurohormonal modulation. There are three major post-synaptic membrane receptors in heart: β_1-adrenoceptors, α_1-adrenoceptors, and muscarinic receptors. The purpose of this study is to (1) determine the effect of activation of these receptors on cardiac contractility and (2) assess the role of Na^+-Ca^{2+} exchange in the inotropic effect mediated by these receptors. Part of the results in this paper have been published previously.[2,3]

[a]This work was supported by Grant R01 HL-33333 and Deutsche Forschungsgemeinschaft. S-S.S. is an Established Investigator of the American Heart Association. R.T.D. is a recipient of a predoctoral fellowship from NIDA Training Grant DA-07232.

[b]Address for correspondence: Department of Pharmacology, University of Rochester, School of Medicine and Dentistry, 601 Elmwood Ave., Rochester, NY 14642.

[c]Present address: Institüt für Pharmakologie und Toxikologie, der Technischen Universität München, München, Germany.

[d]Present address: Department of Medicine, University of Chicago, Chicago, IL 60637.

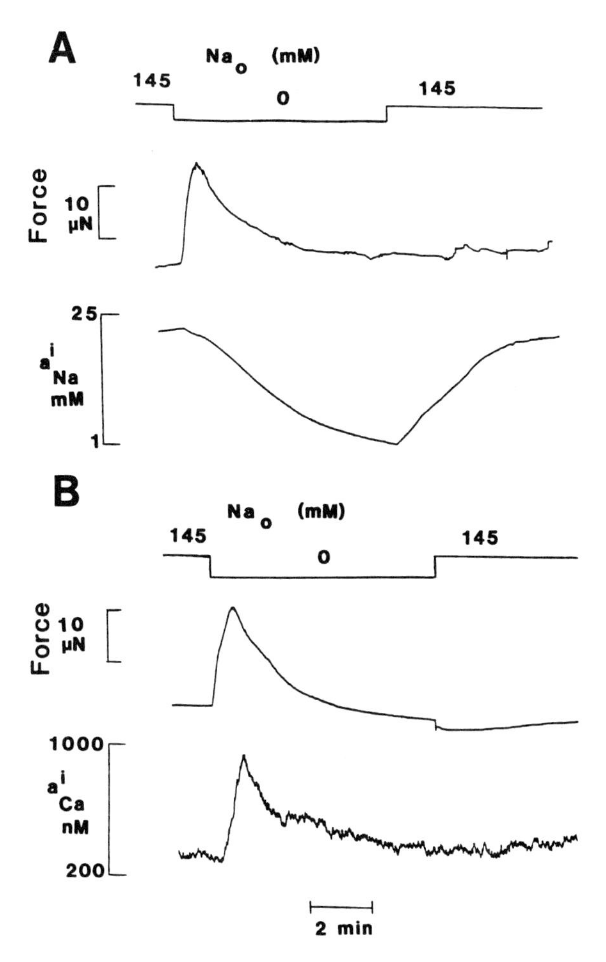

FIGURE 1. (A) The effects of removal of external Na^+ on contractile force and a^i_{Na}. (**B**) The effect of removal of external Na^+ on contractile force and a^i_{Ca}. The preparations (sheep cardiac Purkinje fibers) were exposed to 10 μM strophanthidin to inhibit Na^+-K^+ pump before the removal of external Na^+.

EXPERIMENTAL EVIDENCE FOR Na^+-Ca^{2+} EXCHANGE

The antagonistic interaction between external Na^+ and Ca^{2+} on cardiac contraction were first reported by Ringer more than a century ago.[4] His findings were confirmed by Daly and Clark[5] who concluded (a) that the lack of external K^+ and Na^+, or an excess of Ca^{2+}, increase muscle contraction and (b) that the effects of low external Na^+ showed a striking resemblance to the effects of strophanthidin on the heart.

FIGURE 1 is one of the representative experimental results that support the existence of Na^+-Ca^{2+} exchange in heart. FIGURE 1A shows a simultaneous recording of contractile force and intracellular Na^+ activity (a^i_{Na}) in a sheep cardiac Purkinje fiber. The method of using ion-sensitive microelectrodes to measure a^i_{Na} has been described previously.[6] The fiber was first exposed to 10 μM strophanthidin to raise a^i_{Na} to a new steady-state level of approximately 23 mM. As shown in the figure, when the external Na^+ was removed and replaced by Li^+ at the indicated time, contractile force increased rapidly followed by a gradual decline to the original level. This transient increase in muscle contraction was accompanied by a gradual fall of a^i_{Na} to nearly 0 mM. FIGURE 1B shows that the transient increase in contractile force was due to a transient increase in intracellular Ca^{2+} activity (a^i_{Ca}) as a result of the removal of extracellular Na^+. As recorded by a Ca^{2+}-sensitive microelectrode, removal of extracellular Na^+ caused a^i_{Ca} to increase from around 300 nM up to 1 μM. These data can be interpreted to be due to a Na^+-Ca^{2+} exchange mechanism. When extracellular Na^+ was removed, the electrochemical gradients for Na^+-Ca^{2+} exchange favored the mode of bringing Ca^{2+} into, and Na^+ out of, the cell. This caused a^i_{Na} to fall and a^i_{Ca} to rise. The rise in a^i_{Ca} was then responsible for the increase in muscle contraction. The subsequent decay in a^i_{Ca} (which resulted in the decline of contractile force) could be due to several possible mechanisms such as the decrease in a^i_{Na} (which would reduce activity of Na^+-Ca^{2+} exchange), Ca^{2+} extrusion by other mechanisms (such as plasma membrane Ca^{2+}-ATPase), Ca^{2+} uptake by intracellular buffering systems (such as SR or mitochondria), and binding of Ca^{2+} to various intracellular Ca^{2+}-binding proteins.

FIGURE 2 shows the ability of mitochondria to buffer Ca^{2+} during extracellular Na^+ deprivation. Neonatal rat ventricular myocytes after two weeks of culture were loaded with fura-2 AM. Images of the distribution of intracellular Ca^{2+} concentration were obtained by a fluorescence digital imaging microscope as described previously.[7] FIGURE 2A shows a fluorescent image of two cells excited with light at 360-nm wavelength and emitted at 510 nm. Because 360 nm is the isosbestic point of the excitation spectrum for fura-2, this image reflects the intracellular distribution of fura-2 free acid. As can be seen, fura-2 was highly compartmentalized. Further characterization indicated that the fura-2 was localized significantly in both nuclear and mitochondrial regions. The compartmentation of fura-2 provides an opportunity to measure nuclear and intramitochondrial Ca^{2+} concentrations in living cells. FIGURE 2B shows the spatial distribution of $[Ca^{2+}]_i$ in control solution containing 40 nM ryanodine (to deplete Ca^{2+} stored in the sarcoplasmic reticulum). The rainbow color scale on the right corresponds to various values of $[Ca^{2+}]_i$ that were obtained from an *in vitro* calibration. The image shows that the distribution of $[Ca^{2+}]_i$ was quite heterogeneous. Areas of high fluorescent intensity (green and yellow color) correspond to mitochondrial-dense regions.[8] FIGURE 2C shows the effect of removal of extracellular Na^+ on the spatial distribution of $[Ca^{2+}]_i$. This image was collected 2 min after the removal of extracellular Na^+. Similar to the observation in FIGURE 1B, $[Ca^{2+}]_i$ increased after Na^+ removal. Interestingly, not only the cytosolic Ca^{2+} concentration was increased during Na^+-free exposure, but the mitochondrial Ca^{2+} concentration was also increased as indicated by a higher fluorescent intensity in these regions. This result suggested that mitochondria were able to take up Ca^{2+} from the cytosol during extracellular Na^+ deprivation. FIGURE 2D shows that 16 min after the exposure to Na^+-free solution, cytosolic Ca^{2+} concentration declined to approximately the control value. The mitochondrial area maintained a higher Ca^{2+} concentration than the control level. These results suggest that the mitochondrial Ca^{2+} uptake may contribute to the decline of $[Ca^{2+}]_i$ during Na^+-free exposure.

RECEPTOR-MEDIATED INOTROPIC EFFECTS

The effects of the stimulation of β-adrenoceptors, α_1-adrenoceptors, and low-affinity state muscarinic receptors on contractile force were studied. As shown in FIGURE 3A,

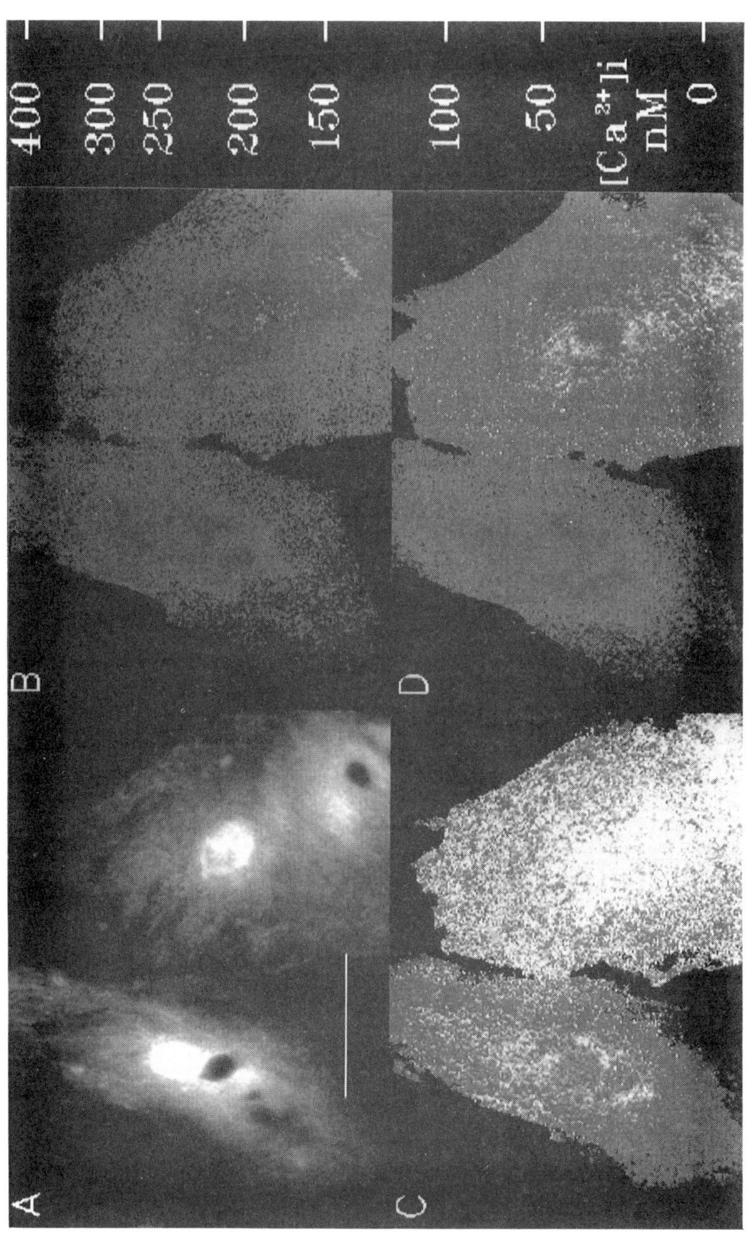

FIGURE 2. The effect of removal of external Na^+ on the spatial distribution of $[Ca^{2+}]_i$ in two-week-old cultured neonatal rat ventricular myocytes. **(A)** Fluorescent image excited at 360 nm. **(B)** Pseudocolor fluorescent ratio image (340/380 nm) of the cells loaded with fura-2AM in normal HEPES-buffered solution. The rainbow color spectrum correlates with the concentration of free Ca^{2+} as obtained from an *in vitro* calibration. **(C)** Pseudocolor fluorescent ratio image of the cells exposed to Na^+-free solution for 2 min, and **(D)** for 16 min.

stimulation of β-adrenoceptors by 2 μM isoproterenol produced a positive inotropic effect in a rat papillary muscle contracting isometrically at a frequency of 0.5 Hz. In addition to the positive inotropic effect, isoproterenol also enhanced the rate of relaxation ("relaxant effect"). FIGURE 3B shows that stimulation of α_1-adrenoceptors by 50 μM phenylephrine in the presence of propranolol (1 μM) produced a biphasic response on the muscle contraction. An initial transient decrease in muscle contraction was followed by a sustained increase in muscle contraction. FIGURE 3C shows that stimulation of low-affinity muscarinic receptors by 300 μM carbachol also produced a positive inotropic effect. The positive inotropic effects observed with α_1-adrenoceptors and

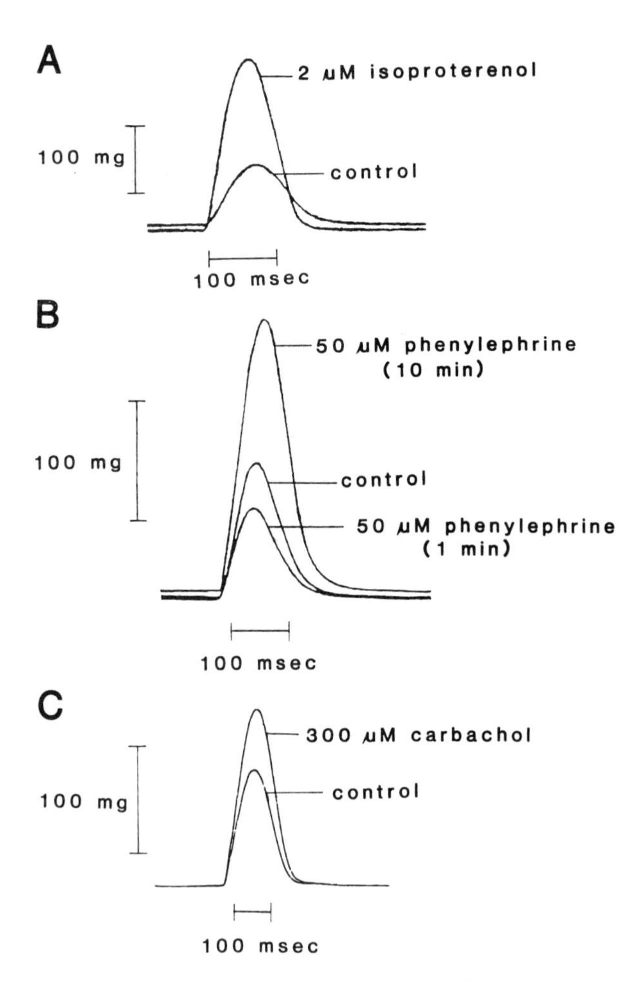

FIGURE 3. Inotropic effects induced by (**A**) isoproterenol (2 μM), (**B**) phenylephrine (50 μM), and (**C**) carbachol (300 μM) in rat papillary muscles obtained from right ventricle.

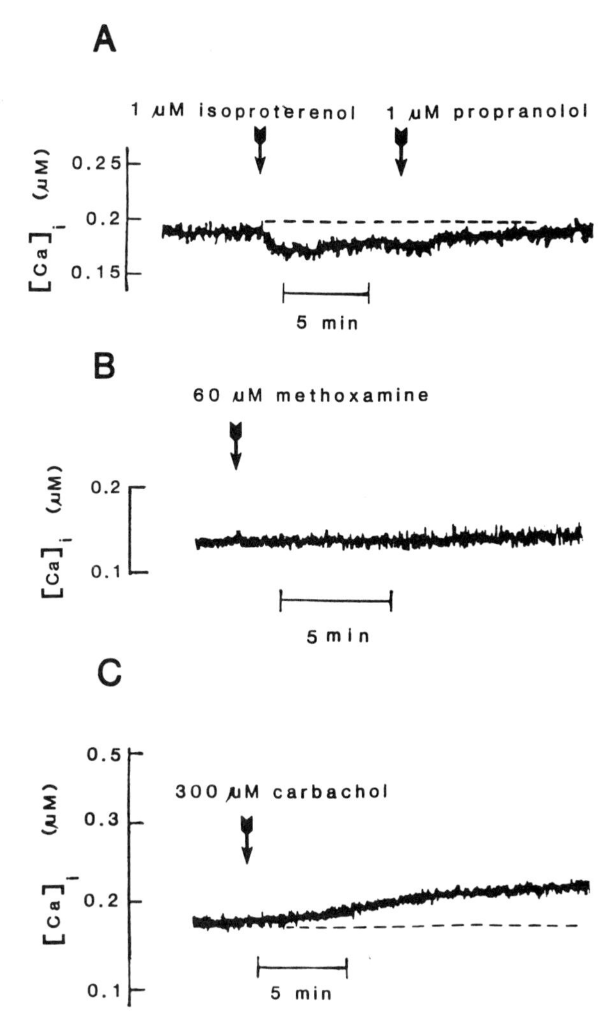

FIGURE 4. (**A**) Effect of isoproterenol (1 μM) on $[Ca^{2+}]_i$ in isolated rat ventricular myocytes (reproduced from Sheu *et al.*[2] with permission). (**B**) Effect of methoxamine (60 μM) on $[Ca^{2+}]_i$ in isolated rat ventricular myocytes. (**C**) Effect of carbachol (300 μM) on $[Ca^{2+}]_i$ in isolated rat ventricular myocytes (reproduced from Korth *et al.*[3] with permission).

low-affinity muscarinic receptors were not accompanied by a relaxant effect similar to the one observed with β-adrenoceptor stimulation.

RECEPTOR-MEDIATED CHANGES IN $[Ca^{2+}]_i$ AND $[Na^{+}]_i$

FIGURE 4A shows the effect of isoproterenol on $[Ca^{2+}]_i$ in quin-2-loaded cells suspended in a modified Krebs-Henseleit solution. $[Ca^{2+}]_i$ decreased from 190 to 169

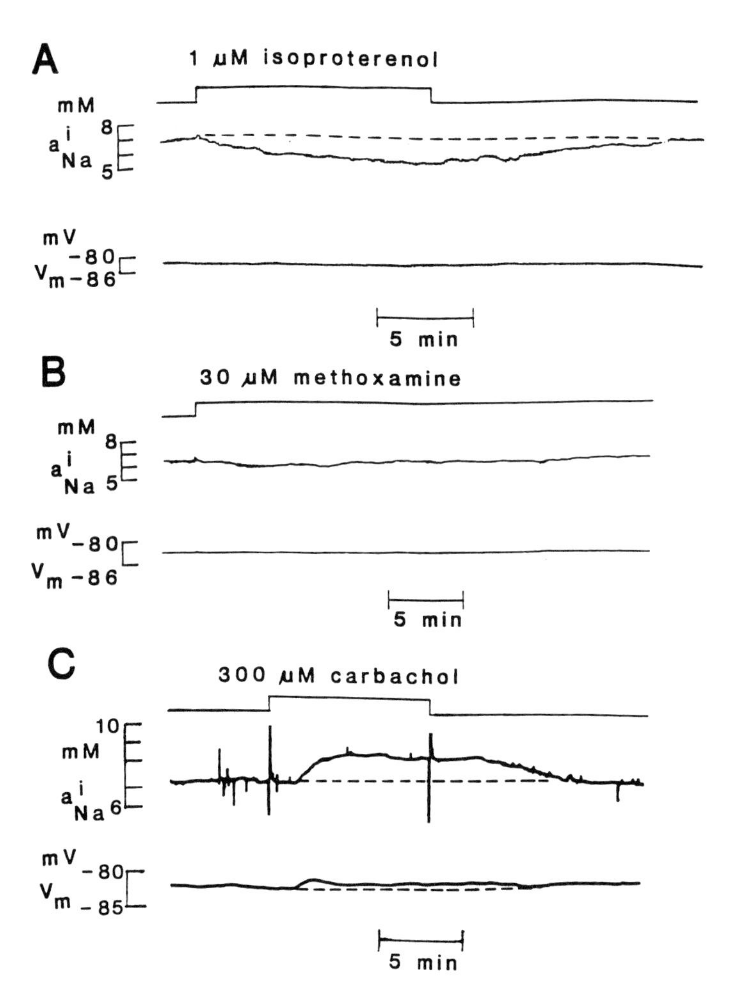

FIGURE 5. **(A)** Effect of isoproterenol (1 μM) on a^i_{Na} in rat papillary muscle from right ventricle. **(B)** Effect of methoxamine (30 μM) on a^i_{Na} in rat papillary muscle from right ventricle. **(C)** Effect of carbachol (300μM) on a^i_{Na} in rat papillary muscle from right ventricle.

nM when 1 μM isoproterenol was added. During continuous exposure of the cells to isoproterenol, there was a slight return of $[Ca^{2+}]_i$ toward the control level, which reached a steady-state value of 178 nM. This decrease in $[Ca^{2+}]_i$ was reversed by 1 μM propranolol, a β-adrenoceptor antagonist. FIGURE 4B shows that 60 μM methoxamine, an α_1-adrenoceptor agonist, failed to produce an effect on $[Ca^{2+}]_i$. FIGURE 4C shows the effect of carbachol (300 μM) on $[Ca^{2+}]_i$. Carbachol caused $[Ca^{2+}]_i$ to increase from a control level of 171 nM to a value of 211 nM. This increase was reversed by 1 μM atropine, a muscarinic receptor antagonist (record not shown).

FIGURE 5A shows the effect of isoproterenol on a^i_{Na} in a rat papillary muscle perfused in a modified Krebs-Henseleit solution. Application of 1 μM isoproterenol caused a^i_{Na} to decrease from 7 mM to 5.5 mM. This decrease in a^i_{Na} was reversible upon washing out of isoproterenol. In nine experiments 1 μM isoproterenol caused an average decrease in a^i_{Na} of 2.4 ± 0.7 mM. FIGURE 5B shows that 30 μM methoxamine failed to produce any effect on a^i_{Na}. FIGURE 5C shows that 300 μM carbachol caused a gradual increase of a^i_{Na} from 7.2 mM to 8.4 mM. This increase in a^i_{Na} was reversible upon washing out of carbachol.

These results show that stimulation of β-adrenoceptors caused a concurrent decrease in a^i_{Na} and $[Ca^{2+}]_i$, while stimulation of α_1-adrenoceptors caused no change in either a^i_{Na} or $[Ca^{2+}]_i$, and stimulation of low-affinity state muscarinic receptor caused a concurrent increase in both a^i_{Na} and $[Ca^{2+}]_i$.

It has been shown that β-adrenoceptor agonists stimulate the Na^+-K^+ pump and thus lead to a decrease in a^i_{Na}.[9] Therefore, it is possible that the reduction of a^i_{Na} by isoproterenol favors the efflux of Ca^{2+} mediated by Na^+-Ca^{2+} exchange resulting in the decrease in $[Ca^{2+}]_i$ and muscle contraction.

The mechanism involved in mediating the positive inotropic effect induced by α_1-adrenoceptor stimulation has yet to be fully elucidated. It apparently does not involve alterations either in resting a^i_{Na} or $[Ca^{2+}]_i$. Endoh and Blinks[12] have suggested that α_1-adrenoceptor stimulation may involve an increase in the sensitivity of the myofilaments to Ca^{2+} as the mechanism for this positive inotropic effect.

Finally, Korth and Kühlkamp[10] were the first to show that choline esters increased a^i_{Na} in guinea pig ventricular muscle. The time course of this increase in a^i_{Na} matches with the time course of the positive inotropic effect. Therefore, they suggested that stimulation of low-affinity state muscarinic receptors first leads to an increase in a^i_{Na} and then to an increase in $[Ca^{2+}]_i$ via Na^+/Ca^{2+} exchange. Consistent with this finding, it has been found that carbachol induces a Na^+-dependent membrane conductance in single guinea pig ventricular myocytes.[11]

RECEPTOR-MEDIATED EFFECT ON RESTING FORCE

The alterations in resting $[Ca^{2+}]_i$ via receptor activation should match with the alterations in resting force. Because under physiological conditions $[Ca^{2+}]_i$ in the quiescent myocardium is below the threshold required for generating resting force, therefore, 22 mM extracellular Ca^{2+} was used to raise resting force. FIGURE 6A shows that resting force increased after raising $[Ca^{2+}]_o$ from 2 to 22 mM and clearly declined upon addition of 1 μM isoproterenol. This finding strongly suggests that the relaxant effect of isoproterenol is due, at least in part, to the decrease in $[Ca^{2+}]_i$. To demonstrate that the carbachol-induced increase in $[Ca^{2+}]_i$ correlates with an increase in the resting force, extracellular Ca^{2+} concentration was raised to 20 mM. Subsequent addition of 300 μM carbachol significantly augmented the resting force as shown in FIGURE 6B.

CONCLUSIONS

The present results demonstrate that Na^+-Ca^{2+} exchange plays an important role in various receptor-mediated inotropic effects. Stimulation of β-adrenoceptors leads to a reduction in a^i_{Na}, possibly by activating the Na^+-K^+ pump. This decrease in a^i_{Na} enhances the Ca^{2+} extrusion, via Na^+-Ca^{2+} exchange, to be at least partially responsible for the relaxant effect.[13] Stimulation of α_1-adrenoceptors produces no change in either a^i_{Na} or $[Ca^{2+}]_i$. It is possible, as suggested by Endoh and Blink,[12] that the positive inotropic effect induced by α_1-adrenoceptors activation is linked to an increased

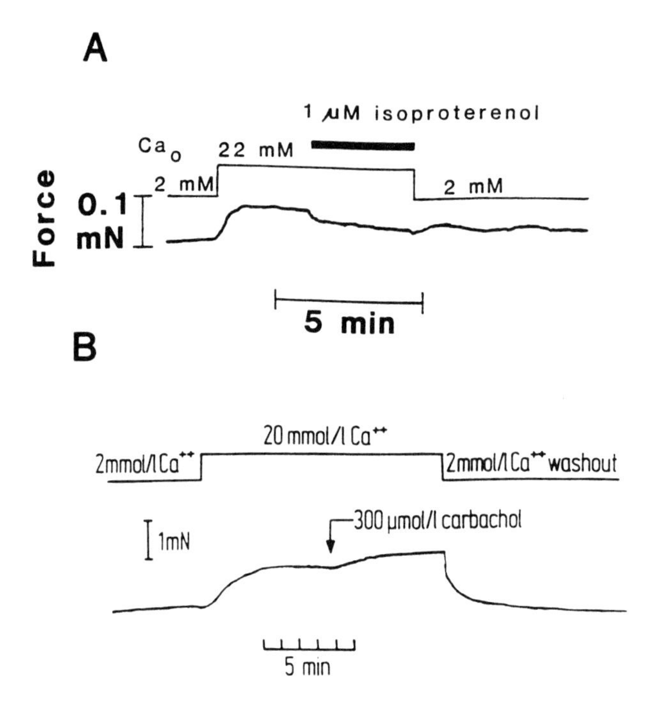

FIGURE 6. (**A**) Isoproterenol (1 μM)-induced relaxation of contracture elicited in a right ventricular rat papillary muscle by increasing extracellular Ca^{2+} from 2 to 22 mM (reproduced from Sheu *et al.*[2] with permission). (**B**) Carbachol (300 μM)-induced increase in resting force in a right ventricular rat papillary muscle (reproduced from Korth *et al.*[3] with permission).

myofilament sensitivity to Ca^{2+}. Finally, stimulation of low-affinity state muscarinic receptors leads to the opening of Na^+-permeable channels that causes an increase in a^i_{Na}. A rise in a^i_{Na} results in an increase in $[Ca^{2+}]_i$ via Na^+-Ca^{2+} exchange and, thereby, augments muscle contraction.

REFERENCES

1. SHEU, S-S. & M. P. BLAUSTEIN. 1991. Sodium/calcium exchange and control of cell calcium and contractility in cardiac vascular smooth muscles. *In* The Heart and Cardiovascular System. H. A. Fozzard, E. Haber, R. B. Jennings, A. M. Katz & H. E. Morgan, Eds. Raven Press. New York. In press.
2. SHEU, S-S., V. K. SHARMA & M. KORTH. 1987. Voltage-dependent effects of isoproterenol on cytosolic Ca concentration in rat heart. Am. J. Physiol. **252:** H697-H703.
3. KORTH, M., V. K. SHARMA & S-S. SHEU. 1988. Stimulation of muscarinic receptors raises free intracellular Ca^{2+} concentration in rat ventricular myocytes. Circ. Res. **62:** 1080-1087.
4. RINGER, S. 1885. Further observations regarding the antagonism between calcium salts and sodium, potassium and ammonium salts. J. Physiol. (*London*) **18:** 425-429.
5. DALY, I. & A. J. CLARK. 1921. The action of ions upon the frog's heart. J. Physiol. (*London*) **54:** 367-383.

6. SHEU, S-S. & H. A. FOZZARD. 1982. Transmembrane Na^+ and Ca^{2+} electrochemical gradients in cardiac muscle and their relationship to force development. J. Gen. Physiol. **80:** 325-351.
7. WILLIFORD, D. J., V. K. SHARMA, M. KORTH & S-S. SHEU. 1990. Spatial heterogeneity of intracellular Ca^{2+} concentration in nonbeating guinea pig ventricular myocytes. Circ. Res. **66:** 234-241.
8. JOU, M-J., L. B. CHEN & S-S. SHEU. 1991. Heterogeneous distribution of the intramitochondrial Ca^{2+} concentration and membrane potential in cultured neonatal rat ventricular myocytes. Biophys. J. **59:** 243a.
9. LEE, C. O. & M. VASSALLE. 1983. Modulation of intracellular Na^+ activity and cardiac force by norepinephrine and Ca^{2+}. Am. J. Physiol. **244:** C110-C114.
10. KORTH, M. & V. KÜHLKAMP. 1985. Muscarinic receptor-mediated increase of intracellular Na^+-ion activity and force contraction. Pflügers Arch. **403:** 266-272.
11. MATSUMOTO, K. & A. J. PAPPANO. 1989. Sodium-dependent membrane current induced by carbachol in single guinea-pig ventricular myocytes. J. Physiol. (*London*) **415:** 487-502.
12. ENDOH, M. & J. R. BLINKS. 1988. Actions of sympathomimetic amines on the Ca^{2+} transients and contractions of rabbit myocardium: Reciprocal changes in myofibrillar responsiveness to Ca^{2+} mediated through α- and β-adrenoceptors. Circ. Res. **62:** 247-265.
13. MORAD, M. 1982. Ionic mechanisms mediating the inotropic and relaxant effects of adrenaline on the heart muscle. *In* Catecholamines in the Non-ischemic and Ischemic Myocardium. R. A. Riemersma & M. A. Oliver, Eds.: 113-135. Elsevier/North-Holland. New York.

The Role of Intracellular Sodium in the Control of Cardiac Contraction[a]

CHIN O. LEE [b]

Department of Physiology
Cornell University Medical College
New York, New York 10021

ALLAN J. LEVI

Department of Physiology
School of Medical Sciences
University of Bristol
Bristol, BS8 1TD, United Kingdom

INTRODUCTION

It is widely recognized that the level of intracellular sodium has a profound influence on the contraction of cardiac muscle.[1] A change in the level of intracellular sodium alters the electrochemical gradient for sodium, and, via sarcolemmal sodium-calcium exchange, this can then lead to a change of intracellular calcium and contraction.

The effect of intracellular sodium on contraction has been examined previously in multicellular cardiac preparations, using ion-selective electrodes to measure sodium. Purkinje fibers were used for virtually all these studies,[2–4] and only a few groups used working ventricular muscle.[5] Experiments using multicellular preparations suffer from a number of limitations. First, the Na-selective electrode measures intracellular sodium in one cell, whereas contraction is measured from the whole preparation. An accurate correlation between sodium and contraction therefore depends on sodium being homogeneous between all cells of the preparation. Second, because Purkinje fibers are specialized for electrical conduction and contract only weakly, their behavior may not necessarily reflect that of ventricular muscle. Third, it is difficult to voltage-clamp multicellular ventricular muscle adequately in order to control membrane potential.

In this study, we have taken a new and timely approach to the investigation of intracellular sodium and contraction in cardiac muscle. We have used the recently developed fluorescent sodium indicator, SBFI,[6] to measure intracellular sodium, and we have performed experiments on single cells isolated from ventricular muscle. Individual myocytes offer significant advantages over multicellular preparations. Intracellular

[a] The Wellcome Trust and the British Heart Foundation are gratefully acknowledged for their financial support. C.O.L. was an Overseas Visiting Fellow of the BHF while this work was performed.

[b] Address for correspondence: Dr. Chin O. Lee, Department of Physiology, Cornell University Medical College, 1300 York Ave., New York, NY 10021.

sodium and contraction can be measured simultaneously in the same cell, and mycoytes can be voltage-clamped with a single electrode.

The first goal of our study was to examine the suitability of the sodium dye for measuring intracellular sodium in cardiac myocytes and to devise a technique for dye calibration. The second goal was to characterize the effect of intracellular sodium on contraction, in cells with action potentials and in voltage-clamped cells. From a technical viewpoint, this is the first study in which intracellular sodium, membrane potential, and contraction have been measured simultaneously in single cardiac cells.

METHODS

Myocyte Isolation

Ventricular myocytes were isolated from rabbit hearts using a combination of enzymic and mechanical dispersion. As the method proved successful and has not been reported before, it is described in the following section. The basic solution (Solution A) contained (in mM): 130 NaCl, 4.5 KCl, 3.5 $MgCl_2$, 0.4 NaH_2PO_4, 5 HEPES, 10 glucose (pH 7.25). Chemicals were either Analar or Aristar grade (BDH) and all solutions were made with deionized water (Mill-Q system, Millipore) and gassed with 100% O_2.

New Zealand white male rabbits (weight 2 to 3.5 kg) were injected intravenously with 5,000 units of heparin. They were killed by cervical dislocation, and their hearts removed rapidly, taking care to get a long length of aorta. The heart was rinsed in Solution A + 750 μM Ca with heparin (10 units ml^{-1}) and the aorta tied onto the glass cannula of the perfusion apparatus. The heart was retrogradely perfused at 37°C with Solution A + 750 μM Ca and surrounded with a warmed water jacket; it soon began to beat regularly and forcefully after perfusion began.

Flow rate was kept constant at 6 ml min^{-1} per gram of heart tissue using a pump (model 502S, Watson-Marlow). Twenty serial measurements showed that heart weight was approximately 0.3% of animal weight, and this allowed the calculation of heart weight for each animal. Aortic pressure was monitored with a pressure transducer to ascertain whether the cannula was in the aorta or ventricle. In 750 μM Ca, perfusion pressure was between 30 and 40 mmHg, and changes of aortic pressure reflected changes of coronary vascular resistance.

After perfusing Solution A + 750 μM Ca for 2 minutes, perfusion was switched to Solution A + 100 μM Na-EGTA for 4 minutes. The heart stopped beating after 10 to 20 seconds, aortic pressure fell transiently and subsequently returned to the initial level. Perfusion was then switched to Solution A + 175 μM Ca, 1 mg ml^{-1} collagenase (Type 1, Worthington) and 0.1 mg ml^{-1} protease (Type XIV, Sigma)—the "enzyme solution." Solution dripping from the heart became viscous after 30 to 40 seconds and perfusion pressure rapidly rose to 60-80 mmHg. After 80 seconds, perfusion was switched to "recirculation" so that effluent draining from the heart was pumped back into the aorta; enzyme solution was then recirculated for a further 10 minutes. Within a few seconds of recirculated effluent reaching the heart, there was a sudden and dramatic fall of aortic pressure to near zero, indicating massive vasodilation of coronary vessels.[7] Aortic pressure remained near zero for the remainder of perfusion.

The heart was removed from the cannula, the atria discarded and the ventricles chopped coarsely by cutting from base to apex 20 times. The ventricles were put into a plastic beaker with 3 ml of recirculated enzyme and 1% bovine serum albumin was added (BSA; Fraction V, Sigma). The tissue was shaken for four periods of 5 minutes at 30 strokes min^{-1} at 37°C while being bubbled with 100% O_2. After each shaking period, the cell suspension was filtered through nylon gauze (200-μm diameter mesh). The cell-containing filtrate was then poured onto 6 ml of Solution A + 750 μM Ca with added 1.5% BSA and centrifuged at 40 *g* for 45 seconds. The supernatant was then replaced with either Solution A + 750 μM Ca or with "KB" solution.[8]

This procedure produced between 40 and 90% Ca-tolerant and rod-shaped cells and the highest percentage of rods usually came from the first two shaking periods. The cells were kept at room temperature until needed and could be used for up to 12 hours after isolation.

Experimental Solutions and Perfusion

Myocytes in the experimental chamber were continuously superfused with Tyrode solution containing (in mM): 130 NaCl, 4.5 KCl, 2 $CaCl_2$, 1 $MgCl_2$, 5 HEPES, 10 glucose, titrated to a pH of 7.4 with 4 moles of NaOH. Strophanthidin (Sigma) was added from a 20 mM stock solution in ethanol; the ethanol concentration in Tyrode never exceeded 0.1%.

The experimental chamber had a volume of 75 μl and a flow rate of 2 to 4 ml min^{-1}. Miniature solenoid valves (LFAA1201618H, Lee Products Ltd) selected the solution entering the chamber, and perfusate within the chamber could be changed within 3 to 5 seconds.[9] Chamber temperature was monitored with a miniature thermistor and maintained constant at 30°C by warming inflowing solution with heating coils. The solution level of the chamber was controlled with a feedback system.[10] The chamber and solenoid valves were mounted on the sliding stage of a Nikon Diaphot microscope that sat on an antivibration table (BT series, Photon Control).

Electrical Recording

Membrane potential was measured with conventional microelectrodes pulled from 1.5-mm o.d. filamented glass (Clark Electromedical Instruments). They were filled with 290 mM KCl and had a resistance between 25 and 40 Mohms. Microelectrode potential was measured with an Axoclamp 2A amplifier (0.1 gain headstage, Axon Instruments Inc.). After compensating for electrode resistance and capacitance, the microelectrode was advanced toward a rod-shaped cell that contracted regularly with external stimulation. Twenty-millisecond, 1-nA hyperpolarizing pulses were passed down the microelectrode, and, as its tip touched the cell, increasing negative voltage deflections were observed. When they reached 30 to 50 mV, the electrode was electrically oscillated for 1 msec (using the "buzz" facility), and this usually impaled it successfully. Cell input resistance and capacitance was measured by applying 10-msec 200-pA hyperpolarizing current pulses. The steady-state potential deflection gave the cell input resistance (R_{CI}), and the time course of potential change was a single exponential with time constant T. Cell capacitance (C) was then calculated from $C = T/R_{CI}$.

Action potentials were stimulated at 1 Hz by 2-msec-duration current pulses passed down the microelectrode, and cells were voltage-clamped using the discontinuous "switch" clamp technique with a sampling rate of 4 kHz. In some experiments, action potential duration (APD) was measured using an electronic meter.[11]

Measurement of Contraction

Contraction (cell shortening) was measured optically with two different methods. The first system used a linear photodiode array as described by Boyett *et al.*[12] The only difference was that we used a 2.5-mm-wide photodiode array (model RL 1024 SAQ, EG&G Reticon). The scanning rate of the array, and therefore the time resolution of the system, was 4 msec. Because the image contrast from a fluorescence objective was often not high enough to get good cell images, we changed to a video edge detector system that used the image from a TV camera to measure cell length.[13] The frame rate of the video system was 50 Hz (UK line frequency) so that it had a time resolution of 25 msec, considerably slower than the array system. This did not cause a problem because rabbit myocyte contractions are relatively slow (time to peak is 200 msec at 30°C), and the edge detector was fast enough to follow them. Results with the video system were identical to those obtained with the array, but the edge detector was much easier to use, its function could be followed on the TV monitor, and cell images could be stored on videotape for later analysis.

Fluorescence Measurement

Myocytes in the experimental chamber were illuminated with ultraviolet light applied via the epifluorescence port of the microscope (FIG. 1). The light source was a 75W zenon lamp (Photon Technology Inc.), and neutral density filters cut down intensity to about 40%. A filter wheel in front of the light was rotated continuously at 20 Hz and alternately selected filters of 340 and 380 nm (Cairn Spectrophotometer System, Cairn Research Ltd., Newnham, Kent). Excitation light was transmitted to the microscope via a flexible fiber optic tube and was then directed onto a cell using a 400-nm dichroic mirror (Nikon) placed beneath the objective (40× oil-immersion phase-contrast fluor objective, NA 1.3; Nikon). Emitted light was collected by the objective and passed to a photomultiplier tube (PMT) mounted on the diascopic port of a beam splitter ("dual CCTV adapter"; Nikon). Light detected by the PMT was restricted to the emission from a single cell by altering a rectangular diaphragm in the emission light path. The interference filter in front of the PMT was centered on 505 nm with a half-bandwidth of 37 nm.

The signal from the PMT was processed with analogue input amplifiers and a "ratio amplifier" that was synchronized to the rotation of the filter wheel (Cairn Research Ltd). The ratio of the light emitted with 340-nm excitation to that emitted with 380-nm excitation (the 340/380 ratio) is a direct indication of the level of intracellular sodium.[6] The ratio amplifier provided a continuous analogue output of the 340/380 ratio, which was filtered with a 4-sec time constant and recorded on a strip chart recorder and magnetic tape for later analysis.

For measuring contraction, cells were illuminated with a second light source (100W halogen) placed above the microscope stage. A red filter with a cut-off below 600 nm was placed in front so as not to interfere with fluorescence measurement. The cell image formed by this light was diverted to the TV camera or array using a 580-nm dichroic mirror mounted in the beam splitter on the side port. The TV camera and linear array were mounted on the episcopic port of the beam splitter.

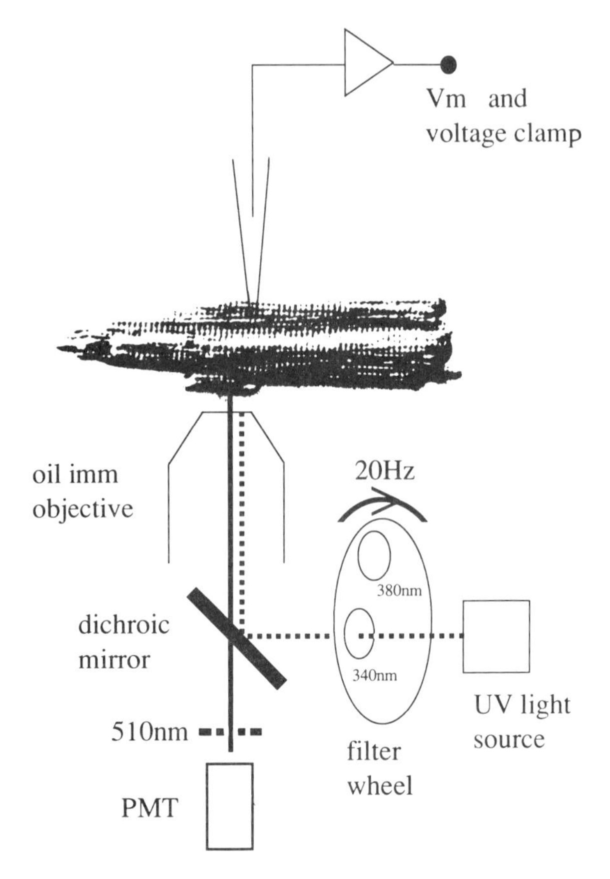

FIGURE 1. Schematic diagram of the experimental apparatus. A myocyte on the microscope stage was illuminated with ultraviolet (UV) excitation light via an oil-immersion (oil imm) objective. A filter wheel in front of the UV light rotated at 20 Hz and alternately placed 340- and 380-nm filters in the light path. Excitation light (indicated by the dotted line) was transmitted to the epifluorescence port of the microscope via a fiber-optic tube. It was then reflected upwards to the cell by a 400-nm dichroic mirror. Emitted light (indicated by the solid line) passed through the 400-nm dichroic mirror to be detected by the photomultiplier (PMT) tube. For simplicity, the fluorescence system is shown without the cell contraction system on this diagram.

Intracellular Loading of SBFI

Freshly isolated cells were suspended in 1 ml of Solution A + 750 μM Ca with added 10 μM SBFI-AM and 4 μl of 25% pluronic acid (both from Molecular Probes). Incubation was continued for 2 hr at room temperature in a dark box to prevent photobleaching. Cells were centrifuged three times after incubation, and each time the supernatant was replaced with fresh solution to remove any residual dye. There was no detectable change in the proportion of rod-shaped cells after dye loading, suggesting that SBFI and pluronic acid are not toxic. After loading, cells were kept in Tyrode for a further 30 min to allow completion of hydrolysis.[14]

Statistics

Numerical data are quoted as mean ± standard error. Unless otherwise stated, all results reported were found in at least four different cells.

RESULTS

Experiments were carried out on a total of 42 cells isolated from 25 hearts. Myocytes had a resting potential of −88 ± 1.09 mV and an action potential duration (measured at −60mV) of 235 ± 20 msec (n = 12). Cell input resistance was 16.5 ± 0.94 Mohms, and average membrane capacitance was 116.3 ± 9.3 pF (n = 15). Any cell that showed signs of Ca_i overload (transient inward currents or aftercontractions) was rejected, and any cell that developed toxicity with strophanthidin was also excluded.

Calibration of SBFI for Intracellular Sodium

Cells were calibrated for intracellular sodium (Na_i) using an *in situ* method. SBFI exhibits a spectral shift when it enters cells,[14,15] and therefore it is not possible to use an *in vitro* calibration technique. We found that a "maximal/minimal ratio" type of calibration was also inappropriate (see Beukelmann & Weir[16] for fura-2) because the 340/380 ratio declined over time with high Na_i. We therefore used an empirical calibration for Na_i based on equilibrating internal and external Na. The calibration solution contained (mM): (NaCl + KCl) 130, EGTA 2, HEPES 10, (pH 7.2), and Na and K were altered so that their total remained 150 mM. Na ionophores were added to increase membrane Na permeability and the Na-K pump was inhibited by strophanthidin to block Na extrusion. The conditions required for optimum equilibration and calibration were critical, and we tested the effect of adding monensin, strophanthidin, and magnesium (Mg) to gramicidin. FIGURE 2A shows an experiment carried out to test the effect of monensin. The cell was initially in Tyrode, and then a calibration solution with 2 mM Na, gramicidin, and strophanthidin was applied. The ratio declined slowly and subsequently increased as external Na was raised. When monensin was added in the presence of 20 mM Na, there was an abrupt increase of ratio and after this the ratio changed more rapidly with external Na. At each level of external Na, the ratio was higher with monensin than in its absence.

FIGURE 2B shows a plot of the 340/380 ratio against external Na for the absence and presence of monensin. Monensin shifted the whole curve upwards. Using the calibration *without* monensin gave an Na_i for this cell of 3.3 mM, whereas the calibration curve obtained *with* monensin gave an Na_i of 1.6 mM. The calibration curve obtained with monensin is likely to give the more correct estimate. The fact that the 340/380 ratio changed more rapidly with monensin suggests that it increased membrane Na permeability, and this should improve equilibration of internal and external Na. The higher ratio observed with monensin at each level of external Na may also reflect better equilibration. Similar experiments were performed to test the effect of strophanthidin and Mg. Strophanthidin was found to be necessary for optimal equilibration. In its absence, the active Na-K pump maintained internal Na lower than external and thus prevented equilibration. An improved equilibration was also obtained by omitting external Mg, as expected from the effect of Mg-free on the Na permeability of Ca channels.[17]

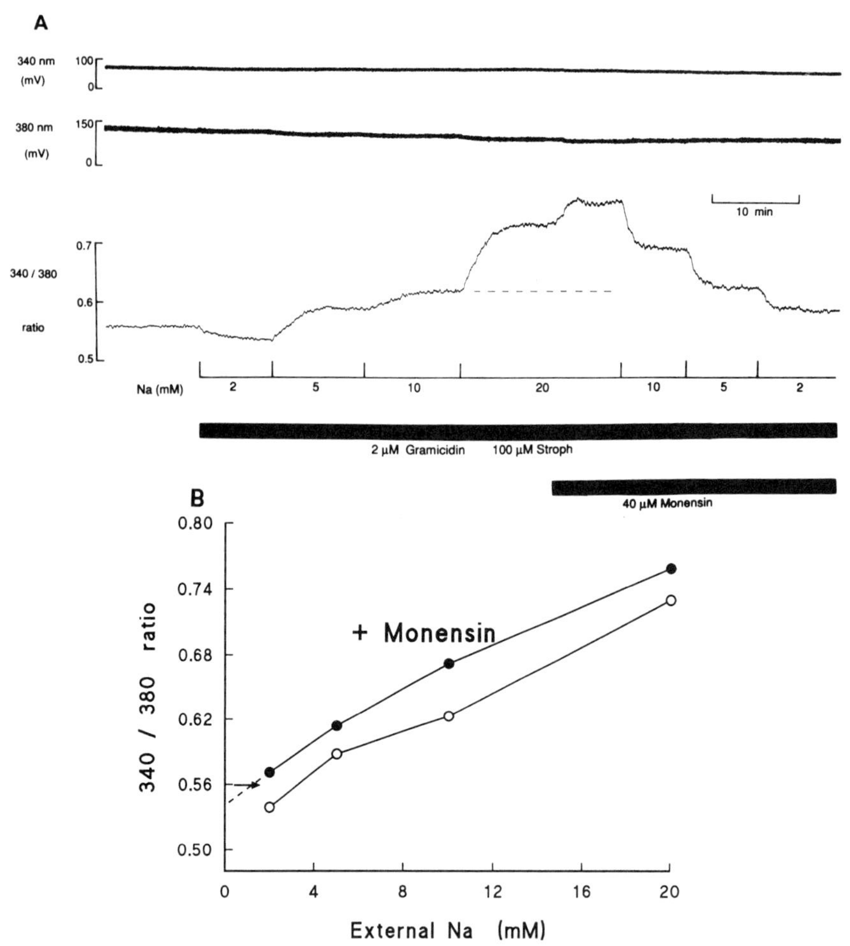

FIGURE 2. The effect of monensin on the calibration of SBFI. **A.** The myocyte was initially in Tyrode solution. A calibration solution containing 2 mM external Na (Na_o), gramicidin and strophanthidin (stroph) was applied, and Na_o was subsequently raised in steps to 20 mM. Monensin was added while Na_o was at 20 mM. **B.** The change of the 340/380 ratio with Na_o, in the presence and absence of monensin. The arrow on the y axis indicates the 340/380 ratio recorded in Tyrode solution at the start of the experiment.

A typical calibration using these conditions is shown in FIGURE 3. Each change of external Na produced a change of ratio, and the calibration curve obtained gave an Na_i in this myocyte of 3.7 mM (FIG. 3B). It was generally found that the 340/380 ratio increased linearly for external Na between 5 and 20 mM, but there was usually an increased slope between 2 and 5 mM. Occasionally, cells went into irreversible contracture during calibration, and, in these cases, dye rapidly leaked out of the cell and calibration was not possible.

Na_i in Quiescent and Stimulated Cells

One of the aims of this study was to determine the level of Na_i in rabbit myocytes. Calibrations such as the one shown in FIGURE 3 were carried out in 24 quiescent cells, and the average level of Na_i was 3.8 ± 0.23 mM (range 1.6 to 6.15 mM). In the experiments described below, we have expressed Na_i as activity (a^i_{Na}) rather than concentration, in order to make the results comparable with Na-selective electrode measurements. An activity coefficient of 0.75 was used to convert Na concentration to activity.

FIGURE 4 shows the typical effect of a change in stimulation rate on a rabbit myocyte. At the start of the record, the cell was stimulated at 1 Hz with 2-msec current pulses down the microelectrode. When stimulation rate was slowed to 0.1 Hz, there

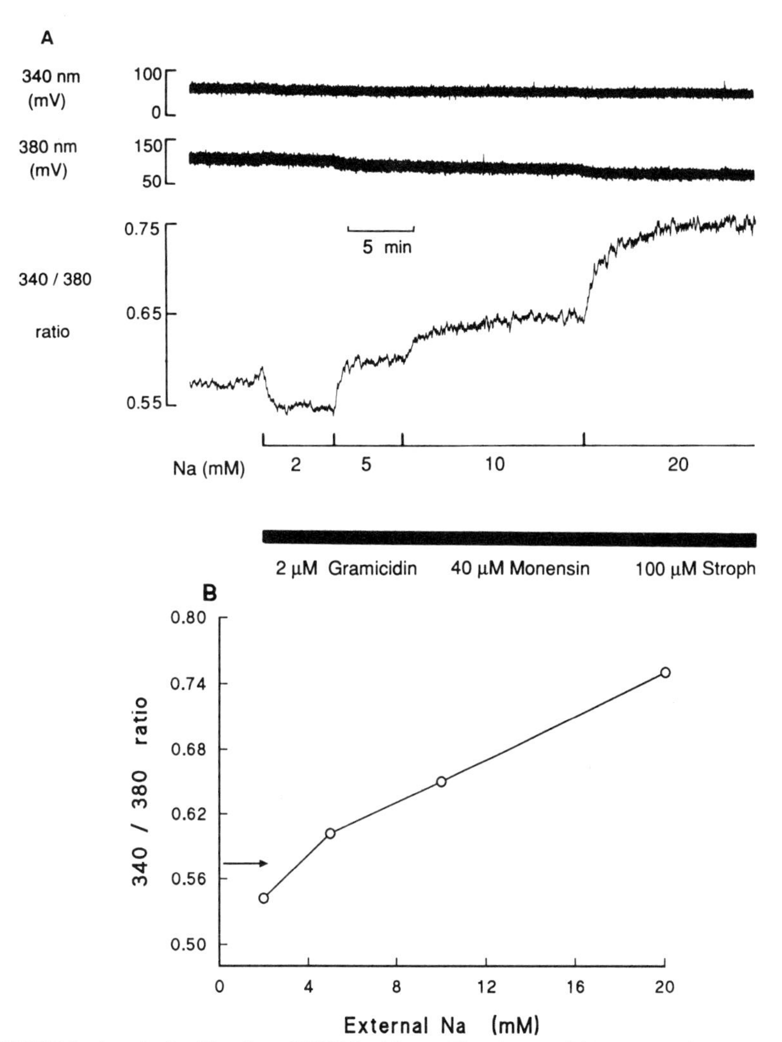

FIGURE 3. A typical calibration of SBFI for Na_i. A. The change of 340-nm and 380-nm signals and the 340/380 ratio when a cell is calibrated using gramicidin, monensin, strophanthidin and Mg-free. The myocyte was in Tyrode at the start of the record. **B.** Plot of 340/380 ratio against Na_o. The arrow on the y-axis indicates the 340/380 ratio in Tyrode at the start of the experiment.

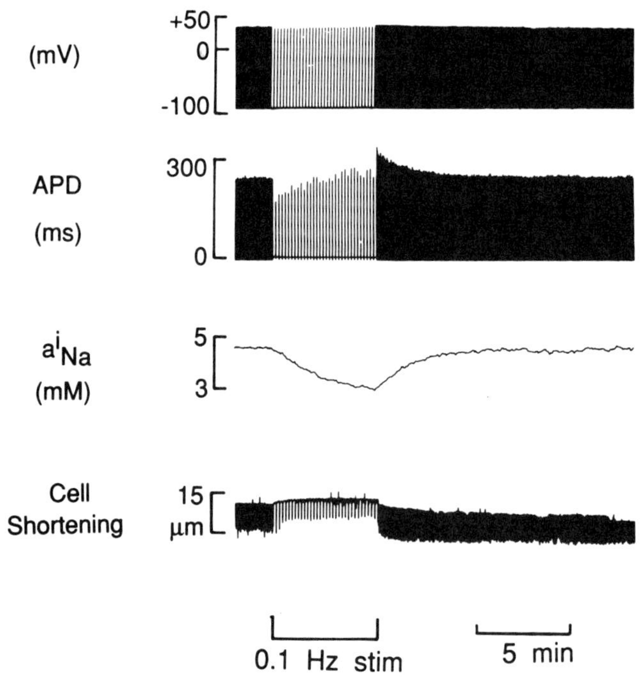

FIGURE 4. The effect of a change in stimulation rate on membrane potential, action potential duration, a^i_{Na}, and contraction. Stimulation rate was slowed from 1 Hz to 0.1 Hz for the period shown. Note that cell contraction (shortening) is shown as a downward deflection in this record.

was a fall of a^i_{Na}, a slow decline of twitch, and a sudden shortening and then slow lengthening of the action potential (AP). All parameters returned to control when 1-Hz stimulation was resumed. There was an average increase of a^i_{Na} by 1.54 ± 0.21 mM when stimulation rate was increased from resting or 0.1 Hz to a level of 1 Hz ($n = 9$ cells). We used the change of a^i_{Na} with stimulation as a routine test to check if the SBFI in a myocyte responded to a^i_{Na}. If there were clear changes of 340/380 ratio with an alteration of stimulation rate, we then continued with the experiment.

The Effect of Strophanthidin on Myocytes with Action Potentials

The first method that we used to modulate a^i_{Na} was to inhibit the Na-K pump with the digitalis analogue strophanthidin. We used strophanthidin because it has a relatively rapid onset and wash-off. FIGURE 5 shows the typical effect of this compound on a myocyte with action potentials. The period of drug exposure and recovery has been divided into four time windows as indicated by the dotted lines. During the first phase just after strophanthidin was applied, there was a depolarization of diastolic potential, a lengthening of the action potential, and an increase of contraction *without* a corresponding rise of a^i_{Na}. The AP then started to shorten during the second phase as a^i_{Na} began to rise and contraction continued to increase. During the third phase, immediately after strophanthidin was washed off, diastolic potential repolarized, the AP shortened markedly, and twitch began to decline at the same time as AP shortening.

However, a^i_{Na} *continued to increase* as the twitch began to fall. At the start of the fourth phase, a^i_{Na} reached a peak value and began to decline along with contraction. Diastolic length also shortened with strophanthidin, and neither it nor a^i_{Na} fully recovered to the initial level. Incomplete recoveries were observed in approximately half the cells tested, but apart from this other features were identical between cells.

This experiment showed clearly that, under these conditions, contraction and a^i_{Na} were dissociated in time. Similar results were found in 12 cells, using a range of strophanthidin doses between 3×10^{-6}M and 2×10^{-5}M. Contraction always reached a peak value before a^i_{Na}, and the mean delay of peak a^i_{Na} behind contraction was 49.8 $\pm$ 3.67 sec.

FIGURE 6A shows that, not only was there a change of AP duration with strophanthidin, there was also a change of AP shape. When the AP initially lengthened, the second half of the AP plateau became more positive, and when the AP shortened on wash-off, its plateau became more negative. FIGURE 6B shows peak contraction plotted against a^i_{Na} for the experiment illustrated in FIGURE 5. As might be expected from the temporal dissociation between contraction and a^i_{Na}, there was also hysteresis between these two variables. As strophanthidin was applied and washed off, the relation moved

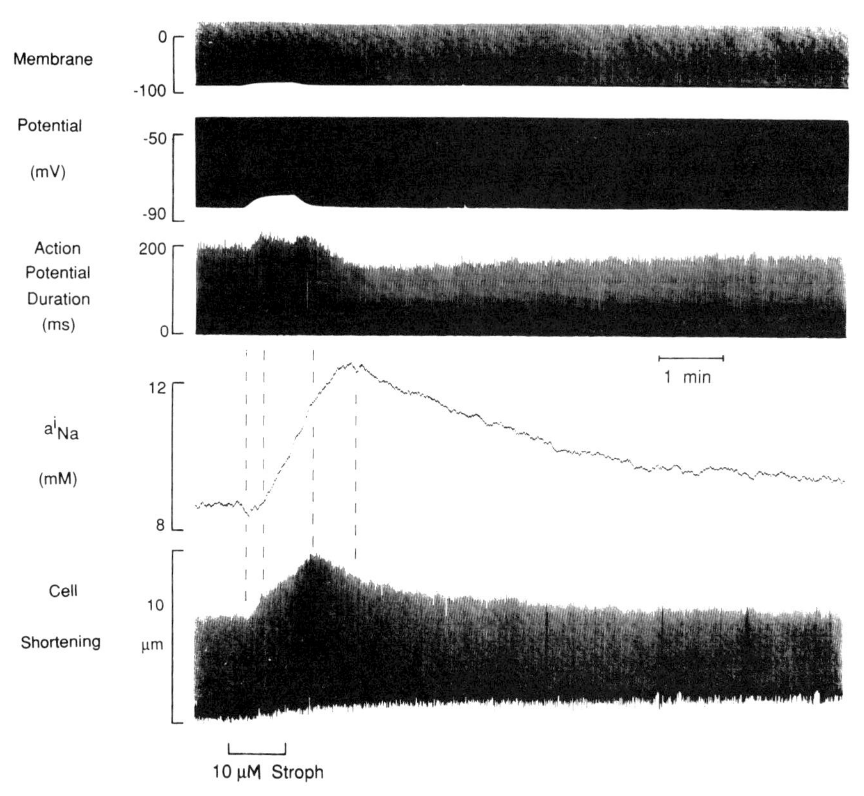

FIGURE 5. The effect of strophanthidin (stroph) on membrane potential, action potential duration, a^i_{Na} and contraction in a myocyte stimulated at 1 Hz with action potentials. There is an obvious lag of the changes in a^i_{Na} behind the changes in contraction.

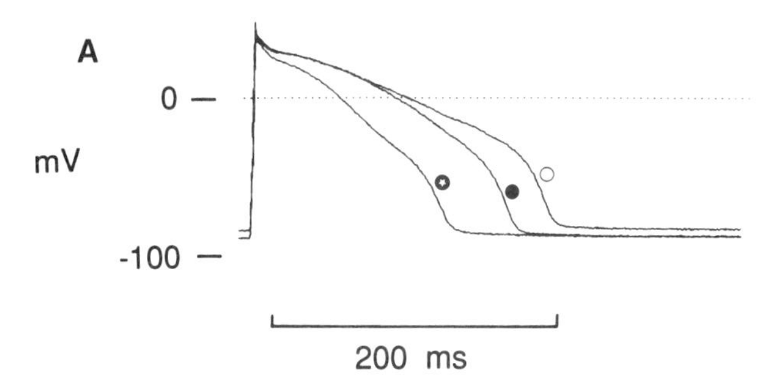

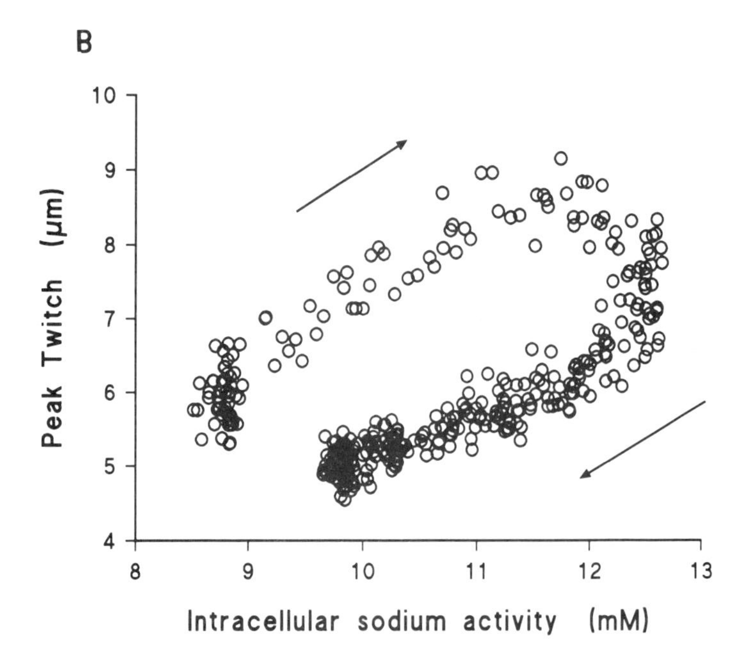

FIGURE 6. A: The changes in action potential shape occurring with strophanthidin. The filled circle indicates the control action potential (AP). The AP denoted by the open circle was recorded just after strophanthidin had been applied and the AP had lengthened. The AP indicated by the star was recorded just after wash-off, when the AP had shortened. **B:** The relation between contraction amplitude and a^i_{Na} for the experiment shown in FIGURE 5.

around the hysteresis loop in a clockwise direction. There was a different contraction-a^i_{Na} relation during the period of increasing contraction compared to when contraction was decreasing. When contraction was increasing, a given level of contraction was related to a lower level of a^i_{Na} than when contraction was declining.

The Effect of Strophanthidin in Voltage-Clamped Myocytes

It is well known that changes of the resting potential and AP can affect contraction.[18] That the alteration in these properties with strophanthidin underlies the observed dissociation between contraction and a^i_{Na} is a strong possibility . Thus, the rapid decline in contraction after strophanthidin removal might be directly caused by the marked AP shortening occurring at the same time. If contraction is affected by changes in the AP as well as by a^i_{Na}, this causes severe problems for evaluating the contraction-a^i_{Na} relation in myocytes with APs.

In order to eliminate the effect of a change in resting and AP with strophanthidin, cells were voltage-clamped. FIGURE 7A shows another exposure to strophanthidin in the *same cell* as FIGURE 5, except that now 200-msec voltage-clamp pulses from -80 to 0 mV were applied at 1 Hz. Pulse duration was altered to give the twitch a similar size as with APs, and strophanthidin was applied for an identical time. Under voltage-clamp conditions, there was no apparent dissociation and no obvious delay of a^i_{Na} behind contraction, in contrast to the situation with APs.

In FIGURE 7B we plotted peak contraction against a^i_{Na} for this experiment. We were rather surprised to find that there was *still* some hysteresis between contraction and a^i_{Na}. Although there was considerably less hysteresis with voltage-clamp than in the same cell with APs, the relation still moved around the hysteresis loop in a clockwise direction as drug exposure and wash-off proceeded. While contraction was increasing, a given level of contraction was associated with a lower level of a^i_{Na} than when contraction was declining.

Removing External K to Inhibit the Na-K Pump

Removal of external K (K_o) is an alternative way to inhibit the Na-K pump and raise a^i_{Na}.[19] Exposures to K-free solution were carried out to assess whether hysteresis was exclusive to strophanthidin, or if it might be a general feature of pump inhibition. In FIGURE 8A, exposure to K-free solution resulted in an increase of contraction and a^i_{Na} with similar time courses. When the pump was reactivated by adding back K_o, contraction and a^i_{Na} reached peak values at the same time and then declined together. In FIGURE 8B, peak contraction is plotted against a^i_{Na} for this experiment. Despite the absence of temporal dissociation, there was *still* hysteresis between contraction and a^i_{Na} and, once again, the relation moved round the loop in a clockwise direction as the experiment proceeded. Similar results were found in six myocytes.

Changes of Ca Current with K-Free Solution

L-type Ca current (I_{Ca}) is thought to be the trigger for intracellular Ca release and contraction in cardiac muscle.[20] If I_{Ca} declines when the pump is reactivated after K-free conditions, this might produce a more rapid decline of contraction than a^i_{Na} and result in hysteresis. The experiment in FIGURE 9 was carried out to determine the effect of K-free solution on I_{Ca}. Holding potential was set at -40 mV to inactivate fast

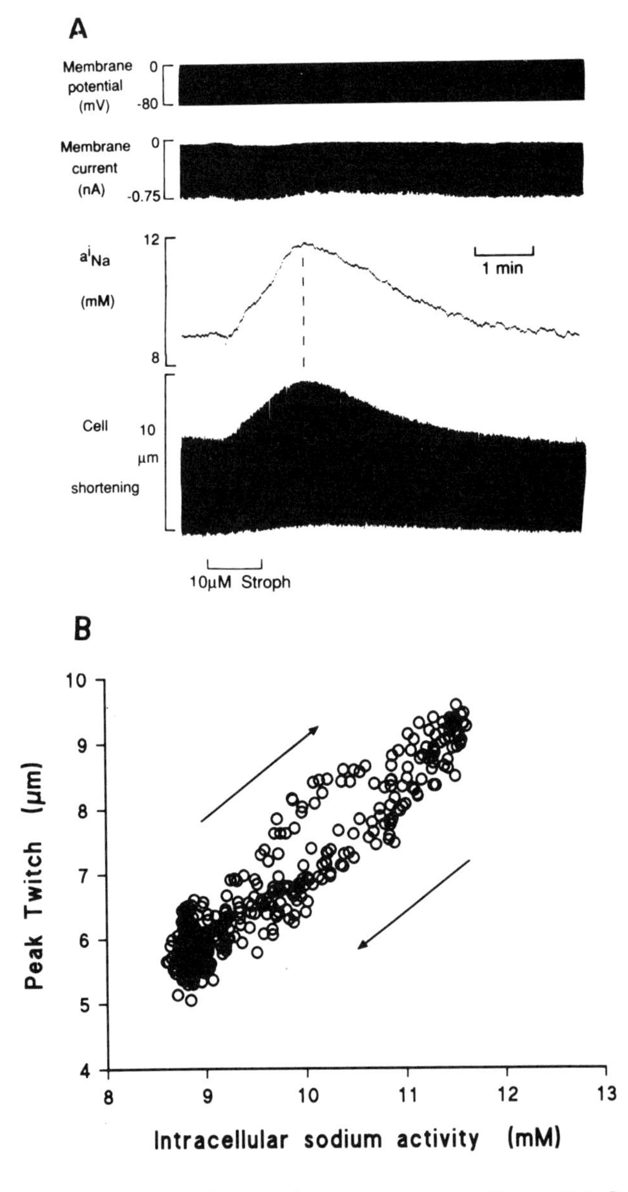

FIGURE 7. A voltage-clamp experiment on the same myocyte as in FIGURE 5. **A.** Strophanthidin was applied while the cell was stimulated with 200-msec voltage-clamp pulses applied at 1 Hz. **B.** The relation between contraction amplitude and a^i_{Na} for this experiment.

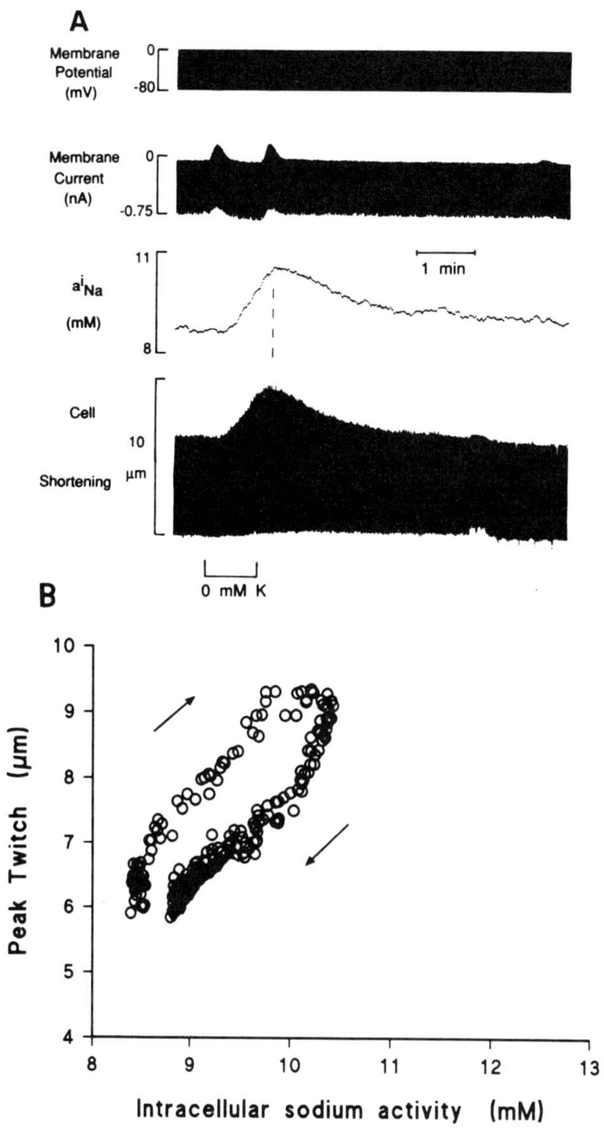

FIGURE 8. The effect of K-free solution on a^i_{Na} and contraction in a voltage-clamped myocyte. **B:** The relation between contraction amplitude and a^i_{Na}.

Na current, and 100-msec pulses to 0 mV were applied at 1 Hz to elicit I_{Ca}. FIGURE 9A shows that contraction and a^i_{Na} increased during K-free exposure, and when the pump was reactivated they declined back to control. A plot of contraction against a^i_{Na} (not shown) revealed the usual type of hysteresis between them. FIGURE 9B shows I_{Ca} elicited at different times. The left panel shows control I_{Ca}, and the right panel shows I_{Ca} at the end of K-free exposure; it is clear that I_{Ca} declined with K-free solution. The

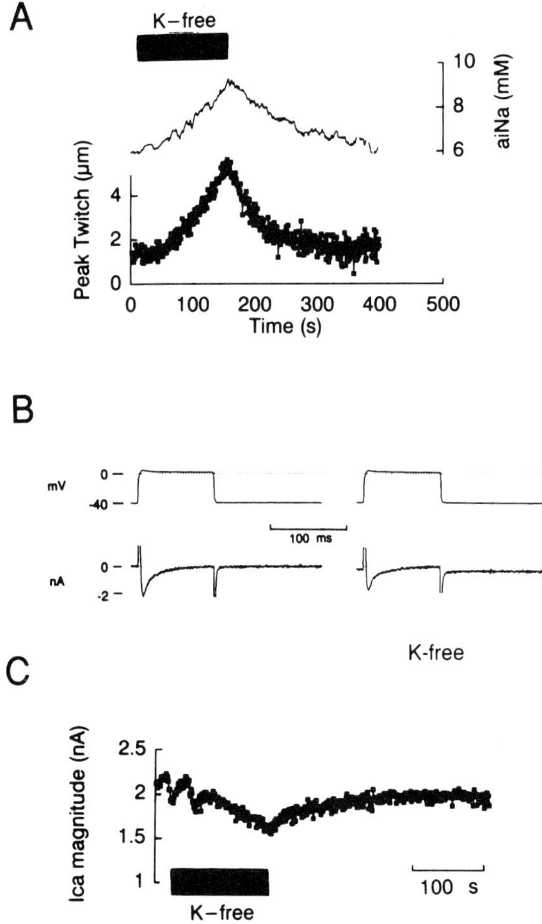

FIGURE 9. The effect of K-free solution on Ca current (I_{Ca}). **A.** Changes of contraction and a^i_{Na} during and exposure to, and recovery from, K-free solution. **B.** Voltage-clamp pulses from −40 to 0 mV to show the effect of K-free on I_{Ca}. The left panel shows a control pulse, and the right panel a pulse recorded at the peak inotropic effect. **C.** The time course of the change in I_{Ca} during and after K-free conditions.

magnitude of I_{Ca} was measured as the difference between peak inward current at the start and steady-state current at the end of the pulse, and FIGURE 9C shows the time course of the change in I_{Ca} amplitude with K-free conditions. There was a monotonic decline of I_{Ca} with exposure to K-free solution, and when K_o was restored it recovered nearly back to the initial level.

DISCUSSION

Suitability of SBFI for Indicating the Na_i of Cardiac Myocytes

Our results show clearly that the fluorescent Na indicator SBFI seems suitable for measuring the Na_i of cardiac myocytes. It responds appropriately to interventions such as a change of stimulation rate and blocking the Na-K pump and gives reasonable values for Na_i in quiescent and stimulated cells. We found an average Na_i in quiescent rabbit cells of 3.8 mM, and this corresponds to an a^i_{Na} of 2.9 mM. This level is broadly in agreement with other values reported for ventricular muscle. Using Na-selective electrodes, Wang *et al.*[5] found an a^i_{Na} of 4 mM in guinea-pig muscle and Desilets and Baumgarten[21] found 8.4 mM in rabbit ventricular myocytes. In view of the fact that electrodes themselves might cause a leak of Na into the cell to raise Na_i, agreement between these and our results is reasonable.

We found that it was necessary to take appropriate precautions in order to calibrate optimally for Na_i. The *in situ* calibration method we used was dependent on a sufficient increase of membrane Na permeability to equilibrate internal and external Na. At the same time, it was important to prevent any dye leaking out of the cell. We have no direct proof that equilibration was accomplished with our procedure, but our results certainly suggest that gramicidin alone was probably insufficient to give complete equilibration. At the very least, equilibration seemed to be much improved by adding monensin and strophanthidin and by removing external Mg. We presume that if Na is not completely equilibrated by these measures, then it must very nearly be so. If monensin and strophanthidin are not added to gramicidin, the Na_i level calculated from the calibration seems to be higher than the real cytoplasmic level. As far as preventing dye leakage from the cell during calibration, as there was no change in the individual 340 and 380 signals when monensin was added, this suggests that it did *not* cause a release of dye from organelles or a loss of dye from the cell.

Changes of Na_i with Stimulation

In agreement with previous studies on multicellular preparations,[3,5] a^i_{Na} rose with stimulation and declined with rest. The changes in a^i_{Na} were monotonic and were of a similar magnitude to that found in previous studies, and this further convinced us that the dye indicated cytoplasmic Na in a realistic manner. When stimulation rate was reduced, twitch declined with a time course similar to a^i_{Na}. This might be due to the fall of a^i_{Na}, which allows an increased Ca extrusion on Na-Ca exchange, resulting in a fall of Ca_i and contraction.

The changes of the AP with an alteration of stimulation rate were interesting. The immediate shortening on reducing stimulation rate to 0.1 Hz might be due to a recovery of transient outward current from inactivation with long rest intervals.[22] The subsequent slow AP lengthening occurred with a time course similar to the decline of Na_i. A reduction of Na_i will reduce outward Na-K pump current and will also reduce any outward current generated by Na-Ca exchange at plateau potentials.[23] A fall of a^i_{Na} might also lead to a reduction of outward Na-activated K current,[24] although there is some doubt as to whether this is significant at low a^i_{Na}. All of these mechanisms will tend to cause a lengthening of the AP as Na_i falls.

The Relation between Contraction and a^i_{Na} with Strophanthidin

In cells with APs, contraction and a^i_{Na} were clearly dissociated in time with strophanthidin, and there was marked hysteresis between them. This is similar to behavior reported in sheep Purkinje fibers by Boyett *et al.*[4] and more recently in guinea-pig myocytes by Boyett and Harrison.[25] However, because membrane potential was not recorded simultaneously in the latter study, it was not possible to discern if changes of the action potential were responsible for the hysteresis they observed.

Temporal dissociation and hysteresis between contraction and a^i_{Na} were dramatically reduced when cells were voltage-clamped, and this is strong evidence that the changes in the AP with strophanthidin were responsible for most of the hysteresis. A longer and more positive depolarization of the membrane is known to increase the contraction of cardiac muscle.[18] Thus, the initial lengthening and more positive plateau of the AP with strophanthidin will increase the contraction associated with a given level of a^i_{Na}. When strophanthidin is washed off, the AP shortening and more negative plateau will reduce the contraction associated with the level of a^i_{Na}. Thus, this mechanism alone could account for temporal dissociation and hysteresis between contraction and a^i_{Na} in myocytes with APs. It is a very likely explanation for much of the hysteresis between contraction and a^i_{Na} that has been reported in previous studies.[4,25]

These experiments also show that if the myocyte AP changes during an intervention, the data will not show the basic contraction-a^i_{Na} relation, because contraction is also influenced by AP changes. In theory, voltage-clamped myocytes *can* reveal the basic contraction-a^i_{Na} relation. It was therefore of considerable surprise that hysteresis between contraction and a^i_{Na} was not completely abolished with voltage-clamp. We have not attempted to investigate the mechanism responsible for this residual hysteresis. One possibility is that a Na gradient might exist from the subsarcolemmal space to the bulk cytoplasm of the cell when Na_i is changing. The potential importance of this becomes clear when one realizes that it is subsarcolemmal Na_i that controls Ca influx via Na-Ca exchange, whereas SBFI probably indicates bulk cytoplasmic Na_i. When the pump is inhibited by strophanthidin, sub-sarcolemmal Na_i might rise faster than bulk cytoplasmic Na, and this will result in a larger Ca influx and contraction for a given bulk Na_i. When the pump is reactivated after strophanthidin removal, high pump activity might reduce subsarcolemmal Na_i below bulk cytoplasmic, resulting in a decline of Ca influx and contraction faster than bulk Na_i. Such a mechanism is capable of producing hysteresis between contraction and a^i_{Na}.

The crucial question is whether such Na gradients exist, but there is little information on which to form an opinion. A recent report using X-ray microprobe analysis[26] suggested that there might be Na gradients within myocytes, but this technique detects total Na (i.e., bound + free) and not the physiologically important quantity of free ionized Na (i.e. Na activity). Estimates of the diffusion coefficient for Na in cytoplasm[27] (about 0.5×10^{-5} cm^2 sec^{-1}) coupled with estimates for passive Na influx[28] (10 pmoles cm^{-2} sec^{-1}) suggest an upper limit for a cytoplasmic Na gradient of 0.2 μM μm^{-1}, which is negligible. However, this diffusion coefficient may not apply to the subsarcolemmal space if there are physical tortuosities that slow diffusion. Slowed diffusion in the subsarcolemmal space might allow larger and significant Na gradients to exist. A clear answer as to whether cytoplasmic Na gradients exist in myocytes will have to wait until high-resolution spatial imaging of Na can be performed.

The Relation between Contraction and a^i_{Na} with K-Free Solution

Although contraction and a^i_{Na} changed approximately in parallel during exposure to K-free solution and on recovery, hysteresis between these variables was still observed. This is further evidence for the existence of hysteresis between contraction and a^i_{Na} and suggests that it is related to pump inhibition rather than being exclusive to strophanthidin. We also recorded changes of I_{Ca} during pump inhibition with K-free solution and showed that I_{Ca} declined as the pump was inhibited and contraction increased. A possible mechanism for the decline of I_{Ca} with pump inhibition is that as Ca_i rose, this caused Ca-induced inactivation of I_{Ca}.[29] In cases where there was an incomplete recovery of I_{Ca} after a period of pump inhibition, it is possible that the reduced I_{Ca} resulted in a smaller contraction relative to the level of a^i_{Na}. This might be a partial explanation for some of the hysteresis observed. However, it is unlikely to be the whole explanation because, in most cases, I_{Ca} recovered fully to the control level. If changes of I_{Ca} cannot account fully for the hysteresis, it is possible that cytoplasmic Na gradients might turn out to be the cause.

SUMMARY

Intracellular sodium was estimated in ventricular myocytes using the new Na-sensitive fluorescent indicator SBFI. Membrane potential and contraction were also measured simultaneously. Using an *in situ* calibration method, we found that intracellular sodium activity (a^i_{Na}) was 2.9 mM in quiescent rabbit cells.

When the digitalis analogue strophanthidin inhibited the Na-K pump of myocytes with action potentials (APs), changes of contraction and a^i_{Na} were dissociated in time. There was also marked hysteresis between contraction and a^i_{Na}. When strophanthidin was applied to the same myocytes under voltage-clamp conditions, temporal dissociation between contraction and a^i_{Na} was dramatically reduced. This suggests that much of the dissociation and hysteresis was due the change in AP shape with strophanthidin.

A small amount of residual hysteresis still existed even with voltage-clamp, and this persisted when the pump was blocked by removal of external potassium as an alternative method. We suggest that a gradient of sodium concentration from the subsarcolemmal space to the bulk cytoplasm might be responsible for hysteresis. Whereas SBFI probably signals the average Na level of the cytoplasm, subsarcolemmal Na may control Ca influx and contraction via Na-Ca exchange.

ACKNOWLEDGMENTS

We would like to thank Roger Thomas, Paul Brooksby, and Jackie Addison for stimulating discussion and gastronomic delights. George Schofield and Spencer Shorte gave us much help with fluorescence and John Bridge provided a video edge detector at extremely short notice. Cliff Christian of Axon Instruments repaired our Axoclamp in an amazingly short time after it blew up! We are also grateful to Dave Clements and Merv Higgins who built many of the mechanical bits, to Jo Howard for much appreciated technical help, and to Margaret Clements for secretarial assistance.

REFERENCES

1. LEE, C. O. 1985. 200 years of digitalis: The merging central role of the sodium ion in the control of cardiac force. Am. J. Physiol. **249:** C367-C378.
2. EISNER, D. A., W. J. LEDERER & R. D. VAUGHAN-JONES. 1981. The dependence of sodium pumping and tension on intracellular sodium activity in voltage-clamped sheep Purkinje fibers. J. Physiol. **317:** 163-187.
3. LEE, C. O. & M. DAGOSTINO. 1982. Effect of strophanthidin on intracellular Na^+ ion activity and twitch tension of constantly driven canine cardiac Purkinje fibers. Biophys. J. **40:** 185-198.
4. BOYETT, M. R., G. HART & A. J. LEVI. 1986. Dissociation between force and intracellular sodium activity with strophanthidin in isolated sheep Purkinje fibers. J. Physiol. **381:** 311-331.
5. WANG, D. Y., S. W. CHAE, Q. Y. GONG & C. O. LEE. 1988. Role of a^i_{Na} in positive force-frequency staircase in guinea-pig papillary muscle. Am. J. Physiol. **255:** C798-C807.

6. MINTA, A. & R. Y. TSIEN. 1989. Fluorescent indicators for cytosolic sodium. J. Biol. Chem. **264:** 19449-19457.
7. LEVI, A. J., S. J. PRICE, S. HALL & A. KAUFMAN. 1990. Changes of coronary vessel resistance during enzymatic isolation of myocytes from guinea-pig hearts. J. Physiol. **425:** 9P.
8. ISENBERG, G. & U. KLOCKNER. 1982. Calcium currents of isolated bovine ventricular myocytes are fast and of large amplitude. Pflugers Arch. **395:** 30-41.
9. LEVI, A. J. 1988. An experimental chamber for isolated cell experiments with minimal temperature gradients. J. Physiol. **401:** 4P.
10. CANNELL, M. B. & W. J. LEDERER. 1986. A novel experimental chamber for single-cell voltage-clamp and patch-clamp applications with low electrical noise and excellent temperature and flow control. Pflugers Arch. **406:** 536-539.
11. KENTISH, J. C. & M. R. BOYETT. 1983. A simple electronic circuit for monitoring changes in the duration of the action potential. Pflugers Arch. **398:** 233-235.
12. BOYETT, M. R., M. MOORE, B. R. JEWELL, R. A. P. MONTGOMERY, M. S. KIRBY & C. H. ORCHARD. 1988. An improved apparatus for the optical recording of contraction of single heart cells. Pflugers Arch. **413:** 197-205.
13. STEADMAN, B. W., K. B. MOORE, K. W. SPITZER & J. H. B. BRIDGE. 1988. A video system for measuring motion in contracting heart cells. IEEE Trans. Biomed. Eng. **35:** 264-272.
14. HAROOTUNIAN, A., J. P. Y. KAO, B. K. ECKERT & R. Y. TSIEN. 1989. Fluorescent ratio imaging of cytosolic free Na in individual fibroblasts and lymphocytes. J. Biol. Chem. **264:** 19458-19467.
15. DONOSO, P., D. A. EISNER & S. C. O'NEILL. 1989. Measurement of intracellular $[Na^+]$ in isolated rat cardiac myocytes using the fluorescent indicator SBFI. J. Physiol. **418:** 48P.
16. BEUKELMANN, D. J. & W. G. WEIR. 1989. Sodium-calcium exchange and guinea-pig cardiac cells: Exchange current and changes in intracellular calcium. J. Physiol. **414:** 499-520.
17. ALMERS, W., E. W. MCCLESKEY & P. T. PALADE. 1984. A non-selective cation conductance in frog muscle membrane blocked by micromolar external calcium ions. J. Physiol. **353:** 565-583.
18. GIBBONS, W. R. & H. A. FOZZARD. 1975. Relationships between voltage and tension in sheep cardiac Purkinje fibers. J. Gen. Physiol. **65:** 345-365.
19. ELLIS, D. 1977. The effects of external cations and ouabain on the intracellular sodium activity of sheep heart Purkinje fibers. J. Physiol. **273:** 211-240.
20. FABIATO, A. 1985. Time and calcium dependence of activation and inactivation of calcium-induced release of calcium from the sarcoplasmic reticulum of a skinned canine cardiac Purkinje cell. J. Gen. Physiol. **85:** 247-289.
21. DESILETS, M. & C. M. BAUMGARTEN. 1986. K^+, Na^+, and Cl^- activities in ventricular myocytes isolated from rabbit heart. Am. J. Physiol. **251:** C197-C208.
22. HIRAOKA, M. & S. KAWANO. 1987. Mechanism of increased amplitude and duration of the plateau with sudden shortening of diastolic intervals in rabbit ventricular cells. Circ. Res. **60:** 14-26
23. LEVI, A. J. 1990. Cardiac glycosides, action potential shortening and electrogenic Na-Ca exchange in isolated guinea-pig cardiac myocytes. J. Physiol. **426:** 15P.
24. KAMEYAMA, M., M. KAKEI, R. SATO, T. SHIBASAKI, H. MATSUDA & H. IRISAWA. 1984. Intracellular Na^+ activates a K^+ channel in mammalian cardiac cells. Nature **309:** 354-356.
25. BOYETT, M. R. & S. M. HARRISON. 1990. The relationship between the intracellular sodium activity (a^i_{Na}) and the strength of contraction of ventricular myocytes isolated from the guinea-pig. J. Physiol. **423:** 60P.
26. ISENBERG, G. & M. F. WENDT-GALLITELLI. 1990. X-ray microprobe analysis of sodium concentration reveals large transverse gradients from the sarcolemma to the centre of voltage-clamped guinea-pig ventricular myocytes. J. Physiol. **420:** 86P.

27. PUSCH, M. & E. NEHER. 1988. Rates of diffusional exchange between small cells and a measuring patch pipette. Pflugers Arch. **411:** 204-211.
28. DEITMER, J. W. & D. ELLIS. 1978. The intracellular sodium activity of cardiac Purkinje fibers during inhibition and re-activation of the Na-K pump. J. Physiol. **284:** 241-259.
29. KOKUBUN, S. & H. IRISAWA. 1984. Effects of various intracellular Ca ion concentrations on the calcium current of guinea-pig single ventricular cells. Jpn. J. Physiol. **34:** 599-611.

Contribution of Sodium-Calcium Exchange to Calcium Regulation in Cardiac Muscle [a]

M. B. CANNELL[b]

Department of Pharmacology & Clinical Pharmacology
St. George's Hospital Medical School
London SW17 0RE, United Kingdom

INTRODUCTION

The existence of a sodium-calcium (Na-Ca) exchange mechanism in the surface membrane of cardiac cells has been appreciated for more than a decade.[1-3] However, detailed information about the role played by the Na-Ca exchanger in calcium homeostasis has not, until recently, been available. While thermodynamic analysis can give information about the direction of transport (assuming that the intra- and extracellular levels of sodium and calcium, the membrane potential, and the stoichiometry are known),[4] such analyses cannot answer questions about the kinetics of exchange and role played by the exchanger in calcium metabolism.

Recently, major advances have been made in isolating currents that arise from the exchanger,[5-7] and it is now clear that current generated by the exchanger will contribute to the time-course of the action potential.[8] Such studies do not provide information about the role of the exchanger in the contraction-relaxation cycle because of the need to manipulate intracellular and/or extracellular ion levels in order to measure the voltage dependence of the exchange under defined (not necessarily physiological) ionic conditions. In particular, the buffering of intracellular calcium ($[Ca^{2+}]_i$) to known levels will preclude studying the role played by the exchanger during the normal calcium transient. In our experiments, the need to buffer $[Ca^{2+}]_i$ was removed by measuring $[Ca^{2+}]_i$ directly with a fluorescent calcium indicator under voltage-clamp conditions. As in other studies,[5-7,9-11] examination of the properties of the exchanger was facilitated by employing pharmacological agents to remove the calcium fluxes due to calcium channels and the sarcoplasmic reticulum. Under these conditions, we have been able to dissect the processes that regulate $[Ca^{2+}]_i$ across the surface membrane and examine the role of the exchanger in calcium homeostasis.[12]

METHODS

The methods used were similar to those published elsewhere.[12,13] Single guinea-pig ventricular myocytes were obtained by enzymatic dissociation.[14] The cells were trans-

[a] This work was supported by grants from the National Institutes of Health (HL39733) and Florida Affiliate of the American Heart Association.

[b] Address for correspondence: Dept. of Pharmacology & Clinical Pharmacology, St. George's Hospital Medical School, Cranmer Terrace, London SW17 0RE, U.K.

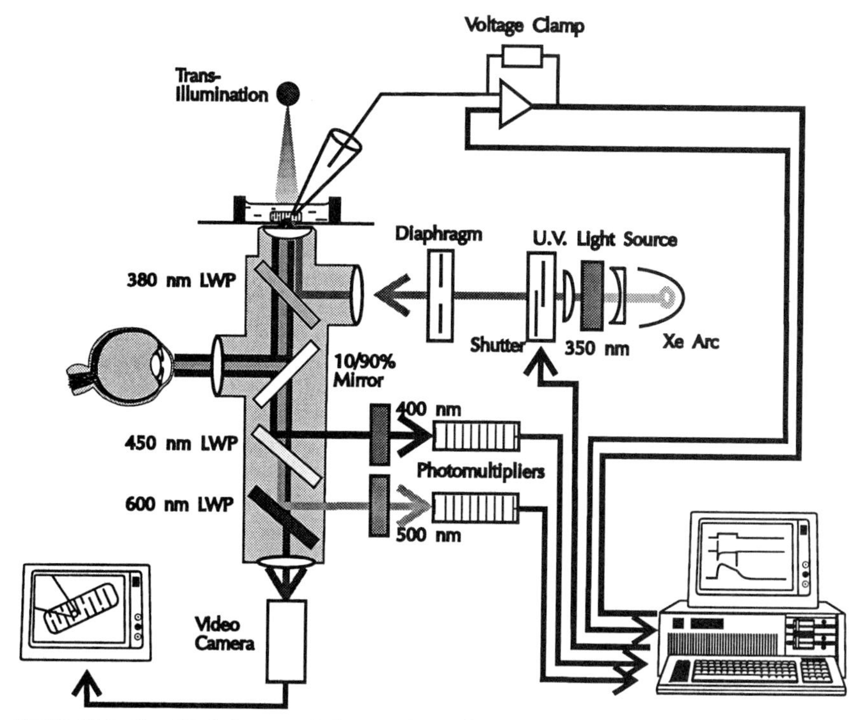

FIGURE 1. Simplified diagram of the experimental apparatus. The cell is trans-illuminated by red light (> 650 nm), which allows recording of cell length and fluorescence at the same time. The cell is also illuminated by ultraviolet (UV) light, and the fluorescence of the intracellular indicator is collected at 400 and 500 nm as illustrated. The cell is voltage-clamped with a single microelectrode, which is also used to dialyze the cell and introduce the calcium indicator (Indo-1) into the myoplasm.

ferred to the experimental chamber[15] mounted on the stage of an inverted microscope (Nikon). Cells were voltage clamped with a single-microelectrode technique,[16] and electrode resistances were typically 1-3 Mohm. The filling solution of the electrode contained (in mM): 135 TEA, 8 NaCl, 3.3 ATP (Mg-salt), 12 phosphocreatine (Tris-salt), creatine phosphokinase (30 U/ml), 30 PIPES, and 30 or 50 μM Indo-1 (K-salt), pH 7.2. The cells were continuously superfused at 35°C with saline containing 140 mM NaCl, 4 mM KCl, 1 mM $MgCl_2$, 2 mM $CaCl_2$, 0.1 mM CaEGTA, 10 mM glucose, 10 mM HEPES, pH 7.4.

When desired, calcium currents were blocked with 20 μM verapamil, sodium currents with 10 μM tetrodotoxin, potassium currents with 4 mM tetraethylammonium, and sarcoplasmic reticulum activity was inhibited by adding 30 μM ryanodine to the bathing solution. Indo-1 fluorescence was recorded at 400- and 500-nm wavelengths while epi-fluorescence illumination at 360 nm was used (see FIG. 1). Rapid solution changes were accomplished by moving a 100-μm bore tube from which the desired solution was freely flowing up to the cell. The tube was then moved away from the cell to allow the control bathing solution flowing through the chamber to return. The timing of the command to change the bathing solution is indicated in the figures.

$[Ca^{2+}]_i$ was calculated from the ratio of Indo-1[17] fluorescence, R = (F400/F500) after subtraction of cell autofluorescence and the equation[17,18] $[Ca^{2+}]_i = K_d(R - R_{min})/(R_{max} - R)$ where R_{min} and K_d were determined *in vitro*[18] and R_{max} determined inside the cell by electrically disrupting the sarcolemma. All signals were digitized (VR-100A CRC Digitizer, Instrutech Corp., Mineola, NY, USA) and recorded on VCR tape for off-line analysis. Voltage-clamp command signals were generated by a microcomputer using custom-written software.

RESULTS

Under "normal" conditions, depolarization of the surface membrane of a cardiac myocyte leads to a rapid increase in $[Ca^{2+}]_i$. FIGURE 2A shows changes in membrane currents and Indo-1 fluorescence evoked by depolarizing the cell from −37 mV to various potentials in the range +17 to +70 mV. The least positive potential is associated with a large phasic inward current that is mainly carried by calcium ions.[19] This influx of calcium triggers a large and rapid increase in $[Ca^{2+}]_i$, which can be seen on the Indo-1 record as a decrease in fluorescence at 505 nm (note that the record has been inverted in the figure). As the depolarizing potential is made more positive, the amplitude of the calcium current decreases, and there is also a decrease in the amplitude of the calcium transient. In addition, an "after transient" develops with repolarization from more positive potentials. These features of the calcium transient and the relationship between the calcium current and the calcium transient have been examined in detail by several groups.[13,20–22] It is thought that the rapid increase in $[Ca^{2+}]_i$ is mainly due to the release of calcium from the sarcoplasmic reticulum triggered by the calcium current and CICR mechanism.[23]

It is clear that the $[Ca^{2+}]_i$ transient is quite rapid, declining to baseline levels in approximately 200 msec following repolarization. (This can be seen more easily by examining the time course of the "after transient.") The time course of the rise in $[Ca^{2+}]_i$ is shown with an expanded time scale in FIGURE 2B. The calcium current peaks about 2 msec after depolarization, and at the same time there is a detectable increase in $[Ca^{2+}]_i$. The change in $[Ca^{2+}]_i$ reaches 90% of its final value about 10 msec after depolarization. Although it is likely that the time course of the change in Indo-1 fluorescence is limited by the kinetics of the dye-calcium reaction, it is clear that the release of calcium (principally from the sarcoplasmic reticulum) is essentially complete in about 10 msec (as expected from $d[Ca^{2+}]_i/dt$ being proportional to the flux of calcium into the myoplasm[24]).

To reveal the contribution of the Na-Ca exchanger, we removed sarcoplasmic reticulum calcium release by adding ryanodine (30 μM),[10,25] and calcium currents were blocked with 20 μM verapamil. Under these conditions, changes in intracellular calcium evoked by depolarization became much slower, and, as illustrated in FIGURES 3 and 4A, take several seconds to reach new steady levels after a change in membrane potential. Two observations support the idea that these changes in $[Ca^{2+}]_i$ are due to the Na-Ca exchanger[12]: (1) Reducing internal sodium by omitting sodium from the electrode-filling solution inhibited these changes in $[Ca^{2+}]_i$. Increasing intracellular sodium (by either inhibiting the sodium pump or increasing the level of sodium in the electrode) led to larger increases in $[Ca^{2+}]_i$ on depolarization (not shown). (2) At the most positive potentials, the driving force for calcium entry should be very small, so that a large increase in intracellular calcium cannot be explained by calcium entry across the surface membrane by a passive process.

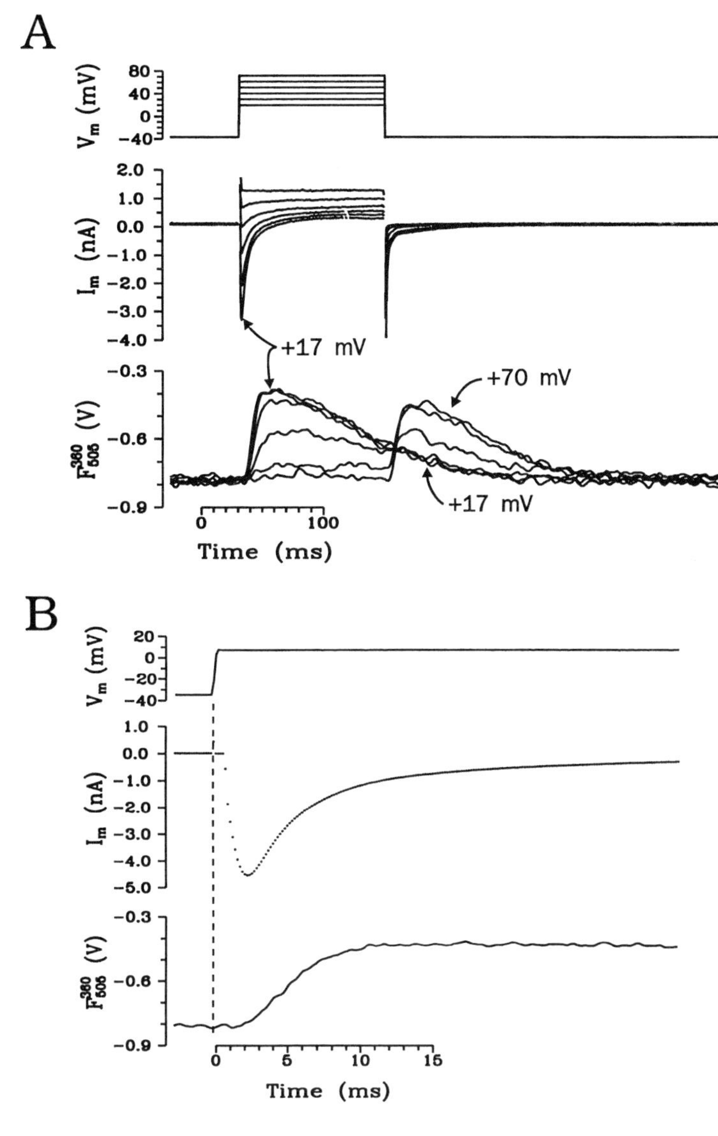

FIGURE 2. (**A**) Voltage-dependence and time course of the $[Ca^{2+}]_i$ transient. Note that at the depolarization, potential becomes more positive, the amplitude of the calcium current decreases, and at the same time there is a decrease in the amplitude of the calcium transient. However, repolarization from very positive potential results in an "after-transient." This after-transient may be triggered by the tail calcium current. (**B**) The time course of the rise of $[Ca^{2+}]_i$ at high time resolution, same experiment as **A**.

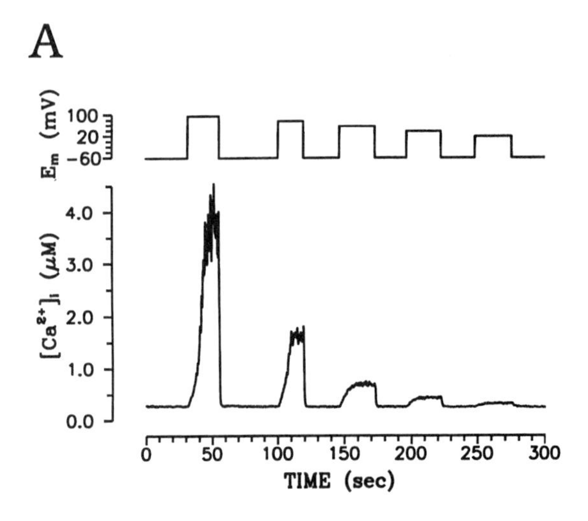

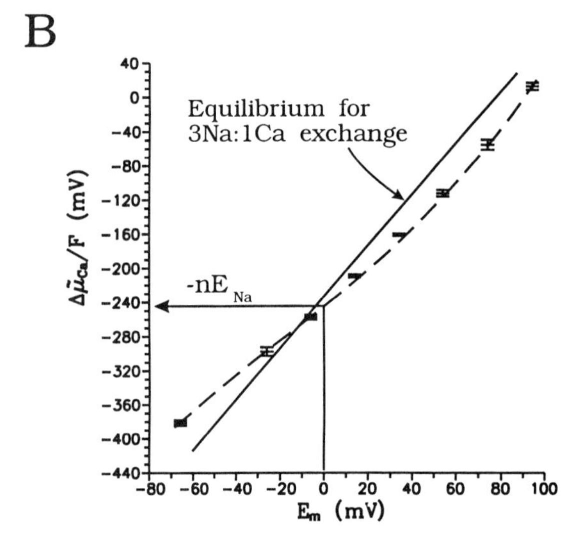

FIGURE 3. (**A**) The effect of depolarization on resting $[Ca^{2+}]_i$. Changes in $[Ca^{2+}]_i$ resulting in depolarization to various potentials in the range +100 to 0 mV are shown. The pulse potentials are separated by 20 mV and the steady level of $[Ca^{2+}]_i$ reached during the pulse has a nearly exponential dependence on membrane potential. The increased noise in the $[Ca^{2+}]_i$ record at higher $[Ca^{2+}]_i$ arises from the nonlinear relationship between fluorescence and $[Ca^{2+}]_i$. Note also that the rate of rise of $[Ca^{2+}]_i$ is slower than the rate of decline of $[Ca^{2+}]_i$. (**B**) Dependence of the electrochemical gradient for calcium entry ($\Delta\mu_{Ca}/F$) on the membrane potential (E_m). This figure shows pooled data from experiments on five different cells, the error bars correspond to 1 SEM. The dashed line is an arbitrary second-order polynomial fitted to the data to enable estimation of the y-intercept of the data, as shown. The solid line indicates the expected relationship between $\Delta\mu_{Ca}/F$ and E_m for a 3Na : 1Ca exchange that is at equilibrium.

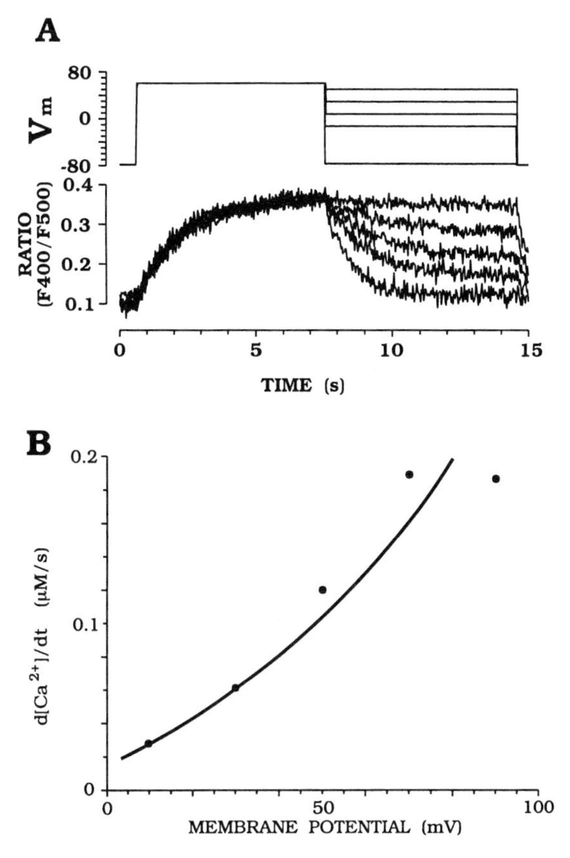

FIGURE 4. (**A**) Changes in $[Ca^{2+}]_i$ when the calcium current is blocked and the sarcoplasmic reticulum is inhibited with ryanodine. Note the time course of rise and fall of calcium takes seconds in these conditions, even though the voltage range is similar to that used in FIGURE 2.(**B**) Voltage dependence of the rate of rise of $[Ca^{2+}]_i$. The rate of rise increases *e*-fold in approximately 60 mV.

The exchanger will attempt to make the electrochemical gradient for calcium ($\Delta\mu_{Ca}$) equal to the sodium electrochemical gradient times the number of sodium ions exchanged per calcium ion ($\Delta\mu_{Ca} = n\Delta\mu_{Na}$) so that $\Delta\mu_{Ca}/F$ measured at steady state should be linearly related to the membrane potential (E_m) with a slope of n and intercept of $-nE_{Na}$. As shown in FIGURE 3B, the intercept of the curve through the data (from five cells) predicts that $-nE_{Na} = -238$ mV. If the level of sodium inside the cell is the same as that in the pipette (8 mM), then F_{Na} would be ~76 mV, which suggests that $n = 3$ (99% confidence limits: $2.9 < n < 3.3$). This data is in good agreement with estimates of the stoichiometry from isolated cardiac sarcolemmal vesicles and electrophysiological experiments.[26–28]

The observed data points fall below the theoretical equilibrium value for a 3Na : 1Ca exchange (shown by the solid line in FIG. 3B) at very positive E_m, an observation that

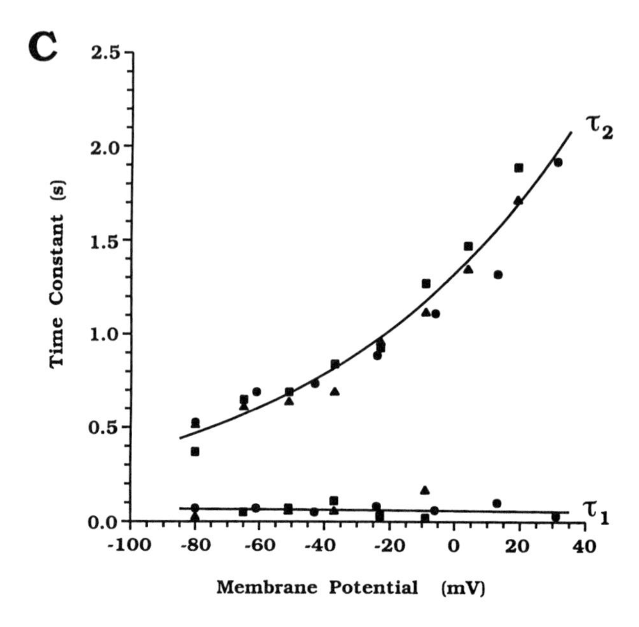

FIGURE 4.*Continued.* **(C)** Voltage dependence of the rate of decline of calcium (the decline of calcium was resolved into two exponential components, $\tau 1$ and $\tau 2$). The more rapid component ($\tau 1$) may be due to residual sarcoplasmic reticulum activity. The slower component ($\tau 2$), which arises from Na-Ca exchange activity, slows *e*-fold with 77 mV of depolarization. (Experimental data replotted from Crespo *et al.*[12])

can be explained by calcium extrusion via a sarcolemmal Ca-ATPase.[29] The Ca-ATPase will tend to oppose calcium entry via the exchanger (there can be no passive calcium transport when $\Delta\mu_{Ca} = 0$) so that the observed level of $[Ca^{2+}]_i$ is less than that expected from the thermodynamics of the Na-Ca exchange reaction. At more negative E_m, the presence of a calcium "leak" (which would be required for the cell to have a stable resting $[Ca^{2+}]_i$) will prevent the exchanger from lowering $[Ca^{2+}]_i$ to the level predicted from the stoichiometry of the exchanger. Although an estimate of the stoichiometry (n) can be made from the slope of the dotted line in FIGURE 3B, it is important to note that the calcium leak and Ca-ATPase fluxes will prevent the exchanger from achieving equilibrium and will result in an underestimate of n (as observed in other studies—for review see Eisner & Lederer[30]) except when the leak flux is equal and opposite to the pump flux. This will occur at the intersection of the solid line in FIGURE 3B with the observed data at about -11 mV (note that our estimate of n was obtained near equilibrium conditions at 0 mV, improving our confidence that $n = 3$). At more negative potentials the leak of calcium into the cell should increase, and the efflux of calcium via Ca-ATPase should decrease (as the decrease in $[Ca^{2+}]_i$ and increase in $\Delta\mu Ca/F$ for calcium entry should inhibit Ca-ATPase). These considerations lead to the conclusion that the exchanger must make a significant contribution to calcium homeostasis at normal diastolic membrane potentials.

The time course of changes in $[Ca^{2+}]_i$ resulting from exchanger activity are illustrated in FIGURE 4. FIGURE 4A shows original records of Indo-1 fluorescence resulting from depolarization to $+60$ mV. It is clear that the rate of rise of $[Ca^{2+}]_i$ is several orders of magnitude slower than is observed when the sarcoplasmic reticulum is active (FIG. 2B). The voltage dependence of the rate of rise of $[Ca^{2+}]_i$ is illustrated in FIGURE 4B and the rate of rise of $[Ca^{2+}]_i$ increases e-fold in about 60 mV. Cells were also depolarized to various potentials from +60 mV to examine the voltage dependence of the rate of decline of $[Ca^{2+}]_i$. Analysis of these data (FIG. 4C) showed that the time course of the decline of $[Ca^{2+}]_i$ consisted of two exponential phases; the first phase was voltage independent and had a mean time constant of 0.06 sec while the time constant of the second phase increased exponentially with voltage. The initial rapid decline in $[Ca^{2+}]_i$ may reflect some residual calcium uptake by the SR that was not blocked by ranodine since, in the absence of ryanodine, $[Ca^{2+}]_i$ declines at a similar rate. The second slower component may reflect the activity of the Na-Ca exchanger, and this component had a time constant of 0.5 sec at −80 mV that increased e-fold in 77 mV (shown by the solid line). Again, these changes are much slower than are observed under control conditions (i.e., in the absence of ryanodine). It is notable that the voltage dependence of the rates of change of $[Ca^{2+}]_i$ measured here are similar to the reported voltage dependence of exchanger currents.[5,6]

To what extent can these thermodynamic and kinetic data be explained by a model for calcium regulation that includes an ohmic calcium leak, a Ca-ATPase, and a Na-Ca exchanger? The simplest formulation for the Na-Ca exchange reaction that was capable of explaining our experimental data was the six-state model illustrated in FIGURE 5A. (Additional considerations that also require such a model include the existence of Ca-Ca exchange in the absence of sodium[31] as well as recent measurements of "gating currents."[32]) The calcium dependence of the rate of calcium extrusion by the Ca-ATPase was taken from the literature, as were the affinities of the exchanger sites for sodium and calcium. The free parameters in the model were further reduced by constraining all ion-association rates to be diffusion limited. The off-rate constants were calculated from the ion-association rate and the affinity constant. With these limitations, the computer had only to parameter fit three membrane-crossing rate constants (the fourth being determined by microreversability) and the amount of exchanger protein to our kinetic and steady-state data.

It was notable that the predicted rate-limiting step(s) for the exchanger reaction were associated with sodium translocation from inside the cell to outside. Calcium translocation was about an order of magnitude faster, and its voltage dependence was undefined by our data (because it was not rate-limiting and was therefore set to zero). The model required the rate-limiting step for sodium translocation to occur over about 35% of the membrane voltage field (to provide the appropriate voltage dependence of the rates of rise and fall of $[Ca^{2+}]_i$). The voltage dependence of calcium fluxes predicted by the model during simulated experiments of the type shown in FIGURE 3 are shown in FIGURE 5B. At diastolic potentials (around −80 mV) about 75% of the leak is balanced by the exchanger. However, the contribution of the Ca-ATPase rises rapidly with depolarization, and this is due to stimulation of the pump by the increase in $[Ca^{2+}]_i$. (It was assumed that the Ca-ATPase rate was not voltage dependent.) FIGURE 5C shows the voltage dependence of steady-state $[Ca^{2+}]_i$, and, as observed experimentally, $[Ca^{2+}]_i$ is relatively voltage insensitive over the range −100 to −30 mV (note that $[Ca^{2+}]_i$ is displayed on a logarithmic scale.) FIGURE 5D illustrates the voltage dependence of calcium fluxes carried by the exchanger as a function of membrane potential at two different levels of $[Ca^{2+}]_i$ that correspond (approximately) to resting and peak levels of $[Ca^{2+}]_i$ during the calcium transient. The rate of calcium influx via the exchanger at +60 mV and $[Ca^{2+}]_i$ of 0.14 μM was approximately 17% of the rate of calcium extrusion by the exchanger at $[Ca^{2+}]_i$ of 1.2 μM at −80 mV, as observed experimentally (see FIG. 4A). In addition, the voltage dependence of calcium influx

A

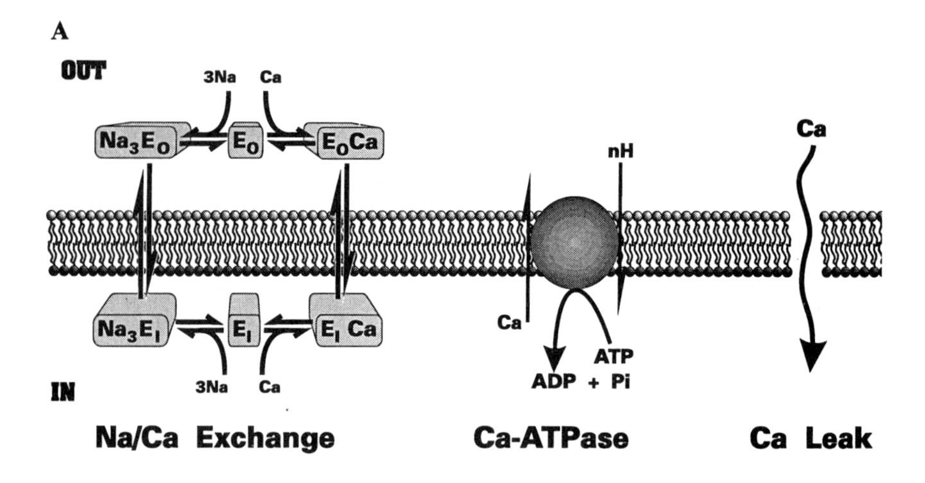

B

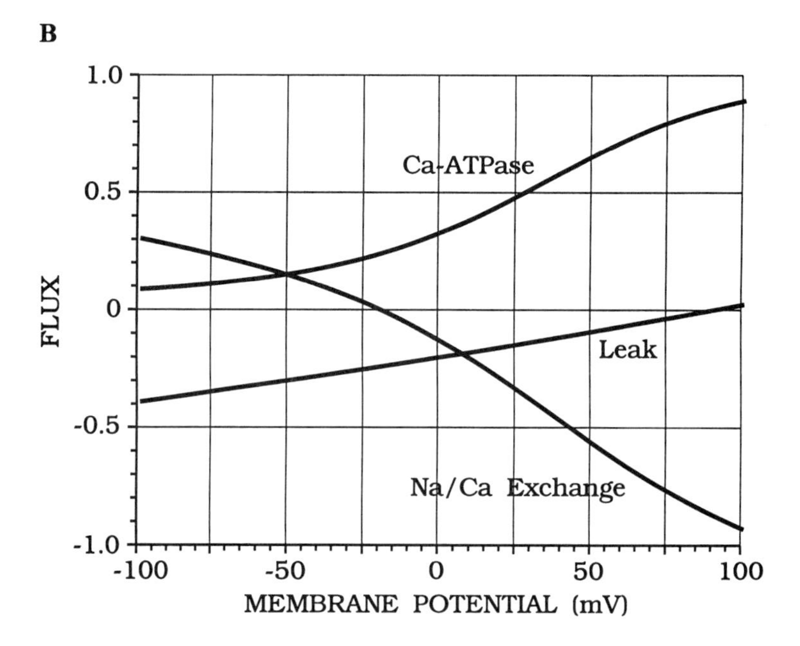

FIGURE 5. Model for calcium homeostasis. (**A**) Schematic representation of the calcium transport processes considered. The exchanger was a six-state model with voltage dependence in the membrane-crossing sodium translocation steps only. The overall rate of exchange was stimulated by calcium as described by Hilgemann. The Ca-ATPase had properties based on those published by other workers.[29] (**B**) Voltage dependence of the fluxes underlying the steady-state voltage-dependence of $[Ca^{2+}]_i$. Note that the leak is ohmic, while the flux due to the Ca-ATPase increases with depolarization due to the increase in $[Ca^{2+}]_i$.

C

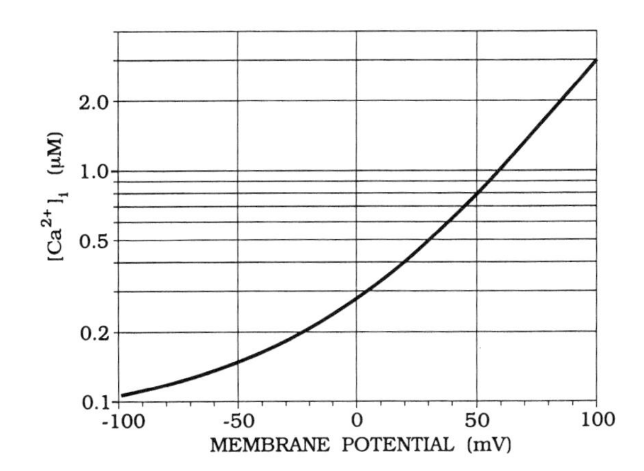

D

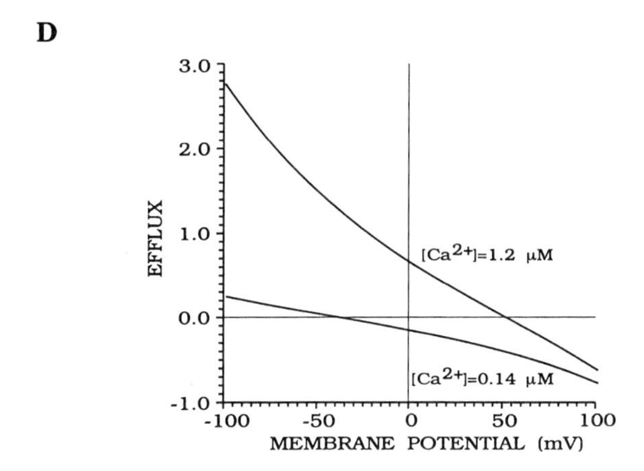

FIGURE 5. *Continued.* (**C**) Voltage dependence of steady-state $[Ca^{2+}]_i$. (**D**) Voltage dependence of the Na-Ca exchanger fluxes at two different levels of $[Ca^{2+}]_i$.

and efflux at the two $[Ca^{2+}]_i$ levels is in good agreement with the data shown in FIGURES 4B and 4C. Overall, the behavior of the model was very similar to that observed experimentally.

Despite the ability of this model to describe calcium regulation under the conditions of our experiments, a remarkable deficiency in the model appeared when attempting to describe the effect of removing external sodium. As shown in FIGURE 5, in the presence of external sodium, the cell is able to extrude a calcium load (introduced by depolarizing the cell in the absence of calcium channel blockers—see FIG. 5 legend) quite rapidly. On inhibiting the Na-Ca exchanger by removing external sodium, $[Ca^{2+}]_i$ increased very slowly. Depolarization in these conditions resulted in a greater increase in $[Ca^{2+}]_i$, and after repolarization, $[Ca^{2+}]_i$ did not return to resting levels. In agreement with these results, recent mechanical experiments have shown that the cell relaxes by less than 15% after 1 sec in the absence of sodium, compared with > 80% in its presence.[33,34] On first sight, these results appear compatible with the model data presented earlier. The inability of the cell to keep $[Ca^{2+}]_i$ at resting levels in the absence of sodium can be explained by the large contribution of exchanger calcium extrusion to resting calcium at negative potentials (*cf.* FIG. 5B). Also, the time course of $[Ca^{2+}]_i$ changes (in response to a calcium load imposed by depolarization-activated calcium channels) are larger and slower because the exchanger can no longer contribute to calcium extrusion. When sodium is replaced, the exchanger then extrudes calcium, returning $[Ca^{2+}]_i$ to resting levels.

FIGURE 6B shows the model simulation of the results shown in FIGURE 6A. Although steady-state levels of $[Ca^{2+}]_i$ are well predicted by the model (as are the kinetics of the change in $[Ca^{2+}]_i$ in the presence of extracellular sodium), when sodium is removed $[Ca^{2+}]_i$ rises too rapidly to its new steady level. In addition, when the cell is depolarized, $[Ca^{2+}]_i$ rises to the experimentally observed level, but declines too rapidly to the steady-state level on repolarization. On returning sodium, the model also predicts a return of $[Ca^{2+}]_i$ to control levels that is too rapid. Thus quite large deficiencies in the model are disclosed by comparing model responses to experimental data (obtained during changes in extracellular sodium). This is despite the fact that the model is able to accurately describe steady levels of $[Ca^{2+}]_i$ observed in the presence and absence of sodium as well as the kinetics of the changes in $[Ca^{2+}]_i$ in the presence of sodium.

DISCUSSION

Regulation of Resting $[Ca^{2+}]_i$ by the Na-Ca Exchange

The data presented here suggests that the exchanger makes a significant contribution to resting $[Ca^{2+}]_i$. While the contribution of the Ca-ATPase has been considered to be more important than the exchanger under *resting* conditions,[29] it has not been straightforward to test this idea directly in intact cells. The modeling suggested that the Ca-ATPase contributes about 25% of the necessary calcium extrusion at rest. When the exchanger was blocked, $[Ca^{2+}]_i$ increased to about 200 nM, a level that stimulates the Ca-ATPase sufficiently to cause calcium extrusion via the Ca-ATPase to increase and equal the leak calcium influx. The voltage dependence of the exchanger coupled with the calcium dependence of the Ca-ATPase ensure that depolarization from −80 mV has only modest effects on resting $[Ca^{2+}]_i$, until about −30 mV, where depolarization starts to increase $[Ca^{2+}]_i$rapidly. A consequence of this observation is that modest diastolic depolarizations will have minimal effects on resting $[Ca^{2+}]_i$.

Contribution of the Exchanger to the Calcium Transient

On depolarization under normal conditions, $[Ca^{2+}]_i$ peaks in less than 20 msec and has a maximum rate of rise of about 150 μM/sec. (This rate may be limited by the dye kinetics.) It should also be noted that this is the rate of rise of *free calcium;* the actual flux is undoubtedly larger as calcium is taken up by buffer systems with the cell. In the presence of ryanodine and calcium-channel blockers, the maximum rate of rise of $[Ca^{2+}]_i$ is reduced by about two orders of magnitude. It therefore seems likely that

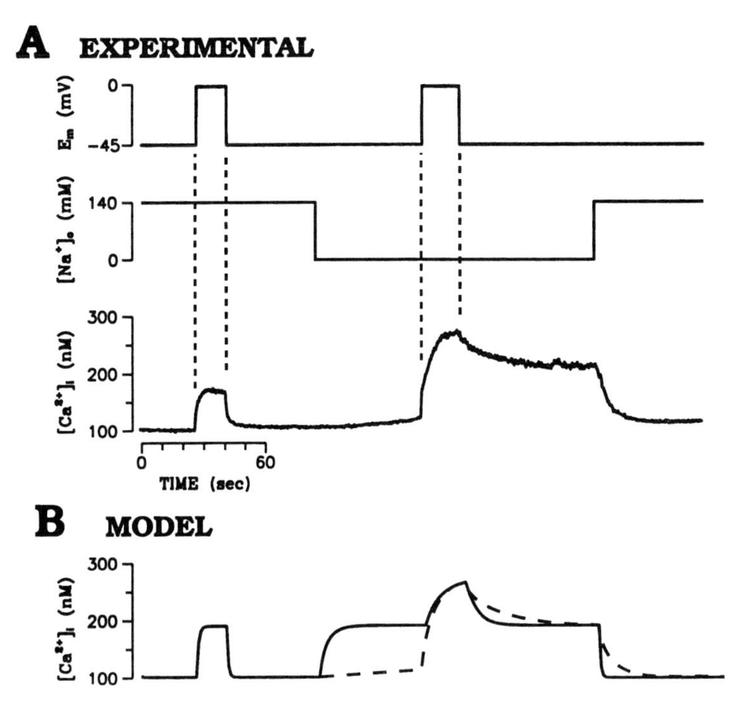

FIGURE 6. Changes in $[Ca^{2+}]_i$ resulting from Na-Ca exchange inhibition in the presence of ryanodine. (**A**) In the absence of calcium-channel antagonists, depolarization resulted in $[Ca^{2+}]_i$ increasing, and this increase was rapidly reversed when the cell was repolarized in the sodium-containing bathing saline. These increases in $[Ca^{2+}]_i$ arise from the activation of calcium channels as the omission of sodium from the pipette dialysis solution prevented calcium entry via the exchanger. When sodium was removed (by isosmotic replacement with lithium) $[Ca^{2+}]_i$ slowly increased. Depolarization in the absence of sodium resulted in a greater increase in $[Ca^{2+}]_i$, and $[Ca^{2+}]_i$ slowly declined to an elevated level after repolarization. Returning sodium to the bathing solution resulted in a rapid decrease in $[Ca^{2+}]_i$. (Data replotted from Crespo *et al.*[12]) (**B**) The model simulation of the experiment shown in **A.** In the presence of sodium the model reproduces the changes in $[Ca^{2+}]_i$ well. Note that on removing sodium, the model predicts too rapid an increase in $[Ca^{2+}]_i$. (The experimental response is shown as the dashed line.) During and after depolarization in sodium-free conditions, the model correctly predicts the steady level of $[Ca^{2+}]_i$, but the timecourse of the change in $[Ca^{2+}]_i$ is too rapid on repolarization (as well as during the return of extracellular sodium).

the exchanger contributes less than 1% of the calcium that activates contraction. In comparison to this meagre contribution to calcium influx, the efflux via the exchanger during relaxation is more substantial. The time constant of the decline of $[Ca^{2+}]_i$ after repolarization to −80 mV under normal conditions is about 100 msec, and in the presence of ryanodine the time constant of the decline of $[Ca^{2+}]_i$ is about 500 msec. Thus the activity of the exchanger can explain as much as 20% of the rate of decline of $[Ca^{2+}]_i$. (The actual contribution will depend on the time course of the decline of membrane potential and $[Ca^{2+}]_i$ also.) Given these estimates for the contribution of the exchanger to calcium movements, it is likely that the activity of the exchanger is directed primarily towards calcium efflux rather than calcium influx throughout the contraction-relaxation cycle. Hence an increase in action potential duration or diastolic depolarization will promote an increase in contractility by decreasing calcium efflux via the exchanger rather than by promoting calcium influx. Interestingly, over the range of plateau potentials (+40 to 0 mV) the membrane attempts to clamp $[Ca^{2+}]_i$ in the range 700 to 300 nM (respectively). Thus at the start of the plateau of the action potential, the surface membrane helps keep $[Ca^{2+}]_i$ elevated to a level where force would be generated, but by the time that the cell is starting to repolarize rapidly (at the end of the plateau), the surface membrane is forcing $[Ca^{2+}]_i$ towards resting levels. Given that the duration of the plateau of the action potential and the calcium transient are both about 200 msec, the voltage dependence of surface membrane $[Ca^{2+}]_i$ regulation matches the time course of the action potential to that of the calcium transient economically.

Modeling Calcium Homeostasis in Cardiac Muscle

Modeling provides a precise mathematical framework in which to test our understanding of calcium regulation. The model presented here, namely, a calcium pump, leak, and Na-Ca exchanger in the sarcolemma, appears reasonable as these are the only processes that have been described that can regulate $[Ca^{2+}]_i$ across the sarcolemma and therefore $[Ca^{2+}]_i$ in the steady state. A similar model has been used by other workers[35] to describe calcium regulation, but it should be noted that the model of the Na-Ca exchange reaction employed here is somewhat more complicated than the single energy barrier model proposed by Mullins[36] as a possible simplification of the true reaction scheme. This choice of model was not arbitrary, but arose from attempts to parameter-fit quite extensive experimental data sets.

Despite the success of the model in describing both the steady levels of $[Ca^{2+}]_i$ and the kinetics of the change in $[Ca^{2+}]_i$ in response to changes in membrane potential, the model predicted changes in $[Ca^{2+}]_i$ in the absence of Na-Ca exchange that were more rapid than those observed experimentally. A simple explanation for this deficiency would be that the buffering power of the cell for calcium is considerably larger in the absence of sodium than in its presence. In connection with this point, it is notable that the largest difference between the model and experimental data is the rate at which $[Ca^{2+}]_i$ rises on removing sodium from the bathing solution. It seems unlikely that even with zero sodium in the pipette that the intracellular sodium concentration would be zero,[37] and if so, the removal of sodium from outside the cell should further increase the rate of rise of $[Ca^{2+}]_i$. However, $[Ca^{2+}]_i$ rises more slowly than expected. Similarly, when sodium is replaced and the exchanger should be stimulated to extrude calcium, we observe that calcium falls more slowly than expected. This occurs despite the fact that the exchanger model correctly predicted the rate of rise and fall of $[Ca^{2+}]_i$ in the presence of sodium (when a similar $[Ca^{2+}]_i$ level was achieved by depolarization).

It is likely that the model is unable to reproduce the changes in $[Ca^{2+}]_i$ (in the absence of sodium) because it does not adequately describe the properties of the intracellular buffers. More specifically, at least a part of the differences between the model and experimental data could be explained by the lack of mitochondrial calcium metabolism in the model (for review see Fry & Harding[38]). On removing external sodium, intracellular sodium will fall and the mitochondrial Na-Ca exchanger will take up calcium. This could then reduce the rate of rise in $[Ca^{2+}]_i$ in these conditions. Similarly, replacement of sodium in the bathing solution will cause intracellular sodium to increase, and this will result in the release of calcium accumulated by the mitochondria in the sodium-free period. If this is the correct explanation for the inability of the model to describe the changes in $[Ca^{2+}]_i$ on removal and replacement of extracellular sodium, then a consequence of adding such an additional buffer system is that the contribution of the sarcolemmal Ca-ATPase to calcium regulation has been overestimated. Such an overestimate will result in $[Ca^{2+}]_i$ declining too rapidly after $[Ca^{2+}]_i$ has been increased in sodium-free conditions (as seen in FIG. 6). A visual estimate of the contribution of mitochondrial calcium uptake can be made by noting the difference between the model predicted and observed time course of change in $[Ca^{2+}]_i$ in response to a calcium load in sodium-free conditions. In FIGURE 6, $d[Ca^{2+}]_i/dt$ was twice that expected, so that the flux due to the Ca-ATPase may have been overestimated twofold. Put another way, the contribution of mitochondrial calcium metabolism to short-term calcium regulation may be as large as that due to the sarcolemmal Ca-ATPase. (It has also been observed that under sodium-free conditions there is a gradual increase in phase-dense particles within the cell that may be the result of mitochondrial swelling in these conditions.[40]) However, as an intracellular store the mitochondria cannot provide long-term calcium homeostasis, a function that is supplied by the Na-Ca exchanger and a sarcolemmal Ca-ATPase in parallel.

CONCLUSION

Under normal conditions the calcium transient is rapid, rising in about 10-20 msec to its peak level and declining to resting levels approximately 200 msec after repolarization. The rapidity of these changes is largely due to the sarcoplasmic reticulum, and calcium influx via the exchanger is too slow at normal levels of intracellular sodium to contribute more than about 1% of the total calcium during the rise of the calcium transient. During repolarization, the exchanger may extrude up to 20% of the calcium released, the majority of the calcium being taken up by the sarcoplasmic reticulum for re-release. Thus the primary role of the exchanger is as a calcium extrusion system. Modeling of experimental data suggested that (1) the rate-limiting step for the Na-Ca exchange reaction occurs after sodium binding and involves a single equivalent charge moving through about 35% of the membrane voltage field; (2) the sarcolemmal Ca-ATPase may provide about 25% of the calcium extrusion from the cell at rest (the remainder being due to the exchanger); and (3) it seems likely that mitochondria contribute to short-term calcium homeostasis, and this contribution may be as large as that due to the sarcolemmal Ca-ATPase.

Computer modeling highlights deficiencies in our understanding of calcium metabolism and the comparison of experimental with simulated data provides powerful insight into the problems of calcium homeostasis in the cardiac cell.

ACKNOWLEDGMENTS

I would like to thank my collaborators Drs. L. Crespo and C. Grantham for their help with experiments.

REFERENCES

1. REUTER, H. & N. SEITZ. 1968. The dependence of calcium efflux from cardiac muscle on temperature and external ion composition. J. Physiol. (London) **195:** 451-470.
2. GLITSCH, H. G., H. REUTER & H. SCHOLZ. 1970. The effect of the internal sodium concentration on calcium fluxes in isolated guinea-pig auricles. J. Physiol. (London) **204:** 25-43.

3. ALLEN, T. J. A., D. NOBLE & H. REUTER, EDS. 1989. Sodium-Calcium Exchange. Oxford University Press. Oxford, England.
4. MULLINS, L. J. 1979. The generation of electric current in cardiac fibers by Na/Ca exchange. Am. J. Physiol. **236:** C103-C110.
5. HUME, J. R. & A. UEHARA. 1986. "Creep currents" in single frog atrial cells may be generated by electrogenic Na/Ca exchange. J. Gen. Physiol. **87:** 857-884.
6. KIMURA, J., S. MIYAMAE & A. NOMA. 1987. Identification of sodium-calcium exchange current in single ventricular cells of guinea-pig. J. Physiol. (London) **384:** 199-222.
7. MECHMAN, S. & L. POTT. 1986. Identification of Na-Ca exchange current in single cardiac myocytes. Nature **319:** 597-598.
8. NOBLE, D. 1986. Sodium-calcium exchange and its role in generating electric current. *In* Cardiac Muscle: The Regulation of Excitation and Contraction. R. D. Nathan, Ed.: 171-200. Academic Press. New York.
9. BARCENAS-RUIZ, L., D. J. BUEKELMANN & W. G. WIER. 1987. Sodium-calcium exchange in heart: Currents and changes in Ca_i. Science **238:** 1720-1722.
10. MARBAN, E. & G. WIER. 1985. Ryanodine as a tool to determine the contributions of calcium entry and calcium release to the calcium transient and contraction of cardiac Purkinje fibers. Circ. Res. **56:** 133-138.
11. CANNELL, M. B., D. A. EISNER, W. J. LEDERER & M. VALDEOLMILLOS. 1986. The effects of membrane potential on intracellular calcium concentration in sheep cardiac Purkinje fibers in Na-free solutions. J. Physiol. **381:** 193-203.
12. CRESPO, L. M., C. J. GRANTHAM, & M. B. CANNELL. 1990. Kinetics, stoichiometry and role of the Na-Ca exchange mechanism in isolated cardiac myocytes. Nature **345:** 618-621.
13. CANNELL, M. B., J. R. BERLIN & W. J. LEDERER. 1987 Effect of membrane potential changes on the calcium transient in single rat cardiac muscle cells. Science **238**:1419-1423.
14. MITRA, R. & M. MORAD. 1985. A uniform enzymatic method for dissociation of myocytes from hearts and stomachs of vertebrates. Am. J. Physiol. **249:** H1056-H1060.
15. CANNELL, M. B. & W. J. LEDERER. 1986. An experimental chamber for single-cell voltage-clamp experiments with temperature and flow control and low electrical noise. Pflügers Arch. **406:** 536-539.
16. HAMILL, O. P., A. MARTY, E. NEHER, B. SAKMANN & F. J. SIGWORTH. 1981. Improved patch-clamp techniques for high-resolution current recording from cells and cell-free membrane patches. Pflügers Arch. **391:** 85-100.
17. GRYNKIEWICZ, G., M. POENIE & R. Y. TSIEN. 1985. A new generation of Ca^{2+} indicators with greatly improved fluorescence properties. J. Biol. Chem. **260:** 3440-3450.
18. WILLIAMS, D. A., K. E. FOGARTY, R. Y. TSIEN & F. S. FAY. 1985. Calcium gradients in single smooth muscle cells revealed by the digital imaging microscope using fura-2. Nature **318:** 558-561.
19. ISENBERG, G. & U. KLOCKNER. 1982. Calcium currents in isolated bovine ventricular myocytes are fast and of large amplitude. Pflügers Arch. **395:** 30-41.
20. BEUCKELMAN, D. J. & W. G. WIER. 1988. Mechanism of release of Ca^{2+} from the sarcoplasmic reticulum of guinea-pig cardiac cells. J. Physiol. **405:** 233-255.
21. CALLEWAERT, G., L. CLEEMAN & M. MORAD. 1988. Epinephrine enhances Ca^{2+} current-regulated Ca^{2+} release and Ca^{2+} reuptake in rat ventricular myocytes. Proc. Natl. Acad. Sci. USA **85:** 2009-2013.
22. LONDON, B. & J. W. KREUGER. 1986. Contraction in voltage-clamped internally perfused single heart cells. J. Gen. Physiol. **88:** 475-505.
23. FABIATO, A. 1985. Simulated calcium current can both cause calcium loading in and trigger calcium release from the sarcoplasmic reticulum of a skinned canine cardiac Purkinje cell. J. Gen. Physiol. **85:** 291-320.
24. MELZER, W., E. RIOS & M. F. SCHNEIDER. 1987. A general procedure for determining the rate of calcium release from the sarcoplasmic reticulum in skeletal muscle fibers. Biophys. J. **51:** 849-864.
25. MEISSNER, G. 1986. Ryanodine activation and inhibition of the Ca^{2+} release channel of sarcoplasmic reticulum. J. Biol. Chem. **261:** 6300-6306.
26. PITTS, B. J. R. 1979. Stoichiometry of sodium-calcium exchange in cardiac sarcolemmal vesicles. J. Biol. Chem. **254:** 6232-6235.

27. REEVES, J. P. & C. C. HALE. 1984. The stoichiometry of the cardiac sodium-calcium exchange system. J. Biol. Chem. **259:** 7733-7739.
28. EHARA, T., S. MATSUOKA & A. NOMA. 1989. Measurement of reversal potential of Na-Ca exchange current in single guinea-pig ventricular cells. J. Physiol. (London) **410:** 227-249.
29. CARONI, P. & E. CARAFOLI. 1981. The Ca^{2+}-pumping ATPase of heart sarcolemma. J. Biol. Chem. **256:** 3263-3270.
30. EISNER, D. A. & W. J. LEDERER. 1985. Na-Ca exchange: Stoichiometry and electrogenicity. Am. J. Physiol. **248:** C189-C202.
31. SLAUGHTER, R. S., J. L. SUTKO & J. P. REEVES. 1983. Equilibrium calcium-calcium exchange in cardiac sarcolemmal vesicles. J. Biol. Chem. **258:** 3183-3190.
32. NIGGLI, E. & W. J. LEDERER. 1991. Molecular operations of the sodium-calcium exchanger revealed by conformation currents. Nature **349:** 621-567.
33. BRIDGE, J. H. B., K. W. SPITZER & P. R. ERSHLER. 1988. Relaxation of isolated ventricular cardiomyocytes by a voltage dependent process. Science **241:** 823-825.
34. BERS, D. M. & J. H. B. BRIDGE. 1989. Relaxation of rabbit ventricular muscle by Na-Ca exchange and sarcoplasmic reticulum pump. Circ. Res. **65:** 334-342.
35. DI FRANCESCO, D. & D. NOBLE. 1985. A model of electrical cardiac electrical activity incorporating ionic pumps and concentration changes. Phil. Trans. R. Soc. B **307:** 353-398.
36. MULLINS, L. J. 1977. A mechanism for Na/Ca transport J. Gen. Physiol. **70:** 681-695.
37. MATHIAS, R. T., I. S. COHEN & C. OLIVA. 1990. Limitations of the whole-cell patch-clamp technique in the control of intracellular concentrations. Biophys. J. **58:** 759.
38. FRY, C. H. & D. P. HARDING. 1989. Homeostasis of calcium ions in calcium-tolerant myocytes. *In* Isolated Adult Cardiomyocytes. Vol. 1: 99-122. H. M. Piper & G. Isenberg, Eds. CRC Press Inc. Boca Raton, FL.
39. HILGEMANN, D. W. 1990. Regulation and deregulation of cardiac Na^{+}-Ca^{2+} exchange in giant excised sarcolemmal membrane patches. Nature **344:** 242-245.
40. GRANTHAM, C. J. & M. B. CANNELL. 1990. Regulation of resting calcium by Na-Ca exchange in guinea pig cardiac myocytes. Biophys. J. **57:** 12a.

The Contribution of Na-Ca Exchange to Relaxation in Mammalian Cardiac Muscle [a]

S. C. O'NEILL, M. VALDEOLMILLOS,[b] C. LAMONT, P. DONOSO,[c] AND D. A. EISNER

Department of Veterinary Preclinical Sciences
University of Liverpool
Liverpool L69 3BX, United Kingdom

INTRODUCTION

It is by now well established that contraction in cardiac muscle is initiated by an increase of intracellular calcium concentration ($[Ca^{2+}]_i$). There are two sources of this Ca^{2+}: (i) from the extracellular fluid via voltage-sensitive Ca^{2+} channels; and (ii) by release from the sarcoplasmic reticulum (SR). Available evidence suggests that in most mammalian cardiac preparations (ii) is quantitatively more important than (i).[1] For the heart to be in a steady state, it is important that, following each beat, the Ca released from the sarcoplasmic reticulum is returned to it (by the Ca-ATPase) and, correspondingly, that Ca which entered from outside the cell is pumped back out of the cell. If more is pumped out than enters, the cell will gradually be depleted of calcium and contraction will cease. Conversely, if less is pumped out than enters, the cell will gain calcium and the toxic manifestations of "calcium overload"[2] will ensue. Therefore the relative calcium-pumping abilities of the SR and surface membrane determine the degree of calcium loading of the SR and thereby the inotropic state of the heart.

Two mechanisms have been described for removing Ca ions from the cell: the Na-Ca exchange[3] (see Eisner & Lederer[4] for review) and the Ca-ATPase.[5] The relative magnitudes of these two processes are still uncertain. In this paper we describe experiments that provide an estimate of the relative calcium pumping by the surface membrane and the SR and, in addition, show the importance of Na-Ca exchange.

METHODS

The experiments were performed on rat ventricular myocytes. The cells were isolated using collagenase and protease by methods published previously.[6] One variation employed was that in some experiments, at the end of the enzymatic digestion, the

[a] The work described in this article was supported by grants from the British Heart Foundation and The Wellcome Trust.

[b] Current address: Departmento de Fisiologia, Facultad de Medicina, Universidad de Alicante, Alicante, Spain.

[c] Current address: Departmento Preclinicas, Facultad de Medicina, Universidad de Chile, Arda Salvador 486, Santiago 9, Chile.

hearts were perfused with a nominally Ca-free solution in which 50 mM taurine replaced 25 mM NaCl. Perfusion with taurine improved the calcium tolerance and the survival of the isolated cells.[7] The cells were loaded with either fura-2 or Indo-1 by loading with the appropriate acetoxymethyl (AM) ester. They were then put in a superfusion bath on the stage of an epifluorescence microscope. A single cell was centered in the field and only the light emitted from that cell was measured. In the majority of experiments Indo-1 was used to measure $[Ca^{2+}]_i$. In these experiments the fluorescence was excited at 340 nm and emission measured at both 400 and 500 nm. In a few experiments fura-2 was used to measure $[Ca^{2+}]_i$. In this case, excitation was at 340 and 380 nm and emission was measured at 500 nm. The apparatus used to measure fluorescence from Indo-1 or fura-2 loaded cells has been described previously.[6,8]

The experimental solution contained (mM): NaCl, 134; KCl, 4; $MgCl_2$, 1; HEPES, 10; $CaCl_2$, 1; glucose, 10; titrated to pH 7.4 with NaOH. In some experiments external Na was removed completely and replaced with Li. All statistics in this paper are presented as mean ± SEM.

In experiments such as that illustrated in FIGURE 4, the total Ca efflux from a Langendorff-perfused rat heart was measured. In these experiments, after reaching a steady-state, external calcium was removed. The calcium concentration in the perfusate leaving the heart was measured and the excess over the contaminant calcium in the fluid entering the heart gave a measure of the net calcium efflux from the heart. The effects of caffeine were studied on this efflux both in Na-containing and in Na-free solutions.

RESULTS

The Contribution of the SR to Ca Removal from the Cytoplasm

In these experiments we have studied the relative extent to which the sarcoplasmic reticulum and the surface membrane contribute to removing Ca^{2+} ions from the cytoplasm. The principle behind the experiments was to add caffeine to make the SR leaky to Ca ions. Under these conditions the SR should not contribute to Ca removal. A similar method has been used before.[9,10] FIGURE 1A shows a record of Indo-1 fluorescence from an isolated rat ventricular myocyte. At the start of the record, the cell was stimulated electrically. This produces the systolic Ca transients shown. Stimulation was then discontinued and, after a pause of a few seconds, caffeine was added. This produced an increase of $[Ca^{2+}]_i$ of comparable magnitude to that produced by electrical stimulation.[8,11] The similarity of the magnitude of the electrically stimulated and the caffeine responses is consistent with caffeine and electrical stimulation releasing comparable quantities of Ca from the SR. One obvious difference between the electrically stimulated and the caffeine responses is that the former relaxes much more quickly than the latter. This is emphasised in FIGURE 1B, which shows the two responses replotted on different time-scales with exponentials fitted through them. In this experiment the rate constants were 4.9 and 0.4 sec^{-1}. Mean values for the decay of the caffeine and electrically stimulated Ca transients are given in TABLE 1.

The Contribution of Na-Ca Exchange to the Removal of Ca^{2+} Ions

This series of experiments was designed to investigate what fraction of the SR-independent fraction of Ca removal could be attributed to Na-Ca exchange. FIGURE 2

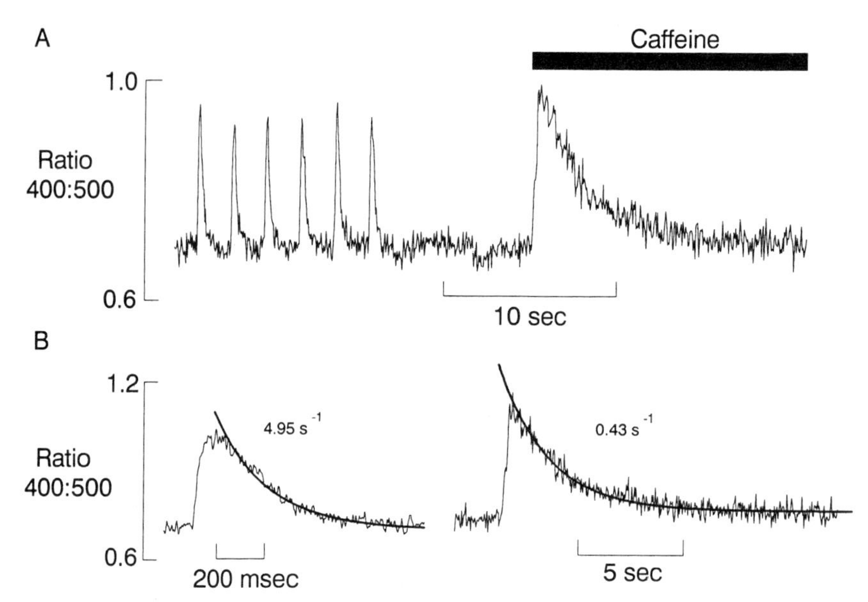

FIGURE 1. Comparison of the electrically stimulated and caffeine responses. (**A**) Original record. The trace shows the ratio of the fluorescence emitted by Indo-1 at 400 nm to that at 500 nm. Six electrically stimulated Ca transients are shown followed by (in the absence of stimulation) the addition of 10 mM caffeine. (**B**) Replotted records of: *left,* the electrically stimulated Ca transient (average of 8); *right,* the caffeine response. Note the difference in time scale. The smooth curves show single exponentials fitted to the data with rate constants: 4.9 sec^{-1}, electrically stimulated; 0.43 sec^{-1}, caffeine.

TABLE 1. Rate Constants for Decay of Electrically Stimulated and Caffeine-Evoked Calcium Transients

		Rate Constant (sec^{-1})	
Isolation Method	*n*	Electrically Stimulated	Caffeine
No taurine	5	3.87 ± 0.37	0.33 ± 0.05
Taurine	6	4.38 ± 0.32	0.89 ± 0.08

NOTE: The rate constants were obtained by least-squares fits. An unpaired *t*-test shows that the caffeine transients from the cells prepared in taurine decay faster than the controls ($p < 0.001$), whereas there is no significant difference between the electrically stimulated transients in the two sets of cells ($p > 0.1$).

shows the effects of adding caffeine (10 mM) under various conditions. The control trace shows that this produced a rapid rise of $[Ca^{2+}]_i$ which decayed quickly. After allowing recovery from the effects of caffeine, external Na was removed in order to inhibit the Na-Ca exchange. The addition of caffeine in the absence of external Na ions produced a rise of $[Ca^{2+}]_i$ that was of comparable magnitude to the control but

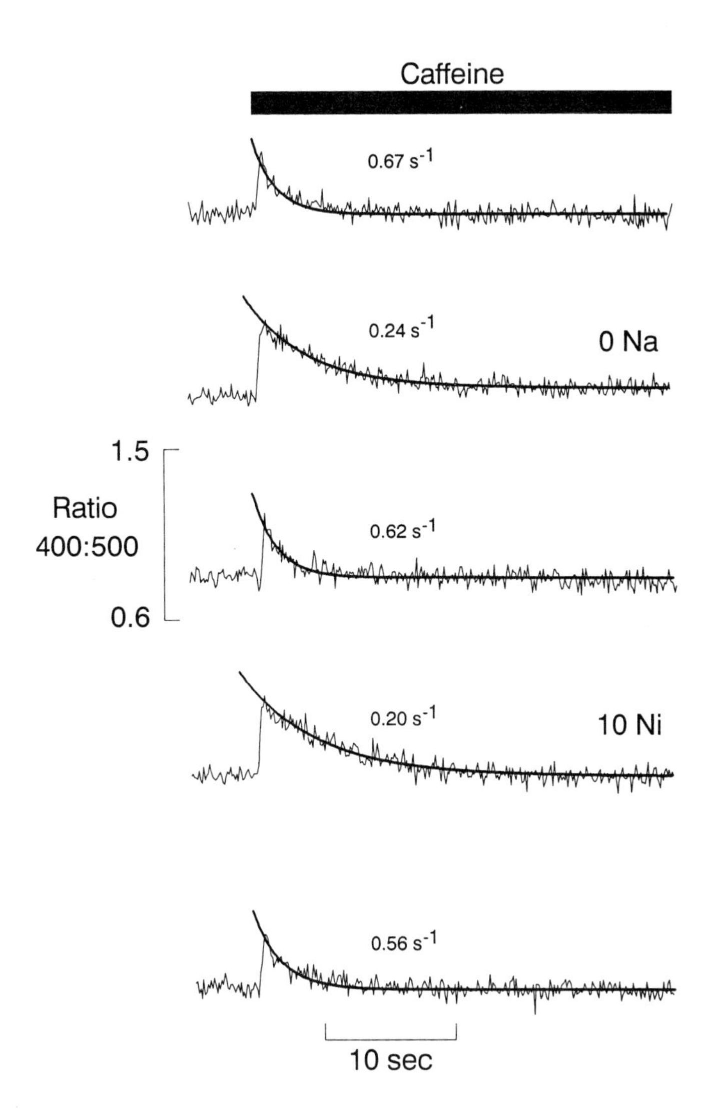

FIGURE 2. The effects of inhibition of Na-Ca exchange on the caffeine response. Caffeine was applied as shown in each panel. Before each application of caffeine, the cell was first exposed for 10 min to the control solution. A Ca-free solution was then applied for 1 min with either Na removed or 10 mM Ni^{2+} added as appropriate. Finally caffeine (10 mM) was added as shown. The continuous curves through the traces are fitted exponentials with rate constants as indicated. The experiments were carried out in calcium-free solutions so that the Na removal did not induce a large Ca entry into the cell. Control experiments (not shown) demonstrated that Ca removal had no effect on the caffeine response.

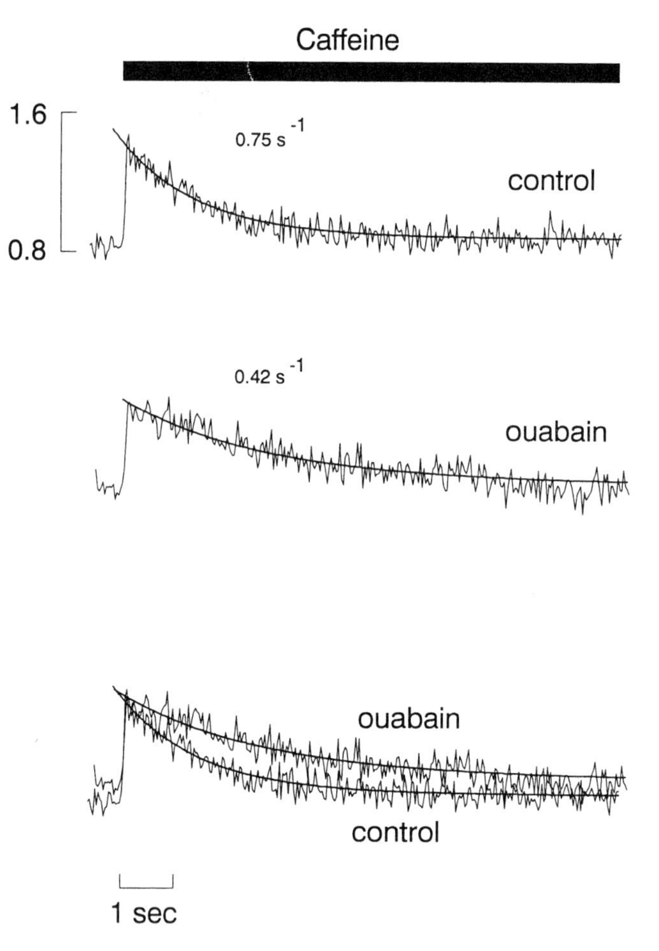

FIGURE 3. The effects of ouabain on the caffeine response. The top two traces show the effects of adding caffeine (10 mM) on $[Ca^{2+}]_i$. The top trace was obtained under control conditions and the lower trace 30 min after adding ouabain (1 mM). The smooth lines through the data are single exponentials with the indicated rate constants. The bottom panel shows the two traces superimposed.

decayed more slowly. The effect of Na removal is completely reversible as shown by the recontrol. The next panel shows the effects of another method of inhibiting Na-Ca exchange—addition of Ni^{2+}ions.[12] Again this slows the rate of fall of $[Ca^{2+}]_i$. On average we found that the rate of decay of the caffeine response was reduced to 0.48 $\pm$ 0.04 (n = 4) of the control rate in a Na-free solution and to 0.50 $\pm$ 0.08 (n = 5) of the control in a solution to which 10 mM Ni^{2+} had been added.

Na removal completely abolishes Na-Ca exchange. We have also investigated the effects of modifying the exchange by elevating the intracellular sodium concentration ($[Na^+]_i$). FIGURE 3 shows the effects of inhibiting the Na-K pump with ouabain. This decreases the rate constant of recovery of $[Ca^{2+}]_i$. It should also be noted that, although ouabain has a significant effect on the rate constant of decay of $[Ca^{2+}]_i$, there is little effect on the magnitude of the increase of $[Ca^{2+}]_i$ produced by caffeine. This contrasts with work on ferret papillary muscles where inhibition of the Na-K pump greatly increased the magnitude of the caffeine response.[11] This suggests that, in the rat, the SR is maximally loaded with calcium before ouabain is added.

The method of cell isolation was also found to have marked effects on the rate constant of decay of the caffeine response. TABLE 1 shows that the caffeine response decays more quickly in cells prepared with taurine in the solution than in cells not exposed to taurine. There are many possible explanations for this observation. However, the fact that the electrically stimulated $[Ca^{2+}]_i$ transient is not significantly affected by taurine suggests that the difference does not arise from, for example, differences in the loading with the calcium indicator. An alternative attractive explanation is based on the observation that taurine-loaded cells have a lower $[Na^+]_i$ than do controls.[13] As shown by the experiment of FIGURE 2, such a decreased $[Na^+]_i$ would account for the acceleration of the fall of $[Ca^{2+}]_i$.

Measurements of Calcium Efflux

The experiments described above show that a significant fraction of the decay of $[Ca^{2+}]_i$ following exposure to caffeine is not mediated by Na-Ca exchange. Two possible mechanisms could account for that fraction of the Ca removal which is not carried out by Na-Ca exchange. (i) Calcium could be pumped out of the cell by a process other than the Na-Ca exchange, and (ii) calcium could be sequestered by intracellular

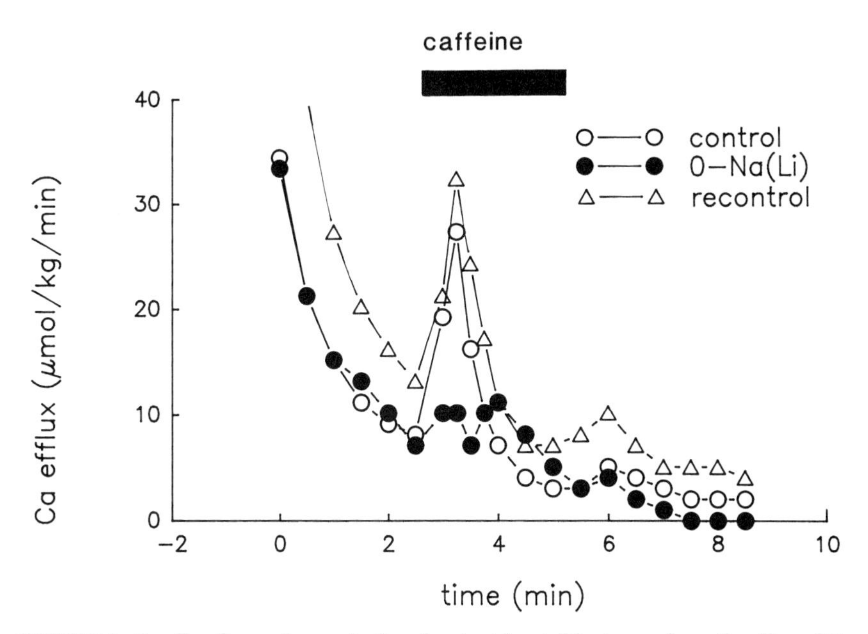

FIGURE 4. Ca efflux from a Langendorff-perfused rat heart. The traces shows the efflux of Ca. At the time indicated caffeine (10 mM) was added. This was done, on the same heart, in the following conditions: (○) control; (●) Na-free; (△) recontrol.

mechanisms. One way to distinguish between these possibilities is by measuring the efflux of Ca^{2+} ions (*cf.* Jundt *et al.*[14] and Barry *et al.*[15]). FIGURE 4 shows the efflux curve from a Langendorff-perfused rat heart. The control trace shows that the addition of caffeine produces a transient increase of Ca efflux. In the Na-free solution the Ca efflux produced by caffeine is not discernable above the background. This result shows therefore (in agreement with some[14] but not all[15] earlier work) that Na removal abolishes the vast majority of the Ca efflux produced by caffeine.

The Effects of Membrane Potential

It is by now well established that Na-Ca exchange in cardiac muscle is voltage dependent. We have therefore looked to see whether the fall of $[Ca^{2+}]_i$ during exposure to caffeine is sensitive to membrane potential. FIGURE 5 shows results from such an experiment. The data show that depolarization does slow Ca extrusion. However the effect is only to halve the rate for a 100-mV depolarization (see discussion).

DISCUSSION

In this paper we have investigated the extent to which various mechanisms contribute to the removal of Ca^{2+} ions from the cytoplasm. In agreement with previous studies[9] the sarcoplasmic reticulum was shown to be the most powerful Ca-removal mechanism. This can be seen by comparing the rates of decay of the electrically stimulated $[Ca^{2+}]_i$ transient with that of the caffeine response. Ca removal in the former is presumably produced by the combined efforts of the SR and other mechanisms while in the latter case only the other mechanisms contribute. If we assume that all the processes are first order, then

$$d[Ca^{2+}]_i/dt = (k_{SR} + k_{other})[Ca^{2+}]_i$$

where k_{SR} and k_{other} are the rate constants for removal of $[Ca^{2+}]_i$ by the SR and the other systems, respectively. It is easy to show that

$$k_{other}/(k_{SR} + k_{other}) = \text{rate of decay of caffeine response/rate of decay of stim response}$$

We found that on average the rate constants of decay of the caffeine and electrically stimulated responses were 0.89 and 4.38 sec^{-1} in taurine-treated and 0.33 and 3.87 in nontaurine-treated cells. From this one can calculate that k_{other} = 0.08 to 0.25 k_{SR} and therefore that the SR accounts for 80 to 90% of the total Ca removal.

As mentioned above, in the steady state, all the Ca that enters the cell must be pumped out across the surface membrane. This predicts that 80 to 90% of the Ca that activates contraction must come from the SR, a potentially testable hypothesis.

The experiments showed that the caffeine response was significantly slowed by either removal of external Na or the addition of Ni^{2+} ions. On average, the response was slowed to about 50% of the control value. The simplest interpretation of this is that Na-Ca exchange contributes about 50% of the SR-independent processes that remove Ca ions from the cytoplasm. It is also noteworthy that ouabain slowed the caffeine response to the same extent as did Na removal and Ni^{2+}. One might expect the rise of $[Na^+]_i$ produced by ouabain to only partially inhibit the Na-K pump in contrast to the other maneuvers that should essentially completely inhibit it. It may, however, have been difficult to resolve this difference on top of the Na-Ca exchange-independent component.

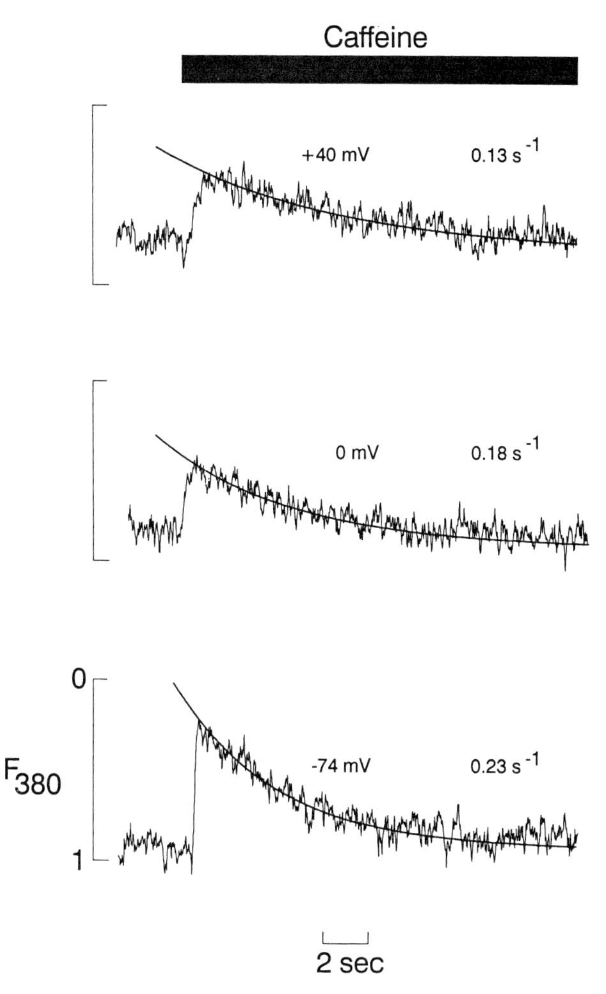

FIGURE 5. The effects of membrane potential on the caffeine response. The traces show the effects of adding caffeine at the potentials indicated: *top,* +40 mV; *middle,* 0 mV; *bottom,* −74 mV. The smooth curves are single exponentials with the rate constants indicated.

The decay of the caffeine response was slowed by depolarization. While this would be expected from a Na-Ca exchange, the magnitude of the effect of membrane potential is less than would be predicted. For example, in measurements in single cardiac cells, the Na-Ca exchange current changes *e*-fold for a 38-mV depolarization.[12] In contrast, in the present work, FIGURE 5 shows that depolarization by more than 100 mV did not quite reduce the rate constant by 50%. This relative lack of effect of membrane potential is, of course, consistent with the hypothesis that the Ca removal represents the sum of a (voltage-sensitive) Na-Ca exchange and another mechanism (presumably voltage-insensitive).

The remaining question concerns the identity of the non-SR and the non-Na-Ca exchange mechanisms that are responsible for the residual Ca removal. There are two possibilities: a surface membrane Ca-ATPase or, alternatively, sequestration in some non-SR. The sarcolemmal Ca-ATPase has been shown to produce a net Ca efflux in intact cardiac preparations.[15] The flux experiments in the present paper showed that most of the Ca efflux was mediated via Na-Ca exchange. This suggests, therefore, that the residual Ca removal from the cytoplasm is mediated via intracellular sequestration. This conclusion is consistent with the results of Kitazawa.[16]

REFERENCES

1. FABIATO, A. 1983. Am. J. Physiol. **245:** 1-14.
2. KASS, R. S., W. J. LEDERER, R. W. TSIEN & R. WEINGART. 1978. J. Physiol. (London) **281:** 187-208.
3. REUTER, H. & N. SEITZ. 1968. J. Physiol. (London) **195:** 451-470.
4. EISNER, D. A. & W. J. LEDERER. 1985. Am. J. Physiol. **248:** 189-202.
5. CARONI, P. & E. CARAFOLI. 1980. Nature **283:** 765-767.
6. EISNER, D. A., C. G. NICHOLS, S. C. O'NEILL, G. L. SMITH & M. VALDEOLMILLOS. 1989. J. Physiol. (London) **411:** 393-418.
7. ISENBERG, G. & U. KLOCKNER. 1982. Pflügers Arch. **395:** 6-18.
8. O'NEILL, S. C., P. DONOSO & D. A. EISNER. 1990. J. Physiol. (London) **425:** 55-70.
9. BERS, D. M. & J. H. BRIDGE. 1989. Circ. Res. **65:** 334-342.
10. CALLEWAERT, G., L. CLEEMANN & M. MORAD. 1989. Am. J. Physiol. **257:** 147-152.
11. SMITH, G. L., M. VALDEOLMILLOS, D. A. EISNER & D. G. ALLEN. 1988. J. Gen. Physiol. **92:** 351-368.
12. KIMURA, J., S. MIYAMAE & A. NOMA. 1987. J. Physiol. (London) **384:** 199-222.
13. CHAPMAN, R. A. & G. C. RODRIGO. 1990. J. Physiol. (London) **426:** 16P.
14. JUNDT, H., H. PORZIG, H. REUTER & J. W. STUCKI. 1975. J. Physiol. (London) **246:** 229-253.
15. BARRY, W. H., C. A. RASMUSSEN, JR., H. ISHIDA & J. H. BRIDGE. 1986. J. Gen. Physiol. **88:** 393-411.
16. KITAZAWA, T. 1988. J. Physiol. (London) **402:** 703-729.

Effects of Hypoxia and Acidification on Myocardial Na and Ca

Role of Na-H and Na-Ca Exchange

S. E. ANDERSON,[a] P. M. CALA,[a] C. STEENBERGEN,[b]
R. E. LONDON,[c] AND E. MURPHY[c]

[a]Department of Human Physiology
University of California
Davis, California 95616

[b]Department of Pathology
Duke University Medical Center
Durham, North Carolina 27710

[c]Laboratory of Molecular Biophysics
National Institute of Environmental Health Sciences
Research Triangle Park, North Carolina 27709

Until recently, increases in cell Na content during ischemic and hypoxic episodes were thought to result from impaired ATP production causing decreased Na-K ATPase activity. We report the results of testing the alternate hypothesis that hypoxia-induced Na uptake is (1) the result of increased entry, as opposed to decreased extrusion, and (2) via Na-H exchange operating in a pH regulatory capacity and that cell Ca accumulation occurs via Na-Ca exchange secondary to collapse of the Na gradient.

METHODS

We used ^{23}Na, ^{19}F, and ^{31}P NMR to measure intracellular Na content (Na_i), free Ca concentration ($[Ca]_i$), pH (pH_i), and high-energy phosphates in isolated rabbit hearts perfused with Krebs-Henseleit solution using a modified Langendorff technique.[1] Perfusates were equilibrated with $95\%O_2/5\%CO_2$ or $95\%N_2/5\%CO_2$ and pH was adjusted to 7.4 ± 0.05 at 25°C. K-free and/or ouabain (1 mM) perfusion was used to inhibit Na efflux from the cells in order to quantify changes in Na uptake. DyTTHA[2] (15 mM) was used to separate intra- and extracellular Na resonances, pH_i was calculated from the calibrated shift in Pi, and $[Ca]_i$ was calculated as $[Ca]_i = K_d$ [Ca-5FBAPTA]/[5FBAPTA] where $K_d = 500$ nM and the ratio of the Ca-bound to Ca-free 5FBAPTA concentrations is equal to the ratio of the bound to free peak areas in the ^{19}F spectrum after loading the myocardial cytoplasm with 5FBAPTA.[3,4]

RESULTS

As shown in FIGURE 1, hypoxia stimulated net Na uptake at a rate > 10 times that of normoxic controls (during the first 12.5 min, Na_i increased from 7.9 ± 5.8 to 34.9 ± 11.0 mEq/kg dry wt [mean ± SD] compared with 11.1 ± 16.3 to 13.6 ± 9.0, respectively). When normoxic hearts were acidified using a 20 mM NH_4Cl prepulse[5] (FIG. 2), pH_i rapidly fell from 7.27 ± 0.24 to 6.63 ± 0.12 but returned to 7.07 ± 0.10 within 20 min, while Na uptake was similar in rate and magnitude to that observed during hypoxia (24.5 ± 13.4 to 132.1 ± 17.7 mEq/kg dry wt). During hypoxia and after NH_4Cl washout, increases in $[Ca]_i$ were similar in time course to those observed for Na_i. Hypoxic Na_i uptake was insensitive to benzamil (50 μM) a selective inhibitor of Na-Ca exchange and conductive Na channels as well as bumetanide (10 μM) a selective inhibitor of Na + K + 2Cl cotransport. On the other hand, during hypoxia and after normoxic acidification, Na uptake as well as pH_i regulation (after NH_4Cl washout) and increases in $[Ca]_i$ were inhibited by amiloride (1mM) and ethylisopropylamiloride (EIPA, 100 μM), a specific inhibitor of Na-H exchange. EIPA and amiloride also decreased changes in coronary resistance and phosphocreatine measured after 60 min of hypoxic perfusion ($p < 0.05$).

These results are consistent with the hypothesis that hypoxia stimulates an increase in myocyte Na uptake (independent of changes in Na-K ATPase activity) and that this increase appears to be via pH regulatory Na-H exchange. Furthermore, changes in $[Ca]_i$ are similar in time course to changes in Na_i and are inhibited by selective inhibition of Na-H exchange, suggesting that hypoxic cell Ca accumulation occurs via Na-Ca exchange after collapse of the membrane Na gradient through Na-H exchange.

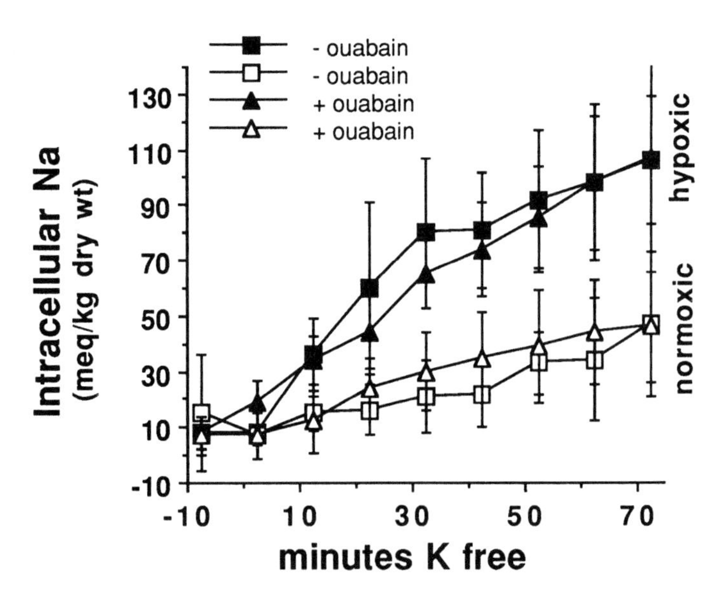

FIGURE 1. Perfused heart intracellular Na (means ± SD) during hypoxia (■, ▲) and normoxia (□,△) with (triangles) and without (squares) 1 mM ouabain. Mean Na uptake during the first 12.5 min of hypoxic K-free perfusion is > 10 times greater than during normoxic K-free perfusion. (Reprinted from Anderson *et al.*[1] with permission from the American Physiological Society).

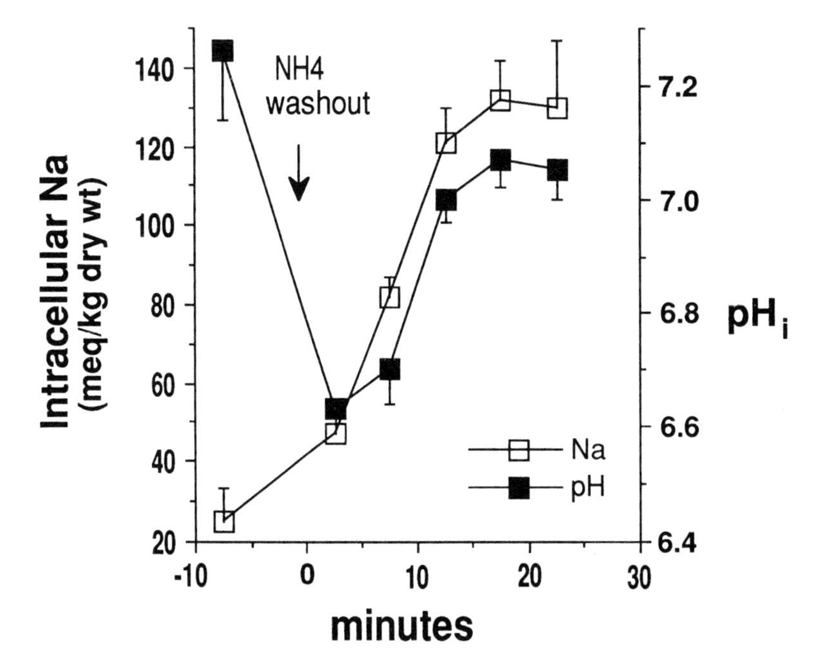

FIGURE 2. Na uptake and intracellular pH (pH_i) regulation after normoxic intracellular acidification. Intracellular Na (□) and pH (■) (means ± SE) are plotted on the left and right ordinates, respectively, versus minutes after NH_4Cl washout. These data support the hypothesis that decreased pH_i stimulates myocyte Na uptake via pH regulatory Na-H exchange. (Reprinted from Anderson *et al.*[1] with permission from the American Physiological Society.)

REFERENCES

1. ANDERSON, S. A., E. MURPHY, C. STEENBERGEN, R. E. LONDON & P. M. CALA. 1990. Am. J. Physiol. **259**(Cell Physiol. 28): C940-C948.
2. PIKE, M. M., J. C. FRAZER, D. F. DEDRICK, J. S. INGWALL, P. D. ALLEN, C. S. SPRINGER, JR. & T. W. SMITH. 1985. Biophys. J. **48:** 159-173.
3. STEENBERGEN, C., E. MURPHY, L. LEVY & R. E. LONDON. 1987. Circ. Res. **60:** 700-707.
4. KIRSCHENLOHR, H. L., J. C. METCALFE, P. G. MORRIS, G. C. RODRIGO & G. A. SMITH. 1988. Proc. Natl. Acad. Sci. USA **85:** 9017-9021.
5. BORON, W. F. & P. DEWEER. 1976. J. Gen. Physiol. **67:** 91-112.

Effects of Transient Changes in Membrane Potential on Twitch Force in Ferret Papillary Muscle

Possible Effects on Na-Ca Exchange

P. ARLOCK,[a] B. WOHLFART,[b] AND
M. I. M. NOBLE[c]

[a]*Department of Animal Physiology*
[b]*Department of Clinical Physiology*
University of Lund
Lund, Sweden

[c]*Academic Unit of Cardiovascular Medicine*
Charing Cross & Westminster Medical School
London, England

Force in relation to membrane potentials was studied in voltage-clamped ferret papillary muscles at 37°C using the single sucrose gap technique. This method allows accurate force measurements during voltage clamp. The preparation was basically stimulated with rectangular voltage pulses at a rate of 1.0 Hz until steady state. The holding membrane potential was −70 mV or −40 mV and the preparation was depolarized to +20 mV for 200 msec. A test clamp denoted "1" consisted of an initial depolarizing step of 20 msec duration to +20 mV in order to activate the initial part of the second inward current ($I_{Ca,f}$)[1] and was followed by a second voltage step to different potentials (−25 mV- +25 mV for 0.02-3.0 sec). Test clamp 1 was followed by a second "2" and a third "3" test clamp pulse of control period amplitude (+20 mV) and duration (200 msec).

RESULTS AND CONCLUSIONS

Variations in test clamp 1 produced different degrees of contractile potentiation of test contractions 2 and 3 (F2 and F3, FIG. 1a) Prolongation of the test clamp caused potentiation or depression of F2 depending on the clamp voltage.[2] These effects were not seen until the preceding clamp pulse was more than approximately 150 msec (i.e., the duration of an action potential). But obviously calcium is entering or leaving the cell during test clamp 1, as is seen on the profound effects of contractility in F2 (and F3 not shown). It is likely at these long durations of the test-clamp potential that the Na-Ca exchange is mainly responsible for the calcium homeostasis. At an estimated clamp voltage of the preceding clamp step (V1) of about −18 mV, there was neither potentiation nor depression of F2. This suggests that at −18 mV there is an equilibrium for the exchanger (FIG. 1b).

There was a straight relation between peak force of contraction F3 and peak force of contraction F2. At Vh −70 mV the F3-F2 curve shifted downwards in a more or less parallel manner compared to the curve obtained at −40 mV. Steady-state force at

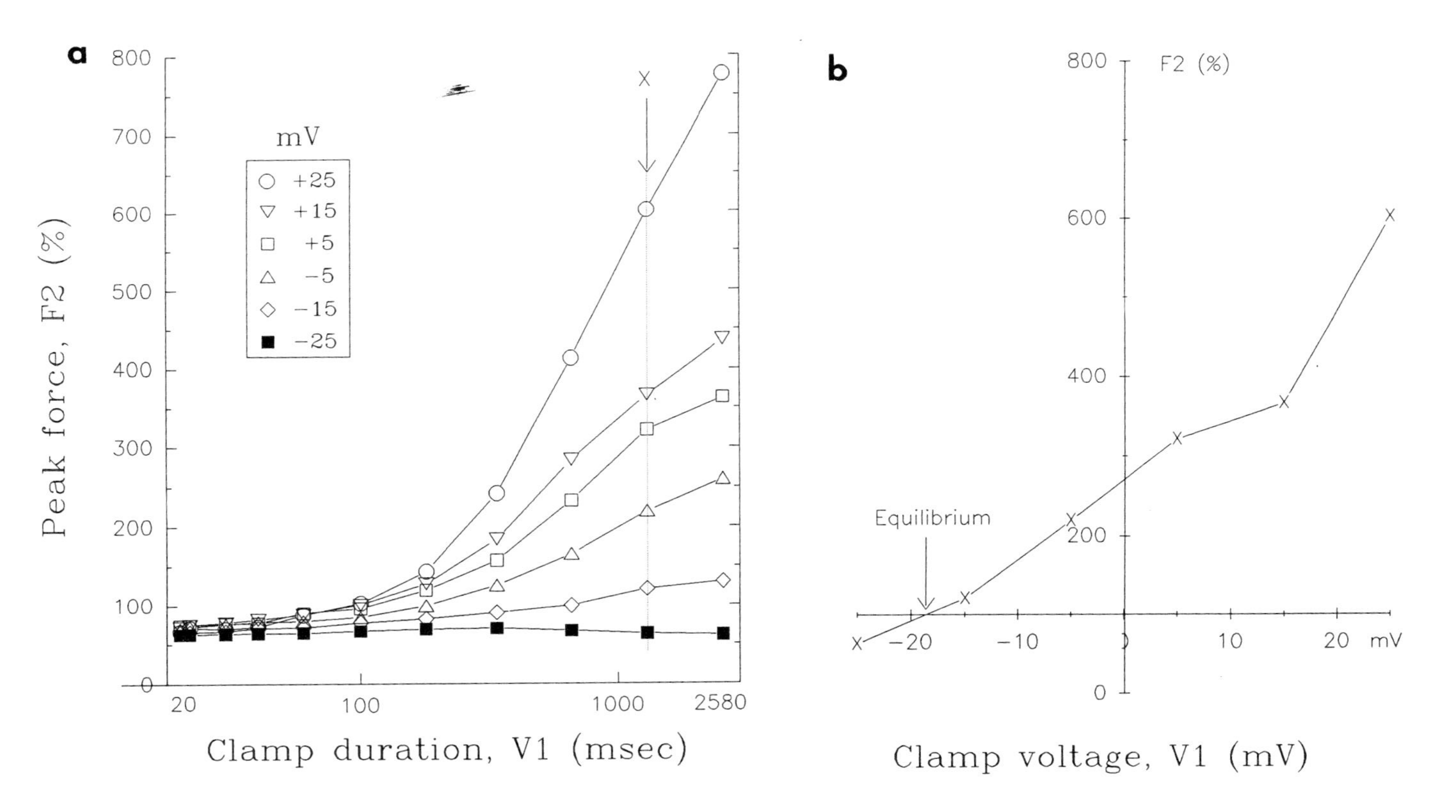

FIGURE 1. Peak force of second test contraction (F2) related to previous cycle voltage-clamp duration/voltage.

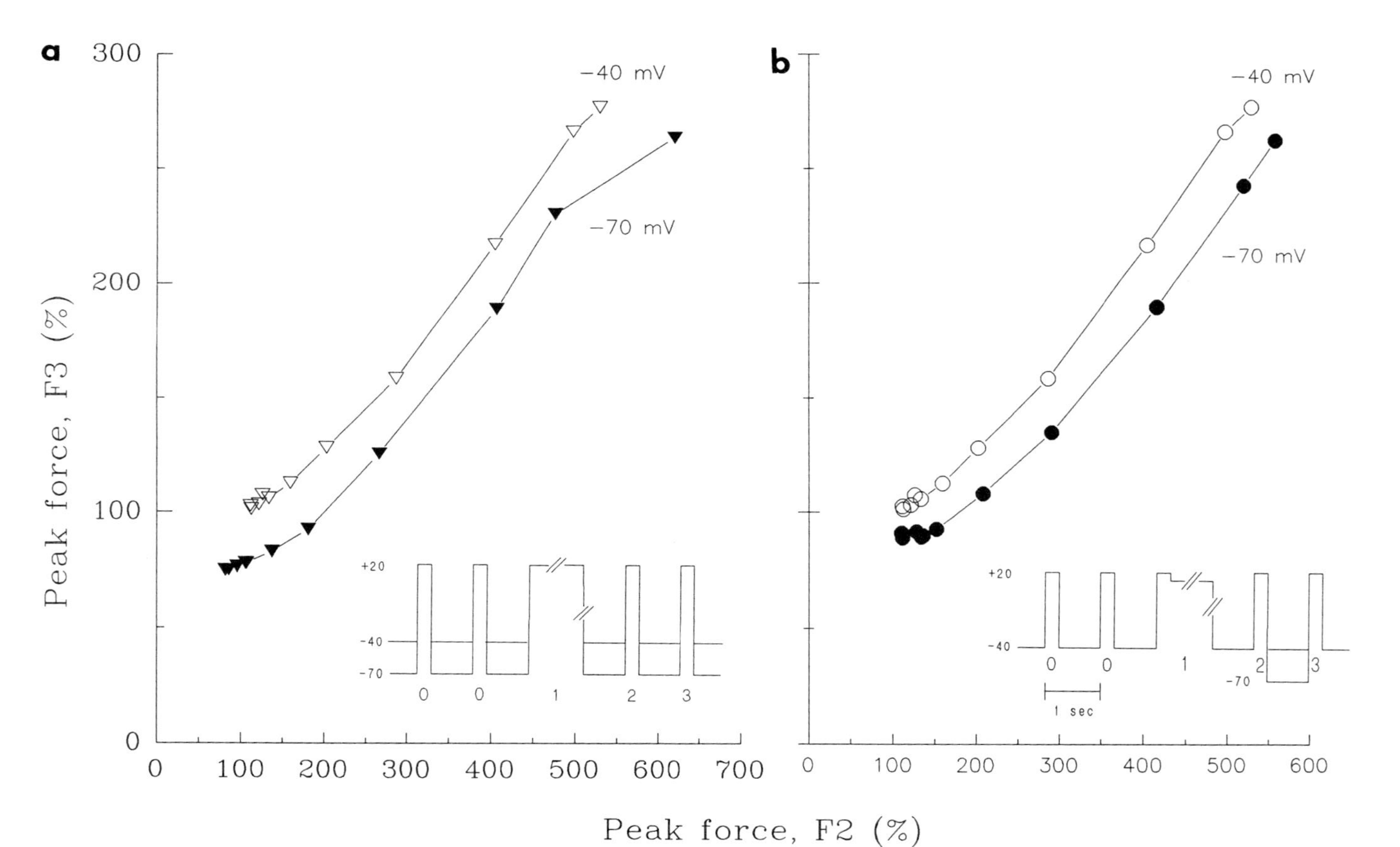

FIGURE 2. F3 versus F2 at two holding potentials and test diastolic potentials.

−70 mV was less than at −40 mV (FIG. 2a). Also at Vh −40 mV when test diastolic interval between test clamp 2 and 3 was −70 mV, the curve shifted downwards compared to a test diastolic interval of −40 mV (FIG. 2b). The straight relation between F3 and F2 could be explained by recirculation of calcium within the cardiac cell being a fraction of the amount of calcium released in the potentiated contractions. The downward shift at a more negative diastolic interval would follow from an increased outflow of calcium via the Na-Ca exchange caused by a greater driving force. The unchanged *slope* and *shape* of the F3-F2 relation suggests that recirculation fraction was unaltered. This means that the recirculation mainly takes place during the time course of an ordinary action potential.

REFERENCES

1. ARLOCK, P. & D. NOBLE. 1985. J. Physiol. **369:** 88P.
2. ARLOCK, P. & B. WOHLFART. 1990. Acta Physiol. Scand. **140:** 63-72.

Contractile Properties of Isolated Ventricular Myocytes from the Spontaneously Hypertensive Rat[a]

PAUL BROOKSBY AND ALLAN LEVI

Department of Physiology
University of Bristol
Bristol, BS8 1TD, United Kingdom

JOHN VANN JONES

Department of Cardiology
Bristol Royal Infirmary
Bristol, BS8 1TD, United Kingdom

Hypertension is a common problem in Western society. It is well known that in hypertension, heart muscle hypertrophies in response to an increased afterload. However, a crucial, as yet unanswered question is whether the individual cells from the hypertrophied myocardium have an increased contraction. It is this question that we set out to investigate in our present study.

We have compared the contractile properties of myocytes isolated from the left ventricle of the spontaneously hypertensive rat (SHR) with those from its normotensive control, the Wistar Kyoto rat (WKY). The SHR is regarded as the best animal model of genetic hypertension in man.[1]

METHODS

The animals (four SHR and four WKY) were compared at around 60 days of age. Ventricular myocytes were isolated enzymatically in the same way for each animal using collagenase and protease.

The resting cell length and width of 20 myocytes from each animal were measured with a graticule at a $\times 450$ magnification. The cells were superfused with Tyrode's solution containing 1.5 mM Ca; temperature was held constant at 32°C. The cells were stimulated at a baseline rate of 1 Hz using platinum field electrodes; stimulation rate was altered to 0.3, 2, and 3 Hz. Contraction was measured optically with a photodiode array.

[a] The support of the Wellcome Trust, British Heart Foundation, and the British Royal Infirmary Research Fund is gratefully acknowledged.

TABLE 1. Contractile and Morphological Characteristics of SHR and WKY Myocytes

Characteristics	SHR (n = 14)	WKY (n = 13)
Cell length (μm)	113.2 (2.5)	94.7 (2.7)
Cell width (μm)	22.6 (0.5)	20.3 (0.7)
Peak contraction (μm)	8.1 (0.7)	4.8 (0.3)
TTP (msec)	68.5 (3.9)	66.7 (3.6)
V_{max} (μm/msec)	205 (21.2)	107 (17.1)
1/2RT (msec)	44.1 (3.7)	48.4 (3.1)

NOTE: Data on contraction is taken from cells stimulated at 1 Hz. Mean (± SEM).

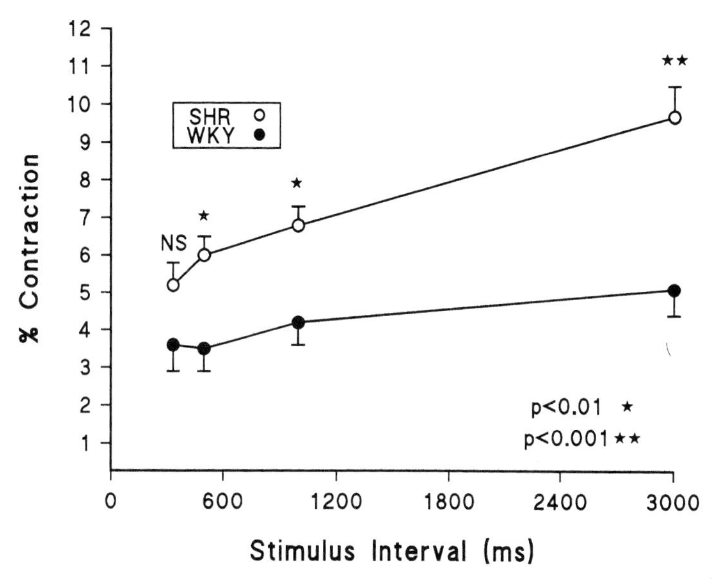

FIGURE 1. The dependence of contraction on stimulus interval for SHR and WKY myocytes. For each interval contraction is larger in the SHR myocytes. Data from the two groups was compared using Student's *t*-test.

RESULTS

The SHR cells were significantly longer and wider than WKY cells (TABLE 1). FIGURE 1 shows a graph of stimulus interval plotted against percent contraction (twitch shortening expressed as a percentage of cell length). It can be seen that with increasing stimulus interval there is an increase in contraction—typical of rat ventricular myocytes. Contraction of SHR myocytes was larger at all frequencies tested.

We analyzed three aspects of the time course of contraction: the time to peak contraction (TTP), the maximal velocity of shortening (V_{max}), and an index of relaxation—the one-half relaxation time (1/2RT), which is the time for the cell to relax from maximal shortening to one-half maximal shortening. As we found no difference in the TTP, this implied that the rate of shortening should be increased in the SHR, indeed this was found to be the case (TABLE 1). We found that the 1/2RT was reduced in the SHR (TABLE 1). These changes in time course were found at all frequencies tested.

CONCLUSION

We have, for the first time, shown that hypertrophied ventricular myocytes from the SHR have a larger contraction in comparison to age-matched normotensive controls, even when the increased cell length is taken into account. We have also shown an increase in the speed of shortening and relaxation, as indicated by the reduced relaxation time in the SHR. At present the cause of this increased contraction in the SHR is unknown. We do, though, have preliminary evidence that it is not due to an increased calcium current. We plan to investigate other possible causes, such as an alteration in the myofilament sensitivity to Ca or increased sarcoplasmic reticulum loading, perhaps due to Na-Ca exchange. Our results indicate that cardiac hypertrophy associated with hypertension, rather than being a pathological process, is, at least in the early stages, a protective adaptation allowing the heart to cope with the increased blood pressure.

REFERENCE

1. FROHLICH, E. D. 1986. J. Hypertension **4**(Suppl. 3): 47-49.

Evidence for the Beat-Dependent Activation of the Na-Ca Exchanger by Intracellular Ca[a]

ROBERT A. HAWORTH, ATILLA B. GOKNUR, AND DOUGLAS R. HUNTER

Department of Anesthesiology
University of Wisconsin
Madison, Wisconsin 53792

The effect of Ca on Na fluxes across the sarcolemma was investigated with isolated adult rat heart cells in suspension using ^{22}Na. The action of Ca was studied under three conditions: after Na loading induced by incubation with ouabain (1 mM) for 30 min at 37°C without Ca, and also under conditions of normal Na with Ca where the effect of electric field stimulation and KCl addition were examined.

When Na-loaded cells were exposed to 0.2 mM Ca, the rate of ^{22}Na uptake and efflux were both strongly stimulated, with no measurable change in total cell Na. The stimulation was prevented by verapamil (10 μM), at this level of Ca, and was insensitive to tetrodotoxin (25 μM). Mn was ineffective at inducing Na fluxes. When Na-loaded cells were exposed to 0.2 mM Ca for 1 min, the subsequent addition of verapamil was unable to reverse the rapid, stimulated rates of Na flux. The stimulated rates were fully reversible by incubation with EGTA for 10 min. By contrast with verapamil, either dichlorobenzamil (100 μM) or ATP depletion (incubation with rotenone [3 μM] plus FCCP [0.2μM] for 8 min) was able to both prevent and reverse the stimulation of Na fluxes by Ca. Addition of Ca to Na-loaded cells did not stimulate ^{86}Rb efflux, and the stimulation of ^{22}Na efflux required extracellular Na. From these results we conclude that entry of Ca into Na-loaded cells through Ca channels results in the activation of Na-Na exchange through the Na-Ca exchanger, and we propose that the extent of this exchange is a measure of the degree of activation of the exchanger by intracellular Ca. This Na-Na exchange activity is completely analogous to that seen in squid axon.[1]

Electrical stimulation of normal cells in suspension in the presence of 1 mM Ca resulted in a large increase of the rate of ^{22}Na uptake and efflux, with no change in total cell Na. The increased flux rates were reversed within seconds of ceasing stimulation. Stimulation had no effect on Na fluxes when Ca was removed or was replaced by Mn. In the absence of stimulation the Ca-dependent rate of ^{22}Na efflux was $< 10\%$ of its value with stimulation. The stimulation-dependent Na fluxes were inhibited by verapamil. They were also partially inhibited by tetrodotoxin, but the inhibition could be overcome by BAY K 8644 (1 μM) or by isoproterenol (1 μM). Ouabain had very little effect on stimulation-dependent Na fluxes. Depolarization of normal cells with KCl (50 mM) also showed a Ca-dependent stimulation of Na fluxes. These were also inhibited by verapamil, but not by tetrodotoxin.

[a]This work was supported by Grant #HL33652 from the National Heart, Lung, and Blood Institute.

We conclude that (1) at rest, the Na-Ca exchanger is almost completely inactive; (2) when stimulated, the exchanger becomes activated directly or indirectly by Ca that enters the cell through Ca channels; and (3) ionic fluxes through the Na-Ca exchanger during the cardiac cycle are much larger than those through channels.

REFERENCE

1. DiPolo, R. & L. Beauge. 1986. Reverse Na/Ca exchange requires internal Ca and/or ATP in squid axons. Biochim. Biophys. Acta **854:** 298-306.

Voltage Dependence and Kinetics of Na-Ca Exchange Tail Current in Rabbit Atrial Myocytes[a]

WON KYUNG HO AND YUNG E. EARM

Department of Physiology
Seoul National University College of Medicine
Seoul 110-460, Republic of Korea

In cardiac cells, sodium-calcium exchange has been considered to be important in regulation of intracellular calcium concentration as a primary mechanism of calcium extrusion. Earm *et al.*[1] recorded the inward current after a short depolarizing pulse and showed that the inward tail current was Na-Ca exchange current activated by intracellular calcium released from sarcoplasmic reticulum. This current contributed to the generation of the late plateau of the action potential in rabbit atrial myocytes. They observed that the voltage dependence of the exchange current did not match that of the calcium current. On the other hand, there are results supporting a simple relation between calcium entry and calcium release.[2,3] Earm and Noble[4] showed the regenerative behavior of calcium release using a single cell version of the rabbit atrial model developed by Hilgemann and Noble[5] and suggested that the buffering property of calcium indicators could modify the intracellular calcium transients. In the present study, we investigated the voltage dependence and kinetic properties of the tail current in different concentrations of intracellular calcium buffer (0.01 to 0.5 mM EGTA or 0.01 to 0.2 mM BAPTA) or at different stimulus intervals (300 msec to 30 sec).

The voltage dependency of the inward current in low concentrations of EGTA or at low-frequency stimulation showed very steep activation; activation occurred from −40 mV and reached a peak level at about −10 to 0 mV. Voltage dependence in high concentrations of buffer or at high-frequency stimulation showed more gradual activation, which is very similar to the activation pattern of the calcium current (FIG. 1).

When EGTA concentration and stimulation frequency were low, the time course of the current decay was slow. Increase of EGTA concentration or stimulation frequency resulted in current decay that was accelerated, and the slow component of the inward current disappeared (FIG. 2).

From these results we could conclude that (1) sarcoplasmic reticulum calcium release is triggered by only a small fraction of the calcium current, and the nonlinear relation between the inward current and the calcium current strongly supports the regenerative Ca release hypothesis of the calcium-induced calcium release (CICR) mechanism, and (2) such regenerative behavior was abolished by a high concentration of Ca buffer or premature stimulation.

[a]This work was supported by a research grant from Seoul National University College of Medicine.

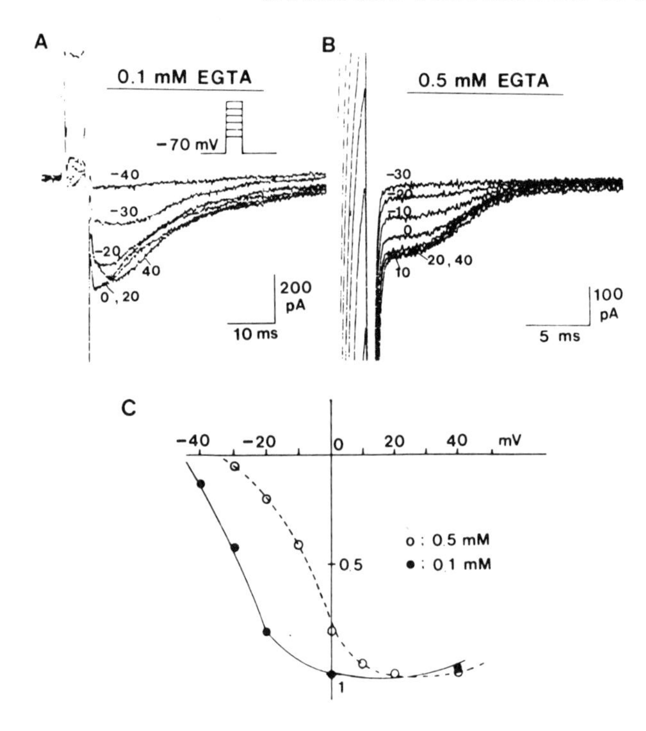

FIGURE 1. Effect of intracellular EGTA concentration on the voltage dependence of the inward tail current. From the holding potential of −70 mV, step-depolarizing pulses of −40, −30, −20, 0, 20, and 40 mV were applied as shown in **A** (inset). Pulse interval was 30 seconds in all experiments if not mentioned. The inward tail currents were activated on repolarization to −70 mV after the depolarizing pulses and superimposed current traces were shown in **A** (in 0.1 mM EGTA) and **B** (in 0.5 mM EGTA). In **C** normalized values of peak current magnitudes were plotted against the potential of the step pulse. The inward current in 0.1 mM EGTA (filled circle) was activated from −40 mV and increased very steeply, then reached a peak at 0 mV. Increase of the EGTA concentration to 0.5 mM shifted the voltage-dependence curve to the right and made a less steep curve (open circles and broken line); 30 μM tetrodotoxin was used to block Na current.

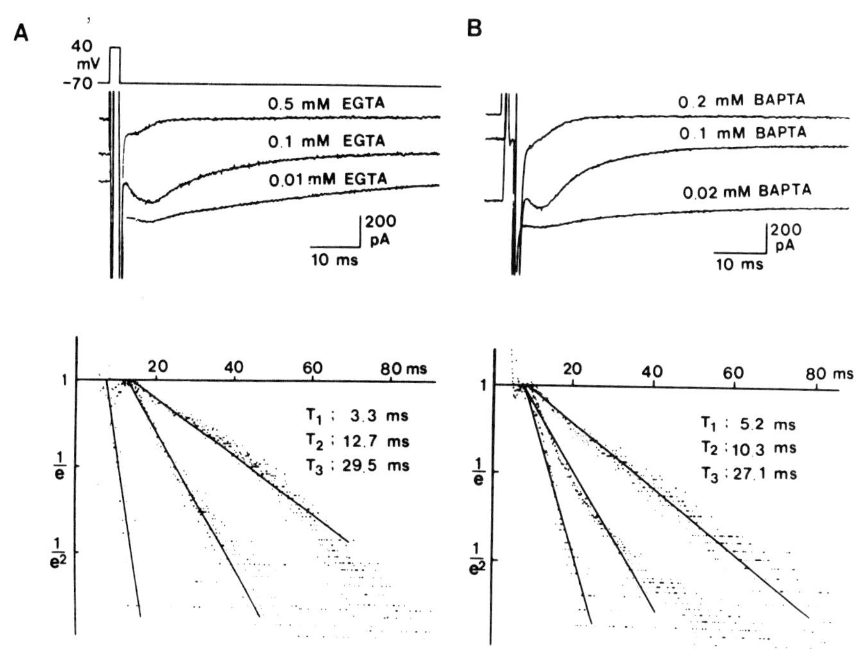

FIGURE 2. Effect of calcium buffer concentration on the time course of the inward current. The inward currents activated by a pulse to +40 mV in 0.5, 0.1, 0.01 mM EGTA are shown in upper panel of **A,** and those in 0.2, 0.1, 0.02 mM BAPTA are shown in **B.** Relaxation of the current becomes faster with increasing buffer concentration. The lower panel represents the semilogarithmic plot of the currents in normalized scale. The time course of the current decay is almost single exponential and straight line is drawn by eye. Time constants of the current decay in 0.5, 0.1, and 0.01 mM EGTA were 3.3, 12.7, and 29.5 msec, and those in 0.2, 0.1, and 0.02 mM BAPTA were 5.2, 10.3, and 27.1 msec, respectively.

REFERENCES

1. Earm, Y. E., W. K. Ho & I. S. So. 1990. Proc. R. Soc. London B **240:** 61-81.
2. Barcenas-Ruiz, L. & W. G. Wier. 1987. Circ. Res. **61:** 148-154.
3. Cleemann, L. & M. Morad. 1991. J. Physiol. **432:** 283-312.
4. Earm, Y. E. & D. Noble. 1990. Proc. R. Soc. London B **240:** 83-96.
5. Hilgemann, D. W. & D. Noble. 1987. Proc. R. Soc. London **B230:** 163-205.

Whole-Cell Current Associated with Na-Ca Exchange in Cultured Chick Cardiac Myocytes[a]

SHI LIU AND JOSEPH R. STIMERS

Department of Medicine
Division of Cardiology
and
Department of Pharmacology and Toxicology
University of Arkansas for Medical Sciences
Little Rock, Arkansas 72231

MELVYN LIEBERMAN

Department of Cell Biology
Division of Physiology
Duke University Medical Center
Durham, North Carolina 27710

We previously demonstrated electrogenic Na-Ca exchange using the ion-selective microelectrode technique.[1] In those experiments, we exposed preparations to 24 mM $[K]_o$ + 0.1 mM ouabain + 1 mM Ba followed by reduction of $[Na]_o$ to 27 mM. Presumably driven by the Ca gradient, intracellular Na activity (a^i_{Na}) rapidly decreased accompanied by a hyperpolarization of membrane potential beyond the equilibrium potentials for K, Na, and Cl. Subsequent restoration of Na_o resulted in an increase in a^i_{Na} and depolarization. Using similar experimental protocols and the patch-clamp technique, we studied the whole-cell current responsible for the previous observations and its ionic dependence. Single myocytes were internally dialyzed with a solution of 24 mM Na, 0.5 mM Ca, 1 mM EGTA, 130 mM K, 2 mM Mg, 35 mM Cl, 111 mM aspartic acid 10 mM (HEPES + Tris), and 2 mM ATP (pH 7.2) and were superfused with a solution of 24 mM K (or K-free), 1 mM Ca, 0.8 mM Mg, 150 mM Cl, 10 mM (HEPES + Tris), 1 mM Ba, 0.1 mM ouabain and 0.1 mM Cd (pH 7.4) to block K inward rectifier, Na-K pump, and Ca currents, respectively. After myocytes were voltage-clamped at −40 mV, reducing $[Na]_o$ from 125 to 27 mM (TMA as substitute) induced an outward current (Na_i-Ca_o exchange) while restoring Na_o induced an inward current (Na_o-Ca_i exchange) (FIG. 1). Na_o-induced currents were almost completely abolished by either Ca-free (+ 1 mM EGTA) (FIG. 1), 1 mM La or 20 mM Mn. Currents sensitive to Mn and La showed strong activation on depolarization (FIG. 2), similar to the Na-Ca exchange current demonstrated previously using single microelectrode-switching voltage clamp.[2] When cells were perfused with either 24 mM $[K]_o$ or K-free solution, varying $[Na]_o$ between 149 mM and 0 mM resulted in an increasing outward current in a concentration-dependent manner. The Hill plot of the change in outward current versus $[Na]_o$ in 1 mM $[Ca]_o$ was linear with a slope of ~2, contrary to our previous finding of slope of 1 under non-voltage-clamped conditions, in which the net Na efflux

[a]Supported by National Institutes of Health Grants HL27105, HL17670, and HL07101.

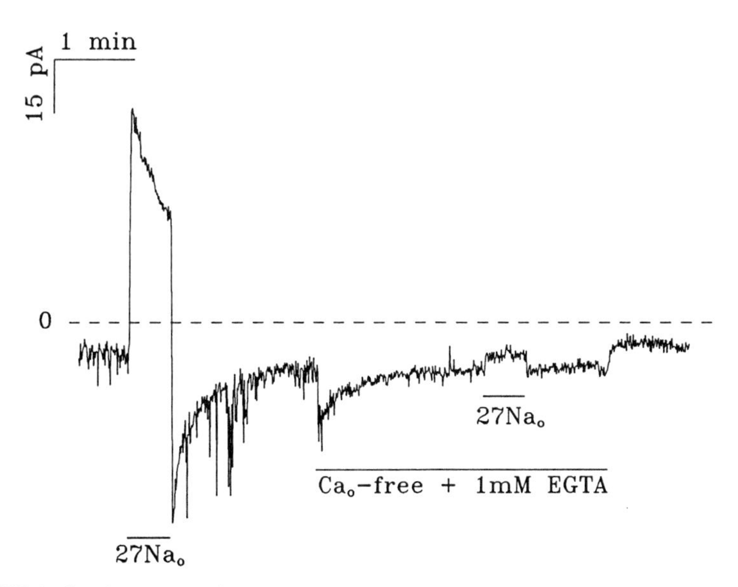

FIGURE 1. Ca$_o$-dependence of Na-Ca exchange current. Cardiac myocyte was voltage-clamped at −40 mV, internally dialyzed with 24 mM Na and 120 nM Ca, and superfused with 125 mM Na and 1 mM Ca in 24 mM K + 1 mM Ba + 0.1 mM ouabain + 0.1 mM Cd. Reduction of $[Na]_o$ to 27 mM induced an outward current that gradually declined (Na$_i$-Ca$_o$ exchange). Restoration of 125 mM $[Na]_o$ induced an inward current that also slowly declined (Na$_o$-Ca$_i$ exchange). Removal of Ca$_o$ (+ 1 mM EGTA) caused a transient inward shift on holding current, consistent with a Na$_o$-Ca$_i$ exchange current. As indicated, Na$_o$-induced currents were almost completely abolished in the absence of Ca$_o$. In addition, restoration of 1 mM $[Ca]_o$ caused an outward shift of holding current, consistent with a Na$_i$-Ca$_o$ exchange current.

was plotted versus $[Na]_o$ in 2.7 mM $[Ca]_o$. The maximal current obtained from the best fit of the Hill equation is 3.2 pA/pF, a value similar to that found in guinea-pig ventricular myocytes.[3] The half-maximal inhibition of this outward current required 82 mM $[Na]_o$. The results suggest that to inhibit the outward current via Na$_i$-Ca$_o$ exchange, two Na$_o$ ions compete with one Ca$_o$ ion for the Ca$_o$-binding site. In K-free, 149 mM $[Na]_o$ solution, increasing $[Ca]_o$ from 0 to 1 mM induced the Na$_i$-Ca$_o$ exchange current of 1.2 ± 0.2 pA/pF (n = 7). When the pipette solution is Ca-free plus 5 mM EGTA, this outward current is reduced to 0.27 ± 0.04 pA/pF (n = 8), consistent with finding that Ca$_i$ presence influences the Na$_i$-Ca$_o$ exchange current.[3] When Na$_o$ was reduced to 27 mM in $[Ca]_o$ ranging from 0 to 4 mM, the outward current was increased. The Hill plot of this low Na$_o$-induced outward current versus $[Ca]_o$ gives a slope of 1, consistent with the concept of one Ca$_o$-binding site for Na$_i$-Ca$_o$ exchange current. The double reciprocal plot reveals a maximal current of 2.9 pA/pF with $K_{0.5}$ of 1.3 mM $[Ca]_o$, a value similar to that reported for guinea-pig ventricular myocytes.[4] In contrast, in 1 mM $[Ca]_o$ solution, restoring $[Na]_o$ from 0 to 149 mM induced an inward current (Na$_o$-Ca$_i$ exchange current) in a concentration-dependent manner. The Hill plot of this Na$_o$-induced current versus $[Na]_o$ gives a Hill coefficient of 1.8 with an apparent $K_{0.5}$ of 50 mM for Na$_o$, in comparison with a Hill coefficient of 3 and $K_{0.5}$ of 88 mM reported for guinea-pig ventricular myocytes.[4] In conclusion, (1) we have identified and characterized a Na-Ca exchange current in cultured chick cardiac

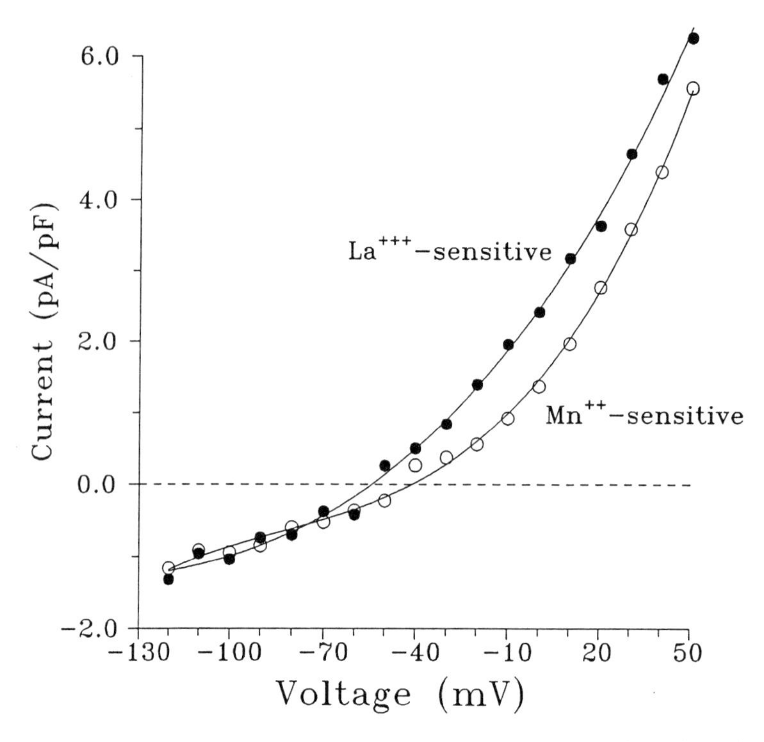

FIGURE 2. The effects of Mn and La on Na-Ca exchange. Cells were voltage-clamped at −40 mV. The same experimental protocols as described in FIGURE 1 were performed in the absence and presence of 20 mM Mn or 1 mM La. The difference current was plotted versus clamp potentials. Both Mn- (open circles) and La-sensitive currents (filled circles) show an outward rectification.

myocytes, with characteristics similar to those found in mammalian myocytes; (2) we have demonstrated that this current could account for our previous findings involving changes in a^{i}_{Na} and membrane potential.

REFERENCES

1. JACOB, R., M. LIEBERMAN & S. LIU. 1987. J. Physiol. **387:** 567-588.
2. STIMERS, J. R. & M. LIEBERMAN. 1988. Biophys. J. **53:** 423a.
3. MIURA, Y. & J. KIMURA. 1989. J. Gen. Physiol. **93:** 1129-1145.
4. KIMURA, J., S. MIYAMAE & A. NOMA. 1987. J. Physiol. **384:** 199-222.

Amiloride Enhances Postischemic Ventricular Recovery During Cardioplegic Arrest

A Possible Role of Na^+-Ca^{2+} Exchange

XUEKUN LIU,[a] RICHARD M. ENGELMAN,[b] JASHIMHA IYENGAR,[a] GERALD A. CORDIS,[a] AND DIPAK K. DAS[a,c]

[a]*Cardiovascular Division*
Department of Surgery
University of Connecticut School of Medicine
Farmington, Connecticut 06030

[b]*Department of Surgery*
Baystate Medical Center
Springfield, Massachusetts 01107

Intracellular Ca^{2+} is of primary importance in the pathogenesis of ischemia and reperfusion injury in the myocardium.[1,2] Although the exact mechanism of intracellular Ca^{2+}accumulation has not been described, several mechanisms have been proposed, including the Ca^{2+} transmembrane inflow through the slow channels[2] and through the permeable membrane caused by myocardial ischemia.[3]Recent studies suggested that massive Ca^{2+} influx may occur as a consequence of the Na^+-Ca^{2+} exchange during reperfusion, which in turn may be caused by accumulation of Na^+ during ischemia.[4] It is known that acidosis induced by anaerobic metabolism leads to enhanced exchange of intracellular H^+ for Na^+. In addition, inhibition of Na^+-K^+-ATPase by energy depletion also is known to cause to the accumulation of intracellular Na^+.

This study proposes to determine if inhibition of H^+-Na^+ exchange by amiloride can lead to myocardial preservation during cardioplegic arrest as seen during open-heart surgery. We used an animal model simulating coronary artery bypass for acute myocardial infarction. The results suggest that amiloride reduces myocardial ischemic-reperfusion injury.

EXPERIMENTAL METHODS

Yorkshire pigs weighing 20-25 kg were anesthetized with an intravenous injection of Nembutal (30 mg/kg), intubated, and ventilated with room air by a Harvard respirator. The chest was opened with a median sternotomy incision. Ater heparinization with sodium heparin (500 units/kg), the animals were placed on cardiopulmonary bypass with a bubble oxygenator and a roller pump. The heart was isolated *in situ* from the

[c]Address for correspondence: Dipak K. Das, Ph.D., Cardiovascular Division, Department of Surgery, University of Connecticut School of Medicine, Farmington, CT 06030.

systemic circulation by cross-clamping the ascending aorta as described previously.[5] The heart was perfused with blood (37°C) at 75 mmHg for 15 min. The left anterior descending artery (LAD) was dissected free. Two pairs of piezoelectric crystals were inserted into the LAD-supplied region and non-LAD-supplied region of LV to measure the myocardial segment shortenings (SS). A balloon catheter filled with saline was also inserted into the LV to measure the isovolumetric LV function as described previously. Coronary blood flow (CBF) was collected by venting the pulmonary artery and quantitated. Creatine kinase activity in the coronary effluent was determined.

After stabilization of the heart, baseline measurements were performed, and the LAD occluded just distal to the first diagonal branch for 60 min at normothermia. The heart was subjected to 60 min of global hypothermic cardioplegic arrest induced by infusion of 50 ml of potassium (35 mEq/L) crystalloid cardioplegic solution (4°C) (CPS), and the heart temperature was maintained at 18 to 20°C. The ligation of the LAD was removed, and a second dose of CPS was given. The experimental group ($n = 6$) received CPS containing 0.25 mM of amiloride. One hour of global reperfusion followed, and myocardial functional and metabolic measurement were determined. At the end of the experiment, transmural myocardial biopsies were taken for assay of high-energy phosphates as described previously.[6]

RESULTS

Effects of Amiloride on Regional Function and Global Function

LAD occlusion resulted in prompt dyskinesia in the LAD distribution. No active contraction was observed during 60 min of LAD occlusion. Restoration of SS in this region 60 min after reperfusion was 10 ± 0.89% and 25 ± 2.0% in control and treated group, respectively ($p < 0.05$) (TABLE 1). Moreover, amiloride treatment significantly improved the LV contractility as indicated by LVDP and dp/dt_{max} during reperfusion ($p < 0.05$ compared with control group). LV end diastolic pressure (LVEDP), which represents the compliance of LV, was also improved by amiloride.

Effects of Amiloride on Coronary Blood Flow and Myocardial Creatine Kinase Release

Coronary blood flow (CBF) did not change significantly during LAD occlusion or after reperfusion without a difference in both groups. The creatine kinase (CK) release was increased during reperfusion in both groups but was significantly reduced by amiloride treatment as shown in TABLE 1.

Effects of Amiloride on Myocardial High-Energy Phosphates

Myocardial high-energy phosphates (ATP and CP) measured at the end of reperfusion are shown in FIGURE 1; amiloride animals maintained a higher level of ATP and CP compared with controls ($p < 0.05$).

TABLE 1. Effects of Amiloride on Left Ventricular Function and Metabolism

	Baseline		LAD 60		R 60	
	Control	Amiloride	Control	Amiloride	Control	Amiloride
LVDP (mmHg)	90 ± 7.2	98 ± 7.8	55 ± 4.4	62 ± 5.3	38 ± 3.1	56 ± 4.5[a]
$LVdp/dt_{max}$ (mmHg/sec)	850 ± 7.0	960 ± 77	630 ± 50	710 ± 57	360 ± 21	710 ± 57[a]
LVEDP (% of baseline)	8.8 ± 0.70	6.8 ± 0.55	10.3 ± 0.82	9.6 ± 0.77	32.8 ± 2.6	14.2 ± 1.14[a]
SS (% of baseline)	100	100	0	0	10 ± 0.89	25 ± 2.0[a]
CBF (ml/min)	107 ± 8.6	101 ± 8.1	102 ± 6.9	92 ± 7.4	117 ± 9.4	110 ± 8.2
CK release (% of baseline)	100	100	112 ± 8.9	106 ± 8.7	243 ± 24	130 ± 11[a]

NOTE: Results are expressed as mean ± SEM of six hearts in each group. Student's *t*-test was performed for statistical analysis. LVDP = left ventricular developed pressure; LV dp/dt_{max} = maximum of first derivative of LV pressure; LVEDP = left ventricular end-diastolic pressure; SS = segment shortening; CBF = coronary blood flow; CK = creatine kinase; LAD 60 = 60 min of LAD occlusion; R 60 = 60 min of reperfusion.

[a] $p < 0.05$ compared with control.

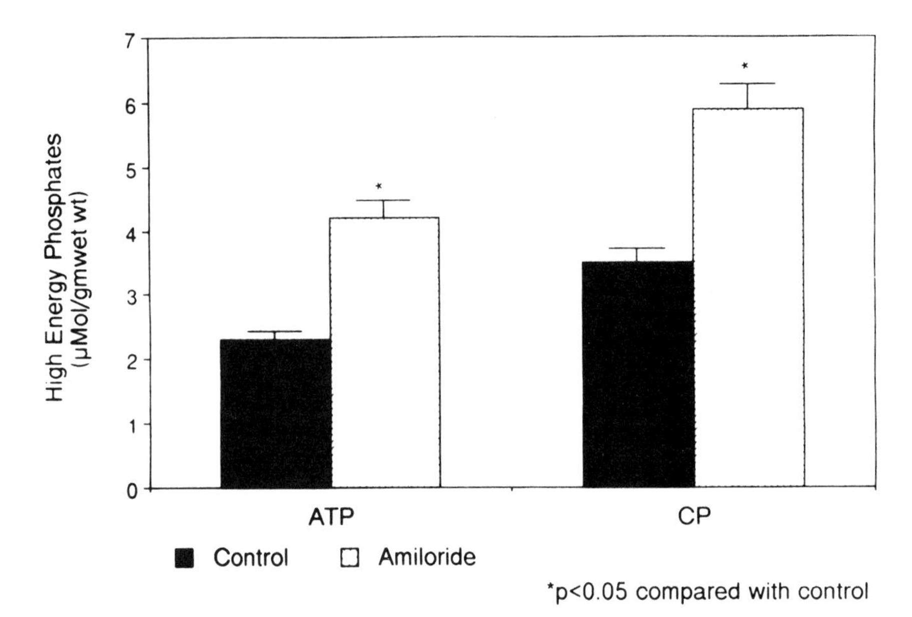

FIGURE 1. Effect of amiloride on myocardial high-energy phosphates. The results are expressed as mean ± SEM of six hearts in each group. Student's *t*-test was performed for statistical analysis. ATP = adenosine triphosphate; CP = creatine phosphate.

DISCUSSION

The present study suggests that inhibition of Na^+-H^+ exchange during ischemia leads to significant myocardial preservation during reperfusion. The better recovery of LV compliance and contractile function, reduction of CK release, and improvement of ischemic myocardial segment shortening all point to the effectiveness of amiloride in reducing myocardial reperfusion injury. The reduction of Na^+ accumulation during ischemia by amiloride and the presumed reduction of Na^+-Ca^{2+} exchange during reperfusion strongly suggest a crucial role of Na^+-H^+ and Na^+-Ca^{2+}exchange for the prevention of intracellular Ca^{2+}accumulation observed during myocardial ischemia and reperfusion.

REFERENCES

1. Nayler, W. G. 1981. Am. J. Pathol. **102:** 262-270.
2. Watts, J. A., G. D. Koch & K. F. LaNoue. 1980. Am. J. Physiol. **238:** H909-H916.
3. Crake, T. & P. A. Poole-Wilson. 1986. J. Mol. Cell. Cardiol. **18**(Suppl IV): 31-36.
4. Tani, M. & J. R. Neely. 1989. Cir. Res. **65:** 1045-1056.
5. Kimura, Y., J. Iyengar, R. M. Engelman & D. K. Das. 1990. J. Cardiovasc. Pharmacol. **16:** 992-999.
6. Cordis, G. A., R. M. Engelman & D. K. Das. 1988. J. Chromatogr. **459:** 229-236.

Late Contraction in Guinea-Pig Ventricular Myocytes Activated by the Na^+-Ca^{2+} Exchange during the Action Potential

K. SCHÜTTLER

Cardiovascular Research Department
CIBA-GEIGY Ltd.
CH-4002 Basel, Switzerland

S. Y. WANG, T. PFEIFER, AND R. MEYER

Second Institute of Physiology
University of Bonn
D-5300 Bonn, Federal Republic of Germany

$[Ca^{2+}]_i$ in myocardial cells is raised by depolarization-dependent entry from the external medium and by Ca^{2+}-induced Ca^{2+} release from the sarcoplasmic reticulum. The aim of this study was to test whether depolarization-dependent entry of Ca^{2+} is able to activate contraction directly in guinea-pig ventricular myocardium.

Experiments were carried out on both isolated ventricular myocytes and papillary muscles at ambient temperature, and the sarcoplasmic reticulum was functionally eliminated by means of two approaches. (1) In isolated ventricular myocytes, membrane potential or membrane currents were measured by patch clamp (whole-cell mode). Simultaneously, unloaded contractions were optically monitored, or intracellular Ca^{2+} transients were recorded by Fura-2 fluorescence ratio (340/380 nm). Sarcoplasmic reticulum was functionally eliminated by incubation of the cells for 90 min in 10 μM ryanodine. During this time the initial portion of contraction was progressively suppressed. A small late contraction, more slowly developing and relaxing, remained and increased with time up to about the magnitude of the normal contraction. The late contraction was found to develop as long as the test pulse or the action potential lasted (FIG. 1A, B). Normal contraction elicited by trains of identical voltage-clamp pulses exhibited a bell-shaped voltage dependence with a maximum around 50 mV (FIG. 1C, solid line). In contrast, late contraction increased with increasing membrane potential over the voltage range from −30 to 90 mV (FIG. 1C, dashed line). The voltage dependence of the late contraction could be fitted by a linear regression ($r^2 = 0.9835$; $s = 0.6458$). (2) From papillary muscle isometric contraction was recorded. A train of normal contractions was interrupted by prolonged periods of rest, lasting 10 to 20 min. This results in a Ca^{2+} depletion of the sarcoplasmic reticulum and thus in an inhibition of Ca^{2+}-induced Ca^{2+} release. Stimulation then elicits so-called rested-state contractions. They were found to be extremely weak and to develop with an increased latency after the stimulus.

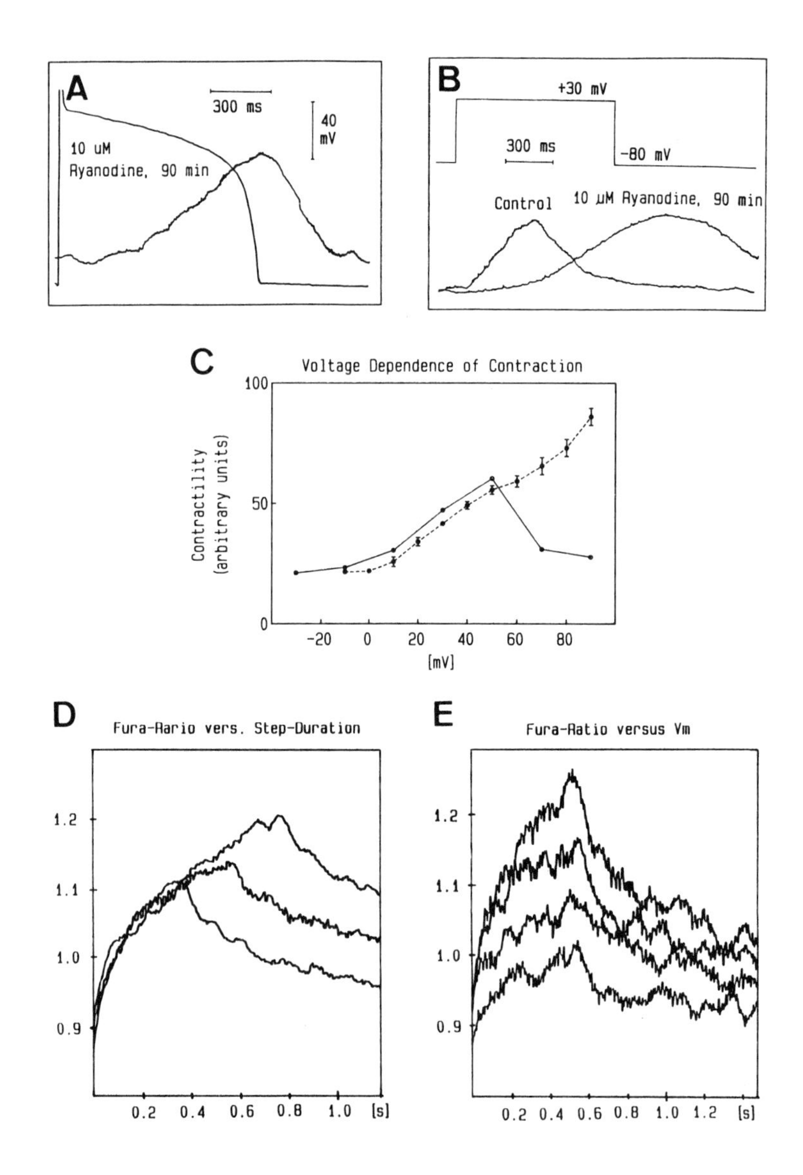

FIGURE 1. Late contraction or Fura-2 fluorescence ratio in isolated ventricular myocytes after application of 10 μM ryanodine. For depolarization action potentials were elicited (**A**) or voltage steps were applied (**B-E**) at a frequency of 0.25 Hz. (**C**) Voltage dependence of the amplitude of the late contraction (dashed line) and of the normal contraction (solid line). The lines connecting the data points were drawn by eye. Holding potential −80 mV. (**D**) Fura-2 fluorescence ratio. Variation of clamp duration (400, 600, 800 msec). (**E**) Increased Fura-2 fluorescence ratio with increased voltage step (from −80 to 0, 20, 40, 80 mV). For further explanation, see text.

Both types of late contraction were increased when the transmembrane Na^+ gradient was reduced either by reducing $[Na^+]_o$ (substitution by choline chloride or LiCl) or by experimental maneuvers expected to increase $[Na^+]_i$ (reduction of $[K^+]_o$ from 5.4 to 0.5 mM; application of ouabain). After application of 30 μM tetrodotoxin, late contraction was reversibly abolished. Also in the presence of 40 μM $NiCl_2$ late contraction was severely suppressed.

Ryanodine treatment inhibited the initial phase of the Ca^{2+}-transient (peak after about 100 msec), and $[Ca^{2+}]_i$ increased permanently during the voltage-clamp pulse. Consequently, prolongation of the clamp pulse resulted in higher $[Ca^{2+}]_i$ values (FIG. 1D; clamp from −80 to 30 mV, duration 400, 600, 800 msec.) Both increasing voltage-clamp potentials (FIG. 1E; clamp from −80 to 0, 20, 40, 80 mV is shown, duration 500 msec) or reduction of $[K^+]_o$ caused a steeper rise in $[Ca^{2+}]_i$.

These properties of both types of late contraction and of $[Ca^{2+}]_i$ are in agreement with the hypothesis that these signals may substantially represent responses to Ca^{2+} directly entering the cell via the Na^+-Ca^{2+} exchange in its reversed mode during excitation or depolarization. Toward the end of the action potential plateau, when the transient Ca^{2+} increase due to the sarcoplasmic reticulum is over, an additional Ca^{2+} influx via the Na^+-Ca^{2+} exchange might contribute to the loading of the sarcoplasmic reticulum for the subsequent beat (as was suggested by the right maximum of the voltage-tension relationship of normal contraction).

Effects of Amiloride Derivatives as Inhibitors of the Na^+-Ca^{2+} Exchange on Mechanical and Electrical Functions of Isolated Cardiac Muscle and Myocytes

E. WETTWER, H. M. HIMMEL, AND U. RAVENS[a]

Pharmakologisches Institut
Universitätsklinikum Essen
D-4300 Essen 1, Federal Republic of Germany

In cardiac sarcolemmal membrane preparations, amiloride derivatives are potent inhibitors of the Na^+-Ca^{2+} exchange transporter with IC_{50} values < 10-50 μM (for review see Ravens & Wettwer[1]). In the whole heart, pharmacological inhibition of the Na^+-Ca^{2+} exchange is expected to increase force of contraction if impairment of Ca^{2+} extrusion from the cell dominates impairment of the Ca^{2+} entry mode. Indeed, both positive and negative inotropic effects of amiloride derivatives have been reported.[2] Because of this controversy, we studied the effects of three amiloride derivatives, 2′,3′-benzobenzamil (BB), 3′,4′-dichlorobenzamil (DCB), and 5-(*N*-4-chlorobenzyl)-2′,4′-dimethylbenzamil (CBDB), on force of contraction (F_c), action potentials (AP), and membrane currents in different preparations from the guinea-pig heart, in order to assess their specificity of action.

METHODS

Isometric F_c was measured in isolated multicellular cardiac muscle preparations (left atria, right papillary muscles, and Langendorff hearts), AP and membrane currents were measured in myocytes (Himmel *et al.*[3] for details of methods). (Amiloride derivatives were a gift from Merck, Sharp & Dohme Research Laboratories, West Point.) The agents were dissolved in DMSO and further diluted with H_2O.

RESULTS AND CONCLUSION

In left atria, BB, DCB, and CBDB reduced F_c (FIG. 1A) by a maximum of 20 ± 8%, but the highest concentrations (10^{-4} M) increased F_c by a maximum of 36 ± 7%. In papillary muscle, the negative inotropic effect of BB (5×10^{-5} M) was accompanied by little change in shape of AP. In the Langendorff heart, BB (10^{-5} M; FIG. 1B) suppressed F_c almost completely and irreversibly, and the spontaneous heart rate

[a] Address for correspondence: Prof. Dr. med. Ursula Ravens, Pharmakologisches Institut, Universitätsklinikum Essen, Hufelandstraße 55, D-4300 Essen 1, Federal Republic of Germany.

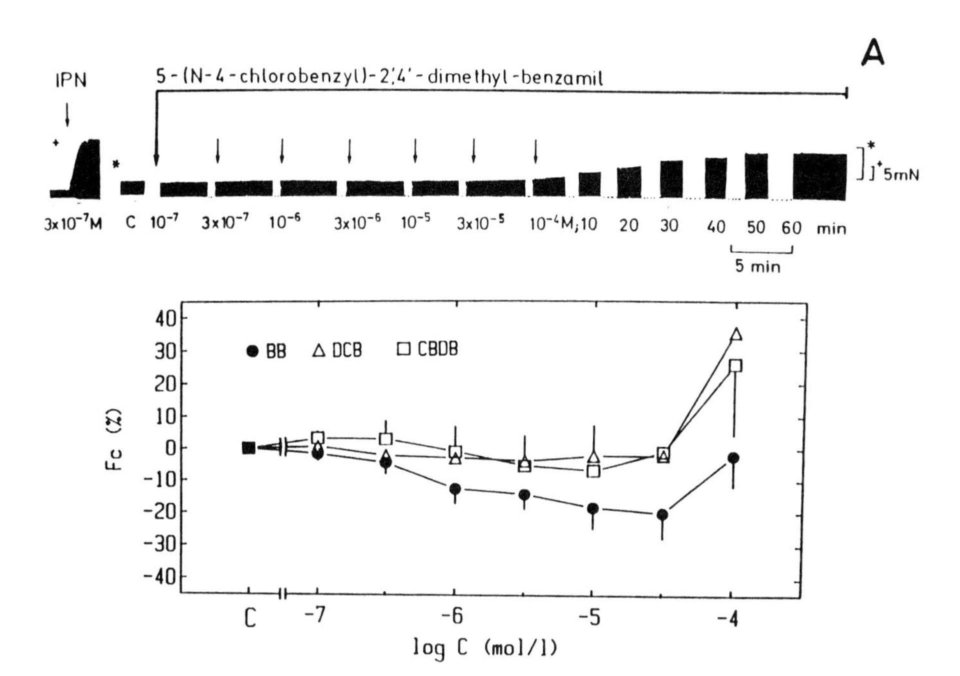

FIGURE 1. Effects of the amiloride derivatives 2′,3′-benzobenzamil (BB), 3′,4′-dichlorobenzamil (DCB), and 5-(*N*-chlorobenzyl)-2′,4′-dimethylbenzamil (CBDB) on force of contraction (F_c) in guinea-pig heart muscle. (**A**) Results from left atria. *Top part:* original tracing of a typical experiment (stimulation frequency 60 $min^{-}1$). After an equilibration period of 60 min, the atrium was exposed to isoprenaline (IPN, 3×10^{-7} M) to test for maximum force increase. After removal of IPN, CBDB was added in cumulatively increasing concentrations. Each concentration was allowed to act for 10 min before the next higher concentration was added with exception of 10^{-4} M, which was allowed to equilibrate for 60 min. The "+" and "*" signs mark different attenuations of the force scale. *Bottom part:* Concentration-response curves for the effects of BB, DCB, and CBDB on F_c of left atria. Mean values ± SEM (n = 4-9) measured after 10 min of exposure to each concentration and expressed as percent change from the control value.

decreased from 135 ± 5 min^{-1} to 59 ± 9 min^{-1}. For a similar reduction in frequency in paced control hearts, F_c decreased by 15.7 ± 13.3%. In single myocytes, the AP amplitude decreased from 122 ± 2 mV to 47 ± 4 mV with BB (5×10^{-5} M). The $Ca^{2+}$and Na^+ currents were inhibited by BB, DCB, and DBDB as illustrated in FIGURE 2 by the results from two experiments with BB. Similar results were obtained with the other amiloride derivatives in two to three cells each.

In conclusion, the three amiloride derivatives have additional effects on membrane currents that may contribute to their inotropic actions. In order to quantify the contribution of the carrier to regulation of cardiac contraction, more specific modulators of the Na^+-Ca^{2+}exchange are required.

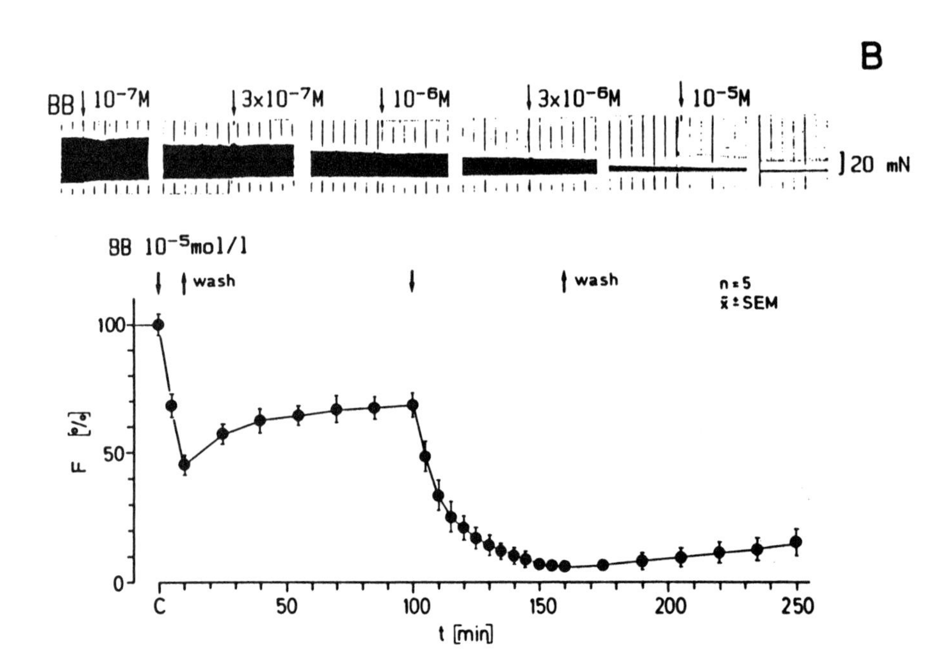

FIGURE 1. *Continued.* (**B**) Results from Langendorff hearts. *Top part:* Original tracing of F_c from an experiment with cumulatively increasing concentrations of BB (10 min at each concentration). *Bottom part:* Time course of F_c expressed in percent of the predrug control value (= 100%) after 10 min of exposure to BB (10^{-5} M), 90 min of wash-out, repeated exposure to BB (10^{-5} M) but for 60 min, and final wash-out for 90 min. Mean values ± SEM from five experiments. Similar results were obtained with DCB and CBDB.

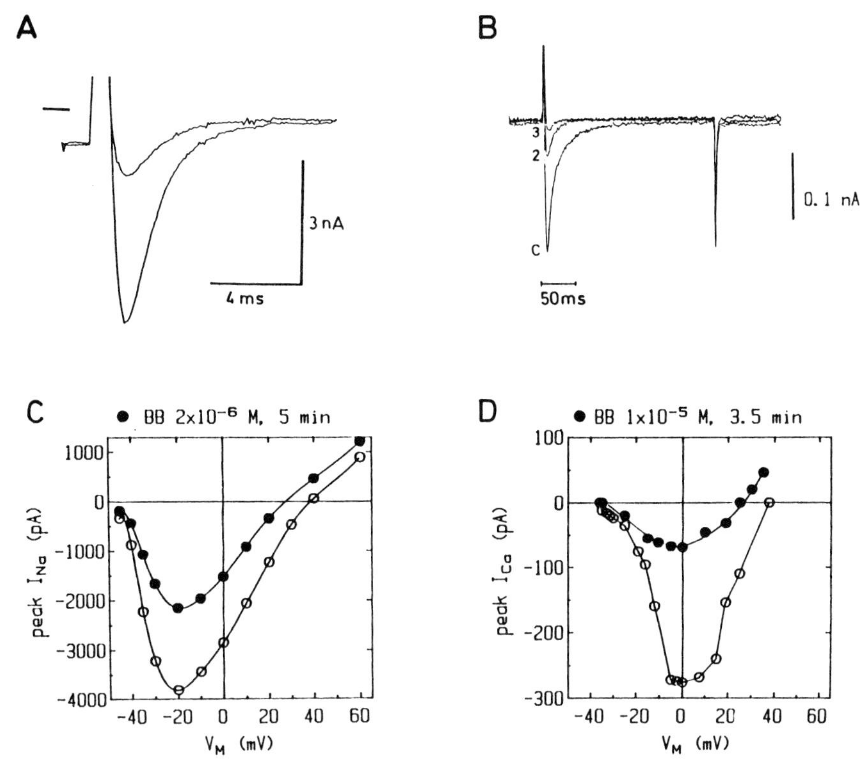

FIGURE 2. Effects of 2′,3′-benzobenzamil (BB) on the sodium current (I_{Na}, A, and C) and the calcium current (I_{Ca}, B, and D) in isolated myocytes of the guinea-pig heart. (**A**) Original tracing of I_{Na} measured in response to a clamp step from a holding potential of −80 mV to −20 mV (single electrode, whole-cell patch-clamp technique; room temperature; low extracellular Na^+ concentration 30 mM). Large and small downward deflections, control I_{Na}, and after 5 min of exposure to 5×10^{-6} M. (**B**) Original tracing of I_{Ca} measured in response to a clamp step from a holding potential of −40 mV (in order to inactivate I_{Na}) to 0 mV under control conditions (c) and after 2 min and 3 min of exposure to BB (5×10^{-5} M) as indicated by the numbers. Room temperature. (**C**) Current-voltage relation of I_{Na} measured in a different cell in the absence (open symbols) and after 5 min of exposure to BB (2×10^{-6} M, closed symbols). (**D**) Current-voltage relation of I_{Ca} measured in a different cell in the absence (open symbols) and after 3.5 min of exposure to BB (1×10^{-5} M, closed symbols). Similar results were obtained with DCB and CBDB in different cells.

REFERENCES

1. Ravens, U. & E. Wettwer. 1989. Modulation of sodium/calcium exchange: A hypothetical positive inotropic mechanism. J. Cardiovasc. Pharmacol. **14**(Suppl. 3): S30-S35.
2. Siegl, P. K. S., E. J. Cragoe, M. J. Trumble & G. J. Kaczarowski. 1984. Inhibition of Na^+/Ca^{2+}exchange in membrane vesicle and papillary muscle preparations from guinea pig heart by analogs of amiloride. Proc. Natl. Acad. Sci. USA **81:** 3238-3242.
3. Himmel, H., H. Glossmann & U. Ravens. 1991. Naftopidil, a new α-adrenoceptor blocking agent with calcium antagonistic properties: Characterization of Ca^{2+} antagonistic effects. J. Cardiovasc. Pharmacol. **17:** 213-221.

Sodium-Calcium Exchange in Smooth and Skeletal Muscles

Introduction to Part VI

The role of Na-Ca exchange in smooth and skeletal muscle is less clear than in cardiac muscle. Exchange activity is observed in t-tubule vesicle preparations from skeletal muscle, and functional effects of exchange activity on skeletal muscle contraction have also been described. Various smooth muscle preparations possess Na-Ca exchange activity, but the precise role of the exchanger in Ca^{2+} homeostasis and tension regulation in these tissues, particularly the resistance arteries of the vascular system, remains an unsettled issue. The complexity of smooth muscle tissue was emphasized, and the roles of Na-Ca exchange in mediating neurotransmitter release and possibly affecting endothelial function in smooth muscle were discussed. Immunocytochemical methods and computer-assisted optical techniques have been utilized to show that the Na-Ca exchange carriers and the Na^+, K^+-ATPase are both distributed on the surface of smooth muscle cells so as to be closely apposed to the underlying sarcoplasmic reticulum.

Sodium-Calcium Exchange in Transverse Tubule Vesicles Isolated from Amphibian Skeletal Muscle[a]

CECILIA HIDALGO,[b,c] FREDY CIFUENTES,[b,c] AND PAULINA DONOSO [b,d,e]

[b]*Centro de Estudios Científicos de Santiago*
and
[c]*Departamento de Fisiología y Biofísica*
[d]*Departamento Preclínicas*
Facultad de Medicina
Universidad de Chile
Santiago, Chile

INTRODUCTION

The Na-Ca exchange transporter of cardiac muscle, a major pathway for calcium fluxes in heart cells, has been extensively studied both in whole cells and in isolated vesicles, as described in this volume. The amino acid sequence and the functional expression of the cardiac exchanger has recently been reported.[1] In contrast, considerably less information exists regarding the properties of the exchanger in skeletal muscle. Only a few reports exist describing sodium stimulation of calcium efflux from muscle fibers[2] or isolated membrane vesicles.[3-5] The contribution of the skeletal Na-Ca exchanger to regulation of calcium fluxes is not known, although studies in intact frog skeletal muscle fibers suggest that the Na-Ca exchanger has a significant role in transporting calcium out of the cell.[2]

This work describes the properties of a sodium-calcium exchanger present in transverse tubule (t-tubule) membrane vesicles isolated from frog skeletal muscle.[5] The exchanger present in t-tubule vesicles shares many properties with the cardiac exchanger, suggesting that it may play a physiological role in calcium homeostasis in skeletal muscle cells.

MATERIALS AND METHODS

Isolation of T-Tubule Membranes

T-tubule membranes were isolated from frog skeletal muscle using the procedure described in detail elsewhere.[6] Briefly, fast muscle was removed from the legs of the frog *Caudiverbera caudiverbera,* minced, homogenized in 0.1 M KCl, 20 mM Tris/

[a]This work was supported by National Institutes of Health Grant GM 35981, by Departamento Técnico de Investigación Grant 2149, and by Fondo Nacional de Ciencia y Tecnología 972-88. F. Cifuentes is the recipient of a FONDECYT Doctoral Fellowship.

[e]Address for correspondence: Dr. Cecilia Hidalgo, C.E.C.S., Casilla 16443, Santiago 9, Chile.

maleate, pH 7.0, and sedimented at 10,000 × g. The resulting supernatant was made 0.6 M in KCl by addition of solid salt and was sedimented at 100,000 × g. The microsomal pellet, containing a mixture of sarcoplasmic reticulum (SR) and t-tubule vesicles, was further fractionated by sedimentation through discontinuous sucrose gradients. The t-tubule vesicles were obtained from the interface of the 25/27.5 (wt/vol) sucrose solutions.

Sodium-Dependent Calcium Influx

Unless otherwise indicated, t-tubule vesicles were equilibrated in a solution containing 100 mM NaCl, 10-20 mM KCl, 1 μM valinomycin, 20 mM Tris-HCl, pH 7.5, for 1.5 hr at 25°C. To initiate calcium uptake, vesicles (usually 5 μl) were placed in the side of a hydrophobic plastic tube[7] and were rapidly mixed at time zero with 0.3 ml of a solution containing 0.1 mM $^{45}CaCl_2$ (10-15 mCi/mmol), variable concentrations of NaCl and KCl as specified in the figure legends, plus 20 mM Tris-HCl, pH 7.5. Osmolarity was kept constant by adding choline chloride. The reaction was stopped by addition of 0.3 ml of ice-cold quench solution A, containing 5 mM $MgCl_2$, 10 mM ethylene glycol bis(β-aminoethyl ether)-*N,N'*-tetraacetic acid (EGTA), 20 mM Tris-HCl, pH 7.5, and 0.5-ml fractions were filtered through either Millipore (HA 0.45 μm) or GF/B filters previously soaked in 0.1 mM $CaCl_2$, 5 mM $MgCl_2$, 20 mM Tris-HCl, pH 7.5. The filters were washed three times with 3 ml of cold quench solution A, dried and counted.

Sodium-Dependent Calcium Efflux

T-tubule vesicles were incubated for two hours with 2 mM $^{45}CaCl_2$ at 25°C, in a solution containing 20 mM K-gluconate, 150 mM Hepes/Tris, pH 7.5, 1 μM valinomycin. The vesicles (5 μl) were mixed at time zero with 0.1 ml of a solution containing 150 mM Na-gluconate, 20 mM K-gluconate, 20 mM Hepes/Tris, pH 7.5, following the procedure described above; 10 mM EGTA was added when indicated in the text. The reaction was stopped by addition of 1 ml cold quench solution B, containing 20 mM K-gluconate, 5 mM $MgCl_2$, 10 mM EGTA, 150 mM Hepes/Tris, pH 7.5, and the filters were processed as described above.

To measure calcium efflux in the absence of external Na, the same procedure was followed except that vesicles equilibrated with $^{45}CaCl_2$ were diluted into a solution containing 150 mM Hepes/Tris, pH 7.5, instead of 150 mM Na-gluconate.

Other Procedures

Binding of [^{3}H]ouabain and [^{3}H]digoxin was done as described in detail previously[8]; binding of [^{3}H]ryanodine was measured as described elsewhere.[9] Protein concentration was determined using a modification of the Lowry method,[10] with bovine serum albumin as standard.

TABLE 1. Characterization of Isolated Transverse Tubule Vesicles

	B_{max} (pmol/mg of protein)
Nitrendipine binding	124 ± 18 (4)
Ouabain binding[a]	215 ± 38 (4)
Ryanodine binding	<0.5
Cholesterol content (μmol/mg protein)	1.13 ± 0.17 (4)
Cholesterol/Phospholipid (molar ratio)	0.55 ± 0.11 (4)

[a] Measured in the presence of detergent.

Materials

Valinomycin (Calbiochem Corp.) was added from a stock solution in dimethylsulfoxide. $^{45}CaCl_2$, [^{3}H]ouabain, [^{3}H]digoxin, and [^{3}H]ryanodine were obtained from New England Nuclear Corp. All other reagents used were of analytical grade.

RESULTS AND DISCUSSION

Properties of the T-Tubule Vesicles

Purity of the Preparation

The t-tubule vesicles used in this work are highly purified, as described in detail elsewhere.[6] They have a high density of dihydropyridine (DHP) receptors (t-tubule membrane markers) and are virtually devoid of ryanodine receptors, markers of the SR junctional membrane (TABLE 1). In addition, the t-tubule vesicles have high cholesterol content and a high density of ouabain-binding sites (TABLE 1), as expected from the fact that t-tubule membranes are rich in cholesterol and possess a similar density of sodium pumps as that of preparations enriched in surface membranes.[6] A recent immunocytochemical study shows that the density of ($Na^+ + K^+$)-ATPase molecules is similar in surface and t-tubule membranes.[11] These results rule out previous findings showing that the sodium pump density is lower in t-tubules than in surface membranes.[12] We have described elsewhere that the isolated t-tubules are not contaminated with light SR, since they lack measurable Ca^{2+}-ATPase activity,[6] nor with surface membranes, since they are devoid of low-affinity binding sites for an ethylenediamino derivative of tetrodotoxin (en-TTX_{II}), a characteristic feature of surface membranes.[8]

Integrity and Sidedness

To determine integrity of the vesicles, we measured ouabain-sensitive (Na^+ + K^+)-ATPase activity and ATP-dependent ouabain binding, both in the absence and the presence of detergent. Only leaky vesicles should display activity, since ouabain

TABLE 2. Characterization of the Integrity and Sidedness of the Isolated Transverse Tubule Vesicles

		B_{max} (pmol/mg of protein)
Total ouabain binding (+ saponin)		237 (100%)
Basal ouabain binding		35 (14.8%)
Sealed vesicles:	**85.2%**	
Total digoxin binding (+ saponin)		220 (100%)
Basal digoxin binding		232 (105%)
Vesicles sealed inside/out:	**100%**	

and ATP act on opposite sides of the vesicles. As illustrated in TABLE 2, in the absence of detergent the maximal ouabain-binding obtained corresponded to only 14.8% of the value measured in the presence of detergent, indicating that 85% of the vesicles are sealed. We routinely measured ouabain binding with and without detergent, and we used only t-tubule preparations that contained at least 85% sealed vesicles.

Vesicle sidedness was determined by measuring the effect of detergent addition on digoxin binding. Since digoxin is a membrane-permeant analogue of ouabain, in the absence of detergent leaky vesicles and vesicles sealed with the ATP-binding site externally located (inside-out vesicles) will bind digoxin; addition of detergent should unmask the binding sites present in the vesicles sealed with the outside-out configuration. The t-tubule vesicles used in this work displayed the same maximal binding of digoxin with and without detergent (TABLE 2) and equal to the maximal value of ouabain binding, indicating that the preparation does not contain vesicles sealed with the outside-out configuration.

In conclusion, the data illustrated in TABLES 1 and 2 show that 85% of the t-tubule vesicles used in this work are sealed, that the sealed vesicles have only the inside-out configuration, and that the preparation is virtually devoid of other membrane contaminants.

Sodium-Dependent Calcium Influx

Equilibrium Determinations

To test whether the t-tubule membranes have Na-Ca exchange activity, we equilibrated the vesicles with 100 mM NaCl; we diluted them in a solution containing 0.1 mM $^{45}CaCl_2$; and we measured the amount of ^{45}Ca taken up by the vesicles as a function of time (FIG. 1). After 4 minutes, the vesicles reached a constant calcium content of 8 nmol/mg of protein. Addition of ionophore A23187 caused a rapid release of ^{45}Ca, indicating that it was present inside the vesicles. Depending on the preparation, after equilibration with 100 mM NaCl maximal ^{45}Ca contents of 7-12 nmol/mg of protein were obtained.

A linear correlation between the concentration of NaCl used to equilibrate the vesicles and the maximal amount of ^{45}Ca accumulated at equilibrium following dilution in 0.1 mM $^{45}CaCl_2$ was found (FIG. 2). Vesicles equilibrated with a sodium-free solution reached a maximal ^{45}Ca content of 2.2 nmol/mg of protein, indicating that in the

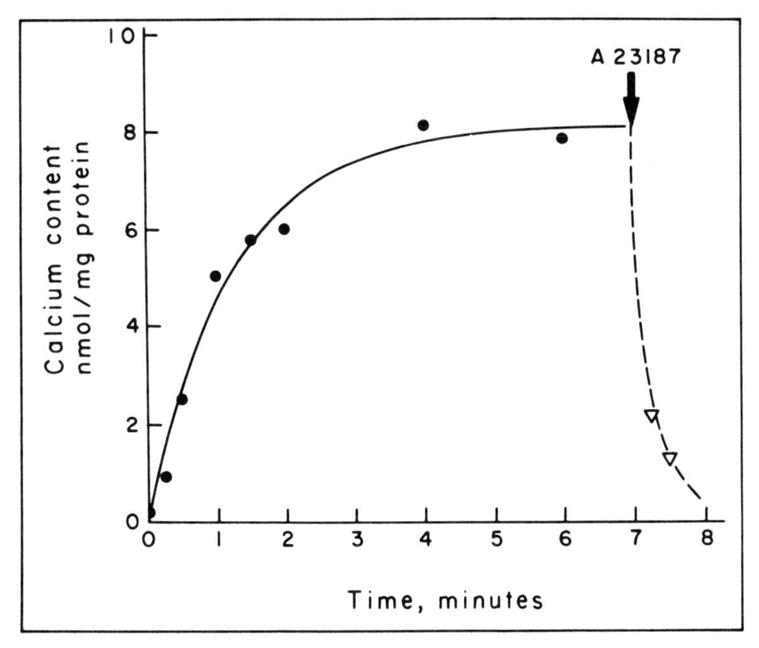

FIGURE 1. Time course of calcium uptake in t-tubule vesicles equilibrated with 100 mM NaCl. Vesicles were incubated in 100 mM NaCl, 20 mM Tris-maleate, pH 7.0, for 90 minutes at 25°C. The reaction was initiated by 60-fold dilution in 0.1 mM $^{45}CaCl_2$, 100 mM KCl, 20 mM Tris-maleate, pH 7.0, and was stopped as described in the text.

absence of sodium the vesicles can take up 20% of the amount of ^{45}Ca accumulated after equilibration in 100 mM NaCl. We have shown elsewhere that this sodium-independent ^{45}Ca accumulation takes place via Ca-Ca exchange[5] and that this exchange is made possible by the high internal calcium content of the isolated vesicles (see below). The linear correlation between the amount of ^{45}Ca accumulated at equilibrium and the concentration of sodium equilibrated inside the vesicles (FIG. 2) indicates that the sodium gradient drives the accumulation of ^{45}Ca. Assuming a 3 : 1 Na-Ca stoichiometry for the exchanger and correcting for the 15% of leaky vesicles maximally present in the preparation, we can estimate that if 100% of the theoretical accumulation of ^{45}Ca had taken place, an intravesicular ^{45}Ca content of 28 nmol/μl would have been obtained in vesicles pre-equilibrated with 100 mM NaCl. If the t-tubule vesicles have an intravesicular volume of 1.5 μl/mg (see below), the measured ^{45}Ca equilibrium content of 11.5 nmol/mg (FIG. 2) is a fraction (28%) of the expected value. We are not taking into account possible effects of membrane potential in this calculation, since it is likely that any transient potentials generated will dissipate in a few minutes after dilution. The existence of alternative pathways for sodium efflux that would partly dissipate the sodium gradient, or of a fraction of vesicles lacking the exchanger (see below), might explain these results. Addition of TTX did not modify the rate of Na-Ca exchange (data not shown), indicating that the Na gradient does not dissipate by sodium efflux through sodium channels, although other routes for sodium efflux might exist.

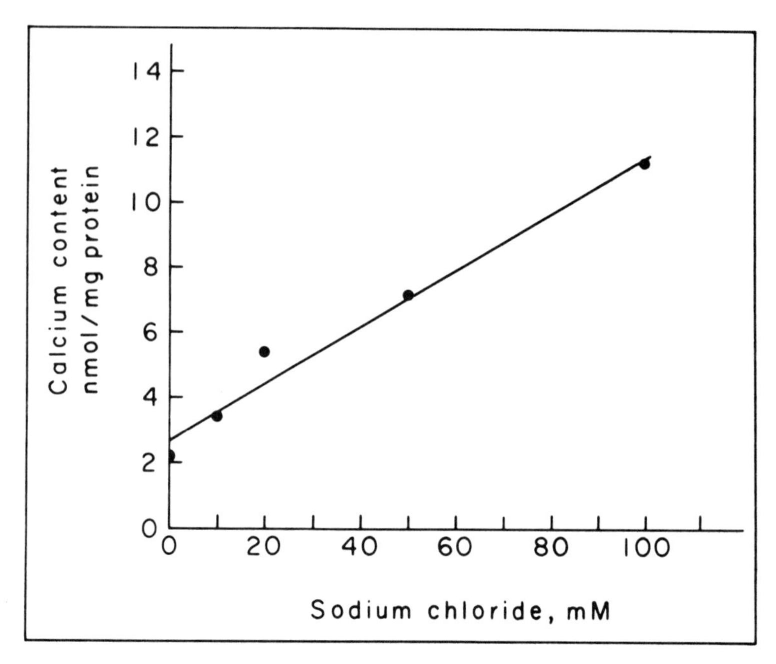

FIGURE 2. Correlation between the concentration of sodium used to pre-equilibrate the vesicles and the amount of calcium accumulated at equilibrium. Vesicles were incubated in solutions containing variable concentrations of NaCl and KCl, so that their sum was equal to 120 mM, 20 mM Tris-HCl, pH 7.5, for 90 minutes at 25°C. The reaction was initiated by 60-fold dilution in 0.1 mM $^{45}CaCl_2$, 120 mM KCl, 20 mM Tris-HCl, pH 7.5, and was stopped as described in the text. The amount of ^{45}Ca accumulated inside the vesicles reached a constant value after 5 minutes; the figure shows these maximal values.

Determination of Initial Rates

To measure initial rates of calcium uptake, we equilibrated the vesicles with 100 mM NaCl and 10 mM KCl, and we measured the amount of ^{45}Ca taken up after 5, 10, and 15 seconds of dilution in a solution containing 0.1 mM $^{45}CaCl_2$ and 100 mM KCl (FIG. 3, open symbols). The data were fitted with a first-order rate equation. It is apparent that the reaction is approximately linear only for the first couple of seconds, so that even at 5 sec there is some deviation (about 10%) from the initial rate value of 2.7 nmol per mg per 5 sec. However, for practical purposes all the values of calcium influx described in this work were measured at 5 seconds.

Electrogenicity

The electrogenic nature of the Na-Ca exchanger present in other cells is well established. If the vesicles are impermeable to other ions that might dissipate electrical potential gradients, ^{45}Ca influx should be inhibited by the positive charge buildup generated outside the vesicles during sodium efflux. To investigate this possibility, we measured ^{45}Ca influx in the presence of the same potassium gradient (K_e = 100 mM,

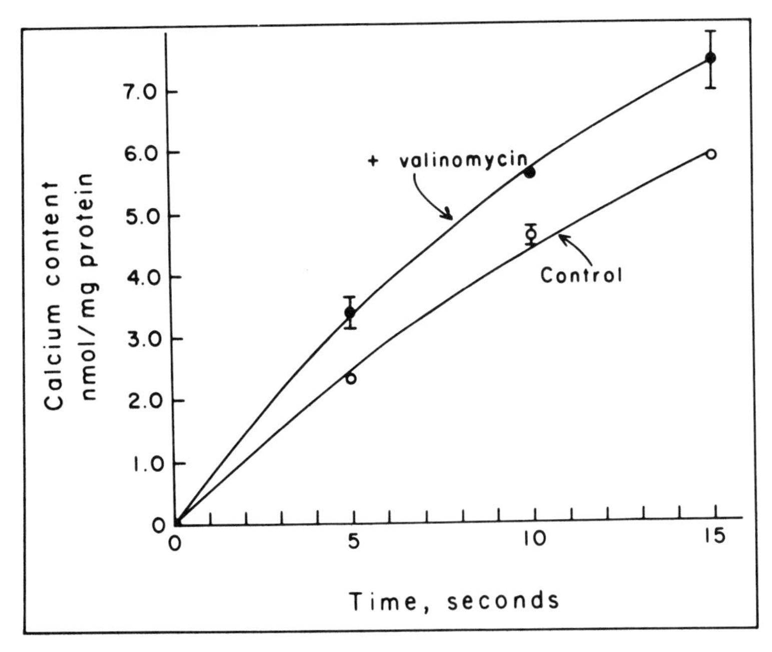

FIGURE 3. Effect of valinomycin on calcium uptake rates. Vesicles were equilibrated with 100 mM NaCl, 10 mM KCl, 20 mM Tris-HCl, pH 7.5, for 90 minutes at 25°C. To initiate the reaction, vesicles were diluted 60-fold in 0.1 mM $^{45}CaCl_2$, 100 mM KCl, 10 mM choline-Cl, 20 mM Tris-HCl, pH 7.5, and the reaction was stopped after a few seconds as described in the text. *Open circles,* no valinomycin; *closed circles,* plus 1 μM valinomycin.

K_i = 10 mM) used above (FIG. 3, open symbols) plus 1 μM valinomycin (FIG. 3, solid symbols). This gradient should generate a potassium diffusion potential equal to −59 mV, outside (cytoplasmic side) minus inside. An initial rate of ^{45}Ca uptake of 3.8 nmol per mg per 5 seconds was obtained in the presence of valinomycin, corresponding to 1.4-fold stimulation of the control rate. This stimulation of calcium influx indicates that the Na-Ca exchange reaction in t-tubule vesicles is electrogenic. We have not studied systematically the potassium permeability of the t-tubule vesicles, but we have observed that even in the absence of valinomycin, they display higher values of ^{45}Ca influx in response to external negative potassium diffusion potentials than to positive potentials (not shown). This finding could explain the rather modest stimulation of calcium influx exerted by valinomycin. Furthermore, if a high chloride permeability existed in the t-tubule vesicles isolated from frog muscle, it would counteract the effect of the positive charge buildup generated by the efflux of sodium and would also rapidly dissipate the potassium diffusion potentials. Either one of these two effects would diminish the effect of valinomycin.

Keeping these limitations in mind, we investigated the effect of imposing potassium diffusion potentials on ^{45}Ca influx (FIG. 4). We found that a potential shift from zero to 59 mV, positive outside, decreased calcium influx by 40% relative to the control kept at zero potential, whereas a 59-mV shift, negative outside, increased calcium influx by 40%. These results confirm the electrogenic nature of the t-tubule exchanger, although more experiments are needed to investigate whether the effect of potential

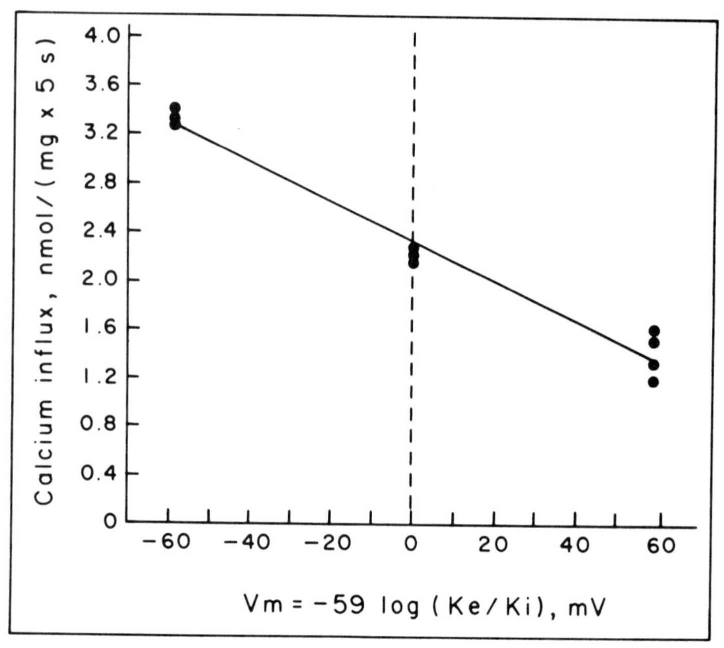

FIGURE 4. Rates of calcium uptake (calcium influx) as a function of the calculated potassium diffusion potentials. Vesicles equilibrated in 100 mM NaCl, 10 mM KCl, 20 mM Tris-HCl, pH 7.5, for 90 minutes at 25°C, were diluted in 0.1 mM $^{45}CaCl_2$ with either 1 mM KCl (V_m = 59 mV), 10 mM KCl (V_m = 0 mV), or 100 mM KCl (V_m = −59 mV). All solutions contained 1 μM valinomycin. Choline chloride was added to maintain osmolarity constant.

would be enhanced by removal of chloride ions. To test this point, experiments are in progress to measure calcium fluxes in solutions where chloride has been replaced by glutamate.

Effect of External Calcium and Internal Sodium on Calcium Influx

To measure the dependence of ^{45}Ca influx on the external calcium concentration, we diluted t-tubule vesicles equilibrated with 100 mM NaCl and 10 mM KCl in solutions containing 100 mM KCl, 1 μM valinomycin, and different ^{45}Ca concentrations. The rate of ^{45}Ca uptake increased hyperbolically with increasing calcium (FIG. 5), yielding an apparent K_m value of 2.7 ± 0.4 μM. This apparent K_m value is in the lower end of K_m values described for the exchanger present in other tissues.

The effect of sodium on ^{45}Ca influx was measured by diluting t-tubule vesicles, equilibrated with solutions containing 5 mM KCl and variable concentrations of NaCl, in a solution containing 0.1 mM $^{45}CaCl_2$, 100 mM KCl, and 1 μM valinomycin. In the absence of sodium, the rate of ^{45}Ca uptake was about one-third of the maximal rate obtained at 120 mM NaCl (FIG. 6, solid symbols). As described elsewhere,[5] this uptake is due to Ca-Ca exchange. Contrary to the clear hyperbolic dependence of the magnitude of the influx with external calcium (FIG. 5), the increase of influx with increasing intravesicular sodium does not follow a hyperbolic function (FIG. 6). If we

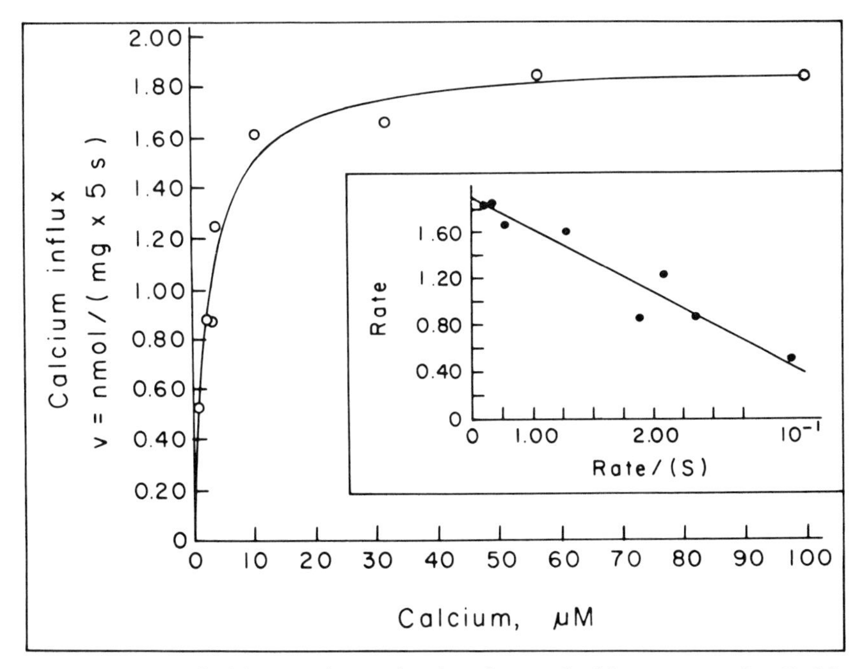

FIGURE 5. Rates of calcium uptake as a function of external calcium concentration. Vesicles were equilibrated in 100 mM NaCl, 10 mM KCl, 20 mM Tris-HCl, pH 7.5, for 90 minutes at 25°C and were diluted in solutions containing variable concentrations of $^{45}CaCl_2$, 100 mM KCl, 10 mM choline chloride, 20 mM Tris-HCl, pH 7.5, 1 μM valinomycin. To obtain solutions of the desired free-calcium concentrations, we used *N*-hydroxyethylethylenediaminetriacetic acid (HEDTA) up to 20 μM and nitrilotriacetic acid (NTA) for higher concentrations. All solutions were checked with a calcium electrode.

assume that increasing intravesicular sodium inhibits Ca-Ca exchange in proportion to the stimulation of the influx due to sodium, and that at 120 mM NaCl the exchange due to Ca-Ca becomes zero, we can generate the points illustrated in FIGURE 6 (open symbols). If we then subtract these calculated values for Ca-Ca exchange from the total values, we obtain the values of influx shown in the inset to FIGURE 6. These corrected values follow reasonably well a theoretical curve generated with a Hill equation with $n = 3$ (FIG. 6, inset) that predicts 50% stimulation of Na-dependent calcium influx at 50 mM intravesicular NaCl.

Calcium Efflux

Passive Equilibration of T-Tubule Vesicles with Calcium

In order to measure the effect of sodium on calcium efflux, we searched for conditions to equilibrate the vesicles with ^{45}Ca. We found that t-tubule vesicles incubated with solutions containing increasing ^{45}Ca concentrations reached equilibrium after 2 hours at 25°C, since their ^{45}Ca contents were the same after 2 and 5 hours of incubation.

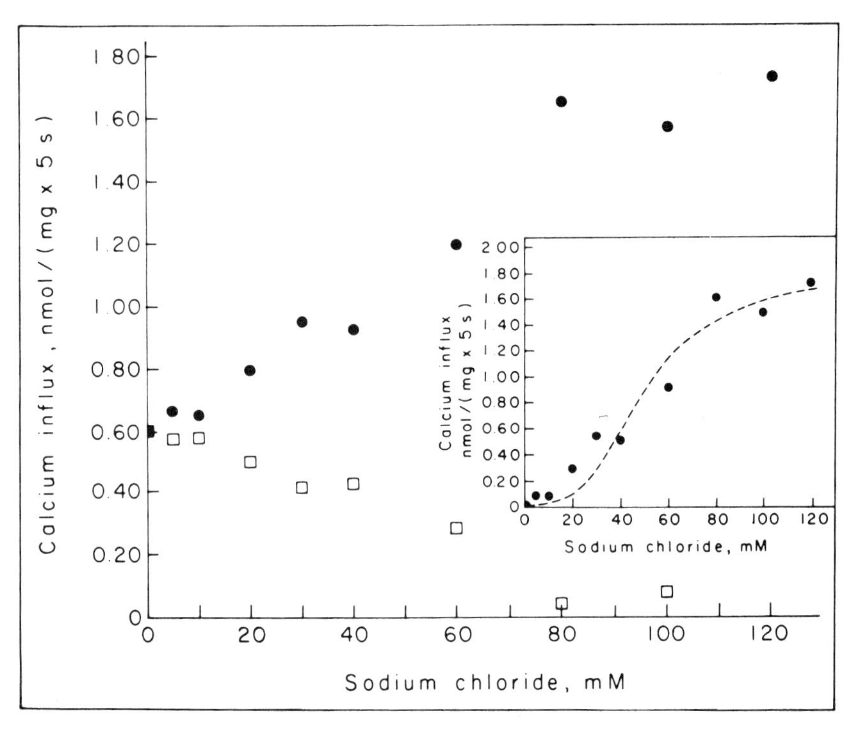

FIGURE 6. Rates of calcium uptake as a function of internal sodium. Vesicles were equilibrated in solutions containing variable NaCl concentrations, 5 mM KCl, variable concentrations of choline chloride (up to 120 mM), 20 mM Tris-HCl, pH 7.5, for 90 minutes at 25°C, and were diluted in solutions containing 0.1 mM $^{45}CaCl_2$, 100 mM KCl, 20 mM choline-Cl, 20 mM Tris-HCl, pH 7.5, 1 μM valinomycin. The total amount of ^{45}Ca accumulated (*solid circles*) were corrected for the accumulation observed in 0 sodium (Ca-Ca exchange) as follows: We assumed that at 120 mM NaCl the exchange due to Ca-Ca becomes zero, and that at intermediate sodium concentrations Ca-Ca exchange decreases in proportion to the stimulation caused by sodium. This calculation generated the points illustrated in FIGURE 6, open squares. Subtraction of the calculated from the experimental values originated the influx values shown in the inset to FIGURE 6. These corrected values follow reasonably well a theoretical curve generated with a Hill equation with $n = 3$ (FIG. 6, inset).

Surprisingly, we observed a nonlinear increase in the amount of ^{45}Ca taken up at equilibrium as a function of external calcium concentration. This nonlinear function represented the sum of a linear component, with a slope of 1.6 nmol per mg of protein per mM, plus a hyperbolic component. The linear component is the sum of the nonspecific calcium binding plus the free calcium equilibrated into the intravesicular space. The measured nonspecific ^{45}Ca binding was <0.1 nmol per mg per mM (see the legend to FIG. 7); thus, by difference, the amount of calcium equilibrated in the vesicles should be about 1.5 nmol per mg per mM, giving an intravesicular volume of 1.5 μl/mg of protein. Subtraction of the total linear component generated a saturable hyperbolic function, corresponding to binding to a single class of sites (FIG. 7). The calculated maximal binding capacity was 70 $\pm$ 2 nmol/mg of protein, with a K_d of 2.3 $\pm$ 0.3 mM. These low-affinity calcium-binding sites are present in the t-tubule vesicular

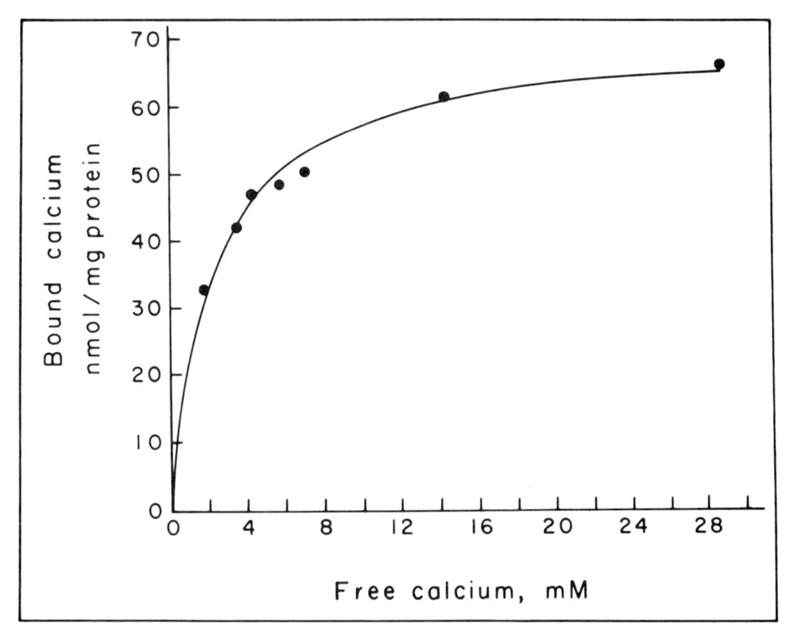

FIGURE 7. Calcium binding to the t-tubule vesicles. Vesicles were incubated for 3 hours at 25°C in solutions containing variable concentrations of $^{45}CaCl_2$, 10 mM K-gluconate, 1 μM valinomycin, 100 mM Hepes-Tris, pH 7.5, and variable concentrations of choline chloride to keep constant osmolarity. A 14 μl fraction (6 μg of protein) of the above solution was diluted in 0.3 ml of a solution containing 5 mM $MgCl_2$, 10 mM EGTA, 1 μM valinomycin, 20 mM Hepes-Tris, pH 7.5 (stopping solution). The vesicles (0.25 ml) were collected on Millipore filters (HA 0.45 μm) and were washed with 4 ml of ice-cold stopping solution. The resulting values of ^{45}Ca accumulated in the vesicles were fitted with a curve equal to the sum of a saturable plus a linear component (1.6 nmol per mg of protein per mM). Subtraction of the linear component yielded the values shown in the figure, that were fitted to a hyperbolic function (B_{max} = 70 ± 2 nmol/ mg; K_d = 2.3 ± 0.3 mM). The linear component represents nonspecific calcium binding plus free calcium equilibrated into the intravesicular space. We measured nonspecific ^{45}Ca binding by first diluting 4 μl (6 μg) of t-tubule vesicles in 0.3 ml of ice-cold stopping solution, adding then 10 μl of the solution containing variable concentrations of ^{45}Ca, filtering 0.25 ml and washing as above. A negligible nonspecific linear component was found, with a slope <0.1 nmol per mg per mM.

lumen, since calcium was lost from the vesicles after addition of the ionophore A23187. Direct determination of the calcium content of the isolated t-tubule vesicles was carried out after reducing the membranes to ashes and measuring calcium by atomic absorption spectroscopy.[5] We found that the isolated vesicles contained 112 ± 39 nmol of calcium per mg of protein (average of eight preparations). Thus, despite the long isolation procedure followed, the t-tubule vesicles retained a very high calcium content, of which at least 62% is bound (FIG. 7) and is available for exchange with external ^{45}Ca. The nature of the low-affinity calcium-binding sites present in the lumen of the t-tubule vesicles remains to be determined. However, the finding that the t-tubules have a high density of low-affinity binding sites in their lumen, which according to the K_d values found (2.3 mM) would be approximately half-saturated in the resting cells, has to be

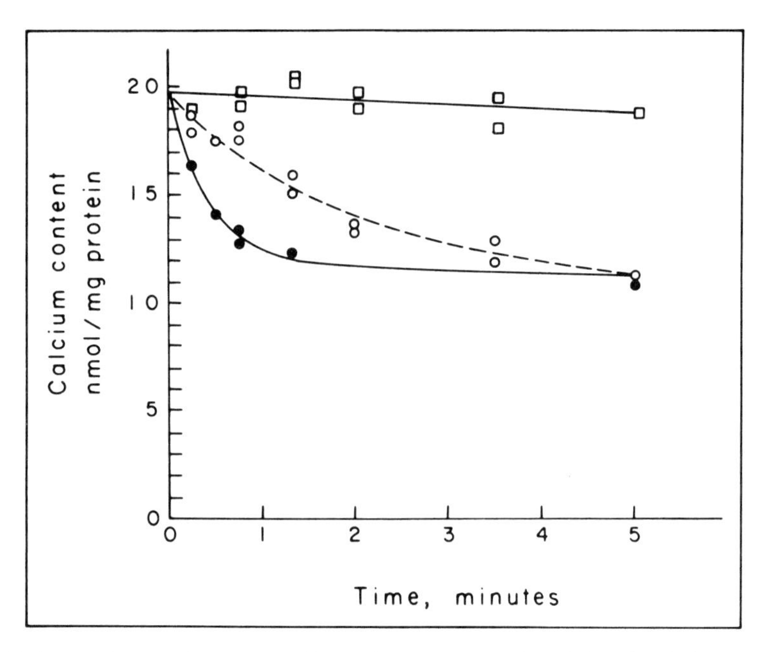

FIGURE 8. Calcium efflux in 150 mM Na-gluconate, 10 mM EGTA (*solid circles*), 150 mM Na-gluconate, 20 μM calcium (*open circles*) or 150 Hepes/Tris, 10 mM EGTA (*open squares*). Only the initial part of the measurements is illustrated, but efflux was determined for longer times that allowed resolution in sodium-containing solutions of a double exponential decay (see FIG. 9), whereas efflux in sodium-free solution decayed with a single exponential. For experimental details, see text.

taken into account when interpreting the effects of modifying extracellular calcium on the physiology of the skeletal muscle cells.

Effect of External Sodium on Calcium Efflux

Vesicles equilibrated with 2 mM $^{45}CaCl_2$ and 10 mM K-gluconate were diluted in solution containing 150 mM Na-gluconate, 10 mM K-gluconate, and 1 μM valinomycin, adding in some experiments 10 mM EGTA to decrease the external calcium concentration below pCa 8; and the amount of ^{45}Ca remaining inside the vesicles was measured as a function of time. As a control, vesicles were diluted in sodium-free solutions containing 10 mM EGTA. Osmolarity was kept constant by addition of Hepes/Tris, pH 7.5, and 1 μM valinomycin was added to keep the potential at 0 mV.

Dilution of t-tubule vesicles equilibrated with ^{45}Ca into a solution containing 150 mM Na-gluconate and 10 mM EGTA produced a calcium efflux too fast to be measured accurately and that was essentially finished before one minute (FIG. 8, solid circles). Dilution into 150 mM Na-gluconate, 20 μM external calcium, without EGTA produced a ^{45}Ca efflux (FIG. 8, open circles) with a rate constant of 0.55 min^{-1} (half time = 1.3 min), and an average value for two different experiments of 0.52 $\pm$ 0.17 min^{-1}. In this t-tubule preparation, 42% of the ^{45}Ca accumulated was rapidly released by

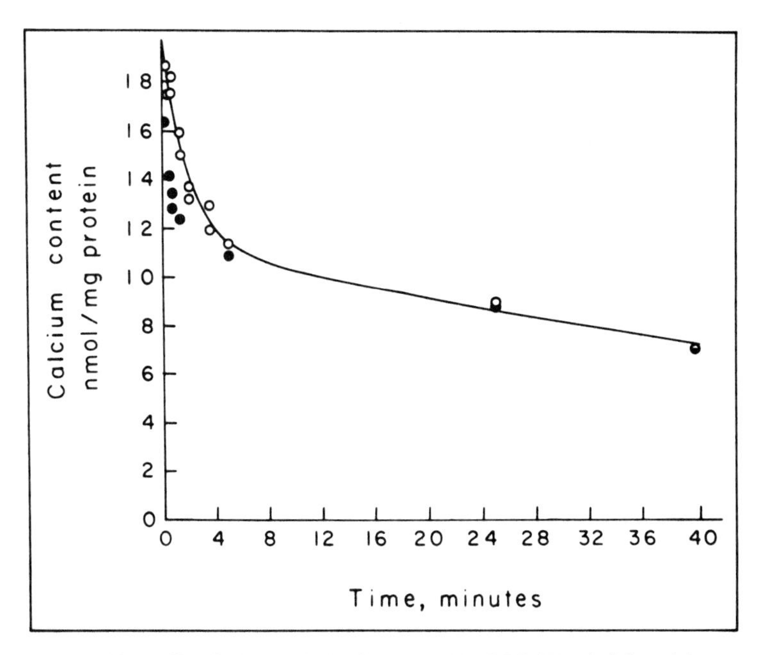

FIGURE 9. Calcium efflux in 150 mM Na-gluconate, 10 mM EGTA (*solid circles*) or 150 mM Na-gluconate, 20 μM calcium (*open circles*). The latter was fitted with a double exponential decay, with rates constants of 0.55 ± 0.24 and 0.011 ± 0.004 min^{-1}. For experimental details, see text.

dilution in 150 mM Na-gluconate, plus or minus 10 mM EGTA. In contrast, dilution in a sodium-free solution containing 10 mM EGTA, despite the large chemical gradient generated for calcium, produced an extremely slow calcium efflux (FIG. 8, open squares). In this case, as determined for times as long as 90 minutes, we found that the ^{45}Ca content decayed with time following a simple exponential, with a rate constant of 0.010 min^{-1} (half-time = 70 min). In contrast, the ^{45}Ca content of vesicles diluted in 150 mM Na-gluconate had a double exponential decay (FIG. 9). The slower exponential had the same rate constant as that determined after dilution into sodium-free solutions, 0.011 min^{-1}, indicating that this component is due to passive calcium efflux from the vesicles.

The rate constant of calcium release in 150 mM Na-gluconate and 10 mM EGTA was determined by measuring the amount of ^{45}Ca remaining in the vesicles at one-second intervals after dilution (FIG. 10). The experimental points were fitted to a single exponential decay plus offset, yielding a rate constant of 7.6 min^{-1} (half-time = 5.5 sec) that gives an initial efflux value of 7.5 nmol per mg of protein per 5 sec. This particular preparation lost 80% of the ^{45}Ca accumulated into the vesicles one minute after dilution in sodium; this was the highest amount of ^{45}Ca release observed. The initial efflux value measured in this experiment was 3.3-fold higher than the influx values measured at zero potential (*cf.* FIG. 4). When measuring efflux under the conditions described in FIGURE 10, the vesicles were effectively equilibrated only with calcium in the inside and were diluted only with sodium on the outside. In the influx measurements, on the other hand, we do not know how much endogenous calcium remained inside the vesicles after equilibration with sodium; this residual calcium might

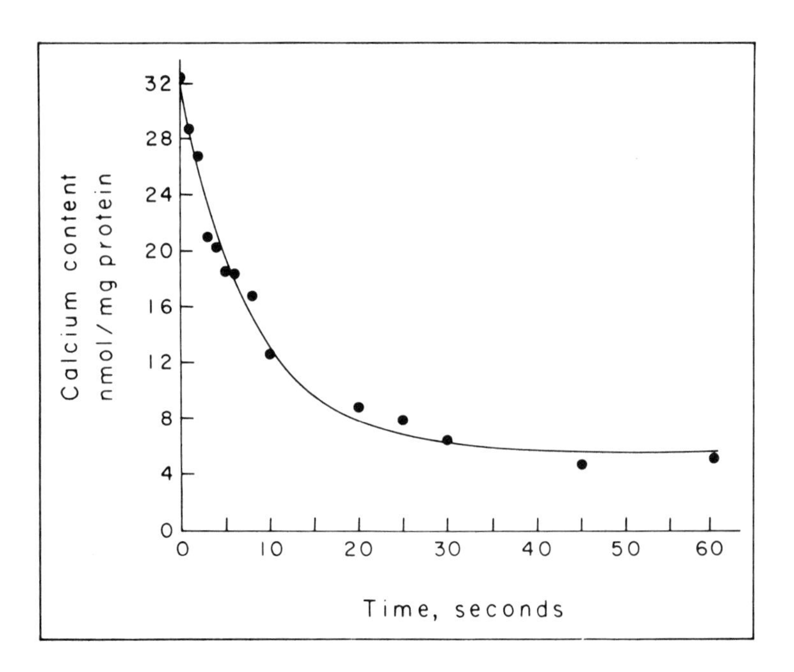

FIGURE 10. Calcium efflux in 150 mM Na-gluconate, 10 mM EGTA. The curve represents a fit of the experimental points to a single exponential decay (time constant = 7.6 min^{-1}) plus offset, since in such short times the second slow exponential decay evident in FIGURE 9 is not resolved. For further experimental details, see text.

compete with internal sodium, thus slowing down the exchange reaction. In fact, we found evidence for competition between sodium and calcium, since addition of 20 μM external calcium to 150 mM Na-gluconate decreased 14-fold the initial efflux values, from 7.5 to 0.5 nmol per mg per 5 seconds (FIGS. 9 and 10).

These combined efflux results indicate that t-tubule vesicles are highly impermeable to calcium, so that even in the presence of a large chemical gradient, very low values of calcium efflux were obtained. Addition of 150 mM sodium, 10 mM EGTA caused a fast calcium efflux mediated by the exchanger; this efflux was decreased when the external solution contained 150 mM sodium and 20 μM calcium. Thus, we failed to detect a stimulation of calcium efflux by addition of calcium to the external side of the vesicles (cytoplasmic surface). Hence, the regulatory calcium site described in the cytoplasmic face of the exchanger of cardiac and nerve cells has been lost in the isolated t-tubule vesicles. Whether this loss is due to proteolysis[13] remains to be investigated. In addition, we found that only a fraction of the calcium equilibrated inside the vesicles, from 40-80%, was released by 150 mM sodium, suggesting that the exchanger is present in a fraction of the vesicles (50-90%, after correcting for the percent leaky vesicles present in the preparation). We are currently investigating the factors responsible for this variability, since we have ruled out significant contamination of the t-tubules with other membrane fractions.

CONCLUSIONS

The results present in this work indicate that t-tubule vesicles have a Na-Ca exchange activity that can couple sodium and calcium fluxes in either direction. The maximal values for calcium efflux were 7.5 nmol per mg per 5 seconds, or 90 nmol per mg per min, and the maximal influx values were 36 nmol per mg per min. These values are several-fold higher than the maximal rates of 7-10 nmol per mg per min, at 25°C, of the calmodulin-stimulated calcium pump present in t-tubule membranes isolated from rabbit skeletal muscle.[14] However, the apparent K_m value for the latter transporter is in the range of 0.3 μM,[14] whereas the affinity of the Na-Ca exchanger for intracellular calcium is close to 3 μM. These results indicate that the exchanger in skeletal muscle cells behaves as in other cells, as a low-affinity, high-capacity transport system to remove internal calcium when its concentration increases to the μM range.

ACKNOWLEDGMENTS

We thank Dr. Marco Tulio Núñez for kindly helping us to fit mathematical functions to the experimental points, Dr. Enrique Jaimovich for many helpful discussions, and Mrs. Mónica Matus for her help in some of the experiments.

REFERENCES

1. NICOLL, D. A., S. LONGONI & K. D. PHILIPSON. 1990. Science **250:** 562-565.
2. CAPUTO, C. & P. BOLAÑOS. 1978. J. Membr. Biol. **41:** 1-14.
3. GILBERT, J. R. & G. MEISSNER. 1982. J. Membr. Biol. **69:** 77-84.
4. MICKELSON, J. R., T. M. BEAUDRY & C. F. LOUIS. 1985. Arch. Biochem. Biophys. **242:** 127-145.
5. DONOSO, P. & C. HIDALGO. 1989. Biochim. Biophys. Acta **978:** 8-16.
6. HIDALGO, C., C. PARRA, G. RIQUELME & E. JAIMOVICH. 1986. Biochim. Biophys. Acta **855:** 79-88.
7. REEVES, J. P. & J. L. SUTKO. 1983. J. Biol. Chem. **258:** 3178-3182.
8. JAIMOVICH, E., P. DONOSO, J. L. LIBERONA & C. HIDALGO. 1986. Biochim. Biophys. Acta **855:** 89-98.
9. BULL, R., J. J. MARENGO, B. SUÁREZ-ISLA, P. DONOSO, J. L. SUTKO & C. HIDALGO. 1989. Biophys. J. **56:** 749-756.
10. HARTREE, E. F. 1972. Anal. Biochem. **48:** 422-427.
11. OHLENDIECK, K., J. M. ERVASTI, J. B. SNOOK & K. P. CAMPBELL. 1991. J. Cell Biol. **112:** 135-148.
12. VENOSA, R. A. 1990. *In* Transduction in Biological Systems. C. Hidalgo, J. Bacigalupo, E. Jaimovich & J. Vergara, Eds.: 275-286. Plenum Press. New York.
13. HILGEMANN, D. W. 1990. Nature **344:** 242-245.
14. HIDALGO, C., M. E. GONZÁLEZ & A. M. GARCÍA. 1986. Biochem. Biophys. Acta **854:** 279-286.

Sodium-Calcium Exchange in Vascular Smooth Muscle

M. J. MULVANY,[a] CHRISTIAN AALKJAER, AND PETER E. JENSEN

Danish Biomembrane Research Centre
and
Department of Pharmacology
Aarhus University
8000 Aarhus C., Denmark

INTRODUCTION

The role of sodium-calcium exchange in vascular smooth muscle has been a controversial and confusing area. On the one hand, as Dr. Blaustein and colleagues showed many years ago[11] and as has been repeated many times since, the tone of large vessels can be dramatically affected by altering the transmembrane sodium gradient. Thus, in aorta and conduit arteries from a number of animal models, reduction of extracellular sodium or increase of intracellular sodium (by inhibiting the sodium pump) causes an increase in tone,[2,3] while Na-Ca exchange has been demonstrated in sarcolemmal vesicles.[4] Measurements of calcium flux have supported the idea that these responses are mediated through Na-Ca exchange.[1] On the other hand, many investigators using more peripheral vessels have been unable to demonstrate large direct effects on smooth muscle tone when the sodium gradient is reduced.[5–8] Nor have they been able to demonstrate sodium-dependent changes in calcium fluxes.[9] We too, in our laboratory, using resistance arteries, have found it difficult to demonstrate increases in tone associated with decreased sodium gradient.[10–12] The aim of this communication is to review old data and present new data concerning the possible role of Na-Ca exchange in controlling the tone of rat mesenteric small arteries.

PREVIOUS EXPERIMENTS

Effect of Low-Sodium and Sodium-Free Solutions on Rat Aorta and Mesenteric Small Arteries

In rat aorta, we,[10,12] like others (as mentioned above), have shown that low-sodium solutions and sodium-free solutions cause an increase in vascular tone and prevent relaxation upon washout of a preconstrictor. In rat mesenteric small arteries, however, neither low-sodium solutions nor sodium-free solutions cause any increase in tone of resting vessels, and in noradrenaline-contracted vessels, reduced extracellular sodium causes relaxation.[11] These experiments do not, therefore, support the hypothesis that

[a] Address for correspondence: Dr. M. J. Mulvany, Department of Pharmacology, Aarhus University, Universitetsparken 240, 8000 Aarhus C., Denmark.

a reduced sodium gradient is important in increasing vascular tone under normal circumstances.

The presence of Na-Ca exchange mechanisms in these vessels can, however, be demonstrated with large reductions in the transmembrane sodium gradient. For example, during relaxation from a potassium-induced tone, the rate of relaxation is reduced when the extracellular sodium is reduced to 25 mmol/L.[12] Furthermore, if intracellular sodium is raised (by exposing vessels to 1 mmol/L ouabain for 1 hr), the vessels respond to low extracellular sodium with a small response (*ca.* 15% of a maximal response), a response that cannot be inhibited by calcium antagonists.[11]

NEW EXPERIMENTS

Methods

Segments of rat mesenteric small arteries were dissected free and mounted on a myograph for measurement of contractility, as in a previously reported procedure.[13] Simultaneous measurements of contractility and cytoplasmic calcium were made in vessels loaded with fura-2 by holding them in PSS (see below) containing 10 μmol/L fura-2AM, 0.5% dimethylsulfoxide, 0.02% Pluronic F-127 (Molecular Probes, Junction City, OR) and 0.1% Cremophor EL for 1 hr at 37°C. Vessels were excited alternately with 340 ± 5 nm and 380 ± 5 nm light, and emitted light was collected by a photomultiplier through a 500-520-nm filter.[14] After the first sequence of stimulations, the vessels were reloaded with fura-2, simultaneously with incubation with ouabain (1 mmol/L). At the end of the experiments, the fluorescence was calibrated.[15] Measurements of ^{22}Na-efflux rate constants were made as described previously,[16] in which vessels were loaded with ^{22}Na and then transferred through a series of washout vials. Simultaneous measurements of contractility and membrane potential were made using intracellular microelectrodes.[17]

Solutions

Solutions were as follows (nmol/L). Physiological saline solution (PSS): NaCl, 119; KCl, 4.6; $NaHCO_3$, 25; $CaCl_2$, 2.5; KH_2PO_4, 1.18; $MgSO_4$, 1.17; ethylenediamenetetraacetic acid (EDTA), 0.026; glucose, 5.5; bubbled with 5% CO_2 in O_2. Hepes buffer: NaCl, 140; KCl, 4.6; $CaCl_2$, 2.5; KH_2PO_4, 1.18; $MgSO_4$, 1.17; EDTA, 0.026; glucose, 5.5; Hepes, 5; bubbled with O_2. Sodium-free solution was prepared as for Hepes buffer but with *N*-methylglucamine substituted for NaCl, and neutralized with HCl (*ca.* 130 mmol/L). All solutions were held at 37°C, pH 7.5.

Effect of Low Na_o on Response to Vasopressin

We have previously demonstrated an inhibitory effect of low-sodium solutions on the response to noradrenaline.[11] To investigate whether this response was agonist-specific, we have investigated the effect of low extracellular sodium on the response to

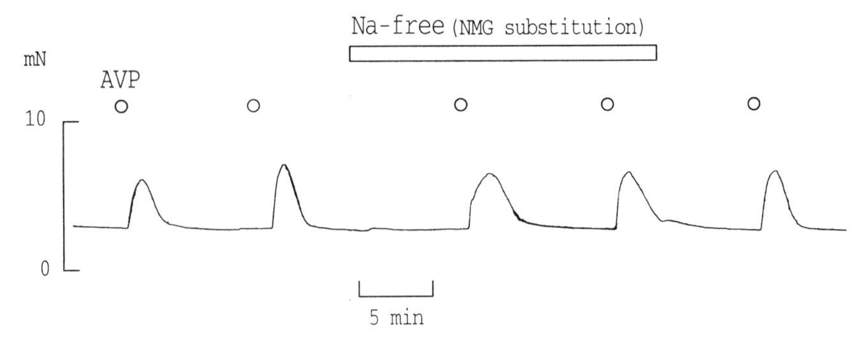

FIGURE 1. Effect of sodium-free Hepes solution, *N*-methylglucamine (NMG) substitution, on response of rat mesenteric small arteries to arginine-vasopressin (AVP). Solution was suffused through myograph chamber (volume, 3 ml) at 8 ml/min. AVP was added as a bolus to the infusion line at circles (maximum concentration in the chamber *ca.* 10^{-7} mol/L). Solutions contained phentolamine (1 μmol/L).

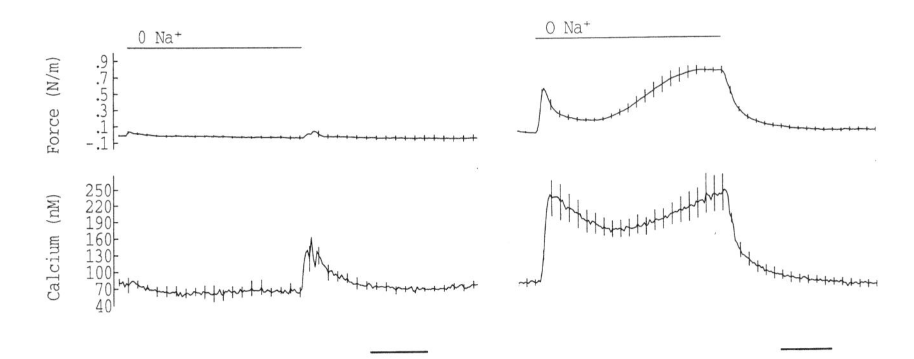

FIGURE 2. Average effect of sodium-free Hepes solution, *N*-methylglucamine substitution (0 Na^+), on the force (*upper curve*) and cytoplasmic calcium (*lower curve*) of three rat mesenteric small arteries before (*left*) and after (*right*) exposure to ouabain (1 mmol/L) for 1 hr. Measurements made at 10-sec intervals; vertical lines show standard error of measurements at 50-sec intervals. Time bars show 5 min.

vasopressin (FIG. 1). Unlike our previous experiments with noradrenaline preconstriction, little depression of the response was seen. However, it is also clear that there was no potentiation of the response, suggesting that also under these conditions, Na-Ca exchange is not playing a dominant role in the control of arterial tone. It should, though, be pointed out that Bova and colleagues,[18] using a similar protocol, were able to demonstrate potentiation of responses to vasopressin by omission of sodium, so that more work is required to clarify this issue.

Response of Sodium-Loaded Vessels to Low Extracellular Sodium

The responses of sodium-loaded vessels to low extracellular sodium was accompanied by an increase in cytoplasmic calcium (FIG. 2), although with normal intracellular

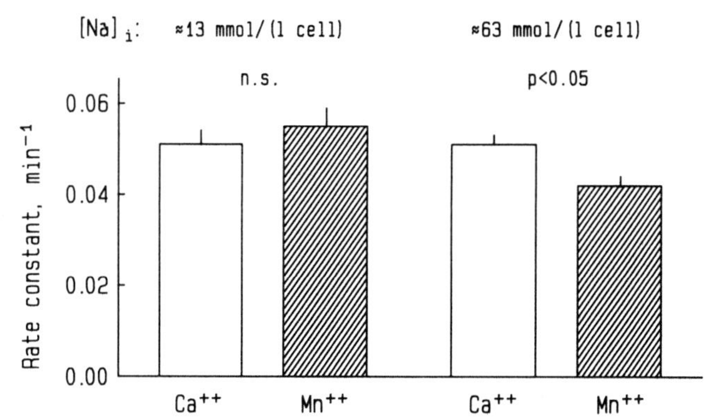

FIGURE 3. Effect of removal of extracellular calcium on the efflux of ^{22}Na from rat mesenteric small arteries. Vessels were loaded with ^{22}Na in PSS and then washed through a series of vials containing PSS in which PO_4^- had been replaced with Cl^- to avoid Mn^{2+} precipitation. Figure shows ^{22}Na-efflux rate constant before (*left*) and 2 hr after (*right*) exposure to ouabain (1 mmol/L). Figures at top show estimated intracellular sodium concentrations. Bars show mean ± standard error (7-13 vessels in each group). Hatched bars show values obtained when Ca^{2+} in the washout solutions was replaced with 1 mmol/L Mn^{2+}.

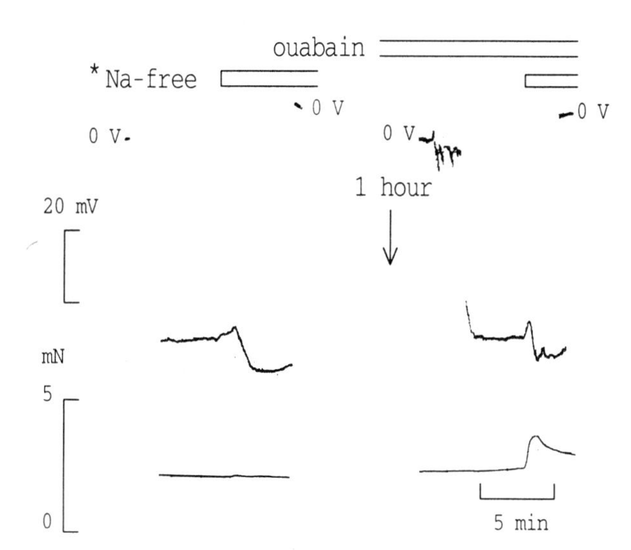

FIGURE 4. Effect of sodium-free Hepes solution, *N*-methylglucamine substitution, on the force (*lower trace*) and membrane potential (*upper trace*) of rat mesenteric small arteries before (*left*) and after (*right*) exposure to ouabain (1 mmol/L) for 1 hr.

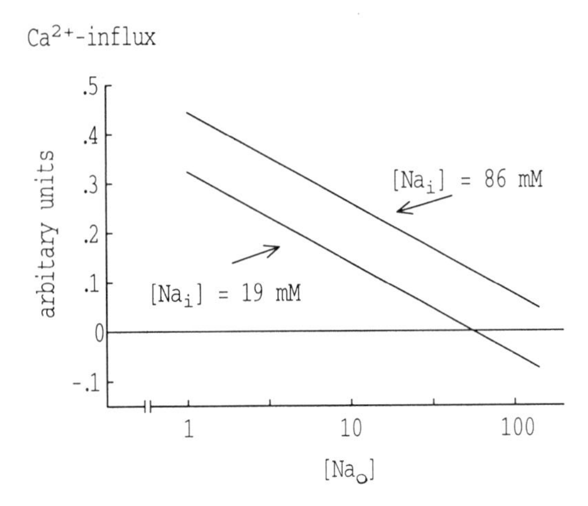

FIGURE 5. Steady-state net fluxes across the plasma membrane through Na-Ca exchange calculated for rat mesenteric small arteries as a function of the extracellular sodium concentration, $[Na_o]$. Flux shown in arbitrary units, given by

$$Ca^{2+}\ \text{influx} = (E_{Ca} - E_m) \cdot z_{Ca} - r \cdot (E_{Na} - E_m) \cdot z_{Na}$$

where $E_{Ca} = (-R \cdot T/(F \cdot z_{Ca})) \cdot \ln([Ca_i]/[Ca_o])$ and $E_{Na} = (-R \cdot T/(F \cdot z_{Na})) \cdot \ln(([Na_i] \cdot \alpha_{Na})/[Na_o])$ Horizontal line indicates no net flux, points above line indicate net influx, points below line indicate net efflux. Other lines show influx for intracellular sodium concentrations of 19 and 86 mmol/(liter cell water), as indicated. E_m, membrane potential (−60 mV[17]); Ca_o, extracellular calcium concentration (2.5 mmol/L; Ca_i, cytoplasmic calcium concentration (80 nmol/L, see FIG. 2); Na_i, intracellular sodium content per volume cell water. α_{Na}, sodium activity coefficient (0.2[20]); Na_o, extracellular sodium concentration (140 mmol/L); r, Na-Ca coupling ratio (3[19]); R, gas constant (8.31 J/(°K·mol)); T, absolute temperature (310°K); $z_{Ca} = 2$, $z_{Na} = 1$; F, Faraday's constant (96,500 coulomb/mol).

sodium, no increase in cytoplasmic calcium was seen. Thus, these experiments support the interpretation (see above) that although Na-Ca exchange may not be of importance under normal conditions, under conditions where intracellular sodium is raised, it can play a role.

Evidence that the contractile response of sodium-loaded rat small arteries to low-sodium solutions is indeed due to sodium-mediated calcium influx was obtained in flux experiments (FIG. 3). Although under normal circumstances, the efflux of ^{22}Na was not affected by the removal of extracellular calcium (by the exchange of Ca^{2+} with 1 mmol/L Mn^{2+}), after vessels had been held in ouabain for 2 hours in order to raise their intracellular sodium to about 63 mmol/(L cell), removal of extracellular calcium reduced the ^{22}Na efflux by about 17% ($p < 0.05$).

Measurements of membrane potential showed that removal of extracellular sodium caused a hyperpolarization of up to 15 mV (FIG. 4). This effectively excludes the possibility that the low-sodium solution could cause the opening of calcium channels, a conclusion consistent with the lack of effect of calcium antagonists on this response mentioned above.

NET CALCIUM FLUXES

These data can be used to calculate the direction of Na-Ca exchange under the different conditions (FIG. 5) on the basis of the classic equilibrium equation.[19] As indicated, reducing extracellular sodium may be expected to cause a sizeable increase in the influx of calcium through Na-Ca exchange and through raising the intracellular sodium a further small increase. However, relating this to the measurements of cytoplasmic calcium (and force) suggests that the other calcium-control systems are able to keep calcium at resting levels despite the calcium influx through Na-Ca exchange, until these become overwhelmed when both extracellular sodium is eliminated and intracellular sodium is raised.

CONCLUSION

The evidence presented here suggests that rat mesenteric small arteries, like other vascular preparations that have been investigated, contain a Na-Ca-exchange mechanism. Under conditions where intracellular sodium is high, this mechanism can play a role in the control of vascular tone. However, under normal circumstances, it does not appear that the Na-Ca-exchange mechanism of rat mesenteric small arteries plays an important role in the control of tone, or cytoplasmic calcium, and that vascular tone is dominated by other calcium homeostatic mechanisms.

REFERENCES

1. REUTER, H., M. P. BLAUSTEIN & G. HAEUSLER. 1973. Na-Ca exchange and tension development in arterial smooth muscle. Phil. Trans. R. Soc. London B **265:** 87-94.
2. BLAUSTEIN, M. P. 1977. Sodium ions, calcium ions, blood pressure regulation, and hypertension: A reassessment and a hypothesis. Am. J. Physiol. **232:** C165-C173.
3. OZAKI, H. & N. URAKAWA. 1981. Involvement of a Na-Ca exchange mechanism in contraction induced by low-Na solution in isolated guinea-pig aorta. Pflügers Arch. **390:** 107-112.
4. MATLIB, M. A., A. SCHWARTZ & Y. YAMORI. 1985. A Na^+-Ca^{2+} exchange process in isolated sarcolemmal membranes of mesenteric arteries from WKY and SHR rats. Am. J. Physiol. **249:** C166-C172.
5. HERMSMEYER, K. 1982. Electrogenic ion pumps and other determinants of membrane potential in vascular muscle. Physiologist **25:** 454-465.
6. VANBREEMEN, C., P. AARONSON & R. LOUTZENHISER. 1979. Sodium-calcium interactions in mammalian smooth muscle. Pharmacol. Rev. **30:** 167-208.
7. KARAKI, H., H. OZAKI & N. URAKAWA. 1978. Effects of ouabain and potassium-free solution on the contraction of isolated blood vessels. Eur. J. Pharmacol. **48:** 439-443.
8. AARHUS, L. L., J. T. SHEPHERD, G. M. TYCE, T. J. VERBEUREN & P. M. VANHOUTTE. 1983. Contractions of canine vascular smooth muscle cells caused by ouabain are due to release of norepinephrine from adrenergic nerve endings. Circ. Res. **52:** 501-507.
9. DROOGMANS, G. & R. CASTEELS. 1979. Sodium and calcium interactions in vascular smooth muscle cells of the rabbit ear artery. J. Gen. Physiol. **74:** 57-70.
10. MULVANY, M. J., H. NILSSON, J. A. FLATMAN & N. KORSGAARD. 1982. Potentiating and depressive effects of ouabain and potassium-free solutions on rat mesenteric resistance vessels. Circ. Res. **51:** 514-524.
11. MULVANY, M. J., C. AALKJAER & T. T. PETERSEN. 1984. Intracellular sodium, membrane potential, and contractility of rat mesenteric small arteries. Circ. Res. **54:** 740-749.
12. PETERSEN, T. T. & M. J. MULVANY. 1984. Effect of sodium gradient on the rate of relaxation of rat mesenteric small arteries from potassium contractures. Blood Vessels **21:** 279-289.
13. MULVANY, M. J. & W. HALPERN. 1977. Contractile properties of small arterial resistance vessels in spontaneously hypertensive and normotensive rats. Circ. Res. **41:** 19-26.
14. GRYNKIEWICZ, G., M. POENI & R. Y. TSIEN. 1985. A new generation of Ca^{2+} indicators with greatly improved fluorescence properties. J. Biol. Chem. **6:** 3440-3450.

15. JENSEN, P. E., H. YAMAGUCHI, C. AALKJAER & M. J. MULVANY. 1991. Intracellular calcium in rat mesenteric small arteries measured by fura-2 and intracellular calcium selective electrodes (abstract). Blood Vessels **28:** 298-299.
16. AALKJAER, C. & M. J. MULVANY. 1985. Effect of ouabain on tone, membrane potential and sodium-efflux compared with ^{3}H-ouabain binding in rat resistance vessels. J. Physiol. **362:** 215-231.
17. MULVANY, M. J., H. NILSSON & J. A. FLATMAN. 1982. Role of membrane potential in the response of rat mesenteric arteries to exogenous noradrenaline stimulation. J. Physiol. **332:** 363-373.
18. BOVA, S., W. F. GOLDMAN, X.-J. YUAN & M. P. BLAUSTEIN. 1990. Influence of sodium gradient on calcium transients and contraction in vascular smooth muscle. Am. J. Physiol. **259:** H409-H423.
19. MULLINS, L. J. 1981. Transport in Heart. Raven Press. New York.
20. AICKIN, C. C. 1987. Investigation of factors affecting the intracellular sodium activity in the smooth muscle of guinea-pig ureter. J. Physiol. **385:** 483-505.

Sodium-Calcium Exchange in Aortic Myocytes and Renal Epithelial Cells

Dependence on Metabolic Energy and Intracellular Sodium[a]

JEFFREY BINGHAM SMITH, RONG-MING LYU, AND LUCINDA SMITH

Department of Pharmacology
Schools of Medicine and Dentistry
University of Alabama at Birmingham
Birmingham, Alabama 35294

The plasma membrane of certain mammalian cells contains a protein called the Na^+-Ca^{2+} exchanger that translocates Na^+ and Ca^{2+} in opposite directions. Although the exchanger was discovered more than twenty years ago,[1,2] its contribution to Ca^{2+} regulation has been difficult to determine because of the lack of a specific inhibitor. Cultured mammalian cells are well suited to assessing the specific contribution of individual transporters to overall Ca^{2+} regulation by hormones and other external stimuli. By determining the effects of a stimulus on free and total Ca^{2+} as well as $^{45}Ca^{2+}$ fluxes, a more reliable picture of Ca^{2+} regulation may be obtained than would be provided by a single approach. To assess the contribution of Na^+-Ca^{2+}exchange to cellular Ca^{2+} regulation, it is necessary to determine the effects of manipulating intra- and extracellular Na^+ concentration. An acute decrease in extracellular Na^+, however, may trigger the production of inositol trisphosphate and the release of stored Ca^{2+}, depending on the particular cell type and mammalian species.[3] External Ca^{2+} and Mg^{2+}potentiate the Ca^{2+}-mobilizing response to an abrupt decrease in external Na^+.[4] Therefore, a combination of experimental approaches is essential for distinguishing changes in cytosolic free Ca^{2+} ($[Ca^{2+}]_i$) that are caused by Ca^{2+} influx via the Na^+-Ca^{2+} exchanger from the $[Ca^{2+}]_i$ changes due to release of stored Ca^{2+} evoked by triggering the Na^+-sensitive receptor. Comprehensive studies of Ca^{2+} regulation indicate that Na^+-Ca^{2+} exchange is plentiful in certain arterial myocytes and renal epithelial cells. The exchanger in arterial myocytes and renal epithelial cells has several kinetic features that are quite similar to those in cardiac myocytes.

[a] This work was supported by Grants DK39258 and HL44408 from the National Institutes of Health.

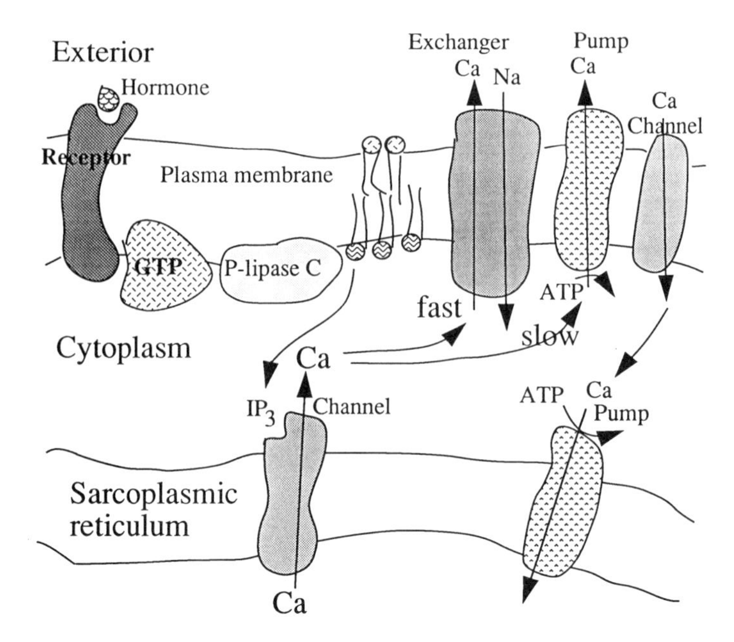

FIGURE 1. Schematic view of the predominant function of Na^+-Ca^{2+} exchange in Ca^{2+} cycling in vascular smooth muscle (modified from Benos, Warnock & Smith).[42] Stimulation of cell-surface receptor by a hormone or neurotransmitter activates phospholipase C (P-lipase C) via a guanosine triphosphate-binding protein (GTP). Phospholipase C produces inositol trisphosphate (IP_3), which opens a Ca^{2+} channel in the sarcoplasmic reticular membrane. $[Ca^{2+}]_i$ rises rapidly and falls rapidly. The rapid fall from the $[Ca^{2+}]_i$ peak is largely caused by net Ca^{2+} efflux via the Na^+-Ca^{2+} exchanger. Because IP_3 keeps the intracellular Ca^{2+} channel in the open state, the sarcoplasmic reticulum does not immediately reaccumulate the lost Ca^{2+}. The sarcoplasmic reticular Ca^{2+}-ATPase slowly reaccumulates Ca^{2+} that leaks back into the cytoplasm via a channel. The Ca^{2+} pump in the plasma membrane causes a slower efflux of Ca^{2+} from the cytoplasm and contributes substantially to the maintenance of basal $[Ca^{2+}]_i$.

ROLE OF NA^+-CA^{2+} EXCHANGE IN VASCULAR SMOOTH MUSCLE

Rat Aortic Myocytes

The primary function of the Na^+-Ca^{2+} exchanger in aortic myocytes, in our view, is to rapidly decrease $[Ca^{2+}]_i$ in stimulated cells as depicted in FIGURE 1. The following evidence supports this view. The stimulation of unidirectional $^{45}Ca^{2+}$ efflux as well as net Ca^{2+} efflux by angiotensin II (ANG) is strongly inhibited by the replacement of external Na^+ with *N*-methyl-D-glucamine (NMG) or choline.[5,6] The replacement of external Na^+ with NMG has no effect on ANG-evoked inositol trisphosphate production,[6] and ANG produces similar peak increases in $[Ca^{2+}]_i$ in the presence and absence of external Na^+.[5] Therefore, the replacement of external Na^+ with NMG probably blocks Ca^{2+} efflux through the plasma membrane rather than affecting Ca^{2+} release from the sarcoplasmic reticulum. The absence of external Na^+ strikingly increases and prolongs the plateau phase of the $[Ca^{2+}]_i$ response to ANG.[5] This potentiation of the plateau phase of the $[Ca^{2+}]_i$ response corroborates the Ca^{2+} flux data and strongly implicates the inhibition of Ca^{2+} efflux as the cause of changes in ANG-evoked Ca^{2+} handling that are produced by external Na^+ substitution. The two known pathways of Ca^{2+}efflux are the Ca^{2+} pump ATPase and the Na^+-Ca^{2+}exchanger. Ca^{2+} efflux via the exchanger depends obligatorily on external Na^+, whereas Ca^{2+} pump activity is not affected by the changes in external Na^+.[7,8] Therefore, it is likely that the inhibition of the Na^+-Ca^{2+} exchanger, which is particularly abundant in these cells,[9,10] accounts for the differences in cell Ca^{2+} handling evoked by ANG in the presence, versus the absence, of external Na^+.

In contrast to the pronounced effects of external Na^+ replacement on Ca^{2+} handling by stimulated cells, total substitution of external Na^+ with NMG or choline has no effect on $[Ca^{2+}]_i$ in nonstimulated cells.[5,11] Because quite sensitive measurements of $[Ca^{2+}]_i$ with fura-2 failed to detect any contribution of exchange activity to the short-term maintenance of basal $[Ca^{2+}]_i$, the exchanger is probably latent in nonstimulated cells. Allosteric activation of the exchanger by an increase in $[Ca^{2+}]_i$ may cause the increase in exchange activity in stimulated cells, although it has not been excluded that the latency is partly due to the low level of $[Ca^{2+}]_i$ relative to the K_m of the exchanger. Thus the selective release of stored Ca^{2+} by ionomycin, a Ca^{2+} ionophore, evokes rapid $^{45}Ca^{2+}$ and net Ca^{2+} efflux that is markedly dependent on the presence of external Na^+.[12] Because ionomycin mimics the effects of ANG on cell Ca^{2+} regulation,[12] an increase in $[Ca^{2+}]_i$ may be sufficient to activate Ca^{2+} efflux via the exchanger. Internal Ca^{2+} allosterically activates the exchanger in squid axon[13] and cardiac myocytes,[14] although the $K_{0.5}$ for activation in single perfused heart cells was 22 nM, which is well below basal $[Ca^{2+}]_i$. The cardiac exchanger was recently cloned and found to have a cluster of basic amino acids interspersed with hydrophobic residues (amino acids 251-270), which is characteristic of calmodulin-binding sites.[15] A synthetic peptide with this sequence binds calmodulin and inhibits exchange activity (K_i of 1.5 μM), apparently by binding to a specific intracellular domain of the exchanger.[16] There are no reports of a direct effect of calmodulin on Na^+-Ca^{2+} exchange, however, and the mechanisms by which ANG and ionomycin activate Ca^{2+}efflux via the exchanger are unknown.

A study of several amiloride congeners that selectively inhibit Na^+-H^+ or Na^+-Ca^{2+} exchange in aortic myocytes supports the view that Na^+-Ca^{2+} exchange is the major pathway of rapid Ca^{2+} efflux from stimulated cells. *N*-(2,4-dimethylbenzyl)amiloride (DMB) at 25-50 μM markedly inhibited ANG-stimulated $^{45}Ca^{2+}$ efflux similarly to the replacement of external Na^+ with NMG.[6] With NMG as the principle external cation, 25 μM DMB caused no further inhibition of ANG-stimulated efflux,[6] as would be expected if DMB and the removal of external Na^+ inhibit the same efflux pathway, namely the exchanger. DMB similarly inhibited net Ca^{2+} efflux evoked by ANG[6] or ionomycin.[12] Because ANG and ionomycin release stored Ca^{2+} by different mechanisms, it is likely that DMB inhibits Na^+-Ca^{2+} exchange rather than the Ca^{2+}-release channel in the sarcoplasmic reticulum. DMB (50 μM) had no effect on the production of [^{3}H]inositol phosphates evoked by ANG.[6] Na^+-H^+ exchange has no major influence on short-term Ca^{2+} regulation in aortic myocytes because complete inhibition of this exchanger with 5-(*N,N*-ethylisopropyl)amiloride had no effect on ANG-evoked $^{45}Ca^{2+}$ or net Ca^{2+}efflux.[6]

The lack of an effect of external Na^+ substitution on $[Ca^{2+}]_i$ and ^{3}[H]inositol phosphate in cells that have not been incubated with ouabain to raise cell Na^+ indicates that lowering external Na^+ does not trigger the release of stored Ca^{2+} in rat aortic myocytes.[6,11] These cells lack the Ca^{2+}-mobilizing receptor that is sensitive to changes in external Na^+.[3]

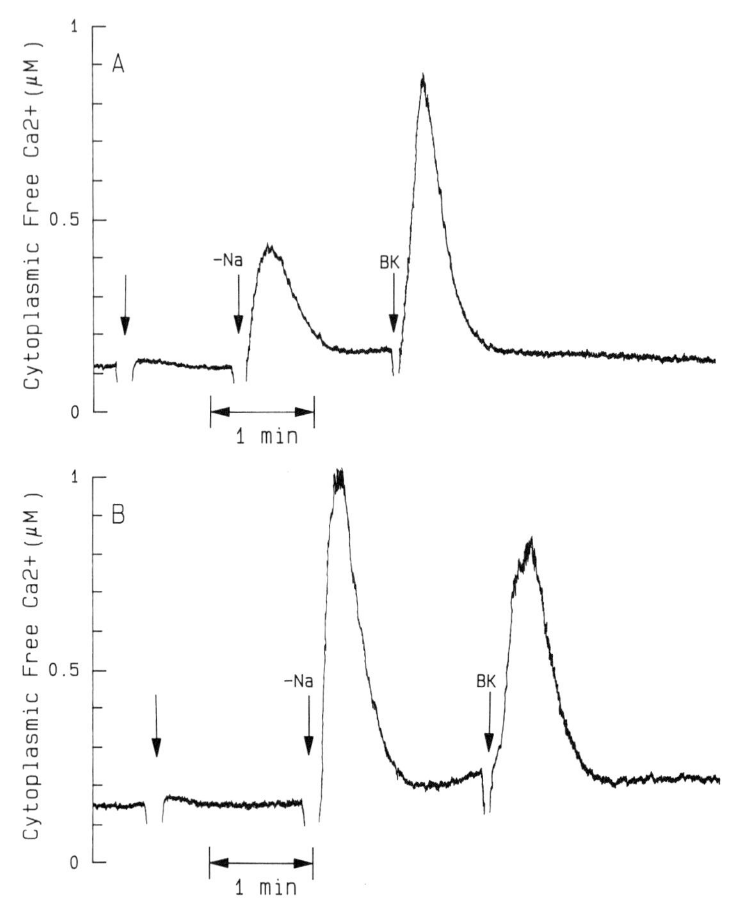

FIGURE 2. $[Ca^{2+}]_i$ spikes evoked by replacing external Na^+ with choline or by adding bradykinin (BK) in human arterial muscle cells with basal Na^+ (panel **A**) or increased Na^+ due to incubation with ouabain (panel **B**) or monensin (panel **C**). At the times indicated by the initial arrows, the solution in the cuvette was aspirated and replaced with fresh physiological salts solution.[11] At the "−Na" arrows, the solution in the cuvette was aspirated and replaced with physiological salts solution containing choline instead of Na^+. At the "BK" arrows, 100 nM bradykinin was added. Myocytes from umbilical arteries were grown on cover glasses, loaded with fura-2, and incubated for 30 min in physiological salts solution containing glucose without (**A**) or with 0.1 mM ouabain (**B**) as described previously. At the "Mon" arrow in panel **C**, 2 μg/ml of the Na^+ ionophore, monensin, was added.

Human Arterial Myocytes

FIGURE 2 shows that replacing external Na^+ with choline or adding bradykinin (BK) evokes an immediate $[Ca^{2+}]_i$ spike in myocytes cultured from human umbilical artery. Raising the concentration of intracellular Na^+ with ouabain (FIG. 2B) or the

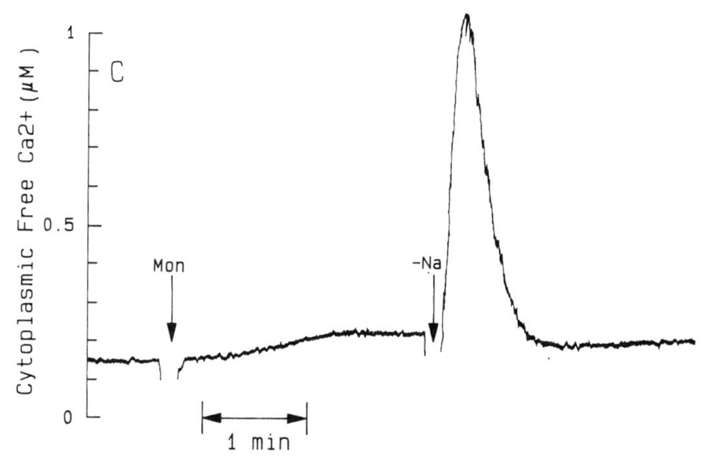

FIGURE 2. *Continued.*

addition of monensin (FIG. 2C) potentiated the effect of external Na^+ replacement on $[Ca^{2+}]_i$. Basal $[Ca^{2+}]_i$ was 127 ± 6 nM (n = 14) and 146 ± 6 nM (n = 14) in control and ouabain-treated cells, respectively. The 30-minute ouabain treatment increased cell Na^+ from ~15 to ~35 mmol/L cell water space. Peak $[Ca^{2+}]_i$ produced by replacing external Na^+ with choline was 363 ± 24 nM (n = 10) and 776 ± 46 nM (n = 14) in control and ouabain-treated cells, respectively. Addition of monensin (FIG. 2C) increased $[Ca^{2+}]_i$ after a short lag from 126 ± 9 to 214 ± 11 nM (n = 8). After the brief incubation with monensin, replacing external Na^+ with choline increased $[Ca^{2+}]_i$ to a peak level of 1001 ± 113 nM (n = 8). Because loading the cells with Na^+ markedly potentiated the rise in $[Ca^{2+}]_i$ produced by external Na^+ substitution, it is likely that the human arterial myocytes have considerable Na^+-Ca^{2+} exchange activity.

To further evaluate the possibility that these cells express the exchanger, we determined the effect of the Na^+ gradient on $^{45}Ca^{2+}$influx. FIGURE 3 shows that replacing external Na^+ with K^+ increased $^{45}Ca^{2+}$ influx only if cell Na^+ had been elevated above the basal level. Additionally, DMB (25 μM), which selectively inhibits exchange activity in rat aortic myocytes[6] and cardiac sarcolemmal vesicles,[17] strongly inhibited $^{45}Ca^{2+}$ uptake by the Na^+-loaded cells similarly to high external Na^+ (FIG. 3A). The combination of DMB and high external Na^+ caused almost the same inhibition of $^{45}Ca^{2+}$ uptake as each produced by itself (FIG. 3A), as would be expected if they inhibited the same uptake pathway, namely Na^+-Ca^{2+} exchange. The concentration of DMB that inhibited Na^+ gradient-dependent $^{45}Ca^{2+}$ uptake by 50% was 1.7 μM (FIG. 3B). DMB is ~6 times more potent as an inhibitor of exchange activity in the human arterial, compared to rat aortic, myocytes.[6] The myocytes cultured from human umbilical artery appear to have about a fifth as much exchange activity as rat aortic myocytes (FIGS. 3 and 4).

Decreasing external Na^+ releases stored Ca^{2+} in human arterial myocytes.[3] Therefore, it is likely that a substantial portion of the increase in $[Ca^{2+}]_i$ produced by replacing external Na^+ with choline in cells with basal Na^+ is caused by the release of stored Ca^{2+} rather than influx via the exchanger. If cells contain both the exchanger and the Na^+-sensitive Ca^{2+}-mobilizing receptor, it is quite difficult to assess the contribution that each makes to an increase in $[Ca^{2+}]_i$. Growth of human skin fibroblasts in

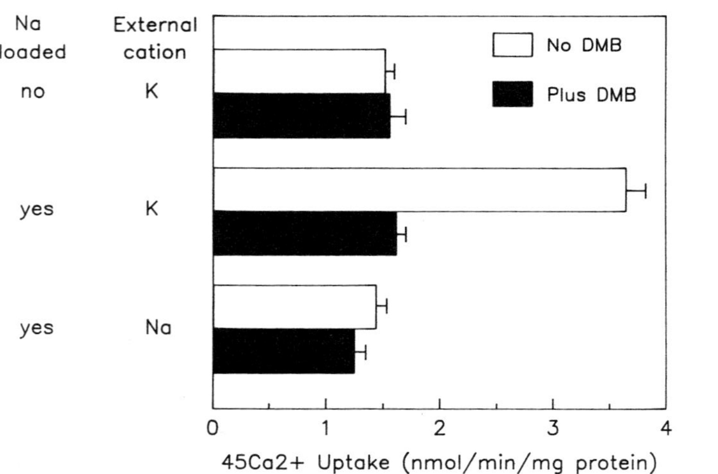

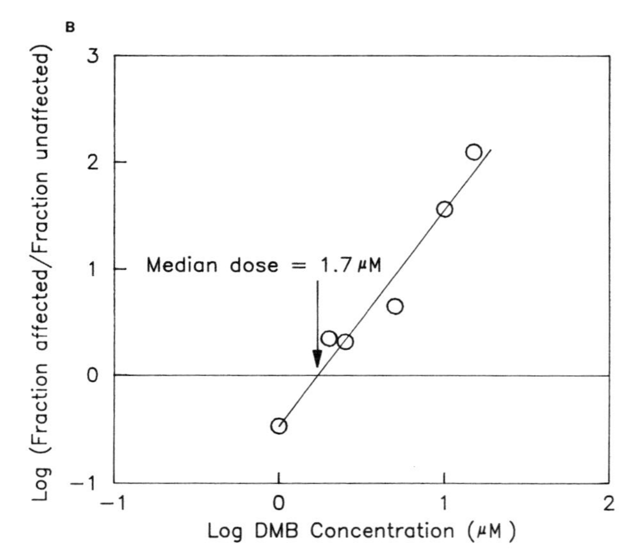

FIGURE 3. Na^+ gradient-dependent $^{45}Ca^{2+}$ influx (**A**) and concentration dependence of the inhibition of Na^+-Ca^{2+} exchange by DMB (**B**) in human arterial muscle cells. Myocytes cultured from human umbilical artery were incubated without or with (Na^+-loaded) 0.1 mM ouabain and 10 μg/ml monensin for 20 min in a physiological salts solution containing glucose and sodium propionate, as described previously.[10] (**A**) $^{45}Ca^{2+}$ uptake was measured after a 1-min incubation in the presence of 140 mM KCl or NaCl as indicated in the presence of 20 mM Hepes/Tris, pH 7.4, 2 μCi $^{45}Ca^{2+}$, 0.1 mM ouabain, and 25 μM DMB as indicated. Values are mean ± SE (n = 3 experiments in duplicate). **B** is a log-log analysis of the inhibition of $^{45}Ca^{2+}$ influx measured in the presence of 140 mM KCl by Na^+-loaded cells.

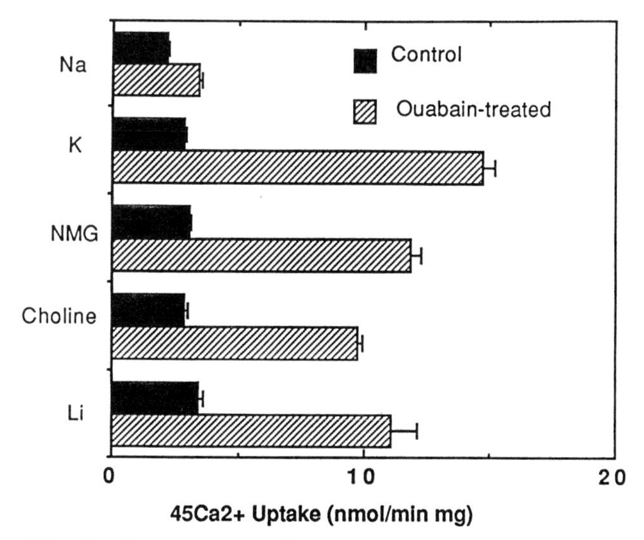

FIGURE 4. Effect of replacing external Na^+ by potassium, NMG, choline, or lithium on $^{45}Ca^{2+}$ influx by rat aortic myocytes with basal or increased intracellular Na^+. The cultures were rinsed with the solution used to measure $^{45}Ca^{2+}$ uptake and incubated for 30 sec in the same solution containing 2 μCi $^{45}Ca^{2+}$. All solutions contained (in mM): 120 of the chloride salt of the indicated cation, 5 KCl, 1 $CaCl_2$, 1 $MgCl_2$, and 20 HEPES/Tris, pH 7.4. To load the indicated cultures with Na^+, 1 mM ouabain was added to the conditioned culture medium 1.5 hours before rinsing the cultures and measuring uptake. Ouabain (1 mM) was also present during assay of uptake by the Na^+-loaded cells. Values are mean ± SE for five experiments on duplicate cultures.

TABLE 1. Comparison of Vascular Tissue Contraction and Cytosolic Free-Ca^{2+} Data in Aortic Myocytes

Condition	Immediate Contraction[a]	Free Ca^{2+}[b]
1. Inhibition of the Na^+ pump	No	No change
2. External Na^+ replacement	No	No change
3. Inhibition of Na^+ pump and external Na^+ replacement	Yes	Increased
4. Addition of NE or ANG and external Na^+ replacement	Yes, potentiated	Increased, potentiated

[a] Rat aorta, Ashida and Blaustein[20] and rat mesenteric artery, Mulvany *et al.*[21]
[b] Rat aortic myoctyes, Smith and Smith[5] and Smith, Zheng, and Smith.[11]

the presence of 100 μM Zn^{2+} selectively desensitizes them to the Ca^{2+}-mobilizing effect of an abrupt drop in external Na^+.[18] Growth in high Zn^{2+} may offer an approach to distinguish the two modes by which a drop in external Na^+ concentration elicits a $[Ca^{2+}]_i$ response.

Agreement between Vascular Tissue Contraction and $[Ca^{2+}]_i$ Data

Regarding the latency of Na^+-Ca^{2+} exchange and its prominent role in Ca^{2+} efflux, there is excellent agreement between the contraction data obtained with isolated vascular tissue and the $[Ca^{2+}]_i$ data from cultured aortic myocytes (TABLE 1). Contractile responses associated with the inhibition of the Na^+ pump are quite small and slow in onset. For example, a 60-minute incubation with enough ouabain to completely inhibit the Na^+ pump increased the tension of helical strips of rabbit aorta by only 2% relative to a high-K^+ contraction.[19] Blockade of the Na^+ pump by superfusion of rat aortic rings with a K^+-free solution slightly increased tension after 30-45 minutes.[20] A 30-minute incubation of aortic myocytes with a saturating concentration of ouabain only slightly increased $[Ca^{2+}]_i$.[11] Inhibition of the Na^+ pump raises cell Na^+, which may inhibit Ca^{2+} efflux via Na^+-Ca^{2+} exchange. However, it is unclear whether these small increases in $[Ca^{2+}]_i$ and contraction are caused by the inhibition of Ca^{2+} efflux via Na^+-Ca^{2+}exchange since a specific inhibitor of the exchanger is not available.

Decreasing external Na^+ from 139 to 7.5 mM increased the tonic tension of bovine tail artery but not rat aorta.[20] Lowering external Na^+ to 25 mM or adding 1 mM ouabain fails to elicit a contractile response from denervated rings of rat mesenteric arteries.[21] Contraction occurs, however, if the rings are incubated with ouabain to raise cell Na^+ and *then* exposed to low external Na^+.[21] $[Ca^{2+}]_i$ experiments with cultured aortic myocytes indicate that there is little if any change in $[Ca^{2+}]_i$ in response to inhibition of the Na^+ pump or abruptly lowering or completely replacing external Na^+ with choline or NMG.[11] Lowering external Na^+ after raising cell Na^+, however, produces a large spike in $[Ca^{2+}]_i$.[11] These observations are consistent with the view that the exchanger is latent in nonstimulated cells and that appreciable Ca^{2+} influx via the exchanger (reverse mode) does not occur even when there is an enormous driving force (e.g., with KCl replacing all external NaCl) unless intracellular Na^+ is raised. The Ca^{2+} influx mode of the exchanger appears to be kinetically limited by the low extent of the saturation of the internal Na^+ site at basal cell Na^+ in rat aortic and mesenteric vessels. The sigmoid dependence of exchange activity on cell Na^+ and the $K_{0.5}$ relative to the normal concentration of intracellular Na^+ supports the view that the Ca^{2+}-influx mode of exchange is kinetically limited in cells with basal Na^+ (see kinetic parameters given below). Additional studies are needed to determine whether the kinetic properties of the Na^+-Ca^{2+} exchanger in vessels from different mammalian species and vascular beds are similar to those of rat aortic myocytes.

Interestingly, lowering external Na^+ potentiates the contractile[20,21] and $[Ca^{2+}]_i$[5] responses to stimuli that trigger the release of Ca^{2+} from the sarcoplasmic reticulum. Additionally, the rate of relaxation of rat aortic rings from caffeine-induced contractions is four times faster at 139 compared to 1 mM external Na^+.[20] These findings are consistent with the view that the exchanger makes a significant contribution to Ca^{2+} efflux in stimulated cells. Hence, the inhibition of Ca^{2+} efflux (forward-mode exchange) in stimulated cells would be expected to potentiate the contractile responses to the stimuli and retard relaxation.

KINETIC FEATURES OF Na^+-Ca^{2+} EXCHANGE IN SMOOTH MUSCLE, RENAL EPITHELIAL, AND HEART CELLS

K_m *and* V_{max} *Values for* Ca^{2+} *Influx Mode of Exchange*

Na^+-Ca^{2+} exchange in smooth muscle, kidney epithelial, and heart cells has remarkably similar kinetic properties (TABLE 2). Most of the kinetic parameters summarized in TABLE 2 were determined from measurements of $^{45}Ca^{2+}$ influx that depended both on raising cell Na^+ *and* replacing external Na^+, which provides a reliable and accurate estimate of exchange activity. FIGURE 4 shows the effects of replacing external Na^+ with K^+, NMG, choline, or Li^+ on $^{45}Ca^{2+}$ influx in aortic myocytes with basal

TABLE 2. Comparison of the Kinetic Properties of the Ca^{2+}-Influx Mode of Na^+-Ca^{2+} Exchange in Renal Epithelial Cells and Aortic and Cardiac Myocytes

	$K_m Ca_o$[a] (μM)	$K_{0.5}$ Na_i (mmol/L)	Hill Coefficient	K_i Mg_o (μM)	V_{max} (nmol/min·mg)	References
Renal epithelial cells	200 ± 20	26 ± 3	3.1 ± 0.2	80 ± 10	~40	30
Smooth muscle cells	100 ± 4	28 ± 4[b]	2.8 ± 0.1[b]	90 ± 7	~30	9-11
Cardiac muscle cells	110 ± 20[c] 140[d]	21 ± 7[d]	2.0 ± 0.2[d]	770 ± 21[e]	~100[c]	28,29

[a] Values were determined in the absence of external Na^+; in the presence of 90 or 140 mM external Na^+ the apparent K_m for external Ca^{2+} was 1.44 ± 0.32 mM and 1.4 mM in aortic and cardiac myocytes, respectively.

[b] Lyu and Smith, unpublished data.

[c] Values are from $^{45}Ca^{2+}$ flux measurements with myocytes cultured from cardiac ventricles from fetal mice.[29]

[d] Values are from Na^+-Ca^{2+} exchange currents of single ventiruclar cells dissociated from guinea pig hearts with collagenase.[28]

[e] The K_i for Mg^{2+} was determined with choline as the principle external cation. The K_i values for the epithelial and smooth muscle cells were obtained with *N*-methyl-D-glucamine as the principle external cation. The concentration of Mg^{2+} that causes 50% inhibition of exchange is the same with choline or *N*-methyl-D-glucamine as the principle external cation, however, the presence of high concentrations of external K^+ decreases the potency of Mg^{2+} by ~7-fold in aortic myocytes and renal epithelial cells.[9,30]

and elevated cell Na^+. Raising cell Na^+ markedly potentiated the increase in $^{45}Ca^{2+}$ influx produced by replacing all external Na^+ with another monovalent cation (FIG. 4).

The V_{max} of exchange in smooth muscle and renal epithelial cells was 30 and 40%, respectively, of the V_{max} in heart cells (TABLE 2). From the V_{max} and a turnover number of 1,000 to 2,000 Ca^{2+} per second,[22,23] it appears that smooth muscle and epithelial cells have approximately 100,000 exchangers per cell. The dependence of exchange activity on internal Na^+ is sigmoidal, with $K_{0.5}$ values between 21 and 28 mM for the three cell types. The stoichiometry of the exchanger is 3 Na^+ per Ca^{2+} in heart cells[24,25] and sarcolemmal vesicles.[26] The Hill coefficients of the Na^+ dependence in smooth muscle and renal epithelial cells are 3.3 and 3.1, respectively, which are consistent with the stoichiometry of the exchanger in heart.

The K_m values for external Ca^{2+} were determined in the presence and absence of external Na^+. In the absence of external Na^+ the K_m values were between 100 and 200 μM for the three cell types. External Na^+ interacts competitively with external Ca^{2+},[27] thus with high concentrations of external Na^+ the apparent K_m for Ca^{2+} increases about 10-fold to 1.4 mM in heart[28] and smooth muscle cells (TABLE 2). It is noteworthy that K_m of the exchanger for internal Ca^{2+} is probably less than 1 μM[5,8]; the value for cardiac ventricular cells is 0.6-0.8 μM.[28] The K_m (0.16 μM) of the Ca^{2+} pump for $[Ca^{2+}]_i$ is consistent with the view that the pump makes a greater contribution than the exchanger to Ca^{2+} efflux in nonstimulated cells.[8]

External Calcium and Magnesium Interact Competitively

External Mg^{2+}, like external Na^+, competitively inhibits Ca^{2+} influx via the exchanger in smooth muscle,[9] heart,[29] and renal epithelial cells.[30] Mg^{2+} is not transported at a detectable rate by the exchanger.[9,30] The K_i values for Mg^{2+} are similar to the K_m values for Ca^{2+} in the smooth muscle and epithelial cells (TABLE 2). The K_i value for the cultured cardiac myocytes is ~9 times greater than the values for the smooth muscle and epithelial cells (TABLE 2). The markedly decreased potency of external Mg^{2+} as an inhibitor of the Ca^{2+}-influx mode of exchange in the heart cells suggests that there are different isoforms of the exchanger in mammalian cells.

High external potassium decreases the potency of Mg^{2+} by seven- to eightfold in the epithelial and smooth muscle cells.[9,11,30] High external K^+ increases the selectivity of the exchanger for Ca^{2+} relative to Mg^{2+} without affecting the apparent affinity of the exchanger for external Ca^{2+}.[30]

Mg^{2+} also competitively inhibits the $^{45}Ca^{2+}$-Ca^{2+} exchange in rod outer segments from vertebrate photoreceptor cells.[31] Potassium (or rubidium) strikingly decreased the potency of Mg^{2+} in the rod outer segments as was just mentioned for the smooth muscle and epithelial cells (TABLE 2). The exchanger in rod outer segments exchanges 4 Na^+ per 1 Ca^{2+} plus 1 K^+.[32,33] The exchanger in the epithelial (Lyu, Smith & Smith, J. Membr. Biol., in press), smooth muscle,[9,11] and heart cells does not appear to require or transport K^+. The Na^+-Ca^{2+} and Na^+-(K^+ + Ca^{2+}) exchangers may have a common structural domain that interacts with K^+, however, because K^+ increases the specificity of the divalent site for Ca^{2+} versus Mg^{2+} in both exchangers.

Divalent metals generally inhibit Na^+-Ca^{2+} exchange. Their inhibitory potency depends on the closeness of the crystal ionic radius to that of Ca^{2+}.[9,34] For example, cadmium, which has an ionic radius that is very close to that of Ca^{2+}, is a much more potent inhibitor than Mg^{2+} or Ba^{2+} whose radii are considerably smaller and larger than that of Ca^{2+}, respectively.

The K_m for Ca^{2+} of the exchanger in squid axons,[35] cardiac,[27] and aortic myocytes[9,28] is about 1000-, 230-, and 200-fold larger in the Ca^{2+} influx compared to the Ca^{2+}-efflux mode. Because the cytoplasmic and external divalent sites have such a marked difference in apparent Ca^{2+} affinity, it is unclear whether intracellular Mg^{2+} in the physiological range (0.5 mM) affects the Ca^{2+} efflux mode of exchange.

DEPENDENCE OF Na^+-Ca^{2+} EXCHANGE ON METABOLIC ENERGY

Energy Dependence in Rat Aortic Myocytes

The Na^+-Ca^{2+} exchanger is not directly coupled to ATP or another high-energy compound because the exchanger is active in dialyzed squid axons[7,36,37] and sarcolemmal vesicles[38] in the absence of a source of metabolic energy. None the less, exchange activity in squid axons,[7,36,37] as well as cardiac[39] and smooth muscle cells,[10] is markedly influenced by ATP or a closely related metabolite. Rather than directly fueling the exchanger, ATP allosterically activates the exchanger in squid axons,[7,36,37] giant sarcolemmal patches from cardiac cells,[39] and apparently cardiac and smooth muscle cells.[10] Three different classes of mitochondrial poisons strongly inhibited Na^+-Ca^{2+} exchange in aortic myocytes.[10] The presence of glucose during the incubation with each mitochondrial poison prevented the inhibition of exchange activity. Glucose yields ATP via glycolysis, thereby preventing the mitochondrial poisons from depleting cellular ATP.[10] The possibility that the mitochondrial poisons directly affected exchange activity is excluded because glucose prevented them from inhibiting exchange. FIGURE 5 shows the time courses of ATP depletion produced by rotenone as well as ATP repletion produced by adding glucose to rotenone-treated aortic myocytes. The time courses for the loss and restoration of Na^+-Ca^{2+} exchange closely parallel cellular ATP levels (FIG. 5). It is notable that about 20% of total exchange activity remains after ATP is virtually eliminated (FIG. 5).

Energy Dependence in Renal Epithelial Cells

TABLE 3 shows the effects of three different classes of mitochondrial poisons and 2-deoxy-D-glucose (DOG), which blocks glycolysis, on exchange activity and ATP levels in renal epithelial cells (LLC-MK_2). The combination of DOG with a metabolic poison was necessary to deplete ATP; glucose prevented ATP depletion (TABLE 3). Virtually eliminating ATP decreased exchange activity by $\sim$80% regardless of the particular mitochondrial poison that was used (TABLE 3). Glucose largely prevented the inhibition of exchange activity, indicating that the poisons do not directly affect the exchanger (TABLES 3 and 4). TABLE 4 shows that removing DOG and dinitrophenol restores cellular ATP and exchange activity. Under the conditions of this experiment the poisons decreased ATP by 79% and exchange activity by 55% (TABLE 4). Removing the poisons for ten minutes increased exchange activity to 87% of that in control cells that were not exposed to the poisons and also increased ATP to $\sim$50% of the level in the control cells (TABLE 4).

A plausible explanation for the energy dependence of Na^+-Ca^{2+}exchange is a phosphorylation reaction catalyzed by a protein kinase. Only hydrolyzable ATP analogues activated exchange in squid axon and ventricular patches, and this activation required Mg^{2+} and Ca^{2+}.[37,39] The energy dependence in the aortic myocytes and renal

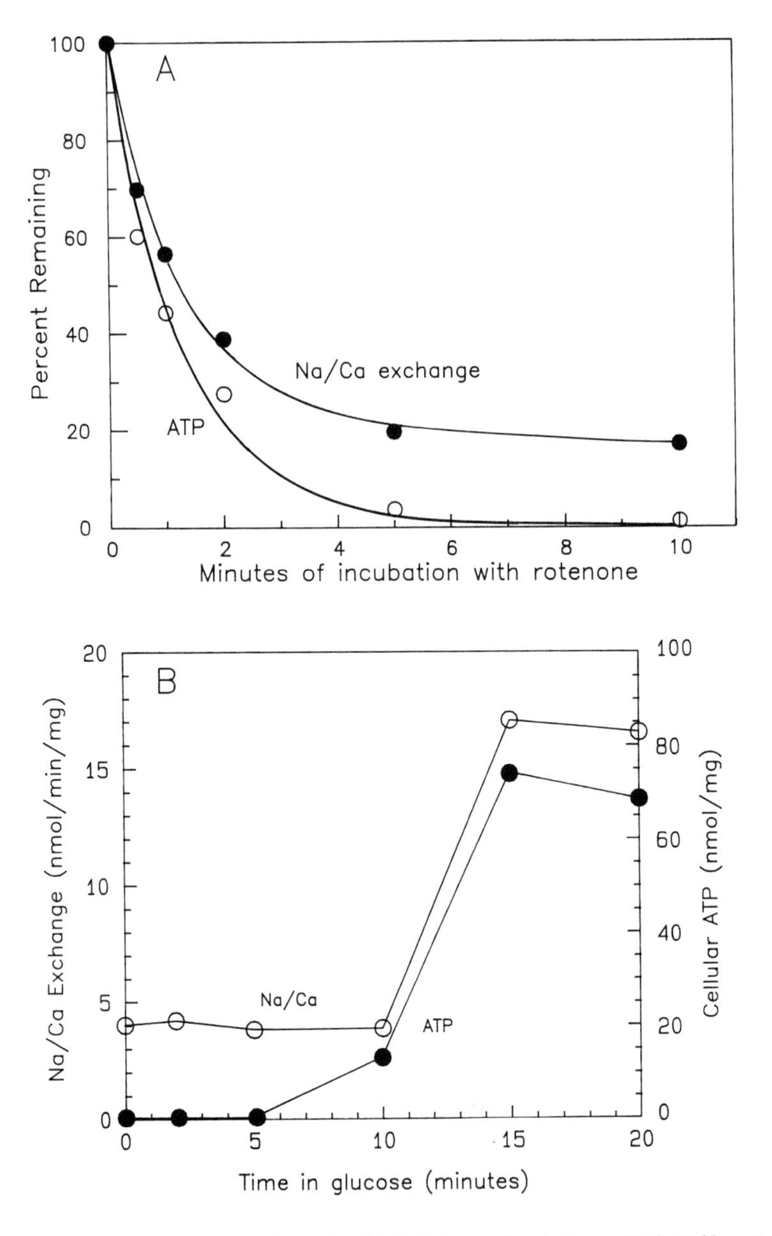

FIGURE 5. Time courses of the effects of ATP depletion and repletion on Na^+-Ca^{2+} exchange in rat aortic myocytes. (After FIG. 3 and 4 in Smith & Smith.[10])

TABLE 3. Effects of Metabolic Poisons on $^{45}Ca^{2+}$ Uptake and Cellular ATP in the Presence and Absence of Glucose

Additions	$^{45}Ca^{2+}$ Uptake % Control		Cellular ATP % Control	
	No Glucose	Plus Glucose	No Glucose	Plus Glucose
None	100 ± 4	99 ± 2	100 ± 4	107 ± 6
DOG	81 ± 2	99 ± 5	34 ± 2	108 ± 5
Oligomycin	105 ± 6	104 ± 2	71 ± 4	98 ± 6
Antimycin A	84 ± 4	84 ± 5	59 ± 4	103 ± 4
Rotenone	84 ± 4	86 ± 4	70 ± 5	95 ± 4
DNP	90 ± 7	90 ± 5	66 ± 9	76 ± 11
DOG + oligomycin	22 ± 2	96 ± 2	3 ± 1	75 ± 11
DOG + antimycin A	17 ± 1	81 ± 5	1 ± 1	94 ± 7
DOG + rotenone	19 ± 1	75 ± 2	2 ± 1	87 ± 8
DOG + DNP	20 ± 2	78 ± 4	2 ± 1	97 ± 6

NOTE: Ouabain (0.1 mM final) was added to the culture medium for 1 hr to load the cells with Na^+. The cultures were rinsed three times with PSS and incubated for 20 min at 37°C with 1 ml of PSS containing 0.1 mM ouabain and the indicated additions. For ATP determinations the medium was aspirated and the cultures were extracted with perchloric acid.[10] For $^{45}Ca^{2+}$ uptake the cultures were incubated for 1 min in 140 mM KCl-Hepes/Tris, pH 7.4, containing 0.1 mM ouabain, 0.25 mM $CaCl_2$, 1 μCi $^{45}Ca^{2+}$. Uptake was also measured in the presence of NaCl instead of KCl. Uptake in the presence of Na^+ or Na^+ plus glucose was 20 ± 2 (n = 8) and 21 ± 1% control (n = 8), respectively. The concentrations of the additions were 2 mM DOG, 2.5 μg/ml oligomycin, 2 μM antimycin A, 2 μM rotenone, 0.2 mM DNP, and 10 mM glucose. Values are means ± SE, n = 8 for "None" and n = 4 for all of the other conditions. For $^{45}Ca^{2+}$ uptake the 100% value was 12.9 (nmol/min·mg cell protein); for ATP the 100% value was 57.6 nmol/mg protein.

epithelial cells may be caused by an ATP-dependent phosphorylation or ATP binding to an allosteric site. The alternative possibility that some changes that are secondary to ATP depletion may have decreased exchange activity should also be kept in mind. However, $^{45}Ca^{2+}$uptake by organelles probably does not contribute significantly to exchange activity assayed by Na^+ gradient-dependent $^{45}Ca^{2+}$influx.[10] First, treating the cells with ruthenium red strongly inhibited mitochondrial $^{45}Ca^{2+}$ uptake, but it had no effect on exchange activity. Second, a prior incubation of the cells with the Na^+-selective ionophore, monensin, strikingly increased exchange activity but strongly inhibited ATP-dependent $^{45}Ca^{2+}$ uptake by organelles as determined after selectively permeabilizing the plasma membrane. Third, ATP-dependent $^{45}Ca^{2+}$ uptake by mitochondria and the sarcoplasmic reticulum in permeabilized cells was very small relative to uptake via the exchanger in intact cells.

It is important to note that energy depletion influences exchange activity assayed under V_{max} conditions in the aortic myocytes and epithelial cells, whereas Mg-ATP increases the apparent affinities of the exchanger for Na^+ and Ca^{2+} in dialyzed squid axons. The type of modulatory effect exerted by ATP on exchange activity and the underlying biochemical mechanism of ATP action apparently differs in squid axon and mammalian cells.

Comparison of Exchange Activity in Intact Myocytes and Purified Membranes

Recent observations indicate that the V_{max} of the exchanger in purified plasma membrane vesicles is considerably lower than the V_{max} in intact aortic myocytes from

TABLE 4. Removal of DOG and DNP Restores Cellular ATP and Na^+-Ca^{2+} Exchange Activity

Incubations with DOG and DNP[a]			$^{45}Ca^{2+}$ Uptake (nmol/min·mg)				
First	Second	Third	K^+	Na^+	Na^+-Ca^{2+} Exchange	% Control	Cellular ATP % Control
−	−	−	12.1 ± 0.2	1.2 ± 0.1	10.8 ± 0.2 (6)	100	100 (6)
−	−	+	6.0 ± 0.1	1.1 ± 0.1	4.9 ± 0.2 (6)	45*	21 (7)*
−	−	+/Glc	12.5 ± 1.2	1.0 ± 0.1	11.6 ± 1.2 (4)	107	86 (4)
+	−	−	10.3 ± 0.1	1.0 ± 0.1	9.4 ± 0.5 (4)	87	49 (4)*
+	−/Glc	−/Glc	10.7 ± 0.4	1.0 ± 0.1	9.7 ± 0.4 (4)	90	96 (4)

[a] The (*) indicates statistical significance at 99% by the Scheffe F test. The "+" or "−" indicates that the incubation was with or without 2 mM DOG and 20 μM DNP, respectively; "Glc" indicates that the cultures were incubated with 10 mM glucose. Ouabain (0.1 mM final) was added to the culture medium 1 hr before the first incubation; ouabain (0.1 mM) was present during all incubations. The cultures were rinsed three times and incubated for 5 min with a physiological saline solution that contained (in mM): 120 NaCl, 20 $NaHCO_3$, 5 KCl, 1 $CaCl_2$, 1 $MgCl_2$, and 20 Hepes/Tris, pH 7.4. Then they were rinsed three times and incubated with the additions indicated for the first, second, and third 5-min incubations. For measuring $^{45}Ca^{2+}$ uptake the cultures were rinsed three times with 140 mM NaCl-Hepes/Tris, pH 7.4, or 140 mM KCl-Hepes/Tris, pH 7.4, before assaying $^{45}Ca^{2+}$ uptake in the presence of Na^+ or K^+, respectively. $^{45}Ca^{2+}$ uptake was measured after a 1-min incubation in 1 ml of 140 mM NaCl- or KCl-Hepes/Tris, pH 7.4, containing 0.25 mM $CaCl_2$, 0.1 mM ouabain, and 1 μCi $^{45}CaCl_2$. For ATP determinations the cultures were rinsed twice with ice-cold saline-Hepes/Tris, pH 7.4, and 1 ml of ice-cold 0.4 M perchloric acid was added to extract ATP.[10] The 100% value for ATP was 25.9 nmol/mg protein. Values are mean ± SE (number of replicate cultures).

which the vesicles were derived (see Lyu, this volume). The V_{max} of exchange activity in the vesicles was only a sixtieth of the value expected on the basis of the activity in the intact cells. Endogenous proteolysis during cell homogenization and membrane purification increases exchange activity, which partially compensates for the larger losses in activity that occur when proteolysis is inhibited. Endogenous proteolysis increases the V_{max} of exchange by 3.4-fold without affecting the K_m for Ca^{2+}. Chymotrypsin activates Na^+-Ca^{2+} exchange in sarcolemmal vesicles from bovine heart[40] and aortic myocytes as well as the exchange current in giant inside-out patches of sarcolemma from guinea pig heart cells.[39] Furthermore, chymotrypsin abolishes the modulatory effects of Ca^{2+} and Mg-ATP on the exchange current.[39] Endogenous proteolysis may also deregulate the exchanger in vesicles from aortic myocytes because vesicles from control and ATP-depleted cells have the same exchange activity even though ATP depletion strongly inhibits exchange in intact cells (Lyu, Reeves & Smith, Biochim. Biophys. Acta., in press).

CONCLUDING REMARKS

Several lines of mammalian cells have no detectable Na^+-Ca^{2+}exchange activity. These include some lines of renal epithelial cells (A6, MDCK, OK, LLC-PK_1) (Lyu, Smith & Smith, J. Membr. Biol., in press), human skin and lung fibroblasts,[3] canine coronary endothelial cells,[41] mouse lung, skin, and amnion fibroblasts.[29] It seems likely that the vast majority of mammalian cell types depend solely on the Ca^{2+}-ATPase for expelling Ca^{2+}. Only certain cell types express the Na^+-Ca^{2+} exchanger. For example, the Na^+-Ca^{2+} exchanger is plentiful in an epithelial line derived from monkey kidney and therefore probably makes a prominent contribution to Ca^{2+} regulation in a specific segment of the nephron. Additionally, the exchanger is present in arterial myocytes from human umbilical arteries, although exchange activity is more abundant in rat aortic myocytes. The plasma membrane Ca^{2+}-ATPase, which appears to be ubiquitous, operates in parallel with the exchanger to expel $Ca^{2+}$from the cell. The ATPase has a 10 to 20 times smaller turnover number than the exchanger. Depleting cellular ATP reversibly inhibits Na^+-Ca^{2+} exchange. Because the exchanger is active in vesicles in the absence of a source of metabolic energy, ATP or a related metabolite probably modulates exchange activity allosterically rather than supplying energy for cation translocation. The exchanger is latent in the unstimulated cell. In the stimulated cell the exchanger constitutes a high-affinity, high-capacity pathway for active Ca^{2+} extrusion that makes a major contribution to decreasing cytosolic free Ca^{2+}. The Na^+-Ca^{2+} exchanger in renal epithelial cells and arterial and cardiac myocytes does not require or transport potassium, in contrast to the exchanger in photoreceptor cells. Studies are currently underway to clarify tissue and cell type-specific expression of the exchanger using cDNA probes.

REFERENCES

1. REUTER, H. & N. SEITZ. 1968. J. Physiol. **195:** 451-470.
2. BAKER, P. F., M. P. BLAUSTEIN, A. L. HODGKIN & R. A. STEINHARDT. 1969. J. Physiol. **200:** 431-458.
3. SMITH, J. B., S. D. DWYER & L. SMITH. 1989. J. Biol. Chem. **264:** 831-837.
4. LYU, R.-M., S. D. DWYER, Y. ZHUANG & J. B. SMITH. 1990. FASEB J. **4:** A1208.
5. SMITH, J. B. & L. SMITH. 1987. J. Biol. Chem. **262:** 17455-17460.
6. SMITH, J. B., R.-M. LYU & L. SMITH. 1991. Biochem. Pharmacol. **41:** 601-609.

7. BAKER, P. F. 1978. Ann. N.Y. Acad. Sci. **307:** 250-268.
8. FURUKAWA, K.-I., Y. TAWADA & M. SHIGEKAWA. 1988. J. Biol. Chem. **263:** 8058-8065.
9. SMITH, J. B., E. J. CRAGOE, JR. & L. SMITH. 1987. J. Biol. Chem. **262:** 11988-11994.
10. SMITH, J. B. & L. SMITH. 1990. Am. J. Physiol. **259:** C302-C309.
11. SMITH, J. B., T. ZHENG & L. SMITH. 1989. Am. J. Physiol. **256:** C147-C154.
12. SMITH, J. B., T. ZHENG & R.-M. LYU. 1989. Cell Calcium **10:** 125-134.
13. DIPOLO, R. & L. BEAUGE. 1986. Biochim. Biophys. Acta **854:** 298-306.
14. KIMURA, J., A. NOMA & H. IRISAWA. 1986. Nature **319:** 596-597.
15. NICOLL, D. A., S. LONGONI & K. D. PHILIPSON. 1990. Science **250:** 562-565.
16. LI, Z., D. A. NICOLL, A. COLLINS, D. W. HILGEMANN, A. G. FILOTEO, J. T. PENNSITON, J. N. WEISS, J. M. TOMICH & K. D. PHILIPSON. 1991. J. Biol. Chem. **266:** 1014-1020.
17. KACZOROWSKI, G. J., F. BARROS, J. K. DETHMERS & M. J. TRUMBLE. 1985. Biochemistry **24:** 1394-1403.
18. ZHUANG, Y., L. SMITH & J. B. SMITH. 1990. J. Cell Biol. **111:** 469a.
19. OZAKI, H., T. KISHIMOTO, H. KARAKI & N. URAKAWA. 1982. Naunyn-Schmiedeberg's Arch. Pharmacol. **321:** 140-144.
20. ASHIDA, T. & M. P. BLAUSTEIN. 1987. J. Physiol. **392:** 617-635.
21. MULVANY, M. J., C. AALKJAER & T. T. PETERSEN. 1984. Circ. Res. **54:** 740-749.
22. CHEON, J. & J. P. REEVES. 1988. J. Biol. Chem. **263:** 2309-2315.
23. NIGGLI, E. & W. J. LEDERER. 1991. Nature **349:** 621-624.
24. CRESPO, L. M., C. J. GRANTHAM & M. B. CANNELL. 1990. Nature **345:** 618-621.
25. BRIDGE, J. H., J. R. SMOLLEY & K. W. SPRITZER. 1990. Science **248:** 376-378.
26. REEVES, J. P. & C. C. HALE. 1984. J. Biol. Chem. **259:** 7733-7739.
27. REEVES, J. P. & J. L. SUTKO. 1983. J. Biol. Chem. **258:** 3178-3182.
28. MIURA, Y. & J. KIMURA. 1989. J. Gen. Physiol. **93:** 1129-1145.
29. WAKABAYASHI, S. & K. GOSHIMA. 1981. Biochim. Biophys. Acta **642:** 158-172.
30. LYU, R.-M., L. SMITH & J. B. SMITH. 1991. Biophys. J. **59:** 137a.
31. SCHNETKAMP, P. P. M. 1980. Biochim. Biophys. Acta **598:** 66-90.
32. SCHNETKAMP, P. P. M., D. K. BASU & R. T. SZERENCSEI. 1989. Am. J. Physiol. **257:** C153-C157.
33. CERVETTO, L., L. LAGNADO, R. J. PERRY, D. W. ROBINSON & P. A. MCNAUGHTON. 1989. Nature **337:** 740-743.
34. TROSPER, T. L. & K. D. PHILIPSON. 1983. Biochim. Biophys. Acta **731:** 63-68.
35. DIPOLO, R. & L. BEAUGE. 1988. Biochim. Biophys. Acta **947:** 549-569.
36. BLAUSTEIN, M. P. 1977. Biophys. J. **20:** 79-111.
37. DIPOLO, R. 1977. J. Gen. Physiol. **69:** 795-813.
38. REEVES, J. P. & J. L. SUTKO. 1979. Proc. Natl. Acad. Sci. USA **76:** 590-594.
39. HILGEMANN, D. W. 1990. Nature **344:** 242-245.
40. PHILIPSON, K. D. & Y. NISHIMOTO. 1982. Am. J. Physiol. **243:** C191-C195.
41. DWYER, S. D., Y. ZHUANG & J. B. SMITH. 1991. Exp. Cell Res. **192:** 22-31.
42. BENOS, D. J., D. G. WARNOCK & J. B. SMITH. 1991. *In* Membrane Transport in Biology. G. H. Giebisch, H. H. Ussing & H. N. Christensen, Eds. Vol. 5, in press. Springer-Verlag. New York.

Evidence for Na-Ca Exchange in Human Resistance Arteries

P. I. AARONSON, L. POSTON, R. G. WOOLFSON,
AND S. V. SMIRNOV

United Medical and Dental Schools
Smooth Muscle Group
Divisions of Pharmacology, Physiology, and Medicine
St. Thomas' Hospital, London SE1 7EH, United Kingdom

INTRODUCTION

The recent discovery that endogenous ouabain may represent a humoral inhibitor of sodium transport[1] and that it may be elevated in experimental hypertension[2] has again stimulated interest in the mechanisms whereby inhibition of active sodium transport in vascular smooth muscle may lead to an increase in tension. Ouabain may increase vascular tone in a number of ways: It may have a direct effect on neuronal transmitter release[3] and reuptake[4]; it may effect an increase in intracellular calcium through depolarization[5]; it may reduce endothelium dependent relaxation[6]; and, as proposed by Blaustein,[7] may lead to increased Ca uptake or decreased Ca extrusion through the membrane Na-Ca exchanger. The last of these possibilities has led to some controversy.

While there is little doubt that Na-Ca exchange is an active pathway in the smooth muscle of large arteries in some species,[8,9] its role in the small arteries that offer greatest resistance to flow and that are involved in the control of blood pressure remains uncertain. In rat and guinea-pig aorta, Na-Ca exchange is well demonstrated by the substantial rise in intracellular calcium and in tension[10] that occurs when the sodium gradient is reduced or reversed. The potentiation of agonist-induced tension[10,11] when the sodium gradient is reduced also provides evidence of elevation of cell calcium through Na-Ca exchange. The rate of relaxation from potassium-induced contraction in large arteries is also reduced when sodium is removed[9] and suggests an important role for this antiporter in the process of relaxation. In the resistance arteries the role of Na-Ca exchange has been questioned by a number of studies.[12-14] In rat and guinea pig resistance arteries, reduction of the sodium gradient either with ouabain or low extracellular sodium, leads to no increase in tone.[12] The presence of Na-Ca exchange in rat mesenteric arteries has been demonstrated only in extreme conditions, for example, when sodium removal and ouabain addition are combined.[15] In that situation a small rise in tension occurs. Removal of sodium and substitution with sucrose has led to a reduction of relaxation from arteries precontracted with potassium, but this effect was not demonstrable when sodium was substituted with choline.[16]

Because the response of different vascular preparations to sodium gradient reduction is clearly heterogeneous,[8] we have recently carried out similar investigations with human arteries.[17] Using a small-vessel myograph, we found that ouabain causes a concentration-dependent increase in tension that is slow in onset. This was likely to be the result of altered Na-Ca exchange, since tension was only partially reduced by the

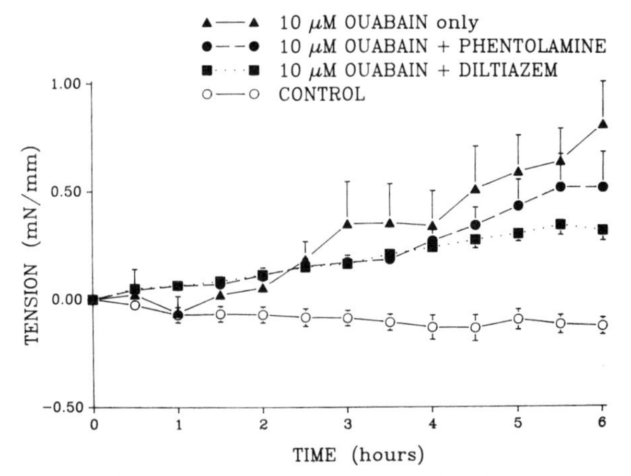

FIGURE 1. Line graphs showing effect of prolonged incubation of human resistance vessels in 10 μM ouabain alone, in 10 μM ouabain with 1 μM phentolamine, in 10 μM ouabain with 1.6 μM diltiazem, or in physiological saline solution (control). Reproduced from Woolfson *et al.*[17] by permission of the American Heart Association.

calcium antagonist diltiazem and not significantly affected by phentolamine (FIG. 1). Prolonged incubation in 25 mM Na substituted with choline also led to a rise in tension. These results suggested therefore that Na-Ca exchange was certainly present in human arteries and could be involved in the development of tension in response to an altered sodium gradient.

Investigations using large arteries have demonstrated a profound effect of a low sodium gradient on subsequent agonist-induced tension.[10,11] This has suggested that the reduced gradient may elevate intracellular calcium, which may be sequestered in intracellular calcium stores. It is apparent therefore that estimation of intracellular calcium as well as tension may help to solve the controversy of the role of Na-Ca exchange in resistance arteries. Using the calcium-sensitive dye, fura 2, and a specially adapted small-vessel myograph, we have carried out investigations into Na-Ca exchange in human resistance arteries.

METHODS

Arteries were obtained from biopsies of subcutaneous fat obtained from the abdominal wall of normotensive patients undergoing elective surgery. The majority of the patients were women having gynecological surgery. The study was approved by the local committee on ethical research and all patients gave informed consent.

The arteries were dissected and mounted on two tungsten wires (40 μm diameter) in the jaws of an automated myograph (Cambustion Ltd., Cambridge, England). One wire jaw was attached to a tension transducer. The other could be moved by a computer-controlled stepper motor. The vessel was subjected to an automated stretch and relaxation procedure to determine the passive tension-circumference characteristics. A best-fit curve including stretch and relaxation curves was then constructed. Using the equation of this curve and Laplace's relationship to estimate transmural pressure, the vessel was then stretched to a circumference 90% of that which it would have at a transmural pressure of 100 mm Hg. This circumference has previously been determined to give maximal active tension. The vessel was then loaded with fura-2 by incubation for 2 hr in 3 μM fura 2-AM with the addition of 0.03% pluronic 127 to aid uptake. The myograph module was then transferred to the stage of an inverting Nikon-Diaphot microscope ($\times$10 Fluar objective). The vessel was illuminated with light of 340, 360, and 380 nm using a rotating filter wheel. Emitted light was collected through a filter (510 nm) and intensity of each wavelength recorded on computer, together with the calcium-sensitive 340/380 ratio (software, Cairn Instruments Ltd.). The digitized output from the tension transducer on the myograph was transferred to a further channel on the same computer for simultaneous recording of tension and the calcium signal.

The vessel was then perfused with PSS (physiological saline solution, 25 mM bicarbonate) prewarmed (37°C) and gassed (95% O_2, 5% CO_2). Vessels were then contracted a total of five times with noradrenaline (NA, 5 μM) in PSS, potassium-substituted PSS (KPSS), and 5 μM NA in KPSS.

INVESTIGATIONS OF NA-CA EXCHANGE

Reversal of Na Gradient

The effect of sodium removal from the extracellular medium on calcium, and tension was investigated by perfusing arteries for 5 minutes with the following sodium-substituted solutions: (a) 116 mM choline chloride, 5 mM Na, 25 mM choline bicarbonate-substituted PSS; (b) 116 mM *N*-methyl glucamine chloride, 2 mM Na, 25 mM choline bicarbonate-substituted PSS, pH 7.4; (c) 116 mM lithium chloride, 5 mM Na, 25 mM choline bicarbonate-substituted PSS. The calcium anatagonist dilitiazem (20 μM) and the α-blocker phentolamine (10 μM) were added to all solutions. In some vessels concentration responses to Na (with appropriate substitution) were carried out.

Effect of Reversal of Sodium Gradient on Relaxation

The effect of sodium removal on intracellular Ca^{2+} and tension recovery from KPSS-induced contraction was investigated. Arteries were precontracted with KPSS (2 min) and relaxation induced in (a) PSS, (b) PSS, $0Ca^{2+}$, (c) PSS $0Ca^{2+}$, 5 mM Na^+. In three experiments relaxation in 5 mM Na^+ medium was investigated in the presence of ryanodine (10 μM).

Role of the Sarcoplasmic Reticulum in Ca_i Response to Reduced Sodium Gradient

Ca bound to the SR was released by quick exposure to caffeine (10 mM, 2 min) and reuptake prevented by subsequent perfusion with ryanodine (10 μM). The arteries were then perfused with choline (or NMG or Li)-substituted PSS + ryanodine. All solutions contained diltiazem (20 μM) and phentolamine (10 μM).

Effect of Sodium Gradient Reversal on AVP-Induced Tension

Arteries were perfused with choline-substituted PSS for 5 minutes, and after a washout period of 15 minutes were submaximally contracted with AVP (500 μunits/ml) in PSS applied directly to the bath. After 10 minutes the solution was replaced with AVP (500 μunits/ml) in choline-substituted PSS (10 min) and then AVP (500 μunits/ ml in PSS). All solutions contained diltiazem (20 μM) and phentolamine (10 μM).

Calibration of [Ca]$_i$

In some experiments, the 340/380 ratio was converted into an estimate of [Ca]$_i$ using the equation presented by Cobbold and Rink.[18] Tissues were incubated in Ca-free PSS containing 10 or 20 mM EGTA. After the 340/380 ratio had decreased to a steady-state value, 10 μM ionomycin was added to the EGTA solution. This resulted in a further decrease in the 340/380 ratio; the subsequent minimum value of this quantity was recorded as R_{min}. Tissues were then returned to normal PSS, resulting in an increase in the 340/380 ratio to a high value which was recorded as R_{max}. Subsequently, tissues were exposed to 10 mM manganese to quench the fura-2 fluorescence, to give an estimate of the tissue autofluorescence. These values were then subtracted from all readings taken during the experiment, and [Ca]$_i$ calculated using a K_d value for the binding of Ca to fura-2 of 224 nM.

RESULTS

The method described enabled calcium and tension to be measured simultaneously in human resistance arteries. The degree of resolution obtained for the calcium signal was sufficiently high to allow the measurement of small alterations in intracellular calcium. (FIG. 2-5). During the run-up procedure, it was routinely observed (see FIG. 1) that intracellular calcium was greater in KPSS than in 5 μM noradrenaline, yet tension development was greater when the arteries were stimulated with noradrenaline.

Reversal of Sodium Gradient

In all arteries investigated (n = 18, mean inner diameter 255 ± SEM 22, i.d. 138-467 μm) the removal of extracellular sodium led to a prompt rise in the calcium-sensitive fura-2 340/380 ratio. In larger vessels (> 350 μm i.d.) this was associated

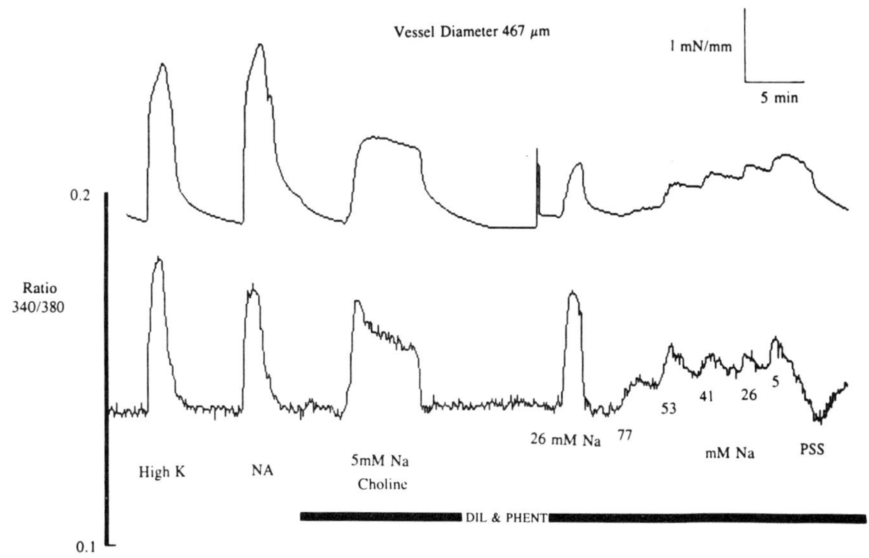

FIGURE 2. Simultaneous measurement of Ca and tension in a human artery (i.d. 467 μm). Responses to KPSS, 5 μM NA, and reduced extracellular sodium (choline substitution) in the presence of 20 μM diltiazem and 10 μM phentolamine.

with an increase in tension (FIG. 2), whereas in the smaller arteries there was little or no increase in tension. The rise in intracellular calcium occurred in the presence of diltiazem and phentolamine and was not affected by the omission of these two agents from the perfusion buffer. The rise in calcium was independent of the sodium substitute used (FIG. 4). It was of interest to note that the change in calcium observed when extracellular sodium (and presumably intracellular sodium) was depleted in a stepwise fashion (FIG. 3) was not as great as that observed when the gradient was abruptly changed (compare responses to 26 mM Na), due perhaps to a progressive discharge of the sodium gradient.

Effect of Reversal of Sodium Gradient on Relaxation

After contraction induced by KPSS, removal of extracellular calcium led to an increase in the rate at which tissues relaxed and $[Ca]_i$ fell (n = 13, mean i.d. 221 ± 13 μm; i.d. 160-308 μm). Removal of sodium at this time retarded the rate of fall of intracellular calcium in 11 of 13 arteries, but had little effect on the rate of relaxation (FIG. 3). In two arteries, pretreatment with ryanodine exaggerated the effect of Na removal on the recovery of $[Ca]_i$ and revealed an effect on relaxation.

Role of the Sarcoplasmic Reticulum in Ca_i Response to Reduced Sodium Gradient

Because the majority of arteries investigated demonstrated a rise in calcium without tension development, it was considered that the measured calcium might be within an

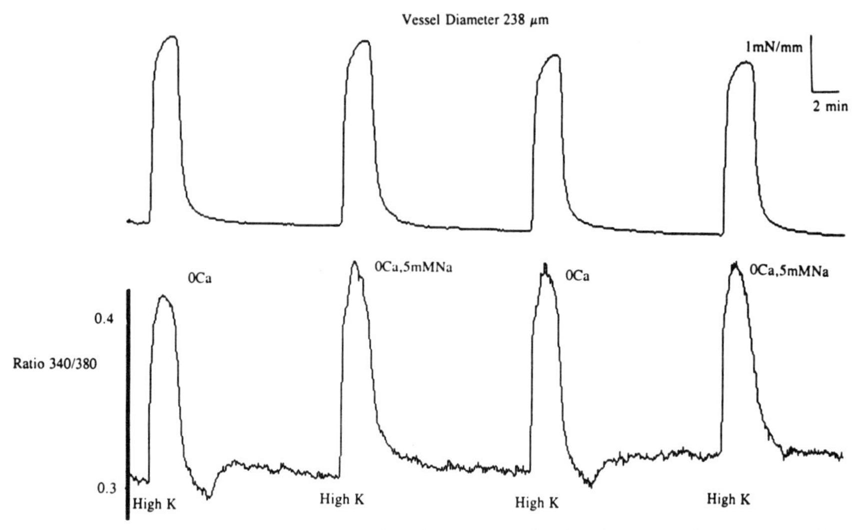

FIGURE 3. Simultaneous measurement of calcium and tension in a human resistance artery (i.d. 238 μm). Response to KPSS and relaxation in the absence of $[Ca]_o$ and in the absence of $[Ca]_o$ + 5 mM Na.

intracellular calcium store. The addition of caffeine to the perfusate led to a transient increase in calcium and tension followed by a prolonged rise in the calcium signal when ryanodine PSS was added. The subsequent removal of extracellular sodium in the presence of ryanodine led to a further rise in calcium, but no rise in tension (n = 3, mean i.d. 187 μm ± 17; i.d. 154-203 μm; FIG. 4).

Effect of Reversal of Sodium Gradient on AVP-Induced Tension

Removal of sodium from the extracellular medium led to a potentiation of AVP-induced tension and to a further rise in the calcium signal (n = 3, mean i.d. 208 ± 27 μm; i.d. 168-258 μm). Return to PSS led to a reversal of the potentiation of both tension and calcium (FIG. 5).

DISCUSSION

This study describes a method for the simultaneous measurement of intracellular calcium and tension in human resistance arteries using a modified automated myograph and inverting fluorescence microscope. Previously, one other similar study had reported simultaneous measurement of calcium and tension in resistance arteries, but using a water-immersion objective and rat mesenteric small arteries.[19] The measured intracellular calcium was of similar magnitude in this study and that of Bukoski *et al.*[19] The apparent increased sensitivity to calcium that occurs in the presence of noradrenaline is in agreement with that demonstrated by Sato and coworkers in rat aortic strips, using fura-2 to estimate intracellular calcium.[20]

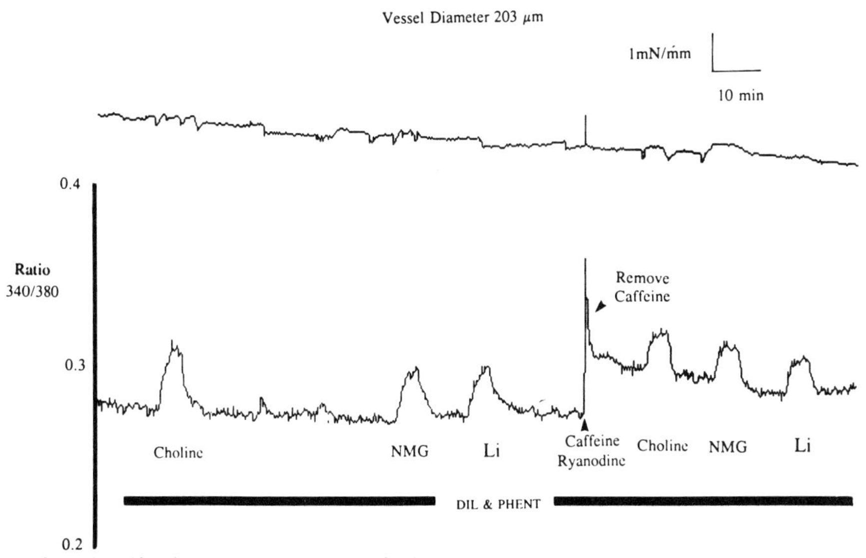

FIGURE 4. Simultaneous measurement of calcium and tension in a human resistance artery (i.d. 203 μm). Demonstration of the effect on Ca release from intracellular stores by caffeine (10 μM) and ryanodine (10 μM) on the low-sodium response; 20 μM diltiazem and 10 μM phentolamine were present throughout.

This study confirms our earlier report that Na-Ca exchange is present in human resistance arteries.[17] In that study in which calcium was not estimated, we demonstrated a slow rise in tension development when the sodium gradient was reversed and also in the presence of ouabain. This contrasted with earlier work in small arteries obtained from the rat and guinea pig in which no rise in tension was observed unless ouabain and low sodium were combined.[15] These somewhat disparate results suggested that, in human resistance arteries, Na-Ca exchange may be of greater importance than originally suggested by animal studies. The measurement of calcium demonstrates that sodium gradient reversal is accompanied by a prompt rise in calcium in the small arteries, but during the time course of this experiment, with no rise in tension. As phentolamine and diltiazem were present, the most likely explanation for the rise in calcium is that Na-Ca exchange is reversed when the sodium gradient favors outward movement of Na. The rise in calcium observed was approximately one-third to one-half of that which occurred upon stimulation with KPSS + 5 μM noradrenaline. In one experiment a concentration-response curve to potassium was carried out. The rise in calcium that occurred in the absence of sodium and without tension development was of equal magnitude to that observed with 30 mM KPSS in which there was tension development. This would suggest that the calcium that enters the artery when external sodium is removed is less available for contraction and is likely to be sequestered in a calcium store. This is in agreement with previous work in the guinea pig tenia coli.[21] The rise in calcium in the absence of sodium was little affected by pretreatment with caffeine or ryanodine or noradrenaline, but tension was still unaffected, suggesting that sequestration of calcium in the stores affected by these three agents is not responsible for the dissociation between the rise in $[Ca]_i$ and tension development. The question then remains as to where the calcium is buffered. It was of interest that in the presence of AVP, sodium removal was accompanied by marked tension development. This might

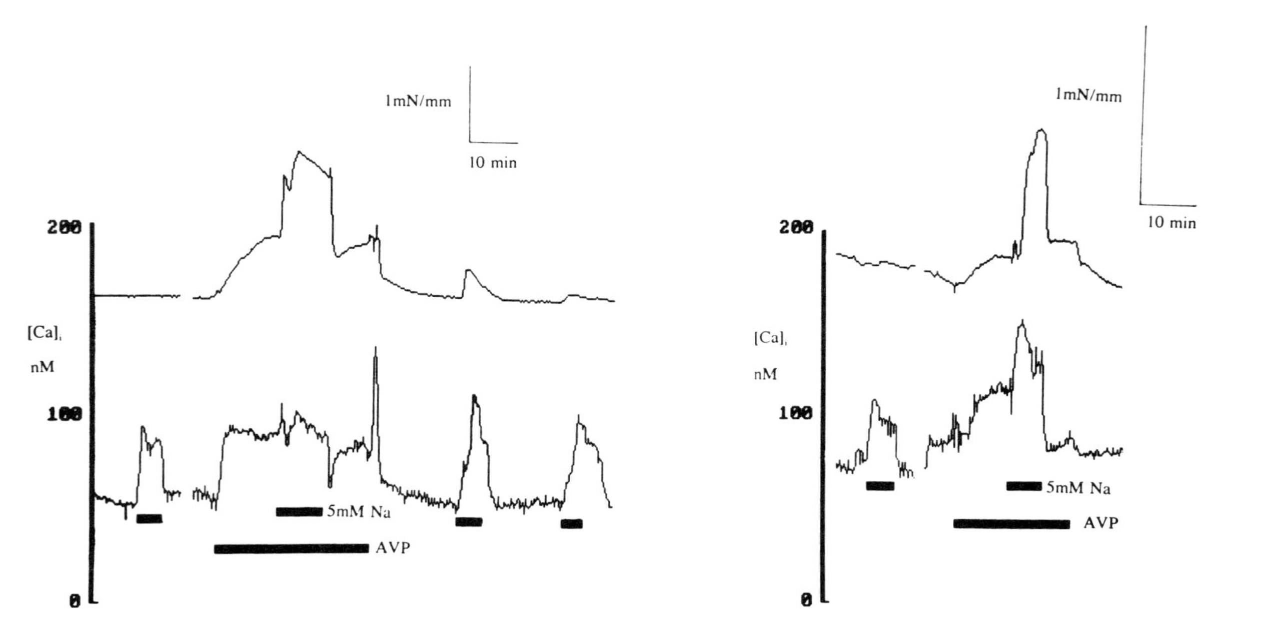

FIGURE 5. Simultaneous measurement of calcium and tension in two human resistance arteries (i.d. 168 μm and i.d. 158 μm). Response to 500 μunits/ml of AVP and effect of reversal of sodium gradient; 20 μM diltiazem and 10 μM phentolamine were present throughout.

suggest that in the presence of AVP the smooth muscle is unable to sequester the Ca^{2+}entering during Na removal. Alternatively AVP may be affecting sensitivity of the myofilaments to Ca^{2+}. Further work with other agonists is underway to determine whether this is a phenomenon associated in general with agonist-induced contraction, or to AVP alone. This will be an important determinant as to the physiological importance of the rise in calcium that occurs when the sodium gradient is compromised, and therefore to the role of Na-Ca exchange in these arteries.

Relaxation from contraction in the vascular smooth muscle of large arteries is dependent upon a Ca ATPase, the activity of the Na-Ca exchanger, and uptake by the SR. In this study the removal of calcium to decrease the inward calcium gradient led to both faster relaxation and a faster fall in the calcium signal after stimulation by KPSS. Removal of both calcium and sodium should demonstrate the role of Na-Ca exchange in the recovery of intracellular calcium after contraction and, indeed, did lead to a marked reduction in recovery of intracellular calcium. This was, however, only occasionally associated with a fall in the rate of tension recovery. In the presence of ryanodine, the effect of sodium removal on calcium recovery was exaggerated and relaxation of tension was clearly impaired. This would suggest that the effect of abolishing Na-Ca exchange-mediated extrusion can normally be compensated for to some extent by buffering of calcium within the SR. However, if both processes are inhibited, the contribution of Na-Ca exchange becomes more evident.

REFERENCES

1. MATTHEWS, W. R., D. W. HARRIS, D. W. DUCHARME, B. S. LUTZE, F. MANDEL, M. A. CLARK, J. H. LUDENS & J. M. HAMLYN. 1990. Endogenous Digitalis Like Factor (EDLF) III. Mass Spectral Characterization (abstract). Hypertension **16(3):** 337.
2. HAMLYN, J. M., J. REGAN, A. LOFTUS, A. ROGOWSKI, R. WHITE, D. HARRIS, D. DUCHARME & J. LUDENS. 1990. Endogenous digitalis-like factor (EDLF) V: Localization, secretion, and levels in DOCA salt hypertension (abstract). Hypertension **16(3):** 338.
3. LORENZ, R. R., D. A. POWIS, P. M. VANHOUTTE & J. T. SHEPHERD. 1980. The effect of acteylstrophanthidin and ouabain on the sympathetic adrenergic neuroeffector junction in canine vascular smooth muscle. Circ. Res. **47:** 845-854.
4. VANHOUTTE, P. M., T. J. VERBEUREN & R. C. WEBB. 1981. Local modulation of the adrenergic neuroeffector interaction in the blood vessel wall. Physiol. Rev. **61:** 151-247.
5. MIKKELSEN, E., K.-E. ANDERSSON & O. F. PEDERSEN. 1979. Effects of digoxin on isolated human peripheral arteries and veins. Acta Pharmacol. Toxicol. **45:** 25-31.
6. WOOLFSON, R. G. & L. POSTON. 1991. The effect of ouabain on endothelium dependent relaxation of human resistance arteries. Hypertension. In press.
7. BLAUSTEIN, M. P. 1977. Sodium ions, calcium ions, blood pressure regulation, and hypertension: A reassessment and a hypothesis. Am. J. Physiol. **223:** C165-C173.
8. OZAKI, H. & N. URAWAKA. 1981. Effects of K free solution on tension development and Na content in vascular smooth muscles isolated from guinea-pig, rat and rabbit. Pflügers Arch. **390:** 107-112.
9. ASHIDA, T. & M. P. BLAUSTEIN. 1987. Regulation of cell calcium and contractility in mammalian arterial smooth muscle: The role of sodium-calcium exchange. J. Physiol. **392:** 617-635.
10. BOYA, S., W. GOLDMAN, X-J. YUAN & M. P. BLAUSTEIN. 1990. Influence of Na^+ gradient on Ca^{2+} transients and contraction in vascular smooth muscle. Am. J. Physiol. **259:** H409-H423.
11. NAPODAMO, R. J., F. S. CALVIA, C. LYONS, J. DESIMONE & H. LYONS. 1962. The reactivity to angiotensin of rabbit aortic strips after either alterations of external sodium environment or direct addition of benzydroflumethazide. Am. Heart J. **64:** 498-502.
12. MULVANY, M. J., C. AALKJAER, H. NILSSON, N. KORSGAARD & T. PETERSEN. 1982. Raised intracellular sodium consequent to sodium-potassium-dependent ATPase inhibi-

tion does not cause myogenic contractions of 150 μm arteries from rat and guinea pig. Clin. Sci. **63:** 45s-48s.
13. MULVANY, M. J., H. NILSSON, J. A. FLATMAN & N. KORSGAARD. 1982. Potentiating and depressive effects of ouabain and potassium free solutions on rat mesenteric resistance vessels. Circ. Res. **51:** 514-524.
14. MULVANY, M. J. 1984. Effect of electrolyte transport on the response of arteriolar smooth muscle. J. Cardiovasc. Pharmacol. **6:** S82-S87.
15. MULVANY, M. J., C. AALKJAER & T. T. PETERSEN. 1984. Intracellular sodium, membrane potential and contractility of rat mesenteric small arteries. Circ. Res. **54:** 740-749.
16. PETERSEN, T. T. & M. J. MULVANY. 1984. Effect of sodium gradient on the rate of relaxation of rat mesenteric small arteries from potassium contractures. Blood Vessels **21:** 279-289.
17. R. G. WOOLFSON, P. J. HILTON & L. POSTON. 1990. Effect of ouabain and low sodium on contractility of human resistance arteries. Hypertension **15:** 583-590.
18. COBBOLD, P. H. & T. J. RINK. 1987. Fluorescence and bioluminescence measurement of cytoplasmic free calcium. Biochem. J. **248:** 313-328.
19. BUKOSKI, R. D., C. BERGMANN, A. GAIRARD & J. C. STOCLET. 1989. Intracellular Ca^{2+} and force determined simultaneously in isolated resistance arteries. Am. J. Physiol. **257:** H1728-H1735.
20. SATO, K. & H. KARAKI. 1988. Changes in cytosolic calcium level in vascular smooth muscle strip measured simultaneously with contraction using fluorescent calcium indicator fura-2. J. Pharmacol. Exp. Ther. **246:** 294-300.
21. VAN BREEMEN, C., P. AARONSON & R. LOUTZENHISER. 1980. The influence of Na on Ca fluxes in the guinea-pig tenia coli. *In* Vascular Neuroeffector Mechanisms. J. A. Bevan, T. Godfraind, R. A. Maxwell & P. M. Vanhoutte, Eds.: 227-236. Raven Press. New York.

Role of Sarcolemmal Membrane Sodium-Calcium Exchange in Vascular Smooth Muscle Tension[a]

M. A. MATLIB

Department of Pharmacology and Cell Biophysics
University of Cincinnati College of Medicine
Cincinnati, Ohio 45267-0575

INTRODUCTION

Cytosolic free Ca^{2+} concentration plays a crucial regulatory role in many cellular processes.[1] A number of Ca^{2+}-transport processes in the cell membrane and in the intracellular organelle membranes regulate cytosolic free-Ca^{2+} concentrations (FIG. 1) One of these processes is Na^+-Ca^{2+} exchange located in the cell membrane, which was first identified in cardiac muscle[2] and nerve.[3–5] Since then this process has been found in the plasma membrane of a number of cell types[6–11] and secretory vesicles.[12] The existence of a similar system in smooth muscle in general and vascular smooth muscle in particular was uncertain until recent years. This article is a brief review of the existence and the potential role of a Na^+-Ca^{2+} exchange system in vascular smooth muscle tension in normal and hypertensive states.

THE EXISTENCE AND THE ROLE OF NA^+-CA^{2+} EXCHANGE IN VASCULAR SMOOTH MUSCLE

Evidence from Studies with Isolated Blood Vessel Segments

Studies with cardiac muscle and nerve membranes established that under normal conditions the Na^+-Ca^{2+} exchange is an active Ca^{2+}-efflux process driven by a Na^+ electrochemical gradient across the membrane (FIG. 1). Thus, this process is directly regulated by internal and external Na^+ concentrations, and by the activity of a Na^+-K^+ pump. Modulation of internal Na^+ concentration by Na^+-K^+ pump activation or inhibition, and by depletion or repletion of external Na^+ concentration, were found useful in the elucidation of the existence and the role of the Na^+-Ca^{2+}exchange process in physiological, pharmacological, and pathological conditions in a number of systems. Studies have been carried out to determine the effects of Na^+ depletion or repletion to modify extracellular Na^+ concentrations, and K^+ depletion or cardiac glycoside treatments to modify intracellular Na^+ concentrations on vascular smooth muscle tension. Reduction of extracellular Na^+ concentration has been shown to induce vasoconstriction.[13-17] It has also been shown that K^+ depletion and digitalis glycoside treatment, which inhibit the Na^+-K^+ pump and increase cytosolic Na^+

[a]The work carried out in this laboratory and discussed here was supported by a grant from the National Institutes of Health (RO1-HL34664).

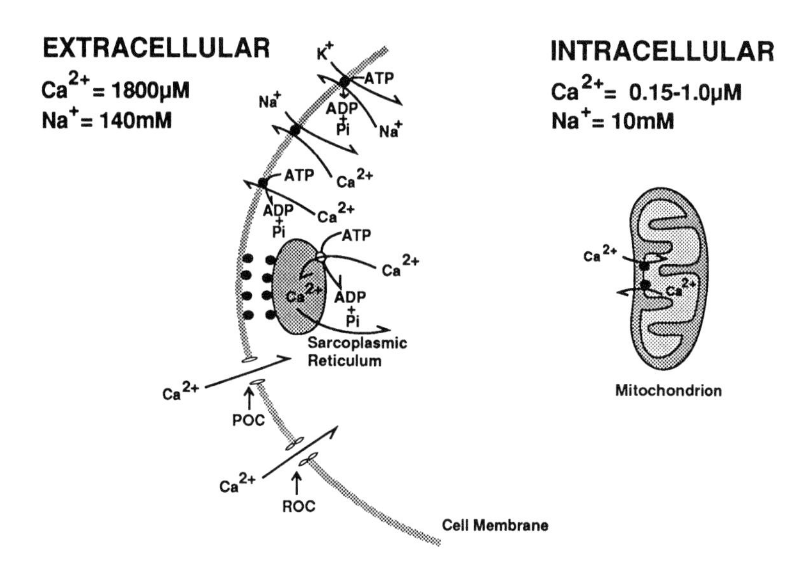

FIGURE 1. Ca^{2+}-transporting systems in a vascular smooth muscle cell. The cell membrane contains a Na^+-Ca^{2+}exchange, an ATP-dependent Ca^{2+} pump, a membrane potential operated Ca^{2+} channel (POC), and a receptor-operated Ca^{2+} channel (ROC) in addition to a Na^+-K^+ pump. The sarcoplasmic reticulum contains an ATP-dependent Ca^{2+} pump and a Ca^{2+} release channel. The mitochondria contain Ca^{2+}uptake and release pathways.

concentration, produce vasoconstriction.[16,17-24] The above conditions have been found to enhance vasoconstriction induced by norepinephrine.[13,18,20-22,24,25] Furthermore, some of the above conditions have also been found to attenuate the rate of relaxation of vascular smooth muscle.[18,21,26,27] The effects of these treatments on vascular smooth muscle tension described above could be or have been explained by the increase in cytosolic Ca^{2+} concentration due to reduction of efflux *via* a Na^+-Ca^{2+} exchange system across the cell membrane.[17,21,23,28] On the basis of this information, Blaustein[29] proposed a hypothesis on the role of Na^+-Ca^{2+} exchange in the regulation of cytosolic free-Ca^{2+} concentrations in vascular smooth muscle in hypertension. However, there were considerable uncertainties at that time with respect to the question of whether or not a specific Na^+-Ca^{2+} exchange process plays any significant role in contractility of smooth muscle in general.[30-32] There have been reports of a contribution from α-adrenergic receptor activation in vascular smooth muscle by catecholamines released from nerve terminals during the manipulations described above to modify cytosolic or extracellular Na^+ concentrations to produce contraction or relaxation.[33-34] Furthermore, studies by Mulvany *et al.*[35,36] indicate that Na^+-Ca^{2+} exchange may play an important role in tension development or maintenance of large arteries, but probably plays a minor role in small arteries and that only under extreme conditions. These studies cast a serious doubt on the existence and the role of a specific Na^+-Ca^{2+} exchange process in vascular smooth muscle, particularly in small or resistant arteries that are involved in the regulation of blood pressure. However, there are a number of pitfalls in identifying Na^+-Ca^{2+} exchange activity in vascular smooth muscle employing intact tissue, which may explain the disparity in the results.[37] The lack of vascular smooth muscle contraction observed in Na^+- or K^+-depleted medium or with cardiac glycoside treatment could be due to sequestration of rising cytosolic Ca^{2+} by the sarcoplasmic reticulum (FIG. 1). Therefore, ryanodine or caffeine has been used to eliminate the involvement of the sarcoplasmic reticulum in Ca^{2+} sequestration. Under these conditions, in K^+- or Na^+-free medium, caffeine or ryanodine produced larger contractions,[37–39] indicating an increase in cytosolic Ca^{2+} concentration. It has also been shown that the rate of relaxation is dependent on the presence of extracellular Na^+ concentration indicating a Na^+-dependent Ca^{2+} efflux presumably *via* a Na^+-Ca^{2+} exchange system. However, there are a number of other pitfalls in this approach. For example, the tissue may contain other cells and nerve terminals that may also contribute to these effects. The contractile state of the tissue may also be affected by changes in cell-to-cell contact, ion binding in the extracellular matrix, surface charge, and other Ca^{2+}-transport processes.

Evidence from Studies with Isolated Cells

In studies with freshly dispersed or cultured vascular smooth muscle cells, some of the pitfalls associated with intact tissues can be avoided. In these studies, ^{45}Ca or ^{22}Na flux across the cell membrane can be measured under conditions where cytosolic or extracellular Na^+ concentrations can be varied, and by measuring the flux of these radioactive ions an estimate of the Na^+-Ca^{2+} exchange activity can be obtained. In cultured rat aortic cells, an extracellular Ca^{2+}-dependent $^{22}Na^+$ efflux and an intracellular Na^+-dependent ^{45}Ca influx attributed to a Na^+-Ca^{2+} exchange process have been observed.[40–43] It has also been observed that Ca^{2+} released from intracellular stores with angiotensin II or ionomycin is exported out of the cells at a much faster rate in the presence than in the absence of extracellular Na^+, which has also been attributed to the activity of a Na^+-Ca^{2+} exchange system across the cell membrane.[41–44] The limitation of these studies is that the cultured cells were quiescent, and therefore the role of the Na^+-Ca^{2+} exchange activity in cell tension could not be assessed.

In other studies, using fura-2 as a Ca^{2+} indicator and digital imaging of the fluorescence of Ca-fura-2 complex in the cell, spatial distribution of intracellular Ca^{2+} concentrations have been measured in freshly dispersed arterial myocytes. In these studies, it has been shown that removal of extracellular K^+ from the superfusion medium, to inhibit Na^+-K^+ pump and load the cells with Na^+, resulted in increased cytosolic free-Ca^{2+} concentration and cell contraction.[45,46] Subsequent reduction of extracellular Na^+ concentration caused further increases in intracellular free-Ca^{2+}concentration and more contraction of the cell. These data indicate that the Na^+-Ca^{2+} exchange process is indeed in the Ca^{2+}-efflux mode of operation under this condition in order to reduce cytosolic Ca^{2+} concentration and cell tension.

Evidence from Studies with Sarcolemmal Vesicles

More direct evidence for the existence of a specific Na^+-Ca^{2+}exchange system in vascular smooth muscle was provided by studies with isolated sarcolemmal membrane vesicles. The advantage of this approach is that some of the pitfalls of the studies with isolated intact tissues and isolated cells can be avoided. For example, the contribution of intracellular organelles and other ion-transport processes can be eliminated, and Na^+-Ca^{2+} exchange can be measured in isolated and purified cell membrane vesicles directly and specifically using radioisotopes of Na^+ and Ca^{2+}. Pursuing this approach, a specific Na^+-Ca^{2+} exchange system has been demonstrated in isolated sarcolemmal membrane vesicles of vascular smooth muscle.[47-54] In all of these studies, evidence has been provided to show that in sarcolemmal membrane vesicles Ca^{2+} uptake or release is dependent on a Na^+ concentration gradient, that is, transport of Ca^{2+} occurs in exchange for Na^+ from the opposite direction.

This approach has been useful in obtaining information on cation specificity, reversibility, and kinetics of the system *in vitro*. It has also been shown that the system is specific for Na^+ and is reversible, that is, Ca^{2+} transport occurs in the direction of the Na^+ concentration gradient on either side of the vesicles.[51] The rate of Ca^{2+} flux has been found to be dependent on a Na^+ concentration gradient across the membrane. The system is highly active at the ratio of a Na^+ concentration gradient (~ 15) that is known to exist in intact cells under physiological conditions. The rate of exchange of Ca^{2+} is saturable with Ca^{2+} or Na^+. The half-maximal rate is obtained at about 40 mM Na^+ and about 2 μM Ca^{2+} when free-Ca^{2+} concentration is maintained with a Ca^{2+}-EGTA buffer system. The maximum velocity (V_{max}) is found to be about 15 nmoles/(min·mg) protein. Comparing the degree of purification (12-fold) of sarcolemmal vesicle preparations from vascular smooth muscle, and cardiac muscle (30-fold), the V_{max} of the system in vascular smooth muscle will be comparable to about 30% of that in the cardiac muscle system.[55] This value is not negligible if one considers that the physiological demand for a slower system in vascular smooth muscle is compatible with its slower rate of development of tension and relaxation when compared to cardiac muscle. However, the possibility exists that the system *in vitro* may not exhibit the same activity as it would *in vivo* because of changes during isolation and purification of the membranes. This is one of the pitfalls of this approach. However, this approach should be useful in obtaining information about the regulation of the exchange system under certain *in vivo* conditions such as phosphorylation-dephosphorylation; membrane polarization; and treatment with endogenous, natural, and synthetic compounds. Furthermore, this approach will be useful in determining the mechanism and to estimate the stoichiometry of the exchange system since this has not yet been determined in the exchange system in vascular smooth muscle.

A survey of the Na^+-Ca^{2+} exchange activity in isolated membrane vesicles of different species and different sizes of blood vessels has shown that the system is kinetically almost identical in the blood vessels so far tested.[54] This study proved that the Na^+-Ca^{2+} exchange system is present in both large and small arteries unlike in studies with intact tissues where the function of the system in small arteries is doubtful.[35,36]

Another pitfall of this approach is that isolated sarcolemmal membranes may be contaminated with synaptic vesicles, which also contain an active Na^+-Ca^{2+} exchange system.[56] To eliminate this problem, sarcolemmal membrane vesicles have been prepared from cultured cells of vascular smooth muscle, and it has been shown that the Na^+-Ca^{2+} exchange system is similar to that in vesicles prepared from the tissue.[53]

Although there are other pitfalls in this approach which have been pointed out by several investigators,[55,57] it produced valuable information. It unequivocally demonstrated the existence of a specific Na^+-Ca^{2+} exchange system in vascular smooth muscle, and it provided information about the ionic specificity, reversibility, and the kinetics of the system *in vitro*.

WHY NA^+-CA^{2+} EXCHANGE IS IMPORTANT IN VASCULAR SMOOTH MUSCLE

Physiological Conditions

Under physiological conditions, a large Na^+ concentration gradient exists across the cell membrane (10 mM inside to 140 mM outside); therefore, the exchange system is expected to efflux Ca^{2+} from the cell in exchange for external Na^+ (FIG. 1). Thus, this system will be important in maintaining low cytosolic free-Ca^{2+} concentration and relaxation. Thus far, there is no report in the literature that provides direct evidence that under physiological conditions the Na^+-Ca^{2+} exchange system can exclusively be involved in the efflux of Ca^{2+} from the cell. An ATP-dependent Ca^{2+}pump (FIG. 1) in the cell membrane[57] may also play an important part in the regulation of cytosolic free-Ca^{2+}concentration in conjunction with the Na^+-Ca^{2+} exchange system. A specific inhibitor of the Na^+-Ca^{2+} exchange system can be useful to determine its contribution vis-à-vis the ATP-dependent pump and the other Ca^{2+}-transport processes under physiological conditions. Unfortunately, there is no specific inhibitor available at the present time. A number of agents inhibit Na^+-Ca^{2+} exchange activity *in vitro,* but they are useless *in vivo* because of their effects on other ion-transport processes in the cell.[58] Thus, the relative contribution of the exchange system, under physiological conditions, in the regulation of cytosolic free-Ca^{2+}concentration and cell tension remains unknown.

Pharmacological Conditions

As discussed in the foregoing sections, an increase in cytosolic Na^+ concentration, either by digitalis glycosides or K^+-free medium to inhibit the Na^+-K^+ pump, increases the cytosolic free-Ca^{2+}concentration in cultured cells and produces contraction in isolated tissues. This is an indirect indication that indeed the Na^+-Ca^{2+}exchange is operating in the Ca^{2+}-efflux mode, which is driven by the Na^+ concentration gradient across the cell membrane, and when this gradient is decreased due to increased cytosolic Na^+ concentration, Ca^{2+} efflux is retarded. Further evidence of this mode of operation is indicated by increased cytosolic free-Ca^{2+}concentrations in cultured cells, and contraction in isolated tissue when extracellular Na^+ is depleted. Depolarization of the cell membrane by these treatments may also affect Na^+-Ca^{2+} exchange activity if the system is electrogenic, that is, more than two Na^+ are exchanged for each Ca^{2+}; however, this remains to be determined.

In addition to the vasoconstricting effect of digitalis glycosides on the isolated blood vessel segments described above, a similar response has been observed in many regional circulations under *in vivo* conditions.[59–63] Although the mechanism of action was not known at that time, it could be attributed to Na^+-K^+-pump inhibition, an increase in cytosolic free-Na^+ concentration, and an increase in cytosolic free-Ca^{2+} concentration due to the attenuation of Na^+-Ca^{2+} exchange activity. Digitalis glycosides are used in the treatment of failing hearts.[64] The therapeutic action is mainly due to a positive inotropic action on cardiac muscle in an identical mechanism as described above.[65] Interestingly, except for a transient increase, no sustained increase in blood pressure has been observed with digitalis glycoside treatment of man or experimental animals. The reasons for the lack of effect on blood pressure may be low plasma digitalis glycoside concentrations and relatively low sensitivity of Na^+-ATPase activity of vascular smooth muscle.

Pathological Conditions

As pharmacological studies indicate, inhibition of the Na^+-Ca^{2+} exchange activity in vascular smooth muscle may lead to increased cytosolic Ca^{2+} concentration resulting in vasoconstriction, increased peripheral resistance, and hypertension. Thus far, there is no report in the literature indicating any intrinsic defect in the Na^+-Ca^{2+} exchange activity of vascular smooth muscle. In Kyoto spontaneously hypertensive rats, a small (statistically not significant) increase in Na^+-Ca^{2+} exchange activity, compared to that in normotensive rats, has been observed in sarcolemmal membrane vesicles isolated from mesenteric artery.[49] However, the exchange system may be involved in the pathogenesis or maintenance of hypertension without being intrinsically defective. Since the exchange activity is regulated by the Na^+ concentration gradient across the cell membrane, a decrease in the Na^+ gradient due to an increase in cytosolic free-Na^+ concentration may lead to a decrease in the Na^+-Ca^{2+} exchange activity and an increase in the cytosolic free-Ca^{2+} concentration. There is evidence that the cytosolic free-Na^+ concentration is increased in smooth muscle cells of hypertensive rats.[66,67] It has also been shown that intracellular free Ca^{2+} is also increased in vascular smooth muscle cells of these rats.[66–68] How can intracellular Na^+ concentration increase in vascular smooth muscle in hypertension? A number of mechanisms, such as passive diffusion, carrier-mediated transport, and inhibition of Na^+-K^+ pump, may lead to an increase in the cytosolic Na^+ concentration. Increased ionic permeability has been shown in aorta of hypertensive rats.[69] Among carrier-mediated Na^+ transport processes, Na^+-H^+ exchange has been shown to be a major pathway for Na^+ influx in vascular smooth muscle.[70] If the Na^+-H^+ exchange activity is increased, it may lead to an increase in the cytosolic Na^+ concentration. In fact, increased Na^+-H^+ exchange activity has been demonstrated in vascular smooth muscle cells of hypertensive rats.[71] However, operation of a Na^+-H^+ exchange may not only modulate the Na^+-Ca^{2+} exchange activity but may also change cytosolic pH, Ca^{2+} binding, and other Ca^{2+}-transport processes. A digitalis glycoside-sensitive Na^+-K^+ pump is another likely Na^+-transport process that may also play an important role in the regulation of cytosolic Na^+ concentration in hypertension. A inhibitor of the Na^+-K^+ pump may increase the cytosolic free-Na^+ concentration and may cause an increase in the cytosolic free-Ca^{2+} concentration due to reduced Na^+-Ca^{2+} exchange activity in vascular smooth muscle in hypertension. A hypothesis along this line has been proposed by Blaustein.[29] This hypothesis was expounded later to explain how this putative endogenous inhibitor may be released in volume-expanded hypertension and its potential action on the Na^+-K^+ pump in a number of locations.[72] A schematic description of this postulation is presented in FIGURE 2. Briefly, a defect in the Na^+ excretion in the kidney and high Na^+ intake result in a slightly positive Na and H_2O balance in the plasma that can produce an increase in the extracellular fluid (ECF) volume. In order to correct this expansion in the ECF volume, endogenous agents are secreted to promote natriuresis and diuresis. These include aldosterone, renin, vasopressin, atrial natriuretic peptide, and a digitalis-like factor among others. The concept of the endogenous digitalis-like factor was first introduced by deWardner and Clarkson.[73] This putative factor, by inhibiting the Na^+-K^+ pump in the renal tubule cells, may inhibit Na^+ reabsorption and may produce natriuresis.[73] This circulating digitalis-like factor in the plasma could conceivably also affect the Na^+-K^+ pump in vascular smooth muscle and nerve terminals (FIG. 2).[72]

A digitalis-like factor has been found to be linked to essential hypertension in man.[74] It was purified from human plasma[75] and was later identified as ouabain, a digitalis glycoside.[76] The origin of this compound in the plasma of undigitalized individuals remains a mystery. The fact that it is in the circulation and linked to essential hypertension indicates that it may be accessible to the Na^+-K^+ pump of vascular smooth muscle and nerve terminals (FIG. 3). As discussed earlier, inhibition of the Na^+-K^+ pump in vascular smooth muscle by ouabain will produce hyperreactivity to agonists, and vasoconstriction. In addition, inhibition of the Na^+-K^+ pump in nerve terminals will also increase axoplasmic Na^+ concentration and Ca^{2+} concentration due to reduced Na^+-Ca^{2+} exchange activity present in this membrane system.[56] Increases in axoplasmic Ca^{2+} will induce release of norepinephrine (NE) by exocytosis. Further-

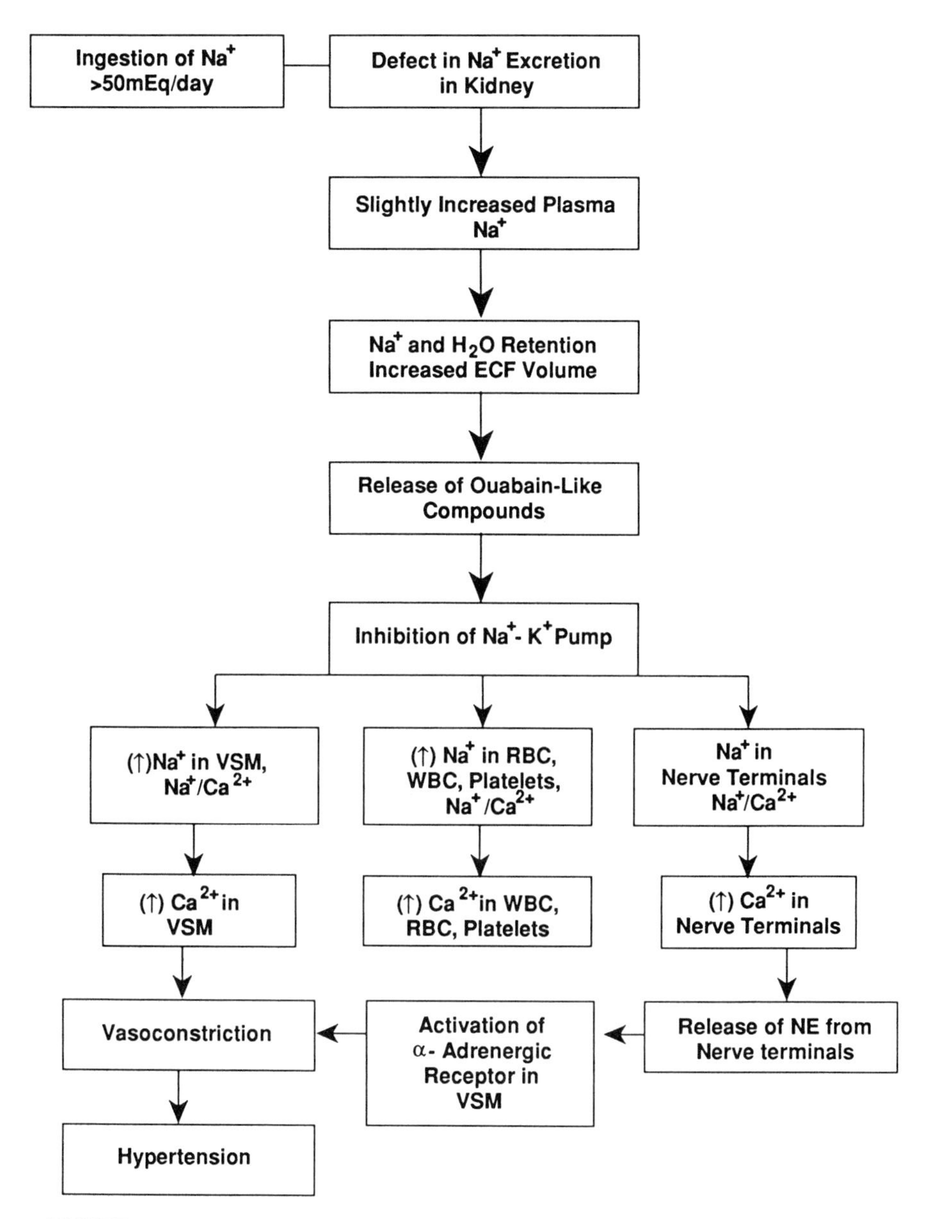

FIGURE 2. A schematic description of the events that may lead to secretion of an endogenous digitalis-like factor and its vasoconstricting effect due to Na^+-K^+ pump inhibition and modulation of Na^+-Ca^{2+} exchange activity in hypertension. Adapted from Blaustein and Hamlyn.[72]

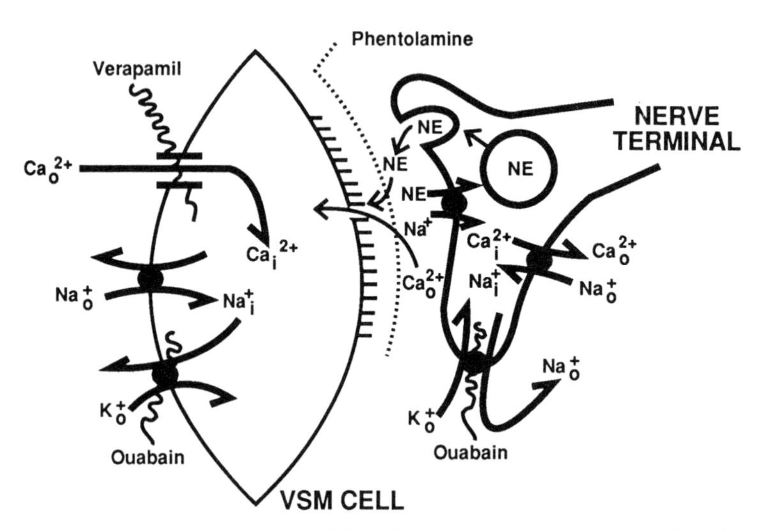

FIGURE 3. Potential role of the Na^+-Ca^{2+}exchange activity of nerve terminals and vascular smooth muscle on cell tension. Inhibition of the Na^+-K^+ pump by ouabain results in increased cytosolic and axoplasmic Na^+ concentrations. In vascular smooth muscle, the decreased Na^+ concentration gradient retards the Ca^{2+} efflux through the Na^+-Ca^{2+} exchange, resulting in increased cytosolic free-Ca^{2+} concentration and vascular smooth muscle (VSM) cell tension. In the nerve terminal, the decreased Na^+ concentration gradient retards Ca^{2+} efflux through Na^+-Ca^{2+} exchange resulting in increased axoplasmic free-Ca^{2+} concentration and the release of norepinephrine (NE) by exocytosis. The decrease in the Na^+ concentration gradient also decreases NE reuptake through a NE-Na^+ cotransport system. The NE at the synaptic cleft activates an α-adrenergic receptor-operated Ca^{2+} channel at the VSM cell surface and the release of Ca^{2+} from an intracellular store (sarcoplasmic reticulum) ensuing further increases in the cytosolic free-Ca^{2+} concentration and more contraction of the cell.

more, increases in axoplasmic Na^+ concentration will decrease NE reuptake since this process is linked to a Na^+ concentration gradient-dependent cotransport mechanism in the nerve membrane. The released NE at the synaptic cleft will activate α-adrenergic receptors in the vascular smooth muscle cell surface resulting in Ca^{2+} entry through the receptor-mediated calcium channel and the release of Ca^{2+} from the sarcoplasmic reticulum.[77] The effects of circulating ouabain or a digitalis-like compound in these two Na^+-K^+ pump systems may bring about vasoconstriction and hypertension as postulated by Blaustein.[72] However, in the past in studies identifying Na^+-Ca^{2+} exchange in vascular smooth muscle, the neural mechanism was avoided by including an α-adrenergic antagonist in the experiment, and it was not part of the evaluation. Therefore, the role of the Na^+-Ca^{2+} exchange of the nerve terminals could not be determined. Postsynaptic α_1-adrenergic antagonists are effective antihypertensive agents, which indicate that a neural mechanism may be operating in hypertension. This mechanism may also involve inhibition of the Na^+-K^+ pump by endogenous ouabain and attenuation of the Na^+-Ca^{2+} exchange activity. However, the relative contribution of the neural and vascular smooth muscle Na^+-Ca^{2+} exchange system to hypertension remains unknown.

A paradox of this mechanism is that digitalization of humans, although producing vascular hyperreactivity to contractile agonists,[78] does not produce hypertension unless there is a renal abnormality.[79] Thus, renal defect appears to be a predisposition to susceptibility to this malaise.

We have carried out experiments on rabbit aortic rings, which are highly innervated, to determine the contribution of the neural Na^+-Ca^{2+} exchange activity in vascular smooth muscle tension. This was accomplished by inhibiting the Na^+-K^+ pump with ouabain to produce contraction in the presence or absence of an α_1-adrenergic antagonist such as phentolamine or prazosin to separate the neural mechanism from that of the vascular smooth muscle (FIG. 3). The results (unpublished observations) suggest that the Na^+-Ca^{2+}exchange activity of the nerve terminal, in addition to that of vascular smooth muscle, may play a significant role in vascular smooth muscle tension. This observation is significant with respect to the mechanism of action of endogenous ouabain or a ouabain-like compound in the development of hypertension and the mechanism of action of antihypertensive α-adrenergic antagonists.

SOME UNRESOLVED ISSUES

It is now quite certain that a Na^+-Ca^{2+} exchange system exists in smooth muscle of large and small arteries.[54] The question that needs to be resolved is why in small arteries the exchange system does not reveal itself completely under conditions similar to large arteries.[35] The reason for this remains to be determined before its importance in vascular smooth muscle tension, particularly hypertension, can be evaluated. Furthermore, there is no direct evidence that the exchange system is in operation under physiological conditions. A specific inhibitor, if found, will be extremely helpful in this respect. The stoichiometry of the Na^+-Ca^{2+} exchange system in vascular smooth muscle is still unknown. If the system exchanges more than two Na^+ for each Ca^{2+}, then it must be electrogenic (current generating) and must be regulated by the membrane potential of the cell. The affinity of the system *in vitro* for Ca^{2+} appears to be outside the physiological range (0.15-1.0 μM) of cytosolic free-Ca^{2+}concentrations. It is possible that the exchange system exists in a state *in vivo* different from the state *in vitro* in isolated membrane vesicles. What causes it to behave differently *in vitro* remains unknown. The molecular composition of Na^+-Ca^{2+}exchange protein in vascular smooth muscle is still unknown. The problem is the paucity of the vascular smooth muscle tissue to isolate and purify the exchange protein. Recently the DNA encoding the cardiac Na^+-Ca^{2+} exchange protein has been cloned.[80] This approach should help elucidate the molecular nature of the exchange protein in vascular smooth muscle. The results may explain similarities and differences in structure and function between the exchange protein of vascular smooth muscle and that of other tissues.

SUMMARY

A body of information obtained by experiments with intact tissues, isolated cells, and sarcolemmal vesicles indicates, beyond a reasonable doubt, that a specific Na^+-Ca^{2+} exchange system exists in vascular smooth muscle. However, its role in the regulation of cytosolic free-Ca^{2+} concentration and cell tension under physiological conditions remains unclear. Under pharmacological conditions in which the Na^+-K^+ pump is inhibited either by digitalis glycosides or K^+-free medium, Na^+-Ca^{2+} exchange may be modulated by increases in cytosolic free Na^+ to increase the cytosolic free-Ca^{2+}concentration and cell tension. Under pathological conditions in which the cytosolic Na^+ concentration is increased as a result of inhibition of the Na^+-K^+ pump by endogenous ouabain or a digitalis-like factor, or activation of the Na^+-H^+ exchange or passive permeability of Na^+, the Na^+-Ca^{2+} exchange activity of vascular smooth muscle and the nerve terminal may play an important role in the development and/or maintenance of hypertension. These and other premises remain to be confirmed or discounted.

ACKNOWLEDGMENTS

I thank Evangeline Motley and Kevin McFarland for their help.

REFERENCES

1. CAMPBELL, A. K. 1983. Intracellular Calcium—Its Universal Role as Regulator. Wiley. New York.
2. REUTER, H. & H. SEITZ. 1968. J. Physiol. (London) **195:** 451-470.
3. BAKER, H. & M. P. BLAUSTEIN. 1968. Biochim. Biophys. Acta **150:** 167-170.
4. BAKER, P. F., M. P. BLAUSTEIN, A. L. HODGKIN & R. A. STEINHARDT. 1969. J. Physiol. (London) **200:** 459-496.
5. BLAUSTEIN, M. P. & A. L. HODGKIN. 1969. J. Physiol. (London) **200:** 497-527.
6. ASHLEY, C. C. & P. C. CALDWELL. 1972. J. Physiol. (London) **242:** 255-272.
7. BERNSTEIN, J. & G. SANTACANA. 1985. Res. Comm. Chem. Pathol. Pharmacol. **47:** 3-34.
8. KACZOROWSKI, G. J., L. COSTELLO, J. DETHMERS, M. J. TRUMBLE & R. L. VANDLEN. 1984. J. Biol. Chem. **259:** 9395-9403.
9. SCHNETKAMP, P. P. M., F. J. M DAEMEN & S. L. BONTING. 1977. Biochim. Biophys. Acta **468:** 259-270.
10. RENGASAMY, A., S. SOURA & H. FEINBERG. 1987. J. Biol. Chem. **258:** 3178-3182.
11. GRUBB, B. R. & P. J. BENTLEY. 1985. Am. J. Physiol. **249**(Reg. Integrat. Comp. Physiol. 8): R172-R178.
12. SAERMARK, T., N. A. THORN & M. GRATZL. 1983. Cell Calcium **4:** 151-170.
13. BOHR, D. F., D. BRODIE & D. CHEN. 1958. Circulation **27:** 746-749.
14. HINKE, J. A. M. & M. L. WILSON. 1962. Am. J. Physiol. **203:** 1161-1166.
15. CORET, I. A. & M. J. HUGHES. 1964. Arch. Intern. Pharmacodyn. **203:** 330-353.
16. TODA, N. 1972. Jpn. J. Pharmacol. **22:** 347-357.
17. REUTER, H., M. P. BLAUSTEIN & G. HAEUSLER. 1973. Phil. Trans. R. Soc. B. **265:** 87-94.
18. LEONARD, E. 1957. Am. J. Physiol. **189:** 185-190.
19. BRIGGS, A. H. & S. SHIBATA. 1966. Proc. Soc. Exp. Biol. Med. **121:** 274-278.
20. MATHEWS, E. K. & M. C. SUTTER. 1967. J. Physiol. Pharmacol. **45:** 509-520.
21. BOHR, D. F., C. SEIDAL & J. SOBIESKI. 1969. Microvasc. Res. **1:** 335-343.
22. BRENDER, D., C. G. STRONG & J. T. SHEPHERD. 1970. Circ. Res. **26:** 647-655.
23. FRIEDMAN, S. M., M. NAKASHIMA, V. PALATY & B. K. WALTERS. 1973. Can. J. Physiol. Pharmacol. **51:** 410-417.
24. BROEKAERT, A. & T. GODFRAIND. 1973. Arch. Intern. Pharmacodyn. **203:** 393-395.
25. DODD, W. A. & E. E. DANIEL. 1960. Circ. Res. **8:** 451-463.
26. BIAMINO, G. & B. JOHANSSON. 1970. Pflügers Arch. **321:** 143-158.
27. MA, T. S. & D. BOSE. 1977. Am. J. Physiol. **232**(Cell Physiol. 1): C59-C66.
28. BLAUSTEIN, M. P. 1974. Rev. Physiol. Biochem. Pharmacol. **70:** 33-82.
29. BLAUSTEIN, M. P. 1977. Am. J. Physiol. **232**(Cell Physiol. 3): C165-C173.
30. VAN BREEMEN, C., P. AARONSON & R. LOUTZENHISER. 1979. Pharmacol. Rev. **30:** 167-208.
31. DROOGMAN, G. & R. CASTEELS. 1979. J. Gen. Physiol. **74:** 57-70.
32. HERMSMEYER, K. 1983. Fed. Proc. **42:** 246-252.
33. KARAKI, H. & N. URAKAWA. 1977. Eur. J. Pharmacol. **43:** 65-72.
34. KARAKI, H., H. OZAKI & N. URAKAWA. 1978. Eur. J. Pharmacol. **48:** 439-443.
35. MULVANY, M. J., C. AALKJAER & T. T. PETERSEN. 1984. Circ. Res. **54:** 740-749.

36. PETERSEN, T. T. & M. J. MULVANY. 1984. Blood Vessels **21:** 279-289.
37. BLAUSTEIN, M. P. 1988. J. Cardiovasc. Pharmacol. **12**(Suppl. 5): S56-S68.
38. BLAUSTEIN, M. P., T. ASHIDA, W. F. GOLDMAN, W. GIL WEIR & J. M. HAMLYN. 1986. Ann. N. Y. Acad. Sci. **188:** 199-216.
39. ASHIDA, T., J. SCHAEFFER, W. F. GOLDMAN, J. B. WADE & M. P. BLAUSTEIN. 1988. Circ. Res. **62:** 854-863.
40. MATLIB, M. A., K. R. WHITMER, E. P. MACCARTHY & B. S. OOI. 1986. J. Hypertens. **4**(Suppl. 5): S222-S223.
41. SMITH, J. B., E. J. CRAGOE & L. SMITH. 1987. J. Biol. Chem. **262:** 11988-11994.
42. NABEL, E. G., B. C. BERK, T. A. BROCK & T. A. SMITH. 1988. Circ. Res. **62;** 486-493.
43. SMITH, J. B. & L. SMITH. 1987. J. Biol. Chem. **262:** 17455-17460.
44. SMITH, J. B., T. ZHENG & R.-M. LYU. 1989. Cell Calc. **10:** 125-134.
45. GOLDMAN, W. F. & M. P. BLAUSTEIN. 1988. J. Cardiovasc. Pharmacol. **12**(Suppl. 5): S13-S19.
46. BOVA, S., W. F. GOLDMAN, X-J. YUAN & M. P. BLAUSTEIN. 1990. Am. J. Physiol. **259**(Heart Cir. Physiol. 28): H409-H423.
47. DANIEL, E. E., A. K. GROVER & C. Y. KWAN. 1982. Fed. Proc. **41:** 2898-2904.
48. MOREL, N. & T. GODFRAIND. 1984. Biochem. J. **218:** 421-427.
49. MATLIB, M. A., A. SCHWARTZ & Y. YAMORI. 1985. Am. J. Physiol. **249**(Cell Physiol. 18): C166-C172.
50. MATLIB, M. A. & J. P. REEVES. 1987. Biochim. Biophys. Acta **904:** 145-148.
51. MATLIB, M. A. 1988. Am. J. Physiol. **255**(Cell Physiol. 24): C323-C330.
52. KAHN, A. M., J. C. ALLEN & H. SHELAT. 1988. Am. J. Physiol. **254**(Cell Physiol. 23): C441-C449.
53. MATLIB, M. A., M. KIHARA, C. FARRELL & R. C. DAGE. 1988. Biochim. Biophys. Acta **939:** 173-177.
54. MATLIB, M. A. 1988. J. Cardiovasc. Pharmacol. **12**(Suppl. 6): S60-S62.
55. REEVES, J. P. 1985. Curr. Top. Membr. Transp. **5:** 77-127.
56. BLAUSTEIN, M. P. & A. C. ECTOR. 1976. Biochim. Biophys. Acta **419:** 295-308.
57. DANIEL, E. E. 1985. Experientia **41:** 905-913.
58. KACZOROWSKI, G. J., M. L. GARCIA, V. F. KING & R. S. SLAUGHTER. 1989. *In* Sodium-Calcium Exchange. T. J. A. Allen, D. Noble & H. Reuter, Eds.: 66-101. Oxford Science Publications. Oxford, U.K.
59. FERRER, M. I., S. E. BRADLEY, H. O. WHEELER, Y. ENSON, R. PREISIG & R. M. HARVEY. 1965. Circulation **32:** 527-537.
60. LEVINSKY, R. A., R. M. LEWIS, T. E. BYNUM & H. G. HANLEY. 1975. Circulation **52:** 130-136.
61. ROSS, J., J. A. WALDHAUSEN & E. BRAUNWALD. 1960. J. Clin. Invest. **39:** 930-936.
62. TREAT, E., H. B. ULANO & E. D. JACOBSON. 1971. J. Pharmacol. Exp. Ther. **179:** 144-148.
63. VATNER, S. F., C. B. HIGGENS, D. FRANKLIN & E. BRAUNWALD. 1971. Circ. Res. **28:** 470-479.
64. HOFFMAN, B. F. & J. BIGGER. 1985. *In* The Pharmacological Basis of Therapeutics. A. G. Gilman, L. S. Goodman, T. W. Rall & F. Murad, Eds.: 716-745. MacMillan Publishing Co. New York.
65. SCHWARTZ, A. 1983. Curr. Top. Membr. Transp. **19:** 825-841.
66. ZIDEK, W., T. KERENYI, H. LOSSE & H. VETTER. 1983. Res. Exp. Med. **183:** 129-132.
67. LOOSE, H., W. ZIDEK & H. VETTER. 1984. J. Cardiovasc. Pharmacol. **6**(Suppl. 1): S32-S34.
68. SUGIYAMA, T., M. YOSHIZUMI, F. TAKAKU, H. UREBE, M. TSUKA-KOSHI, T. KASUYA & Y. YAZAKI. 1986. Biochem. Biophys. Res. Commun. **141:** 340-345.
69. JONES, A. W. 1974. Fed. Proc. **33:** 133-137.
70. LITTLE, P. J., E. J. CRAGOE & A. BOBIK. 1986. Am. J. Physiol. **251**(Cell Physiol. 20): C707-C712.
71. BERK, B. C., G. VALLENGA, A. J. MUSLIN, H. M. GORDON, M. CANESSA & R. W. ALEXANDER. 1989. J. Ciin. Invest. **83:** 822-829.
72. BLAUSTEIN, M. P. & J. M. HAMLYN. 1984. Am. J. Med. **77:** 45-59.
73. DEWARDNER, H. E. & E. M. CLARKSON. 1985. Physiol. Rev. **65:** 685-759.

74. Hamlyn, J. M., R. Schaeffer, P. D. Levinson, B. P. Hamilton, A. A. Kowarski & M. P. Blaustein. 1982. Nature **300:** 650-652.
75. Hamlyn, J. M., D. W. Harris & J. H. Ludens. 1989. J. Biol. Chem. **264:** 7395-7404.
76. Hamlyn, J. M., M. P. Blaustein, S. Bova, D. W. Ducharme, D. W. Harris, F. Mandel, W. R. Mathews & J. H. Ludens. 1991. Proc. Natl. Acad. Sci. USA, in press.
77. Bolton, T. 1986. Br. Med. Bull. **42:** 421-429.
78. Guthrie, G. P. 1984. J. Clin. Endocrinol. Metab. **58:** 76-80.
79. Fujimura, A., A. Ebara & O. Yaoka. 1984. Jpn. J. Hypertens. **7**(1): 49.
80. Nicoll, D. A., S. Longoni & K. D. Philipson. 1990. Science **250:** 562-565.

Role of Na^+-Ca^{2+} Exchanger in β-Adrenergic Relaxation of Single Smooth Muscle Cells[a]

EDWIN D. W. MOORE, KEVIN E. FOGARTY, AND FREDRIC S. FAY

Department of Physiology/Biomedical Imaging Group
University of Massachusetts Medical Center
Worcester, Massachusetts 01605

β-Adrenergic agents cause smooth muscle of many tissues to relax. Despite the importance of this process for the function of many of our organs, we still do not fully understand the mechanism underlying this basic physiological response. Considerable progress has been recently achieved by analyzing the response to β-adrenergic stimulation at the single-cell level, using isolated smooth muscle cells.[1-3] Our own work on smooth muscle has focused on single cells enzymatically isolated from the stomach of the toad *Bufo marinus.* These cells are relatively large and retain their structural,[4] physiological,[5] pharmacological,[6] and biochemical[2] properties following isolation, thereby facilitating studies at the single-cell level. They have been the subject of intense study for about 20 years and the knowledge base that has been developed is extremely powerful both for planning and interpreting studies. Finally, mechanisms uncovered in this cell system have in general been shown to be present in many other smooth muscle cell types, making it a generally useful model.

Previous studies directed at understanding the events underlying β-adrenergic relaxation of these smooth muscle cells have focused on changes in radioactive ion fluxes,[7,8] changes in membrane electrical properties,[3] as well as changes in the 3′5′-cyclic adenosine monophosphate (cAMP)-dependent protein kinase pathway[1,2] that accompanies the inhibitory effect of β-adrenergic agents like isoproterenol on contractility.[1] Those studies led to the view of β-adrenergic relaxation of smooth muscle shown in FIGURE 1. According to the model, isoproterenol (ISO), a β-adrenergic agonist, acts via a rise in cAMP and activation of cAMP-dependent protein kinase to stimulate the Na^+-K^+ pump. The consequent steepening of the transmembrane $[Na^+]$ gradient in turn would be expected to place the Na^+-Ca^{2+} exchange into Ca^{2+}-extrusion mode, thereby diminishing $[Ca^{2+}]$ changes and contractile force induced by excitatory stimuli. In addition ISO also hyperpolarizes the membrane through an increase in potassium conductance,[3] and this too would be expected to diminish contractile responsiveness by decreasing the extent of depolarization and consequently decreasing the activation of Ca^{2+} channels by excitatory stimuli.[9] The notion that the Na^+-K^+ pump is stimulated by ISO has rested principally on radioactive ion-flux studies that revealed that ISO stimulated $^{42}K^+$ influx and $^{24}Na^+$ efflux and that these effects could be mimicked by dibutyryl cAMP and blocked by ouabain.[2,7] Furthermore, the notion that stimulation

[a]Supported in part by a grant from the National Institutes of Health (HL 14523) and by a grant from the National Science Foundation (DIR-8720188). Ed Moore was supported by a Neuromuscular Disease Research Fellowship from the Muscular Dystrophy Association.

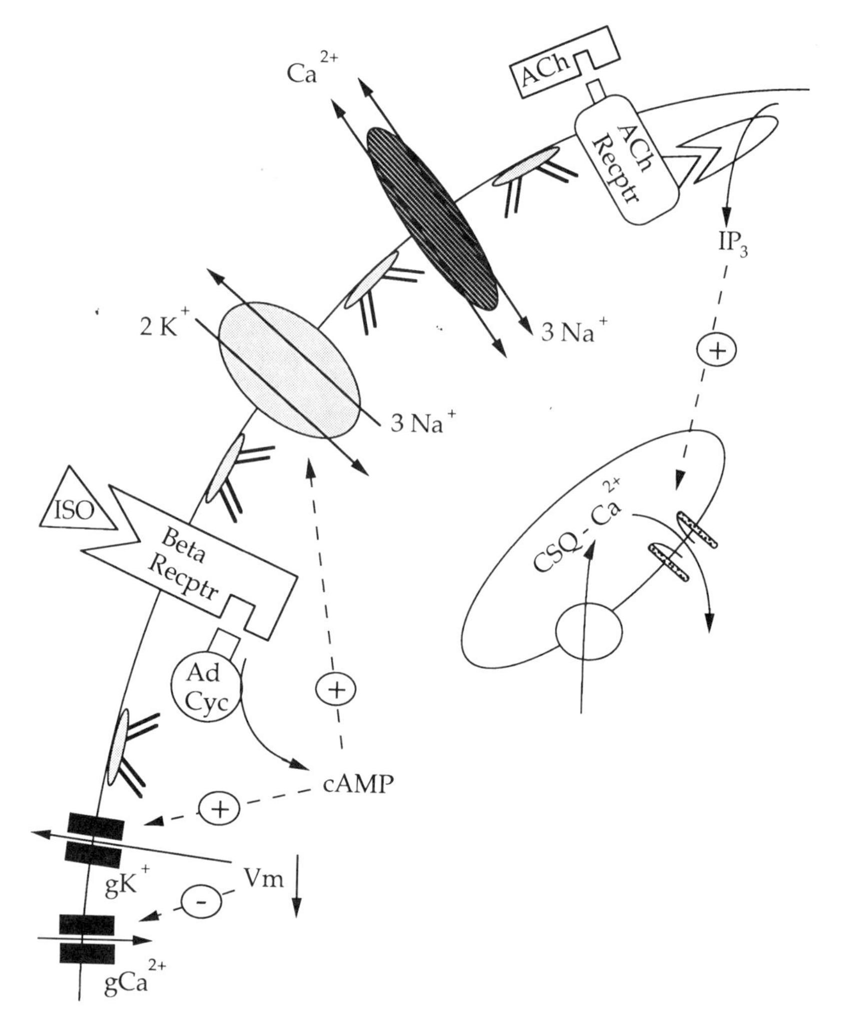

FIGURE 1. Model for the proposed mechanism of action of ISO in smooth muscle cells. ISO, through the β-adrenergic receptor, activates adenylate cyclase and increases cAMP production. cAMP directly or through cyclic AMP dependent protein kinase increases g_{K+} which hyperpolarizes the cell and, in addition, stimulates the Na^+-K^+ pump. Na^+-K^+-pump stimulation both hyperpolarizes the cell and decreases $[Na^+]_i$ and these effects place the Na^+-Ca^{2+} exchanger into Ca^{2+} extrusion mode. Contraction is therefore inhibited in response to agents, which, like acetylcholine (ACh), release Ca^{2+} from intracellular stores.

of the Na^+-K^+ pump was linked to decreased contractility because of Ca^{2+} extrusion through the Na^+-Ca^{2+} exchange rested on observations that ^{45}Ca efflux was transiently stimulated by ISO and this effect could be mimicked by dibutyryl-cAMP and blocked by ouabain as well as by a reduction in extracellular $[Na^+]$.[8] While the model represented a reasonable way to explain the kinetic data, the methods available in the early 1980s and late 70s never allowed us to determine if ISO caused $[Na^+]$ to fall—a key feature linking stimulation of the Na^+-K^+ pump and Na^+-Ca^{2+} exchange by ISO. With the recent development of sodium-sensitive fluorescent ratiometric indicators[10] it became possible to directly test this prediction.

We have thus utilized the Na^+-sensitive dye SBFI to measure $[Na^+]$ in single isolated smooth muscle cells. The excitation and emission spectra of SBFI are virtually identical to the spectra of the Ca^{2+}-sensitive fluorescent dye fura-2. The peak of the excitation spectrum of SBFI shifts to shorter wavelengths when the dye binds Na^+. The dye was loaded into the smooth muscle cells by incubating the cells with the cell-permeant acetoxymethylester derivative of SBFI. Because properties of the dye inside the cell appeared to be slightly different from that of the dye in free solution, at the end of each experiment for each cell we calibrated the dye by exposing the cell to solutions containing varying concentrations of Na^+ and K^+ in the presence of monensin and nigericin. Measurements of the ratio of fluorescence of cells loaded with SBFI revealed that resting $[Na^+]_i$ was uniform throughout the cell. The fluorescence ratio at rest, when referenced to changes in the ratio of the dye at various $[Na^+]$ in the presence of ionophores, averaged 12.5 ± 4.2 mM. When a cell was superfused with solution containing 100 μM ISO, $[Na^+]_i$ decreased on average by 7 mM in 3 sec. Up to three identical responses have been observed in the same cell when it was repetitively challenged with ISO at 5-minute intervals. In order to determine if the fall in $[Na^+]_i$ induced by ISO is due to stimulation of the Na^+-K^+ pump cells were preincubated in K^+-free medium for 5 min to inhibit the pump. As expected, preincubation in K^+-free medium caused $[Na^+]_i$ to rise by about 10 mM. When such cells were subsequently challenged with ISO, $[Na^+]_i$ did not decrease. Re-incubation in medium replete with normal levels of K^+ allowed $[Na^+]_i$ to return to normal resting levels and the $[Na^+]_i$ again decreased when the cell was challenged with ISO. As the effect of ISO on contractility appeared to be mediated by interaction with a β-receptor and consequent rise in cAMP,[1,2] we examined the effect of ISO on $[Na^+]_i$ to determine whether it was similarly mediated. We found that the response to ISO was fully blocked by the β-adrenergic antagonist pindolol at concentrations as low as 10 nM. Furthermore, we found that forskolin (30 μM) and 8-bromo-cyclic AMP (5 mM) both of which would be expected to elevate intracellular cAMP levels triggered a fall in $[Na^+]_i$ when applied to a smooth muscle cell. The magnitude of the changes in $[Na^+]_i$ induced by these agents was similar to that observed in response to ISO as shown in FIGURE 2. Thus, direct measurements of $[Na^+]_i$ during the action of isoproterenol on single smooth muscle cells reveals events in strong support of our hypothesis for the mechanism of ISO action on contractility in smooth muscle. ISO stimulates the Na^+-K^+ pump, and a decrease in $[Na^+]_i$ is indeed observed. This supports the notion that through these effects ISO acts to enhance the ability of the Na^+-Ca^{2+} exchanger to drive Ca^{2+} from the cell.

We were puzzled, however, by additional observations that suggested to us that this scheme is perhaps too simple to explain how ISO relaxes smooth muscle. Several observations have prompted this reassessment. For one, when we analyze the processes dominating Ca^{2+} removal from the cell following brief voltage activation we find that Ca^{2+} removal from the cytoplasm is dominated by a process that is largely voltage insensitive and has a very high affinity for Ca^{2+}—its apparent K_m is at most a few hundred nM,[11] hardly hallmarks of Na^+-Ca^{2+} exchange. Thus acute removal of Ca^{2+} from the cytoplasm appears more likely to be driven by Ca^{2+} pumps rather than the Na^+-Ca^{2+} exchanger. Yet we know that ISO stimulates the Na^+-K^+ pump and accelerates unidirectional ^{45}Ca efflux in a manner expected for Na^+-Ca^{2+} exchange and that ISO markedly attenuates the contractile response of these cells. The inhibition of contractile responsiveness is especially potent for agents such as acetylcholine that act principally by releasing Ca^{2+} from internal stores.[12] Thus it seems as if ISO's effect

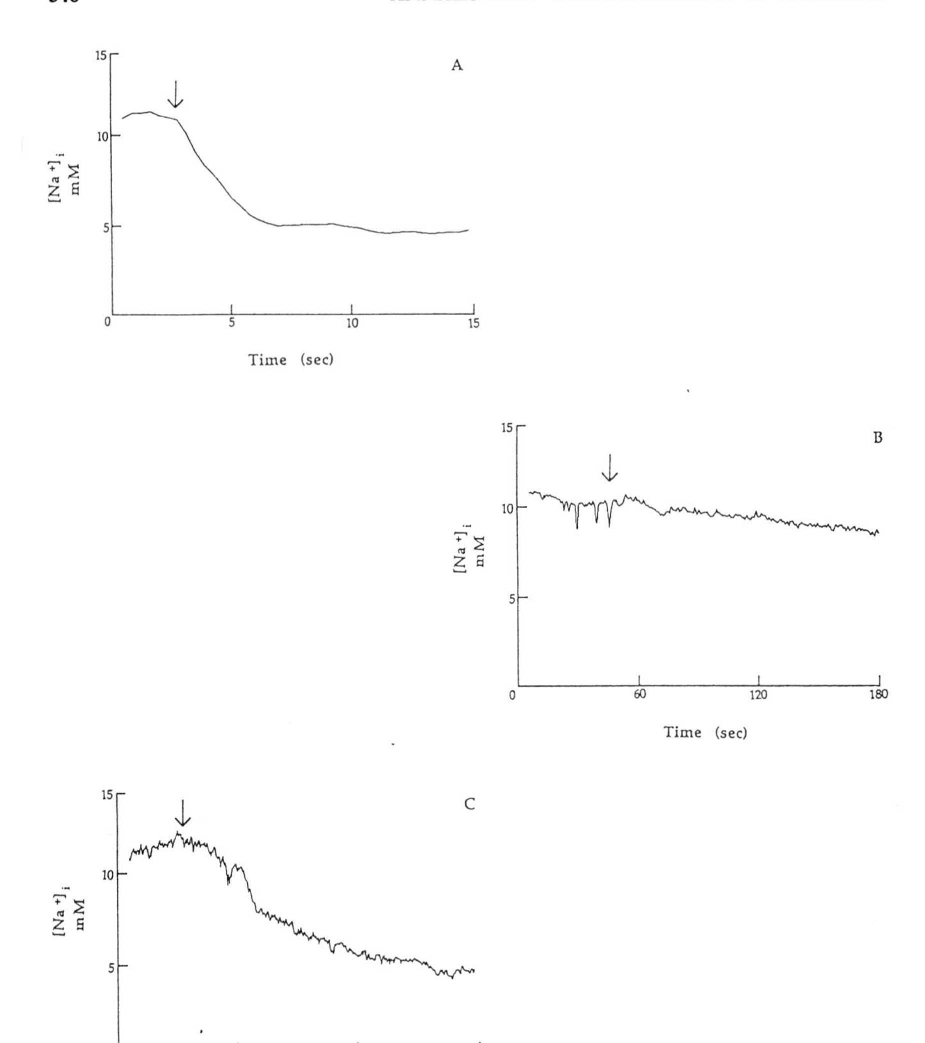

FIGURE 2. Records of $[Na^+]_i$ recorded on a dual-wavelength microfluorimeter. **A.** Isolated cells were maintained in amphibian physiological salt solution (APS) of the following composition (all concentrations are mM) 109 NaCl, 2 KCl, 0.56 Na_2HPO_4, 0.14 NaH_2PO_4, 20 $NaHCO_3$, 1.8 $CaCl_2$, 0.98 $MgSO_4$, 11.1 glucose, equilibrated with 95% O_2, 5% CO_2, pH 7.4. At the point marked by the arrow the cell was superfused with APS containing 100 μM ISO. $[Na^+]_i$ concentrations were recorded every 10 msec., the trace has been averaged at 50-msec intervals to reduce noise. **B.** After the response in A was recorded, the shutter was closed and the cell was superfused with fresh APS for 3 min, followed by APS with 10 nM pindolol for 2 minutes. At the arrow, 100 μM ISO was reintroduced in the continued presence of pindolol. Note the different time scales in A & B. **C.** Five minutes after pindolol and ISO were removed from the cell with fresh APS, the cell was superfused, at the arrow, with 5 mM 8-bromo-cAMP.

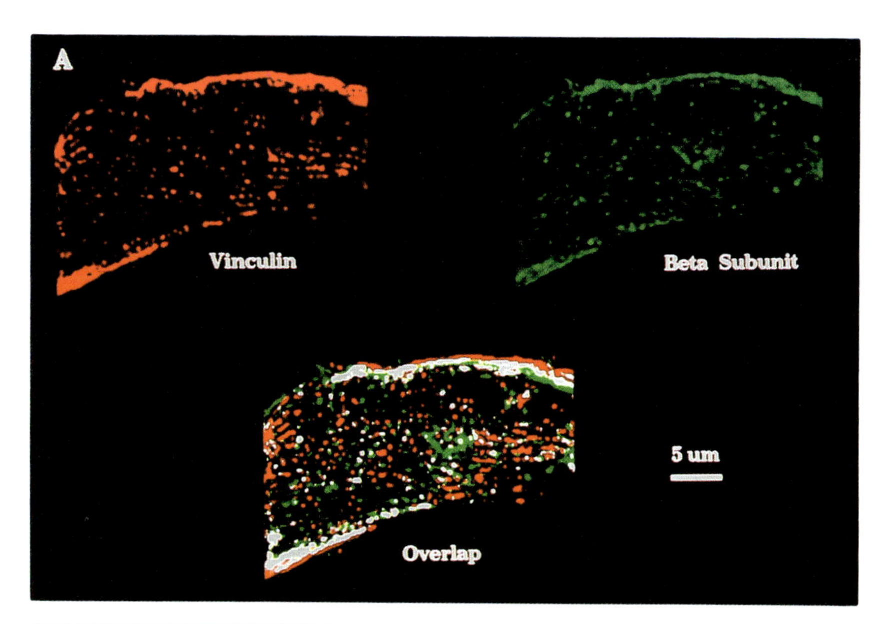

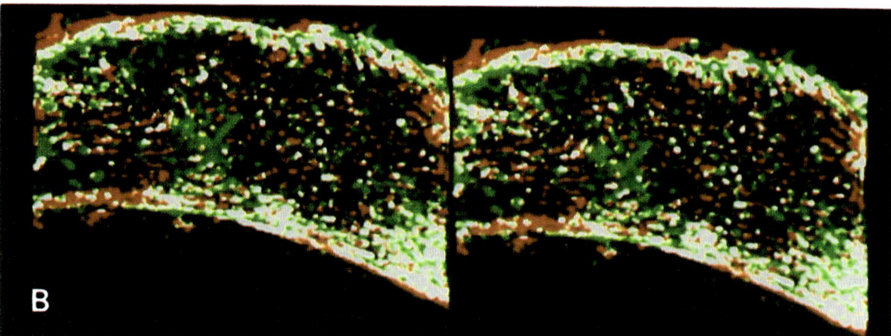

FIGURE 3. Reconstructed images of a single smooth muscle cell which was fixed, and dual-labeled (using indirect immunofluorescence), showing the distribution of vinculin Texas Red), the β-subunit of the Na^+-K^+ pump (fluorescein) and a combined image that shows the extent of overlap or colocalization between the two. In the combined image, white pixels are those where both vinculin and the β-subunit are located in the same pixel (pixel dimensions = 0.15 μm^2). **B.** Stereo pair (12 ° rotation between images) of the combined image presented in A.

on the Na^+-K^+ pump and effects on Na^+-Ca^{2+} exchange are to diminish Ca^{2+} releasable from internal stores.

One possible way of reconciling these observations might be if the Na^+-Ca^{2+} exchanger in these smooth muscle cells acted principally to remove Ca^{2+} from the sarcoplasmic reticulum and not directly on that in the cytoplasm. This might come about if the Na^+-Ca^{2+} exchanger somehow had preferred access to the sarcoplasmic reticulum. It has long been recognized that the membrane of smooth muscle is made up of two general domains[4,13]—one contains numerous caveolae in regions in close apposition to elements of the sarcoplasmic reticulum, and the other region of the membrane contains dense amorphous plaques that are known to be enriched in proteins anchoring elements of the contractile apparatus.[14] The caveolae-enriched regions run for some distance, alternating with regions devoid of these caveolae.[13] Thus the membrane of smooth muscle appears to be divided into two macrodomains—one is rich in caveolae in close apposition to elements of the sarcoplasmic reticulum, and the other region is rich in proteins anchoring the contractile machinery. Where are the proteins involved in ion regulation?

In order to obtain insights into this issue, we have begun to investigate the distribution in single isolated smooth muscle cells of the Na^+-K^+ pump, the Na^+-Ca^{2+} exchanger, calsequestrin (a marker for the sarcoplasmic reticulum[15]), and vinculin (a marker for regions of the attachment of the contractile machinery to the cell membrane[16]). The distribution of these molecules with respect to one another has been investigated by fluorescence immunocytochemistry. In order to obtain a high-resolution view of the distribution of these molecules with respect to one another, we stained cells with antibodies specific for each of these four proteins, and then collected a series of images at intervals through focus in such cells. Because the fluorescent microscope has a very large depth of field, light from all planes are superimposed in any given image plane, making analysis of such images virtually impossible.[17] We thus applied to this series of optical sections a constrained deconvolution technique based on regularization theory that effectively reverses the distortion introduced by the optics. The restored image provides a more reliable estimate of the distribution in 3D of labeled molecules within the cell[18] as shown in FIGURE 3.

The results to date suggest that α and β subunits of the Na^+-K^+ pump as well as the Na^+-Ca^{2+} exchanger are all in the same macrodomain on the cell membrane. This domain is distinct from that occupied by vinculin, which presumably marks the regions devoid of caveolae where the contractile machinery attaches to the surface of the cell. The region rich in Na^+-K^+ pump and Na^+-Ca^{2+} exchanger appears to exist close to regions occupied by the sarcoplasmic reticulum as the patterns of patches of calsequestrin appear to be closely associated with the plasma membrane ion pumps and exchangers.

These results raise the possibility that the Na^+-Ca^{2+} exchanger may indeed have preferred access, by virtue of its distribution, to Ca^{2+} in the sarcoplasmic reticulum. Furthermore, the similarity in the distribution of the Na^+-K^+ pump and Na^+-Ca^{2+} exchanger may well facilitate the coupling of activity of the Na^+-K^+ pump and Na^+-Ca^{2+} exchanger, as they appear to share the same domain in the cytoplasm of these smooth muscle cells.

REFERENCES

1. HONEYMAN, T. W., P. MERRIAM & F. S. FAY. 1978. Effect of isoproterenol on cyclic AMP levels and contractility of isolated smooth muscle cells. J. Mol. Pharm. **14:** 86-98.
2. SCHEID, C. R., T. W. HONEYMAN & F. S. FAY. 1979. Mechanism of β-adrenergic induced relaxation of smooth muscle. Nature **277:** 32-36.
3. YAMAGUCHI, H., T. W. HONEYMAN & F. S. FAY. 1988. β-Adrenergic actions on membrane electrical properties of dissociated smooth muscle cells. Am. J. Physiol. Cell **254:** C423-C431.
4. FAY, F. S. & C. M. DELISE. 1973. Contraction of isolated smooth muscle cells—structural changes. Proc. Natl. Acad. Sci. USA **70:** 641-645.
5. FAY, F. S. 1977. Isometric contractile properties of single isolated smooth muscle cells. Nature **265:** 553-556.
6. FAY, F. S. & J. J. SINGER. 1977. Characteristics of responses of isolated smooth muscle cells to cholinergic drugs. Am. J. Physiol. **232:** C144-C154.

7. SCHEID, C. R. & F. S. FAY. 1984. β-Adrenergic stimulation of ^{42}K influx in isolated smooth muscle cells. Am. J. Physiol., Cell **246:** C415-C421.
8. SCHEID, C. R. & F. S. FAY. 1984. β-Adrenergic effects on transmembrane ^{45}Ca fluxes in isolated smooth muscle cells. Am. J. Physiol., Cell **246:** C431-C438.
9. WALSH, J. V., JR. & J. J. SINGER. 1981. Voltage clamp of single freshly dissociated smooth muscle cells: Current-voltage relationships for three currents. Pflügers Arch. **390:** 207-210.
10. MINTA, A. & R. W. TSIEN. 1989. Fluorescent indicators for cytosolic sodium. J. Biol. Chem. **264:** 19449-19457.
11. BECKER, P. L., J. V. WALSH, JR., J. J. SINGER & F. S. FAY. 1989. Regulation of calcium concentration in voltage-clamped single smooth muscle cells. Science **244:** 211-214.
12. WILLIAMS, D. A. & F. S. FAY. 1986. Calcium transients and resting levels in isolated smooth muscle cells as monitored with quin-2. Am. J. Physiol. **250** (Cell Physiol. 19): C779-791.
13. GABELLA, G. 1984. Structural apparatus for force transmission in smooth muscles. Physiol. Rev. **64:** 455-477.
14. COOKE, P. H., G. KARGACIN, R. CRAIG, K. FOGARTY, F. S. FAY & S. HAGEN. 1987. Molecular structure and organization of filaments in single, skinned smooth muscle cells. *In* Regulation and Contraction of Smooth Muscle. A. P. Somlyo & M. Siegman, Eds. Alan R. Liss, Inc. New York.
15. WUYTACK, F., L. RAEYMAEKERS, J. VERBIST, L. P. JONES & R. CASTEELS. 1987. Smooth-muscle endoplasmic reticulum contains a cardiac-like form of calsequestrin. Biochim. Biophys. Acta **899:** 151-158.
16. DRENCKHAHN, D. & H. G. MANNHERZ. 1983. Distribution of actin and the actin associated proteins myosin, tropomyosin, alpha-actinin, vinculin, and villin in rat and bovine exocrine glands. Eur. J. Cell Biol. **30:** 167-176.
17. FAY, F. S., W. CARRINGTON & K. FOGARTY. 1989. 3D molecular distribution in single cells analyzed using the digital imaging microscope. J. Microsc. **153** (pt. 2): 133-149.
18. CARRINGTON, W., K. E. FOGARTY, L. M. LIFSHITZ & F. S. FAY. 1989. Three dimensional imaging on confocal and wide field microscopes. *In* Handbook of Biological Confocal Microscopy. J. Pawley, Ed. Plenum Press. New York and London.

A Comparison of Free Intracellular Calcium and Magnesium Levels in the Vascular Smooth Muscle and Striated Muscle Cells of the Spontaneously Hypertensive and Wistar Kyoto Normotensive Rat

M. AMEEN

MRC Unit
and
University Department of Clinical Pharmacology
Radcliffe Infirmary
Oxford, England

J. E. DAVIES AND L. L. NG

Department of Pharmacology
Leicester Royal Infirmary
Leicester, England

Abnormalities in the cellular handling of Ca^{2+} (see McCarron[1]) and Mg^{2+} (see Altura *et al.*[2]) have been postulated to play an important role in the pathogenesis of hypertension. Observed increases in intracellular Ca^{2+} concentration in the platelets of hypertensive humans have not been confirmed in the platelets of spontaneously hypertensive (SHR) rats compared to the Wistar Kyoto normotensive (WKY) rat. Furthermore the inaccessibility of vascular smooth muscle cells (VSM) has limited experiments to platelets and red cells. We now present the $[Ca^{2+}]_i$ and $[Mg^{2+}]_i$ in VCM and striated muscle (SM) from SHR and WKY rat cells in culture.

METHODS

VSM cells were cultured from eight 13-week-old SHR or WKY rat aortas. A collagenase/trypsin cocktail, 1 g/L of each enzyme in phosphate-buffered saline, was used to dissociate the tissue. SM cultures were set up from 1-g pieces of hindquarter skeletal muscle. The cells were cultured in Ham's F-12 medium supplemented with 10% fetal calf serum. The cultures were freed from contamination with fibroblasts by plating on gelatin-coated plates and by using described methods of clonal analysis.[3] Briefly, single clonal colonies of about 1000 muscle cells were harvested, groups of three were pooled, and grown to only about 50% confluence (to prevent initiation of myogenesis) in Ham's F-12 medium with 0.5% chicken extract and 20% fetal calf serum, and then, either used or frozen for long-term storage. Simultaneously, some cells derived from the same clones were tested for their myogenic potentials.

Confluent cells grown on cover slips were loaded with 10 μmol/L of fura-2 acetoxymethyl ester for Ca^{2+} measurements or 10 μmol/L furaptra-AM for Mg^{2+} measurement in RPMI 1640 (15 mmol/L HEPES and 0.1% fetal calf serum) for 1 hr at 37°C. Then, they were washed with RPMI 1640 and left in medium for 30 minutes, followed by measurement of fluorescence. The excitation wavelengths for fura-2 were 340 and 380 nm, and for furaptra they were 335 and 370 nm. Emission was measured at 510 nm for both dyes. Emission values were corrected for dye leakage by the measurement of fluorescence in the buffer and for autofluorescence. For reading the resting intracellular concentrations, the cover slips were mounted on clamps at an angle of 30° to the incident beam of light and immersed in buffer (NaCl, 140; KCl, 5; $CaCl_2$, 1.8; $MgSO_4$, 0.8; glucose, 5; and HEPES 15 [mmol/L], pH 7.4) solution in a quartz cuvette. Free $[Ca^{2+}]_i$ and $[Mg^{2+}]_i$ were calculated using the previously outlined[4] R_{max} and R_{min} calibration technique. For fura-2, resting 340/380 ratio was measured for the cells. Then, 2 μmol/L of 4 bromo A23187 was added to the NaCl buffer, and R_{max} values were measured. Then, 10 mmol/L *N,N,N',N'*-tetraacetic acid (EGTA) with Tris base was added to the NaCl buffer. This resulted in an increase in the pH of the buffer to 8.5, enabling the measurement of R_{min}. For the calibration of Mg^{2+}, R_{max} was measured following the lysis of the cells with 0.1% Triton X-100. The following equation was used to calculate intracellular concentrations

$$Ca^{2+} \text{ or } Mg^{2+} = K_d \times \frac{(R - R_{min}) \times S_F}{R_{max} - R) \times S_B}$$

where S_F was the fluorescence at zero concentration of Ca^{2+} or Mg^{2+} at 380 nm and 370 nm, respectively. S_B was the fluorescence at full saturation. R was the fluorescence at the two emission wavelengths at resting pH, and, the values of K_d that we used were 224 nmol/L for fura-2 and 1.5 mmol/L for furaptra.

RESULTS

The free $[Ca^{2+}]_i$ in SM cells from SHR rats was significantly higher than those from WKY rats, see FIGURE 1 (SHR 199.0 ± 51.6 vs. WKY 146.7 ± 38.4 nmol L^{-1}, $p < 0.01$). VSMC from both rat strains showed similar free $[Ca^{2+}]_i$ (SHR 92.9 ± 25.2 vs. WKY 79.2 ± 45.2 nmol L^{-1}). These results are similar to those reported by Nabika *et al.*[5] The free $[Mg^{2+}]_i$ in both SM cells and VSMC from SHR was significantly lower than levels in the cells from WKY rats (FIG. 2; SM cells from SHR 0.406 ± 0.064 vs. WKY 0.535 ± 0.061 mmol L, $p < 0.001$; VSMC from SHR 0.397 ± 0.069 vs. WKY 0.520 ± 0.105 mmol L, $p < 0.002$). The free Ca^{2+}/Mg^{2+} ratios ($\times 10^{-4}$) were therefore significantly higher in both SM cells (SHR 4.93 ± 0.88 vs. WKY 2.73 ± 0.47, $p < 0.005$) and for aortic VSMC from hypertensive rats (SHR 2.23 ± 0.45 vs. WKY 1.27 ± 0.46, $p < 0.02$).

DISCUSSION AND CONCLUSION

Previous work on raised $[Ca^{2+}]_i$ in the platelets from hypertensive patients and SHR rats has been referred to above. There have been a few reports of altered $[Ca^{2+}]_i$ in VSMC from different vascular trees. In general, these reports showed elevated $[Ca^{2+}]_i$ in SHR VSMC from venous vessels, mesenteric artery, and aorta. In VSMC

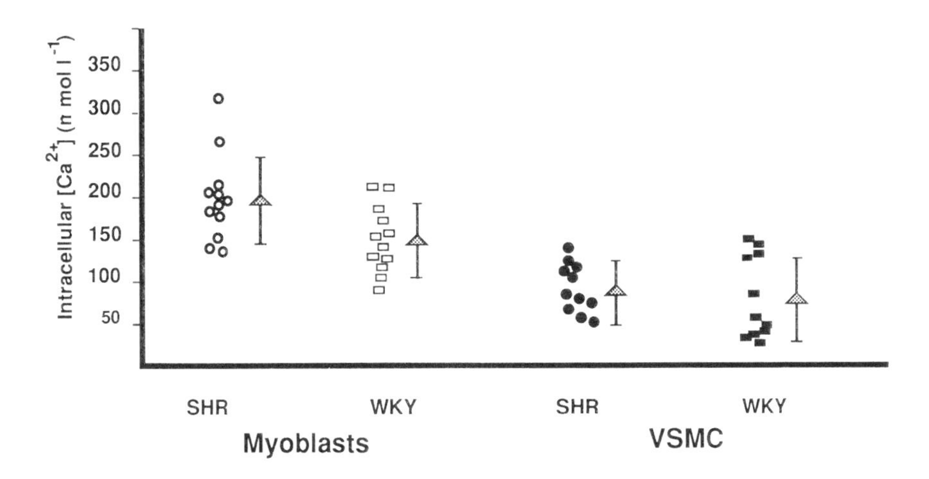

FIGURE 1. Intracellular free Ca^{2+} in vascular smooth muscle and striated muscle cells of spontaneously hypertensive and Wistar Kyoto normotensive rats in culture. Values of Ca^{2+} have been corrected for autofluorescence. For SHR and WKY striated muscle cells, $n = 12$, and $n = 11$ for SHR and WKY vascular smooth muscle cells. Statistical analysis was performed using an OXSTAT statistics package (Microsoft Corporation). Values are shown as ± SD.

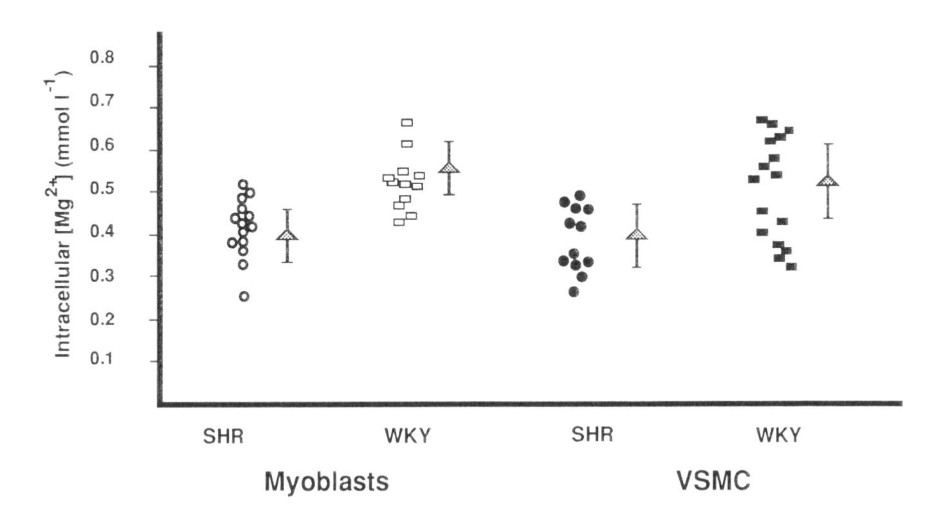

FIGURE 2. Intracellular free Mg^{2+} in vascular smooth muscle and striated muscle cells of spontaneously hypertensive and Wistar Kyoto normotensive rats in culture. Values of Mg^{2+} have been corrected for autofluorescence. For SHR vascular smooth muscle and WKY striated muscle cells, $n = 12$; $n = 16$ for WKY vascular smooth muscle cells; and $n = 14$ for SHR and striated muscle cells. Statistical analysis was performed using an OXSTAT statistics package (Microsoft Corporation). Values are shown as ± SD.

stimulated with various agonists, there is agreement that $[Ca^{2+}]_i$ rises to a greater degree in cells from SHR rats, suggesting a possible abnormality of $[Ca^{2+}]_i$ handling in these cells. Therefore, our results raise the possibility that the abnormality of $[Ca^{2+}]_i$ handling may not be confined to aortic VSMC. This work was done on cultured cells in between passages 10-15. Recently Buloski[6] has reported that differences in VSMC $[Ca^{2+}]_i$ arose out of serial passaging before disappearing in later passages.

Research into the role of cellular magnesium in hypertension is at an early stage. We have found that $[Mg^{2+}]_i$ in both aortic VSMC and SM cells in SHR was lower than WKY rats. It has been shown that a deficiency of Mg^{2+} could contribute to increased vascular tone, and that Mg^{2+} can act as a natural Ca^{2+}antagonist.[2]

In conclusion, in this work, we have described, in cultured cells that were not subject to a high-pressure load, changes that would lead to a raised intracellular Ca^{2+}/Mg^{2+} ratio in aortic VSMC and striated muscle cells. If such changes are found in resistance vessels, they may lead to an increased basal tone and a possible elevation of the blood pressure. Furthermore, the changes that we find in VSMC and striated muscle cells from the SHR rat may represent an intrinsic abnormality in the handling of these divalent ions by this strain of rat.

REFERENCES

1. McCARRON, D. A. 1985. Hypertension **7:** 607-627.
2. ALTURA, B. M., B. T. ALTURA, A. GEBREWOLD, H. ISING & T. GUNTHER. 1984. Science **223:** 1315-1317.
3. BLAU, H. M. & C. WEBSTER. 1981. Proc. Natl. Acad. Sci. USA **78:** 5623-5627.
4. GRYNKIEWICZ, G., M. POENIE & R. T. TSIEN. 1985. J. Biol. Chem. **260:** 3440-3450.
5. NABIKA, T., P. P. VELLETRI, M. BEAVEN, J. ENDO & W. LOVENBERG. 1985. Life Sci. **7:** 579-584.
6. BUKOSKI, R. D. 1990. J. Hypertens. **8:** 37-43.

Na-Ca Exchange Studies in Frog Phasic Muscle Cells[a]

E. CASTILLO,[b] H. GONZALEZ-SERRATOS,[b]
H. RASGADO-FLORES,[c] AND M. ROZYCKA [b]

[b]*Department of Biophysics*
School of Medicine
University of Maryland
Baltimore, Maryland 21201

[c]*Department of Physiology and Biophysics*
University of Health Sciences/The Chicago Medical School
North Chicago, Illinois 60064

A net Ca^{2+} influx exists during contraction and at rest in skeletal muscle cells,[1-3] but since the sarcoplasmic reticulum (SR) has a limited capacity to store Ca^{2+}, a homeostatic mechanism must exist that regulates intracellular calcium concentration ($[Ca^{2+}]_i$), otherwise contractures may occur. Intra- and extracellular Na^+ concentrations affect Ca^{2+} fluxes,[4,5] which has been interpreted as evidence of a Na-Ca exchange mechanism.[6] Recently, transverse tubule vesicles isolated from frog skeletal muscle were found to exhibit substantial Na-Ca exchange activity[7,8] and a Na-Ca exchanger has been proposed to explain contractility changes due to extracellular Na^+ concentrations ($[Na^+]_o$) modifications[9,10] or mechanical triggering.[11] We have found that post-fatigue caffeine contractures were larger than pre-fatigue ones and proposed that this is due to activity of this exchanger.[12] But the Na-Ca exchanger has been poorly described and no clear functional role has been attributed to it. The present experiments were undertaken to further characterize this mechanism and its functional role in skeletal muscle. The experiments were performed on single muscle fibers isolated from the semitendinosus muscle of the frog *Rana pipiens.* The dissection, isolation of the fibers, and the experimental set-up have been described previously.[13]

TWITCH TENSION

The fibers were stimulated at 0.008 Hz to avoid the staircase effect. A decrease in $[Na^+]_o$ enhanced twitch tension (up to 43%) with an increase of time to peak (up to

[a]Supported by National Institutes of Health Grant RO1-NS17048 (to H. G.-S.) and a grant from the American Heart Association (to H. R.-F.). H. Rasgado-Flores is an established investigator of the American Heart Association.

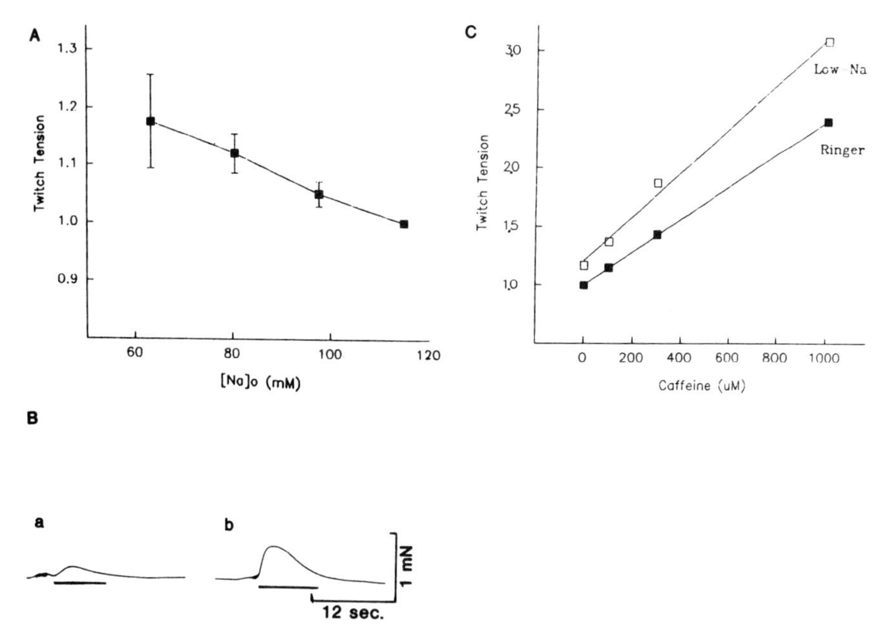

FIGURE 1. Effect of lowering extracellular sodium concentration on peak twitch tension, caffeine potentiation, and K contractures. (**A**) relationship between peak fractional twitch tension normalized to force obtained with 114.4 mM $[Na^+]_o$ and $[Na^+]_o$ in a single muscle cell. Each point is the average measurement of 14 different fibers, the bars are SEM. (**B**) Effect of $[Na^+]_o$ on brief 30 mM K contractures. The fiber was exposed to the 30 mM K solution during the times indicated by the lines under each record. The contractures were done in the presence of 100 mM (*a*) and (*b*) of 10 mM $[Na^+]_o$. The KCl product was kept constant in all the solutions. Contracture *a* was done first and *b* second after the fiber was allowed to rest for 30 min. (**C**) Relationship between peak fractional twitch tension normalized to force obtained in 114.5 mM $[Na^+]_o$ and caffeine concentrations in the presence of 114.5 mM $[Na^+]_o$ (*filled symbols*) and of 63 mM $[Na^+]_o$ (*open symbols*). The initial twitch tension potentiation in low $[Na^+]_o$ and 0 caffeine is less than the twitch potentiation in low Na and 1 mM caffeine concentration.

4.1 msec) and no change in twitch duration. As shown in FIGURE 1A, the potentiation was inversely proportional to $[Na^+]_o$ until the action potential failed below 63.25 mM $[Na^+]_o$. Low $[Na^+]_o$ enhanced post-tetanic potentiation but proportionally smaller than the low $[Na^+]_o$ pretetanic potentiation.

POTASSIUM CONTRACTURES

Studies on K contractures permitted assessment of the effect of $[Na^+]_o < 55$ mM. FIGURE 1B shows two 30 mM K contractures done in 100 mM (a) and 10 mM $[Na^+]_o$ (b). The contracture in low $[Na^+]_o$ was three times larger than in high $[Na^+]_o$. Uninterrupted contractures elicited with other K concentrations (not shown) in low $[Na^+]_o$ were larger and longer lasting than in high $[Na^+]_o$. The tension-[K] relationship in low $[Na]_o$ was steeper with no change in mechanical threshold and with a similar maximal tension to that in high $[Na^+]_o$.

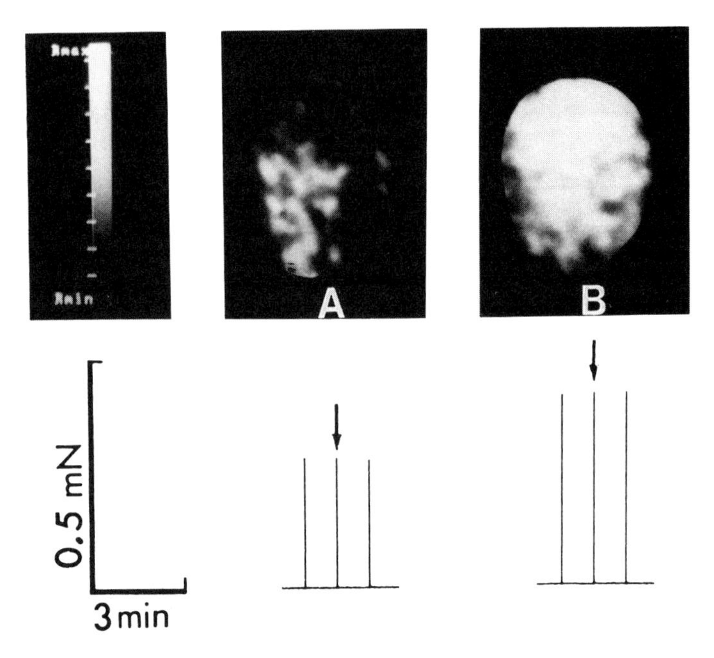

FIGURE 2. Cross-section images of Ca^{2+}release during single twitches in 114.5 mM $[Na^+]_o$ (**A**) and 80.1 mM $[Na^+]_o$ (**B**). The corresponding tension recordings are shown under the images. The arrows indicate the twitch during which the images were taken. The fiber was loaded with FLUO-3 (AM form), and the scale to the left is a relative scale of $[Ca^{2+}]$ where R_{max} represents the maximal and R_{min} the minimal Ca^{2+} concentrations.

CAFFEINE EXPOSURE

To assess if $[Na^+]_o$ could affect the SR Ca^{2+}pool, we studied twitch potentiations due to $[Na^+]_o$ on low-concentration (< 2 mM) caffeine concentration-induced potentiation. FIGURE 1C shows that the slope of the twitch tension-caffeine concentration relationship was steeper in the low $[Na^+]_o$ than in high $[Na^+]_o$. This indicates that the total caffeine potentiation is larger in low $[Na^+]_o$. The difference in the slopes reflects the extra Ca^{2+} available in the muscles in low $[Na^+]_o$.

INTRACELLULAR Ca^{2+} RELEASE

Cross section images of Ca^{2+} release (FIG. 2) showed that low $[Na^+]_o$, resulted in an increase of intracellular Ca^{2+} release that was proportional to the decrease in $[Na^+]_o$.

EFFECT OF DICHLOROBENZAMIL

We tested to see if dichlorobenzamil (DCB, a Na-Ca exchanger inhibitor)[14] could block the exchanger in skeletal muscle. We measured the effect of 1 μM DCB on twitch tension. A summary of four experiments showed that DCB increased twitch tension by 67 ± 20%.

CONCLUSIONS

Potentiation of twitch tension, caffeine contractures, and K contractures by low $[Na^+]_o$ can all be explained by inhibition of a Na-Ca exchanger operating in the Ca^{2+} efflux/Na^+ influx mode. This conclusion is further supported by the potentiating effect of DCB on twitch tension. Assuming a $[Na^+]_o$ = 113.5 mM, $[Na^+]_i$ = 6.2[15] mM, $[Ca^{2+}]_o$ = 1.8 mM, and $[Ca^{2+}]_i$ = 0.1 μM, the equilibrium potential of the exchanger[16] would be −27 mV. Under resting conditions, with a membrane potential of −79 mV,[15] the exchanger would be expected to operate in the Ca^{2+}-efflux mode, extruding the excess Ca^{2+} that constantly leaks into the cell. The Na-Ca exchanger therefore contributes to maintain $[Ca^{2+}]$ in the steady state.

REFERENCES

1. BIANCHI, C. P. & A. M. SHANES. 1959. Calcium influx in skeletal muscle at rest, during activity, and during potassium contracture. J. Gen. Physiol. **42:** 803-815.
2. CURTIS, B. A. 1966. Ca fluxes in single twitch muscle fibers. J. Gen. Physiol. **50:** 255-267.
3. CURTIS, B. A. 1970. Calcium efflux from frog twitch muscle fibers. J. Gen. Physiol. **55:** 243-253.
4. COSMOS, E. E. & E. J. HARRIS. 1961. *In vitro* studies of the gain and exchange of calcium in frog skeletal muscle. J. Gen. Physiol. **44:** 1121-1130.
5. CAPUTO, C. & P. BOLAÑOS. 1978. Effect of external sodium and calcium on calcium efflux in frog striated muscle. J. Membr. Biol. **41:** 1-14.
6. BLAUSTEIN, M. P. 1974. The interrelationship between sodium and calcium fluxes across cell membranes. Rev. Physiol. Biochem. Pharmacol. **70:** 32-82.
7. DONOSO, P. & C. HIDALGO. 1989. Sodium-calcium exchange in transverse tubules isolated from frog skeletal muscle. Biochim. Biophys. Acta **978:** 8-16.
8. HIDALGO, C., F. CIFUENTES & P. DONOSO. 1991. Sodium-calcium exchange in transverse tubule vesicles isolated from amphibian skeletal muscle. Ann. N.Y. Acad Sci. This volume.
9. KAWATA, H. & N. FUJISHIRO. 1988. Effects of external sodium removal on the contracture of frog skeletal muscle. Jpn. J. Physiol. **38:** 33-46.
10. NOIREAUD, J. & C. LEOTY. 1988. Effect of external sodium substitution on potassium contractures of mammalian muscles: Possible involvement of sarcolemma-bound calcium and Na-Ca^{2+}exchange. Q. J. Exp. Physiol. **73:** 233-236.
11. CURTIS, B. A. 1988. Na/Ca exchange and excitation-contraction coupling in frog fast fibers. J. Musc. Res. Cell Motil. **9:** 416-427.
12. GARCIA, M. C., H. GONZALEZ-SERRATOS, J. P. MORGAN, C. PERREAULT & M. ROZYCKA. 1991. Differential activation of myofibrils during fatigue in phasic skeletal muscle cells. J. Musc. Res. Cell Motil. In press.
13. GONZALEZ-SERRATOS, H. 1975. Graded activation of myofibrils and the effect of diameter on tension development during contractures in isolated skeletal muscle fibres. J. Physiol. **253:** 321-329.
14. FELIN, C., V. BARBRYP, P. VIGNE, O. CHASSANDE, E. J. CRAGOE, JR. & M. LAZDUNESKY. 1988. Amiloride and its analogs as tools to inhibit Na^+ transport via the Na^+ channel, the Na/H^+ + antiport and Na^+/Ca^{2+} exchanger. Biochemistry **70:** 1285-1290.
15. RASGADO-FLORES, H. & M. BLAUSTEIN. 1987. Na/Ca exchange in barnacle muscle cells has a stoichiometry of 3 Na^+/1 Ca^{2+}. Am. J. Physiol. **252**(Cell Physiol. 21): C499-C504.
16. ALVAREZ-LEEFMANS, F. J., S. M. GAMIÑO, F. GIRALDEZ & H. GONZALEZ-SERRATOS. 1986. Intracellular free magnesium in frog skeletal muscle fibres measured with ion-selective micro-electrodes. J. Physiol. **378:** 461-483.

Contribution of Na^+-Dependent and ATP-Dependent Ca^{2+} Transport to Smooth Muscle Calcium Homeostasis[a]

R. A. COONEY, T. W. HONEYMAN, AND C. R. SCHEID[b]

Department of Physiology
University of Massachusetts Medical School
Worcester, Massachusetts 01655

Controversy has been longstanding as to the importance of Na^+-Ca^{2+} exchange in smooth muscle calcium homeostasis.[1] Many investigators discounted the importance of this transport system because of its low calcium affinity and assumed that steady-state calcium regulation is dominated by ATP-dependent Ca^{2+}transport. Few studies have measured Na^+-dependent Ca^{2+} flux at submicromolar calcium levels, however. Thus the present studies examined the characteristics of Na^+-dependent and ATP-dependent calcium uptake in plasma membrane vesicles derived from pig stomach muscle.

Membrane vesicles enriched 15-20-fold for plasma membrane markers were prepared from pig gastric smooth muscle as described previously.[2] Na^+-Ca^{2+} exchange activity was assessed after 1, 5, or 15 seconds using a standard filtration assay in which Na^+-loaded vesicles (preequilibrated with 130 mM NaCl) were diluted either into Na^+-containing buffer (no gradient) or into K^+-containing buffer (with gradient). Studies were carried out at 12°C and Na^+-dependent uptake was defined as the difference in uptake in the presence versus the absence of a transmembrane Na^+ gradient.

Uptake of label occurred rapidly, plateauing within 5-10 seconds and the Na^+-dependent component was relatively small, ~15-20% of the total uptake. However, this component appeared to reflect true vesicular uptake since it was inhibited by 20 μM dichlorobenzamil (~65% reduction, $p < 0.01$) and by 1 μM monensin (~100% reduction, $p < 0.001$), and since the label was released on reversal of the transmembrane Na^+ gradient.

Na-dependent Ca^{2+} uptake was observed over a wide range of Ca^{2+} concentrations. The rate of Na^+-dependent Ca^{2+} accumulation increased with increasing Ca^{2+} levels between 1 and 100 μM, and the apparent $K_{m(Ca)}$ was 50 μM. Between 100 and 1000 nM Ca^{2+}, Na^+-dependent Ca^{2+} uptake showed little Ca^{2+}dependency. However, the uptake observed between pCa 6.8 and 5.0 was appreciable, averaging ~6 nmoles/(min·mg protein) (FIG. 1). This transport was similar in magnitude to that observed for ATP-dependent Ca^{2+}uptake; and as evident in FIGURE 2, Na^+-dependent uptake constituted a significant portion of total calcium transport over a wide range of calcium levels. These findings, like those reported recently for bovine aortic smooth muscle,[3]

[a] Supported by a grant from the National Institutes of Health (HL41188).

[b] Address for correspondence: Cheryl R. Scheid, Ph.D., Dept. of Physiology, S4-209, University of Massachusetts Medical School, 55 Lake Avenue N., Worcester, MA 01655.

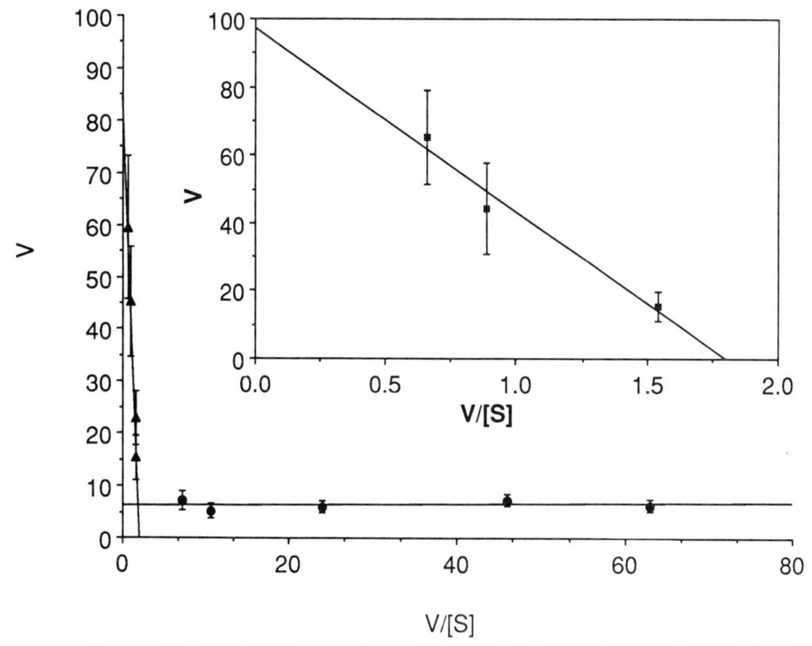

FIGURE 1. Calcium dependency of Na^+-Ca^{2+}exchange in pig gastric smooth muscle. Plot of the initial velocity of uptake V (measured at 1 sec at 12°C) versus V/[S], where [S] = calcium concentration in μM. Na^+-dependent uptake exhibits two components. Uptake at free calcium concentrations between 100 and 1000 nM averaged ~6 nmoles/(mg·min) irrespective of the calcium concentration. At free calcium concentrations > 1 μM, the rate of Na^+-dependent uptake increased markedly. *Inset:* expanded plot showing the calcium dependence of uptake rate between pCa 5 and 4. The apparent $K_{m(Ca)}$ obtained from this plot was 48 μM and the apparent V_{max} was 90 nmoles/(mg·min) under these conditions. Values are means ± standard errors of the mean for 6-16 determinations.

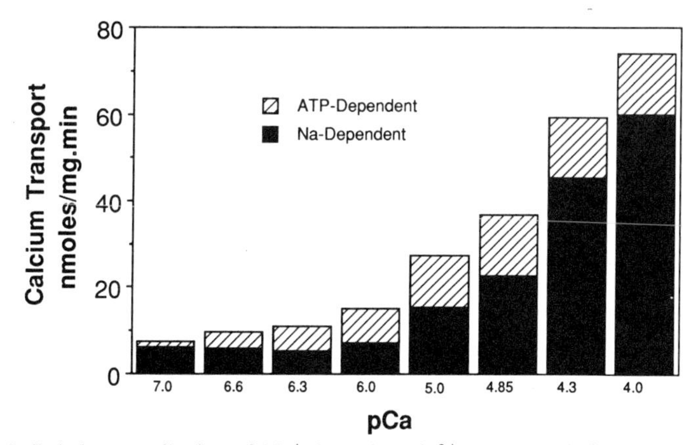

FIGURE 2. Relative contribution of Na^+-dependent Ca^{2+} transport (*solid bars*) and ATP-dependent calcium transport (*hatched bars*) to net calcium transport as a function of extravesicular calcium concentrations. Na^+-dependent transport rate was assessed at 12 °C after 1 sec of uptake. ATP-dependent transport was assessed at room temperature (22 °C) after 1 min of uptake.

suggest that Na^{+}-Ca^{2+} exchange may play a significant role in smooth muscle calcium handling.

REFERENCES

1. BLAUSTEIN, M. J. 1988. Cardiovasc. Pharmacol. **12**(Suppl 5): S56-S68.
2. LUCCHESI, P. A., R. A. COONEY, C. MANGSEN-BAKER, T. W. HONEYMAN & C. R. SCHEID. 1988. Am. J. Physiol. **255:** C226-C236.
3. SLAUGHTER, R. S., J. L. SHEVELL, J. P. FELIX, M. L. GARCIA & G. J. KACZOROWSKI. 1989. Biochemistry **28:** 3995-4002.

Free Cytosolic Calcium Regulation via Na^+-Ca^{2+} Exchange in Cultured Vascular Smooth Muscle Cells[a]

MARYJO GODINICH, MICHAEL S. LAPOINTE, AND DANIEL C. BATLLE[b]

Department of Medicine
Northwestern University Medical School
and
VA Lakeside Medical Center
Chicago, Illinois 60614

Recent studies have presented functional evidence for the existence of a Na^+-Ca^{2+} exchanger in cultured smooth muscle cells.[1-9] The contribution of the Na^+-Ca^{2+} exchanger to the maintenance of resting iCa^{2+} and its mode of operation under physiological conditions, however, remain to be well defined. We have utilized cultured aortic vascular smooth muscle cells (VSMC) to investigate the impact of changes in the transmembrane Na^+ gradient on free cytosolic Ca^{2+} (iCa^{2+}) and to investigate whether the changes observed correspond to the forward or the reverse mode of operation of the Na^+-Ca^{2+} exchanger.[9] Our data suggest an important role for sodium-dependent iCa^{2+} regulation via the Na^+-Ca^{2+} exchanger acting as a Ca^{2+}-efflux pathway (forward mode) in the recovery from an abrupt iCa^{2+} elevation. The operation of this exchanger as a Ca^{2+} influx pathway (reverse mode) could also be easily demonstrated when intracellular sodium was previously increased by addition of ouabain to inhibit sodium efflux via the Na^+-K^+ ATPase.

METHODS AND RESULTS

Experiments were performed on confluent subcultures of VMSC fed with serum-free media 16 to 36 hours before the study. VSMC were loaded with fura-2 AM (2-4 μM) for 30 minutes at 37°C. Fura-2 fluorescence was measured using a Perkin-Elmer spectrofluorometer (LS-5 model) connected to an IBM PC-XT that was programed to alternate rapidly between the two excitation wavelengths (340 and 380 nm) while keeping the emission wavelength (510 nm) constant. Free cytosolic Ca^{2+} was calculated as previously described.[9,10] The basic assay solution used to superfuse the coverslips had the following composition (mM): NaCl 133.8, KCl 4.7, $CaCl_2$ 1.25, $MgCl_2$ 1.25, Na_2HPO_4 0.97, NaH_2PO_4 0.23, glucose 3, HEPES 5 (pH 7.4). When ouabain (2 mM) was used, cells were superfused for 30 minutes with this agent before and during the

[a]This study was supported, in part, by a Veterans Administration Merit Review Award (Dr. Batlle). Dr. Daniel C. Batlle is a member of the Feinberg Cardiovascular Research Institute of Northwestern University.

[b]Address for correspondence: Daniel C. Batlle, Northwestern University Medical School, Department of Medicine, 303 E. Chicago Avenue, Chicago, Il 60614.

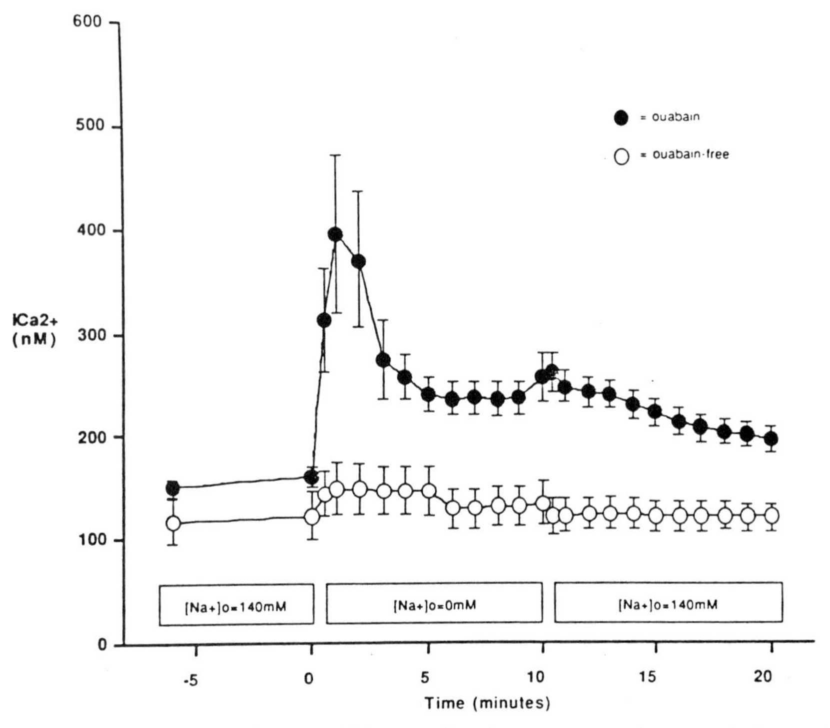

FIGURE 1. Summary data (mean ± SE) comparing the responses to the removal of external Na^+ in VSMC pretreated with ouabain (closed symbols, $n = 14$) versus cells not pretreated with ouabain (open symbols, $n = 19$). A rapid and significant increase in $iCa^{2+}$was seen in both conditions, although the iCa^{2+} increment was much larger ($p < 0.01$) in cells pretreated with ouabain. After the initial surge, a new steady-state iCa^{2+} is reached after about 5 minutes of superfusion in the absence of external Na^+, but iCa^{2+} recovery is not complete. Reintroduction of Na^+ to the superfusate resulted in a further decrease of iCa^{2+}, although in cells treated with ouabain it remained significantly higher than baseline iCa^{2+} (see text). (Reprinted from Batlle *et al.*[9] with permission from the *American Journal of Physiology [Cell].*)

subsequent experimental maneuver. Incubation of this duration with ouabain has been shown to increase intracellular Na^+ (iNa^+) from 8 to about 30 mM.[11] In other experiments ionomycin (10^{-6} M) was utilized to rapidly increase iCa^{2+}. This increase is largely due to release of IP_3-sensitive Ca^{2+} stores and also due to external calcium influx.[12] Student's *t*-test (unpaired *t*-test) was used to compare differences between experimental groups. Changes within the same experimental group were analyzed using the paired *t*-test.

When Na^+ was removed from the superfusate (choline replaced), there was a rapid increase in iCa^{2+} followed by a gradual return toward baseline. In cells pre-exposed to ouabain for 30 minutes, removal of $[Na^+]_o$ resulted in a marked and more rapid increase in iCa^{2+}. In a series of 14 experiments, the peak increment in iCa^{2+} from baseline (Δ iCa^{2+} 36 ± 9 nM) was observed within 4 minutes of exposure to a Na^+-free solution (from 124 ± 23 to 150 ± 25 nM, $p < 0.01$, FIG. 1).[1] The increment in iCa^{2+} from baseline was greater in ouabain-pretreated cells than in cells not pretreated with ouabain

(ΔiCa^{2+} 236 $\pm$ 75 versus 36 $\pm$ 9 nM, respectively, $p < 0.025$).[9] In both groups, iCa^{2+}declined over the next five minutes of exposure to a Na^+-free solution, until a new steady state was reached. Throughout the exposure to the Na^+-free solution, iCa^{2+} remained significantly higher in ouabain-treated cells than for the group not treated with ouabain. An additional fall in iCa^{2+} was seen upon re-exposure to a Na^+-containing solution (140 mM) in both experimental groups (FIG. 1).[9]

The source of the iCa^{2+} increase observed when Na^+ was removed from the superfusate was investigated by superfusing cells in a nominally Ca^{2+}-free solution with EGTA (250 mM) added to chelate any existing external Ca^{2+}. After pretreatment with ouabain for 30 minutes, exposure to a Na^+-free superfusate completely failed to increase iCa^{2+} under these conditions (data not shown).

In cells superfused with a standard sodium-containing solution, addition of ionomycin (10^{-6}M) produced a rapid increase in iCa^{2+} within the first 60 seconds of exposure (from 130 $\pm$ 19 to 271 $\pm$ 47 nM, $p < 0.005$, $n = 15$).[9] The increase was followed by a rapid decline and achievement of a new steady state well above the baseline iCa^{2+}. When ionomycin was removed from the external solution, there was only a slight reduction in iCa^{2+} (from 221 $\pm$ 31 to 199 $\pm$ 28 nM, $p < 0.05$, $n = 15$). In experiments where cells superfused with the standard sodium-containing solution were switched to a Na^+-free, ionomycin-containing solution, there was a marked increase in iCa^{2+} (from 149 $\pm$ 13 to 520 $\pm$ 78 nM, $p < 0.005$, $n = 7$) (FIG. 2). The peak increment from baseline (Δ iCa^{2+} 259 $\pm$ 80 nM) was more pronounced than that seen in the Na^+-containing counterparts (peak Δ iCa^{2+} 144 $\pm$ 38 nM, $n = 15$, $p < 0.05$). After the initial rapid decrease in iCa^{2+}, cells superfused with ionomycin in the absence of external Na^+ displayed a fall in iCa^{2+} and then achieved a new steady state with iCa^{2+} ~350 nM (FIG. 2). Reexposure to a Na^+-containing solution, still containing ionomycin, resulted in a marked decrease in iCa^{2+} within 60 seconds of exposure (from 349 $\pm$ 36 to 225 $\pm$ 11 nM, $p < 0.005$, $n = 7$). Under these conditions, iCa^{2+} continued to decline to a new steady state so that recovery was almost complete five minutes after the readdition of external Na^+ (FIG. 2).

DISCUSSION

The iCa^{2+} changes observed in response to Na^+ removal indicate that in VSMC the Na^+-Ca^{2+} antiporter can operate in the reverse mode (that is, as a Ca^{2+}-influx pathway) when the intracellular sodium concentration is high. In contrast, the Na^+-Ca^{2+} exchanger operates in the forward mode (that is, as a Ca^{2+}-efflux pathway) when iCa^{2+} is elevated. The latter mode of operation, in conjunction with a Na^+-independent mechanism of Ca^{2+}-efflux (probably the plasma membrane Ca^{2+}ATPase) and other processes leading to Ca^{2+} buffering and reuptake in internal stores, allows for iCa^{2+} recovery toward normal after iCa^{2+} has been abruptly increased to levels prevailing in stimulated cells (FIG. 2).

The reverse mode of operation of the Na^+-Ca^{2+} exchanger is clearly shown by the fact that when cells are Na^+-loaded with ouabain, the rise in iCa^{2+} is much greater than that seen in the absence of ouabain (FIG. 1). This observation does not permit the conclusion, however, that the antiporter is actively functioning in the reverse mode under "resting" conditions (i.e., normal iNa^+ and normal iCa^{2+}), or when cells are stimulated. Indeed, using ionomycin to effect a rise in iCa^{2+} to a value that may prevail in a stimulated cell under physiological conditions (about 400 mM), we were able to provide clear evidence for the forward mode of operation of the Na^+-Ca^{2+}exchanger in VSMC. (FIG. 2).[9] This is clear from our finding that reintroducing Na^+ to the

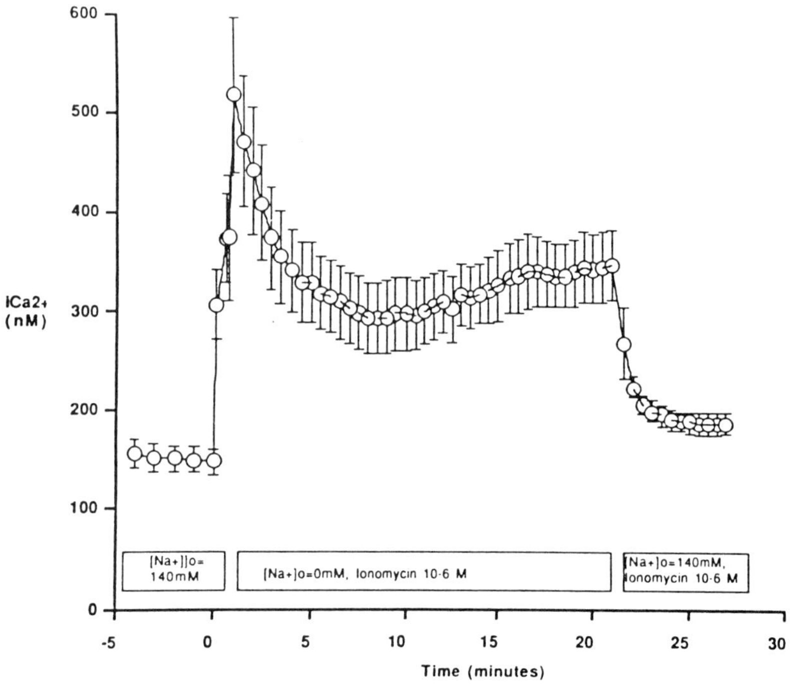

FIGURE 2. Summary data (mean ± SE) showing the effect of ionomycin on iCa^{2+} in the absence of external sodium. After recording a stable baseline iCa^{2+}, cells were superfused with a Na^+-free, ionomycin (10^{-6} M) containing solution. This maneuver produced a larger increase followed by rapid but incomplete recovery with achievement of a new steady-state ($iCa^{2+} \approx 300$ nM). Readdition of sodium in the presence of ionomycin resulted in a marked decrease in iCa^{2+}. Even after Na^+ addition, however, iCa^{2+}remained significantly above baseline ($p < 0.001$). (Reprinted from Batlle *et al.*[9] with permission from the *American Journal of Physiology [Cell].*)

external perfusate, in the continued presence of ionomycin, resulted in a rapid return of iCa^{2+} toward baseline values. The Na-dependent fall in iCa^{2+} that we ascribe to Na^+-Ca^{2+} exchanger operating in the forward mode was clearly additive to the initial decline, which is sodium-independent and likely mediated, at least in part, by the action of the plasma membrane Ca^{2+}-ATPase, an enzyme with a high specific activity in smooth muscle.[13,14] However, the fact that Ca^{2+} remains well above baseline even when the Ca^{2+} pump is activated by the iCa^{2+} surge suggests that this pump alone is unable to reestablish baseline iCa^{2+}. That reintroduction of Na^+ greatly contributes to iCa^{2+} recovery strongly suggests that the Na^+-Ca^{2+} exchanger plays an important role in Ca^{2+} extrusion, and thus regulation of iCa^{2+} in VSMC.

In summary, we present evidence that shows that a sodium-dependent mechanism, presumably the Na^+-Ca^{2+} exchanger, brings the iCa^{2+} of an activated VSMC toward its resting level. Sodium-dependent iCa^{2+} regulation in VSMC via a Na^+-Ca^{2+} exchanger operating as a Ca^{2+}-efflux mechanism (forward-mode operation) may thus be the physiologic role of the Na^+-Ca^{2+} exchanger, since elevations in Ca^{2+}similar to those produced by ionomycin also are attained with hormonal stimulation.

REFERENCES

1. GROVER, A. K., C. Y. KWAN, P. K. RANGACHARI *et al.* 1983. Na^+/Ca^{2+} exchange in smooth muscle plasma membrane-enriched fraction. Am. J. Physiol. **244:** C158-C165.
2. BLAUSTEIN, M. P., T. ASHIDA, W. F. GOLDMAN, W. G. WIER & J. M. HAMLYN. 1986. Sodium/calcium exchange in vascular smooth muscle: A link between sodium metabolism and hypertension. Ann. N.Y. Acad. Sci. **488:** 199-216.
3. PRITCHARD, K. & C. C. ASHLEY. 1987. Evidence for Na^+/Ca^{2+} exchange in isolated smooth muscle cells: A fura-2 study. Pflügers Arch. **410:** 401-407.
4. SMITH, J. B. & L. SMITH. 1987. Extracellular Na^+ dependence of changes in free Ca^{2+}, $^{45}Ca^{2+}$ efflux and total cell Ca^{2+}produced by angiotensin II in cultured arterial muscle cells. J. Biol. Chem. **262:** 17455-17460.
5. SMITH, J. B., E. J. CRAGOE & L. SMITH. 1987. Na^+/Ca^{2+} antiporter in cultured arterial smooth muscle cells. J. Biol. Chem. **262:** 11988-11994.
6. SMITH, J. B., T. ZHENG & R.-M. LYU. 1989. Ionomycin releases calcium from the sarcoplasmic reticulum and activates Na^+/Ca^{2+} exchange in vascular smooth muscle cells. Cell Calc. **10:** 125-134.
7. SMITH, J. B., T. ZHENG & L. SMITH. 1989. Relationship between cytosolic free Ca^{2+} and Na^+/Ca^{2+} exchange in aortic smooth muscle cells. Am. J. Physiol. **256:** C147-C154.
8. VIGNE, P., J.-P. BREITTMAYER, D. DUVAL *et. al.* 1988. The Na^+/Ca^{2+} antiporter in aortic smooth muscle cells. J. Biol. Chem. **263:** 8078-8083.
9. BATLLE, D. C., M. J. GODINICH, M. S. LAPOINTE, E. MUÑOZ, F. CANONE & N. MEHRING. 1991. Extracellular sodium-dependency of free cytosolic calcium regulation in aortic vascular smooth muscle cells. Am. J. Physiol. (Cell) **261:** L845-856.
10. GRYNKIEWICZ, G., M. POENIE & R. Y. TSIEN. 1985. A new generation of Ca^{2+} indicators with greatly improved fluorescence properties. J. Biol. Chem. **260:** 3440-3450.
11. GUTTERMAN, C. & D. BATLLE. 1990. Measurements of free cytosolic sodium (iNa^+) in cell suspensions using a novel fluorescent probe SBFI (Abstract). Am. J. Hypertens. **3:** 57A.
12. KAUFFMAN, R. F., R. W. TAYLOR & D. R. PFEIFFER. 1980. Cation transport and specificity of ionomycin. J. Biol. Chem. **255:** 2735-2739.
13. VAN BREEMEN, C. 1989. Cellular mechanisms regulating $[Ca^{2+}]_i$ in smooth muscle. Annu. Rev. Physiol. **51:** 315-329.
14. WUYTACK, F., L. RAEYMAEKERS & R. CASTEELS. 1985. The Ca^{2+} transport ATPases in smooth muscle. Experientia **41:** 900-904.

Sodium-Calcium Exchange in Bovine Aortic Endothelial Cells

BARBARA A. HANSEN,[a] DANIEL C. BATLLE,[b] AND MARTHA E. O'DONNELL[c,d]

[a]*Department of Cell Biology and Anatomy*
The Chicago Medical School
North Chicago, Illinois 60064

[b]*Department of Medicine Section of Nephrology/Hypertension*
Northwestern University Medical School
Chicago, Illinois 60611

[c]*Department of Human Physiology*
School of Medicine
University of California
Davis, California 95616

Regulation of intracellular calcium (Ca_i) is an important function for all cells, given the central role this ion plays in signal transduction. In endothelial cells, stimuli such as bradykinin and histamine increase Ca_i both by release from the endoplasmic reticulum (ER) and by a mechanism dependent on extracellular Ca.[1,2] A portion of the Ca that accumulates in the cytoplasm exits from the cell rather than being returned directly to the ER. Before a second response, the ER Ca pool is refilled, necessitating a Ca-influx pathway that may be independent of receptor stimulation.[2] One mechanism that is, in principle, capable of mediating both net influx and efflux is Na-Ca exchange, although it is not known whether endothelial cells express this activity. The data from this study demonstrates the presence of Na-Ca exchange in bovine aortic endothelial cells (BAEC) as a Na-dependent ^{45}Ca influx.

RESULTS AND DISCUSSION

Endothelial cells exhibit a Na-dependent ^{45}Ca uptake. BAEC have a basal rate of ^{45}Ca uptake in the first 0.5 min after exposure to 140 mM Na (Na-BSS, legend FIG. 1) of 0.68 $\pm$ 0.09 μmole Ca/g protein (FIG. 1). After treatment with ouabain to elevate internal Na (Na_i), uptake is increased to 0.99 $\pm$ 0.15 μmole Ca/g protein $\times$ 0.5 min, and when ouabain-treated cells are transferred to Na-free BSS, uptake increases further to 2.05 $\pm$ 0.20 μmole Ca/g protein $\times$ 0.5 min. The rate of ^{45}Ca uptake is a function of extracellular Na (Na_o), decreasing with increasing Na_o as expected with Na-Ca exchange activity. As a control, cells treated with ouabain in Na-free Li-BSS show no enhanced rate of ^{45}Ca uptake; Li does not substitute for Na in the Na-Ca exchangers characterized in other cells.[3]

The Na content of resting BAEC measured by atomic absorption is 89 $\pm$ 2 μmole/g protein. Using a cell water content of 3.7 μl/mg protein,[4] this yields a calculated Na_i

[d]To whom correspondence should be addressed.

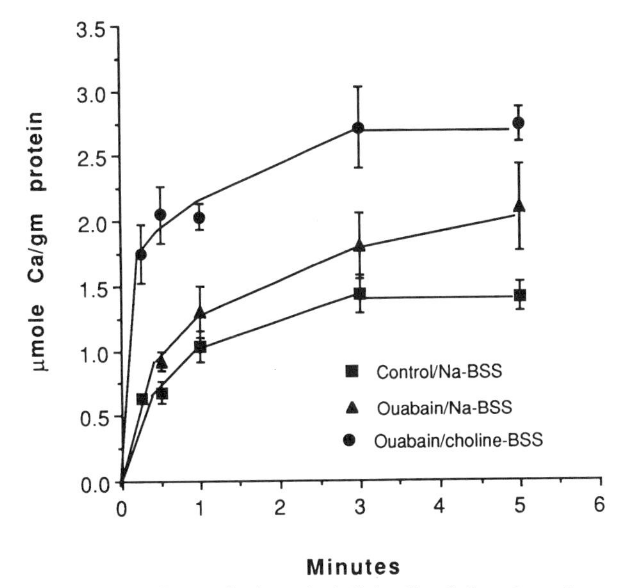

FIGURE 1. Na-dependent Ca uptake in endothelial cells. Cultured confluent monolayers of BAEC were pre-incubated for 1 hour at 37 °C in Na-BSS (in mM: 140 NaCl, 5 KCl, 1.8 $CaCl_2$, 0.4 $MgSO_4$, 5.6 glucose, 20 HEPES/Tris, pH 7.45) without (control) or with (Na-loaded) 1 mM ouabain + 10 μM bumetanide. At time = 0, the Na-BSS was aspirated and replaced with either Na-BSS or choline-BSS (NaCl replaced with 140 choline Cl) containing 1.5 μCi ^{45}Ca. Ouabain and bumetanide were present in the assay buffer of the Na-loaded experiments. At the times indicated on the graph, the assay buffer was aspirated and the monolayers rapidly rinsed with ice-cold 0.1 M $MgCl_2$. The dishes were allowed to dry, and the monolayers were extracted with 0.2% sodium dodecyl sulfate, and counted for radioactivity. Protein content was determined using separate monolayers from the same experiment.[12]

of 24 mM. This value is higher than that measured in most muscle cells, but is comparable to results obtained by others in endothelial cells,[4] in fibroblasts,[5] and in neutrophils[6] and may be typical for nonexcitable cells. Treatment of BAEC with 1 mM ouabain for 1 hour elevates Na_i to 146 ± 10 μmole/g protein (39 mM). If ouabain treatment is carried out in EGTA-Ca BSS (600 nM Ca) to prevent Na_i loss through the exchanger, Na_i reaches 509 μmole/g protein (137 mM) providing further evidence of Na-Ca exchange in these cells.

The ability of 3,4-dichlorobenzamil (DCB), a characterized inhibitor of Na-Ca exchange in other cells, to inhibit the Na-dependent ^{45}Ca uptake in BAEC was determined. DCB inhibited the uptake in a dose-dependent manner with a K_i of approximately 80 μM.

Transfer of BAEC to Na-free BSS might be expected to increase, at least transiently, the Ca_i. Using fura-2 to monitor Ca_i, we were able to detect small but reproducible increases in resting Ca_i upon exposure to choline-BSS (FIG. 2). The increase in ouabain-treated cells was not different than that in untreated controls, although ouabain treatment alone reproducibly elevated Ca_i from 88 ± 10 to 104 ± 13 nM (n = 8), consistent with the action of ouabain on resting Ca_i in cardiac myocytes. Luckhoff *et al.* have noted that the mechanism that allows Ca refilling of the ER after bradykinin stimulation requires Ca_o, but results in only a small increase in Ca_i during the refilling period unless

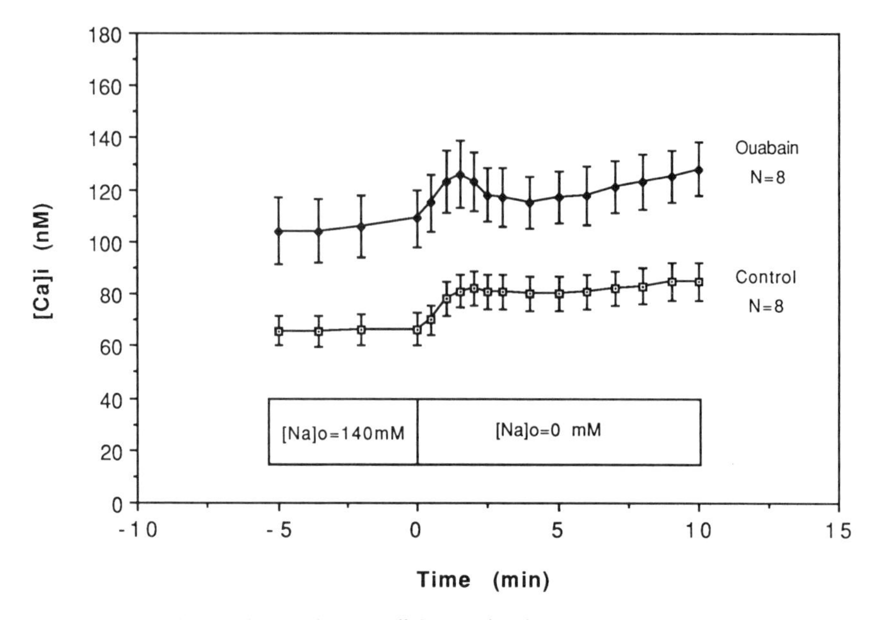

FIGURE 2. Na-free medium evokes a small Ca transient in BAEC. Endothelial cells were grown to confluency on glass coverslips in D-MEM with 10% FCS. This medium was replaced with serum-free medium 16-36 hours before the experiment. BAEC were loaded with fura-2 AM in serum-free medium (4 μM) for 30 minutes at 37 °C. The cells were washed three times with Na-BSS and allowed to stand for at least 30 minutes before beginning the experiment. The coverslip containing the fura-2 loaded cells was placed in a cuvette maintained at 37 °C. Cells were perfused at 2 ml/min with prewarmed Na-BSS and effluent continuously withdrawn with a peristaltic pump. Under these conditions, the contribution of extracellular fura-2 to the measured fluorescence was found to be negligible. Fura-2 fluorescence was measured using a Perkin-Elmer model LS-5 spectrofluorimeter programmed to rapidly alternate the excitation wavelength between 340 and 380 nm (5-nm slit width), while recording the emission intensity at 510 nm (10-nm slit width). Free cytosolic Ca^{2+} was calculated according to the following formula of Grynkiewicz *et al.*[13]

$$Ca_i = [(R - R_{min})\, S_f K_d]/[(R_{max} - R)\, S_b]$$

where K_d, the dissociation constant of fura-2-Ca^{2+} was assumed to be 224 nM. Determination of the calibration values, R_{max} (6.98), R_{min} (0.79) and S_f/S_b (4.19) has been described previously.[14]

the accompanying uptake by the ER is blocked.[7] It was concluded that Ca uptake by the ER in endothelial cells is able to buffer a moderate influx of Ca. ER uptake may explain the minimal effect Ca influx evoked by Na-free medium has on the Ca_i observed in our experiments.

Our results indicate that endothelial cells have Na-Ca exchange activity that might be expected to play a role in Ca translocation across the plasma membrane. Using the values 140 mM Na_o, 24 mM Na_i, 1.8 mM Ca_o, 88 nM Ca_i, $E_{Na\text{-}Ca}$ is calculated to be -127 mV. Under these conditions, and with an E_m of -55 to -67 mV in the resting endothelial cell,[8,9] the exchanger would operate in a net Ca-influx mode. From this calculation, one can predict that the exchanger (1) contributes to Ca influx following agonist stimulation; (2) along with a Ca/ATPase, contributes to Ca efflux at the peak of the response when Ca_i approaches 1 μM, and the membrane may be slightly hyperpolarized due to activation of Ca-activated K channels[10,11]; and (3) contributes to the subsequent influx necessary to refill the ER.

REFERENCES

1. SCHILLING, W. P., A. K. RITCHIE, L. T. NAVARRO & S. G. ESKIN. 1988. Am. J. Physiol. **255:** H219-H227.
2. HALLAM, T. J., R. JACOB & J. E. MERRITT. 1989. Biochem. J. **259:** 125-129.
3. VIGNE, P. A., J.-P. BREITTMAYER, D. DUVAL, C. FRELIN & M. LAZDUNSKI. 1988. J. Biol. Chem. **263:** 8078-8083.
4. DWYER, S. D., Y. ZHUANG & J. B. SMITH. 1991. Exp. Cell Res. **192:** 22-31.
5. SMITH, J. B., S. D. DWYER & L. SMITH. 1989. J. Biol. Chem. **264:** 831-837.
6. DALE, W. E. & L. SIMCHOWITZ. 1991. The role of Na^+/Ca^{2+} exchange in human neutrophil function. Ann. N.Y. Acad. Sci. This volume.
7. LUCKHOFF, A. & R. BUSSE. 1990. FEBS Lett. **276:** 108-110.
8. LASKEY, R. E., D. J. ADAMS, A. JOHNS, G. M. RUBANY & C. VAN BREEMAN. 1990. J. Biol. Chem. **265:** 2613-2619.
9. JOHNS, A., T. W. LATEGAN, N. J. LODGE, U. S. RYAN, C. VAN BREEMAN & D. J. ADAMS. 1987. Tissue Cell **19:** 733-745.
10. COLDEN-STANFIELD, M., W. P. SCHILLING, A. K. RITCHIE, S. G. ESKIN, L. T. NAVARRO & D. L. KUNZE. 1987. Circ. Res. **61:** 632-640.
11. BUSSE, R., H. FICHTNER, A. LUCKHOFF & M. KOHLHARDT. 1988. Am. J. Physiol. **255:** H965-H969.
12. LOWRY, D. H., N. J. ROSEBROUGH, A. L. FARR & R. J. RANDALL. 1951. J. Biol. Chem. **193:** 265-275.
13. GRYNKIEWICZ, G., M. POENIE & R. Y. TSIEN. 1985. J. Biol. Chem. **260:** 3440-3450.
14. MORGAN-BOYD, R., J. M. STEWART, R. J. VAVREK & A. HASSID. 1987. Am. J. Physiol. **253:** C588-C598.

α-Adrenoceptor Agonist-Induced Stimulation of Na-Ca Exchange in Rabbit Abdominal Aorta[a]

M. A. KHOYI, R. A. BJUR, AND D. P. WESTFALL

Department of Pharmacology
University of Nevada School of Medicine
Reno, Nevada 89557

The Na-Ca antiporter couples translocation of Ca^{2+} with the movement of Na^+ in the opposite direction. In many tissues, the Na-Ca antiporter has been found to be electrogenic and voltage sensitive[1-3] and to exchange three Na ions for each Ca ion transported across the cellular membrane.[4] Therefore, in depolarized cells, the antiporter can exchange Ca^{2+}_o with Na^+_i. Although the activity of the Na-Ca antiporter is known to be influenced by $[Ca^{2+}_i]$, ATP, and certain exogenous agents,[3,5,6] there has been as yet no demonstration of hormonal or neuronal regulation of Na-Ca exchange in smooth muscle. This may be due to alterations in the function of the antiporter during the preparation of isolated cells or membranal vesicles.[6,7] The present investigation was designed (a) to measure Na-Ca exchange in pieces of tissue instead of isolated cells or membranal vesicles, (b) to determine whether agonists stimulate Na^+_i-Ca^{2+}_o exchange, and (c) whether such stimulation is sensitive to changes in cell membrane potential.

METHODS

Experiments were carried out on rings of rabbit abdominal aorta. Na^+_i-Ca^{2+}_o exchange was measured by the method of Vigne *et al.*[8] Na^+-loading solution had the following composition (mM): NaCl 65, KCl 80, $MgCl_2$ 1, $CaCl_2$ 1.8, EGTA 2, HEPES-Tris 5, glucose 10, nifedipine 0.01, ouabain 1, and gramicidin 0.001. For measuring Na^+_i-Ca^{2+}_o exchange, aortic rings were transferred to a solution similar to Na-loading solution except that EGTA and gramicidin were omitted, LiCl or *N*-methyl-*d*-glucamine replaced NaCl and 0.01% BSA was added to quench the gramicidin remaining in the tissue.

RESULTS AND CONCLUSION

As shown in TABLE 1, the rate of Ca^{2+} uptake is significantly enhanced in aortic rings that have been preloaded with Na^+ compared to Na^+-depleted tissues. Ca^{2+} uptake into Na^+-loaded tissues was inhibited by benzamil or incorporation of Na^+, but not Li^+, in the assay medium. The addition of norepinephrine, phenylephrine, or methoxamine significantly enhanced Ca^{2+} uptake into Na^+-loaded tissues but not into

[a] Supported by National Institutes of Health Grant HL38126 and by the American Heart Association.

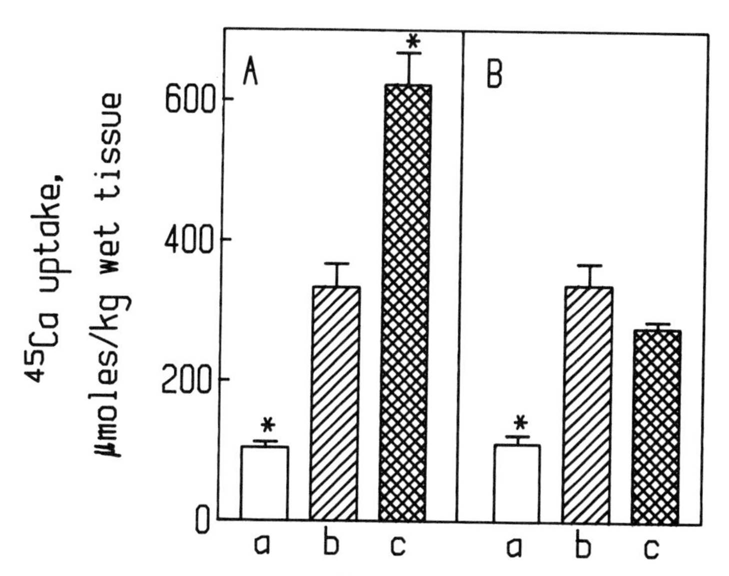

FIGURE 1. Effect of the concentration of K^+ in the assay medium on the stimulatory effect of norepinephrine on Na_i^+-Ca_o^{2+} exchange. Experiments were carried out as described in TABLE 1. After equilibration, aortic rings were exposed to 1.8 mM $^{45}Ca^{2+}$-containing medium for 8 min. The concentration of K^+ in this medium was 80 mM (**A**) or 5 mM (**B**). (a) Na^+-free tissues, (b) Na^+-loaded tissues, (c) Na^+-loaded tissues exposed to norepinephrine (10 μM) during the uptake period. Note that Na_i^+-independent and Na_i^+-dependent $^{45}Ca^{2+}$ uptake in unstimulated tissue is not affected by the concentration of K^+ in the assay medium. However, the effect of norepinephrine is blocked by reducing the concentration of K^+ to 5 mM (Li^+ substitution).

* $p < 0.05$ when compared to Na^+-loaded tissue of the group.

TABLE 1. Na_i^+-Dependent $^{45}Ca^{2+}$ Uptake into Rabbit Abdominal Aortic Rings[a]

	Ca^{2+} 1.8 mM		Ca^{2+} 0.1 mM	
	Na-free	Na-loaded	Na-free	Na-loaded
Control	96.4 ± 4.20	247.0 ± 28.1[b]	23.2 ± 2.1	75.1 ± 6.42[b]
Benzamil			ND[d]	
(100 μm)	83.7 ± 5.96	154.0 ± 12.6[b,c]	ND	
NE (10 μM)	113.8 ± 6.47	634.0 ± 26.2[b,c]	ND	76.0 ± 4.00
Na^+ 65 mM	85.6 ± 3.26	206.6 ± 33.8[b]	20.5 ± 1.1	28.5 ± 2.26[b,c]
Na^+ + NE	ND	468.0 ± 38.7[c]	ND	25.1 ± 0.90[c]

[a] Rabbit abdominal aortic rings were preincubated in Na-free or 65 mM Na^+-loading solution for 2 hr. Tissues were then transferred to Na^+-free assay medium containing 0.1 or 1.8 mM $^{45}Ca^{2+}$ for 8 min. Tissues were then washed in ice-cold La^{3+} solution ($MgCl_2$ 93 mM, $LaCl_3$ 10 mM, HEPES-Tris 10 mM, KCl 5 mM, glucose 5 mM) for 50 min, blotted, weighed, and radioactivity determined by liquid scintillation spectrometry.

[b] $p < 0.05$ when compared with appropriate Na^+-free tissue.

[c] $p < 0.05$ when compared with control group.

[d] ND = not determined.

Na-depleted tissues. These agonists had no significant effect on Na_i^+-Ca_o^{2+} exchange in the presence of phentolamine or when the concentration of Ca^{2+} in the assay medium was reduced to 0.1 mM. Also, changing the concentration of K^+ in the assay medium from 80 mM to 5 mM (*N*-methyl-*d*-glucamine or Li^+ substitution) completely blocked the stimulatory effect of norepinephrine on Na_i^+-Ca_o^{2+} exchange (FIG. 1). These results indicate that, in depolarized aorta of rabbit, Na_i^+-Ca_o^{2+} exchange can be stimulated by α-adrenoceptor agonists, and this stimulatory effect is sensitive to the concentrations of Ca^{2+} and K^+ in the assay medium.

REFERENCES

1. HUME, J. R. & A. UEHARA. 1986. J. Gen. Physiol. **87:** 857-884.
2. KIMURA, J., S. MIYAMAE & A. NOMA. 1987. J. Physiol. **384:** 199-222.
3. PHILIPSON, K. D. 1985. Ann. Rev. Physiol. **47:** 561-571.
4. RASGADO-FLORES, H., E. M. SANTIAGO & M. P. BLAUSTEIN. 1989. J. Gen. Physiol. **93:** 1219-1241.
5. DIPOLO, R. & L. BEAUGE. 1990. Calcium transport in excitable cells. *In* Intracellular Calcium Regulation. F. Bronner, Ed.: 381-413. Alan R. Liss, Inc. New York.
6. REEVES, J. P. 1990. Sodium-calcium exchange. *In* Intracellular Calcium Regulation. F. Bronner, Ed.: 305-347. Alan R. Liss, Inc. New York.
7. PHILIPSON, K. D. 1990. The cardiac Na^+-Ca^{2+} exchanger. *In* Calcium and the Heart. G. A. Langer, Ed.: 85-108. Raven Press. New York.
8. VIGNE, P., J. P. BREITTMAYER, D. DUVAL, C. FRELIN & M. LAZDUNSKI. 1988. J. Biol. Chem. **263:** 8078-8083.

Sodium Withdrawal Contractures in Tonic Skeletal Muscle Fibers of the Frog[a]

J. MUÑIZ, M. HUERTA, J. L. MARIN, AND C. VÁSQUEZ

Centro Universitario de Investigaciones Biomedicas
Universidad de Colima
Colima, Mexico 28000

Skeletal muscle possesses two different fibers: twitch and tonic or slow. Tonic fibers, in contrast to twitch fibers, can maintain tension during prolonged depolarization.[1,2] The sodium-calcium exchange system was investigated by contractile response to alteration of the extracellular sodium concentration in the tonic muscle fibers of the frog. Isometric tension was recorded from tonic bundles of cruralis muscle of *Rana pipiens.* Normal solution was (mM): NaCl 117.5, KCl 2.5, $CaCl_2$ 1.8; pH was adjusted to 7.4 with imidazole-chloride. Na^+-free solutions were prepared by replacing NaCl with an osmotically equivalent amount of Tris-Cl, TMA-Cl, TEA-Cl or *N*-methyl-glucamine-Cl. In some experiments 15 μM of veratridine or 1 μM of strophantidin was added to normal solution. When Ca^{2+} was omitted from the solutions, it was replaced by 0.5 mM $NiCl_2$.[3] All experimental solutions contained *d*-tubocurarine (50 μM) to prevent a possible acetylcholine-like effect.[2] Experiments were done at room temperature (20-22 °C). Contractures were evoked in solutions where the extracellular sodium was withdrawn. FIGURE 1 is a representative result of the effect of sodium-withdrawal solution containing *d*-tubocurarine. The initial tension developed slowly, reaching a maximum to establish a plateau. On return to normal solution, there was a second rise in tension that relaxed spontaneously. This second response was somewhat variable and depends upon the compounds that substitute NaCl. Similar results were obtained when sodium was substituted with Tris, TMA, or glucamine. The replacement of Na^+ by TEA^+ did not affect the resting potential. It was -77 ± 8 mV ($n = 10$) in normal solution and -72 ± 6 mV ($n = 9$) in TEA-Cl solution. In the presence of *d*-tubocurarine, the amplitude of sodium-withdrawal contracture was diminished, and its time course was modified. However, when Na^+ was replaced by TEA^+ the sodium-withdrawal contractures were not modified by the *d*-tubocurarine. Contracture induced in Na^+ withdrawal are highly dependent on external Ca^{2+} concentration. If external calcium was omitted from the solution, the tension of sodium-withdrawal contracture was greatly reduced (FIG. 2, striped bar) as compared to control (FIG. 2, left empty bar), and it was re-established once external calcium was restored (FIG. 2, right empty bar). On the other hand, veratridine and strophantidin tend to increase the intracellular sodium concentration. If the bundles were soaked in normal solution with these drugs, added 20 min before the contracture, the tension was increased. The present results

[a] This work was partially supported by Grants C89-01-0150 and C90-07-0248 from SEP-SECIC, Mexico and P228CCOX881176, D111-901366, and P228CCX881177/891870 from CONACyT, Mexico.

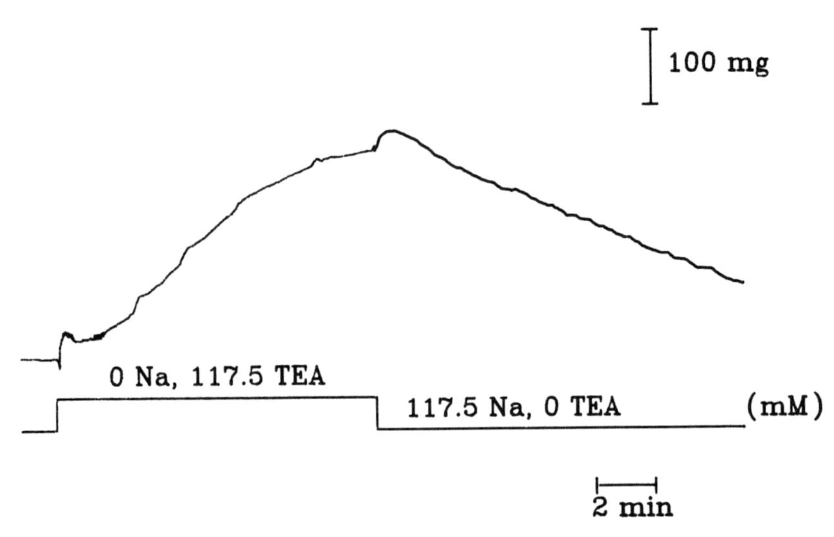

FIGURE 1. Sodium withdrawal contracture in tonic skeletal muscle fibers. The step indicates the beginning and end of application of Na^+-free solution. The solutions contain *d*-tubocurarine (50 μM). The diameter of bundle was less than 500 μM.

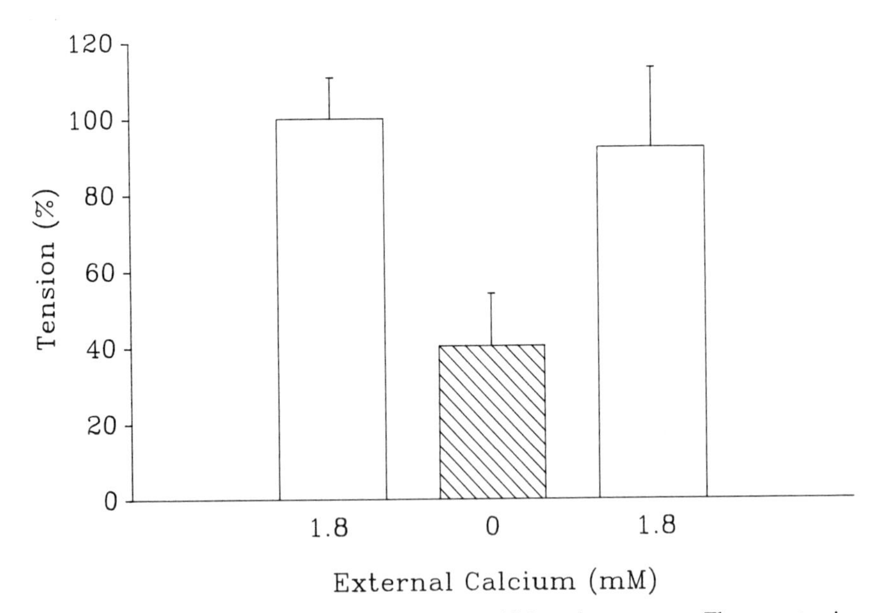

FIGURE 2. Effect of Ca^{2+}-free solution on sodium withdrawal contracture. The mean tension of the initial control sodium withdrawal contractures elicited in the presence of 1.8 mM Ca^{2+} was taken as 100% (left empty bar). Data correspond to the mean ± SE ($n = 4$). The difference between initial control contractures and contractures in Ca^{2+}-free solution (striped bar) were significant ($p < 0.005$, Student's *t*-test). This effect was reversible (right empty bar). The diameter of bundles was less than 500 μm.

suggest that the membrane of tonic skeletal muscle fibers possesses the sodium-calcium exchange system.

REFERENCES

1. KUFFLER, S. W. & E. M. VAUGHAN-WILLIAMS. 1953. J. Physiol. **121:** 318-340.
2. HUERTA, M., J. MUÑIZ & E. STEFANI. 1986. J. Physiol. **376:** 219-230.
3. HILLE, B., A. M. WOODHULL & B. I. SHAPIRO. 1975. Phil. Trans. R. Soc. London B. **270:** 301-318.

Sodium-Calcium Exchange in Other Tissues

Introduction to Part VII

Much work has been done on the role of plasmalemmal Na-Ca exchange in excitable cells, including neurons and all types of muscle, as reflected by most of the preceding articles. However, the exchanger has also been observed in many other types of cells, as discussed in this section.

Na-Ca exchange activity has been observed in basolateral cell membranes of intestinal as well as renal epithelial cells; in the kidney it has been identified in proximal and distal tubule cells as well as in cortical collecting tubule cells. In the proximal tubule it probably does not play a direct role in net transepithelial Ca^{2+} transport because most of the reabsorbed Ca^{2+} moves through the paracellular pathway. However, the exchanger may well play a role in regulating proximal tubular Na^+ reabsorption by modulating $[Ca^{2+}]_i$, which, in turn, modulates Na^+ entry via amiloride-sensitive, Ca^{2+}-regulated apical Na^+ channels.

The exchanger also appears to be prominent in various secretory cells including adrenal medullary cells, pancreatic B cells, platelets, and neutrophils. In all of these cells, cytosolic Ca^{2+} plays a key role in secretion. Therefore, modulation of cytosolic and sequestered Ca^{2+} as well as mediation of Ca^{2+} extrusion, by the exchanger, appear to be important for the normal function of these cells. Nevertheless, detailed information on the specific contribution of the exchanger to secretory activities, comparable to the information available concerning cardiac function, is lacking. Also, although there is no information available about the molecular structure of the exchanger in these cells (except for the kidney), the kinetic properties suggest that the exchangers in the epithelial and secretory cells are likely to be homologous to the exchanger in cardiac sarcolemma rather than the one in rod photoreceptor outer segments.

Human erythrocytes do not have a Na-Ca exchanger: They have a Na^+ pump to regulate $[Na^+]_i$ and cell volume and an ATP-driven Ca^{2+} pump to regulate $[Ca^{2+}]_i$. Also, red cells do not have intracellular stores of Ca^{2+} that require modulation. However, dog and ferret red cells, which do not possess Na^+ pumps, do have prominent Na-Ca exchangers. In these cells the exchanger apparently functions in the Ca^{2+}-entry mode: Na^+ is extruded (to help regulate cell volume), and the entering Ca^{2+} is extruded via the ATP-driven Ca^{2+} pump.

These observations on the distribution of the exchanger and its physiological role in various types of cells foster the view that human erythrocytes, and perhaps other erythrocytes that possess Na^+ pumps and do not have Ca^{2+} channels, may be exceptional. Most types of cells other than erythrocytes may require Na-Ca exchangers to modulate intracellular Ca^{2+} stores as well as to extrude Ca^{2+} when $[Ca^{2+}]_i$ rises as a result of cell activation.

The Role of Na-Ca Exchange in Renal Epithelia

An Overview

E. E. WINDHAGER, G. FRINDT, AND S. MILOVANOVIC

Department of Physiology and Biophysics
Cornell University Medical College
New York, New York 10021

The first indication for the presence of Na-Ca exchange in renal epithelia was obtained by Blaustein[1] in studies on kidney slices in which Ca efflux was reduced upon lowering the Na concentration of the medium. Subsequent work aimed at detecting Na-Ca exchange in renal tubules falls into three main categories. First, indirect approaches, in which as in Blaustein's study, the Na gradient across the basolateral membrane was manipulated. The effect of this maneuver was then examined either in terms of Ca-dependent changes in transepithelial net transport of Na, Ca, and water, or intracellular Na or Ca ion concentrations. Second, a more direct approach has used isotopic fluxes of Na and Ca in membrane vesicles under a variety of conditions. A third approach consists of searching for kidney-derived mRNA that codes for Na-Ca exchange. Unfortunately, attempts to measure Na-Ca exchange currents by whole-cell patch-clamping of cells of the mammalian cortical collecting tubule (G. Frindt & L. Palmer, unpublished) were unsuccessful, probably because the density of the exchangers in the plasma membranes of these cells is too low to permit detection of such a current with presently available recording techniques.

Inherent in all the indirect studies of the effect of alterations in the basolateral Na gradient on epithelial net transport was the assumption that an elevation of intracellular [Ca^{2+}] inhibits the transepithelial movement of Na and water. This assumption was indeed confirmed in a large number of publications[2–9] in which Ca ionophores, or other compounds known to raise cell [Ca^{2+}], were used. A similar inhibition of net reabsorption was also observed when the Na gradient across the basolateral cell membrane had been reduced, suggesting that cytosolic [Ca^{2+}] had risen due to the operation of a Na-Ca exchange process. For example, Grinstein and Erlij[10] observed an inhibition of the short circuit current in frog skins when Na was absent in the serosal medium. This inhibition was prevented, however, when Ca had been removed from the inner bathing solution. Similar findings have been obtained in studies on the urinary bladder of the toad,[11,18] the proximal tubule of mammalian kidneys,[7,41] and the cortical collecting tubule of the rabbit kidney.[9]

More recently, the effect of changes in the electrochemical potential gradient across the basolateral membrane on intracellular [Ca^{2+}] has been examined using either intracellular ion-selective microelectrodes[13–15] or various indicators such as aequorin[16,21] or the fluorescent dyes quin-2[17,19] or fura-2.[20–22] The large size of proximal tubule cells in amphibian kidneys makes it possible to use Ca^{2+}-selective microelectrodes.[13–15]

TABLE 1. Intracellular Na^+ and Ca^{2+} Activities in Perfused Proximal Tubules of *Necturus* Kidney

	a^i_{Na} (nM)	a^i_{Ca} (nM)
Control	12.8	82
Reduced Basolateral Na Gradient		
Low peritubular $[Na^+]$	8.2	585
Ouabain	70.1	835
Zero $[K^+]_{bath}$	51.5	237
Luminal gramicidin	41.8	617
Increased Basolateral Na Gradient		
Low luminal $[Na^+]$	7.8	77
Zero luminal organic solutes	8.3	61

TABLE 1 summarizes measurements of cytosolic calcium (a^i_{Ca}) and sodium (a^i_{Na}) ion activities in *Necturus* proximal tubules during different experimental maneuvers aimed at altering the electrochemical Na gradient. Four different experimental conditions were examined in which the peritubular Na gradient was decreased. In the first case this was achieved by lowering peritubular [Na]. In the second and third, the Na pump was inhibited, whereas in the fourth condition the rate of Na entry across the apical membrane was increased by gramicidin. In all these cases, a^i_{Ca} rose as expected from the operation of a Na-Ca exchange mechanism. The fact that a^i_{Ca}, obtained at nominally zero [K] in the peritubular fluid, was lower than when ouabain was applied is a reflection of the high degree of ion permeability of the proximal tubule that permits K ions to diffuse from the lumen into the peritubular space. When the Na gradient across the basolateral membrane was increased by reducing the rate of Na entry across the apical membrane, a^i_{Ca} was diminished. The case where a luminal perfusate of low $[Na^+]$ was used indicates that Na-Ca exchange does not occur in the apical cell membrane. If Na-Ca exchange were present in the brush border, lowering of [Na] in the lumen should elevate a^i_{Ca} in a fashion similar to that when [Na] is reduced in the peritubular fluid.

FIGURE 1 illustrates the relationship between a^i_{Ca} and the electrochemical potential gradient for Na ions across the basolateral cell membrane, based upon the data shown in TABLE 1 and simultaneously measured membrane voltages.[14,15] The actual level of a^i_{Ca} is largely determined by the electrochemical Na gradient. This implies that Na-Ca exchange dominates over other transfer modes in setting the level of a^i_{Ca} under conditions where the Na gradient was altered.

Similar conclusions have been reached in studies using suspensions of rat proximal tubules by Dominguez and his collaborators.[21] When extracellular [Na] was reduced to 15 mM, intracellular $[Ca^{2+}]$, measured with aequorin, increased by about 100%. In addition, these authors also measured unidirectional Ca fluxes. They concluded that both a decrease of Ca efflux and an increase in Ca uptake contribute to the rise in intracellular $[Ca^{2+}]$ that occurs in response to a decrease in bath [Na].

The renal connecting tubule, which is located between the distal and the collecting tubules of the mammalian nephron, also changes its intracellular $[Ca^{2+}]$ in response to a lowering of the peritubular Na gradient. FIGURE 2, taken from the work of Bourdeau and Lau[22] shows cytosolic $[Ca^{2+}]$ in isolated, perfused connecting tubules during control conditions, during deletion of peritubular [Na], and after restitution of

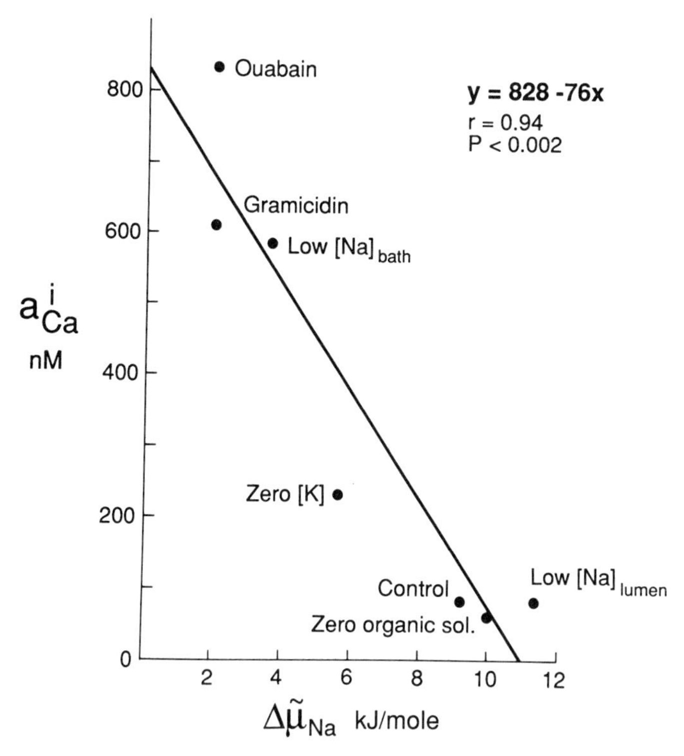

FIGURE 1. Relationship between a^i_{Ca} and the electrochemical potential gradient for Na^+ across the basolateral cell membrane of *Necturus* proximal tubules (modified from Yang *et al.*[15]).

150 mM Na. As replacement of Na, either mannitol (upper diagram) or choline or TEA (lower diagram) were used. A very rapid and marked increase in cytosolic $[Ca^{2+}]$ was observed upon deletion of Na from the bathing solution, irrespective of the nature of the Na substitute. In other experiments, Bourdeau and Lau[22] have found that this response does not occur when bath $[Ca^{2+}]$ was clamped to 0.1 micromolar. The same study failed to detect any evidence for the existence of Na-Ca exchange in the apical cell membrane.

A number of studies have addressed the question whether Na-Ca exchange also occurs in the mammalian cortical collecting tubule (CCT). In nonperfused cortical and outer medullary collecting tubules of adrenalectomized rats, Taniguchi *et al.*[20] have shown that after Na loading by ouabain treatment the cells exhibit a marked rise in cytosolic $[Ca^{2+}]$ when the extracellular [Na] is suddenly reduced. This response depended on the presence of extracellular Ca. In a study by Silver, Frindt, and Palmer (unpublished observations), surgically opened cortical collecting tubules showed an increase in cytosolic $[Ca^{2+}]$ upon exposure to 10^{-4} M ouabain. FIGURE 3 shows Fura-2 measurements of cytosolic $[Ca^{2+}]$ under control conditions (left column) and after 10 minutes of ouabain treatment (right column). Ouabain raised $[Ca^{2+}]$ from 166 to 189 nanomolar, corresponding to an 18% increase. Thus, both studies are consistent with the existence of a Na-Ca exchange mechanism in mammalian collecting tubules.

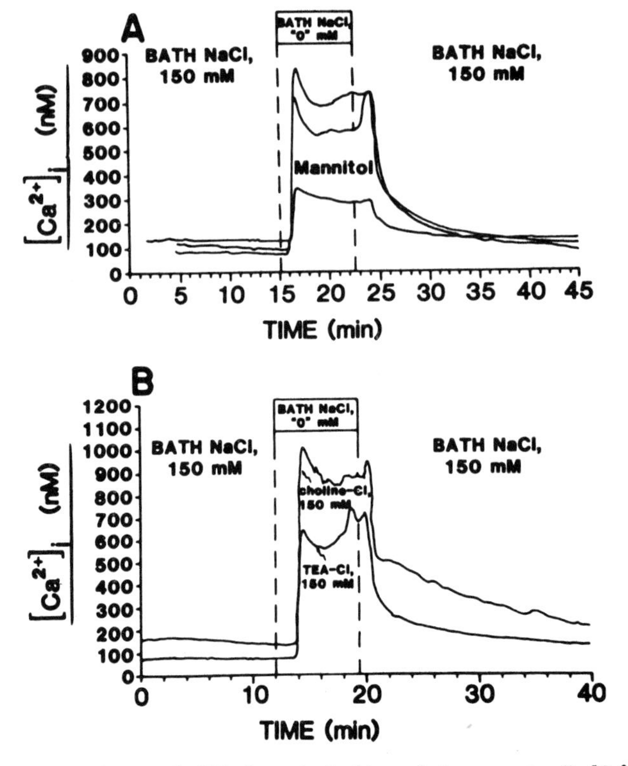

FIGURE 2. Effect of removal of Na from the bathing solution on cytosolic [Ca^{2+}] in isolated, perfused connecting tubules of the rabbit kidney. *Upper diagram:* mannitol replacement. *Lower diagram:* Choline or TEA replacement of Na. (From Bourdeau & Lau[22]; used with permission.)

Early and convincing evidence for the existence of Na-Ca exchange in the basolateral membranes of renal epithelia was obtained by Gmaj *et al.*[23] and later confirmed and extended by many investigators.[24–29] The key observations on inside-out basolateral membrane vesicles, prepared from kidney cortex, were that Ca efflux from Ca-loaded vesicles was enhanced by an inwardly directed Na gradient.[12,23] This effect was stimulated by inside-negative membrane potentials, indicating electrogenicity of Na-Ca exchange.[26] Furthermore, Ca uptake against a Ca gradient was obtained upon reversal of the Na gradient ($Na_{in} > Na_{out}$) across the vesicle membrane.[25–29] Kinetic studies in which the Na-Ca exchanger had been partially purified and incorporated into liposomes[30] are consistent with a stoichiometry of a $3Na^+/1\ Ca^{2+}$.

The existence of Na-Ca exchange in renal tissue can also be demonstrated by molecular biological techniques. For this purpose and also to obtain information on the molecular nature of renal Na-Ca exchange, we attempted to express kidney-derived mRNA coding for Na-Ca exchange in *Xenopus laevis* oocytes.[31] Oocytes were injected with either 50 nl of water (controls) or 50 ng of kidney mRNA (total or size-fractionated). Na-Ca exchange activity was determined three days after injection by measuring rates of oocyte Ca uptake, which were dependent on outwardly directed Na gradients. Na concentration gradients were manipulated by Na-loading the oocytes in

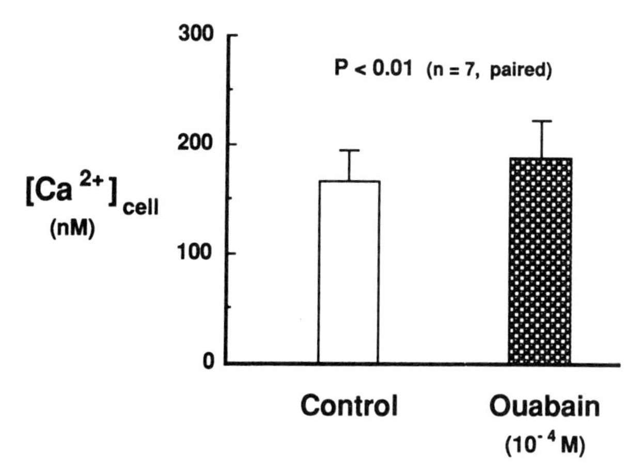

FIGURE 3. Effect of ouabain on cytosolic [Ca^{2+}] in surgically opened cortical collecting tubules (R. Silver and G. Frindt, unpublished).

the presence of nystatin. They were then exposed to either zero or 90 mM Na in the bathing media. In Na-free solutions, Na was replaced by 90 mM K.

FIGURE 4 shows Ca uptake by either water-injected oocytes (the two columns on the left) or whole rat kidney mRNA-injected oocytes (shown in the two columns on the right side). Hatched columns represent Ca uptake of oocytes exposed to K media, whereas open columns indicate data from oocytes exposed to the Na medium. Na gradient-dependent (Na-GD) Ca uptake was calculated as the difference between Ca uptake in K medium and Na medium (the difference between hatched and open columns). For five batches of mRNA-injected oocytes (right side of the slide) the Na-GD Ca uptake was 6.6 picomoles per oocyte per 30 min. This is significantly higher than the value of 3.4 picomoles/(oocyte · 30 min) in water-injected oocytes ($p < 0.001$). These data are consistent with the oocyte expression of a renal Na-Ca exchanger.

In order to examine a possible contribution of Ca influx through Ca channels to the measured Ca uptake, we used the verapamil analogue D 600 or nifedipine. Both Ca channel blockers did not alter Na gradient-dependent Ca uptake in mRNA-injected oocytes. We also examined the effect of 0.1 mM lanthanum, a potent, although not a specific, inhibitor of Na-Ca exchange.[32] In mRNA-injected oocytes, La^{2+} reduced Na gradient-dependent Ca uptake by more than 90%.

Na-Ca exchange, in the reversed mode, depends on the presence of a low concentration of a monovalent cation in the extracellular solution. This has been shown in a variety of cells, including squid axon,[33] photoreceptors,[34] and ventricular myocytes.[35] To test whether the renal Na-Ca exchanger expressed in oocytes shows this requirement, we have measured Ca uptake in Na-loaded oocytes, in the presence and absence of extracellular K. In these experiments (FIG. 5), Ca uptake was measured in the presence of 10 micromolar Ca. Oocytes were exposed to media containing either 90 mM Na (represented by open columns), 90 mM choline (hatched columns), or 85 mM choline plus 5 mM K (stippled columns). On the left side of the slide, Ca uptake by water-injected oocytes is shown. There was no detectable Na gradient-dependent Ca uptake. On the right side, Ca uptake by rat kidney mRNA-injected oocytes is shown. Exposure to choline alone (hatched column) increased Ca uptake significantly. A further increase in Ca uptake was observed upon exposure to choline plus 5 mM K

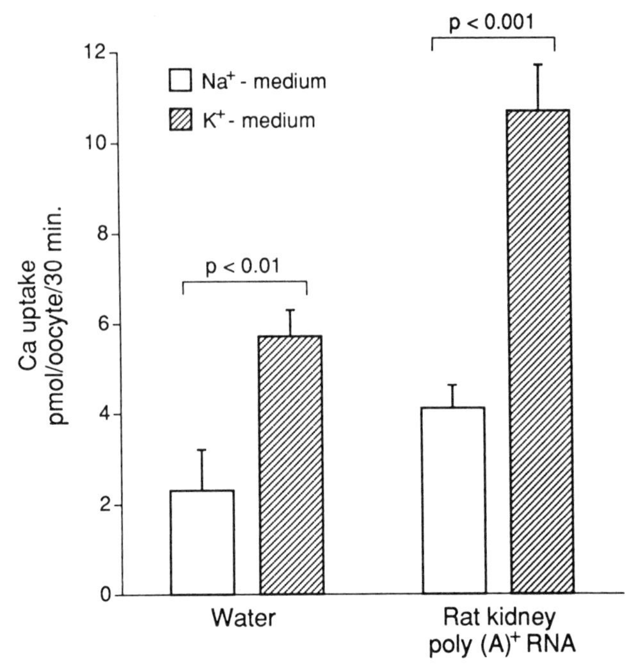

FIGURE 4. Expression of rat kidney Na^+-Ca^{2+} exchanger in *Xenopus* oocytes. Oocytes were injected with either 50 nl water or 50 nl poly(A)+RNA (1 μg/μl). Three days later the cells were loaded with Na by exposure to nystatin and Ca uptake was measured in either Na or K medium (5-9 oocytes per group). The bars represent the mean (±SEM) Ca^{2+} uptake of five different experiments. (From Milovanovic *et al.*[31]; used with permission.)

(stippled column), indicating that the expressed Na-Ca exchange was stimulated by the presence of extracellular K. In photoreceptors,[34] Ca efflux is stimulated by the lowering of extracellular K, presumably by increasing the K gradient across the cell membrane. In photoreceptors,[34] a stoichiometry of 4 Na^+-Ca^{2+}-K^+ has been proposed. The present results on oocytes are consistent with this view, but we cannot exclude other explanations, such as the possibility that K acts as a cofactor for the binding of Ca to the exchanger.[36] A putative stoichiometry of 4 Na^+-Ca^{2+}-K^+ is of special interest for kidney function because the reversal potential for such an exchanger is significantly more positive than the membrane potential, indicating the carrier would function in the forward mode under physiological conditions. More work is required to elucidate this problem.

To estimate the approximate size of the message coding for renal Na-Ca exchange, a size fractionation of rat kidney mRNA was performed, using a sucrose density gradient (FIG. 6). Fifty nanograms of different pools of fractionated mRNA, designated RK 2 to RK 5, were injected into oocytes. The upper part represents a sucrose density gradient profile. The lower part of the slide shows that the highest Na-gradient dependent Ca uptake was observed in oocytes injected with mRNA pool number RK 3, corresponding to 18 to 23 S. This indicates that the mRNA encoding Na-Ca exchange activity is approximately 3-4 kilobases in size. The estimate of length represents only an approximation, since relatively poor enrichment of specific mRNA in

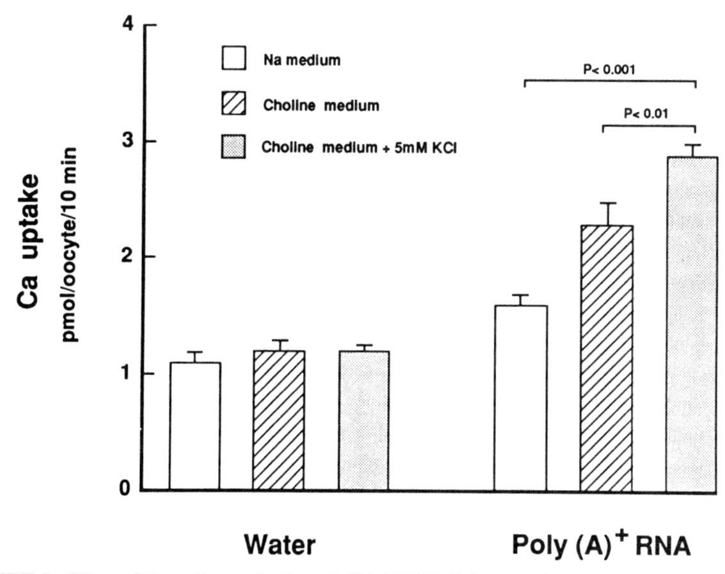

FIGURE 5. Effect of K on Ca uptake by poly(A)+RNA-injected, Na-loaded oocytes. Ca uptake was measured in the presence of 10 μM Ca^{2+} in either Na medium, choline medium, or choline medium plus 5 mM K (7-16 oocytes per group). Each bar represents the mean (±SEM) of five different experiments. (From Milovanovic *et al.*[31]; used with permission.)

RK 3 was obtained when compared to total kidney mRNA (shown in the black column on the left). This low degree of expressed activity may be attributable to partial degradation of mRNA during fractionation, which may greatly alter the mobility of RNA in the gradient. It should be noted, however, that a similar size was proposed for rabbit heart[37] and chick heart[38] exchanger based on sucrose density gradient centrifugation.

FIGURE 7 shows a Northern blot analysis of rat kidney mRNA (5 μg) in the left lane and total heart mRNA (30 μg), used as a positive control in the right lane. The blot was probed with 3′(1.5 kb) and 5′(1 kb) cDNA coding for the rat heart Na-Ca exchanger, kindly provided by Ken Philipson. The probe hybridized with 7 kb and, to a lesser extent, with 4 kb rat kidney mRNA. At the present time we do not know whether the appearance of two bands is due to degradation or to the existence of two different genes or to different RNA-splicing products. Further work is needed for the detailed characterization of the mRNA, but the Northern blot analysis clearly indicates the presence of kidney mRNA encoding renal Na-Ca exchange. We can assume, therefore, a certain degree of homology between cardiac and renal Na-Ca exchangers.

Regarding the physiological role of Na-Ca exchange in renal tubules, our knowledge is incomplete due to the lack of a specific inhibitor. It is possible that the magnitude of Ca extrusion via the exchanger is negligible and that the ATP-driven Ca pump in the basolateral membrane effects most if not all of this transport of Ca. This conclusion rests mainly on kinetic studies on basolateral membrane vesicles and on the calculated reversal potential of the exchanger. Measured levels of cytosolic [Na^+], [Ca^{2+}] and membrane voltage indicate that at physiologically prevailing conditions, the Na-Ca exchanger idles.[39] Furthermore, under conditions simulating physiological conditions,

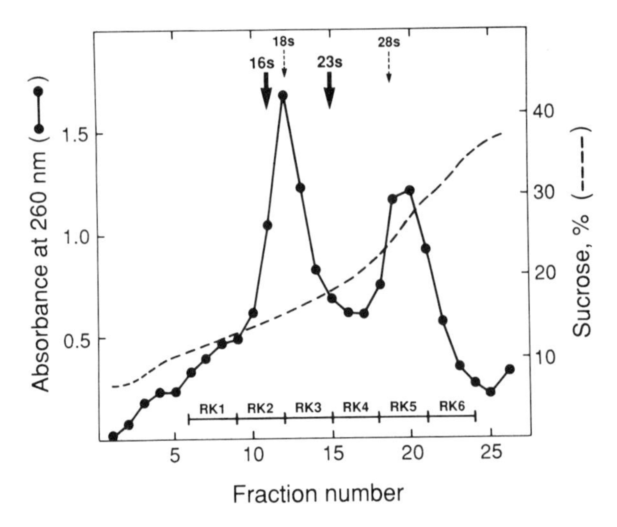

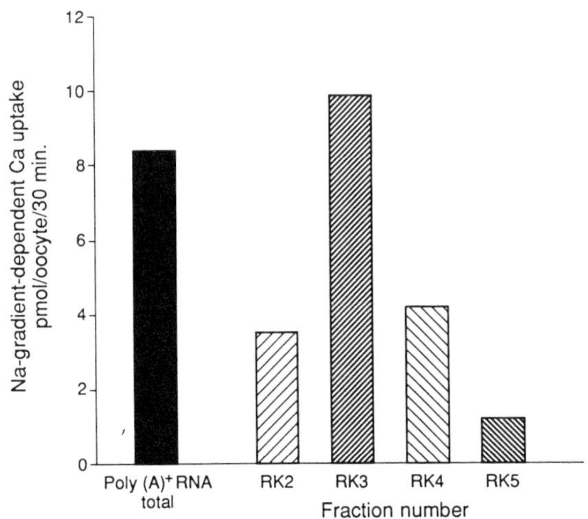

FIGURE 6. Sucrose density gradient fractionation of rat kidney poly(A)+RNA. *Upper diagram:* 270-microgram poly(A)+RNA was fractionated over a sucrose density gradient. The line with solid circles represents the absorbance at 260 nm, and the dotted line, sucrose concentration. Downward arrows indicate the position of 16 S and 23 S. Ribosomal RNA from *E. coli* was used as a marker after fractionation over a similar sucrose gradient. Fractions indicated (RK 1 to RK 6) were separately pooled and the RNA was precipitated and redissolved in water. (From Milovanovic *et al.*[31]; used with permission.)

Lower diagram: expression of Na-gradient-dependent Ca uptake in oocytes after injection of size-selected mRNA (fractions RK2 to RK5; 50 ng mRNA/ooctye). Each bar represents Na gradient-dependent Ca uptake in K medium minus uptake in Na medium. Peak activity of Na-Ca exchange was observed with injection of fraction RK3.

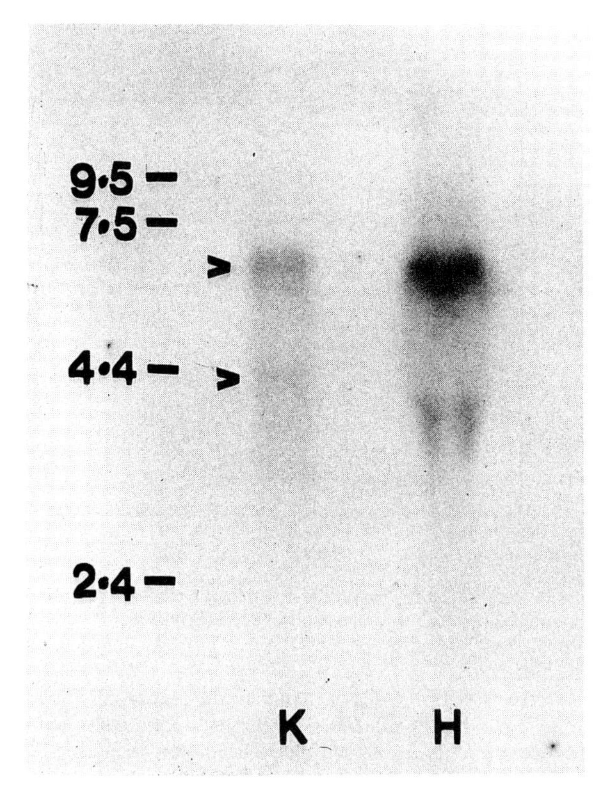

FIGURE 7. Northern blot analysis of rat kidney mRNA (left lane) and total heart mRNA (right lane) (S. Milovanovic, N. Yan, S. Tate, G. Frindt, K. Philipson & E. E. Windhager, unpublished observations).

the estimated K_m and V_{max}, both corrected for membrane vesiculation and phosphorylation, yield an estimated Ca efflux via the Na-Ca exchanger that is only some 10% of that mediated by the ATP-driven Ca pump.[39] It should be emphasized that these estimates were obtained in membrane vesicles and must be reexamined in intact cells. Some of the studies on whole tubules[40] suggest that presently unknown intracellular factors may enhance Ca extrusion via the Na-Ca exchanger beyond the levels calculated from vesicle studies.

Regarding the role of Na-Ca exchange in renal net transport of calcium, a number of facts are clearly established. In work on proximal tubules, Ullrich *et al.*[41] have demonstrated a significant inhibition of active transepithelial Ca transport during peritubular capillary perfusion with fluid containing either ouabain or low [Na]. Also in distal convoluted tubules and connecting tubules,[40] active net Ca reabsorption is inhibited by the same maneuvers. In addition, the normally observed PTH-induced stimulation of Ca transport is abolished when peritubular [Na] is reduced. This finding suggests that an undisturbed operation of the Na-Ca exchange process is a prerequisite for the calcium-retaining effect of this hormone.

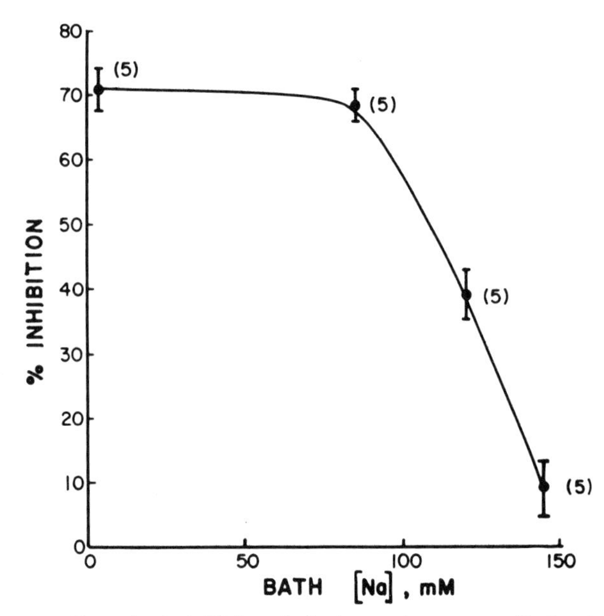

FIGURE 8. Effect of lowering bath [Na] on the hydrosmotic response of isolated perfused rabbit cortical collecting tubules to antidiuretic hormone. (From Frindt *et al.*[12]; used with permission.)

With respect to the role of Na-Ca exchange in the regulation of water transport by renal tubules, the most recent studies (see Breyer[42] for review and Craven & De Rubertis[43]) on collecting tubules concur that experimental maneuvers believed to cause steady-state increases in cytosolic [Ca^{2+}] inhibit the hydroosmotic response to vasopressin and cyclic AMP. The mechanism by which cytosolic Ca^{2+} exerts this effect is multifactorial and beyond the scope of the present overview. However, a significant role of Na-Ca exchange in determining the magnitude of the water permeability response to ADH of the mammalian cortical collecting tubule has been demonstrated.[12] FIGURE 8 shows that a reduction in peritubular [Na] inhibits this response. A pathophysiological role of this effect may be predicted from these results because the steepest increase in inhibition is observed in a range of peritubular Na concentrations that may be encountered in various diseases.

The influence of Na-Ca exchange on renal transport of sodium has been alluded to in the previous discussion of evidence for the existence of the exchanger in renal tubules. Several investigators[10,44–47] have implicated Na-Ca exchange in the hypothesis of a negative-feedback mechanism for Na transport. According to this view, any primary increase in apical Na entry or any decrease in Na pump activity will raise cytosolic [Na^+], which in turn via Na-Ca exchange causes cell [Ca^{2+}] to rise, an effect that inhibits apical Na permeability. In the kidney, this hypothesis has been examined in the cortical collecting tubule and a significant reduction in net transport of Na was found when peritubular [Na] was reduced or when ionomycin was added to the bathing solution.[9]

The view that increased levels of cell [Ca^{2+}] inhibit the rate of apical Na entry was further examined in patch-clamp studies of apical Na channels in cortical collecting tubules.[45] For this purpose, CCTs were split open with sharpened needles under microscopic observation to provide access for the patch-clamp pipette to the luminal cell

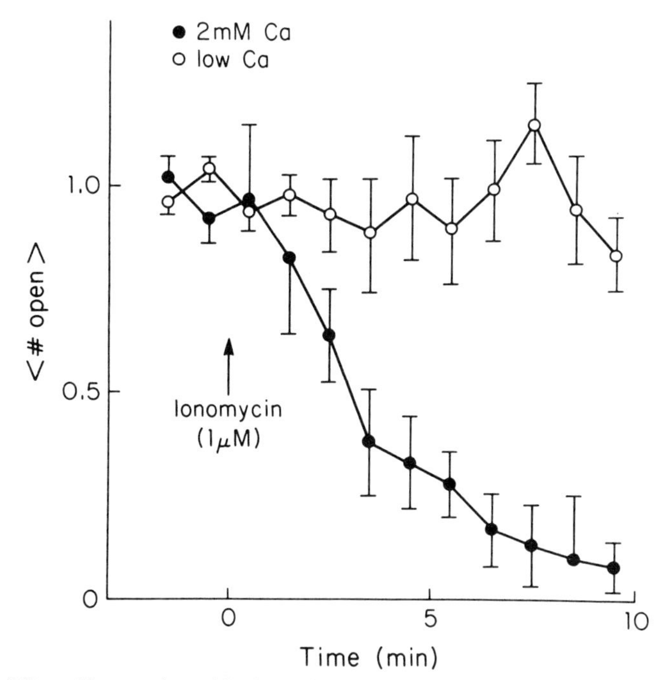

FIGURE 9. Effect of ionomycin on Na-channel activity in cell-attached patches. (From Palmer & Frindt[45]; used with permission.)

membrane. Na channel activity was examined in inside-out and cell-attached apical membrane patches. In a series of experiments with inside-out patches, Na channel activity was measured at different Ca^{2+} concentrations in the fluid bathing the cytoplasmic side of the membrane patches. Contrary to expectations, despite variations in $[Ca^{2+}]$ from 10^{-8} to 10^{-5} M, there was no change in Na-channel activity.[45] This finding was different from the results of Garty and Asher[46] obtained in membrane vesicles from permeabilized toad bladder epithelia. They found that increasing the $[Ca^{2+}]$ in the same range reduced the amiloride-sensitive Na flux. Similar conclusions had previously been reached by Chase and Al-Awqati.[47] Garty and Asher[46] had observed that the inhibitory effect of Ca^{2+} was temperature dependent. It seemed possible, therefore, that a Ca^{2+}-induced inhibition of Na channel activity might be due to biochemical reactions requiring cytoplasmic components not present in the excised inside-out patch preparation. To examine this possibility, the effect of increasing cytoplasmic $[Ca^{2+}]$ on Na channel activity was measured in cell-attached patches.[45]

FIGURE 9 summarizes some of the results obtained in the presence and absence of ionomycin.[45] The mean number of open Na channels in two groups of tubules is plotted as a function of time. The black circles refer to tubules exposed to 2 mM Ca and the open circles to tubules bathed in fluid containing Ca at a concentration $< 10^{-8}$ M. After a control period, ionomycin was added to both groups. Upon addition of ionomycin, Na channel activity at normal extracellular Ca dropped sharply after a brief delay of 2-3 minutes, reaching almost complete inactivation after 5 minutes. In contrast, there was no inactivation of Na channels when the extracellular Ca concentration was below the

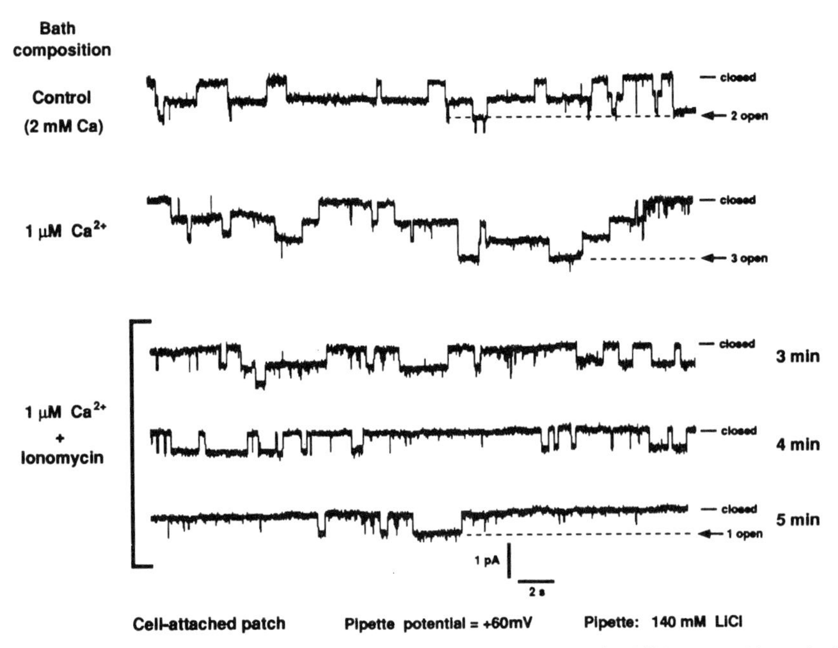

FIGURE 10. Effect of ionomycin on Na-channel activity at 1 μM [Ca^{2+}] in the bathing solution (G. Frindt & L. Palmer, unpublished).

presumed intracellular level. These data demonstrate a Ca-mediated inhibition of apical Na channels.

To examine the effect of less extreme increases in cytosolic [Ca^{2+}], Na-channel activity was measured in cell-attached patches in tubules exposed to 1 μM [Ca^{2+}] in the presence and absence of 5 μM ionomycin. The seal of the pipette to the membrane was established in control solution in 2 mM Ca^{2+}. A representative experiment showing patch-clamp recording obtained from a single cell-attached patch is shown in FIGURE 10. The upper trace shows the current transitions caused by the opening of two channels. Then, when extracellular Ca was reduced by three orders of magnitude, presumably also causing a reduction in cytosolic [Ca^{2+}], on the average, the number of open Na channels increased. The lower three traces show the effect of the addition of ionomycin to 1 micromolar [Ca^{2+}], presumably resulting in an elevation of cytosolic [Ca^{2+}] but to no more than 10^{-6} M. The three lower traces were obtained 3, 4, and 5 minutes after ionomycin addition. Na channel activity was evidently reduced, and, again, a delay of 2 to 3 minutes was observed, supporting the view that Ca ions inhibit Na-channel activity indirectly by an enzymatic reaction. The increase in cytosolic [Ca^{2+}] required to effect a marked inhibition of Na channel activity lies clearly within the range that might be expected to occur under physiological or pathophysiological conditions and during alterations in Na-Ca exchange activity.

The conclusion that cytosolic Ca ions may act indirectly on apical Na channels is also supported by studies of Ling and Eaton,[48] who measured Na-channel activity in apical membranes of A6 cells, a cell line homologous to the collecting tubule. They found that a reduction in intracellular [Na^+], caused by the addition of amiloride to the luminal fluid, resulted in an increase in the open probability of the Na channels.

This is consistent with the view that due to the operation of Na-Ca exchange, changes in cytosolic Na were followed by parallel changes in intracellular [Ca^{2+}] and that, in turn, the reduced level of [Ca^{2+}] stimulated the activity of the Na channels. Then, upon the addition of the phorbol ester PMA, the activity of Na channels was completely abolished. This last observation suggests that increased cytosolic [Ca^{2+}] may inhibit Na-channel activity by activation of PKC, resulting in a phosphorylation reaction. It is unknown at present whether this phosphorylation occurs in the Na channel itself or a channel-associated protein.

It should be pointed out that phosphorylation of proteins is by no means the only possible manner by which changes in cytosolic [Ca^{2+}] may influence apical Na permeability. Other biochemical reactions, such as methylation of proteins or changes in the arachidonic acid metabolism of the cell, must be examined.

Na-Ca exchange may also influence apical Na-channel activity indirectly by secondary changes in intracellular pH, which in turn can alter the activity of Na channels. For example, in a study by Borle *et al.*[49] on MDCK cells, a reduction of peritubular [Ca^{2+}] to zero caused a 71% fall in cell [Ca^{2+}] and a rise in cell [Na^+] by 56%, attributable to Na-Ca exchange. Cell [Na^+] had risen, sufficiently to drive Na^+ out and H^+ into the cells, causing intracellular acidification. Because there is a nearly linear relationship between cytosolic pH and the mean number of open Na channels in apical membranes of the CCT between pH 6.4 and 7.4, it is possible that the tendency for increasing P_{Na} by the low cell [Ca^{2+}] is partially opposed by the tendency of the lowered cell pH to inhibit Na-channel activity.

In summary, drastic experimental manipulations of Na-Ca-exchange activity lead to marked changes in epithelial transport. However, its physiological role remains controversial. We know that the intracellular regulation of ion transport is mediated by a multifactorial system of interacting signals, such as transport-related changes in cytosolic [Ca^{2+}], pH, cell volume, and membrane voltage, as well as changes in cell metabolism. It will be necessary to examine the time course of changes in these parameters during experimental conditions that mimic physiological alterations in ion transport. One might hope that with growing knowledge of the molecular structure of the Na-Ca exchanger, a specific inhibitor may eventually become available. Its use in studies of epithelial transport would greatly enhance our knowledge of the physiological role of Na-Ca exchange in kidney function.

REFERENCES

1. BLAUSTEIN, M. P. 1974. The interrelationship between sodium and calcium fluxes across cell membranes. Rev. Physiol. Biochem. Pharmacol. **70:** 33-82.
2. TAYLOR, A. 1975. Effect of quinidine on the action of vasopressin. Fed. Proc. **34:** 285.
3. WIESMAN, W., S. SINHA & S. KLAHR. 1977. Effects of ionophore A23187 on baseline and vasopressin-stimulated sodium transport in the toad bladder. J. Clin. Invest. **59:** 418-425.
4. ARRUDA, J. A. L. & S. SABATINI. 1980. Effect of quinidine on Na, H^+, and water transport by the turtle and toad bladders. J. Membr. Biol. **55:** 141-147.
5. CHASE, H. S. & Q. AL-AWQATI. 1981. Regulation of sodium permeability of the luminal border of toad bladder by intracellular sodium and calcium. J. Gen. Physiol. **77:** 693-712.
6. ERLIJ, D., L. GERSTEN, G. STERBA & H. F. SCHOEN. 1986. Role of prostaglandin release in the response of tight epithelia to Ca^{2+}ionophores. Am. J. Physiol. **240:** F558-F568.
7. FRIEDMAN, P. A., J. S. FIGUEIREDO, T. MAACK & E. E. WINDHAGER. 1981. Sodium-calcium interactions in the renal proximal convoluted tubule of the rabbit. Am. J. Physiol. **240**(Renal Fluid Electrolyte Physiol. 9): F558-F568.
8. JONES, S. M., G. FRINDT & E. E. WINDHAGER. 1988. Effect of peritubular [Ca] or ionomycin on hydrosmotic response of CCTs to ADH or cAMP. Am. J. Physiol. **254**(Renal Fluid Electrolyte Physiol. 23): F240-253.

9. FRINDT, G. & E. E. WINDHAGER. 1990. Ca^{2+}-dependent inhibition of sodium transport in rabbit cortical collecting tubules. Am. J. Physiol. **258**(Renal Fluid Electrolyte Physiol. 27): F568-F582.
10. GRINSTEIN, S. & D. ERLIJ. 1978. Intracellular calcium and the regulation of sodium transport in the frog skin. Proc. R. Soc. London B Biol. Sci. **202:** 353-360.
11. TAYLOR, A., E. EICH, M. PEARL, S. S. BREM & E. Q. PEEPER. 1987. Cytosolic calcium and the action of asopressin in toad urinary bladder. Am. J. Physiol. **252**(Renal Fluid Electrolyte Physiol. 21): F1028-F1041.
12. FRINDT, G., E. E. WINDHAGER & A. TAYLOR. 1982. Hydrosmotic response of collecting tubules to ADH and cAMP at reduced peritubular sodium. Am. J. Physiol. **243**(Renal Fluid Electrolyte Physiol. 12): F505-F513.
13. LEE, C. O., A. TAYLOR & E. E. WINDHAGER. 1980. Cytosolic calcium ion activity in epithelial cells of *Necturus* kidney. Nature **287:** 859-861.
14. LORENZEN, M., C. O. LEE & E. E. WINDHAGER. 1984. Cytosolic Ca^{2+} and Na^{+} activities in perfused proximal tubules of *Necturus* kidney. Am. J. Physiol. **247**(Renal Fluid Electrolyte Physiol. 16): F93-F102.
15. YANG, J. M., C. O. LEE & E. E. WINDHAGER. 1988. Regulation of cytosolic free calcium in isolated perfused proximal tubules of *Necturus.* Am. J. Physiol.(Renal Fluid Electrolyte Physiol. 24): F787-F799.
16. SNOWDOWNE, K. W. & A. B. BORLE. 1985. Effect of low extracellular sodium on cytosolic ionized calcium. Na^{+}-Ca^{2+} exchange as a major calcium influx pathway in kidney cells. J. Biol. Chem. **260:** 14998-15007.
17. CRUTCH, B. & A. TAYLOR. 1983. Measurement of cytosolic free Ca^{2+} concentration in epithelial cells of toad urinary bladder. J. Physiol. **345:** 109-127.
18. WONG, S. M. E. & H. S. CHASE. 1986. Role of intracellular calcium in intracellular volume regulation. Am. J. Physiol. **250**(Cell Physiol. 19): C841-C852.
19. JACOBS, W. R. & L. J. MANDEL. 1987. Fluorescent measurements of intracellular free calcium in isolated toad urinary bladder cells. J. Membr. Biol. **97:** 53-62.
20. TANIGUCHI, S., J. MARCHETTI & F. MOREL. 1989. Na/Ca exchangers in collecting cells of rat kidney. Pflueger's Arch. **415:** 191-197.
21. DOMINGUEZ, J. H., J. K. ROTHROCK, W. L. MACIAS & J. PRICE. 1989. Na^{+} electrochemical gradient and Na^{+}-Ca^{2+} exchange in rat proximal tubule. Am. J. Physiol. **257**(Renal Fluid Electrolyte Physiol. 26): F531-F538.
22. BOURDEAU, J. E. & K. LAU. 1990. Basolateral cell membrane Ca-Na exchange in single rabbit connecting tubules. Am. J. Physiol. **258**(Renal Fluid Electrolyte Physiol. 27): F1497-F1503.
23. GMAJ, P., H. MURER & R. KINNE. 1979. Calcium ion transport across plasma membranes isolated from rat kidney cortex. Biochem. J. **178:** 549-557.
24. VAN HEESWIJK, M. P. E., J. A. M. GEERTSEN & C. H. VAN OS. 1984. Kinetic properties of the ATP-dependent Ca^{2+} pump and the Na^{+}/Ca^{2+} exchange system in basolateral membranes from rat kidney cortex. J. Membr. Biol. **79:** 19-31.
25. JAYAKUMAR, A., L. CHENG, C. T. LIANG & B. SACKTOR. 1984. Sodium gradient-dependent calcium uptake in renal basolateral membrane vesicles. J. Biol. Chem. **259:** 10827-10833.
26. SCOBLE, J. E., S. MILLS & K. A. HRUSKA. 1985. Calcium transport in canine renal basolateral membrane vesicles. J. Clin. Invest. **75:** 1096-1105.
27. GMAJ, P., M. ZURINI, H. MURER & E. CARAFOLI. 1983. A high-affinity calmodulin Ca pump in the baso-lateral plasma membranes of kidney cortex. Eur. J. Biochem. **136:** 71-76.
28. DE SMEDT, H., J. B. PARYS, R. BORGRAEF & F. WUYTACK. 1981. Calmodulin stimulation of renal (Ca^{2+} + Mg^{2+})-ATPase. FEBS Lett. **131:** 60-62.
29. GMAJ, P., K. MALMSTROM, J. BIBER, M. AMSTUTZ, W. GHIJSEN & H. MURER. 1985. Renal proximal tubular transport of calcium and inorganic phosphate studies with isolated vesicles. Mol. Physiol. **8:** 59-76.
30. TALOR, Z. & J. A. L. ARRUDA. 1985. Partial purification and reconstitution of renal basolateral Na^{+}-Ca^{2+} exchanger into liposomes. J. Biol. Chem. **260:** 15473-15476.
31. MILOVANOVIC, S., G. FRINDT, S. TATE & E. E. WINDHAGER. 1991. The expression of renal Na^{+}/Ca^{2+} exchange activity in *Xenopus laevis* oocytes. Am. J. Physiol. **261**(Renal Fluid Electrolyte Physiol. 30): F207-F212.

32. SIMCHOWITZ, L. & E. J. CRAGOE, JR. NA^{+}-CA^{2+} exchange in human neutrophils. Am. J. Physiol. **254**(Cell Physiol. 23): C150-C164.
33. BLAUSTEIN, M. P. & A. L. HODGKIN. 1969. The effect of cyanide on the efflux of calcium from squid axon. J. Physiol. **200:** 497-527.
34. CERVETTO, L., L. LAGNADO, R. J. PERRY, D. W. ROBINSON & P. A. MCNAUGHTON. 1989. Extrusion of calcium from rod outer segments is driven by both sodium and potassium gradients. Nature **337:** 740-743.
35. GADSBY, D. C., M. NAKAO, M. NODA & R. N. SHEPHERD. 1988. [Na] dependence of outward Na/Ca exchange current in guinea-pig ventricular myocytes. J. Physiol. **407:** 135P.
36. LAGNADO, L. & P. A. MCNAUGHTON. 1989. The sodium-calcium exchange in photoreceptors. *In* Sodium-Calcium Exchange. T. J. A. Allen, D. Noble & H. Reuter, Eds.: 261-297. Oxford University Press. New York.
37. LONGONI, S. M. J., T. COADY, T. IKEDA & K. D. PHILIPSON. 1988. Expression of cardiac sacrolemmal Na^{+}-Ca^{2+} exchange activity in *Xenopus laevis* oocytes. Am. J. Physiol. **255**(Cell Physiol. 24): C870-C873.
38. SIGEL, E., R. BAUR, H. PORZIG & H. REUTER. 1988. mRNA-induced expression of the cardiac Na^{+}-Ca^{2+} exchanger in *Xenopus* oocytes. J. Biol. Chem.: 14614-14616.
39. GMAJ, P. & H. MURER. 1988. Calcium transport mechanisms in epithelial cell membranes. Min. Electrol. Metab. **14:** 22-30.
40. SHIMIZU, T., K. YOSHITOMI, M. NAKAMURA & M. IMAI. 1990. Effects of PTH, calcitonin, and cAMP on calcium transport in rabbit distal nephron segments. Am. J. Physiol. **259**(Renal Fluid Electrolyte Physiol. 28): F408-F414.
41. ULLRICH, K. J., G. RUMRICH & S. KLOSS. 1976. Active Ca^{2+} reabsorption in the proximal tubule of rat kidney. Pflueger's Arch. **364:** 223-228.
42. BREYER, M. D. 1991. Regulation of water and salt transport in collecting duct through calcium-dependent signaling mechanisms. Am. J. Physiol. **260**(Renal Fluid Electrolyte Physiol. 29): F1-F11.
43. CRAVEN, P. A. & R. DERUBERTIS. 1991. Effects of extracellular sodium on cytosolic calcium, PGE_2 and cAMP in papillary collecting tubule cells. Kidney Int. **39:** 591-597.
44. TAYLOR, A. 1989. The role of sodium-calcium exchange in sodium transporting epithelia. *In* Sodium-Calcium Exchange. T. J. A. Allen, D. Noble & H. Reuter, Eds.: 298-323. Oxford University Press. New York.
45. PALMER, L. G. & G. FRINDT. 1987. Effects of cell Ca and pH on Na channels from rat cortical collecting tubule. Am. J. Physiol. **253**(Renal Fluid Electrolyte Physiol 22): F333-339.
46. GARTY, H. & C. ASHER. 1985. Ca^{2+}-dependent, temperature-sensitive regulation of Na^{+} channels in tight epithelia. J. Biol. Chem. **260:** 8330-8335.
47. CHASE, H. S. & Q. AL-AWQATI. 1983. Calcium reduces the sodium permeability of luminal membrane vesicles from toad bladder. Studies using a fast reaction apparatus. J. Gen. Physiol. **81:** 643-666.
48. LING, B. N. & D. C. EATON. 1989. Effects of luminal Na^{+} on single Na^{+} channels in A6 cells, a regulatory role for protein kinase C. Am. J. Physiol. **256**(Renal Fluid Electrolyte Physiol. 25): F1094-F1103.
49. BORLE, A. B. C. J. BORLE, P. DOBRANSKY, A. M. GORECKA-TISERA, C. BENDER & K. SWAIN. 1990. Effects of low extracellular Ca^{2+} on cytosolic free Ca^{2+}, Na^{+}, and pH of MDCK cells. Am. Physiol. **259**(Cell Physiol. 28): C19-C25.

Calcium Extrusion by the Sodium-Calcium Exchanger of the Human Platelet[a]

DUNCAN H. HAYNES, PETER A. VALANT, AND PHILIP N. ADJEI

Department of Molecular and Cellular Pharmacology
University of Miami School of Medicine
Miami, Florida 33101

INTRODUCTION

The human platelet plays an essential role in the control of bleeding. During vascular damage, interaction of receptors on the platelet surface with collagen, thrombin, or ADP triggers Ca^{2+} influx and Ca^{2+} release from intracellular stores (dense tubules, DT). The resulting increase in cytoplasmic Ca^{2+} activity ($[Ca^{2+}]_{cyt}$) leads to exocytotic release of activating factors, shape change, and "stickiness" resulting in platelet aggregation. In contrast to the majority of excitable cells discussed at this symposium, the platelet does not undergo cycles of activation and quiescence. Cell activation is a terminal, one-time event in the life of a platelet, since the processes of release of exocytotic granules, shape change, and appearance of "stickiness" are essentially irreversible. Therefore, the platelet must remain correctly quiescent until a proper stimulus is presented. Platelet Ca^{2+} must be tightly regulated to avoid spurious activation and intravascular coagulation. The resting $[Ca^{2+}]_{cyt}$ = 100 nM.[1] This is accomplished by competition between the extrusion and passive leakage processes depicted in FIGURE 1.

The plasma membrane of the platelet (PM) contains a Cd^{2+}-sensitive, verapamil-insensitive channel that allows passive inward leakage of Ca^{2+}.[2] This is opposed by an extrusion system consisting of a Ca^{2+}-ATPase pump and a Na^+-Ca^{2+} exchanger.[3] The experimentation characterizing the pump and exchanger was based on a protocol characterizing unopposed Ca^{2+} extrusion from platelets overloaded with the high-affinity fluorescent Ca^{2+} indicator, quin-2. The Ca^{2+} dependence of the rate of extrusion ($V_{extrusion}$) was shown to follow the following equation:

$$V_{extrusion} = \frac{V_{max,1} * [Ca^{2+}]_{cyt}^{1.7}}{(K_{m,1}^{1.7} + [Ca^{2+}]_{cyt}^{1.7})} + k_{linear} * [Ca^{2+}]_{cyt} \qquad (1)$$

where $[Ca^{2+}]_{cyt}$ is the concentration of free Ca^{2+} in the cytoplasm. The first term describes the action of a Ca^{2+}-ATPase extrusion pump. Our *in situ* analysis[3] showed it has a K_m = 80 ± 10 nM, a V_{max} = 98 ± 8 μM/min and a 1.7-power dependence on $[Ca^{2+}]_{cyt}$. Thus, the pump makes its major contribution for $[Ca^{2+}]_{cyt} \leq 400$ nM. At $[Ca^{2+}]_{cyt}$ = 110 nM the pump is continuously working at 63% of its V_{max}. The

[a] This work was supported by United States Public Health Service Grants GM 23990 and HL 382288 and by a grant from the American Heart Association, Florida Affiliate.

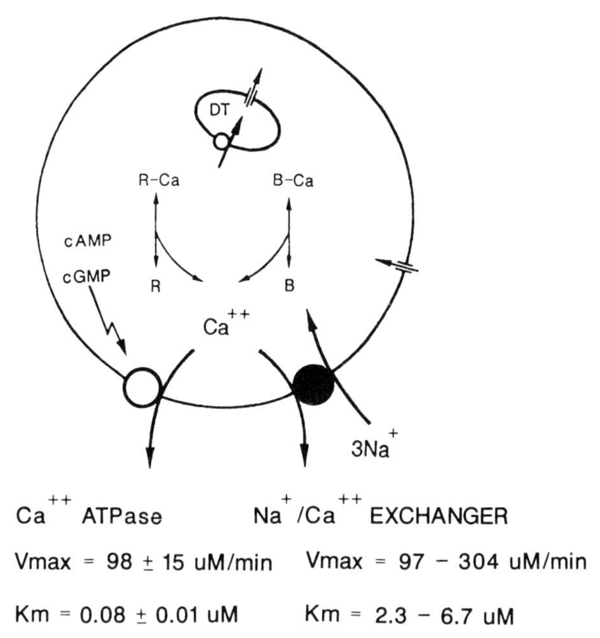

FIGURE 1. A schematic showing the two major processes responsible for the extrusion of Ca^{2+} across the plasma membrane, the Ca^{2+}-ATPase and the Na^+-Ca^{2+} exchanger. The activity of these is opposed by the inward passive leakage of Ca^{2+} across the plasmalemma. Both cAMP and cGMP stimulate the Ca^{2+}-ATPase.[10] Within the platelet the dense tubules (DT) are able to sequester cytoplasmic Ca^{2+} via a dense tubular Ca^{2+} pump. The activity of the latter is opposed by passive Ca^{2+} leakage from the dense tubules. The letter B denotes intrinsic cytoplasmic Ca^{2+}-binding sites, while R denotes rhod-2, the Ca^{2+}-sensitive fluorescent probe used to measure the cytoplasmic Ca^{2+} activity.

second term is due to the Na^+-Ca^{2+} exchanger, which makes increasing contributions as $[Ca^{2+}]_{cyt}$ is raised above 400 nM. This component was identified as a Na^+-Ca^{2+} exchanger by the Na^+ dependence of the kinetics of Ca^{2+} extrusion in the $[Ca^{2+}]_{cyt} \geq$ 400 nM range.[3] In the $[Ca^{2+}]_{cyt}$ range accessible by quin-2 (≤ 1.5 μM), it appeared to have a linear dependence on $[Ca^{2+}]_{cyt}$. However, further characterization of the Na^+-Ca^{2+} exchanger at higher values of $[Ca^{2+}]_{cyt}$ was limited by the high affinity of quin-2 for Ca^{2+} (K_d = 115 nM).

The earliest evidence for the presence of a Na^+-Ca^{2+} exchanger in the human platelet was provided by Rengasamy *et al.*[4] These investigators showed that plasma membrane vesicles isolated from human platelets exhibited saturable $^{45}Ca^{2+}$ uptake in exchange for intravesicular Na^+. The K_m for Ca^{2+} was 22 μM. They also demonstrated Na^+ dependence of Ca^{2+} efflux from Ca^{2+}-loaded vesicles and electrogenicity of the Na^+-Ca^{2+} exchange process. This was followed by our observation of a Na^+-dependent component of Ca^{2+} extrusion[3] as described in the preceding paragraph. Further evidence for an exchanger in intact platelets came from Schaeffer and Blaustein[5] who showed that substitution of extracellular Na^+ with iso-osmolar sucrose induced a rapid increase in $[Ca^{2+}]_{cyt}$ suggestive of a Na^+-Ca^{2+} exchanger working in reverse. A similar finding was made in our laboratory: Addition of monensin, an ionophore that allows

Na^+ entry into the cell by the mechanism of Na^+-H^+ exchange, increases resting $[Ca^{2+}]_{cyt}$.[6]

The present communication describes more recent *in situ* characterization of the Na^+-Ca^{2+} exchanger via the use of rhod-2, a fluorescent Ca^{2+} indicator. This indicator has a fivefold lower affinity (K_d = 500 nM) for Ca^{2+} than quin-2, thus making Ca^{2+} extrusion processes at higher $[Ca^{2+}]_{cyt}$ more experimentally accessible. As a result, we have been able to characterize the kinetics of the Na^+-Ca^{2+} exchanger *in situ* more accurately, and show that it also has saturation kinetics.

MATERIALS AND METHODS

Chemicals

All chemicals used were reagent grade.

Platelet Isolation

Platelets were isolated as previously described[3] with the following modification: The 4-(2-hydroxyethyl)-1-piperazine ethanesulfonic acid (HEPES) concentration of the isolation medium and of all other media was increased from 2.5 to 25 mM to improve their pH-buffering capacity.

Rhod-2 Loading and Fluorometry

Washed suspensions of 2×10^8 platelets/ml were loaded with rhod-2/AM for 90 minutes at room temperature. The usual loading concentration of rhod-2/AM was 12 μM. The platelet suspensions were then centrifuged at $400 \times g$ and the pellets resuspended in an aliquot of Na^+ Tyrode at pH 7.35. The suspension was stored in the dark at room temperature and its platelet concentration determined turbidimetrically with periodic calibration against a Coulter counter.[3] Aliquots (40-100 μl) of the suspension were then introduced into various media after pre-equilibration of the media to 37°C. The final volume was 2.0 ml yielding a final concentration of 1.6×10^7 platelets/ml.

The instruments and techniques used in fluorescence measurements have been previously described.[3] For measurements of rhod-2 fluorescence, the excitation wavelength was 553 nm and emission wavelength was 576 nm, with excitation and emission slits set at a width of 12 nm.

Rhod-2 as an Indicator of Free Ca^{2+}

To characterize the behavior of rhod-2, as commercially available, titrations of the dye with increasing levels of free Ca^{2+} were performed in the presence of 2 mM EGTA. The unesterified form of the dye shows two affinities for Ca^{2+} and shows two or more peaks on HPLC. Calcium titrations of lysates of rhod-2-loaded cells also show two affinities (high: K_d = 474 ± 119 nM and low: K_d = 0.509 ± 0.146 mM), with the fluorescence amplitudes in a ratio of 30 : 70. We are in the process of characterizing the dye more closely.

The relationship between the fluorescence of the rhod-2-containing mixture and Ca^{2+} activity can be described by the following equation:

$$F = F_{min} + \left((F_{max} - F_{min}) * \left(\frac{X_1 * [Ca^{2+}]}{(K_{d,1} + [Ca^{2+}])} + \frac{(1 - X_1) * [Ca^{2+}]}{(K_{d,2} + [Ca^{2+}])} \right) \right) \quad (2)$$

where F is the observed fluorescence, F_{min} is the minimal fluorescence observed in the absence of Ca^{2+}, F_{max} is the maximal fluorescence observed when the dye is saturated with Ca^{2+}, X_1 is fluorescence contributed by the high-affinity form of the dye, $K_{d,1}$ is its dissociation constant, and $(1 - X_1)$ and $K_{d,2}$ are the corresponding values for the low-affinity form.

The activity of cytoplasmic Ca^{2+} ($[Ca^{2+}]_{cyt}$) could not be determined by the simple procedure given for quin-2 by Tsien *et al.*[7] First, the relatively high rates of leakage necessitated corrections for the amount of dye in the external medium. This was determined from the instantaneous changes in fluorescence resulting from the addition of saturating Ca^{2+} (2 mM) when external Ca^{2+} ($[Ca^{2+}]_o$) = 0 or from the addition of EGTA when $[Ca^{2+}]_o$ = 2.0 mM. Second, the two affinities of the dye necessitated the use of Eq. 2 for both the cytoplasmic and extracellular compartment. In practice we have found that $[Ca^{2+}]_{cyt}$ never reached levels sufficient for the low-affinity form of the dye to contribute to the measured fluorescence. Thus, under our conditions only the first term in Eq. 2 is pertinent. The value of $[Ca^{2+}]_{cyt}$ can be calculated from

$$[Ca^{2+}]_{cyt} = K_{d,1} * [\alpha/(1 - \alpha)] \quad (3)$$

where α is the degree of saturation of the high-affinity form and is given by:

$$\alpha = (F - F_{min})/(X_1 * (F_{max} - F_{min})) \quad (4)$$

using Eq. 2 and the values of its constants given earlier. In all determinations, the K_d of rhod-2 for Ca^{2+} was assumed to be 500 nM.

Protocol for Measuring the Rate of Ca^{2+} Extrusion (Method I)

This protocol is based on method developed with quin-2.[3] In the latter, $[Ca^{2+}]_{cyt}$ was increased to 1.5 μM using the Ca^{2+} ionophore ionomycin, followed by removal of external Ca^{2+} using EGTA. This experimentation yielded progress curves for the decrease in quin-2 fluorescence corresponding to active extrusion of Ca^{2+} from the cytoplasm. FIGURE 2 shows application of the technique to rhod-2. Following the addition of 0.1 mM EGTA to chelate contaminating ambient Ca^{2+}, 2 mM Ca^{2+} is added. After a period that allows $[Ca^{2+}]_{cyt}$ to reach its steady-state value of 100 nM (state A), 1.25 μM ionomycin is added to flood the cytoplasm with Ca^{2+}. Ionomycin catalyzes both Ca^{2+} influx from the external medium and Ca^{2+} leakage from DT (state B). This concentration of ionomycin has been shown to be sufficient to short-circuit active dense tubular uptake of cytoplasmic Ca^{2+}.[3] When the high-affinity form of the dye approaches saturation at $[Ca^{2+}]_{cyt} \simeq 10$ μM, sufficient EGTA is added to lower $[Ca^{2+}]_o$ to 10 μM. This halts further influx and allows the extrusion processes to proceed essentially unopposed (state C). The value of $[Ca^{2+}]_o$ = 10 μM was chosen to (a) eliminate the extracellular contribution of the low-affinity form and (b) to guarantee saturation of the extracellular high-affinity form. The latter is of particular advantage since it ensures that any leakage that might occur during the Ca^{2+}-extrusion process will not be counted as Ca^{2+} extrusion.

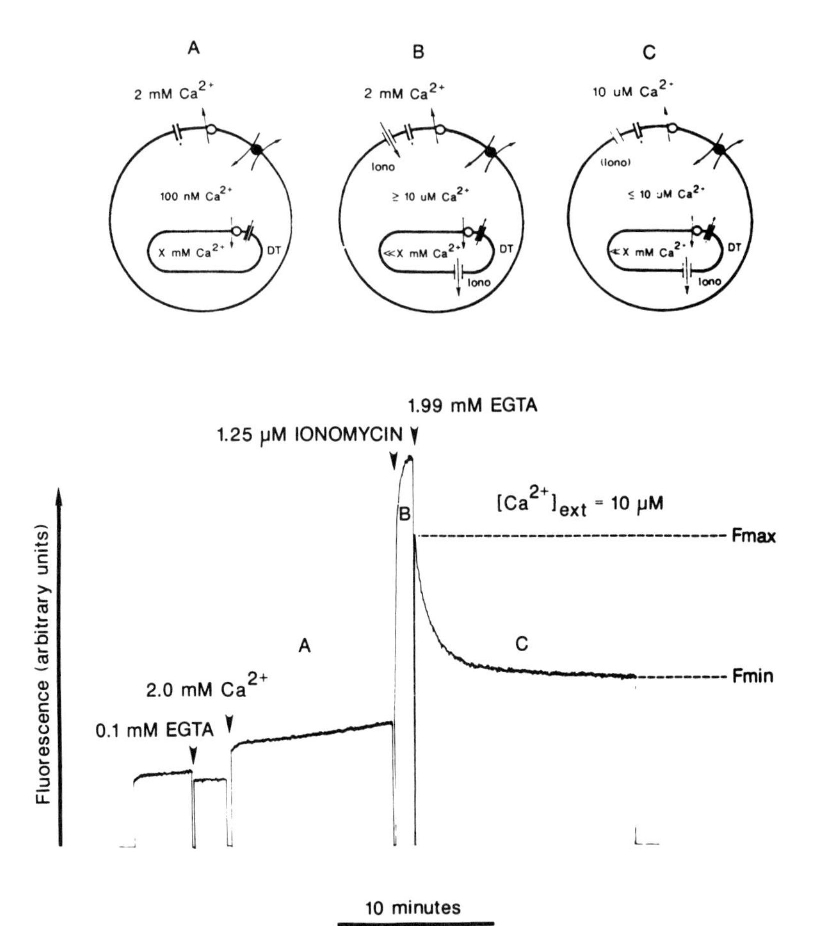

FIGURE 2. Example of the protocol used for measuring rates of active net Ca^{2+} extrusion (Method I). The composition of the Na^+ Tyrode used for platelet isolation, dye preincubation, and short-term storage had the following composition: 135 mM NaCl, 2.7 mM KCl, 0.36 mM NaH_2PO_4, 25 mM HEPES, 11.9 mM $NaHCO_3$ and 10 mM D-glucose. For purposes of experimentation, $NaHCO_3$ was omitted from the medium and replaced by equimolar NaCl. Stock solutions of rhod-2/AM were prepared in DMSO. Stock solutions of ionomycin were prepared in ethanol.

The platelets had been preincubated with 12 μM rhod-2/AM. The experiment was performed in the presence of external Na^+, as shown above, or in its absence in other experiments. Upon the addition of EGTA, the instantaneous drop in fluorescence reflects external rhod-2 that has leaked from the platelets. The slopes of the progress curve of decreasing fluorescence were determined for 0.1, 0.25, 0.5, 1.0, 2.5, 5.0, and 7.5 μM cytoplasmic Ca^{2+}. The absolute rates of fluorescence decrease were converted into mmol Ca^{2+}/liter cell volume per minute by multiplication by $[\text{rhod-2}]_{cyt}/(F_{max} - F_{min})_{cyt}$ using the high-affinity signal only and assuming quantum yields of both forms of the dye to be equal. Abbreviations are defined as follows: Iono, ionomycin; DT, dense tubules.

Method I allows determination of $V_{extrusion}$ as a function of $[Ca^{2+}]_{cyt}$ and thus provides a means for testing Eq. 1. The slopes of the progress curve for Ca^{2+} extrusion are measured at points corresponding to various values of $[Ca^{2+}]_{cyt}$ determined by Eq. 4, and plots of absolutes rates of decreasing fluorescence ($-dF/dt$) versus $[Ca^{2+}]_{cyt}$ are constructed. The rates are then normalized relative to ($F_{max} - F_{min}$) and the product of the normalized rates and cellular dye concentration allows expression of the absolute rates in terms of mmol Ca^{2+}/liter cell volume per minute. A platelet volume of 10 femtoliters is used in the calculation.

During the extrusion process, intrinsic cytoplasmic binding sites will compete with the rhod-2-Ca^{2+} complexes as a source of Ca^{2+} ions destined for export. This can result in underestimation of the true rate of Ca^{2+} extrusion.[3] The higher the degree of loading with indicator dye, the higher will be the probability that the exported Ca^{2+} is released from the Ca^{2+}-indicator complex rather than from an intrinsic binding site; as a result the measured efflux more closely approximates the actual rate of extrusion. Presently, even at the highest degrees of loading of the high-affinity dye (0.6 mM), the dye concentration was insufficient to outweigh the cytoplasmic binding sites. As a result it was not possible to determine to what extent the rates at high values of $[Ca^{2+}]_{cyt}$ were underestimated. However, it was possible to estimate a lower limit for the concentration of low-affinity intrinsic Ca^{2+}-binding sites: 1-2 mmol per liter cell volume.

In addition there is some uncertainty about the actual values of cytoplasmic rhod-2. As discussed above, the dye is a mixture of at least two forms with fluorescence amplitude ratios of 30 : 70 between the high-affinity and low-affinity forms. In the present experimentation, conditions are designed to elicit the behavior of the high-affinity form only (cf. FIG. 2). Present calculations of absolute rates of Ca^{2+} movement across the PM are based on the assumption that the molar ratio of high-affinity to low-affinity dye is also 30 : 70. Thus the *absolute* values of V_{max} for the exchanger may have to be adjusted somewhat if subsequent experimentation shows that the two forms of the indicator have different quantum yields.

Ionomycin Short-Circuit Protocol (Method II)

The main advantage of this method is that the form of the Ca^{2+} dependence of extrusion rates that one obtains is not influenced by the presence of intrinsic cytoplasmic binding sites. The method is illustrated in FIGURE 3. Small concentrations of ionomycin are insufficient to overwhelm the pump and exchanger. When low concentrations of ionomycin are added, a steady state is set up between the processes of passive influx and active extrusion ($V_{influx} = V_{extrusion}$). At steady state $[Ca^{2+}]_{cyt}$ becomes time invariant (cf. FIG. 3). The value of V_{influx} ($= V_{extrusion}$) is given by[b]:

$$V_{influx} = k_{leak} * [Ca^{2+}]_o + k_{iono} * [Iono] * [Ca^{2+}]_o \quad (5)$$

Thus under steady-state conditions $V_{extrusion}$ ($= V_{influx}$) is a linear function of the ionomycin concentration. The proportionality constant k_{iono} is determined by measuring V_{influx} at ionomycin concentrations sufficiently high that the extrusion system is overwhelmed and a steady state is *not* achieved. A knowledge of k_{iono} in *absolute* terms

[b] The Ca^{2+} dependence of the rate of ionomycin-mediated Ca^{2+} influx displays saturation kinetics with a K_m for Ca^{2+} of 7.7 mM.[3] However, when $[Ca^{2+}]_o$ is maintained constant at 2 mM, the dependence becomes a linear function of $[Ca^{2+}]_o$, as shown above.

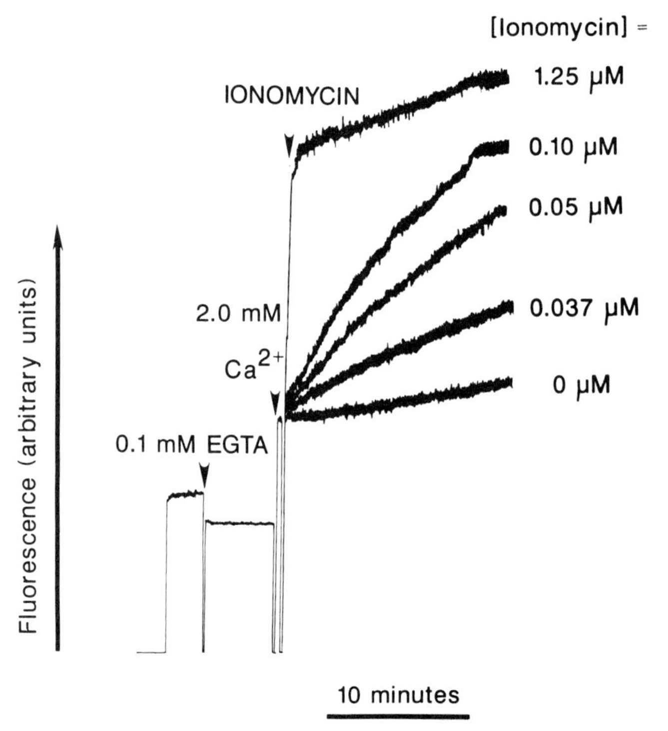

FIGURE 3. Example of the protocol for used in the ionomycin short-circuit method (Method II). The platelets had been preincubated with 12 μM rhod-2/AM. The experiments were performed in the presence of external Na^+ (in HCO_3^--free medium; see legend for FIG. 2 for composition). The initial absolute rates of fluorescence increase were converted into mmol Ca^{2+}/liter cell volume per minute as explained in the legend for FIGURE 2.

enables us to calculate the ionomycin contribution to V_{influx} in absolute terms (μmol Ca^{2+}/liter cell volume per min). The value of k_{leak} is known independently.[3] Thus we are able to determine $V_{extrusion}$ vs. $[Ca^{2+}]_{cyt}$ in steady state and test the dependence given in Eq. 1.

RESULTS

FIGURE 4 shows the Ca^{2+} dependence of the rate of Ca^{2+} extrusion in both the presence and absence of external Na^+. The experimentation was performed using the Ca^{2+}-extrusion protocol (Method I). By this method rates are obtained from the slopes of the progress curve for Ca^{2+} extrusion. The figure shows a large Na^+ dependence of the rates for the 0.25-7.5 μM range. At $[Ca^{2+}]_{cyt}$ = 7.5 μM the measured rates of extrusion in the presence of Na^+ are about twice as large as those observed in its absence (substitution with K^+, choline$^+$, and *N*-methyl-D-glucamine$^+$). The difference is due to Na^+-Ca^{2+} exchange. The measured rates in the absence of Na^+ and in the presence of Na^+ with $[Ca^{2+}]_{cyt} \leq 0.3$ μM are in agreement with values obtained for

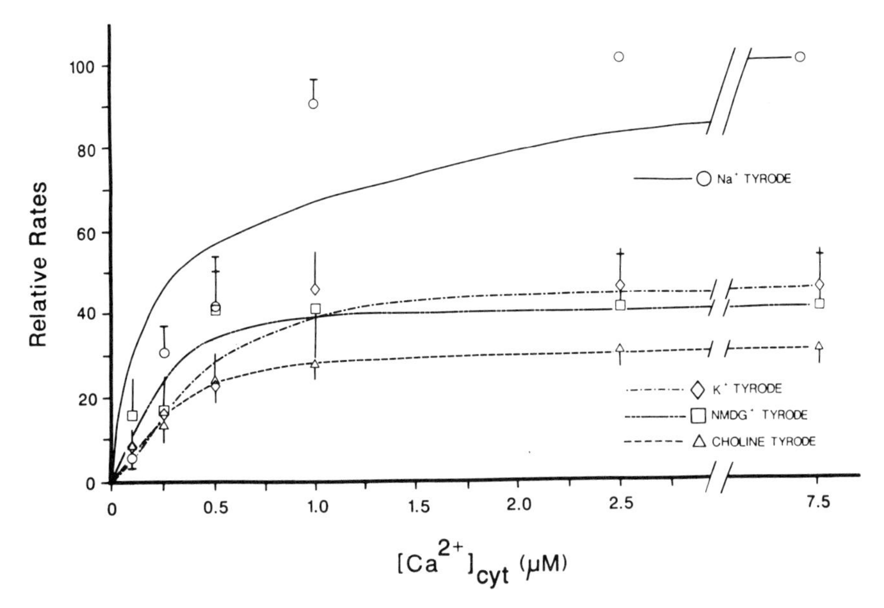

FIGURE 4. Dependence of the rate of Ca^{2+} extrusion on $[Ca^{2+}]_{cyt}$ in the presence and absence of external Na^+. All media were HCO_3^--free (see legend for FIG. 2 for composition). Na^+-free media were prepared by iso-osmolar substitution of KCl, choline chloride, or *N*-methyl-D-glucamine (NMDG) for NaCl. For each platelet preparation, the absolute rates of fluorescence decrease at various values of $[Ca^{2+}]_{cyt}$ were normalized relative to the maximal rate observed in the presence of external Na^+ and were then averaged. The normalized rates (ordinate) have been plotted against $[Ca^{2+}]_{cyt}$ (abscissa). The data shown represent the means ± SE of three experiments. The standard deviation in 100% values of V_{max} for the curve obtained in Na^+ Tyrode was 57%.

the Ca^{2+}-ATPase pump using the quin-2 method. The data of FIGURE 4 are fit with the equation:

$$V = \frac{V_{max,1} * [Ca^{2+}]_{cyt}^{1.7}}{K_{m,1}^{1.7} + [Ca^{2+}]_{cyt}^{1.7}} + \frac{V_{max,2} * [Ca^{2+}]_{cyt}}{K_{m,2} + [Ca^{2+}]_{cyt}} \quad (6)$$

where the first term refers to the pump and the second term to the exchanger. The best fit was obtained by setting $K_{m,1} = 0.08$ μM and $V_{max,1} = 39.1\%$ and fitting for $V_{max,2}$ and $K_{m,2}$. The value of 39.1% for $V_{max,1}$ was the mean of the best fit values to the first term of the above equation using data obtained in choline⁺, NMDG⁺, and K⁺ Tyrode. The values of the fitted parameters in all the different media are summarized in TABLE 1.

The Ca^{2+} dependence presented above was confirmed in experimentation using the ionomycin short-circuit method (Method II). As described in the previous section, this technique uses varied concentrations of ionomycin to set varying values of $[Ca^{2+}]_{cyt}$ in a steady state resulting from competition between the influx and extrusion processes. FIGURE 5A shows the relationship between the steady-state $[Ca^{2+}]_{cyt}$ and the ionomycin concentration giving rise to that value. In FIGURE 5B these data are converted into a $V_{extrusion}$ versus $[Ca^{2+}]_{cyt}$ as described in the METHODS. The dependence is quite similar to that observed by the extrusion method as described in FIGURE 4. The experiment

TABLE 1. Summary of Values for Kinetic Parameters for Ca^{2+}-Extrusion Systems

Condition	Pump		Exchanger	
	$V_{max,1}$ (μM/min)	$K_{m,1}$ (μM)	$V_{max,2}$ (μM/min)	$K_{m,2}$ (μM)
		Method I		
Na^+ Tyrode	83 ± 59 (98[b])	0.08 ± 0.01	97 ± 63	2.3 ± 1.2
$Choline^+$ Tyrode	60 ± 24	0.25 ± 0.02	—	—
$NMDG^+$ Tyrode	102 ± 102	0.20 ± 0.05	—	—
K^+ Tyrode	87 ± 38	0.35 ± 0.06	—	—
		Method II		
Na^+ Tyrode	113 ± 16	0.07 ± 0.01	304 ± 38	6.7 ± 2.0

NOTE: The above parameters are defined in Eq. 6. All values are expressed as the mean ± SD (n = 3).

[b] Johansson & Haynes.[3]

was repeated and average V_{max} and K_m data are given in TABLE 1. We consider the V_{max} value obtained by this method more accurate, since the result does not require knowledge of the intrinsic buffering capacity of the cell.

Influx through the Exchanger with High $[Na^+]_{cyt}$

FIGURE 6 shows a typical experiment in which we examined the effect of monensin (an ionophore that catalyzes Na^+ for H^+ exchange) on resting $[Ca^{2+}]_{cyt}$. The experiment was performed with platelets that had been preincubated with 20 μM quin-2/AM (i.e., subjected to quin-2 overload). Following the addition of 2 mM Ca^{2+} and the ensuing rise in $[Ca^{2+}]_{cyt}$ to a new steady state, 20 μM monensin was added to the platelet suspension in the presence of external Na^+. In the particular experiment shown, this resulted in an increase in $[Ca^{2+}]_{cyt}$ from 192 nM to 362 nM. The results from several such experiments gave an average increase in $[Ca^{2+}]_{cyt}$ of 186 ± 62 nM (mean ± SD; n = 6). The above results show the activity of the Na^+-Ca^{2+} exchanger working in its reverse mode. Monensin causes an initial increase in cytoplasmic Na^+ by virtue of the large inward Na^+ gradient and its high selectivity for Na^+. In some instances the dose of monensin added is sufficient to overwhelm the Na^+-K^+ ATPase as evidenced by the failure of monensin to elicit a significantly greater response in the presence of 0.2 mM ouabain (data not shown). In other instances a rapid rise in Na^+ activity is followed by a more gradual return to basal levels monitored by the Na^+-sensitive indicator, SBFI (unpublished observations). The rise in cytoplasmic Na^+ activity causes the Na^+-Ca^{2+} exchanger to operate in reverse mode, at near-maximal rate.

It should be pointed out that in addition to enhancing cytoplasmic levels of Ca^{2+}, monensin causes a sustained increase in cytoplasmic pH of 0.3 to 0.5 units (unpublished observations). Appropriate control experiments were performed to rule out a pH effect on the affinity of quin-2 for Ca^{2+}. Preliminary experiments also indicate that cytoplasmic alkalinization resulting from the addition of 25 mM ammonium chloride neither releases dense tubular Ca^{2+} (monitored with chlortetracycline) nor increases $[Ca^{2+}]_{cyt}$.

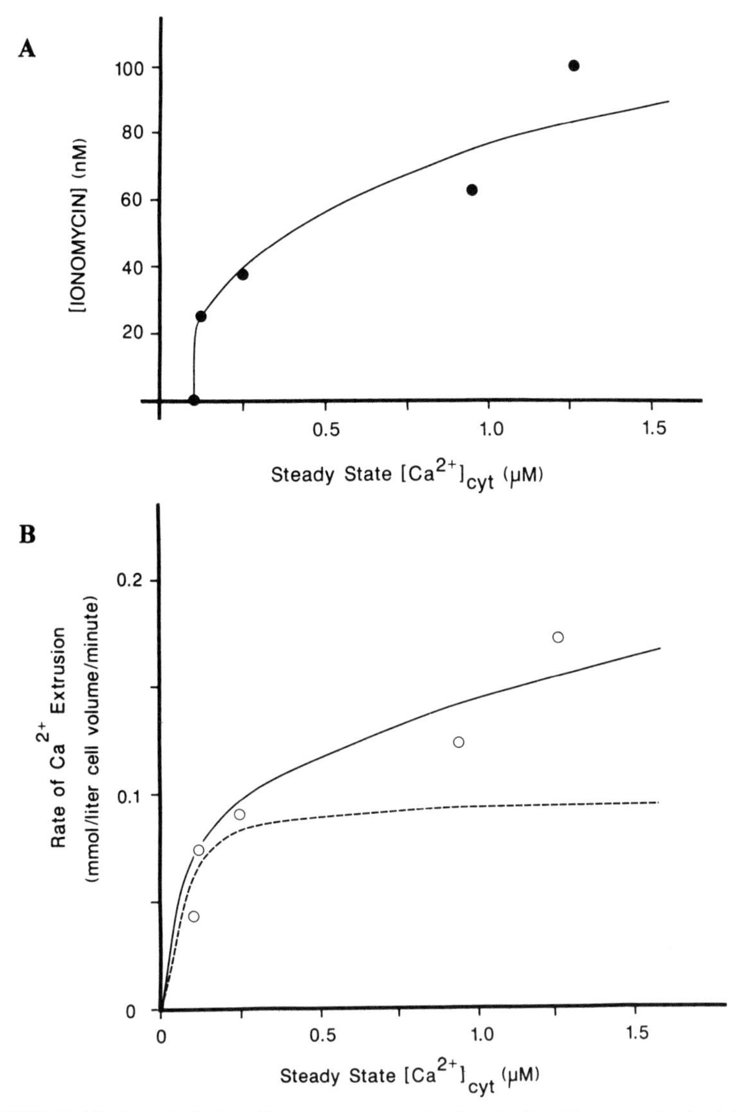

FIGURE 5. **(A)** A typical plot of ionomycin concentration (ordinate) versus steady-state levels of cytoplasmic Ca^{2+} (abscissa). The cytoplasmic Ca^{2+} activity was determined as described in the methods. **(B)** Kinetics of Ca^{2+} extrusion as a function of $[Ca^{2+}]_{cyt}$ determined by the ionomycin short-circuit method. A typical experiment done in Na^+ Tyrode is shown. The solid curve shows the rate of extrusion contributed by activity of both the Na^+-Ca^{2+} exchanger and the Ca^{2+}-ATPase. The dashed line shows the contribution of the Ca^{2+}-ATPase by itself. The latter was calculated from previously determined values of K_m and V_{max} and the first term in Eq. 1.[3] The contribution of passive inward leakage in this experiment was 0.04 mmol/liter cell volume per minute.

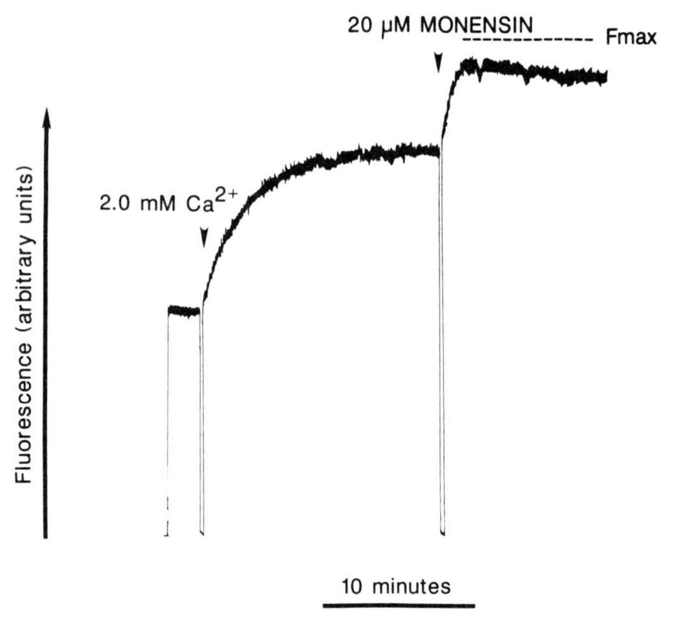

FIGURE 6. An experiment showing a large effect of monensin on resting levels of $[Ca^{2+}]_{cyt}$ in quin-2-overloaded platelets. Platelets that had been preincubated with 20 μM quin-2 (quin-2 overload) were resuspended in Na^+ Tyrode at pH 7.4 and 2 mM Ca^{2+} was added. At steady state (resting $[Ca^{2+}]_{cyt}$), 20 μM monensin was added and the maximal increase in fluorescence was measured. Ionomycin (1 μM) was finally added to obtain F_{max}. F_{min} was obtained in parallel experiments in which the platelets were lysed with 40 μM digitonin in the presence of 2 mM EGTA. In the experiment shown, $[Ca^{2+}]_{cyt}$ rose from 192 to 362 nM.

DISCUSSION

The present data improve the quantitation of kinetics of the Na^+-Ca^{2+} exchanger in platelets. The use of rhod-2, although presenting more technical difficulties than quin-2, has allowed us to measure Ca^{2+} extrusion in a fivefold higher $[Ca^{2+}]_{cyt}$ range. At $[Ca^{2+}]_{cyt} = 7.5$ μM, the contribution of the exchanger to Ca^{2+} removal is 1-1.5-fold as large as that of the pump. We have shown that the exchanger has a K_m of 2.3-6.7 μM and a V_{max} of 97-304 μmol/liter cell volume per minute. We believe the higher value of V_{max} (obtained by Method II) to be more reliable. When the exchanger operates in its reverse mode, the resulting rise in $[Ca^{2+}]_{cyt}$ is to 300 nM. This indicates that rate of the exchanger in reverse corresponds to the pump working at 90% of V_{max}. The reverse operation of the exchanger may prove important to activation with certain agonists. The discussion below will consider the relative contribution of the pump and exchanger to the total Ca^{2+}-extrusion process.

Both our data and Eq. 6 predict that under resting conditions the pump will make the major contribution to Ca^{2+} extrusion. In a previous communication,[3] we showed that in the resting condition the pump works at 63% of its V_{max}. Expressed differently, the value of $k_{leak}*[Ca^{2+}]_o = 0.63\ V_{max,pump}$. We suggest that this circumstance is not an accident, but rather a design of nature with the purpose of controlling $[Ca^{2+}]_{cyt}$ within narrow limits (approx. 110 nM). The energetic cost of this system of regulation is low, corresponding to the cost of exporting 62 μmol Ca^{2+}/liter cytoplasm per min. This would correspond to an ATP expenditure of approximately 1% of its cytoplasmic concentration per minute. The functional cost of loss of regulation is high: We have shown that increases of platelet $[Ca^{2+}]_{cyt}$ by 61%, as seen in arterial thrombosis, are associated with elevated dense tubular Ca^{2+}, lowered activation thresholds, and increased rates of aggregation.[8,9] Normalization of platelet $[Ca^{2+}]_{cyt}$ and dense tubular Ca^{2+} by treatment with Ca^{2+}-channel-blocking drugs is accompanied by normalization of platelet function.

It is reasonable to ask why the platelet contains a Na^+-Ca^{2+} exchanger if Ca^{2+}-ATPase pump activity is sufficient to maintain resting $[Ca^{2+}]_{cyt}$ values. We speculate that the exchanger serves as an additional check against spurious activation due to Ca^{2+} influx through transient breaks in the plasma membrane and other random stimuli experienced during the platelet's short lifetime in the circulation. The exchanger lends the platelet the ability to clear Ca^{2+} more rapidly. Although additional removal capacity could just as easily be obtained by doubling the number of copies of the Ca^{2+}-ATPase pump, this would require an increase in the rate of the passive leak to maintain proper resting $[Ca^{2+}]_{cyt}$. The Na^+-Ca^{2+} exchanger can provide insurance against high $[Ca^{2+}]_{cyt}$ without making a substantial contribution to the rate of Ca^{2+} removal in the resting state. This is due to its much larger K_m and first-power dependence on $[Ca^{2+}]_{cyt}$. On the other hand, the Ca^{2+}-ATPase pump has a K_m close to the $[Ca^{2+}]_{cyt}$ value that it is responsible for maintaining, and has a 1.7-power dependence on $[Ca^{2+}]_{cyt}$. This guarantees that the Ca^{2+} pump is capable of a very dynamic compensatory response to small changes $[Ca^{2+}]_{cyt}$. This compensatory response is subject to further regulation, as evidenced by stimulation of the Ca^{2+}-ATPase by increased cytoplasmic levels of cAMP and cGMP, which raises its V_{max}.[10]

We believe that a second function of the Na^+-Ca^{2+} exchanger is to provide an active counter-challenge to agonist-triggered Ca^{2+} influx and dense tubular Ca^{2+} release processes. If these are adequate and sustained, both the pump and the exchanger will be overwhelmed and the platelet will be irrevocably committed to aggregate. Thus the concerted action of the pump and exchanger is an important determinant of whether a marginal stimulus will be sufficient to cause platelets to aggregate.

REFERENCES

1. Rink, T. J., S. W. Smith & R. Y. Tsien. 1982. FEBS Lett. **148:** 21-26.
2. Jy, W. & D. H. Haynes. 1987. Biochim. Biophys. Acta **929:** 88-102.
3. Johansson, J. S. & D. H. Haynes. 1988. J. Membr. Biol. **104:** 147-163.
4. Rengasamy, A., S. Soura & H. Feinberg. 1987. Thromb. Haemostas. **57:** 337-340.
5. Schaeffer, J. & M. P. Blaustein. 1989. Cell Calc. **10:** 101-113.
6. Johansson, J. S. 1990. Thesis dissertation, University of Miami, Miami, FL.
7. Tsien, R. Y., T. Pozzan & T. J. Rink. 1982. J. Cell Biol. **94:** 325-334.
8. Jy, W., Y. S. Ahn, N. Shanbaky, L. F. Fernandez, W. J. Harrington & D. H. Haynes. 1987. Circ. Res. **60:** 346-355.
9. Shanbaky, N. M., Y. Ahn, W. Jy, W. Harrington, L. Fernandez & D. H. Haynes. 1987. Thromb. Haemostas. **57:** 1-10.
10. Johansson, J., J. Tao, W. Jy & D. H. Haynes. 1991. Biophys. J. **59:** 336a.

Kinetic Models of Na-Ca Exchange in Ferret Red Blood Cells

Interaction of Intracellular Na, Extracellular Ca, Cd, and Mn[a]

M. A. MILANICK [b] AND M. D. S. FRAME [c]

University of Missouri
Department of Physiology
School of Medicine
University of Missouri-Columbia
Columbia, Missouri 65212

INTRODUCTION

The ferret red cell provides an interesting mammalian model system for the study of the kinetics of Na-Ca exchange. In this paper we present some of the unusual features of the ferret red cell and the unusual role Na-Ca exchange plays in this cell type. A background discussion on the kinetic models (ping-pong vs. sequential) and on the use of alternative substrates is presented. This is followed by a summary of our findings in ferret red cells: The Na-Ca exchanger follows sequential kinetics, and Mn and Cd are alternative substrates for Ca on this exchanger. Finally the implications of these results are discussed in terms of a model we find intriguing and plausible: that the reaction sequence of the exchanger involved Ca binding, translocation, and then the binding of at least 1 Na ion, followed by Ca dissociation, the binding of the remaining Na ions, and translocation and dissociation of the Na ions to the other side.

FERRET RED CELLS

The primary function of the Na-Ca in ferret red blood cells is different from that described in other cell types.[1] In most other cell types, the primary role of the Na-Ca exchanger is to regulate intracellular Ca levels.[2-4] Net Ca fluxes on the Na-Ca exchanger result in net Na movements; it is the responsibility of the Na pump to control the intracellular Na levels. Changes in intracellular Na often lead to changes of H^+ through the action of Na-H^+ or Na/HCO_3^- exchangers. These changes in intracellular Na and H^+ are thought to be of less importance in controlling cell function than changes of cell Ca.

[a] This work was supported by grants from the American Heart Association, Missouri Affiliate, and from the National Institutes of Health, DK37512.

[b] Address for correspondence: Department of Physiology, MA415 Medical Science Building, University of Missouri-Columbia, Columbia, MO 65212.

[c] Current address: Department of Biophysics, 601 Elmwood Avenue, University of Rochester Medical Center, Rochester, NY 14642.

In contrast, in ferret red cells, the Na-Ca exchanger functions to regulate intracellular Na levels.[1] It is the responsibility of the Ca pump to maintain low Ca levels. The V_{max} of the Ca pump in ferret red cells is probably 10 times the V_{max} of the Na-Ca exchanger.

In ferret red cells, the physiological operation of Na-Ca exchange is Ca influx-Na efflux, the so-called "reverse" model. This is because the driving forces for Na-Ca exchange are "reversed" in ferret red cells. The intracellular Na concentration of ferret red cells is unusual compared to other cells, being approximately 150 mmoles/liter cell water.[5] The membrane potential of human, and presumably ferret, red cells is −10 mV. The membrane potential predicts a Nernstian concentration of 220 mmol Na/liter cell water, thus there is a small electrochemical gradient for Na entry. In contrast, there is a large electrochemical gradient for Ca entry, similar to that found in most cells. Intracellular free Ca is on the order of 100 nM,[6] extracellular free Ca is slightly greater than 1 mM. So the Ca gradient drives net Na efflux. This net Na efflux is critical to maintain normal cell volume since these cells have no Na-K pump. Presumably, in the absence of Ca, the cells would gain Na and lyse. This has not been directly tested in ferret red cells, but Parker has shown that extracellular Ca does regulate cell volume in dog red cells.[7] Dog red cells are similar to ferret red cells in that both lack a Na-K pump and have Na-Ca exchange.[8] Red cells of all species of animals in the order *Carnivora* probably have high-Na red cells; the high Na content has been measured in bear, cat, and seal (Willis *et al.*[9] and J. F. Hoffman, unpublished) as well as dog and ferret.

For the analysis of kinetics, Na-Ca exchange activity is conveniently measured as Ca influx from a Na-free solution into cells in which Ca efflux on the Ca pump is prevented by preincubation of the cells in vanadate. Ferret red cells offer some unusual advantages as a model system for the study of Na-Ca exchange. A large quantity of uniform cells is available. The intracellular volume is considerably larger than for vesicles so the intracellular ion contents can be maintained for relatively long times. The cell volume is easily measured. The absence of cellular organelles, of Na-K pumps, and Ca channels reduces the number of potential complications. The inability of direct electrical measurements is a major disadvantage.

KINETICS

Ping-pong versus Sequential

Before discussing the different kinetic models, we will review our motivation for determining the kinetic mechanism. (1) The kinetic mechanism can often be used to place physical constraints on the model. For example, the anion transport system of red blood cells exchanges Cl for Cl in a 1 : 1 stoichiometry following ping-pong kinetics. This implies (see below) that there is only one Cl transport site and thus, in the simple case, there is no conformation of the protein in which two Cl ions are bound (at least to the transport site). (2) Knowledge of the kinetic mechanism allows one to model the response of a cell to stimulus. Detailed models of cardiac function require knowledge of how the exchanger responds to changes of intracellular Ca and membrane potential (4,10). This requires a kinetic model that provides the quantitative relationship between flux, ion concentrations, and membrane potential.

The constraints necessary to develop a model of the physical mechanism and the constraints necessary to develop a model of the cell physiology may be different. For example, the physical mechanism of the Na/K pump appears to be ping-pong; that is,

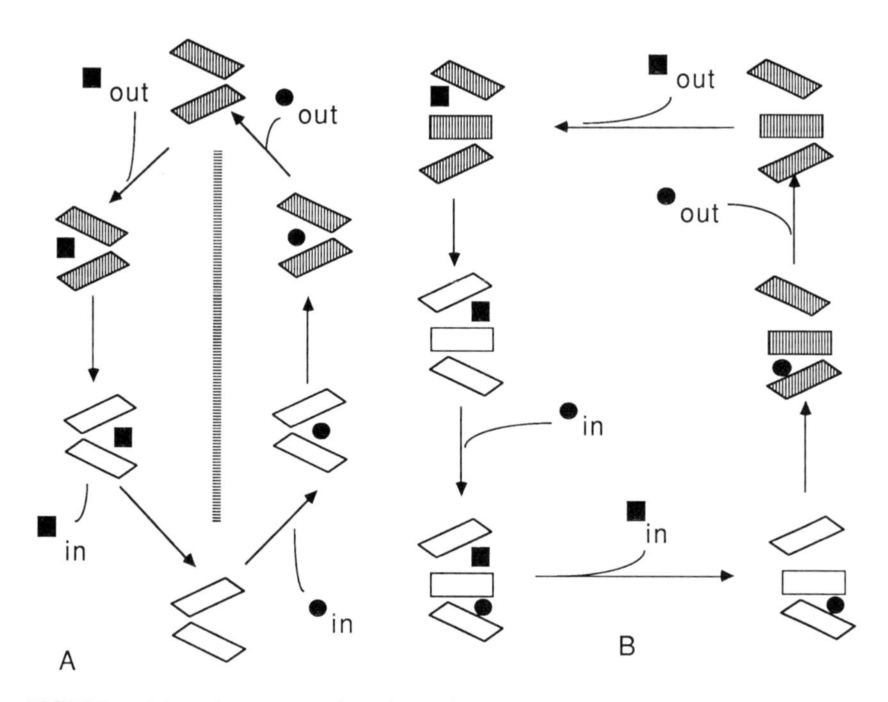

FIGURE 1. Schematic representation of two different kinetic models for the exchange of an extracellular Ca ion (square) for intracellular Na ions (circle). *Left* (A): In the ping-pong model extracellular Ca binds, is transported, and is then released to the cytosol. Then intracellular Na ions bind, are transported, and are released to the exosol. The Na steps and the Ca steps are separate as there is no conformation that has both Na ions and Ca ions bound. The dashed line in the middle represents the slippage pathway. If the rate of interconversion of the empty-inward-facing conformation to empty-outward-facing conformation is non-zero, then the exchange of Ca for Na is not completely coupled, that is, there will be some Ca influx in the absence of Na. See text for further details.

Right (B): In this sequential model, an unusual ordering of the binding and translocation steps is presented. In this model, extracellular Ca binds and is translocated. Then the intracellular Na ions bind, and this is followed by release of Ca. Then the Na ions are translocated and released to the exosol. An essential feature of all sequential models is that there is at least one conformation that has both Na ions and Ca ions bound. This conformation occurs just after intracellular Na ions bind and before Ca dissociates to the cytosol in this particular model.

there is no conformation of the pump where both Na and K are bound to transport sites at the same time. Nevertheless, the kinetic equation of the Na pump is sequential (because of slippage, or imperfect coupling, see below), and this type of equation would be more appropriate in cell physiological modeling. We will use the terms ping-pong or sequential *mechanism* when referring to the physical model and ping-pong or sequential *kinetics* when referring to the mathematical behavior.

Early mechanisms for exchange systems included two classes of models, ping-pong or consecutive mechanisms and sequential or simultaneous mechanisms.[11-14] In a ping-pong mechanism (FIG. 1) the protein binds an extracellular calcium ion, transports it to the inside and unloads the calcium. In order to undergo another cycle of Ca transport, the protein must reset. To reset, the protein binds intracellular Na ions, transports them to the outside, and unloads the Na. A significant feature of this model is that the protein alternatively binds Ca or Na and there is no intermediate in which both Na and Ca are bound to transport sites. In contrast, in a sequential mechanism the protein has at least one conformation in which both Na and Ca are bound. In an early model of a sequential mechanism, intracellular Na ions and extracellular Ca ions bound to the protein, then, following a conformational change, Na ions were released to the exosol and Ca ions to the cytosol.

The determination of the kinetic mechanism can be an attempt to determine which conformational *states* of the protein are allowed and which are forbidden. Some physical implications can be drawn from such results. Additional kinetic information is provided by determining which conformational *changes* are allowed and which are forbidden; that is, what are the possible paths between states of the protein. For example, in a fully coupled ping-pong mechanism for Na-Ca exchange, the resetting of the transport site requires either Na or Ca bound to the site. That is, after Ca unloads to the inside, the protein cannot undergo a conformational change that results in the empty transport site having access to the outside. If that conformational change were allowed, then there would be some Ca influx without Na efflux; that is, the system is not fully coupled or the system allows some slippage (dashed line in FIG. 1A). This uncoupling occurs for the Na-K pump,[15] but not for the anion exchanger.[12] Uncoupling of Na-Ca exchange is undesirable under most physiological conditions. However, the determination of a kinetic mechanism requires the study of the protein under nonphysiological conditions, particularly at low internal and external substrate concentrations. Slippage cannot, *a priori,* be excluded under these conditions. Because the conditions are nonphysiological, there would be no selective pressure against slippage. That is, there is no selection against a protein that allowed the conformational change from empty internal site to empty external site. While an uncoupled ping-pong system is conceptually similar to a ping-pong system, it is mathematically more similar to a sequential system.

Initially, on physical grounds, the sequential model was not favored because it was difficult to conceive of protein models that allowed two ions of like charge to be transported past each other within the bounds of a protein molecule. While not compelling, this argument does have a certain appeal. However, it is clear that there are classes of sequential models that avoid this difficulty. For a three Na to one Ca exchange system, it may be that before the Ca ion dissociates the three Na ions bind (FIG. 1). Indeed, the binding of Na may promote the dissociation of Ca. In other words, Na binding decreases the affinity for Ca. In this case, the enzyme still has a conformation (ECaNa) in which at least one Na ion and a Ca ion are bound to transport sites, but there is no puzzle about moving one ion past the other. This model raises a different concern. How long must the Ca be bound after the Na binds to be sequential? If the Ca ion dissociates instanteously after Na binds, one probably has ping-pong kinetics, since in this case the lifetime of the ECaNa is very short.

The problem of the lifetime of $ECaNa_{(in)}$ in distinguishing between ping-pong and sequential is the mirror image of the problem of the lifetime of $_{(out)}NaCaE$ in distinguishing between competition and noncompetition between Ca and Na ions.[1] The defining feature of inhibitors that are not competitive is that there exists a conformation that has both the substrate and the inhibitor bound; whereas for the competitive inhibitor, by definition, the binding of substrate and inhibitor are mutually exclusive. This concept is usually straightforward for single-substrate systems. For example, in the anion-exchange mechanism it is clear that Cl and iodide compete. This not only means that the protein does not have both Cl and I bound, but that the conformation that has just released the Cl very rapidly becomes the conformation that is just about to bind the I. (A quantitative analysis of how fast this rate must be is presented in Milanick.[1]) This seems reasonable because of the similarity of Cl and I. But the situation may be more complicated in the case of Na-Ca exchange because the conformation that has just finished releasing three Na ions may not be the exact same conformation that will bind the Ca. In this case, Na and Ca will compete if the conformational change from the form that binds Na to the form that binds Ca is very rapid compared to the rate of Na binding.[1]

The Kinetic Mechanism May Depend upon Transporter Isoform

While it might be assumed that a particular transport gene family would follow the same kinetics in different cell types, that does not appear to be the case. The Cl/HCO_3 exchange system of mature red cells (AE1) follows ping-pong kinetics[12]; a different, though very similar protein (AE2), follows sequential kinetics in HL-60 cells, a tumor cell line of blood cell progenitors.[16] In the study of HL-60 cells, Restrepo *et al.* have clearly shown that the sequential kinetics is due to a two-site mechanism and is not due to slippage. They used two different approaches: (i) the standard V_{max}/K_m approach and direct measurements that place upper bounds on the amount of slippage[16]; and (ii) a novel approach through the use of a competitive inhibitor to distinguish directly between ping-pong with slippage models and sequential models.[17] If two similar proteins can have different mechanisms for transport, then it is important to determine the mechanism of Na-Ca exchange in a variety of systems. Because two proteins with similar amino-acid sequences have two physically distinct mechanisms, then caution should be exercised (1) when using similarities in sequence to suggest similarities in mechanism, and (2) when suggesting that two physical models are completely distinct. As indicated above, the relative lifetime of the tertiary complex of Na, Ca, and exchanger is critical for whether the process is ping-pong or sequential. In the case of the anion exchanger, it has been suggested that a sequential mechanism becomes ping-pong if one of the ions remains tightly bound to one of the two transport sites.[12] In both of these examples, a change in one rate constant could change the mechanism from ping-pong to sequential.

Alternative Substrates as Kinetic Tools

Steady-state kinetic information about a transport system can often be augmented by study of the properties of alternative substrates in addition to the physiological substrates. Li does not appear to substitute for Na on the Na-Ca exchanger. Sr does appear to substitute for Ca with very similar properties.[18] We studied Mn and Cd to determine if they were alternative substrates and if their properties were sufficiently different to exploit previous approaches that utilized alternative substrates.

In other transport systems, alternative substrates have been used to determine which steps are altered by a particular modulation. Consider two examples: (1) Papain treatment inhibits (but does not abolish) Cl-Cl exchange in red cells. The exchanger mediates 1 : 1 exchange of Cl, and neither the rate constant for Cl efflux nor Cl influx is known. Thus papain could inhibit the efflux half-cycle or the influx half-cycle. Jennings and Adams have shown that the papain treatment inhibited the efflux half-cycle by examining the effect of papain treatment on Cl efflux/sulfate influx.[19] Sulfate transport is 1000 times slower than Cl transport at neutral pH. Thus during Cl efflux/sulfate influx, the sulfate half-cycle must be rate limiting. Under these conditions, papain did not inhibit Cl-sulfate exchange. Therefore, the inhibition due to papain treatment of Cl-Cl exchange is presumably due to a decrease of the Cl efflux half-cycle. Decreasing the Cl-efflux rate 10-fold with papain treatment would have little effect on

Cl-sulfate exchange because the sulfate-influx step would still be 100 times slower and remain the rate-limiting step. (2) Alternative substrates have also been useful in probing the partial reactions of the Na-K pump. Under some conditions, Li is an alternative substrate for K; Li transport is faster than K. K-K exchange is accelerated by ATP and by Pi; presumably ATP accelerates the influx steps and Pi the efflux steps. Most studies of steady-state K-K exchange fluxes probably cannot confirm nor exclude this model. However, studies of Li-K exchange allow one to separately examine the effects of ATP when influx is slow (Li efflux/K influx) and the effects of Pi when efflux is slow (K efflux/Li influx).[20]

The success of these studies in exploiting the properties of alternative substrates encouraged us to determine if Mn and Cd were alternative substrates for Ca in the ferret red blood cell. For example, ATP stimulates Na-Ca exchange.[21] This modulation may be direct or indirect. In either case, if we had found that Mn was transported more slowly than Ca, then we could have examined the effect of ATP on Na-Mn and Na-Ca exchange. This would have determined if the ATP stimulation were due to an increase in the rate of the divalent cation-dependent steps or the rate of the Na steps during the exchange cycle. In fact, though Cd and Mn are alternative Ca substrates, the properties of Na-Cd and Na-Mn exchange are too similar to allow this type of analysis. Nevertheless, there are other important advantages of using Cd and Mn to answer particular questions; these are discussed below.

If one step of the overall cycle is clearly slower than the rest, then that is the rate-limiting step. In steady state, the rate of transition between conformations is the same; the rate is the product of the concentration of that conformation and the rate constant of the conformational change. Thus the conformational intermediate with the highest concentration is the form just before the rate-limiting step. For example, under some conditions, the slow step in the Na-K cycle is the release of K from E_2PK; under these conditions, most of the pump is in the E_2PK conformation.[22] At low substrate levels, the empty transporter is a predominate enzyme form and the rate-limiting step is clearly substrate addition; addition of substrate increases the rate. At high substrate concentration, that is, when additional increases in substrate do not increase the overall cycle rate, the association of substrate with transporter is not rate limiting. Under these conditions, the rate-limiting step could be substrate translocation or dissociation (or perhaps association of the other substrate, if not at a saturating concentration.)

Competitive inhibitors and alternative substrates can provide information about the substrate-binding site. For example, a divalent cation-binding site that contained a sulfhydryl group would be expected to bind Cd much better than Ca, since sulfur is a better ligand for Cd than Ca. Cd is a potent inhibitor of the plasma membrane Ca pump; the K_m for Ca is more than 100 times larger than the K_i for Cd. This result supports the conclusion that the Ca pump Ca-binding site contains an important sulfhydryl group.[23] A popular notion is that the general nature of many Ca-binding sites is conserved; this would be supported if there were also an important sulfhydryl group in the Ca site of the Na-Ca exchanger.

Not only do binding of different ligands allow comparison between different types of transport proteins (Na-Ca exchanger and Ca pump), but they may provide important information about isoforms for the same transport protein. If one purifies Na-Ca exchangers from different tissues and measures transport function in a common lipid environment, the K_d values should be the same if the proteins are identical. However, when the protein is assayed in its native environment, other factors can affect the comparison of K_d values. This is particularly a problem if the local surface charge is tissue dependent. However, variations in surface charge should not affect the ratio of K_i (Cd or Mn) to K_m (Ca), since surface charge should affect Cd, Mn, and Ca similarly (see Frame & Milanick[30]). K_i and K_m are defined by the standard competitive inhibition equation, $v = V_{max} Ca/[K_m \times (1 + Mn/K_i) + Ca]$. Differences in the ratio of K_i to K_m may reflect differences in the chemical nature of the binding site; in general, however, the translocation rate may influence K_m.[12]

RESULTS

Ping-pong versus Sequential

Ping-pong and sequential mechanisms behave differently at low substrate concentrations but have similar properties at high substrate concentrations. We determined the kinetic mechanism of Na-Ca exchange in ferret red cells by a modification of the standard technique of examining the ratio of V_{max}/K_m for Ca as a function of intracellular Na. The novel part of our approach was to measure Ca uptake at very low Ca as a direct estimate of V_{max}/K_m, since the Michealis-Menten equation reduces to $v = V_{max} Ca/K_m$ when $Ca << K_m$.[24] We found that the Ca influx at low Ca (50 to 100 nM free Ca when 4 μM Ca and 17 μM EGTA was added to the flux solution) was three- to fivefold lower when the intracellular Na concentration was low compared to when the intracellular Na concentration was high. To be sure that the Ca concentration was well below K_m, the Ca influx was also determined at a high Ca concentration (170 μM). The Ca influx at 50 nM free Ca was less than 1/20th of the Ca influx at 170 μM Ca, thus the ratio: [flux (at 50 nM free Ca)/Ca] provides a good estimate of V_{max}/K_m.[24] A ping-pong mechanism predicts that the flux at very low Ca is independent of Na, and our result clearly contradicts that prediction. The reason a coupled ping-pong mechanism predicts Ca influx independent of Na is that, for any given range of intracellular Na, it is always possible to find concentrations of Ca low enough that the Ca influx half-cycle is rate-limiting for the overall cycle. Under these conditions, an increase in Na will not affect the transport rate. In other words, for this Ca concentration, Na is at a saturating concentration in this range (see Milanick[24] for a more complete intuitive explanation).

We tested whether the experimental procedures could have resulted in the Na dependence of the Ca influx. We found similar results whether the intracellular Na concentration was varied using the *p*-chloromercuribenzene sulphonate technique in intact cells or by using the ghost procedure. Na was substituted with either Li or choline, and similar results were obtained. Potential Ca backflux was prevented by the inclusion of EGTA in the ghosts. We determined that red cells did not have a submembrane space that prevents rapid equilibration of free Ca, that is, a Ca pool.[24] A Ca pool that excluded EGTA and allowed Ca backflux in low-Na cells but not in high-Na cells would have provided an alternative explanation of our results. We conclude that the Na-Ca exchanger follows sequential kinetics, because the Ca influx, at very low Ca, is Na dependent. Sequential kinetics has been observed in a number of other cell types (see, e.g., References 11, 13, 14, 25, 26), however ping-pong kinetic have been observed in some other recent studies (see, e.g., References 27, 28, and this volume). While the *kinetics* are sequential, our data are not complete enough to conclude that the *mechanism* is sequential. In particular, we cannot rule out a ping-pong mechanism with variable stoichiometry, slippage being the extreme case of variable stoichiometry, that is, 0 Na to 1 Ca.[24]

The three most straightforward approaches for distinguishing ping-pong with slippage from sequential mechanisms rely upon the use of inhibitors. (1) One could use an inhibitor, peptide, or antibody that only blocked Na-Ca exchange to determine if any Ca flux in the absence of Na was mediated by the same protein. This approach has

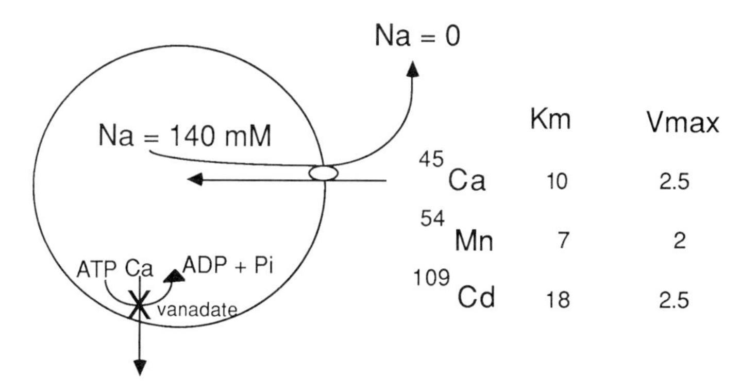

FIGURE 2. Schematic summary of the experimental conditions and kinetic parameters for Na-Ca, Na-Mn, and Na-Cd exchange in ferret red blood cells. For the measurement of Ca influx, the cells are preloaded with vanadate to inhibit Ca efflux by the Ca pump. This step is not required for Mn and Cd, since they are not transported by the Ca pump, but is often included so that direct comparisons can be made with the Ca measurements; vanadate does alter the rate of Na-Mn and Na-Cd exchange.[2,3,21,31] Representative K_m (in μM) and V_{max} values (in mmoles/liter packed cells × hour) are shown on the right. The approximate conversion factor from mmoles per liter packed cells × hour to nmoles per mg membrane protein per minute is 3. Modified from data presented in Frame and Milanick.[30]

been used to place an upper bound on the uncoupled Cl flux mediated by the anion exchanger of HL-60 cells.[16] (2) One could use a noncompetitive inhibitor and the ion dependencies to determine the mechanism as Sachs has done for the Na-K pump.[15] (3) One could also use the substrate dependence of inhibition by a competitive inhibitor to distinguish between the two mechanisms. This approach has been recently used by Restrepo and Knauf studying the anion exchanger in HL-60 cells.[17] We are currently determining which approach is most practical in ferret red cells.

Mn and Cd Fluxes in Ferret Red Blood Cells

Many properties of Mn and Cd influx are similar to Ca influx in ferret red cells. We measured the uptake of the radioisotopes 45 Ca, 54 Mn, or 109 Cd into ferret red cells as previously described[29,30] (FIG. 2). All three ions have a hyperbolic concentration dependence on the extracellular divalent cation concentration. Indeed, the K_m and V_{max} values are similar. There is some day-to-day variability in these parameters; for all three cations the range for K_m values is generally 5 to 20 μM and for V_{max} values is usually 0.5 to 2 mmoles per (liter packed cell × hour). Extracellular Na inhibits the influx of all three divalent cations. Intracellular Na stimulates the influx. Intracellular Na is normally high in ferret red cells (150 mM) so low-Na cells still have about 5 to 20 mM Na. Even this decrease in intracellular Na resulted in a 3- to 10-fold decrease in divalent cation influx. This Na dependence is consistent with Na-divalent cation exchange, though clearly we have not proven an obligatory exchange for Mn and Cd.

The best evidence that Mn and Cd share the same transport mechanism as Ca is that Mn and Cd competitively inhibit Ca influx. Ca inhibits Mn and Cd influx, though we have not yet determined if the inhibition is competitive, as predicted by our model.

Mg is a weak inhibitor of Ca, Mn, or Cd influx. Thus it seems likely that Mn is being transported by a Na-Ca mechanism and not by a Na-Mg mechanism.

One important difference between Ca influx and the influx of Mn and Cd is that the influx of Ca is only observable in ferret red cells when the Ca pump is blocked; in many studies the Ca pump was blocked by vanadate preloading the cells. In other studies, ghosts were resealed containing high concentrations of EGTA.[24] ATP depletion of ghosts was not a fruitful technique: Na-Ca exchange activity was decreased in the ghosts depleted of ATP.[21] All three of these factors, vanadate, EGTA, and ATP, are known to alter the Na-Ca exchanger in other cell types.[2,3,31] In the case of ATP in ferret red cells, we cannot rule out the possibility that the ATP stimulation of Na-Ca exchange is due to the decrease in free-Mg concentration, that is, that intracellular Mg inhibits Na-Ca exchange in ferret red blood cells. However, in contrast to Ca influx, Cd and Mn influxes were observable in "native" red cells, that is, without vanadate preloading.

The similarity of the properties of Mn, Cd, and Ca transport in red cells makes the type of kinetic analysis used in other systems with alternative substrates difficult. The similarity does, however, provide three important pieces of information. (1) Mn and Cd fluxes can provide us information about the Na-Ca exchanger without concern for the possible operation of the Ca pump. (2) The similarity of the V_{max} values places constraints on the rate-limiting steps. (3) The similarity of the K_m values suggests that there is not an important sulfhydryl group in the divalent cation-binding site.

The apparent affinity of Na-Ca exchangers in different tissues would provide some information about how similar the divalent cation-binding sites are. In order to avoid potential tissue differences in surface charge, we have compared the ratio of K_i/K_m (see above). The ratio in red cells is close to one, that is, the K_i values for Mn and Cd are similar to the K_m values for Ca. Studies in other tissues reported IC_{50} values and not K_i values. When a K_i value is calculated from the IC_{50}, with the reasonable assumption that Cd and Mn compete with Ca in these other tissues, we find that the K_i/K_m ratio is also close to unity in bovine cardiac vesicles and human neutrophils.[30,32,33] In contrast, the K_i/K_m ratio for Cd is much less than one in vascular smooth muscle, consistent with the notion of an important thiol group in the Ca-binding site in this tissue.[34] These comparisons suggest that the divalent cation-binding site may be functionally conserved in blood cells and cardiac tissue. It will be interesting to determine if this reflects a structural similarity.

A MODEL

The K_m or K_i values for Ca, Mn, and Cd for the extracellular transport site are certainly within an order of magnitude. This suggests that the extracellular binding site of ferret red cell Na-Ca exchanger does *not* contain a *functionally important* thiol group. In contrast, Cd had a much greater affinity than Ca on the Ca pump, suggesting an important thiol group in the Ca site of the Ca pump. Thus Ca-binding sites of these two proteins are structurally different, in contrast to the idea that membrane transport proteins that bind similar ions might have the same structural features.

An important thiol group is thought to be part of the Na-binding site of the cardiac sarcolemmal vesicle Na-Ca exchanger.[35] If the red cell Na-Ca exchanger also has an important thiol group as part of the Na-binding site, and no important thiol group in the Ca site, then the Na and Ca sites are distinct. This would be consistent with sequential kinetics and not easily reconciled with a ping-pong mechanism.

The Rate-Limiting Step

The V_{max} values for Ca, Mn, and Cd influx were very similar. This places definite constraints on the rate-limiting step for the exchange cycle,[30] either (1) the divalent cation translocation or dissociation step is rate limiting and the same rate for all three ions, or (2) the Na-translocation step (or dissociation step) is rate limiting.

We discuss one model that fits our data since we find it intriguing and it is similar to models discussed for Na-Ca exchange in other tissues.[26] Consider a sequential model in which the divalent cation binds and is translocated, but before dissociation, the Na ions bind (compare FIG. 1). One of the Na sites may contain a thiol group, as in cardiac tissue,[32] but the divalent cation binds to a site that does not contain a thiol group. The binding of Na causes the divalent cation to dissociate, followed by Na translocation and dissociation. One of these two latter steps is the rate-limiting step in this particular model.

This model fits with the data we have presented:

1. It is sequential since there is a conformation in which both Na and Ca are bound.
2. The Ca site does not contain an important thiol group.
3. Differences in the rate of Ca, Mn, and Cd translocation probably exist, but would not be observable because the steady-state V_{max} reflects primarily the slow rate of the Na step.

SUMMARY

The kinetic equation that best describes the intracellular Na dependence of Ca influx into ferret red cells is sequential; whether this implies that there is a conformation of the protein that has both Na and Ca ions bound remains to be determined. Cd and Mn substitute very well for Ca on the exchanger in ferret red cells; this suggests that the Ca-binding site does not contain an important thiol and that the one of the Na steps may be rate limiting.

REFERENCES

1. MILANICK, M. A. 1989. Na/Ca exchange in ferret red blood cells. Am. J. Physiol. **256:** C390-C398.
2. BLAUSTEIN, M. P. 1984. The energetics and kinetics of sodium-calcium exchange in barnacle muscles, squid axons and mammalian heart: The role of ATP. *In* Electrogenic Transport: Fundamental Principles and Physiological Implications. M. P. Blaustein & M. Lieberman, Eds.: 129-147. Raven Press. New York.
3. DIPOLO, R. & L. BEAUGE. 1990. Calcium transport in excitable cells. *In* Intracellular Calcium Regulation.: 381-413. Alan R. Liss, Inc. New York.
4. NOBLE, D. 1987. Experimental and theoretical work on excitation and excitation-contraction coupling in the heart. Experientia **43:** 1146-1150.
5. FLATMAN, P. W. & P. L. R. ANDREWS. 1983. Cation and ATP content of ferret red cells. Comp. Biochem. Physiol. **74:** 939-943.
6. LEW, V. L. & J. GARCIA-SANCHO. 1989. Measurement and control of intracellular Ca in intact red cells. Methods Enzymol. **173:** 100-112.
7. PARKER, J. C., H. J. GITELMAN, P. S. GLOSSON & D. L. LEONARD. 1975. Role of calcium in volume regulation by dog red blood cells. J. Gen. Physiol. **64:** 84-96.
8. PARKER, J. C. 1989. Sodium-calcium and sodium-proton exchangers in red blood cells. Methods Enzymol. **173:** 292-300.

9. WILLIS, J. S., R. A. NELSON, C. GORDON, P. VILARO & Z. H. ZHAO. 1990. Membrane transport of sodium ions in erythrocytes of the American black bear, *Ursus americanus.* Comp. Biochem. Physiol. **96A:** 91-96.
10. HILGEMANN, D. W. & D. NOBLE. 1987. Excitation-contraction coupling and extracellular calcium transients in rabbit atrium: reconstruction of basic cellular mechanisms. Proc. R. Soc. London **230B:** 163-205.
11. BLAUSTEIN, M. P. 1977. Effects of internal and external cations and of ATP on sodium-calcium and calcium-calcium exchange in squid axons. Biophys. J. **20:** 79-111.
12. FROHLICH, O. & R. B. GUNN. 1986. Erythrocyte anion transport: The kinetics of a single-site obligatory exchange system. Biochim. Biophys. Acta **864:** 169-194.
13. HILGEMANN, D. W. 1988. Numerical approximations of sodium-calcium exchange. Prog. Biophys. Molec. Biol. **51:** 1-45.
14. LAUGER, P. 1987. Voltage dependence of sodium-calcium exchange: Predictions from kinetic models. J. Membr. Biol. **99:** 1-11.
15. SACHS, J. R. 1980. The order of release of sodium and addition of potassium in the sodium-potassium pump reaction mechanism. J. Physiol. (London) **302:** 219-240.
16. RESTREPO, D., D. J. KOZODY, L. J. SPINELLI & P. A. KNAUF. 1989. Cl-Cl exchange in promyelocytic HL-60 cells follows simultaneous rather than ping-pong kinetics. Am. J. Physiol. **257:** C520-527.
17. RESTREPO, D. & P. A. KNAUF. 1991. The anion exchanger of HL-60 cells is simultaneous: A novel method to eliminate ping-pong with slippage kinetics. Red Cell Club meeting, Biophys. Soc. 1991.
18. RASGADO-FLORES, H., S. SANCHEZ-ARMASS, M. P. BLAUSTEIN & D. A. NACHSHEN. 1987. Strontium, barium, and manganese metabolism in isolated presynaptic nerve terminals. Am. J. Physiol. **252:** C604-C610.
19. JENNINGS, M. L. & M. F. ADAMS. 1981. Modification by papain of the structure and function of band 3, the erythrocyte anion transport protein. Biochemistry **20:** 7118-7123.
20. KENNEY, L. J. & J. H. KAPLAN. 1988. The vectorial effect of ligands on the occluded intermediate in red cell sodium pump transport. Prog. Clin. Bio. Res. **268A:** 525-530.
21. FRAME, M. D. S. & M. A. MILANICK. 1990. ATP increases Na/Ca exchange and Na/Mn exchange in resealed ferret red blood cell ghosts. Biophys. J. **57:** 180a.
22. FORBUSH, B., III. 1987. Rapid release of ^{42}K and ^{86}Rb from an occluded state of the Na,K-pump in the presence of ATP or ADP. J. Biol. Chem. **262:** 11104-11115.
23. VERBOST, P. M., G. FLIK, P. K. T. PANG, R. A. C. LOCK & S. E. W. BONGA. 1989. Cadmium inhibition of the erythrocyte Ca pump. J. Biol. Chem. **264:** 5613-5615.
24. MILANICK, M. A. 1991. Na/Ca exchange: Evidence against a ping-pong mechanism and against a Ca pool in ferret red blood cells. Am. J. Physiol. **261:** C185-C193.
25. HODGKIN, A. L., B. J. NUNN. 1987. The effect of ions on sodium-calcium exchange in salamander rods. J. Physiol. **391:** 371-398.
26. REEVES, J. P. 1985. The sarcolemmal sodium-calcium exchange system. Curr. Top. Membr. Trans. **25:** 77-127.
27. LI, J. & J. KIMURA. 1990. Translocation mechanism of Na-Ca exchange in single cardiac cells of guinea pig. J. Gen. Physiol. **96:** 777-788.
28. HILGEMAN, D. W. & G. A. NAGEL. 1991. Steady-state ion dependencies of chymotrypsin-deregulated cardiac Na/Ca exchange current conform to consecutive mechanism with voltage dependence in the sodium translocation pathway. Biophys. J. **59:** 137a.
29. FRAME, M. D. 1990. The influence of manganese and cadmium on the Na/Ca exchange system in ferret red cells. Ph.D. thesis. University of Missouri-Columbia, Columbia, MO.
30. FRAME, M. D. S. & M. A. MILANICK. 1991. Mn and Cd transport on the Na/Ca exchanger of ferret red blood cells. Am. J. Physiol. **261:** C467-C475.
31. DIPOLO, R. & L. BEAUGE. 1983. The calcium pump and sodium-calcium exchange in squid axons. Ann. Rev. Physiol. **45:** 313-324.
32. TROSPER, T. L. & K. D. PHILIPSON. 1983. Effects of divalent and trivalent cations on Na-Ca exchange in cardiac sarcolemmal vesicles. Biochim. Biophys. Acta **731:** 63-68.

33. SIMCHOWITZ, L. & E. J. CRAGOE. 1988. Na-Ca exchange in human neutrophils. Am. J. Physiol. **254:** C150-C164.
34. SMITH, J. B., E. J. CRAGOE & L. SMITH. 1987. Na/Ca antiport in cultured arterial smooth muscle cells. J. Biol. Chem. **262:** 11988-11994.
35. PIERCE, G. N., R. WARD & K. D. PHILIPSON. 1986. Role for sulfur-containing groups in the Na-Ca exchange of cardiac sarcolemmal vesicles. J. Membr. Biol. **94:** 217-225.

The Role of Na^+-Ca^{2+} Exchange in Human Neutrophil Function[a]

WILLIAM E. DALE AND LOUIS SIMCHOWITZ[b]

Department of Medicine
Veterans Administration Medical Center
and
Department of Medicine
Department of Cell Biology and Physiology
Washington University School of Medicine
St. Louis, Missouri 63110

INTRODUCTION

Cytoplasmic calcium concentration ($[Ca^{2+}]_i$) plays a critical role in regulation of normal function in most cell types and the human neutrophil is no exception. Although a rise in $[Ca^{2+}]_i$ is rarely sufficient for full activation of the granulocytic cell response, the importance of calcium transients in the neutrophil is undeniable. Previous reports have shown that various stimuli produce changes in $[Ca^{2+}]_i$ that accompany chemotaxis,[1] phagocytosis,[1,2] degranulation,[3] and free radical generation during the respiratory burst.[4,5] Although information from early experiments in neutrophils focused on the existence of intracellular reservoirs of Ca^{2+},[6] more recent data implicate Ca^{2+} influx as an important means of regulating ($[Ca^{2+}]_i$).[7] In other cell types, multiple pathways may exist for transporting Ca^{2+} across the plasma membrane. However, a carrier-mediated exchange of external Ca^{2+} for internal Na^+ appears to be the primary pathway for Ca^{2+} entry into the resting neutrophil. This review will discuss our present understanding of this Na^+-Ca^{2+}-exchange mechanism. We will examine the kinetic properties and activation of the counter-transport system and will follow this with a discussion of the role of Na^+-Ca^{2+} exchange in mediating neutrophil superoxide generation, chemotaxis, and phagocytosis.

Na^+-Ca^{2+} EXCHANGE IN RESTING CELLS

Recently, our laboratory reported the existence of a Na^+-Ca^{2+} exchange mechanism in human neutrophils[7] and described some of its basic biochemical properties. It is clear that the neutrophil plasma membrane Na^+-Ca^{2+} exchanger displays many features reminiscent of similar counter-transport systems in muscle and nerve.[8–10] On the basis of one-way $^{45}Ca^{2+}$ influx studies,[7] we found the carrier to be sensitive to benzamil (apparent K_i ~75 μM), a benzyl derivative of amiloride. Phenamil and 2′,4′-dichlorobenzamil (analogues that bear a substituent on the terminal nitrogen atom

[a] This work was supported by the Department of Veterans Affairs and by National Institutes of Health Grant GM-38094.

[b] Address for correspondence: Dr. Louis Simchowitz (151-JC), V.A. Medical Center, 915 North Grand Avenue, St. Louis, MO 63106.

of the guanidino group of amiloride) were equipotent to benzamil while amiloride itself was inactive. Notably, none of these compounds had any significant effect on the amiloride-sensitive Na^+-H^+ exchanger over the same concentration range.[11] The dose-response curves for benzamil inhibition of $^{45}Ca^{2+}$ influx over a wide range of external Ca^{2+} show that the interaction is noncompetitive in nature. On the other hand, $^{45}Ca^{2+}$ influx was resistant to the "Ca^{2+} channel blockers" nifedipine, verapamil, and diltiazem.

Not surprisingly, internal Na^+ trans-stimulated $^{45}Ca^{2+}$ influx. The half-saturation constant for Na^+ was ~26 mM, a value that coincides with the normal resting content of steady-state cells.[12] Of note, the relationship of the intracellular Na^+ concentration to the rate of $^{45}Ca^{2+}$ influx followed a Hill equation with a Hill coefficient of 2.6 ± 0.5. This value implies an exchange stoichiometry of ~3 Na^+ : 1 Ca^{2+}, which is consistent with that found in other cell types.[13,14] In keeping with the electrogenic nature of the transporter, $^{45}Ca^{2+}$ influx was strongly voltage dependent: Membrane depolarization from ~ −60 to 0 mV enhanced Ca^{2+} uptake by a factor of 2.3. An analysis of the substrate selectivity of the external translocation site of the carrier based on competition kinetics with Ca^{2+} yielded the following ion sequence of decreasing affinities: $La^{3+} > Cd^{2+} > Mn^{2+} \sim Sr^{2+} > Ba^{2+} > Co^{2+} \sim Ni^{2+} > Mg^{2+}$. The K_m for Ca^{2+} in an otherwise substrate-depleted, Na^+-free medium was ~150 μM. Essentially all of the $^{45}Ca^{2+}$ influx into steady-state cells (~6 μmol/liter of cell water·min) could be accounted for via this route, indicating that a benzamil-sensitive Na^+-Ca^{2+} exchange represents the predominant (if not the exclusive) pathway for the influx of Ca^{2+} ions in human neutrophils. These results are compatible with the apparent lack of classical voltage-dependent Ca^{2+} channels in these cells.[15,16] From an analysis of the electrochemical driving forces and given a 3 : 1 coupling ratio, it could be demonstrated that the net flux is in the direction of Ca^{2+} entry in return for Na^+ efflux under normal steady-state conditions.[7] Therefore, in contrast to excitable cells[8-10] where Na^+-Ca^{2+} exchange mediates net Ca^{2+} extrusion ("forward mode"), the physiologically relevant transport occurs in the opposite direction in human neutrophils ("reverse mode") to promote the uptake of Ca^{2+} into the cell.

As shown in FIGURE 1, we also observed the transport rate of the Na^+-Ca^{2+} exchanger to be stimulated ~20-fold by exposing cells to the chemotactic factor *N*-formyl-methionyl-leucyl-phenylalanine (FMLP). This initial observation provided the impetus for examining the biological significance of this carrier for intracellular Ca^{2+} homeostasis and for the regulation of human neutrophil function. The following sections of this chapter are devoted to outlining our current thoughts regarding the role of Na^+-Ca^{2+} exchange in three different functional expressions of cell activation: superoxide radical (O_2^-) generation, chemotaxis, and phagocytosis.

SUPEROXIDE RADICAL PRODUCTION

The role of the Na^+-Ca^{2+} counter-transport system in regulating the respiratory burst of neutrophils in response to activation by the chemotactic tripeptide FMLP was explored.[17] For these studies, O_2^- release was assessed by the reduction of ferricytochrome *c* according to well-established protocols.[12,17] In the presence of 1 mM Ca^{2+}, a variety of di- and trivalent cations suppressed the generation of O_2^- radicals in a series of decreasing efficacy: $La^{3+} \sim Zn^{2+} >> Sr^{2+} \sim Cd^{2+} > Ba^{2+} > Co^{2+} > Ni^{2+} \sim Mg^{2+}$.[17] This sequence is similar to their rank order of activity in inhibiting $^{45}Ca^{2+}$ influx via Na^+-Ca^{2+} counter-transport.[7] Benzamil, phenamil, and 2′,4′-dichlorobenzamil, analogues of amiloride that selectively block Na^+-Ca^{2+} exchange

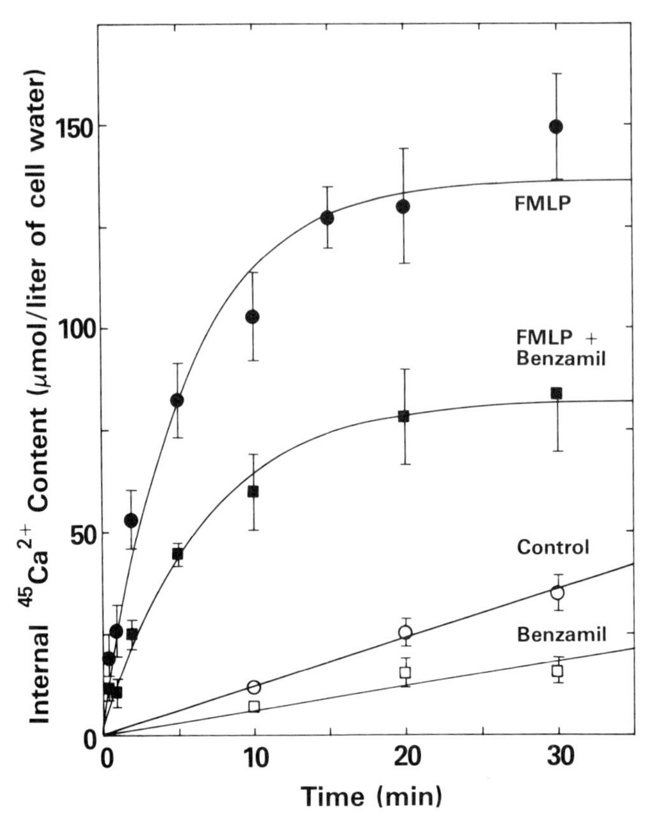

FIGURE 1. Time course of FMLP-induced $^{45}Ca^{2+}$ influx into human neutrophils: effect of benzamil. Neutrophils were stimulated at time zero by the addition of 100 nM FMLP in the presence or absence of 100 μM benzamil. Controls with and without drug were also included. The medium contained 5 mM K^+, 140 mM Na^+, 0.1 mM Ca^{2+}, and 0.5 mM Mg^{2+}. Reaction mixtures were incubated at 37°C, and at stated times samples were taken and analyzed for their $^{45}Ca^{2+}$ contents. Results were taken from four experiments. The upper two curves are single exponential fits to data: for FMLP, initial influx rate = 24.9 ± 4.7 μmol/liter·min and final uptake = 136.9 ± 8.9 μmol/liter of cell water; for FMLP + benzamil, initial influx rate = 12.4 ± 3.4 μmol/liter·min and final uptake = 82.7 ± 7.8 μmol/liter of cell water. The lower two curves are straight line fits with slopes of 1.20 ± 0.10 and 0.609 ± 0.089 μmol/liter·min for control and 100 μM benzamil, respectively. The figure has been reproduced courtesy of the *American Journal of Physiology: Cell Physiology.*[7]

in neutrophils with no activity against Na^+-H^+ exchange,[11] likewise suppressed the release of O_2^- with apparent K_i values of ~30 μM (FIG. 2). The effect of the cations was competitive with Ca^{2+}, while the interaction between the benzamil derivatives and Ca^{2+} appeared to be noncompetitive in nature as was the case in resting cells.[7] Both the divalent cations and benzamil also inhibited the transient FMLP-induced rise in cytoplasmic Ca^{2+} as monitored by fura-2 fluorescence (FIG. 3). In medium containing 1 mM Ca^{2+}, these agents reduced peak cytosolic Ca^{2+} levels after FMLP stimulation to values seen in the absence of extracellular Ca^{2+}. For a given cation, there was a

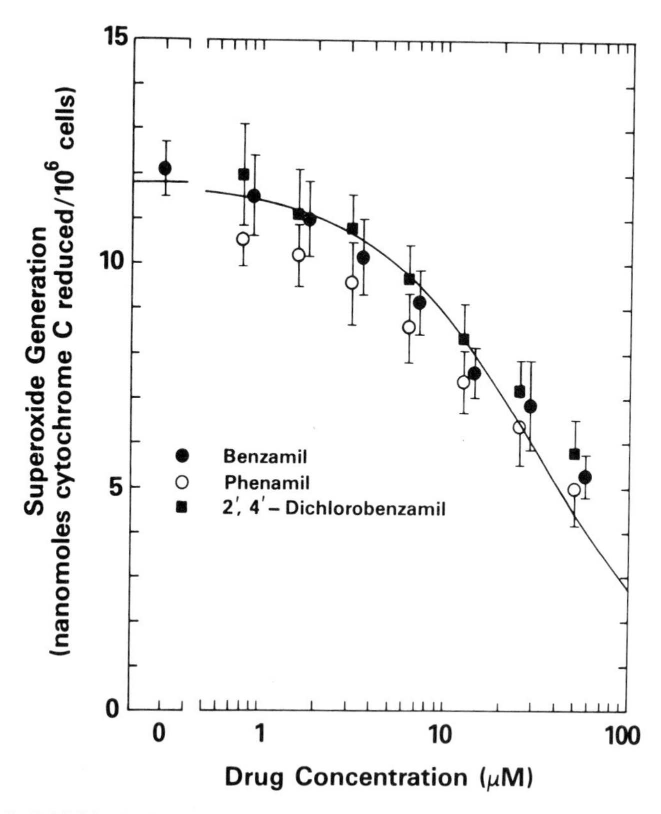

FIGURE 2. Inhibition by benzamil and related compounds of FMLP-stimulated O_2^- production by human neutrophils. The release of O_2^- from cells exposed to 100 nM FMLP was measured in the presence of varying concentrations (0-50 μM) of benzamil, phenamil, or 2′,4′-dichlorobenzamil. The medium contained 1 mM Ca^{2+} and 140 mM Na^+. Results are from three to four experiments for each drug. The combined data points have been fit to a Michaelis-Menten inhibition equation that yielded a K_i of 30.0 ± 4.4 μM. Fits to the individual sets of data gave apparent K_i values of 29.3 ± 6.6, 25.6 ± 7.5, and 34.8 ± 8.9 μM for benzamil, phenamil, and 2′,4′-dichlorobenzamil, respectively. The figure has been reproduced courtesy of the *Journal of Biological Chemistry.*[17]

strong correlation between its ability (as measured by the apparent K_i) to inhibit resting Na^+-Ca^{2+} exchange, FMLP-stimulated rise in cytosolic Ca^{2+}, and FMLP-induced O_2^- production.

These results are compatible with the hypothesis that the influx of Ca^{2+} via Na^+-Ca^{2+} exchange contributes to the transient elevation in intracellular free Ca^{2+} that is observed following FMLP stimulation. Certain polyvalent cations block the entry of Ca^{2+} ions by competing for binding to the external translocation site on the exchange carrier, while benzamil acts by suppressing the maximal transport rate.[7] Additionally, the importance of Na^+-Ca^{2+} exchange in modulating O_2^- production was demonstrated in other ways. Internal Na^+ depletion (i.e., removal of the exchange partner for external Ca^{2+}) abolished $^{45}Ca^{2+}$ uptake and O_2^- generation and also

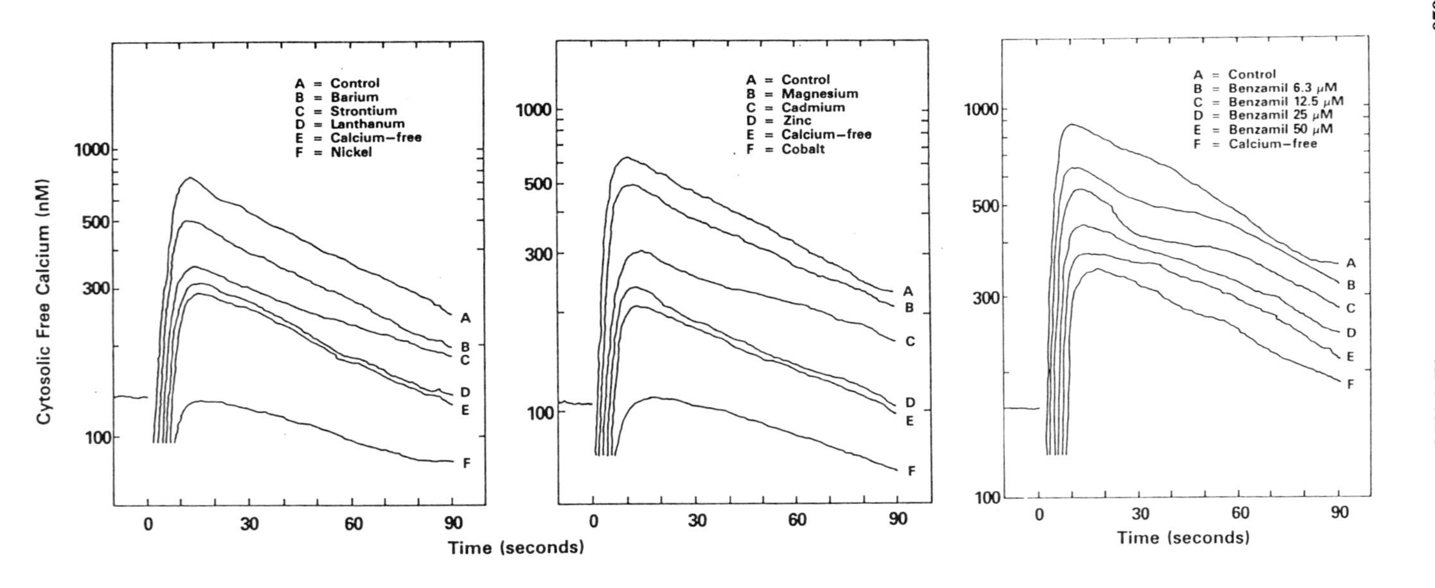

FIGURE 3. Effect of benzamil as well as various di- and trivalent cations on the level of cytosolic free Ca^{2+} in FMLP-stimulated neutrophils. The intracellular Ca^{2+} transients were monitored by the fluorescence of fura-2. The cells were loaded with the probe as described[17] and then resuspended in the standard 1 mM Ca^{2+} medium, but lacking Mg^{2+}. As indicated, stated concentrations of the different cations (100 or 1000 μM) were added to the medium. Controls were performed in the presence or absence (Ca^{2+}-free) of 1 mM Ca^{2+}. The labels describe the conditions under which the experiments took place. At zero time, FMLP was added to a final concentration of 100 nM and the changes in fluorescence followed for ~2 min. The fluorescence readings were calibrated to cytoplasmic Ca^{2+}concentrations according to the procedure of Grynkiewicz *et al.*[35] *Left Panel:* Tracings for 100 μM La^{3+} and 1000 μM of either Sr^{2+}, Ba^{2+}, or Ni^{2+}. *Middle Panel:* Tracings for 100 μM Zn^{2+} and 1000 μM of either Cd^{2+}, Mg^{2+}, or Co^{2+}. *Right Panel:* Dose-dependence of benzamil (0-50 μM). Reprinted courtesy of the *Journal of Biological Chemistry.*[17]

markedly suppressed the peak cytosolic Ca^{2+} level attained after FMLP stimulation. In contrast, internal Na^+ loading increased the rate of Na^+-Ca^{2+} exchange and caused enhanced O_2^- release that coincided with higher than normal peak intracellular Ca^{2+} levels.[17] These results emphasize that Na^+-Ca^{2+} exchange, through its effects on cytoplasmic Ca^{2+}, plays a major regulatory role in activation of the respiratory burst in chemotactic factor-stimulated neutrophils. Although the exact role that intracellular Ca^{2+} transients play in signal transduction is not yet clear, the evidence indicates that elevations of cytosolic Ca^{2+} probably serve a priming function and potentiate responses to other stimuli.[18,19]

The importance of Na^+-Ca^{2+} exchange in FMLP-induced O_2^- generation also explains the pH-dependence of this response.[12] The relationships to pH_o of $^{45}Ca^{2+}$ influx rate via Na^+-Ca^{2+} exchange[7] and the rate of O_2^- production by FMLP-activated neutrophils[12] are identical: Alkalinization enhances, while acidification suppresses, both rates with the points falling along a titration curve (pK′ ~7.10). In this context, the potential interaction between the Na^+-H^+ and Na^+-Ca^{2+} counter-transport systems in chemotactic factor-activated human neutrophils should be emphasized. FMLP-induced Na^+-H^+ exchange may serve to facilitate the entry of Ca^{2+} through the Na^+-Ca^{2+} exchange carrier by two means: (1) the rise in internal Na^+ stimulates Ca^{2+} influx and (2) the intracellular alkalinization due to the Na^+-H^+ exchanger potentiates the rate of Ca^{2+} influx.

CHEMOTAXIS

The effect of a series of di- and trivalent cations on the locomotor response of human neutrophils to 10 nM FMLP was also investigated.[20] Here, migration was assessed by the leading-front method. The cations inhibited FMLP-stimulated chemotaxis in the rank order: $Ni^{2+} \sim Co^{2+} > Sr^{2+} > Zn^{2+} > Mn^{2+} \sim La^{3+} > Cd^{2+} \sim Ba^{2+} >> Mg^{2+}$. Moreover, neither benzamil nor phenamil, at concentrations of 100 μM, significantly altered chemotaxis by themselves. The lack of effect of the two drugs that cause ~70% inhibition of $^{45}Ca^{2+}$ influx through Na^+-Ca^{2+} exchange[7] strongly implies that this influx pathway is not required for optimal chemotaxis. The simultaneous addition of benzamil did, however, prevent the suppressive effects of each of the polyvalent cations on motility (TABLE 1). Clearly, then, these cations must enter the cell in order to produce a detrimental effect on the neutrophil's motile apparatus. The ability of Ni^{2+}, Co^{2+}, and Mn^{2+} to permeate across the cell membrane can also be indirectly detected by the quenching of fura-2 fluorescence[21,22] (FIG. 3). The ion-selectivity sequence noted above and the total lack of activity of benzamil on chemotaxis are strikingly different than for O_2^- generation, thereby implying that the two functional expressions must lie along separate transductional pathways.

In view of the crucial role of cytoskeletal elements and the contractile machinery in all forms of cell motility,[23,24] we postulated that once inside the cell, these cations might inhibit chemotaxis by interfering with actin polymerization. This idea was particularly appealing because a number of investigators had previously reported that exposure of neutrophils to FMLP leads to an increase in actin polymerization as manifested by a rise in the content of fibrillar or F-actin and a corresponding fall in the amount of monomeric, globular or G-actin.[25–29] In confirmation of this work, we observed that FMLP evokes a rapid and dramatic shift in the physical state of actin[20] as monitored by the fluorescence of rhodamine-phalloidin, which binds only to F-actin (FIG. 4). As shown, 100 μM Ni^{2+} abolished the FMLP-induced changes in actin polymerization. Moreover, when the dose-dependencies of the various polyvalent cations on the increase

TABLE 1. Ability of Benzamil to Reverse the Inhibitory Effect of Polyvalent Cations on FMLP-Induced Chemotaxis in Human Neutrophils[a]

Added Cation (μM)	Migration (μm) When Stimulated with: Medium	FMLP 10 nM	FMLP 10 nM + Benzamil 100 μM
None	35.9 ± 1.8[b]	75.0 ± 4.7	74.0 ± 5.2
Ni^{2+} 20		34.3 ± 1.7	68.7 ± 1.8
Co^{2+} 20		40.8 ± 3.9	67.1 ± 5.3
Sr^{2+} 40		37.8 ± 3.5	78.3 ± 1.6
La^{3+} 250		32.0 ± 3.9	74.0 ± 3.5
Cd^{2+} 500		35.5 ± 5.5	66.4 ± 3.1
Ba^{2+} 800		39.7 ± 2.8	81.4 ± 7.8

[a] Aliquots of neutrophil suspensions were exposed to 10 nM FMLP (chemotaxis) in the presence or absence of benzamil and various polyvalent cations. Controls (random motility) were performed in the absence of FMLP. The media contained 1 mM Ca^{2+} to which was added, at stated concentrations, either Ni^{2+}, Co^{2+}, Sr^{2+}, La^{3+}, Cd^{2+}, or Ba^{2+}, alone or in combination with benzamil. Results of three experiments are expressed as the mean ± SEM of the distance (μm) that the leading front of cells had migrated into the filter at 30 min. Data have been taken from a previous article[20] courtesy of the *Journal of Biological Chemistry.*

[b] In the absence of FMLP stimulation, none of the cations altered neutrophil migration into the filter. The given value represents the combined data under all control conditions.

in F-actin content induced by FMLP were examined, the findings agreed well with the profile of inhibition against chemotaxis.[20] As with chemotaxis, benzamil exhibited a protective effect, completely overcoming the inhibitory action of the polyvalent cations on actin polymerization (FIG. 4). Though the participation of a benzamil-sensitive Na^+-Ca^{2+} exchange is apparently not required for chemotaxis, these results indicate that a number of foreign ions can gain access to the cell interior through it. Upon entry into the cytosol, they then interfere with the formation of filaments from actin monomers.

PHAGOCYTOSIS

Recently, our laboratory conducted a series of experiments devoted to an analysis of the potential role of Na^+-Ca^{2+} exchange in neutrophil phagocytosis. For these studies, we utilized a conventional model in which sheep red blood cells (SRBC) are opsonized with a polyclonal IgG antiserum obtained from rabbits immunized with SRBC.[30] Neutrophils attached to a plastic slide are overlaid with these opsonized SRBC in the standard medium for 30 min at 37°C. After lysis of the extracellular SRBC with NH_4Cl, the adherent neutrophil monolayer was fixed and stained. Results are expressed as the phagocytic index, which is the number of ingested SRBC per 100 neutrophils. The attachment and engulfment of SRBC take place through the binding of immune complexes on the erythrocyte surface to neutrophil plasma membrane F_c receptors. This assay is generally termed "unstimulated phagocytosis." In other instances, the activating agents FMLP or phorbol dibutyrate (PDBu) were added to the neutrophils along with SRBC, a process referred to as "stimulated phagocytosis."[31,32]

TABLE 2 shows the dependence of unstimulated phagocytosis on extracellular Ca^{2+}. In agreement with previous reports,[33,34] there is an absolute requirement for external

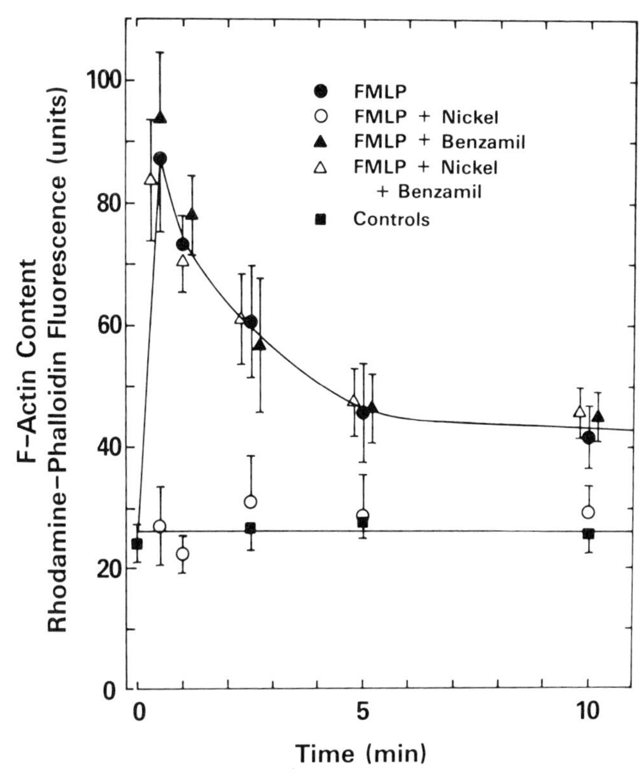

FIGURE 4. Time course of changes in the cellular content of F-actin in human neutrophils induced by FMLP: effect of Ni^{2+} and benzamil. Neutrophils were first allowed to adhere to glass cover slips and then exposed at zero-time to 10 nM FMLP, 10 nM FMLP + 100 μM Ni^{2+}, 100 μM Ni^{2+}, or medium alone (control). Studies were also conducted in the presence or absence of 100 μM benzamil. The bathing solutions each contained 1 mM Ca^{2+}. At stated times, the reactions were terminated by the addition of 2% formaldehyde. The fixed cell preparations were next incubated with 33 nM rhodamine-phalloidin, washed, and examined in a spectrofluorometer. The polymerization of actin, as assessed by the F-actin content, was quantitated from the fluorescence readings. The curve through the data points for FMLP, FMLP + benzamil, and FMLP + Ni^{2+} + benzamil was drawn by eye. In the absence of FMLP stimulation, the fluorescence values under all treatment conditions were indistinguishable: The points shown represent the combined data. A horizontal line has been drawn at 26.1, the average fluorescence level of the controls. Results have been taken from four experiments. The figure has been reproduced with the permission of the *Journal of Biological Chemistry.*[20]

Ca^{2+}: In the presence of 2 mM EGTA or EDTA, the engulfment of opsonized SRBC is essentially abolished. This finding is consistent with a requirement for a Ca^{2+} influx from the exterior (although alternative explanations are possible). Notably, 40 μM benzamil has a striking inhibitory effect on phagocytosis (TABLE 2) which supports the hypothesis that the putative Ca^{2+} influx is occurring via Na^+-Ca^{2+} exchange. FIGURE 5 displays the dose-dependence of the inhibitory effect of benzamil on unstimulated phagocytosis (K_i = 6.0 ± 1.2 μM). This K_i value is, however, ~5-10-fold lower than that for inhibition of $^{45}Ca^{2+}$ influx into resting cells[7] or FMLP-stimulated O_2^-

TABLE 2. Effect of Extracellular Ca^{2+} and Benzamil on the Phagocytosis of IgG-Coated Sheep Erythrocytes by Human Neutrophils

Stimulus	$[Ca^{2+}]_o$ (mM)	Additions	Phagocytic Index
Medium	1	—	96 ± 15
	1	Benzamil (40 μM)	11 ± 3
	—	—	8 ± 2
FMLP (10 nM)	1	—	256 ± 38
	1	Benzamil (40 μM)	19 ± 4
	—	—	24 ± 5
PDBu (60 nM)	1	—	293 ± 54
	1	Benzamil (40 μM)	271 ± 40
	—	—	260 ± 46

NOTE: IgC-coated SRBC were sedimented on top of a monolayer of adherent neutrophils in medium alone (unstimulated phagocytosis) or in the presence of 10 and 60 nM, respectively, of the activating agents FMLP and PDBu (stimulated phagocytosis). Experiments were also performed with and without 40 μM benzamil in Ca^{2+}-containing medium (1 mM Ca^{2+}, 0.5 mM Mg^{2+}) and under Ca^{2+}-free conditions (2 mM EGTA, 0.5 mM Mg^{2+}). The phagocytic index (i.e., the number of ingested SRBC/100 neutrophils) was determined after 30 min at 37°C. Results represent the means ± SEM of four separate experiments, each performed in duplicate.

release.[17] The differential sensitivities of these processes could relate to changes in binding affinity of the carrier due to the various methods of activation, to different extents of the Ca^{2+} influx contribution to the functional responses, or to other compensating factors such as the release of Ca^{2+} from internal stores. Alternatively, the block of phagocytosis might represent some nonselective effect having nothing to do with Na^+-Ca^{2+} exchange.

Evidence supporting the notion that suppression of phagocytic capacity by benzamil is truly related to prevention of the requisite influx of Ca^{2+} via Na^+-Ca^{2+} exchange is provided in TABLE 3. It can be seen that increasing the extracellular Ca^{2+} level can overcome inhibition by benzamil: Raising external Ca^{2+} from 1 to 16 mM reduces the extent of inhibition of unstimulated phagocytosis by benzamil (15 μM) from 78 to 9%. Further evidence against a nonspecific effect of benzamil is illustrated in FIGURE 5 where the effects of the drug on unstimulated as well as on FMLP (10 nM)- and PDBu (60 nM)-stimulated phagocytosis are compared. It is important to emphasize that FMLP-activated phagocytosis, like unstimulated phagocytosis, displays an obligatory requirement for external Ca^{2+}, whereas PDBu-stimulated phagocytosis does not (TABLE 2). In fact, PDBu-stimulated phagocytosis proceeds quite normally in the absence of all Ca^{2+} in the bathing medium. FIGURE 5 shows that the inhibitory potency of benzamil on FMLP-stimulated and unstimulated phagocytosis is comparable (apparent $K_i = 6.8 \pm 0.9$ and 6.0 ± 1.2 μM, respectively). In contrast, over the concentration range 10-80 μM, benzamil has no effect on PDBu-induced phagocytosis. These findings correlate with the different external Ca^{2+}-dependencies of the FMLP- and PDBu-stimulated responses. These results fit into a convenient model that unstimulated as well as FMLP-stimulated phagocytosis is dependent on a Ca^{2+} influx from the exterior via a benzamil-sensitive Na^+-Ca^{2+} exchange mechanism. On the other hand, PDBu-stimulated phagocytosis lacks this Ca^{2+} requirement and so inhibition of Na^+-Ca^{2+} exchange by benzamil would not be expected to appreciably influence the response.

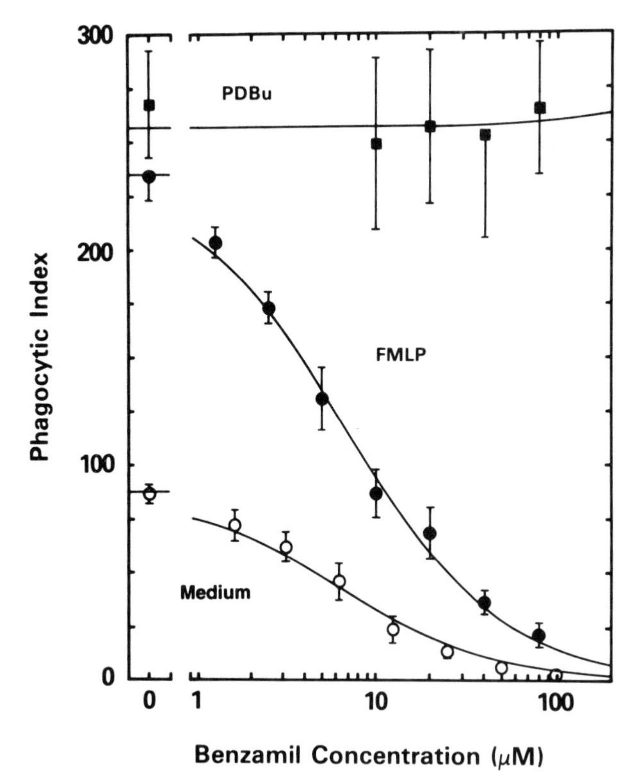

FIGURE 5. Dose-dependence of benzamil on unstimulated, FMLP-, or PDBu-stimulated phagocytosis by human neutrophils. Experiments were conducted in medium containing 1 mM Ca^{2+} and varying concentrations of benzamil (0-100 μM). FMLP (10 nM) or PDBu (60 nM) was added to the reaction mixtures concomitant with addition of the opsonized SRBC. Results have been taken from four to five experiments for each condition. The two lower curves are least-squares fits to Michaelis-Menten inhibition equations yielding apparent K_i values of 6.0 ± 1.2 and 6.8 ± 0.9 μM, respectively, for the medium (unstimulated) and FMLP (stimulated phagocytosis) data sets. The data points for PDBu were fit to a straight line with a slope of 0.030 ± 0.681, which could not be distinguished from zero.

The apparent K_i values for a number of other benzamil analogues are listed in TABLE 4. The structure-activity relationships are roughly similar to those originally reported for inhibition of $^{45}Ca^{2+}$ influx into resting neutrophils.[7] Benzamil and phenamil displayed comparable activity (K_i values of 6.0 and 7.2 μM, respectively). 2′,4′-Dichlorobenzamil (which was equipotent to benzamil in the $^{45}Ca^{2+}$ influx assay[7]) and analogue OOOO were somewhat more potent, exhibiting apparent K_i values of 1.4 and 2.3 μM. Other substituted derivatives such as MMMM, NNNN, and KKKK were considerably less potent.

The influence of a number of divalent cations on F_c-mediated phagocytosis is presented in TABLE 5. It can be seen that the rank order of efficacy in blocking phagocytosis bears only a very superficial resemblance to that for inhibition of FMLP-induced chemotaxis. Again, Ni^{2+} and Co^{2+} are among the most effective cations. However, Ba^{2+} and Cd^{2+} display comparatively greater activity toward phagocytosis,

TABLE 3. Raising Extracellular Ca^{2+} Overcomes the Suppressive Effect of Benzamil on Unstimulated Phagocytosis in Human Neutrophils

$[Ca^{2+}]_o$ (mM)	Conditions	Phagocytic Index
1	Medium	120 ± 16
1	Benzamil (15 μM)	27 ± 6
16	Medium	95 ± 21
16	Benzamil (15 μM)	86 ± 12

NOTE: Opsonized SRBC were added to an adherent neutrophil monolayer in the absence of FMLP and PDBu. The medium contained either 1 or 16 mM Ca^{2+}, with or without 15 μM benzamil. After a 30-min incubation, the phagocytic index (i.e., the number of ingested SRBC/100 neutrophils) was determined. Results represent the means ± SEM of five experiments.

while Sr^{2+} and Mn^{2+} exhibit considerably less. (The decidedly increased relative potency of Ba^{2+} could be due to its ability to block K^+ channels in neutrophils, thereby causing the cells to depolarize. This idea may have some validity because depolarization in high-K^+ medium suppresses phagocytosis by ~40% [L. Simchowitz, unpublished observations].) However, the apparent K_i values for inhibition of phagocytosis by each of the polyvalent cations is at least one and sometimes two orders of magnitude greater than for chemotaxis. The underlying basis for these observations remain unresolved, but could be related to the same factors that have been invoked to explain the markedly enhanced sensitivity of phagocytosis to benzamil as discussed above. Additionally, the activity sequence might reflect a combination of the two different factors that determine the nearly opposite rank orders of the cations for inhibition of FMLP-stimulated O_2^- generation and chemotaxis, that is, prevention of Ca^{2+} influx and interference with the polymerization of actin.[17,20]

DISCUSSION

Chemotaxis requires a rather sophisticated temporal and spatial recognition apparatus to sense and respond to an ever-changing chemical gradient of the stimulus.[36–38] Likewise, phagocytosis involves the dynamic interplay and ordered sequence of a number of discrete events, including particle attachment and engulfment. In other well-studied models of cell motility such as amoeboid motion and echinoderm fertilization,[23,39] these related phenomena are known to involve complicated sol:gel transformations and reversible cycles of polymerization and depolymerization of cytoskeletal and contractile elements. In all of these preparations, changes in cytosolic free Ca^{2+} levels have been implicated as playing a major role in the cascade of events.

In all probability, at least some of the actions of Ca^{2+} are mediated through effects on cytoskeletal elements and the contractile apparatus of the neutrophil. As with other cells,[23,39] chemotaxis by neutrophils requires changes in the organization and physical state of actin. A number of studies have already documented the Ca^{2+}-dependence of actin polymerization and sol:gel transformations in amoeboid movements and in gamete activation.[23,39] There is little doubt that these complex interactions are of fundamental importance in understanding all aspects of cell motility.

In examining the effects of a variety of di- and trivalent cations on FMLP-induced functional responses, we have found that optimal superoxide radical production is

TABLE 4. Ability of Benzamil and Its Analogues to Block Phagocytosis in Human Neutrophils

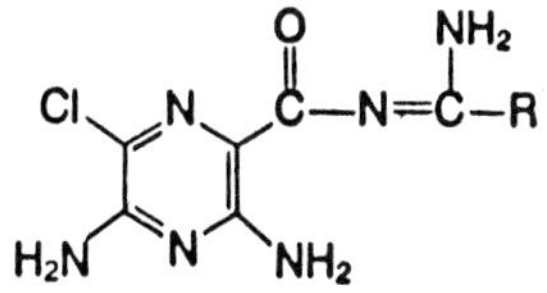

Code Name or Letters	R =	Apparent K_i (μM)
2′,4′-Dichlorobenzamil	-NHCH$_2$(C$_6$H$_3$)Cl$_2$ 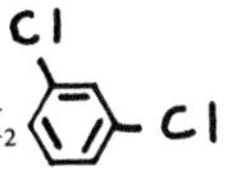	1.4 ± 0.3
OOOO	-NHCH$_2$(C$_6$H$_3$)(CF$_3$)$_2$ 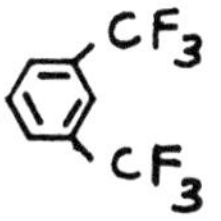	2.3 ± 0.6
Benzamil	-NHCH$_2$(C$_6$H$_5$) 	6.0 ± 1.2
Phenamil	-NH  (C$_6$H$_5$)	7.2 ± 2.1
MMMM	-NH  (fluorenyl)	17.7 ± 5.9
NNNN	-NH(CH$_2$)$_{11}$CH$_3$	29.7 ± 12.0
KKKK	-NHCH$_2$CH$_2$(C$_6$H$_4$)-OH 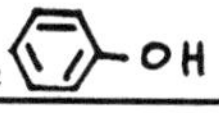	36.2 ± 7.9

NOTE: See legend to FIGURE 5 where the effect of the drugs on unstimulated phagocytosis was evaluated. The apparent K_i values are expressed as the means ± SEM of three to four experiments.

dependent on the uptake of Ca^{2+} from the bathing medium.[17] This influx takes place through Na^+-Ca^{2+} exchange and may be blocked by benzamil and by a series of di- and trivalent cations that prevent the entry of Ca^{2+} by competing with Ca^{2+} at the external translocation site of the carrier. Thus, Na^+-Ca^{2+} exchange seems to play an important role in FMLP-stimulated O_2^- generation as the major influx pathway for extracellular Ca^{2+}, which, upon entry into the cytosol, presumably triggers subsequent steps in the activation cascade.

TABLE 5. Ability of Polyvalent Cations to Suppress Phagocytosis in Human Neutrophils

Cation	Apparent K_i (mM)
Ni^{2+}	1.1 ± 0.4
Ba^{2+}	1.2 ± 0.2
Co^{2+}	1.4 ± 0.3
Cd^{2+}	1.7 ± 0.6
Mn^{2+}	6.1 ± 1.7
Sr^{2+}	8.6 ± 2.9

NOTE: The effect of various cations on unstimulated phagocytosis was examined in medium containing 1 mM Ca^{2+}. The apparent K_i values (means ± SEM) were determined from the results of three to four experiments. The extent of inhibition by La^{3+} was 13% at 0.1 mM, the highest concentration tested.

The situation is rather different for FMLP-stimulated chemotaxis. The lack of effect of benzamil on the migratory behavior of neutrophils emphasizes that a Ca^{2+} influx via Na^+-Ca^{2+} exchange seems to play little or no role. Presumably, the release of Ca^{2+} from internal stores is adequate to sustain chemotaxis. Alternatively, the normal resting level of cytosolic Ca^{2+} may be sufficient for a nominal chemotactic response and no increase is needed. However, optimal chemotaxis could still be dependent on a very small amount of Ca^{2+} influx from the exterior. Because benzamil only partially blocks Na^+-Ca^{2+} exchange, the requirement for a Ca^{2+} influx could still be met by the residual carrier activity during the course of the assay.

On the other hand, F_c-mediated phagocytosis seems to be exquisitely dependent on an influx of Ca^{2+} via Na^+-Ca^{2+} exchange as reflected in the marked sensitivity of the process to low doses of benzamil and related analogues. It is enigmatic as to why phagocytosis is benzamil-sensitive while chemotaxis is not. Furthermore, the identity of the putative Ca^{2+}-sensitive step(s) in phagocytosis remains a mystery. Perhaps this is related to the fact that intracellular pH (and in particular an intracellular alkalinization by way of FMLP-activated Na^+-H^+ exchange) plays a regulatory role in chemotaxis[40] but appears to be inconsequential for the phagocytosis of IgG-coated SRBC (unpublished observations). These points leave little doubt that we are dealing with three very different phenomena and that O_2^- generation, chemotaxis, and phagocytosis lie along separate activation schemes.

ACKNOWLEDGMENTS

We acknowledge the expert technical assistance of Margaret A. Foy and Jacquelyn T. Engle and the secretarial skills of Annette Irving. We also thank Drs. Paul De Weer, Albert Roos, Eric Brown, and Carlos Rosales for their helpful discussions while the work was in progress.

REFERENCES

1. SAWYER, D. W., J. A. SULLIVAN & G. L. MANDELL. 1985. Intracellular free calcium localization in neutrophils during phagocytosis. Science **230:** 663-666.

2. LEW, D. P., T. ANDERSSON, J. HED, F. DI VIRGILIO, T. POZZAN & O. STENDAHL. 1985. Ca^{2+}-dependent and Ca^{2+}-independent phagocytosis in human neutrophils. Nature (London) **315:** 509-511.
3. LEW, D. P., A. MONOD, F. A. WALDVOGEL, B. DEWALD, M. BAGGIOLINI & T. POZZAN. 1986. Quantitative analysis of the cytosolic free calcium dependency of exocytosis from three subcellular compartments in intact human neutrophils. J. Cell Biol. **102:** 2197-2204.
4. ROMEO, D., G. ZABUCCHI, N. MIANI & F. ROSSI. 1975. Ion movement across leukocyte plasma membrane and excitation of their metabolism. Nature (London) **253:** 542-544.
5. BECKER, E. L., M. SIGMAN & J. M. OLIVER. 1979. Superoxide production induced in rabbit polymorphonuclear leukocytes by synthetic chemotactic peptides and A23187. Am. J. Pathol. **95:** 81-98.
6. COOKE, E., F. A. AL-MOHANNA & M. B. HALLET. 1989. Ca^{2+}-dependent and independent mechanisms in neutrophil activation; roles of kinase C, diacylglycerol, and unidentified intracellular messengers. *In* The Neutrophil: Cellular Biochemistry and Physiology. M. B. Hallet, Ed.: 219-241. CRC Press. Boca Raton, FL.
7. SIMCHOWITZ, L. & E. J. CRAGOE, JR. 1988. Na^{+}-Ca^{2+} exchange in human neutrophils. Am. J. Physiol. **254**(Cell Physiol. 23): C150-C164.
8. BLAUSTEIN, M. P. & M. T. NELSON. 1982. Sodium-calcium exchange: Its role in the regulation of cell calcium. *In* Membrane Transport of Calcium. E. Carafoli, Ed.: 217-236. Academic Press. New York.
9. DIPOLO, R. & L. BEAUGE. 1983. The calcium pump and sodium-calcium exchange in squid axons. Annu. Rev. Physiol. **45:** 313-324.
10. PHILIPSON, K. D. 1985. Sodium-calcium exchange in plasma membrane vesicles. Annu. Rev. Physiol. **47:** 561-571.
11. SIMCHOWITZ, L. & E. J. CRAGOE, JR. 1986. Inhibition of chemotactic factor-activated Na^{+}/H^{+} exchange in human neutrophils by analogues of amiloride: Structure-activity relationships in the amiloride series. Mol. Pharmacol. **30:** 112-120.
12. SIMCHOWITZ, L. 1985. Intracellular pH modulates the generation of superoxide radicals by human neutrophils. J. Clin. Invest. **76:** 1079-1089.
13. REEVES, J. P. & C. C. HALE. 1984. The stoichiometry of the cardiac sodium-calcium exchange system. J. Biol. Chem. **259:** 7733-7739.
14. EISNER, D. A. & W. J. LEDERER. 1985. Na-Ca exchange: Stoichiometry and electrogenicity. Am. J. Phyisol. **248**(Cell Physiol. 17): C189-C202.
15. VON TSCHARNER, V., B. PROD'HOM, M. BAGGIOLINI & H. REUTER. 1986. Ion channels in human neutrophils activated by a rise in free cytosolic calcium concentration. Nature **324:** 369-372.
16. KRAUSE, K.-H. & M. J. WELSH. 1990. Voltage-dependent and Ca^{2+}-activated ion channels in human neutrophils. J. Clin. Invest. **85:** 491-498.
17. SIMCHOWITZ, L., M. A. FOY & E. J. CRAGOE, JR. 1990. A role for Na^{+}/Ca^{2+} exchange in the generation of superoxide radicals by human neutrophils. J. Biol. Chem. **265:** 13449-13456.
18. GRINSTEIN, S. & W. FURUYA. 1988. Receptor-mediated activation of electropermeabilized neutrophils: Evidence for a Ca^{2+} and protein kinase C-independent signalling pathway. J. Biol. Chem. **263:** 1779-1783.
19. NASMITH, P. & S. GRINSTEIN. 1989. Diacylglycerol kinase inhibitors R59022 and dioctanoylethyleneglycol potentiate the respiratory burst of neutrophils by raising cytosolic Ca^{2+}. Biochem. Biophys. Res. Commun. **161:** 95-100.
20. SIMCHOWITZ, L. & E. J. CRAGOE, JR. 1990. Polyvalent cations inhibit human neutrophil chemotaxis by interfering with the polymerization of actin. J. Biol. Chem. **265:** 13457-13463.
21. ARSLAN, P., F. DI VIRGILIO, M. BELTRAME, R. Y. TSIEN & T. POZZAN. 1985. Cytosolic Ca^{2+} homeostasis in Ehrlich and Yoshida carcinomas: A new, membrane-permeant chelator of heavy metals reveals that these ascites tumor cell lines have normal cytosolic free Ca^{2+}. J. Biol. Chem. **260:** 2719-2727.
22. ANDERSSON, T., C. DAHLGREN, T. POZZAN, O. STENDAHL & P. D. LEW. 1986. Characterization of fMet-Leu-Phe receptor-mediated Ca^{2+} influx across the plasma membrane of human neutrophils. Mol. Pharmacol. **30:** 437-443.

23. Taylor, D. L. & J. S. Condeelis. 1979. Cytoplasmic structure and contractility in amoeboid cells. Int. Rev. Cytol. **56:** 57-144.
24. Korn, E. D. 1982. Actin polymerization and its regulation by proteins from nonmuscle cells. Physiol. Rev. **62:** 672-737.
25. White, J. R., P. H. Naccache & R. I. Sha'afi. 1983. Stimulation by chemotactic factor of actin association with the cytoskeleton in rabbit neutrophils. Effects of calcium and cytochalasin B. J. Biol. Chem. **258:** 14041-14047.
26. Howard, T. H. & W. H. Meyer. 1984. Chemotactic peptide modulation of actin assembly and locomotion in neutrophils. J. Cell Biol. **98:** 1265-1271.
27. Wallace, P. J., R. P. Wersto, C. H. Packman & M. A. Lichtman. 1984. Chemotactic peptide-induced changes in neutrophil actin conformation. J. Cell Biol. **99:** 1060-1065.
28. Sklar, L. A., G. M. Omann & R. G. Painter. 1985. Relationship of actin polymerization and depolymerization to light scattering in human neutrophils: Dependence on receptor occupancy and intracellular Ca^{2+}. J. Cell Biol. **101:** 1161-1166.
29. Howard, T. H. & D. Wang. 1987. Calcium ionophore, phorbol ester, and chemotactic peptide-induced cytoskeleton reorganization in human neutrophils. J. Clin. Invest. **79:** 1359-1364.
30. Brown, E. J., J. F. Bohnsack & H. D. Gresham. 1988. Mechanism of inhibition of immunoglobulin G-mediated phagocytosis by monoclonal antibodies that recognize the Mac-1 antigen. J. Clin. Invest. **81:** 365-375.
31. Gresham, H. D., J. A. McGarr, P. G. Shackelford & E. J. Brown. 1988. Studies on the molecular mechanisms of human F_c receptor-mediated phagocytosis. Amplification of ingestion is dependent on the generation of reactive oxygen metabolites and is deficient in polymorphonuclear leukocytes from patients with chronic granulomatous disease. J. Clin. Invest. **82:** 1192-1201.
32. Gresham, H. D., A. Zheleznyak, J. S. Mormol & E. J. Brown. 1990. Studies on the molecular mechanisms of human neutrophil F_c receptor-mediated phagocytosis: Evidence that a distinct pathway for activation of the respiratory burst results in reactive oxygen metabolite-dependent amplification of ingestion. J. Biol. Chem. **265:** 7819-7826.
33. Silverstein, S. C., S. Greenberg, F. Di Virgilio & T. H. Steinberg. 1989. Phagocytosis. *In* Fundamental Immunology, 2nd Edition. W. E. Paul, Ed.: 703-720. Raven Press. New York.
34. Lew, D. P., T. Andersson, J. Hed, F. Di Virgilio, T. Pozzan & O. Stendahl. 1985. Ca^{2+}-dependent and Ca^{2+}-independent phagocytosis in human neutrophils. Nature **315:** 509-511.
35. Grynkiewicz, G., M. Poenie & R. Y. Tsien. 1985. A new generation of Ca^{2+} indicators with greatly improved fluorescence properties. J. Biol. Chem. **260:** 3440-3450.
36. Malech, H. L., R. K. Root & J. I. Gallin. 1977. Structural analysis of human neutrophil migration. J. Cell Biol. **75:** 666-693.
37. Zigmond, S. H. & S. J. Sullivan. 1979. Sensory adaptation of leukocytes to chemotactic peptides. J. Cell Biol. **82:** 517-527.
38. Keller, H. U. 1983. Motility, cell shape and locomotion of neutrophil granulocytes. Cell Motil. **3:** 47-60.
39. Tilney, N. G., D. P. Kiehart, C. Sardet & M. Tilney. 1978. Polymerization of actin. IV. Role of Ca^{2+} and H^+ in the assembly of actin and in membrane fusion in the acrosomal reaction of echinoderm sperm. J. Cell Biol. **77:** 536-550.
40. Simchowitz, L. & E. J. Cragoe, Jr. 1986. Regulation of human neutrophil chemotaxis by intracellular pH. J. Biol. Chem. **261:** 6492-6500.

Norepinephrine and Catecholamine Release from Peripheral Sympathetic Nerves and Chromaffin Cells Maintained in Primary Tissue Culture

The Role of Sodium-Calcium Exchange

TAMÁS L. TÖRÖK[a]

Department of Pharmacodynamics
Semmelweis University of Medicine
H-1445 Budapest, Nagyvárad-tér 4
P.O.B.: 370, Hungary

Intracellular free Ca^{2+} ions play a fundamental role in "excitation-secretion coupling."[1,2] Under steady-state conditions the Na^+ gradient, created by the Na^+ pump (Na^+,K^+-activated ATPase),[3] is partly responsible for the low level of Ca^{2+} inside ($\sim 10^{-7}$ M) through the Na^+-Ca^{2+}-exchange.[4–8] Na^+-Ca^{2+} exchange, like the Na^+ pump, is electrogenic and voltage-sensitive.[6,9–16].

Inhibition of the Na^+ pump either by cardiac glycosides or by removal of K^+ from the external medium lead to a rise in internal Na^+. The Na^+ gained inside enhances the influx of Ca^{w+} through the reverse Na^+-Ca^{2+} exchange.[6,17–20] Because the exchanger is voltage dependent, depolarization, which is characteristic for Na^+-pump inhibition,[21-32] itself may increase the influx of Ca^{2+} through reverse Na^+-Ca^{2+} exchange.[20,33–36]

A further characteristic of $[Na^+]_i$-$[Ca^{2+}]_o$ exchange is that, like the Na^+ pump,[24,25,37–40] it can be activated by external monovalent cations such as K^+, Rb^+, or Li^+.[6,10,13,15,41] Thus when the Na^+ pump is reactivated, for example, by readmission of K^+ to Na^+-enriched and K^+-depleted preparations, reverse Na^+-Ca^{2+} exchange is possibly also transiently activated resulting a transient net Ca^{2+} gain inside in smooth muscle of guinea-pig taenia caeci without producing tension response.[42]

In this paper, peripheral sympathetic nerves in isolated pulmonary artery are compared with primary cultured bovine chromaffin cells.

In the isolated main pulmonary artery of the rabbit inhibition of Na^+-pump by removal of K^+ from the external medium increased the release of labeled norepineph-

[a]Present address: Department of Pharmacology and Therapeutics, University of Manitoba, Faculty of Medicine, 770 Bannatyne Avenue, Winnipeg, Manitoba, Canada R3E OW3.

rine (NE)[43] (FIG. 1). During the first period of K^+-free perfusion, the release of [^{3}H]NE was absolutely dependent on the presence of external Ca^{2+}; that is, in Ca^{2+}-free (+1 mM EGTA) solution, transmitter release was abolished. When Ca^{2+} was readmitted to K^+-free solution, the release of neurotransmitter was increased again (FIG. 1), and the rate of release was directly proportional to the preceeding perfusion period with K^+-free solution, being greater for longer exposure times (FIG. 2).

In Na^+-pump-inhibited arteries, Ca^{2+}-removal only transiently inhibited the release of [^{3}H]NE, since after a delay lasting about 90-120 min, the release of neurotransmitter started to increase again (FIG. 3). This is consistent with the $[Na^+]_i$-caused Ca^{2+} release from internal stores.[27,42–47] FIGURE 3 also shows that after reactivation of the Na^+ pump by readmission of K^+ to K^+-free solution, the subsequently readmitted Ca^{2+} failed to produce transmitter release (see also TABLE 1). A likely explanation is that the Na^+ gradient is re-established on Na^+-pump reactivation, therefore Ca^{2+} cannot enter into the nerves through reverse Na^+-Ca^{2+} exchange. In order to study further the internal Na^+ requirement for transmitter release, low external Na^+ (26.2 mM) was used (FIG. 4). In these experiments 113 mM external Na^+ was replaced by Li^+ during the second hour of K^+-free perfusion. In Li^+ solution, the release of [^{3}H]NE slightly increased. When Ca^{2+} was readmitted in low Na^+-containing solution, the release of transmitter was significantly less than in normal Na^+-containing medium. The release decreased from 11.0 ± 2.0 (n = 4) to 0.9 ± 0.1 (n = 5) Δpmol/6 min (see also FIG.

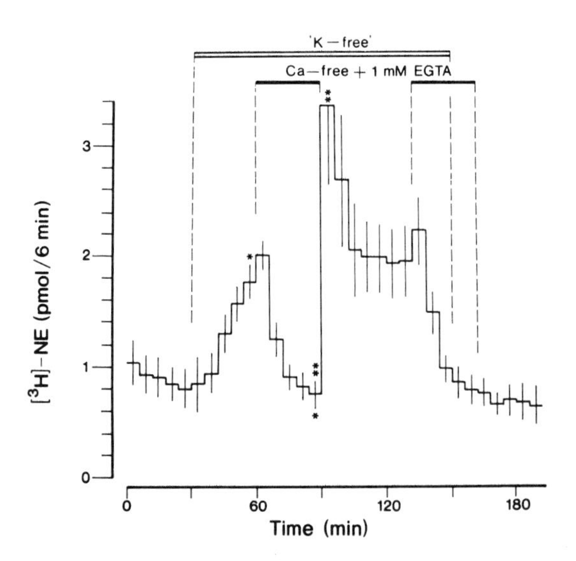

FIGURE 1. The effect of K^+-free and Ca^{2+}-free (+ 1 mM EGTA) solution on spontaneous [^{3}H]NE release from rabbit pulmonary arteries in the presence of neuronal (cocaine, 3×10^{-5} M) and extraneuronal (corticosterone, 5×10^{-5} M) uptake blockers. The release of [^{3}H]NE was calculated in pmol/6 min (*ordinate*) and was plotted against time (min; *abscissa*). K^+-removal increased the [^{3}H]NE release, an action antagonized by removal of Ca^{2+} from the external medium. On Ca^{2+} readmission, the [^{3}H]NE release increased in K^+-free solution (60 min Na^+ loading). After 120 min K^+-free perfusion, K^+ (5.9 mM) was readmitted in Ca^{2+}-free solution. The subsequently readmitted Ca^{2+} failed to produce transmitter release. Means ±SEM of four identical experiments are given. Significant differences between values: *$p < 0.01$; **$p < 0.02$. (Reprinted from Magyar *et al.*[43] with permission.)

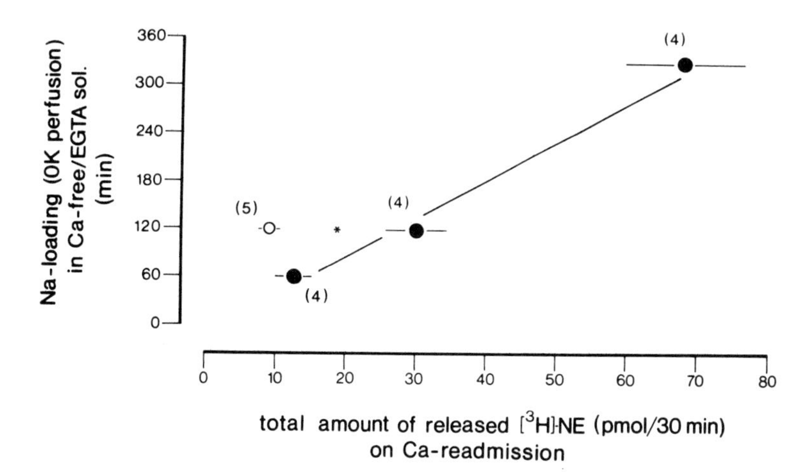

FIGURE 2. Correlation between Na^+-loading periods (K^+-free perfusion) in Ca^{2+}-free (+1 mM EGTA) solution (*ordinate*; min), and the release of [^{3}H]NE in response to Ca^{2+}-readmission (*abscissa*; pmol/30 min). The readmitted Ca^{2+}-induced [^{3}H]NE release was directly related to the preceeding Na^+-loading periods (60, 120, and 330 min). Open circle: 60 min before Ca^{2+}-readmission, 113 mM Li^+ was substituted for Na^+ in K^+-free solution. Note that Li^+ substitution significantly inhibited the release of [^{3}H]NE on Ca^{2+} readmission (*$p < 0.005$). Means ±SEM of four or five identical experiments are given (previously unpublished data of T. L. Török).

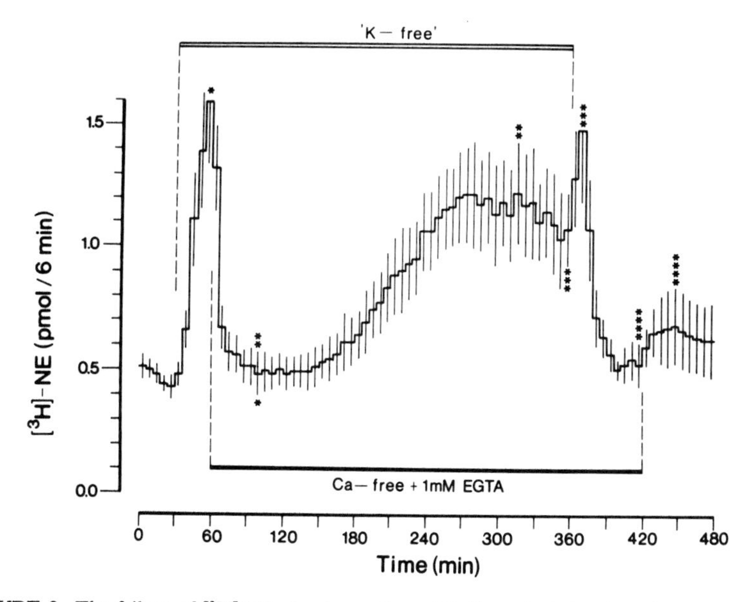

FIGURE 3. The failure of [^{3}H]NE-releasing action of Ca^{2+} readmission in pulmonary arteries after reactivation of Na^+ pump. Readmission of K^+ in Ca^{2+}-free solution after 330 min K^+-free perfusion abolished the release of [^{3}H]NE after a small transient increase. Later on (60 min), the readmitted Ca^{2+} was ineffective in releasing [^{3}H]NE. Means ±SEM of four identical experiments are given. Differences between values: *$p < 0.01$; **$p < 0.02$; ***$p > 0.2$; ****$p > 0.4$. (Reprinted from Magyar *et al.*[43] with permission.)

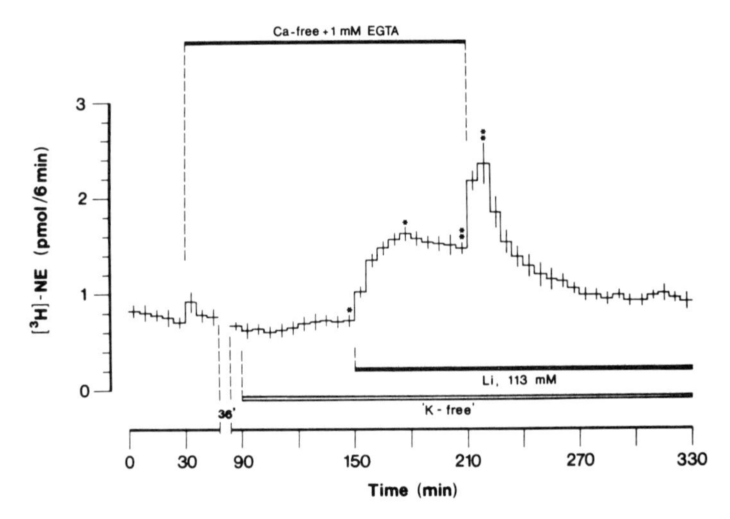

FIGURE 4. The inhibitory effect of Li^+ (113 mM) on [^{3}H]NE release induced by readmission of Ca^{2+} to K^+-free solution. After 60 min Ca^{2+}-free (+1 mM EGTA) perfusion, K^+ (5.9 mM) was also removed in normal Na^+ (139.2 mM)-containing medium. Sixty minutes later external Na^+ was reduced to 26.2 mM. Note that the release of [^{3}H]NE slightly increased in Li^+ solution. Later on (60 min), the readmitted Ca^{2+} slightly increased the release of [^{3}H]NE. Means ±SEM of five experiments. Significant differences between values: $*p < 0.001$; $**p < 0.05$. (Previously unpublished data of T. L. Török.)

2, open symbol). This suggests that internal Li^+ cannot substitute for internal Na^+ in activating the exchanger.

When Ca^{2+} and K^+ were readmitted together to Na^+-loaded arteries the release of [^{3}H]NE transiently increased (FIG. 5), and it was significantly less than when Ca^{2+} was readmitted alone (TABLE 1, compare treatment 1 with 3). A similar phenomenon was observed on cholinergic nerves[48] and Purkinje fibers.[49] Different alkali metal ions (Rb^+, Cs^+, and Li^+) could substitute for K^+ in inhibiting the transmitter release on Ca^{2+}-readmission[42,43] (TABLE 1). The calculated Na^+-pump activation values were as follows: Rb^+, 1.1; K^+, 1.0; Cs^+, 0.32; Li^+, 0.19.

The chromaffin cells of adrenal medulla are a highly specialized form of postganglionic sympathetic neurons. Previously it was shown in bovine chromaffin cells maintained in primary tissue culture, that internal Na^+ and external Ca^{2+} are required for catecholamine (CA) release induced by ouabain.[50] FIGURE 6 shows that ouabain (10^{-4}M) in Ca^{2+}-free (+1 mM EGTA) solution was ineffective in releasing CAs during a 60-min incubation period. On Ca^{2+} readmission however the CA-release markedly increased. The CA-releasing action of ouabain was external Na^+-dependent since in choline solution (5 mM Na^+), the readmitted Ca^{2+} failed to induce significant release of CAs (FIG. 6). In these experiments TTX (10^{-6}M) and ouabain (10^{-4}M) were present throughout to block ion fluxes through the Na^+ channel and Na^+ pump, respectively. The CA release from Na^+-loaded (10^{-4}M ouabain in Ca^{2+}-free/EGTA solution for 60 min) cells was directly proportional to the concentration of Ca^{2+} readmitted (FIG. 7), and further increased when Ca^{2+} was readmitted in low Na^+ (5 mM)-containing

TABLE 1. [^{3}H]Norepinephrine Release in Response to Readmission of 2.5 mM Ca^{2+} after 330 Min Na^+ Loading (K^+-Free Perfusion) Period and in Response to Readmission of Different Alkali Metal Ions (5.9 mM) in the Presence and Absence of External Ca^{2+}

Treatment	[^{3}H]NE Release (Δpmol/6 min)[a]	p (significance)	Ratio
1. Readmission of Ca^{2+} in K^+-free solution[b]	19.9 ± 2.9 $p < 0.001$ (4)		
2. Readmission of K^+ in Ca^{2+}-free solution[b]	0.4 ± 0.2 $p > 0.2$ (4)	2/1 $p < 0.001$	
3. Readmission of K^+ and Ca^{2+} together[b]	3.0 ± 0.7 $p < 0.02$ (4)	3/1 $p < 0.01$ 3/2 $p < 0.01$	3/4 1.1 3/5 0.32 3/6 0.19
4. Readmission of Rb^+ and Ca^{2+} together[b]	2.8 ± 0.9 $p < 0.02$ (4)	4/3 $p > 0.8$	
5. Readmission of Cs^+ and Ca^{2+} together[b]	9.3 ± 1.2 $p < 0.001$ (4)	5/3 $p < 0.01$	
6. Readmission of Li^+ and Ca^{2+}together[b]	15.6 ± 3.8 $p < 0.001$ (4)	6/3 $p < 0.02$	

[a] Peak release minus the last resting output obtained before treatment (cocaine, 3 × 10^{-5} M and corticosterone, 5 × 10^{-5} M are present throughout). Means ± SEM are given; number of experiments in parentheses.

[b] 330-min pre-perfusion with K^+-free and 300 min with Ca^{2+}-free (+ 1 mM EGTA) solution before Ca^{2+} and/or alkali metal ionreadmission.

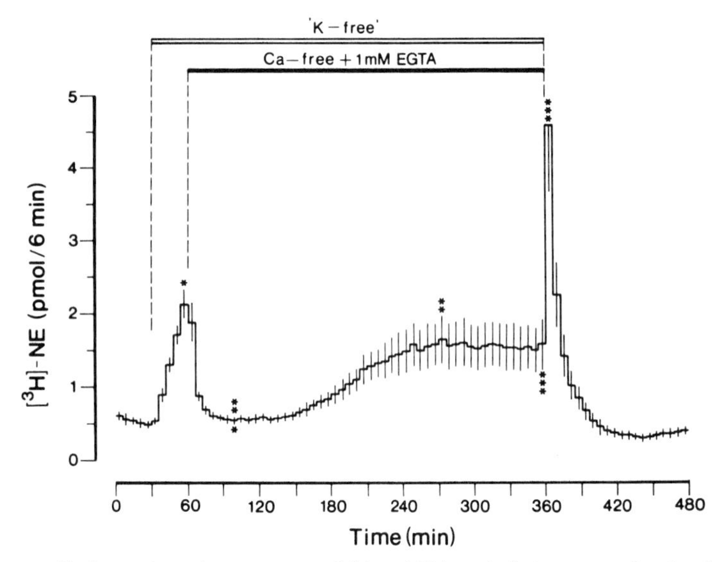

FIGURE 5. [^{3}H]NE release in response to Ca^{2+} and K^+ readmission to arteries that had been pre-perfused with K^+- and Ca^{2+}-free solution. When Ca^{2+} and K^+ were readmitted together, the release of [^{3}H]NE transiently increased. Means ± SEM of four identical experiments are given. Significant differences between values: *$p < 0.001$; **$p < 0.05$; ***$p < 0.02$. (Reprinted from Magyar *et al.*[43] with permission.)

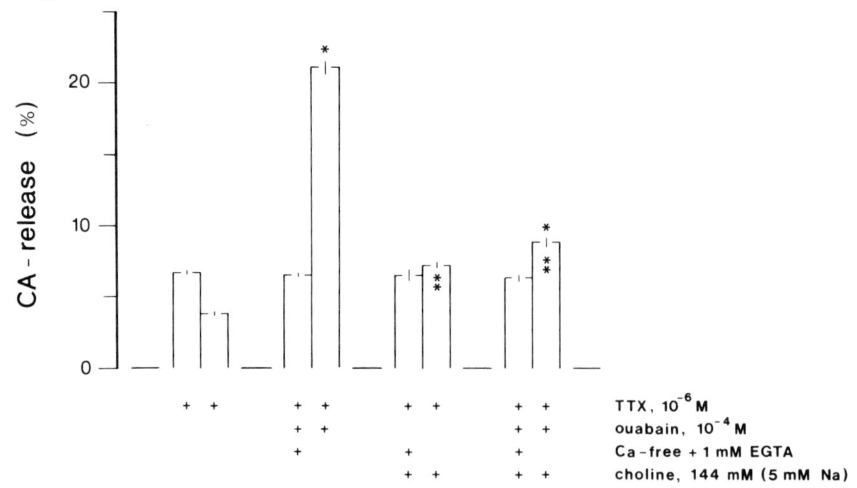

FIGURE 6. External Ca^{2+}- and Na^+-dependent CA releasing action of ouabain (10^{-4} M) in bovine chromaffin cells maintained in primary tissue culture. In this figure (and also in FIGS. 8-11) each pair of columns shows data from two sequential 60-min incubation periods: The first column shows CA release over the period 0-60 min; the second column shows the release of CAs from the same cells over the period 60-120 min. CA release was expressed as a percentage of the total initial cell CA content. Ouabain failed to produce CA release in Ca^{2+}-free (+1 mM EGTA) solution (first column of second pairs of data). Readmission of Ca^{2+} to ouabain-treated cells markedly increased the release of CAs, but only if normal external Na^+ (149 mM) was present during the first 60-min incubation period (compare second columns of second and fourth pairs of data). Note, that TTX (10^{-6} M) was present throughout. Further note that Ca^{2+} was readmitted in the presence of ouabain. Means ± SEM of six determinations are given. Significant differences between values: *$p < 0.0005$; **$p < 0.0025$. (Previously unpublished data of T. L. Török.)

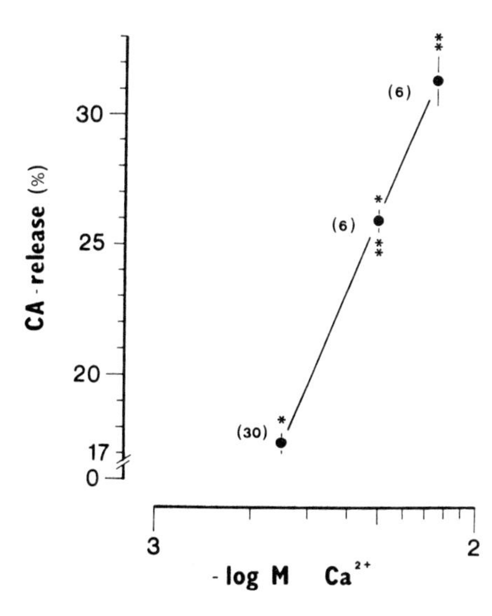

FIGURE 7. Correlation between Na^+-loading (10^{-4} M ouabain in Ca^{2+}-free (+1 mM EGTA) solution for 60 min) and the subsequently readmitted Ca^{2+} (2.5, 5.0 and 7.5 mM)-induced CA release. *Ordinate:* CA release (%); *abscissa:* −log M concentrations of Ca^{2+}. Note that TTX (10^{-6} M) and ouabain (10^{-4} M) were present throughout. Means ± SEM are given. Number of determinations in parentheses. Significant differences between values: $*p \ll 0.005$; $**p < 0.0005$. (Previously unpublished figure of T. L. Török.)

solution (Li^+, choline$^+$, Tris$^+$, or sucrose substitution), though Li^+ was less effective than the other Na^+ substitutes.[50] These results are consistent with Na^+-Ca^{2+} competition at the external site of the exchanger[6,51] and, also, that Li^+ can substitute to some extent for Na^+ in Na^+-Ca^{2+} competition.[6]

In chromaffin cells, as in squid axon[6,10,11,15,52–54] external monovalent cations activate the $[Na^+]_i/[Ca^{2+}]_o$-exchange.[42,50] This was evidenced by the finding that removal of K^+ from the external medium significantly inhibited the CA release evoked by readmission of Ca^{2+} to ouabain-treated cells.[50] On the other hand a small increase of external K^+ (10.8 mM; two times of normal) further potentiated the CA release on Ca^{2+} readmission (FIG. 8). External K^+ was also required to increase the CA release from Na^+-loaded cells when the Ca^{2+} binding was increased either by readmission of elevated Ca^{2+} (7.5 mM, FIG. 9), or by readmission of normal Ca^{2+} (2.5 mM) in low Na^+ (5 mM)-containing medium (Li^+, FIG. 10 or choline$^+$, FIG. 11 substitution).

Different alkali metal ions (Rb^+, Cs^+, and Li^+) could substitute for K^+ in enhancing the CA release from Na^+-loaded cells on Ca^{2+} readmission (TABLE 2). The calculated $[Na^+]_i$-$[Ca^{2+}]_o$ exchange activation values were as follows: Rb^+, 1.22; K^+, 1.0; Cs^+, 0.33; Li^+, 0.19.

In conclusion, the $[Na^+]_i$-$[Ca^{2+}]_o$ exchange plays a significant role in the "excitation-secretion coupling" of peripheral sympathetic nerves and adrenal medullary chromaffin cells. Recently it has been shown that there is amino-acid sequence similarity between a region of the alpha-subunit of Na^+, K^+-ATPase[55] and Na^+-Ca^{2+}-exchange[56] proteins. This similar structural (and perhaps functional) region of the two proteins may be identical with the external activation sites of $[Na^+]_i$-$[K^+]_o$-exchange pump and $[Na^+]_i$-$[Ca^{2+}]_o$ exchange, since activation of these sites by monovalent cations is very similar in sympathetic nerves and chromaffin cells.

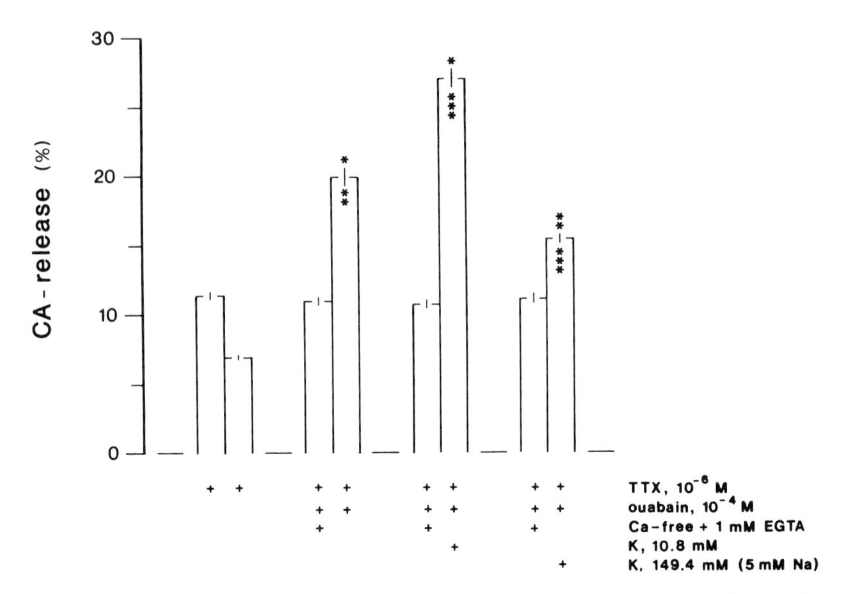

FIGURE 8. The effect of different K^+-concentrations on CA release induced by Ca^{2+} readmission to Na^+-loaded cells. See legend to FIGURE 6 for general description of experiment. Cells were preincubated with normal K^+ (5.4 mM), and ouabain-containing Ca^{2+}-free solution for 60 min (TTX was present). Subsequently Ca^{2+} was readmitted in the presence of normal K^+ (second column of second pair of data), slightly elevated K^+ (10.8 mM; second column of third) and excess K^+ (149.4 mM; second column of fourth pair of data) containing medium. Note, that TTX and ouabain were present throughout. Further note that slightly elevated K^+ (10.8 mM; two times normal) potentiated the readmitted Ca^{2+}-induced CA release, while excess K^+ (149.4 mM) had the opposite effect. Means ±SEM of six determinations are given. Significant differences between values: *$p < 0.005$; **$p < 0.0125$; ***$p < 0.0005$. (Previously unpublished figure of T. L. Török.)

TABLE 2. The Effect of Different Alkali Metal Ions on CA Release Evoked by Readmission of Ca^{2+} to Na^+-Loaded Chromaffin Cells

	CA Release (%)[a]			
	10^{-4}M Ouabain in Ca^{2+}-Free Solution (1)	Readmission of Ca^{2+} (2)	(Δ%)	Ratio
5.4 mM Rb^+	9.5 ± 0.3	20.7 ± 0.5	11.2	1.22
5.4 mM K^+	9.4 ± 0.3	18.6 ± 0.4	9.2	
5.4 mM Cs^+	8.4 ± 0.3	11.1 ± 0.3	2.7	0.33
5.4 mM K^+	8.3 ± 0.2	16.6 ± 0.4	8.3	
5.4 mM Li^+	8.5 ± 0.3	10.4 ± 0.2	1.9	0.19
5.4 mM K^+	8.7 ± 0.2	18.5 ± 0.3	9.8	

[a] The cells were loaded with Na^+ by using ouabain (10^{-4}) for 60 min in TTX (10^{-6}M)-containing Ca^{2+}-free (+ 1 mM EGTA) solution in the presence of 5.4 mM K^+.[1] Subsequently, 2.5 mM Ca^{2+} was readmitted for 60 min in the presence of ouabain and TTX either in K^+ or in Rb^+, Cs^+, and Li^+ (5.4 mM)-containing solution.[2] Means ± SEM of six determinations.

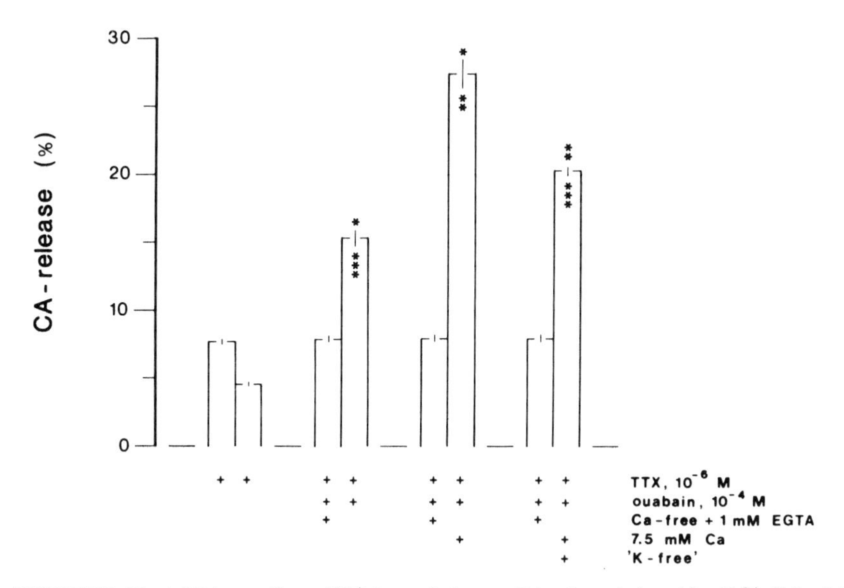

FIGURE 9. The inhibitory effect of K^+-free solution on CA release induced by Ca^{2+} (7.5 mM) readmission to Na^+-loaded cells. Cells were loaded with Na^+ by using ouabain (10^{-4} M) in Ca^{2+}-free solution for 60 min in the presence of 5.4 mM K^+ (TTX was present). Subsequently, normal Ca^{2+} (second column of second pair of data), and elevated Ca^{2+} (7.5 mM) was readmitted in the presence (second column of third pair of data) and in the absence (second column of fourth pair of data) of external K^+. Note that TTX and ouabain were present. Further note that 7.5 mM Ca^{2+} further enhanced the CA release from Na^+-loaded cells, and this was inhibited by K^+removal. Means ±SEM of six determinations are given. Significant differences between values: *$p \ll 0.0005$; **$p < 0.0005$; ***$p < 0.0005$. (Previously unpublished figure of T. L. Török.)

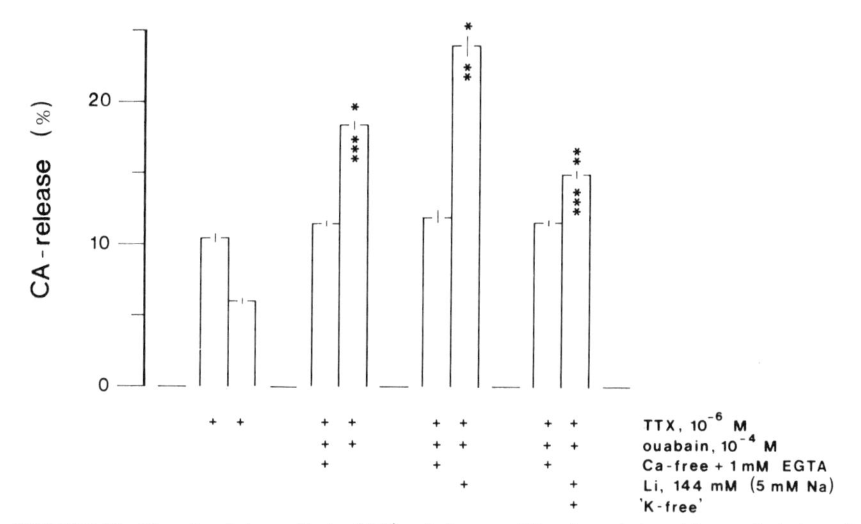

FIGURE 10. The stimulatory effect of Li^+ solution on CA release induced by readmission of Ca^{2+} to Na^+-loaded cells and its external K^+ dependence. Cells were loaded with Na^+ by using ouabain in Ca^{2+}-free solution, containing normal Na^+ (149 mM) and K^+ (5.4 mM). Subsequently Ca^{2+} was readmitted in the presence of normal Na^+ (second column of second pair of data) and low Na^+ (5.0 mM)-containing solution in the presence (second column of third pair of data) and in the absence (second column of fourth pair of data) of external K^+. Note that TTX and ouabain were present. Further note that in Li^+ solution the release of CAs further increased, but this was inhibited by K^+ removal. Means ±SEM of six determinations are given. Significant differences between values: *$p < 0.005$; **$p \ll 0.0005$; ***$p < 0.0005$. (Previously unpublished figure of T. L. Török.)

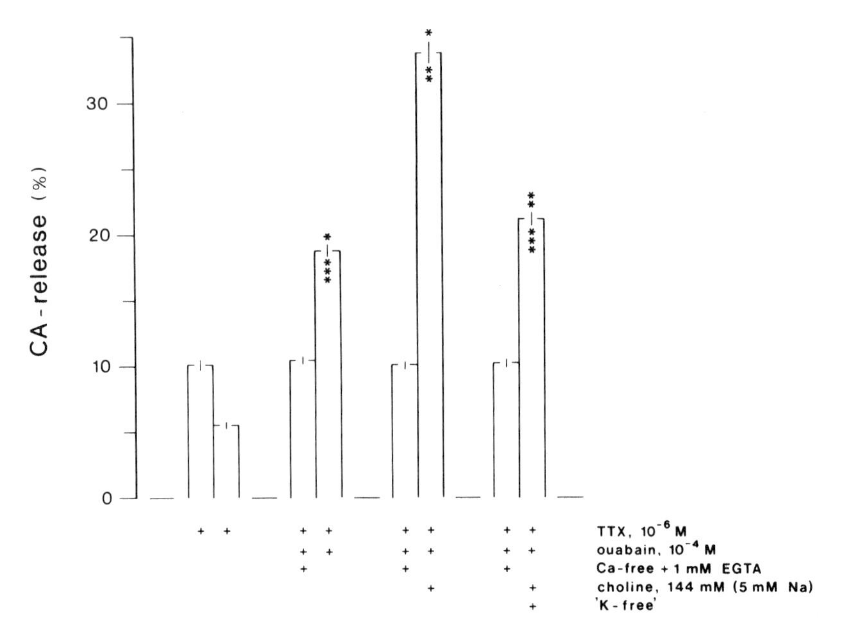

FIGURE 11. As in FIGURE 10, except that choline substitution was used. Note that choline was more effective than Li^+ in enhancing the CA release from Na^+-loaded cells on Ca^{2+} readmission (compare with FIG. 10). Further note that K^+ removal inhibited the CA release in choline solution. Means ± SEM of six determinations. Significant differences between values: *$p < 0.0005$; **$p < 0.0005$; ***$p < 0.025$. (Previously unpublished figure of T.L. Török.)

ACKNOWLEDGMENTS

I thank Ms. Carroll Powney for secretarial assistance and Professors F. S. LaBella and C. V. Greenway for reading the manuscript.

REFERENCES

1. KATZ, B. 1969. The Release of Neural Transmitter Substances. University Press. Liverpool, U.K.
2. BAKER, P. F. 1972. *In* Progress in Biophysics and Molecular Biology. F. A. V. Butler & K. D. Noble, Eds. Vol. **24:** 177-223. Pergamon Press. New York.
3. SKOU, J. C. 1965. Physiol. Rev. **45:** 596-617.
4. BAKER, P. F. & M. P. BLAUSTEIN. 1968. Biochem. Biophys. Acta **150:** 167-170.
5. REUTER, H. & N. SEITZ. 1968. J. Physiol. London **195:** 451-470.
6. BAKER, P. F., M. P. BLAUSTEIN, A. L. HODGKIN & R. A. STEINHARDT. 1969. J. Physiol. London **200:** 459-496.
7. BLAUSTEIN, M. P. & A. L. HODGKIN. 1969. J. Physiol. London **200:** 497-527.
8. GLITSCH, H. G., H. REUTER & H. SCHOLTZ. 1970. J. Physiol. London **209:** 25-43.
9. BLAUSTEIN, M. P. 1974. Rev. Physiol. Biochem. Pharmacol. **70:** 33-81.
10. BLAUSTEIN, M. P. 1977. Biophys. J. **20:** 79-111.

11. BLAUSTEIN, M. P., J. M. RUSSELL & P. DE WEER. 1974. J. Supramol. Struct. **2:** 558-581.
12. MULLINS, L. J. & F. J. BRINLEY, JR. 1975. J. Gen. Physiol. **65:** 135-152.
13. BAKER, P. F. & R. DIPOLO. 1984. *In* Current Topics in Membrane Transport: The Squid Axon. P. F. Baker, Ed. Vol. **22:** 195-247. Academic Press. New York, London.
14. DIPOLO, R., F. BEZANILLA, F. CAPUTO & H. ROJAS. 1985. J. Gen. Physiol. **86:** 457-478.
15. ALLEN, T. J. A. & P. F. BAKER. 1986. J. Physiol. London **378:** 53-76.
16. BLAUSTEIN, M. P. 1988. *In* Handbook of Experimental Pharmacology. Vol. **83:** 275-304. P. F. Baker, Ed. Springer Verlag. Berlin, Heidelberg.
17. SHEU, S.-S. & H. A. FOZZARD. 1982. J. Gen. Physiol. **80:** 325-351.
18. BLAUSTEIN, M. P. & J. M. HAMLYN. 1983. Ann. Int. Med. **89(5):** (Part 2) 785-792.
19. EISNER, D. A., W. J. LEDERER & R. D. VAUGHAN-JONES. 1983. J. Physiol. London **335:** 443-470.
20. MULLINS, L. J., J. REQUENA & J. WHITTEMBURY. 1985. Proc. Natl. Acad. Sci. USA **82:** 1847-1851.
21. GLYNN, I. M. 1964. Pharmacol. Rev. **16:** 381-407.
22. KERKUT, G. A. & R. C. THOMAS. 1965. Comp. Biochem. Physiol. **14:** 167-183.
23. BAKER, P. F. 1966. Endeavour **25:** 166-172.
24. RANG, H. P. & J. M. RITCHIE. 1968. J. Physiol. London **196:** 183-221.
25. BAKER, P. F., M. P. BLAUSTEIN, R. D. KEYNES, J. MANIL, T. I. SHAW & R. A. STEINHARDT. 1969. J. Physiol. London **200:** 459-496.
26. THOMAS, R. C. 1972. Physiol. Rev. **52:** 563-594.
27. BAKER, P. F. & A. C. CRAWFORD. 1975. J. Physiol., London **247:** 209-226.
28. GLYNN, I. M. & S. J. D. KARLISH. 1975. Annu. Rev. Physiol. **37:** 13-55.
29. AKERA, T. & T. M. BRODY. 1985. Trends Pharmacol. Sci. **6:** 156-159.
30. GLITSCH, H. G. 1982. Annu. Rev. Physiol. **44:** 389-400.
31. GADSBY, D. 1984. Annu. Rev. Biophys. Bioeng. **13:** 373-398.
32. BLAUSTEIN, M. P. 1985. Trends Pharmacol. Sci. **6:** 289-292.
33. BAKER, P. F. & P. A. MCNAUGHTON. 1976. J. Physiol., London **259:** 103-144.
34. MULLINS, L. J. & J. REQUENA. 1981. J. Gen. Physiol. **78:** 683-700.
35. BAKER, P. F. 1986. *In* Calcium and the Cell.: 73-92. Wiley. Chichester, U.K.
36. REEVES, J. P. 1990. *In* Intracellular Calcium Regulation.: 305-347. Alan R. Liss, Inc. New York.
37. BAKER, P. F. & C. M. CONNELLY. 1966. J. Physiol., London **185:** 270-297.
38. SJODIN, R. A. & L. A. BEAUGÉ. 1967. Curr. Mod. Biol. **1:** 105-115.
39. SJODIN, R. A. & L. A. BEAUGÉ. 1968. J. Gen. Physiol. **52:** 389-407.
40. BEAUGÉ, L. A. 1984. *In* Current Topics in Membrane Transport: The Squid Axon. P. F. Baker, Ed. Vol. **22:** 131-175.
41. CONDRESCU, M., H. ROJAS, A. GERARDI, R. DIPOLO & L. BEAUGÉ. 1990. Biochim. Biophys. Acta **1024:** 198-202.
42. TÖRÖK, T. L. 1989. Progr. Neurobiol. **32:** 11-76.
43. MAGYAR, K., T. T. NGUYEN, T. L. TÖRÖK & P. T. TÓTH. 1987. J. Physiol., London **393:** 29-42.
44. CARAFOLI, E. & M. CROMPTON. 1978. Curr. Top. Memb. Transp. **10:** 151-216.
45. SCHOFFELMEER, A. N. M. & A. H. MULDER. 1983. J. Neurochem. **40:** 615-621.
46. TÖRÖK, T. L., Z. S. SALAMON, T. T. NGUYEN & K. MAGYAR. 1984. Q. J. Exp. Physiol. **69:** 841-865.
47. TÖRÖK, T. L. & K. MAGYAR. 1986. Q. J. Exp. Physiol. **71:** 105-114.
48. VIZI, E. S. 1977. J. Physiol. London **267:** 261-280.
49. CHAPMAN, R. A., H. A. FOZZARD, I. R. FRIEDLANDER & C. T. JANUARY. 1985. J. Physiol. London **364:** 78P.
50. TÖRÖK, T. L. & D. A. POWIS. 1990. Exp. Physiol. **75:** 573-586.
51. REEVES, J. P. & J. L. SUTKO. 1983. J. Biol. Chem. **258:** 3178-3182.
52. DIPOLO, R., H. ROJAS & L. A. BEAUGÉ. 1982. Cell Calc. **3:** 19-41.
53. DIPOLO, R. & L. A. BEAUGÉ. 1984. J. Gen. Physiol. **84:** 895-914.
54. DIPOLO, R. & L. A. BEAUGÉ. 1990. J. Gen. Physiol. **95:** 819-835.
55. SHULL, G. E., A. SCHWARTZ & J. B. LINGREL. 1985. Nature **316:** 691-695.
56. NICOLL, D. A., S. LONGONI & K. D. PHILIPSON. 1990. Science **250:** 562-565.

Sodium-Calcium Exchange in the Pancreatic B Cell[a]

ANDRÉ HERCHUELZ AND
PIERRE-OLIVIER PLASMAN

Laboratoire de Pharmacodynamie et de Thérapeutique
Université Libre de Bruxelles
Faculté de Médecine
B-1000 Bruxelles, Belgium

INTRODUCTION

It is unanimously accepted that calcium represents a key factor in the process of insulin release from the pancreatic B cell of the islet of Langerhans. When the B cell is exposed to glucose, the main physiological stimulus of insulin release, a complex series of events is initiated that culminate in a rise in cytosolic free-calcium concentration ($[Ca^{2+}]_i$).[1–4] Key events in this sequence can be summarized as follows. Through its metabolism, glucose closes ATP-sensitive potassium (K^+) channels.[5–6] The resulting decrease in K^+ conductance leads to membrane depolarization that opens voltage-sensitive Ca^{2+} channels.[7] It is well established that the opening of the latter channels is a major contributing factor to the rise in $[Ca^{2+}]_i$ induced by glucose in the pancreatic B cell.[1–4,8,9]

While the mechanism of Ca^{2+} entry into the B cell has been the subject of intensive investigation for many years, the mechanisms of Ca^{2+} extrusion from the cell have been more neglected and remain to be investigated in detail. Like most other cells, the B cell is thought to be equipped with a double system responsible for Ca^{2+} extrusion: a Ca^{2+}-ATPase[10–11] and a Na-Ca-exchange transport system. However, the relative contribution of the two processes to Ca^{2+} extrusion remains to be determined. It is the aim of this paper to review our present knowledge concerning Na-Ca exchange in the pancreatic B cell.

NA-CA EXCHANGE

Initial Observations

The first direct evidence for the existence of a Na-Ca exchanger in the pancreatic B cell was obtained more than a decade ago.[12–14] Thus, extracellular sodium ($Na^+{}_o$) removal was found to induce a fourfold increase in ^{45}Ca uptake by pancreatic islets[12] and to markedly decrease ^{45}Ca outflow from preloaded islets, perfused in the absence of extracellular Ca^{2+}($Ca^{2+}{}_o$).[13–16] However, on the basis of indirect evidence,[17] the existence of a Na-Ca exchanger, similar to that described in squid axons by Baker and Blaustein,[18] was postulated in the B cell already in 1968. Indeed, factors known to

[a] This work was supported in part by grants from the Belgian Foundation for Scientific Medical Research.

cause a cellular accumulation of Na^+ (e.g., ouabain and the absence of extracellular K^+_o) were observed to initiate, or at least to enhance, insulin release from the B cell. Likewise, Na^+_o removal induced an initial enhancement of insulin release. (For a review of these reports of indirect evidence, see Malaisse.[19])

Forward Na-Ca Exchange

Forward Na-Ca exchange or Na^+_o-dependent Ca^{2+}_i efflux has been investigated by monitoring ^{45}Ca outflow from prelabeled rat[15] or mouse[16] pancreatic islets. The process was studied in the absence of Ca^{2+}_o to suppress Ca-Ca exchange. Although the stoichiometry of the latter exchange has never been established, it is a common finding that the stimulation of Ca^{2+} entry into the islet cells, by whatever agent or condition, leads to an increase in ^{45}Ca outflow from perfused islets.[20] In the absence of Ca^{2+}_o, the isosmotical replacement of Na^+_o by choline,[15-16] lithium,[15-16] or K^+ (see below) salts or by sucrose[15-16] provokes a marked decrease in ^{45}Ca outflow (FIG. 1). The magnitude of the decrease in outflow averages 60 and 70%,[15,21] and the phenomenon is rapidly reversible.[15-16,21] The half-maximal activation of ^{45}Ca outflow by the reintroduction of Na^+_o is detected around 45 mM.[21] The reduction of the Na^+_o concentration to 12 mM by isosmotic replacement of Na^+ with choline also reduces ^{45}Ca outflow from insulin-releasing cells of a clonal cell line.[22] Further evidence for the existence of a forward Na-Ca exchange in the islet cells is that the dissipation of the Na^+ gradient across the B cell membrane by ouabain, which inhibits the Na^+ pump, also resulted in a decrease in ^{45}Ca outflow (followed by a secondary rise) from islets perfused in the absence of Ca^{2+}_o.[23] In two further studies, however, no initial decrease in ^{45}Ca outflow was observed, or at least considered significant, in response to ouabain[24] or the absence of K^+_o.[25]

Reverse Na-Ca Exchange

In early studies, Na^+_o removal or a lowering of its concentration was found to increase ^{45}Ca uptake.[12,15,26,27] Most of these measurements were carried out over long periods of time (20 min to 120 min[12,15,26]) or used choline salts for isosmotic replacement of Na^+.[12,26,27] Hence, part of the observed effects could have been due to cholinergic effects. In one of these studies, no effect of Na^+ removal was observed over a short-term incubation period.[26]

Early studies also evidenced a rapid and important increase in ^{45}Ca outflow from prelabeled islets perfused in the presence of Ca^{2+}_o, in response to the lowering of $[Na^+]_o$.[15,16] This increase in ^{45}Ca outflow has been interpreted within the frame of a process of Ca-Ca exchange induced by Ca^{2+} entry into the cell in response to Na^+_o removal.

Recently, reverse Na-Ca exchange was reinvestigated and examined in much more detail.[28] Na^+_i-dependent $^{45}Ca^{2+}$ uptake was measured in dissociated rat pancreatic islet cells over short incubation time periods (15 sec to 8 min), Na^+_o being replaced by sucrose.[28]

Removal of Na^+_o induces a dose-dependent increase in ^{45}Ca uptake (FIG. 2). In the complete absence of Na^+_o, ^{45}Ca uptake measured over 5 min is stimulated by about 615%. The uptake is biphasic with an initial and fast rate that is linear over the first 30 sec. This phase is followed by a slower one that is linear up to the eighth min (FIG. 3). The rate of uptake during the fast phase is 1,100 fmol $Ca^{2+} \cdot min^{-1} \cdot 1000\ cells^{-1}$, a value equivalent to that recorded in response to 50 mM K^+ over the same period of

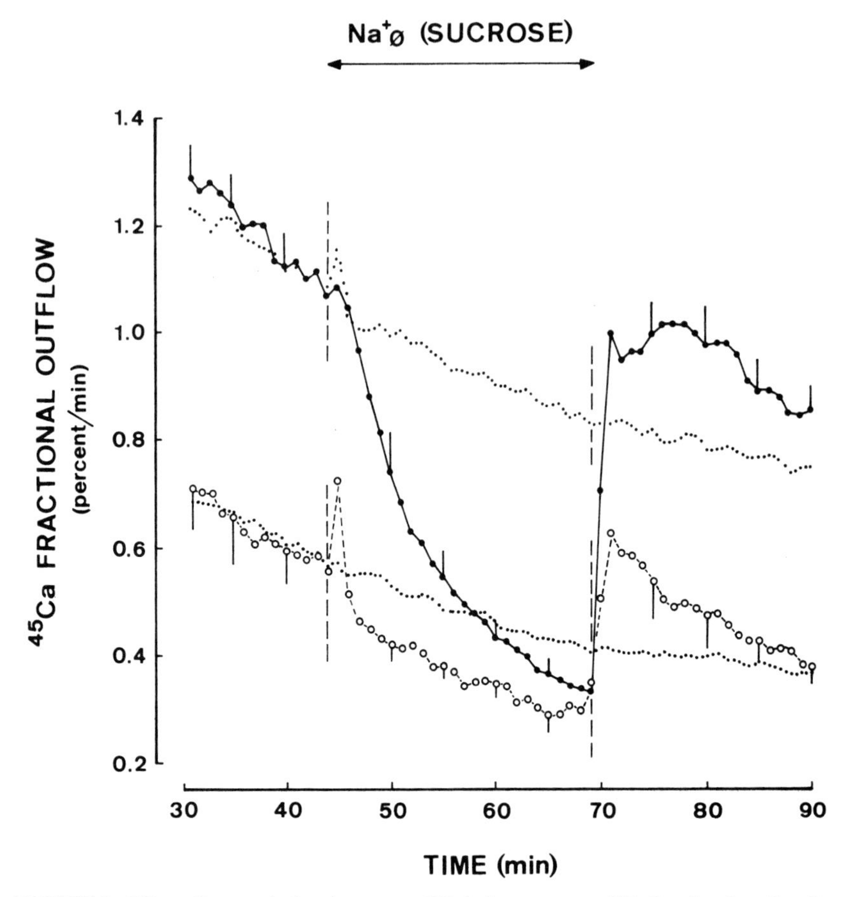

FIGURE 1. Effect of isosmotical replacement of $Na^{+}{}_{o}$ by sucrose on ^{45}Ca fractional outflow from pancreatic islets perfused in the absence of $Ca^{2+}{}_{o}$. The upper curve (●——●) was carried out in the absence of glucose. The lower curve (○- - - -○) was carried out in the presence of glucose (16.7 mM). Stippled lines are control experiments in which $Na^{+}{}_{o}$ was maintained at 139 mM throughout either in the absence of glucose (upper stippled line) or in the presence of 16.7 mM glucose (lower stippled line). Mean ± SEM of at least four experiments. (From A. Herchuelz, unpublished data.)

time (30 sec). The slower second phase probably results from the fall in the $[Na^{+}]_i$ concentration that must occur with increasing time of exposure to low-Na^{+} media. An increase in ^{45}Ca uptake is also observed when $Na^{+}{}_{o}$ is replaced by choline (in the presence of atropine) or lithium chloride instead of sucrose but the uptake is of lower magnitude (63 and 61% of the effect of sucrose).[28]

The changes in $[Ca^{2+}]_i$ induced by $Na^{+}{}_{o}$ removal have been examined in single pancreatic B cells using fura-2 and dual-wavelength microfluorimetry.[29] In about 50% of the cells examined, isosmotic replacement of $Na^{+}{}_{o}$ by sucrose induces a biphasic increase in $[Ca^{2+}]_i$, consisting of an initial phase lasting 10 to 25 sec, followed by a sustained plateau (FIG. 4).

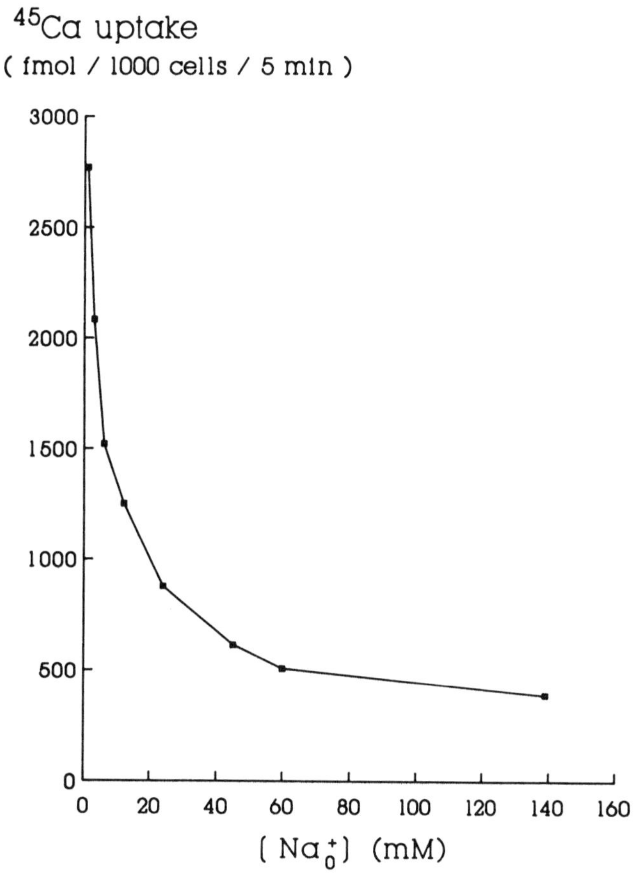

FIGURE 2. Effect of various concentrations of Na^+_o on ^{45}Ca uptake by isolated islet cells. Na^+ was isosmotically replaced by sucrose. Mean values refer to at least 14 individual samples in each case. (From P.-O. Plasman *et al.*[28] by permission of the *American Journal of Physiology.*)

It was verified that the ^{45}Ca uptake induced by Na^+ removal was indeed mediated by Na-Ca exchange. The uptake is unaffected by selective blockers of the voltage-sensitive Ca^{2+} and Na^+ channels, namely by nifedipine and tetrodotoxin.[28] In contrast, the uptake is reduced by known inhibitors of the process, namely by lanthanum, and 3′-4′-dichlorobenzamil (ED_{50}:14 μM). Last, a rise in $[Na^+]_i$ induced by ouabain enhances the uptake of ^{45}Ca due to the removal of Na^+ by 21 $\pm$ 8%.[28]

In a preliminary study, Na^+_i-dependent Ca^{2+} uptake was measured in plasma membrane vesicles of an insulin-releasing tumoral cell line.[30] Na^+_o removal increased ^{45}Ca uptake and decreased ^{45}Ca outflow from the vesicles. One weakness of the study is that Na^+_o was replaced by K^+ (160 mM). Indeed, high K^+ may depolarize the plasma membrane and stimulate Ca^{2+} uptake by itself. On the other hand, K^+ may substitute for Na^+ in the exchange process,[31] alter the kinetics of the exchanger,[32] and even be cotransported with Ca^{2+} by the exchanger.[33] Last, data obtained in tumoral

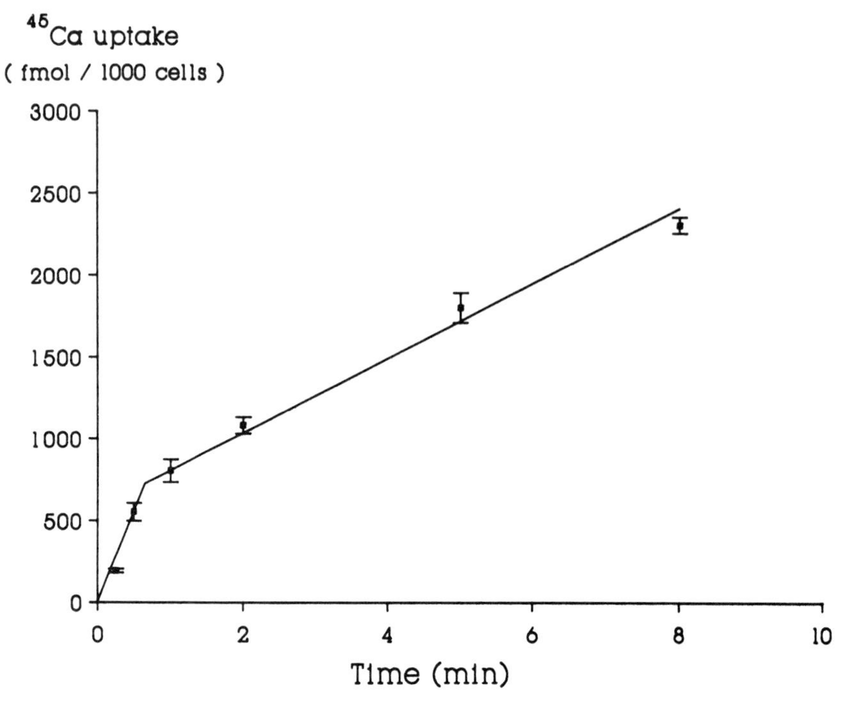

FIGURE 3. Effect of isosmotical replacement of Na^+_o by sucrose on ^{45}Ca uptake by isolated islet cells over different periods of time. Lines were drawn using least-squares analysis. Best fit was obtained by decomposing the curve into two straight lines with the cutting point between 30 sec and 1 min. Mean values ± SE refer to at least 10 individual samples in each case. (From P.-O. Plasman *et al.*[28] by permission of the *American Journal of Physiology.*)

cell lines must be examined with caution since these cells may differ considerably from normal B cells.[34]

In summary, the B cell is equipped with a process of Na-Ca exchange that appears to display quite a large capacity. In the following pages, the regulation of the process and its role in the process of insulin release will be examined.

REGULATION OF NA-CA EXCHANGE

Effects of Glucose

Effect of Glucose on Forward Na-Ca Exchange

Glucose, like Na^+_o removal, reduces ^{45}Ca outflow from islets perfused in the absence of Ca^{2+}_o.[35–37] Because the effect of glucose and Na^+_o removal were initially observed not to be additive, it has been proposed that glucose may inhibit forward Na-Ca exchange.[15] Three other studies provided slightly different results, so that the latter view has been challenged. In the first of these studies, Na^+_o deprivation inhibited ^{45}Ca outflow to a greater extent than glucose, although the effect of the sugar was suppressed

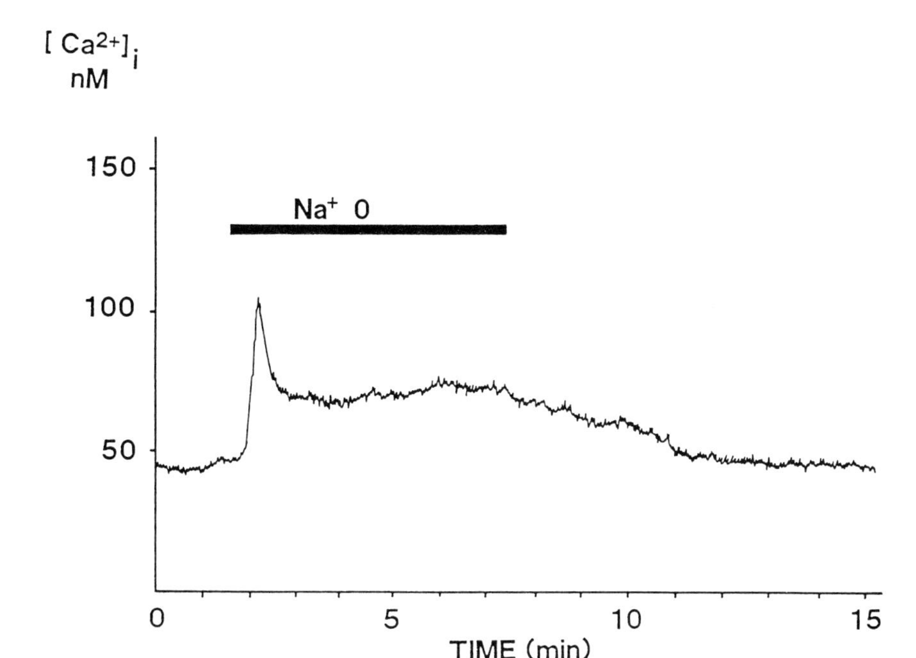

FIGURE 4. Effect of isosmotical replacement of Na^+_o by sucrose on $[Ca^{2+}]_i$ in single isolated pancreatic B cell. $[Ca^{2+}]_i$ was measured using fura-2 and dual-wavelength microfluorimetry. This trace is representative of 20 experiments. (From A. Herchuelz, unpublished data.)

TABLE 1. Effect of Na^+_o Removal[a] and Glucose on ^{45}Ca-Outflow[b] Rate from Perfused Islets

Na^+ (mM)	Na^+ Substitute (mM)	Glucose (mM)	^{45}Ca-Outflow Rate (%/min) min 64-68
139	—	—	0.87 ± 0.02 (5)
0	Choline	—	0.34 ± 0.01 (4)
0	K^+	—	0.32 ± 0.04 (4)
0	Sucrose	—	0.32 ± 0.01 (4)
139	—	16.7	0.43 ± 0.02 (4)

[a] Na^+_o was isosmotically replaced by choline or K^+ salts or by sucrose.

[b] ^{45}Ca outflow was expressed as a fractional outflow rate (percent of content lost per min at the time indicated). In parentheses, number of experiments.

in the absence of Na^+_o.[16] In the second study, the effect of glucose was only reduced by half in a Na^+-depleted medium.[38] In the last, the inhibitory effect of glucose was claimed to be intact in a Na^+-free medium, provided K^+ salts were used as substitutes for Na^+ salts.[39] The results of these studies are based on the short-term effects of glucose and Na^+ removal on ^{45}Ca efflux (about 30 min). Since the dynamics of the effect of glucose and Na^+ removal may not be the same, we re-examined the ^{45}Ca outflow rate from islets exposed for about 65 min to the two experimental conditions. The absolute effect of glucose averages 80% of that of Na^+_o removal (TABLE 1). On the other hand, the effect of Na^+_o replacement by choline or K^+ salts or by sucrose, computed after 30 min of exposure, (FIG. 1) is reduced by about 76% in the presence of glucose, while the effect of glucose, also computed after 30 min, is reduced by about 74% in the absence of Na^+_o whether replaced by choline or K^+ salts or by sucrose. Thus, we confirm that the effect of glucose is not completely suppressed in the absence of Na^+_o.[38] Nevertheless, it averages 80% of the effect of Na^+ removal and is inhibited by about 75% and not 50% in the absence of Na^+_o. On the other hand, we do not confirm that the effect of glucose is preserved when Na^+_o is replaced by K^+.[39] In fact the effect of glucose is reduced by 76% with K^+ salts, as with other substitutes, but the remaining effect has been taken erroneously as a maintained effect, because the authors used a mode of expression of their data that does not allow comparison of absolute effects.[39]

One major argument that has been used to reject the view that glucose may inhibit Na-Ca exchange was the finding that glucose decreased rather than increased $[Ca^{2+}]_i$ as measured in Ca^{2+}-deprived islet cells.[40] Indeed, the argument was advanced that, if the sugar inhibited Na-Ca exchange, it should increase $[Ca^{2+}]_i$.[40] A decrease in $[Ca^{2+}]_i$ induced by glucose (measured using the Ca^{2+} indicator fura-2) was confirmed in two other studies carried out in cell suspensions (either as an initial decrease in the presence of Ca^{2+}_o,[41–42] or as a sustained decrease in its absence).[42] However, this decrease in $[Ca^{2+}]_i$ was observed in only one out of four studies conducted in single B cells.[43–46] In the three negative studies, the effect of glucose was either never seen[46] or seen only occasionally[44–45] in cells with reduced viability, most probably as the result of fuel depletion.[44] Since the measurement of $[Ca^{2+}]_i$ with fura-2 in cell suspensions is the subject of more artifacts than in single perfused cells using dual-wavelength microfluorimetry,[47] it is most probable that glucose does not decrease $[Ca^{2+}]_i$ in healthy (not fuel-depleted) pancreatic B cells.[46] Incidentally, the claim that glucose should increase $[Ca^{2+}]_i$ in the absence of Ca^+_o if the sugar inhibits Na-Ca exchange is probably not valid. Indeed, by analogy with the situation found in other cells, it is probable that the B cell Na-Ca exchanger displays a high capacity but a low affinity for Ca^{2+}; that is, the process is not active at very low $[Ca^{2+}]_i$ for example, in the absence of Ca^{2+}_o.[48–49] In agreement with such a view, Na^+_o removal does not affect $[Ca^{2+}]_i$ in single B cells exposed to a Ca^{2+}-depleted medium[29] (see also below). Would that imply that the decrease in ^{45}Ca outflow induced by glucose does not result from an inhibition of the process of Na-Ca exchange located at the level of the B cell plasma membrane and responsible for Ca^{2+} extrusion from the B cell? Indeed, the inhibitory effect of glucose on ^{45}Ca outflow is rather obvious in the absence of Ca^{2+}_o, namely, at a low $[Ca^{2+}]_i$. In our view, the answer to the question is not necessarily affirmative in view of the high sensitivity of the radioisotopic method in detecting Ca^{2+} movements even when not attended by changes in $[Ca^{2+}]_i$.[46] However, as an alternative, it has been proposed that the inhibitory effect of glucose on ^{45}Ca outflow could result from the inhibition of a nonelectrogenic Na-Ca exchange localized at the level of the B cell mitochondria[50] (see below), a view compatible with a previous suggestion of an increased uptake at the organelle level.[16] This alternative hypothesis, however, remains to be fully substantiated.

Incidentally, other insulin secretagogues that are metabolized by the pancreatic B cell also reduce ^{45}Ca outflow from perfused islets (e.g. 2-ketoisocaproate, leucine, and glyceraldehyde).[51-53] Hence, these secretagogues may inhibit Na-Ca exchange like glucose, depending on the extent of their metabolism in the islets.

Effect of Glucose on Reverse Na-Ca Exchange

The effect of glucose on reverse Na-Ca exchange has been studied by measuring Na^{+}_{i}-dependent Ca^{2+} uptake by dissociated pancreatic islet cells incubated in the presence of nifedipine.[28] Nifedipine was used to block the effect of glucose to increase ^{45}Ca uptake by opening voltage-sensitive Ca^{2+} channels. Glucose increased ^{45}Ca uptake due to Na^{+} removal. However, at variance with the effect of glucose on voltage-sensitive Ca^{2+} channels, the effect of the sugar on Na-Ca exchange was already maximal at a low concentration (2.8 mM). This effect could be mediated by an increase in intracellular ATP (or cytoplasmic ATP/ADP ratio) with subsequent phosphorylation of the carrier.[28] First, the effect of the sugar to increase cellular ATP (or cytoplasmic ATP/ADP ratio) displays a biphasic pattern with a first increase in cytosolic ATP when the concentration of the sugar is raised from 0 to 1.4 or 2.8 mM.[54] Second, ATP has been shown to activate the exchanger in nerve and cardiac tissue via a phosphorylation step[48,55] mediated in the latter tissue by a calmodulin-dependent protein kinase.[55] Although the existence of the latter kinase in insulin-producing cells has been demonstrated,[56] the view that it could mediate the observed effect of glucose remains to be fully substantiated. In favor of this view, however, is the profound inhibitory effect exerted by metabolic inhibitors on Na-Ca exchange (see below). Incidentally, a stimulatory effect of glucose on reverse Na-Ca exchange is not incompatible with an inhibitory effect on forward Na-Ca exchange. For instance, in nerve cells, membrane depolarization inhibits forward Na-Ca exchange while stimulating reverse Na-Ca exchange.

Effects of pH

Effect of pH on Forward Na-Ca Exchange

Some evidence suggests that intracellular pH (pH_i) may inhibit forward Na-Ca exchange. Indeed, a rise in PCO_2 and HCO_3^- concentration of the extracellular medium that may decrease pH_i from 7.1 to 6.4 was found to decrease ^{45}Ca outflow from perifused islets.[57] The latter data are difficult to interpret because control experiments were carried out in the absence of Na^{+}_{o} but the presence of Ca^{2+}_{o}, that is, under conditions that stimulate Ca-Ca exchange.[15] A rise in extracellular pH (pH_o) also increased, while a decrease in pH_o reduced, ^{45}Ca outflow from perfused islets.[58] In this study, no control experiments were carried out in the absence of Na^{+}_{o}. In a subsequent study, an increase in pH_i by 0.51 units, induced by the removal of extracellular bicarbonate and CO_2, resulted in a sustained increase in ^{45}Ca outflow that was slightly reduced in the absence of Ca^{2+}_{o} but completely abolished in the absence of both Ca^{2+}_{o} and Na^{+}_{o}.[59] The increase in ^{45}Ca outflow was also inhibited by glucose.[59] Taken as a whole, the latter data indeed indicate that a rise in pH_i increases ^{45}Ca outflow by a process that is not Ca-Ca exchange but that can be inhibited by the absence of Na^{+}_{o} or the presence of glucose (presumably Na-Ca exchange).

Effect of pH on Reverse Na-Ca Exchange

The effect of pH_o and pH_i on reverse Na-Ca exchange was recently investigated in pancreatic islet cells by measuring Na^{+}_{i}-dependent Ca^{2+} uptake in the presence of nifedipine over 1-min periods.[60] pH_o markedly affected reverse Na-Ca exchange. Indeed, at a pH_o of 6.0, Na-Ca exchange was reduced by about 65% while at a pH_o of 9, it was increased by about 70%.[60] pH_i affected reverse Na-Ca exchange even more

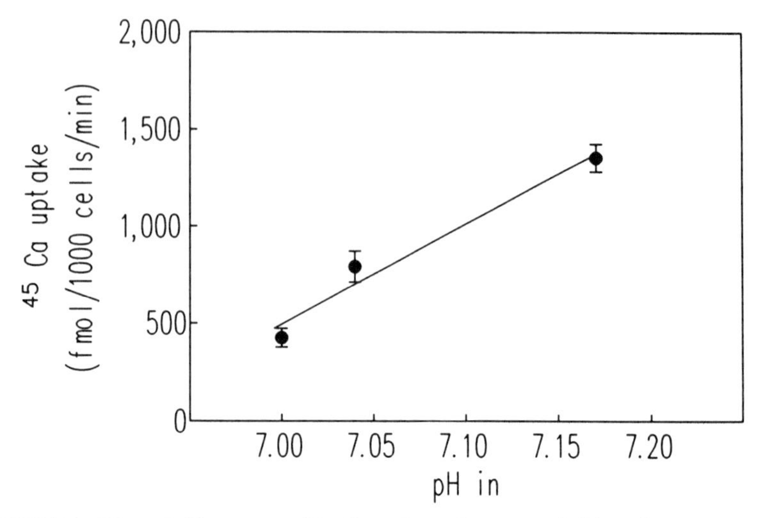

FIGURE 5. Effect of pH_i on reverse Na-Ca exchange by pancreatic islet cells. pH_i was modified by using Na acetate and NH_4Cl. The data are corrected for basal uptake observed in the presence of 139 mM Na^+ at the various pHs investigated, so that only Na^+_i-dependent Ca^{2+} movements are reported. (From A. Herchuelz, unpublished data.) Mean ±SEM of 28-48 individual samples in each case.

markedly than pH_o.[60] pH_i was altered by using Na acetate (10 mM) and NH_4Cl (10 mM). FIGURE 5 shows that a decrease in pH_i from 7.04 to 7.0 (Na acetate) reduced reverse Na-Ca exchange by about 45% while an increase in pH_i from 7.04 to 7.17 (NH_4Cl) enhanced Na-Ca exchange by about 70%. Thus the pH_i appears as a potent modulator of Na-Ca exchange in the pancreatic B cell and changes in pH_i may represent a mechanism by which glucose influences the carrier. Indeed the sugar is known to induce a sustained increase in pH_i.

Effects of Membrane Depolarization

Membrane depolarization as mediated by a rise in K^+_o concentration does not appear to reduce ^{45}Ca outflow from perifused islets and hence to affect forward Na-Ca exchange.[15] In the presence of nifedipine, a rise in K^+_o concentration (25 and 50 mM) provokes a dose-related decrease of the uptake of ^{45}Ca induced by Na^+_o removal in rat pancreatic islet cells (reverse Na-Ca exchange).[28] However, when $[Na^+]_i$ is raised (using ouabain and glucose), membrane depolarization induces a dose-related increase in ^{45}Ca uptake.[28] In chick myocardial cells also, the influence of membrane potential is observed only under raised $[Na^+]_i$,[61] a finding in agreement with the knowledge that Na^+ loading increases the voltage sensitivity of Na-Ca exchange.[62] The electrogenicity of the exchanger was examined in insulinoma cells by measuring the uptake of the potential-sensitive probe, tetraphenylphosphonium ($[^3H]TPP$). Na^+_o removal increased $[^3H]TPP$ uptake.[30] Unfortunately, Na^+_o was replaced by K^+, a condition that reduce the significance of the observation (see above). Thus, the Na-Ca exchanger of the pancreatic B cell is influenced by membrane potential and is probably also electrogenic. Because glucose induces rhythmic changes in membrane potential, it is not inconceivable that the sugar could increase Ca^{2+} inflow into the B cell by reverse Na-Ca exchange during the depolarization phases of the electrical activity.

Effects of High-Energy Substrates

It was suggested that the stimulatory effect of glucose on reverse Na-Ca exchange could result from an increase in intracellular ATP concentration with subsequent phosphorylation of the carrier.[28] To further substantiate this view, we have examined the effects on Na-Ca exchange of three types of drugs able to markedly decrease the ATP concentration and ATP/ADP ratio in the islets (oligomycin, an ATP synthase inhibitor; antimycin A, rotenone, and KCN, three electron transfer inhibitors; and 3′,4′-dinitrophenol, an uncoupler of oxidative phosphorylation).[64] All five drugs reduced reverse Na-Ca exchange by about 25 to 50% without affecting basal uptake measured in the presence of Na^{+}_{o} (139 mM).[60] The similarity in the action of the five different inhibitors on Na-Ca exchange and basal uptake almost exclude any mechanism other than depletion of the cells in high-energy substrates (e.g. ATP), in mediating the inhibition of Na-Ca exchange.[60] The inhibitory effect of the metabolic poisons was not reversed in the presence of glucose, at variance with the situation found in smooth muscle.[65] However, our results were not unexpected since, in islet cells, glucose oxidation is almost completely abolished by these drugs.[64]

These data suggest that reverse Na-Ca exchange may represent a potential target for ATP generated in the B cell by glucose and other nutrients.[60] Thus, one may speculate that glucose, through its metabolism, could activate reverse Na-Ca exchange and, by doing so, favor Ca^{2+} entry into the B cell by another means than by voltage-sensitive Ca^{2+} channels.

Effects of Protein Kinase C Activation

The involvement of protein kinase C (PKC) in glucose-induced insulin secretion is unclear.[66] However, PKC activation was recently suggested to favor Ca^{2+} outflow from the B cell.[67] We have examined the effect of PKC activation on reverse Na-Ca exchange by using the phorbol ester TPA (12-O-tetradecanoyl-phorbol-13-acetate). TPA failed to affect Na-Ca exchange.[60] Thus, the exchanger does not appear to be a target for PKC, at least when working in its reverse mode.[60]

Dependence on K^{+}_{o}

In rod outer segments, K^+ was recently proposed to be cotransported with Ca^{2+} by the exchanger.[33,68] This cotransport of K^+ was suggested to allow a reduction of $[Ca^{2+}]_i$ to much lower values than previously supposed and to be a general phenomenon.[33] In the B cell, however, reverse Na-Ca exchange was totally independent of the presence of K^+ in the extracellular medium.[60]

Effects of Drugs

As mentioned above, tetrodotoxin and nifedipine do not affect Na-Ca exchange in islet cells, at variance with lanthanum and 3′,4′-dichlorobenzamil (DCB).[28]

Recently, we have used DCB to further investigate the effect of glucose on forward Na-Ca exchange in islet cells.[50] Unfortunately, the drug was of poor specificity and potently inhibited both K^+ channels (probably ATP-dependent) and voltage-sensitive Ca^{2+} channels,[50] as in some other tissues.[69] However, in the absence of $Ca^{2+}{}_o$, none of the actions exerted by glucose on ^{45}Ca outflow is thought to result from changes in K^+ permeability and of course Ca^{2+} inflow.[2] In the absence of $Ca^{2+}{}_o$, DCB, like glucose, reduced basal ^{45}Ca outflow, the effects of the sugar and the drug being not additive. Furthermore, this effect of DCB and its ability to impair the inhibitory effect of glucose were reproduced by the absence of $Na^+{}_o$ and disappeared under the latter experimental condition. Therefore, these data reinforce the view that a significant part of the inhibitory effect of glucose on ^{45}Ca outflow indeed results from an inhibition of Na-Ca exchange.[50] Since DCB inhibits both Na-Ca exchange at the level of the plasma membrane and the mitochondria,[69] the latter data do not distinguish between these two potential sites of action of glucose.

Other drugs that were observed to alter Na-Ca exchange directly include amiloride and quinine.[70–71] Amiloride, however, is far less potent than its derivative DCB (IC_{50}: about 1 mM, unpublished results). Quinine (100 μM) was considered to inhibit Na-Ca exchange since the drug reduces ^{45}Ca outflow from pancreatic islets like glucose and $Na^+{}_o$ removal, and that the three effects are not additive.[71]

ROLE OF NA-CA EXCHANGE

Because the B cell is electrically excitable, Na-Ca exchange may participate to both Ca^{2+} extrusion from the cell and Ca^{2+} inflow from the extracellular medium.[72] By analogy with the situation found in other cells, Ca^{2+} extrusion would occur during repolarization phases of the electrical activity while Ca^{2+} inflow would occur during depolarization phases.[72] Up until now, no evidence has been obtained to suggest that Ca^{2+} may enter into the B cell by reverse Na-Ca exchange in response to a physiological stimulus.

However, both direct and indirect evidence exists to suggest that forward Na-Ca exchange may participate in Ca^{2+} extrusion from the B cell and modulate $[Ca^{2+}{}_i]$.[29,73] The indirect evidence is that the absence of $Na^+{}_o$ or the presence of ouabain may reduce ^{45}Ca outflow induced in Ca^{2+}-deprived islets by the release of intracellular Ca^{2+} while simultaneously stimulating insulin release.[73]

To provide more direct evidence for a role of Na-Ca exchange in Ca^{2+} extrusion from the B cell, we have examined the effect of $Na^+{}_o$ removal on $[Ca^{2+}]_i$ in single pancreatic B cells.[29] As mentioned above, $Na^+{}_o$ removal induced an important increase in $[Ca^{2+}]_i$ in the presence of $Ca^{2+}{}_o$ (FIG. 4). However, in the absence of $Ca^{2+}{}_o$, $Na^+{}_o$ failed to affect $[Ca^{2+}]_i$. Next, Ca^{2+} was released from the endoplasmic reticulum (ER) by using thapsigargin, an inhibitor of the ER Ca^{2+}-ATPase,[75] and caffeine, which sensitizes the Ca^{2+}-induced Ca^{2+} release channels of the ER to resting Ca^{2+} levels.[76] In the absence of $Na^+{}_o$, thapsigargin (1 μM) induced a sustained increase in $[Ca^{2+}{}_i]$ compared to a transient effect in the presence of $Na^+{}_o$. Furthermore, normalizing the $[Na^+]_o$ induced a rapid drop of $[Ca^{2+}]_i$ in cells exposed to the absence of $Na^+{}_o$ for

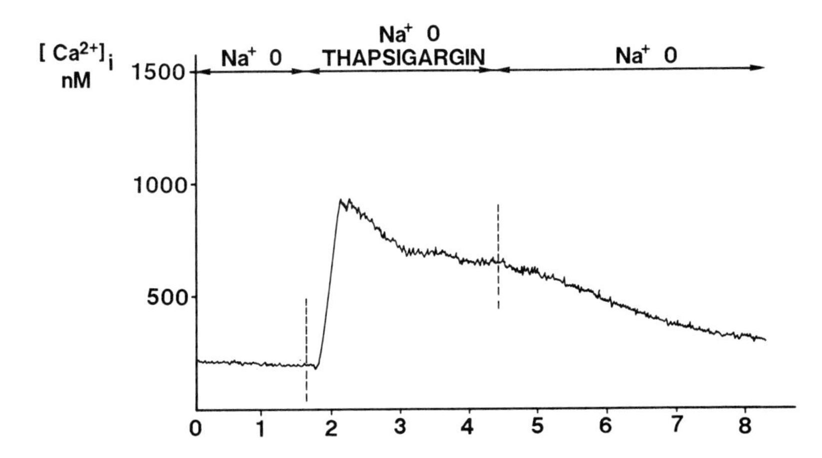

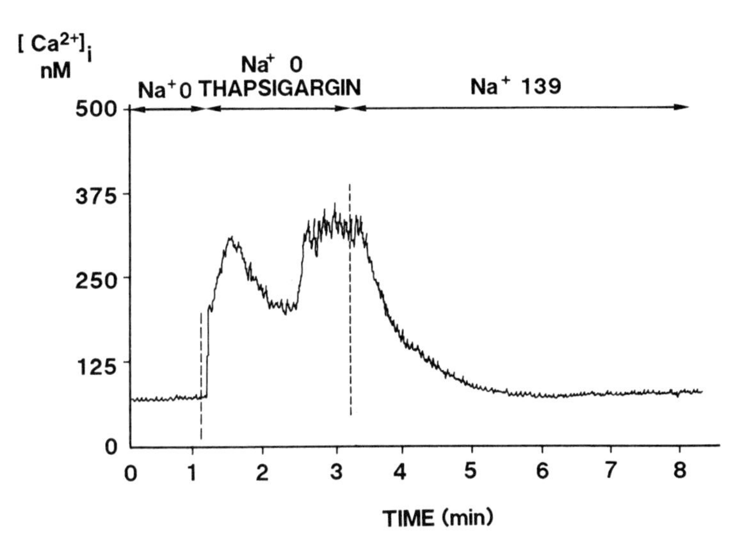

FIGURE 6. Effect of thapsigargin (1 μM) on $[Ca^{2+}]_i$ in single isolated pancreatic B cells perfused in the presence of extracellular Ca^{2+} but the absence of $Na^+{}_o$ (replaced by sucrose). In the lower panel, $Na^+{}_o$ was reintroduced (139 mM) during the last third of the trace. In the upper panel, $Na^+{}_o$ was maintained at a low value throughout the experiment. Notice that in the upper panel, thapsigargin removal is followed by a slow return of $[Ca^{2+}]_i$ to normal values. In the lower panel, $Na^+{}_o$ reintroduction at the time of thapsigargin removal was followed by a rapid drop in $[Ca^{2+}]_i$. Each trace is representative of at least six experiments. $[Ca^{2+}]_i$ was measured as in FIGURE 4. (From A. Herchuelz, unpublished data.)

about 20 min and to the presence of thapsigargin for about 3 min[29] (FIG. 6). A similar picture was observed when caffeine (10 mM) was used to increase $[Ca^{2+}]_i$.

These data provide the first direct evidence that Na-Ca exchange, working in its Ca^{2+}-efflux mode, regulates $[Ca^{2+}]_i$ in the pancreatic B cell. In view of the absence of an effect of $Na^+{}_o$ removal on $[Ca^{2+}]_i$ in islets incubated in the absence of $Ca^{2+}{}_o$, for example, in islet cells displaying a $[Ca^{2+}]_i$ below the basal level (100 nM), the B cell Na-Ca exchange appears as a low-affinity Ca^{2+} system designed to extrude Ca^{2+} when its intracellular concentration is raised above basal levels, for instance when the B cell is stimulated by an insulin secretagogue.

MITOCHONDRIAL NA-CA EXCHANGE

A distinct nonelectrogenic Na-Ca exchange is present in mitochondria derived from some electrically excitable cells.[76] This Na-Ca exchange is probably present in pancreatic islets, as suggested by both direct and indirect evidence.

Thus, a rise in $[Na^+]_i$ induced either by the removal of $K^+{}_o$ or the addition of ouabain provokes, after an initial decrease, an important rise in ^{45}Ca outflow from perifused islets.[23,25] These increases are not suppressed in the absence of $Ca^{2+}{}_o$, a finding that confirms that they do not result from an increase in Ca^{2+} inflow. Ca^{2+} movements in isolated mitochondria from rat insulinoma cells have been studied by monitoring Ca^{2+} concentration in their surrounding medium. Na^+ activated Ca^{2+} efflux with a K_m of 4 mM and had a maximal effect at 10 mM.[77]

The existence of this second Na-Ca exchange mechanism certainly complicates the interpretation of ^{45}Ca-outflow experiments in which $Na^+{}_o$ is lowered. Indeed, the lowering of $[Na^+]_o$ may lead to $[Na^+]_i$ lowering and possibly to a decrease in ^{45}Ca outflow.[16] Nevertheless, this view remains to be fully substantiated.

CONCLUSIONS

The pancreatic B cell is equipped with a process of Na-Ca exchange localized at the level of the plasma membrane and responsible for Ca^{2+} extrusion from the cell. From preliminary findings, the process appears to be a low-affinity Ca^{2+}-pumping system designed to extrude Ca^{2+} from the cell when its intracellular concentration is raised above basal levels. The process does not cotransport K^+.

In its Ca^{2+}-efflux mode, the process is stimulated by alkaline pH_i and inhibited by acidic pH_i and probably also by glucose. In its Ca^{2+}-influx mode, the process is inhibited by acidic pH_i and stimulated by alkaline pH_i, membrane depolarization, intracellular ATP, and glucose. It is unaffected by protein kinase C activators.

By analogy with the situation found in other tissues,[63] it is not inconceivable that the exchanger, in addition to its participation in Ca^{2+} extrusion from the cell, also carries Ca^{2+} inflow in the cell by reverse Na-Ca exchange. This view, however, remains to be demonstrated. Glucose could activate reverse Na-Ca exchange by altering the membrane potential, the pH_i, and the intracellular concentration of ATP.

REFERENCES

1. HELLMAN, B., E. GYLFE, P. BERGSTEN, H. JOHANSSON & N. WESSLÉN. 1988. *In* Pathogenesis of Non-Insulin Dependent Diabetes Mellitus. V. Grill & S. Efendic, Eds.: 39-60. Raven Press. New York.
2. HERCHUELZ, A. & W. J. MALAISSE. 1982. *In* The Role of Calcium in Biological Systems. L. J. Anghileri & A. M. Tuffet-Anghileri, Eds. Vol **III:** 17-32. CRC Press. Boca Raton, FL.
3. PRENTKI, M. & F. M. MATSCHINSKY. 1987. Physiol. Rev. **67:** 1185-1248.
4. WOLLHEIM, C. B. & G. W. G. SHARP. 1981. Physiol. Rev. **61:** 914-973.

5. COOK, D. L. & C. N. HALES. 1984. Nature **311:** 271-273.
6. ASHCROFT, F. M., D. E. HARRISON & S. J. H. ASHCROFT. 1984. Nature **312:** 446-448.
7. HENQUIN, J. C. 1978. Nature **271:** 271-273.
8. MEISSNER, H. P. & M. PREISSLER. 1979. *In* Treatment of Early Diabetes. R. A. Camerini Davalos & B. Hanover, Eds.: 97-107. Plenum. New York.
9. ATWATER, I., P. CARROLL & M. X. LI. 1989. *In* Insulin Secretion. B. Draznin, M. S. Shlomo & L. R. Derek, Eds.: 49-68. Alan R. Liss, Inc. New York,
10. FORMBY, B., K. CAPITO, J. EGEBERG & C. J. HEDESKOV. 1976. Am. J. Physiol. **230:** 441-448.
11. HOENIG, M., J. L. RODERICK & D. C. FERGUSON. 1990. Biochim. Biophys. Acta **1022:** 333-338.
12. DONATSCH, P., D. A. LOWE, B. P. RICHARDSON & P. TAYLOR. 1977. J. Physiol. **267:** 357-376.
13. HERCHUELZ, A. & W. J. MALAISSE. 1979. Diabetes **28:** 371.
14. HELLMAN, B., H. ABRAHAMSSON, T. ANDERSSON, P.-O. BERGGREN, P. FLATT & E. GYLFE. 1979. *In* Proceedings of the 10th Congress of the IDF. W. K. Waldhäusl, Ed.: 160-165. Excerpta Medica. Amsterdam.
15. HERCHUELZ, A., A. SENER & W. J. MALAISSE. 1980. J. Membr. Biol. **57:** 1-12.
16. HELLMAN, B., T. ANDERSSON, P.-O. BERGGREN & P. RORSMAN. 1980. Biochem. Med. **24:** 143-152.
17. HALES, C. N. & R. D. G. MILNER. 1968. J. Physiol. **199:** 177-187.
18. BAKER, P. F., M. P. BLAUSTEIN, A. L. HODGKIN & R. A. STEINHARDT. 1967. J. Physiol. London **192:** 43P-44P.
19. MALAISSE, W. J. 1975. *In* Handbook of Experimental Pharmacology. G. V. R. Born, O. Eichler, A. Farah, H. Herken & A. D. Welch, Eds. Vol. **XXXII**/2: 145-155. Springer-Verlag. Berlin, Heidelberg, New York.
20. HERCHUELZ, A., E. COUTURIER & W. J. MALAISSE. 1980. Am. J. Physiol. **238:** E96-E103.
21. HENQUIN, J. C. 1979. J. Physiol. **296:** 103.
22. ABRAHAMSSON, H., P.-O. BERGGREN & B. HELLMAN. 1984. Am. J. Physiol. **247:** E719-E725.
23. SIEGEL, E. G., C. B. WOLLHEIM, A. E. RENOLD & G. W. G. SHARP. 1980. J. Clin. Invest. **66:** 996-1003.
24. LEBRUN, P., W. J. MALAISSE & A. HERCHUELZ. 1983. J. Membr. Biol. **74:** 67-73.
25. HERCHUELZ, A. & W. J. MALAISSE. 1980. J. Physiol. **302:** 263-280.
26. HELLMAN, B., J. SEHLIN & I.-B. TÄLJEDAL. 1978. Pflügers Arch. **378:** 93-97.
27. HELLMAN, B., J. SEHLIN & I.-B. TÄLJEDAL. 1971. Am. J. Physiol. **221:** 1795-1801.
28. PLASMAN, P.-O., P. LEBRUN & A. HERCHUELZ. 1990. Am. J. Physiol. **259:** E844-E850.
29. HERCHUELZ, A. 1991. Diabetologia **2:** A73.
30. HOENIG, M., L. H. CULBERSON, C. A. WHEELER & D. C. FERGUSON. 1990. Diabetologia **33:** A104.
31. BAKER, P. F., M. P. BLAUSTEIN, A. L. HODGKIN & R. A. STEINHARDT. 1969. J. Physiol. **200:** 431-458.
32. KACZOROWSKI, G. J., M. L. GARCIA, V. F. KING & R. S. SLAUGHTER. 1989. *In* Sodium-Calcium Exchange. T. J. F. Allen, D. Noble & H. Reuter, Eds.: 66-101. Oxford University Press. Oxford, U.K.
33. CERVETTO, L., L. LAGNADO, R. J. PERRY, D. W. ROBINSON & P. A. MCNAUGHTON. 1989. Nature **337:** 740-743.
34. MALAISSE, W. J. 1990. *In* Molecular Biology of Islets of Langerhans. H. Okamoto, Ed.: 315-339. University Press. Cambridge, U.K.
35. MALAISSE, W. J., G. R. BRISSON & L. E. BAIRD. 1973. Am. J. Physiol. **224:** 389-394.
36. HERCHUELZ, A. & W. J. MALAISSE. 1978. J. Physiol. **283:** 409-424.
37. GYLFE, E., A. BUITRAGO, P.-O. BERGGREN, K. HAMMARSTRÖM & B. HELLMAN. 1978. Am. J. Physiol. **235:** E191-E196.
38. HENQUIN, J. C., R. DE MIGUEL, M. G. GARRINO, M. HERMANS & M. NENQUIN. 1985. FEBS Lett. **187:** 177-181.
39. HELLMAN, B. & E. GYLFE. 1984. Biochim. Biophys. Acta **770:** 136-141.
40. RORSMAN, P., H. ABRAHAMSON, E. GYLFE & B. HELLMAN. 1984. FEBS Lett. **170:** 196-200.

41. NILSSON, T., P. ARKHAMMAR & P.-O. BERGGREN. 1988. Biochem. Biophys. Res. Commun. **153:** 984-991.
42. GOBBE, P. & A. HERCHUELZ. 1989. Res. Commun. Chem. Pathol. Pharmacol. **63:** 231-247.
43. GRAPENGIESSER, E., E. GYLFE & B. HELLMAN. 1989. Exp. Clin. Endocrinol. **93:** 321-327.
44. PRALONG, W.-F., C. BARTLEY & C. B. WOLLHEIM. 1990. EMBO J. **9:** 53-60.
45. WANG, J.-L. & M. L. MCDANIEL. 1990. Biochem. Biophys. Res. Commun. **166:** 813-818.
46. HERCHUELZ, A., R. POCHET, C. PASTIELS & A. VAN PRAET. 1991. Cell Calcium **12:** 577-586.
47. MALGAROLI, A., D. MILANI, J. MELDOLESI & T. POZZAN. 1987. J. Cell Biol. **105:** 2145-2155.
48. DIPOLO, R. & L. BEAUGÉ. 1988. Biochim. Biophys. Acta **947:** 549-569.
49. CARAFOLI, E. 1988. Methods Enzymol. **157:** 2-11.
50. PLASMAN, P.-O., P. LEBRUN, E. J. CRAGOE & A. HERCHUELZ. 1991. Biochem. Pharmacol. **41:** 1759-1768.
51. HUTTON, J. C., A. SENER, A. HERCHUELZ, I. ATWATER, S. KAWAZU, A. C. BOSCHERO, G. SOMERS, G. DEVIS & W. J. MALAISSE. 1990. Endocrinology **106:** 203-219.
52. MALAISSE, W. J., J. C. HUTTON, A. R. CARPINELLI, A. HERCHUELZ & A. SENER. 1980. Diabetes **29:** 431-437.
53. MALAISSE, W. J., A. HERCHUELZ, J. LEVY, A. SENER, D. G. PIPELEERS, G. DEVIS, G. SOMERS & E. VAN OBBERGHEN. 1976. Mol. Cell. Endocrinol. **4:** 1-12.
54. MALAISSE, W. J. & A. SENER. 1987. Biochim. Biophys. Acta **927:** 190-195.
55. CARONI, P. & E. CARAFOLI. 1983. Eur. J. Biochem. **132:** 451-460.
56. SCHUBART, V. K., N. FLEISCHER & J. ERLICHMAN. 1980. J. Biol. Chem. **255:** 4120-4124.
57. CARPINELLI, A. R., A. SENER, A. HERCHUELZ & W. J. MALAISSE. 1980. Metabolism **29:** 540-545.
58. HUTTON, J. C., A. SENER, A. HERCHUELZ, I. VALVERDE, A. C. BOSCHERO & W. J. MALAISSE. 1980. Horm. Metab. Res. **12:** 294-299.
59. LEBRUN, P., W. J. MALAISSE & A. HERCHUELZ. 1982. Biochim. Biophys. Acta **721:** 357-365.
60. PLASMAN, P.-O. & A. HERCHUELZ. 1991. Diabetologia (Sup.2): A82.
61. LEE, H.-C. & W. T. CLUSIN. 1987. Biophys. J. **51:** 169-176.
62. DIPOLO, R., F. BEZANILLA, C. CAPUTO & H. ROJAS. 1985. J. Gen. Physiol. **86:** 457-478.
63. SHEU, S.-S. & M. P. BLAUSTEIN. 1986. *In* The Heart and the Cardiovascular System. H. A. Fozzard, Eds.: 509-535. Raven. New York.
64. MALAISSE, W. J., J. C. HUTTON, S. KAWAZU, A. HERCHUELZ, I. VALVERDE & A. SENER. 1979. Diabetologia **16:** 331-341.
65. SMITH, J. B. & L. SMITH. 1990. Am. J. Physiol. **259:** C302-C309.
66. WOLLHEIM, C. B. & R. REGAZZI. 1990. FEBS Lett. **268:** 376-380.
67. BERGGREN, P.-O., P. ARKHAMMAR & T. NILSSON. 1989. Biochem. Biophys. Res. Commun. **165:** 416-421.
68. SCHNETKAMP, P. P. M., K. B. DEBESH & R. T. SZERENCSEI. 1989. Am. J. Physiol. **257:** C153-C157.
69. KACZOROWSKI, G. J., M. L. GARCIA, V. F. KING & R. S. SLAUGHTER. 1989. *In* Sodium-Calcium Exchange. T. J. F. Allen, D. Noble & H. Reuter, Eds.: 66-101. Oxford University Press. Oxford, U.K.
70. LEBRUN, P., E. VAN GANSE, M. JUVENT, M. DELEERS & A. HERCHUELZ. 1986. Biochim. Biophys. Acta **886:** 448-456.
71. HERCHUELZ, A., P. LEBRUN, A. CARPINELLI, N. THONNART, A. SENER & W. J. MALAISSE. 1981. Biochim. Biophys. Acta **640:** 16-30.
72. BLAUSTEIN, M. P. 1988. J. Cardiovasc. Pharmacol. **12:** S56-S68.
73. JANJIC, D. & C. B. WOLLHEIM. 1983. Am. J. Physiol.: E222-E229.
74. THASTRUP, O., P. J. CULLEN, B. K. DROBAK, M. R. HANLEY & A. P. DAWSON. 1990. Proc. Natl. Acad. Sci. USA **87:** 2466-2470.
75. ENDO, M. 1985. Cur. Top. Membr. Transp. **25:** 181-230.
76. KACZOROWSKI, G. J., R. S. SLAUGHTER, V. F. KING & M. L. GARCIA. 1989. Biochim. Biophys. Acta **988:** 287-302.
77. PRENTKI, M., D. JANJIC & C. B. WOLLHEIM. 1983. J. Biol. Chem. **258:** 7597-7602.

Characterization of Na^+-Ca^{2+} Exchange in the Beta Cell

M. HOENIG, L. H. CULBERSON, C. A. WHEELER, AND D. C. FERGUSON

Department of Physiology and Pharmacology
College of Veterinary Medicine
University of Georgia
Athens, Georgia 30602

Na^+-Ca^{2+} exchange plays an important role in the regulation of cytosolic free-Ca^{2+} concentrations ($[Ca^{2+}]_i$) in many systems. Few studies have examined Na^+-Ca^{2+} exchange in islets. Removal of Na^+ from the medium and ouabain treatment resulted in increased $^{45}Ca^{2+}$ retention[1-3] and insulin release.[2-4] It has been proposed that glucose interferes with Na^+-Ca^{2+} exchange, because it was found that Na^+ removal during glucose stimulation had little effect on the $^{45}Ca^{2+}$ efflux rate from islets.[5] However, in other studies it was demonstrated that glucose inhibited $^{45}Ca^{2+}$ efflux from islet cells by a different mechanism than an inhibition of the Na^+-Ca^{2+} countertransport.[6,7] Henquin *et al.*[6] found that in the presence of Na^+, glucose inhibited the Ca^{2+}- efflux rate by approximately 50%. When choline was substituted for Na, the efflux rate was decreased, and addition of glucose led to a further decrease. Similarly, other studies also showed that glucose inhibits $^{45}Ca^{2+}$ efflux in sodium-free medium.[7] Recently, Plasman *et al.*[8] demonstrated that glucose increased the uptake of Ca in sodium-free medium. The effect of glucose was maximal at concentrations that do not lead to insulin release. Because nifedipine blocked the effects of glucose on Ca uptake in the presence of extracellular Na, it was concluded that glucose stimulates reverse Na-Ca exchange.

In the present studies plasma membrane vesicles from a glucose-responsive insulinoma exhibited properties consistent with the presence of a membrane Na^+-Ca^{2+} exchange. The exchange was rapid and was dependent on the external Ca^{2+} concentration (K_m = 4.1 ± 1.1 μM; V_{max} = 6.9 ± 2.1 nmoles/(mg protein · 5 sec); n = 7). External Na^+ inhibited the uptake in a dose-dependent manner (IC_{50} = 15 mM). The exchange process was reversible. Dissipation of the Na gradient by 10 μM monensin decreased Na^+-Ca^{2+} exchange from 0.74 ± 0.17 nmoles/(mg protein · sec) to 0.11 ± 0.05 nmoles/(mg protein · sec), (n = 4). Exchange was not influenced by veratridine, tetrodotoxin, or ouabain. No effect was seen using the calcium-channel blockers nitrendipine or nifedipine. Glucose (30 mM) had no direct effect on Na^+-Ca^{2+} exchange, while glyceraldehyde (20 mM) inhibited it by 50% and dihydroxyacetone (10 mM) by 30%. Na^+-induced efflux of calcium was seen in Ca^{2+}-loaded vesicles and was half-maximal at $[Na^+]$ of 11.1 ± 0.75 mM. The curve describing the dependence of Ca^{2+} efflux on $[Na^+]$ was sigmoidal (FIG. 1) with a Hill coefficient close to 3 (2.7 ± 0.07, n = 5), indicating that activation of Ca^{2+} release involves a minimum of three sites that interact cooperatively. Similarly, the electrogenicity of this exchange was demonstrated using the lipophilic cation tetraphenylphosphonium ([^{3}H]TPP), a membrane potential-sensitive probe. In the presence of 40 μM Ca, the uptake of [^{3}H]TPP uptake was rapid (maximal at 10 sec) but transient while the uptake in the absence of Ca was constant (FIG. 2). The electrogenicity of the process indicates that the Na^+-Ca^{2+} exchange produces a membrane current that will affect the membrane potential and potential-sensitive cellular processes and also that the ion transport by the exchanger depends on the membrane potential.[9] The Na^+-Ca^{2+} exchange in similar membrane preparations has a 20-fold higher capacity and a K_m (Ca), which is two orders of magnitude greater than that of the Ca^{2+} pump.[10] It would then be predicted in the beta cell that resting cytosolic free-Ca^{2+} concentrations of approximately 100 nm[11] may be regulated by both Ca^{2+}-pump and Na^+-Ca^{2+} exchange; however, Na^+-Ca^{2+} exchange becomes the dominant system in the regulation of cytosolic Ca when intracellular Na^+ concentrations increase or when high amounts of Ca^{2+} need to be transported rapidly across the cell membrane.

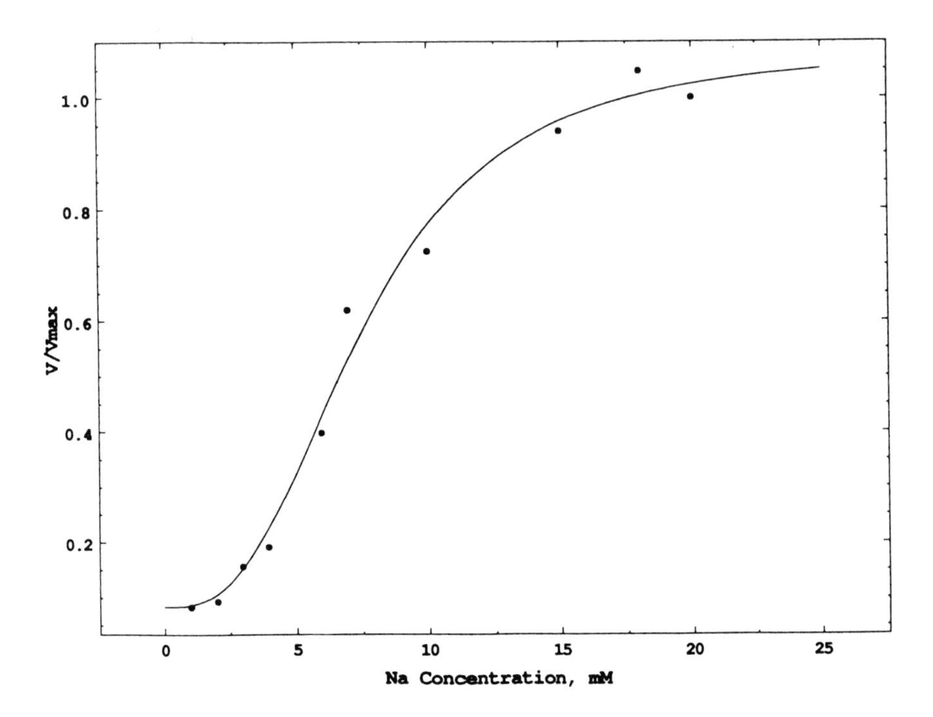

FIGURE 1. Na^+-dependent efflux of Ca^{2+} from insulinoma membrane vesicles. Membrane vesicles were pre-equilibrated with 40 μM $^{45}Ca^{2+}$ for 30 min at 37 °C and then diluted into medium containing various concentrations of NaCL and KCl. The Na^+ concentrations shown are final concentrations after addition of efflux medium. KCl was used to maintain osmolarity as the Na concentrations were varied. A representative experiment is shown. The data from five experiments were analyzed and fit to the Hill equation as described in MATERIALS and METHODS. The Hill coefficient was 2.7 $\pm$ 0.07.

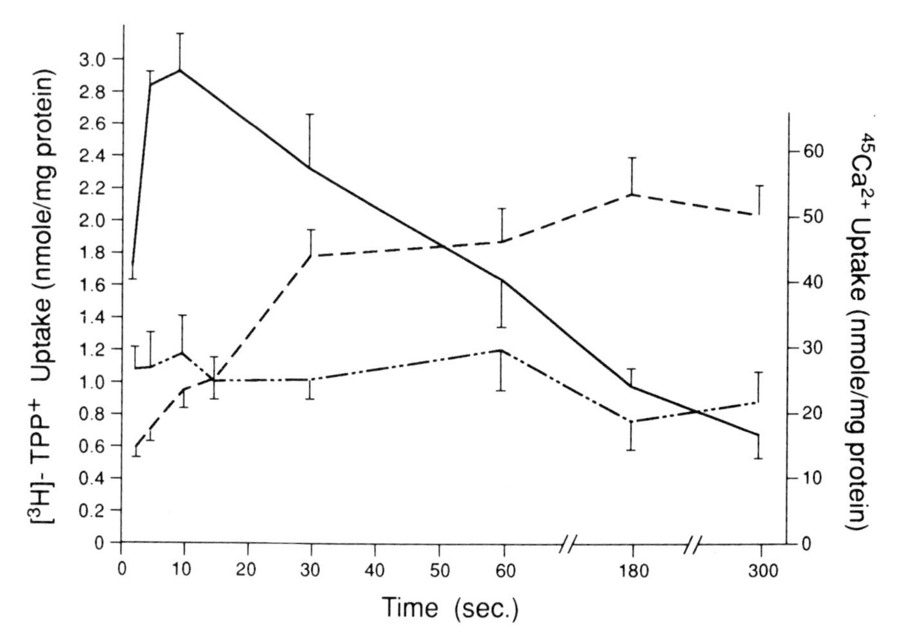

FIGURE 2. Uptake of $[^3H]TPP^+$ and $^{45}Ca^{2+}$ by insulinoma membrane vesicles. Membrane vesicles were pre-equilibrated with 160 mM NaCl medium. Then they were diluted 1 : 30 into medium containing 160 mM KCl and 1 μCi $[^3H]TPP^+$ (specific activity: 39.5 Ci/mmol) with either 40 μM $CaCl_2$(—) or 100 μM EGTA (—-·-—) or 40 μM $^{45}Ca^{2+}$ (---). Uptake was terminated at 2, 5, 10, 15, 20, 30, 60, 180, and 300 sec as described in MATERIALS and METHODS. The mean of eight experiments is shown.

REFERENCES

1. HELLMAN, B., J. SEHLIN, & I.-B. TALJEDAHL. 1979. Pflügers Arch. **378:** 93-97.
2. PHANG, W., L. DOMBOSKI & Y. KRAUSZ, *et al.* 1984. Am. J. Physiol. **247:** E701-708.
3. LAMBERT, A. E., J. C. HENQUIN & P. MALVAUX. 1974. Horm. Metab. Res. **6:** 470-475.
4. SIEGEL, E. G., C. B. WOLLHEIM & A. E. RENOLD, *et al.* 1980. J. Clin. Invest. **66:** 996-1003.
5. HERCHUELZ, A., A. SENER & W. J. MALAISSE. 1980. J. Membr. Biol. **57:** 1-12.
6. HENQUIN, J. C., R. DE MIGUEL & M. G. GARRINO, *et al.* 1985. FEBS Lett. **187:** 177-181.
7. HELMAN, B. & E. GYLFE. 1984. Biochim. Biophys. Acta **770:** 136-141.
8. PLASMAN, P. O., P. LEBRUN & A. HERCHUELZ. 1990. Am. J. Physiol. **259:** E844-850.
9. MULLINS, L. J. 1979. Am. J. Physiol. **236:** C103-110.
10. HOENIG, M., L. H. CULBERSON & D. C. FERGUSON. Endocrinology **128:** 1381-1384.
11. HOENIG, M. & G. W. G. SHARP. 1986. Endocrinology **119:** 2502-2507.

Evidence for Sodium-Calcium Exchange in Rodent Osteoblasts

NANCY S. KRIEGER[a]

Departments of Medicine and Pharmacology
University of Rochester
Rochester, New York 14642

We have previously presented evidence that suggested that hormone-stimulated bone resorption requires Na-Ca exchange. This was based on the inhibitory effects of ionophores[1] and drugs such as dichlorobenzamil[2,3] on stimulated Ca release from bone in organ culture. However, in the organ culture system it was not possible to determine whether Na-Ca exchange was involved in the final osteoclast-mediated release of calcium from bone or in an initiating step occurring in the osteoblast, the direct target cell for calcemic hormones such as parathyroid hormone (PTH) and 1,25-dihydroxyvitamin D_3.[4] In order to test this more directly, we have begun to characterize Na-dependent Ca transport in UMR-106 cells, a rat osteoblastic osteosarcoma cell line, as well as in primary cells isolated from neonatal mouse calvaria, using the Ca-sensitive fluorescent dye, fura-2.[5]

UMR-106 rat osteosarcoma cells are a clonal line that has differentiated properties of mature osteoblasts.[6] These were compared with primary cells isolated by sequential collagenase digestion of neonatal mouse calvariae.[7] Cells were loaded with 3 μM fura-2-AM at 37°C for 40 min in a HEPES buffer, pH 7.3, containing 140 mM NaCl, no Ca in the absence or presence of 0.3 mM ouabain. Cells were then washed, resuspended in the original buffer, and kept on ice. Immediately before measuring fluorescence an aliquot of cells was centrifuged and resuspended in the appropriate buffer ($\pm$ NaCl, $\pm$ 0.3 mM ouabain). Fluorescence was measured at excitation wavelengths of 340 nm and 380 nm and an emission wavelength of 505 nm in a stirred cuvette at 22°C in an Alphascan spectrofluorimeter (Photon Technology Int.), which is capable of rapid dual-wavelength measurements. Calibration of maximal and minimum Ca binding was done by lysing cells with 0.25% NP-40 and quenching with 10 mM EGTA, pH 8.5, respectively.

If the external Na concentration was decreased to 2 mM while maintaining isotonicity with choline chloride, UMR-106 cells that were Na-loaded as a result of ouabain inhibition of the Na,K-ATPase, exhibited significantly greater uptake of Ca than control cells (FIG. 1). The mean change in intracellular Ca was 30% greater in Na-loaded cells (TABLE 1). The Na-dependent increase depended on the presence of external Ca and the increase was maximal within $\sim$150 sec after addition of Ca to the cuvette. Comparable results were observed with the primary bone cells.

We should now be able to determine whether this osteoblastic Na-Ca exchange transport is directly involved in the effect of PTH on the osteoblast.

[a] Address for correspondence: Nancy S. Krieger, Ph.D., Department of Medicine, University of Rochester School of Medicine and Dentistry, 601 Elmwood Ave., Box MED, Rochester, NY 14642.

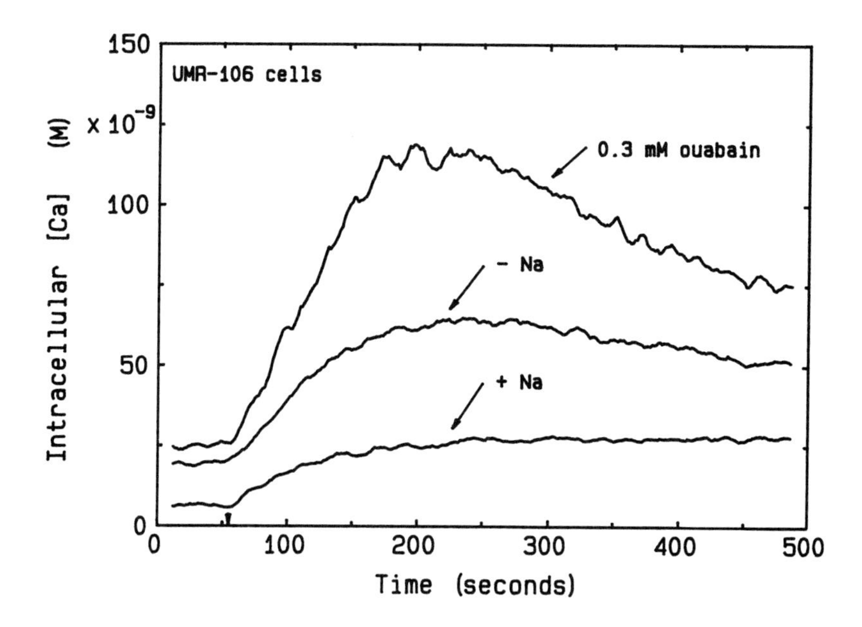

FIGURE 1. Sodium-dependent calcium influx in UMR osteosarcoma cells. UMR-106 cells were loaded with fura-2 in buffer containing 140 mM NaCl, and no Ca, with or without 3 mM ouabain. Cells were washed and resuspended in the original buffer (+ Na), buffer containing 2 mM NaCl (− Na), or buffer containing 2 mM NaCl and 0.3 mM ouabain (ouabain). 1.5 mM $CaCl_2$ was added to the cuvette at 60 sec. Data from a representative experiment are presented as the calculated intracellular Ca levels obtained from measuring the ratio of excitation fluorescence at 340 nm and 380 nm.[5]

TABLE 1. Effect of Decreasing Extracellular [Na] on Intracellular [Ca] Measurements in UMR-106 Osteosarcoma Cells

Treatment	n	Change in $[Ca]_i$ (nM)
Control	18	20.1 ± 0.7
−Na buffer	22	46.7 ± 2.3[a]
Na loaded	24	60.6 ± 2.1[a,b]

NOTE: Peak change in intracellular [Ca] was measured 125-160 sec after addition of 1.5 mM $CaCl_2$ in each experiment. Control = cells loaded with fura-2 in the presence of 140 mM NaCl and resuspended in the same buffer; −Na = cells loaded in the presence of 140 mM NaCl and then resuspended in buffer containing 2 mM NaCl; Na loaded = cells loaded in the presence of 140 mM NaCl and 0.3 mM ouabain and then resuspended in buffer containing 2 mM NaCl and 0.3 mM ouabain. Data are the means ± SE for replicate experiments; n = number of determinations per group.

[a] $p < 0.001$ compared to + Na control.
[b] $p < 0.015$ compared to −Na buffer.

REFERENCES

1. KRIEGER, N. S. & A. H. TASHJIAN, JR. 1980. Nature **287:** 843-845.
2. KRIEGER, N. S. & P. H. STERN. 1982. Am. J. Physiol. **243:** E499-E504.
3. KRIEGER, N. S. & S. G. KIM. 1988. Endocrinology **122:** 415-420.
4. RODAN, G. A. & T. J. MARTIN. 1981. Calc. Tiss. Int. **33:** 349-351.
5. GRYNKIEWICZ, G., M. POENIE & R. Y. TSIEN. 1985. J. Biol. Chem. **260:** 3440-3450.
6. PARTRIDGE, N. C., D. ALCORN, V. P. MICHELANGELI, G. RYAN & T. J. MARTIN. 1983. Cancer Res. **43:** 4308-4314.
7. HEFLEY, T. J., P. H. STERN & J. S. BRAND. 1983. Exp. Cell Res. **149:** 227-236.

Extracellular Na^+ Removal Stimulates Chorionic Gonadotropin and Placental Lactogen Release by Human Placental Explants

P. LEBRUN,[a] B. POLLIOTTI,[b] C. ROBYN,[b] AND S. MEURIS [b]

[a]*Laboratory of Pharmacology and*
[b]*Human Reproduction Research Unit*
Faculty of Medicine
Free University of Brussels
B-1000 Brussels, Belgium

Ionic movements leading to the secretion of human chorionic gonadotropin (hCG) and placental lactogen (hPL) by placenta remain poorly understood. Recent findings, however, indicate that calcium plays a central role in the secretory process of these two proteic hormones.[1] Because sodium is another important cation involved in ionic homeostasis of living cells,[2] we examined to what extent the omission of extracellular Na^+ affects the release of hCG and hPL from placental explants.

METHODS

Placentas were obtained after vaginal delivery from normal-term pregnancies. Explants were placed into vials and physiological media changed every 30 min. Hormonal variations, as assessed by radioimmunoassay, were expressed in percent and corresponded to the ratio between the releases measured at 180 and 150 min.

RESULTS

Isosmotical replacement of extracellular Na^+ (Na_o^+) by sucrose, K^+, or choline increased the hCG and hPL release from placental explants. The effect of Na_o^+ removal

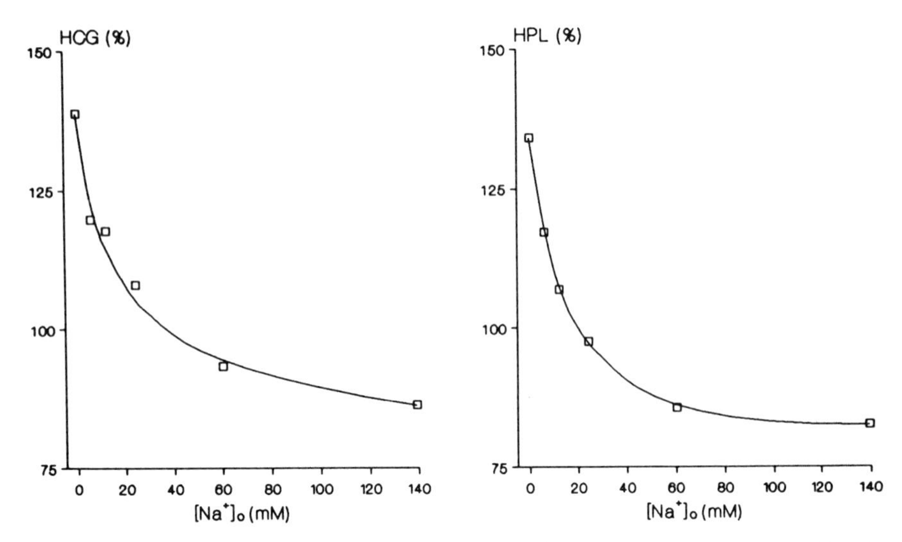

FIGURE 1. Effect of different concentrations of extracellular Na^+ ($[Na^+]_o$) on the hCG (*left panel*) and hPL (*right panel*) release from placental explants. Na_o^+ was isosmotically replaced by choline. The media containing choline chloride were supplemented with 10 μM atropine sulfate.

on the release of both hormones was concentration dependent (FIG. 1) and was inhibited in the absence of extracellular Ca^{2+}. To further characterize the mechanism(s) by which Na_o^+ omission stimulated the hCG and hPL release, the effect of various drugs was examined. Nifedipine (20 μM) and D600 (50 μM), two blockers of the voltage-sensitive Ca^{2+} channels, as well as tetrodotoxin (5 μM), a specific Na^+-channel blocker, did not modify the stimulatory effects of Na_o^+ omission. By contrast, amiloride (2 mM), Mg^{2+} (10 mM) and Sr^{2+} (10 mM) reduced the increase in hCG and hPL release provoked by Na_o^+ removal. Amiloride, Mg^{2+}, and Sr^{2+} have previously been shown to affect the Na^+-Ca^{2+} exchange.[3,4] Last, the secretory responses to Na_o^+ omission were enhanced in the presence of ouabain (2 mM), an inhibitor of the Na^+, K^+-ATPase.

DISCUSSION

These observations indicate that Na_o^+ omission provokes a Ca^{2+}-dependent stimulation of hCG and hPL release. The pharmacological analysis of the secretory effects of Na_o^+ deprivation suggests the existence in placental cells of a process of sodium-calcium exchange.

REFERENCES

1. POLLIOTTI, B., S. MEURIS, P. LEBRUN & C. ROBYN. 1990. Placenta **11:** 181-190.
2. BAKER, P. F. 1986. J. Cardiovasc. Pharmacol. **8:** S25-S32.
3. TROSPER, T. L. & K. D. PHILIPSON. 1983. Biochim. Biophys Acta **731:** 62-68.
4. BENOS, D. J. 1982. Am. J. Physiol. **242:** C131-C145.

Catecholamine Release from Adrenal Gland Evoked by Lithium

A Consequence of $[Li]_i$-$[Ca]_o$ Counter-Transport Mechanism?[a]

F. DE ABAJO, M. A. SERRANO-CASTRO, AND
P. SÁNCHEZ-GARCÍA[b]

Department of Pharmacology and Therapeutics
Medical School UAM
28029-Madrid, Spain

Lithium, a monovalent cation that shares many features with Na, gradually accumulates inside the cells, probably because it is a poor substrate for the Na pump.[1] Previous work from our laboratory has demonstrated that an increase in $[Na]_i$ is a critical factor for the calcium-dependent catecholamine secretion evoked by ouabain in the cat adrenal gland.[2] On these bases we thought that cat adrenal glands perfused with Li-containing Krebs would behave as ouabain-treated glands. Therefore the experiments to be described here were undertaken in order to explore this possibility.

METHODS

Both cat adrenal glands were isolated and prepared for retrograde perfusion with normal Krebs-bicarbonate solution at room temperature as previously described.[2] Calcium-free solutions were made up simply by deleting $CaCl_2$ from the perfusion medium. When the Na concentration was reduced to 25 mM, the osmolarity of the solution was maintained with equivalent amounts of sucrose (sucrose-Krebs) or LiCl (Li-Krebs). When $[Na]_o$ was decreased to less than 25 mM, Tris-hydroxymethyl-amino methane (10 mM), instead of bicarbonate, was used as a buffer. Potassium-rich solutions were prepared by the addition of KCl and the reduction of similar amounts of NaCl to maintain isotonicity. The catecholamine content of the samples was determined according to a conventional photofluorimetric method.

RESULTS AND DISCUSSION

Replacement of Na (119 mM) by Li in the normal-Krebs used to perfuse the gland evoked a progressive increase of spontaneous catecholamine release; this secretion reached a maximum (0.26 ± 0.04 μg/2 min; $n = 16$) within 45 minutes and was

[a] This work was supported by grants from CICYT and FISS (Spain).

[b] Address for correspondence: Dr. Pedro Sánchez-García, M.D., Prof. and Chairman, Dept. of Pharmacology and Therapeutics, Medical School UAM, C/ Arzobispo Morcillo, 4, 28029-Madrid, Spain.

TABLE 1. The Effect of Replacement of Li (119 mM) by Na or Sucrose in the Perfusion Fluid on Catecholamine (CA) Release in the Cat Adrenal Gland Induced by Li-Krebs[a]

n	CA Release after 45-min Perfusion with Li-Krebs (Control)	Li Substitution in the Perfusion Fluid		CA Release: Increase Over the Control
		by Na (2 min)	by Sucrose (2 min)	
7	0.30 ± 0.01	Yes	---	0.06 ± 0.01
10	0.22 ± 0.01	---	Yes	1.06 ± 0.3

[a] Results are expressed as CA μg/2 min ± SE mean; *n* = number of experiments.

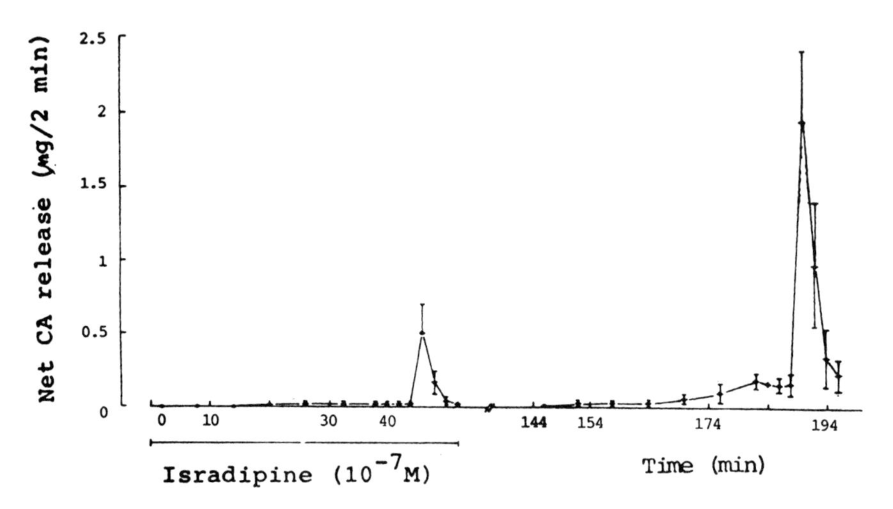

FIGURE 1. The effect of isradipine (10^{-7} M) on the catecholamine (CA) release evoked by Li-Krebs and by the replacement of Li (119 mM) by sucrose for 2 min at 46 and 190 min in the perfused cat adrenal gland. Data on this figure correspond to two consecutive Li-Krebs perfusion periods separated by a 90-min wash-out in normal Krebs. Each point represent the mean from three experiments and vertical lines show SE mean.

calcium dependent; such a secretory profile is exquisitely similar to that produced by ouabain in the same model. Replacement of Li by sucrose, but not by Na, in glands preperfused 45 minutes with Li-Krebs induced a sharp secretory response (TABLE 1). In ouabain (10^{-4} M, 10 min) pretreated glands perfused 30 minutes with normal-Krebs, partial replacement of Na by sucrose but not by Li induces a sharp increase in the catecholamine output. Reintroduction of Ca (2.5 mM; 2 min) in glands preperfused 30 min with Ca-free, Mg-containing Li-Krebs evoked a great increase in catecholamine release (7.2 ± 0.7 μg/8 min; *n* = 28). This secretory response is directly dependent of $[Li]_o$ and the length of the Li loading period.[3] That effect was not seen when the glands were perfused with Ca-free normal, choline- or sucrose-Krebs. Previous data from our laboratory demonstrated that ouabain (10^{-4} M) greatly potentiated the catecholamine secretory response induced by K (17.7 mM).[4] A similar degree of K^+ potentiation was found in glands perfused with Li-Krebs.[5]

Finally, it is important to note that the Li-evoked secretion and the secretion induced by Li replacement by sucrose is partially blocked by the calcium-blocking agent isradipine ((+)PN-200-110, 10^{-7} M) (FIG. 1).

In summary our results demonstrate that Li enhances the rate of catecholamine release by its accumulation in the chromaffin cells and suggest that a dual mechanism is probably involved: (1) chromaffin cell depolarization and subsequent opening of voltage-operated calcium channels and (2) activation of a $[Li]_i$-$[Ca]_o$ counter-transport system.

REFERENCES

1. KEYNES, R. D. & R. C. SWAN. 1959. The permeability of frog muscle fibres to lithium ions. J. Physiol. (London) **147:** 626-638.
2. GARCIA, A. G., M. HERNANDEZ, J. F. HORGA & P. SANCHEZ-GARCIA. 1980. On the release of catecholamines and dopamine beta-hydroxylase evoked by ouabain in the perfused cat adrenal gland. Br. J. Pharmacol. **68:** 571-583.
3. ABAJO, F. J., M. A. S. CASTRO, B. GARIJO & P. SANCHEZ-GARCIA. 1987. Catecholamine release evoked by lithium from the perfused adrenal gland of the cat. Br. J. Pharmacol. **91:** 539-546.
4. GARCIA, A. G., E. GARCIA-LOPEZ, J. F. HORGA, S. M. KIRPEKAR, C. MONTIEL & P. SANCHEZ-GARCIA. 1981. Potentiation of K-evoked catecholamine release in the cat adrenal gland treated with ouabain. Br. J. Pharmacol. **74:** 637-680.
5. ABAJO, F. J., M. A. S. CASTRO & P. SANCHEZ-GARCIA. 1989. The key role of sodium in the ouabain-mediated potentiation of potassium-evoked catecholamine release in cat adrenal glands. Br. J. Pharmacol. **98:** 455-462.

Index of Contributors